Genes, Brain, and Emotions

Series in Affective Science

Series Editors: Richard J. Davidson and Klaus Scherer

The Evolution of Emotional Communication
Eckart Altenmüller, Sabine Schmidt, and Elke Zimmermann (eds)

The Neuropsychology of Emotion
John C. Borod

Persons, Situation, and Emotions: An Ecological Approach
Herman Brandstätter and Andrzej Eliasz

Handbook of Emotion Elicitation and Assessment
James A. Coan and John J.B. Ellen (eds)

Anxiety, Depression, and Emotion
Richard J. Davidson

What the Face Reveals: Basic and Applied Studies of Spontaneous Expression Using the Facial Action Coding System (FACS) 2e
Paul Ekman and Erika L. Rosenberg (eds)

The Nature of Emotion: Fundamental Questions
Paul Ekman and Richard J. Davidson

The Psychology of Gratitude
Robert A. Emmons and Michael E. McCullough (eds)

Who Needs Emotions? The Brain Meets the Robot
Jean-Marc Fellous and Michael A. Arbib (eds)

Emotions in Psychopathology: Theory and Research
William F. Flack and James D. Laird (eds)

The Nature of Emotion: Fundamental Questions, 2e
Edited by Andrew S. Fox, Regina C. Lapate, Alexander J. Shackman, and Richard J. Davidson

Shame: Interpersonal Behaviour, Psychopathology, and Culture
Paul Gilbert and Bernice Andrews (eds)

Pleasures of the Brain
Martin L. Kringelbach and Kent C. Berridge

Infant Chimpanzee and Human Child: A Classic 1935 Comparative Study of Ape Emotions and Intelligence
N.N. Ladygina-Kohts (deceased) and Frans B.M. de Waal (eds), Boris Vekker (translator)

Feelings: The Perception of Self
James D. Laird

Cognitive Neuroscience of Emotions
Richard D. Lane and Lynn Nadel (eds)

The Development of Social Engagement: Neurobiological Perspectives
Peter J. Marshall and Nathan A. Fox (eds)

Science of Emotional Intelligence: Knowns and Unknowns
Gerald Matthews, Moshe Zeidner, and Richard D. Roberts (eds)

Expectancy and Emotion
Maria Miceli and Cristiano Castelfranchi

Affective Neuroscience: The Foundations of Human and Animal Emotions
Jaak Panskepp

Nonverbal Behaviour in Clinical Settings
Pierre Philippot, Robert S. Feldman, and Erik J. Coats (eds)

Emotion in Memory and Development: Biological, Cognitive, and Social Considerations
Jodi Quas and Robyn Fivush (eds)

Memory and Emotion
Daniel Reisberg and Paula Hertel (eds)

Emotion Explained
Edmund T. Rolls

Emotion, Social Relationships, and Health
Carol D. Ryff and Burton Singer (eds)

Oxford Companion to Emotion and the Affective Sciences
David Sander and Klaus Scherer

A Blueprint for Affective Computing: A Sourcebook and Manual
Klaus R. Scherer, Tanja Bänzinger, and Etienne Roesch

Appraisal Processes in Emotion: Theory, Methods, Research
K. Scherer, A. Schorr, and T. Johnstone (eds)

Bodily Sensibility: Intelligent Action
Jay Schulkin

Boo! Culture, Experience, and the Startle Reflex
Ronald C. Simons

Thinking and Feeling: Contemporary Philosophers on Emotions
Robert C. Solomon

Collective Emotions
Christian von Scheve and Mikko Salmela (eds)

Genes, Brain, and Emotions
Interdisciplinary and Translational Perspectives

Edited by
Andrei C Miu
Judith R Homberg
and Klaus-Peter Lesch

OXFORD
UNIVERSITY PRESS

OXFORD
UNIVERSITY PRESS

Great Clarendon Street, Oxford, OX2 6DP,
United Kingdom

Oxford University Press is a department of the University of Oxford. It furthers the University's objective of excellence in research, scholarship, and education by publishing worldwide. Oxford is a registered trade mark of Oxford University Press in the UK and in certain other countries

First Edition Published in 2019

Impression: 1

Published in the United States of America by Oxford University Press
198 Madison Avenue, New York, NY 10016, United States of America

British Library Cataloguing in Publication Data

Data available

Library of Congress Control Number: 2019930282

ISBN 978–0–19–879301–4

Printed and bound by
CPI Group (UK) Ltd, Croydon, CR0 4YY

Contents

Part 3 **Biological mechanisms**

Part 4 **Disorders and therapy**

Contributors

Rebecca Alexander
Research School of Psychology, ANU College of Health and Medicine, Australian National University, Australia; Neuroscience Research Australia (NeuRA), Australia

Bradley M Avery
Department of Psychology, University of North Carolina at Greensboro, USA

Catherine Belzung
UMR 1253, iBrain, Université de Tours, Inserm, Tours, France

Mirela I Bîlc
Cognitive Neuroscience Laboratory, Department of Psychology, Babeş-Bolyai University, Romania

Dorret I Boomsma
Department of Biological Psychology, Vrije Universiteit Amsterdam, The Netherlands

Thomas J Bouchard Jr
Department of Psychology, University of Minnesota, USA

Turhan Canli
Department of Psychology
Stony Brook University, USA

Danielle C Cath
UMC Groningen, Department of Psychiatry and Rijksuniversiteit Groningen, The Netherlands

Airiss R Chan
Centre for Addiction and Mental Health, University of Toronto, Canada

Sven Cichon
Department of Biomedicine, University of Basel, Switzerland
and Institute of Neuroscience and Medicine (INM-1), Research Center Jülich, Germany

Jonathan RI Coleman
King's College
London, Institute of Psychiatry, Psychology and Neuroscience, UK; Institute of Neuroscience and Medicine (INM-1), Research Center Jülich, Germany

Annette Conzelmann
Department of Child and Adolescent Psychiatry, Psychosomatics and Psychotherapy, University of Tübingen, Germany

John F Cryan
Department of Anatomy and Neuroscience, University College Cork, Ireland

Harriet de Wit
Department of Psychiatry and Behavioural Neuroscience, University of Chicago, USA

Hanneke Den Ouden
Donders Institute for Brain, Cognition and Behaviour, Centre for Cognitive Neuroimaging, Radboud University Nijmegen, The Netherlands

Mana R Ehlers
Department of Psychology, University of British Columbia, Canada

Thalia C Eley
King's College
London, Institute of Psychiatry, Psychology and Neuroscience, UK

Daniela Felice
Department of Anatomy and Neuroscience, University College Cork, Ireland

Megan Flom
Department of Psychological and Brain Sciences, Boston University, USA

Emily Funsten
Program in Neuroscience, Middlebury College, USA

Justine Megan Gatt
School of Psychology, University of New South Wales, Australia; Neuroscience Research Australia (NeuRA), Australia

Ina Giegling
Department of Psychiatry, Psychotherapy and Psychosomatics, Martin Luther University Halle-Wittenberg, Germany

Ilona Gorbovskaya
Centre for Addiction and Mental Health, Canada

Irving I Gottesman (deceased),
Department of Psychology, University of Minnesota, USA

Joshua C Gray
Department of Medical and Clinical Psychology, Uniformed Services University, USA

Anand Gururajan
Department of Anatomy and Neuroscience, University College Cork, Ireland

Waylon J Hastings
Department of Biobehavioral Health, Pennsylvania State University, USA

Lisa Heltzel
Donders Institute for Brain, Cognition, and Behaviour, Radboud University Medical Centre, The Netherlands

Judith R Homberg
Donders Institute for Brain, Cognition, and Behaviour, Radboud University Medical Centre, The Netherlands

Wendy Johnson
Department of Psychology, University of Edinburgh, UK

Joan Kaufman
Center for Child and Family Traumatic Stress, Kennedy Krieger Institute, USA, and Department of Psychiatry and Behavioral Sciences, Johns Hopkins School of Medicine, USA

Richard S Lee
Department of Psychiatry and Behavioral Sciences, Johns Hopkins School of Medicine, USA

Klaus-Peter Lesch
Division of Molecular Psychiatry, Center of Mental Health, University of Würzburg, Germany

Kathryn J Lester
School of Psychology, University of Sussex, UK

Tina B Lonsdorf
Department of Systems Neuroscience, University Medical Center Hamburg-Eppendorf, Hamburg, Germany

James MacKillop
Peter Boris Centre for Addictions Research, McMaster University, Canada

Andrei C Miu
Cognitive Neuroscience Laboratory, Department of Psychology, Babeş-Bolyai University, Romania

Janitza L Montalvo-Ortiz
Department of Psychiatry, Yale University School of Medicine, USA

Erik M Mueller
Department of Psychology, University of Marburg, Germany

Thomas W Mühleisen
Institute of Neuroscience and Medicine (INM-1), Research Centre Jülich, Germany; C. and O. Vogt Institute for Brain Research, Medical Faculty, Heinrich Heine University Düsseldorf, Germany

Daniel J Müller
Department of Psychiatry, University of Toronto, Canada

Olivia F O'Leary
Department of Anatomy and Neuroscience, University College Cork, Ireland

Abraham A Palmer
Department of Psychiatry, University of California San Diego, USA

Clarissa C Parker
Department of Psychology and Program in Neuroscience, Middlebury College, USA

Paul Pauli
Department of Biological Psychology, Clinical Psychology and Psychotherapy, University of Würzburg, Germany

Lukas Pezawas
Department of Psychiatry and Psychotherapy, Medical University of Vienna, Austria

Alexander Prehn-Kristensen
Department of Child and Adolescent Psychiatry and Psychotherapy, Center for Integrative Psychiatry, School of Medicine, Christian Albrechts University, Kiel, Germany

Ulrich Rabl
Department of Psychiatry and Psychotherapy, Medical University of Vienna, Austria

Tobias Renner
Department of Child and Adolescent Psychiatry, Psychosomatics and Psychotherapy, University of Tübingen, Germany

Karin Roelofs
Donders Institute for Brain, Cognition and Behaviour, Centre for Cognitive Neuroimaging, Radboud University Nijmegen, The Netherlands

Dan Rujescu
Department of Psychiatry, Psychotherapy and Psychosomatics Martin Luther University Halle-Wittenberg, Germany

Sandra Sanchez-Roige
Department of Psychiatry, University of California San Diego, USA

Vaibhav Sapuram
Department of Psychology, University of North Carolina at Greensboro, USA

Kimberly J Saudino
Department of Psychological & Brain Sciences, Boston University, USA

Idan Shalev
Department of Biobehavioral Health, Pennsylvania State University, USA

Kayvon Sharif
Program in Neuroscience, Middlebury College, USA

Dirk JA Smit
Department of Psychiatry, Amsterdam University Medical Center at Meibergdreef, The Netherlands

Celine L St Pierre
Department of Genetics, Washington University in St. Louis, USA

Alexandre Surget
UMR 1253, iBrain, Université de Tours, Inserm, Tours, France

Rebecca M Todd
Department of Psychology, University of British Columbia, Canada

Jenny van Dongen
Department of Biological Psychology, Vrije Universiteit Amsterdam, The Netherlands

Tatjana van Strien
Department of Clinical Psychology, Radboud University Nijmegen, The Netherlands

Inge Volman
FMRIB Centre, University of Oxford, Nuffield Department of Clinical Neurosciences, John Radcliffe Hospital, UK

Suzanne Vrshek-Schallhorn
Department of Psychology, University of North Carolina at Greensboro, USA

Nuno R Zilhão
Vrije Universiteit Amsterdam, The Netherlands

Chapter 1

Genes, brain and emotions: Introduction

Andrei C Miu, Judith R Homberg, and Klaus-Peter Lesch

Emotions reflect our relation with the world and play an essential role in our ability to adapt to environmental challenges. They arise from the dynamic interpretation of events as opportunities and threats, and by rapidly changing our cognitive and somatic activity, they prepare us to respond to these events. Not surprisingly, research in recent decades has offered extensive evidence for the involvement of emotions in almost every aspect of our cognitive performance and behavior, and has highlighted the multiple implications of emotions for social functioning, happiness, and health. In addition, evidence has also emphasized that, depending on factors such as the level of attentional engagement with the environment, the way events are construed, and the abilities to regulate emotional responses, the consequences of emotions can be adaptive or maladaptive. Indeed, emotional difficulties have been described in most neuropsychiatric disorders, and reducing emotional symptoms is an important aim of most clinical interventions in psychopathology. These issues show the breadth and coverage of affective science, a field that has grown tremendously in the last few decades.

A major subject, which has been approached relatively recently in affective science, is concerned with the contribution of genetic differences to emotions. To what extent do individual differences in emotional responses depend on genes? What are the biological mechanism by which genes influence emotions? How do genes contribute to the escalation of emotional problems into clinical disorders? Addressing these questions has involved multiple methods in order to allow, on the one hand, the reliable assessment of emotional phenotypes at subjective, cognitive, and behavioral levels, and on the other hand, the investigation of genetic differences. From its inception, this topic has developed at the intersection of psychology and behavioral and molecular (epi)genetics. Other fields have joined this interdisciplinary venture more recently. Standing out among these fields, cognitive neuroscience has opened the possibility of investigating the neural mechanisms of emotion, which lie intermediate between genes and emotional experience. Neural endophenotypes of emotion have been examined using non-invasive functional neuroimaging and electrophysiological methods in humans, and neuropsychological, pharmacological or, more recently, optogenetic methods in experimental models. Importantly, experimental models have also permitted genetic manipulations. The wide array of methods and the multiple levels of analysis have rapidly pushed the boundaries of research into the genetics of emotion. As is often the case with maturing fields of investigation, the research questions have become more specialized, the theories have grown increasingly interdisciplinary, and the toolbox that researchers need to test these theories has enlarged.

The aim of this volume is to offer a comprehensive account of current research on the genetics of emotion. Major methodological approaches are reviewed in the first part, including the two traditional "workhorses" in the field, twin studies and gene–environment interaction studies, and the more recently developed epigenetic modification assays, genome-wide association studies, and optogenetic methods. Parts 2 and 3 focus on a variety of cognitive and biological mechanisms that may be viewed as intermediate phenotypes in the pathways between genes and emotional experience. Part 2 includes chapters on fear conditioning, emotional action control, emotion regulation, emotional memory, and emotion-linked decision making. Part 3 reviews work on neural phenotypes examined using functional neuroimaging, electroencephalography, telomere length assays, and psychophysiological measures. Finally, Part 4 of the volume focuses on the genetics of emotional dysregulation in neuropsychiatric disorders, and genetic contributions to the response to psychological, behavioral, and pharmacological interventions. The chapters in this part offer introductions to aspects such as the shared genetics between personality and psychopathology, genetic influences on emotional resilience, and genetic moderators of child maltreatment in psychopathology, as well as genetic contributions to post-traumatic stress disorder, impulsivity, anger, and aggression as risk factors for suicide, emotional eating, and obsessive–compulsive disorder, and Tourette's syndrome. Two chapters in the fourth part of the volume review work in therapy genetics and pharmacogenetics, respectively, with a focus on anxiety and affective disorders.

As shown by the variety of topics covered in this volume, the research to date offers a rich framework for assessing emotional phenotypes and their neural mechanisms, and interpreting the complex gene–environment and gene–development interactions. The genetics of emotion field has rapidly developed and, due to the increasing availability of genomic methods, it has begun its transition to the genome-wide level of analysis. With this challenge ahead, the field needs to take stock of the wealth of available methods, empirical findings, and theories that have accumulated to date. Taking investigations to the genomic level and improving translational research will advance our understanding of the genetics of behavior, including emotions. We hope this volume will provide a guide to this interdisciplinary field and will prove useful to researchers and students alike.

We thank all the authors who have contributed to this volume and Oxford University Press for supporting this project.

Part 1

Methods and approaches

Chapter 2

Twin studies of emotion

Megan Flom and Kimberly J Saudino

2.1 Introduction to genetic and environmental influences on emotions

People are different in their experience and expression of emotion, and as indicated in other chapters in this volume, these differences are related to a variety of outcomes. This, then, begs the question as to what factors explain these differences. The twin design can address this question at the level of etiology by exploring the relative contributions of genetic and environmental influences on individual differences in emotion, and links between emotion and outcomes. After providing an overview of the twin methodology, the present chapter will summarize findings from twin studies that inform on basic emotions (e.g. fear, anger, sadness, happiness, and emotionality), complex emotions, emotion processes, and mechanisms underlying the relations between emotion and psychopathology.

2.2 The twin design

The classic twin design compares the resemblances of identical (monozygotic, MZ) twins with fraternal (dizygotic, DZ) twins to decompose the phenotypic (i.e. observed) variance of a behavior or trait into genetic, shared, and nonshared environmental components. *Heritability*, the genetic effect size, is the proportion of phenotypic variance that can be attributed to genetic factors. If genetic influences are important to a behavior, then MZ twins who share 100% of their genes should be more similar for that behavior than DZ twins who share, on average, 50% of their segregating genes. *Shared environmental variance* is familial resemblance that is not explained by genetic variance and comprises environmental influences that are shared by family members (e.g. socioeconomic status (SES), neighborhood, friends, or even such things as the number of TVs or books in the house). If relevant to the behavior under study, shared environmental influences will make family members more alike. *Nonshared environmental variance* includes environmental influences that are unique to each individual in a family. These unique environmental influences operate to make members of the same family different from one another. Possible sources of nonshared environmental variance include differential parental treatment; relationships with friends, peers, and teachers; and non-systematic factors such as accidents, illness, and measurement error (1).

2.2.1 How can twins inform research on emotions?

The twin design can be applied to the study of emotion in a variety of ways. First, and most basically, it can provide estimates of the extent to which individual differences in emotions and related constructs are due to genetic and environmental influences (e.g. how heritable is sadness, or emotion regulation?). Second, twins can be used to explore genetic and environmental links between different emotions, thus providing information on the etiology underlying the structure of emotions. Third, when applied to longitudinal data, twin methodology can identify genetic and environmental sources of stability and change in emotions across age. Finally, it can examine sources of genetic and environmental overlap between emotions and psychopathology (e.g. do fear and anxiety share common genes?). Taken together, these applications of the twin design to the study of emotion have implications for intervention and molecular genetic work.

2.3 Basic emotions

Perhaps the largest contribution twin studies have made to our understanding of emotion is through the study of temperament. The conceptualization of temperament as individual differences in the expression of primary emotions is widely accepted by temperament researchers, and often includes emotion-related traits such as anger, fear, sadness, happiness, and emotionality (2, 3). Although twin studies on emotion have now expanded beyond the context of temperament, the study of temperament has been the foundation for such work and hence will be the focus of much of the research reviewed on basic emotions.

2.3.1 Genetic and environmental influences on basic emotions

With few exceptions, twin studies find that basic emotions are heritable. The specific etiological pattern does, however, vary somewhat by emotion domain and measurement method. Anger, fear, and sadness are heritable in infancy and childhood when assessed via parent report (4–8) and observers (6, 9–11). Overall, heritabilities range from 0.50 to 0.70 indicating that genetic factors explain somewhere between 50 and 70% of variation in these emotions. In most cases, the remaining variance is due to the nonshared environment, but there have been some studies that also find modest shared environmental influences on anger and sadness in childhood (6, 8, 10, 12, 13).

Although observer- and parent-report measures of behavioral sadness are genetically influenced, this was not the case in a functional Magnetic Resonance Imaging (fMRI) study of neural activation associated with sadness. In a sample of eight-year-old twins who were shown sad film clips, there was no familial resemblance in neural activation in brain regions associated with sadness for either MZ or DZ twins (14). Hence, variation in neural activation was influenced only by the nonshared environment. While it is possible that this neural activation is due solely to individual-specific experiences, the lack of genetic influences may also be due to the small sample size (104 twin pairs) or measurement error.

Anger and fear are also genetically influenced in adults (15–19). Heritability estimates generally fall within a similar range of magnitude as for younger twins. Sadness as a discrete emotion has not been investigated in adult twins, most likely because the focus is often on the more clinical expression of the emotion (e.g. depression) and/or as a component of the personality trait of neuroticism.

These discrete basic emotions are often subsumed under broader concepts of negative and positive emotionality. In children, adolescents, and adults, negative emotionality is moderately to highly heritable (e.g. 0.40–0.64), with remaining variance explained by the nonshared environment (11, 20–23). Positive emotionality—or as it is sometimes referred, positive affect—tends to be less heritable than negative emotionality (11, 13, 24, 25). In fact, positive affect and related behaviors (e.g. smiling, interest in others) display little or no genetic influences and moderate shared environmental influences during infancy and early childhood (11, 13, 25, 26). However, in middle childhood and beyond, positive emotionality is moderately to highly heritable (0.33–0.79) and the influence of shared environments is negligible (27–32). It is possible that shared environmental influences in early, but not later, childhood reflects the influence of maternal personality and attachment security on positive affect in young children (11).

Interestingly, while measures of overall negative and positive affect are substantially heritable past middle childhood, *momentary* measures of these two emotion dimensions (i.e. assessed using the experience sampling method (ESM)) have demonstrated little to no genetic effects in adults (33 ,34). This raises the possibility that general perceptions of negative or positive affect assessed retrospectively have different etiologies than the immediate emotions tapped by momentary affect measures (34). However, as Menne-Lothmann and colleagues note, it is also possible that these results may reflect noise in the dynamic nature of ESM data. More research is needed. Overall, with few exceptions, familial resemblance in basic emotions is nearly always a result of shared genes and not shared environments, particularly in adults. Cross-cultural twin studies find a similar pattern across multiple countries (Germany and Poland) and multiple ages (adolescence through old age) (35). Despite widespread support for the substantial heritability of basic emotions, the environment is still relevant to variation in discrete emotions. Notably, twin studies not only inform on magnitude of genetic effects, they are also able to inform on environmental influences to individual differences in emotions. As reviewed earlier, it is those environments that differ *within* families (i.e. nonshared environments), not between families (i.e. shared environments), that are of key importance for most basic emotions. An implication of this is that researchers exploring how specific environments influence emotions might do well to focus on experiences unique to each member of a family.

In addition to informing about the main effects of the environment, twin studies can also explore interactive effects of genes and environments. Although not well studied, there is some evidence that the environment may moderate genetic and environmental influences on basic emotions. For example, the heritability of negative emotionality was higher for children who experienced poorer quality home environments (36). In other words, under a favorable environment genetic factors may play a smaller role in explaining individual differences in negative emotionality.

2.3.2 **Using twins to understand development of basic emotions**

Emotions are not a static construct, they change across age. Behavioral genetics addresses the question of developmental change in two ways: (i) differential heritability (i.e. whether the magnitude of genetic effects differs across age); and (ii) genetic and environmental contributions to rank-order continuity and change (i.e. the extent to which the same genetic and environmental effects operate across age). With respect to differential heritability, genetic effects for positive emotionality tend to increase across age (27, 29, 37), but this is not the case for negative emotionality or related discrete emotions, such as fear, anger, and sadness (13, 20, 22). This evidence for differential heritability for positive, but not negative, emotionality does not mean that there is change in the genes that influence positive emotionality but not negative emotionality, only that there is a difference in the *relative contributions* of genetic factors on individual differences in positive emotionality across age. Even when the same genes operate at different ages, their relative influence on individual differences (i.e. heritability estimates) may differ. Similarly, even when estimates of heritability are similar across development, as is the case for negative emotionality, the genes that influence these emotions may differ from one age to the next. Thus, comparing heritability estimates cannot inform about sources of continuity and change and does not address developmental processes.

By exploring genetic and environmental contributions to rank-order stability, twin studies can address the extent to which there are common and unique genetic and environmental effects across age, thereby informing about underlying developmental processes. For example, to what extent do genetic effects persist from one age to the next (i.e. stability) and to what extent are there age-specific genetic effects (i.e. change)? Longitudinal twin analyses suggest that stability in basic emotions is almost always due to genetic factors, whereas change is most often a result of both genetic and nonshared environmental influences (10, 20, 21, 31, 38). These findings have implications for emotion researchers in other areas. First, the fact that nonshared environmental influences tend to be age-specific and do not persist across age suggests that emotion researchers interested in uncovering specific environments impacting emotions need to consider that the environments that operate at one age will most likely differ from those that operate at another. Second, for researchers interested in identifying specific genes associated with emotions, those emotions that are highly heritable and have high genetic stability across age will make promising candidates for molecular genetic work.

2.3.3 Etiologic links between basic emotions

Twin studies can also inform about genetic and environmental overlap between different emotions, which can elucidate the underlying structure of broad emotion domains such as negative emotionality. Phenotypic factor analyses suggest three underlying components (anger, fear, and sadness) of negative emotionality (39), but twin studies can provide additional information by telling us *why* these three basic emotions are related. Although this has not been extensively studied, common genetic effects seem to be the mechanism that primarily explains the underlying structure of parent-reported negative emotionality in middle childhood (12). The results are more mixed for observer-rated negative emotionality. When assessed via observation in structured episodes, anger and sadness were significantly associated, and again, this covariation was due to overlapping genetic effects. However, observed fear was not correlated with either anger or sadness, and thus had no genetic or environmental effects in common (12). One reason for the different etiologic pattern may be because the type of fear assessed in the observation episodes is different than what parents see in more naturalistic situations. That said, even for parent reports, it is fear that demonstrates the most genetic independence from other components of negative emotionality (11, 12). Thus, despite support for an overall negative emotionality construct, there are important distinctions between fear, anger, and sadness that should not be ignored.

When exploring the structure of several different self-report measures of fear and fearlessness in adults, both phenotypic and genetically informed biometric factor analyses suggest a bifactor model with a higher-order broad, bipolar fear/fearlessness factor, and three independent second-order subfactors of distress, stimulation seeking, and sociability (17). The biometric analyses of variation in this bifactor revealed genetic and nonshared environmental contributions to individual differences in the higher-order fear/fearlessness factor. This consistency across phenotypic and biometric factor analyses provides support for the validity of a fear/fearlessness factor in terms of an underlying genetic and environmental architecture.

Despite the abundance of twin research on negative and positive affect/emotionality, less is known about the etiologic overlap between the two domains. Although it may seem intuitive that they are flip sides of the same coin, the two are often treated as distinct dimensions (40) and, perhaps because of this, to our knowledge, only one twin study has explored the etiologic underpinnings common to both. In middle childhood, genetic and nonshared environmental effects on positive and negative affect overlap, though not entirely (27). While there are genetic and environmental influences unique to each domain, the finding of a common genetic component could mean that there may be a genetic liability to a general tendency to experience or display emotions. More research is needed particularly at other ages to further investigate this possibility. Nonetheless, the modest association between positive and negative affect, as well as overlapping and unique etiologic effects, suggest that there is value to treating positive and negative emotionality as related, but distinct, constructs when examining the relation to psychopathology and other outcomes.

2.4 **Complex emotions**

Unlike basic emotions, which are more automatic in nature, complex emotions are higher-order emotions that require more cognitive processing. Although there are many complex emotions, they have not gained much attention in the twin literature. Two possible exceptions to this are self-conscious shyness and neuroticism. Self-conscious shyness (e.g. feelings of embarrassment), which emerges later than the more basic emotion of fearful shyness, is also highly heritable, with genetic factors explaining as much as 90% of the variation (5). However, despite both being highly heritable, self-conscious shyness and fearful shyness are not associated, which means that they have no common genetic or environmental underpinnings (5).

Neuroticism, "the subjective and stable tendency towards different states of negative affect" (41) encompasses not only the basic set of emotions that comprise negative affect but also the emotional processes of reacting to experiences. Variance in neuroticism is explained largely by genetic and nonshared environmental effects (42–44), a pattern that has been consistently found across multiple countries (e.g. United States, Australia, the Netherlands, Russia). In fact, estimates of genetic and environmental variance explaining individual differences in neuroticism were highly congruent across twin studies from Canada, Germany, and Japan (45).

More recently, researchers have also started to explore genetic and environmental influences on neuroticism as a *state,* rather than as a trait, which has been the primary focus of behavioral genetic work. When longitudinal trait–state models are applied to studies of neuroticism, trait variance (i.e. stable individual differences) is more genetically influenced than the state component (i.e. occasion-specific variance) (46, 47). Long-term stability is due to both genetic and nonshared environmental influences. In contrast, occasion-specific state variance is mainly influenced by the nonshared environment, which tends to increase over time. The decreases in heritability and increased importance of the environment across age, in addition to increased stability in neuroticism across age, is consistent with the notion of social selection and social influence. That is, people select environments that correlate with their levels of neuroticism and these environments, in turn, produce experiences that influence neuroticism (46, 47). This is related to the concept of an active gene–environment correlation wherein one's genetically influenced traits, neuroticism in this case, influences their selection of environments that they experience, thus providing support for the transaction between genetics and the environment in neuroticism (46, 47).

Underlying associations between neuroticism as a trait and other affective state measures have also been explored. Research finds that neuroticism and momentary positive affect (a state-based emotion) are negatively associated and that this arises because of overlapping nonshared environments (48). Neuroticism and momentary negative affect, on the other hand, are positively associated due to common genetic and nonshared environmental effects. This suggests that neuroticism may be a useful indicator of environmental risk for decreased daily positive affect and an environmental and genetic indicator of risk for increased daily negative affect (48).

2.5 **Emotion processes**

Behavior genetic research on emotion processes has largely focused on emotion regulation and related concepts, and to a lesser extent, the neurobiology of emotion processing. Emotion regulation is an important emotion process that encompasses the regulation of internal affect, external expressive behaviors, and emotion-related physiological reactions, attentional processes, cognition and motivation (49, 50). This process is crucial to adaptive functioning across the lifespan, yet has not often been studied using behavior genetic methodology (51). In a study of infant twins, there were modest genetic contributions to variations in infant gaze aversion (an emotion regulation tactic used by infants) to an unfamiliar model but not to a familiar model (i.e. mother) (52). Hence, nonshared environmental influences explained most of the variance in this emotion regulation strategy as it is just emerging. Genes, however, may play a larger role as emotion regulation develops. For example, in toddlerhood, genetic and nonshared influences accounted for 43% and 48%, respectively, of the variance in individual differences in emotion regulation observed within the lab (51). Genetic influences have also been found on related constructs such as effortful control (53, 54), physiological regulation (55), anger control (19), and affective intensity and lability (56). There are also hints of gene-by-environment interaction as indicated by the finding that cortisol reactivity (a potential physiological marker of emotion regulation) in infants was influenced by genetic and nonshared environmental factors under conditions of low familial adversity, but by only the environment (both shared and nonshared) under conditions of high familial adversity (57). In other words, when infants experience more "risky" environments, the impact of genes on physiological stress regulation is overshadowed by the environment. This research raises the possibility of finding gene-by-environment interactions in more traditional measures of emotion regulation.

The two twin studies that have explored emotional processes related to reward experience (i.e. the degree to which positive affect increases in response to pleasurable events) and stress sensitivity/perception (i.e. the degree to which negative affect increases in response to unpleasurable events) have mixed findings. When assessed by questionnaires tapping general tendencies related to these two processes, genetic factors explained roughly half of the variation (58). However, ESM measures of reward experience and stress sensitivity showed no familial resemblance (34). The authors suggest that the small sample may have resulted in an underpowered test of genetic effects. However, the fact that there was so little co-twin resemblance for either MZ or DZ twins raises the possibility of measurement issues. Ten assessments per day over five days may not be enough to capture reward experience and stress sensitivity fully. It may also be that ESM measures get at different aspects of these two constructs. More research using ESM methodology in larger samples over more extended periods of time is needed to inform about these possibilities.

As previously indicated, negative emotionality as a basic emotion category is genetically influenced. Negative urgency, an emotion *process* that reflects the tendency to engage in rash action in response to negative affect, is also genetically influenced, though less so than negative affect as an emotion. Specifically, in adults, genetic factors explained roughly 35% of the variation in negative urgency, with the nonshared environment explaining remaining variance. Overlap between negative urgency and negative affect was primarily due to genetic effects, although there were also genetic factors unique to each (59). Thus, the two are related, but distinct, constructs at the level of underlying etiology.

Neurobiological methods are starting to be applied to the study of emotion processing in twins. A common assumption of researchers interested in emotion processing is that variation in neural responses to emotional stimuli (e.g. response to affective stimuli such as faces and complex emotional scenes) is heritable, but this has only recently been examined empirically. When investigated in adult twins using affect-modulated event-related potentials, P300 neural responses to complex emotional scenes displayed modest to moderate genetic influences (heritabilities ranging from 0.22 to 0.30) with, once again, the nonshared environment explaining most of the remaining variance (60). Genetic influences have also been found for neural responses to emotionally salient facial stimuli (61). By integrating twin methodology and neurobiology, this research strengthens the notion of a neurobiological marker of emotional neural responsivity.

Although not a traditional twin study, neural and behavioral differences between MZ twins have also been studied. By examining differences between MZ twins, it is possible to get a better understanding of nonshared environmental influences on neurobiological measures. For example, behavioral, psychophysiological, and fMRI measures were assessed in a pair of MZ twins with amygdala dysfunction (62). Despite having the same pathology, one twin was more behaviorally and psychophysiologically "normal" (e.g. had a normal-sized social network, and an acoustic startle response). When examining neurological functioning using fMRI, the more "normal" twin showed responses to fearful faces in areas of the brain relevant to the cortical mirror-neuron systems, thereby highlighting the potential of this system to compensate adaptively for amygdala dysfunction in regard to social information. This finding based on a single pair of twins offers novel support for targeting mirror-neuron systems to help compensate for impaired emotion processing.

2.6 Links between emotion and psychopathology

There is an abundance of evidence showing genetic contributions to emotion-related psychopathology such as depression (63), anxiety (64), fears and phobias (65), and childhood behavior problems (66, 67). In addition to simply estimating genetic and environmental variances of emotion-related psychopathology, behavior genetic methods can also be used to explore genetic and environmental sources of *covariance* between emotion and psychopathology. That is, to what extent are there common genetic and environmental influences across the two domains?

Research looking at the overlap between negative emotionality and psychopathology has examined the genetic and environmental links between negative emotionality/affect and major depressive disorder and conduct disorder (23); internalizing and externalizing problems (68–70); attention and aggression problems (71); a general internalizing/externalizing bifactor (72); depressive symptoms, attachment-related anxiety and attachment-related avoidance in romantic relationships (73); and dysregulated eating (59). With regard to specific negative emotions, research has looked at links between anger and aggression (4), conduct problems (9), stressful life events (19), and borderline disorder (16). Fear and sadness have been etiologically linked with anxiety (7); and neuroticism with anxiety (74, 75), depression (76, 77), perceived stress (78), and obsessive compulsive symptoms (79). The overall pattern of results is similar across studies. That is, the links between negative emotion and psychopathology are almost always primarily due to genetic factors. When the environment does contribute to links between negative emotionality and psychopathology, it tends to be modest and of the nonshared variety. Common genetic influences suggest pleiotropic effects on psychopathology and negative emotion across development and highlights the potential of negative emotionality as a broad vulnerability factor that underlies various dimensions of psychopathology. As such, this research provides novel support for the view of psychopathology as a continuum. Furthermore, the genetic overlap between negative emotionality and dimensions of psychopathology, the high heritability of negative emotionality, and its relation to psychopathology in the general population, makes it a possible endophenotype for molecular research (22). Indeed, the utility of using temperament rather than psychopathology as the target behavior or trait in molecular genetic work has been previously suggested (80).

The pattern of genetic overlap between negative emotionality and psychopathology is so pervasive that it is most interesting when etiologic links between the two are *not* explained by genetic factors. This is the case with fear and sadness, and separation anxiety. Although genetic factors explained the association between over-anxious symptoms and both fear and sadness in infancy and early childhood, separation anxiety was entirely environmentally linked with fear and sadness (7). In other words, the shared and nonshared environments explained the phenotypic association between separation anxiety and fear and sadness, suggesting that intervening on separation anxiety in infancy and early childhood may be a way to begin addressing nascent emotional problems. This research also highlights that similar dimensions of psychopathology can have different underlying etiologic associations with emotion.

While most research on links between emotion and psychopathology has focused on negative emotionality, this does not provide a complete picture of how psychopathology maps onto emotion. This is particularly true given that we know negative and positive emotionality have, at least to a certain extent, unique etiologic underpinnings. By investigating positive emotionality in this context, research can provide novel implications for how we conceptualize and intervene on different dimensions of psychopathology. To our knowledge, only two twin studies have looked at links between positive affect or subjective well-being with psychopathology. In both cases, there was a negative association suggesting that positive emotionality may buffer against psychopathology. A benefit of using twins is the ability to elucidate the extent to which this protective aspect of positive emotionality is due to protective genes or environments. In contrast to research finding genetic links between negative emotionality and behavior problems, in early childhood positive affect showed only environmental links with internalizing problems (81). That is, it was common shared and nonshared environmental factors that increased positive affect and decreased internalizing problems. Although in early childhood the association between positive affect and psychopathology is due to the environment, in adolescence the overlap between subjective well-being and psychopathology was mainly genetic (82). Together these studies suggest that while positive emotionality may serve as a buffer for psychopathology, the mechanisms by which this works may differ across age.

2.7 **Conclusion and future directions**

Twin studies clearly indicate the importance of both genes and the environment in emotion across development. Nearly all emotions and emotion processes are heritable, and overlap amongst emotions and between emotions and psychopathology is primarily a result of shared genetic factors. Future work should continue exploring the etiologic overlap between different discrete and complex emotions to improve understanding of underlying structures and interrelations amongst them. Additional research on genetic and environmental links between psychopathology and emotion, particularly emotion processes and positive emotionality, is also needed.

The application of the twin design to the study of temperament has provided us with a foundation for understanding the etiology of emotion, and has since been expanded to inform on complex emotions, emotion processes, and links between emotion and psychopathology. However, there are a number of gaps in the literature. The bulk of twin research on emotion has relied on questionnaires. More lab-based methods with large samples are required to improve the definition of phenotypes. A greater focus on behavioral and neurological measurement in twin studies would allow for a more complex understanding of the genetic and environmental etiology of the trait versus state aspects of emotion and emotion processes, as well as increase our understanding of genetic liabilities to emotion-related neurobiological processes. In other words, using these lab-based behavioral and neurological approaches is crucial in moving beyond a questionnaire-focused research of emotion to one concentrating on more emotional complexities (e.g. emotion regulation; emotion processing). This will be particularly relevant to our understanding of the overlap between emotion and psychopathology, as well as the overlap with neurobiology. In fact, the necessity and vast potential of a thoughtful approach to integrating behavioral genetics and neuroscience has been advocated (83). Despite this, few twin studies have taken a neurobiological approach. This is most likely due to the high cost of the lab-based twin studies, as well as the lack of cohesion between an individual differences approach (twin studies) and the, until recently, primarily group differences approach of neurobiology. Future research should make a concerted effort to merge the two perspectives. Researchers in both realms will benefit.

References

1. **Plomin R, Chipuer HM, Neiderhiser JM.** Behavioral genetic evidence for the importance of nonshared environment. In: Hetherington EM, Reiss D, Plomin R, Hetherington EM, Reiss D, Plomin R (eds). Separate Social Worlds of Siblings: The Impact of Nonshared Environment on Development. Hillsdale, NJ: Lawrence Erlbaum Associates, Inc; 1994, pp. 1–31.
2. **Goldsmith HH, Campos JJ.** Toward a theory of infant temperament. In: Emde RN, Harmon RJ (eds). The Development of Attachment and Affiliative Systems [Internet]. Boston, MA: Springer; 1982 [cited 2016 Sep 7], pp. 161–93. Available from: <http://link.springer.com/chapter/10.1007/978-1-4684-4076-8_13>
3. **Shiner RL, Buss KA, McClowry SG, Putnam SP, Saudino KJ, Zentner M.** What is temperament now? Assessing progress in temperament research on the twenty-fifth anniversary of Goldsmith et al. (1987). Child Dev Perspect. 2012 Dec;**6**(4):436–44.
4. **Deater-Deckard K, Beekman C, Wang Z, Kim J, Petrill S, Thompson L**, et al. Approach/positive anticipation, frustration/anger, and overt aggression in childhood. J Pers. 2010 Jun;**78**(3):991–1010.
5. **Eggum-Wilkens ND, Lemery-Chalfant K, Aksan N, Goldsmith HH.** Self-conscious shyness: Growth during toddlerhood, strong role of genetics, and no prediction from fearful shyness. Infancy. 2015 Mar;**20**(2):160–88.
6. **Emde RN, Robinson JL, Corley RP, Nikkari D, Zahn-Waxler C.** Reactions to restraint and anger-related expressions during the second year. In: Emde RN, Hewitt JK, Emde RN, Hewitt JK (eds). Infancy to Early Childhood: Genetic and Environmental Influences on Developmental Change. New York, NY: Oxford University Press; 2001, pp. 127–40.
7. **Goldsmith HH, Lemery KS.** Linking temperamental fearfulness and anxiety symptoms: A behavior-genetic perspective. Biol Psychiatry. 2000 Dec 15;**48**(12):1199–209.
8. **Mullineaux PY, Deater-Deckard K, Petrill SA, Thompson LA, DeThorne LS.** Temperament in middle childhood: A behavioral genetic analysis of fathers' and mothers' reports. J Res Personal. 2009 Oct;**43**(5):737–46.
9. **Deater-Deckard K, Petrill SA, Thompson LA.** Anger/frustration, task persistence, and conduct problems in childhood: A behavioral genetic analysis. J Child Psychol Psychiatry. 2007 Jan;**48**(1):80–7.
10. **Gagne JR, Goldsmith HH.** A longitudinal analysis of anger and inhibitory control in twins from 12 to 36 months of age. Dev Sci. 2011 Jan;**14**(1):112–24.
11. **Goldsmith HH, Lemery KS, Buss KA, Campos JJ.** Genetic analyses of focal aspects of infant temperament. Dev Psychol. 1999 Jul;**35**(4):972–85.
12. **Clifford S, Lemery-Chalfant K, Goldsmith HH.** The Unique and Shared Genetic and Environmental Contributions to Fear, Anger, and Sadness in Childhood. Child Dev. 2015 Sep 1;**86**(5):1538–56.
13. **Goldsmith HH, Buss KA, Lemery KS.** Toddler and childhood temperament: Expanded content, stronger genetic evidence, new evidence for the importance of environment. Dev Psychol. 1997 Nov;**33**(6):891–905.
14. **Côté C, Beauregard M, Girard A, Mensour B, Mancini-Marïe A, Pérusse D.** Individual variation in neural correlates of sadness in children: A twin fMRI study. Hum Brain Mapp. 2007 Jun 1;**28**(6):482–7.
15. **Cates DS, Houston BK, Vavak CR, Crawford MH, Uttley M.** Heritability of hostility-related emotions, attitudes, and behaviors. J Behav Med. 1993;**16**(3):237–56.
16. **Distel MA, Roeling MP, Tielbeek JJ, van Toor D, Derom CA, Trull TJ**, et al. The covariation of trait anger and borderline personality: A bivariate twin-siblings study. J Abnorm Psychol. 2012 May;**121**(2):458–66.
17. **Kramer MD, Patrick CJ, Krueger RF, Gasperi M.** Delineating physiologic defensive reactivity in the domain of self-report: Phenotypic and etiologic structure of dispositional fear. Psychol Med. 2012 Jun;**42**(6):1305–20.
18. **Plomin R, Pedersen NL, McClearn GE, Nesselroade JR, Bergeman CS.** EAS temperaments during the last half of the life span: Twins reared apart and twins reared together. Psychol Aging. 1988 Mar;**3**(1):43–50.

19. **Wang X, Trivedi R, Treiber F, Snieder H.** Genetic and environmental influences on anger expression, John Henryism, and stressful life events: The Georgia Cardiovascular Twin Study. Psychosom Med. 2005 Jan;**67**(1):16–23.
20. **Blonigen DM, Carlson MD, Hicks BM, Krueger RF, Iacono WG.** Stability and change in personality traits from late adolescence to early adulthood: A longitudinal twin study. J Pers. 2008 Apr;**76**(2):229–66.
21. **Ganiban JM, Saudino KJ, Ulbricht J, Neiderhiser JM, Reiss D.** Stability and change in temperament during adolescence. J Pers Soc Psychol. 2008 Jul;**95**(1):222–36.
22. **Singh AL, Waldman ID.** The etiology of associations between negative emotionality and childhood externalizing disorders. J Abnorm Psychol. 2010 May;**119**(2):376–88.
23. **Tackett JL, Waldman ID, Van Hulle CA, Lahey BB.** Shared Genetic Influences on Negative Emotionality and Major Depression/Conduct Disorder Comorbidity. J Am Acad Child Adolesc Psychiatry. 2011 Aug;**50**(8):818–27.
24. **Emde RN, Plomin R, Robinson J, Corley R, DeFries J, Fulker DW**, et al. Temperament, emotion, and cognition at fourteen months: The MacArthur Longitudinal Twin Study. Child Dev. 1992 Dec;**63**(6):1437–55.
25. **Saudino KJ.** Special Article: Behavioral Genetics and Child Temperament. Dev Behav Pediatr. 2005 Jun;**26**(3):214–23.
26. **Cohen DJ, Dibble E, Grawe JM.** Fathers' and mothers' perceptions of children's personality. Arch Gen Psychiatry. 1977 Apr;**34**(4):480–7.
27. **Allan NP, Mikolajewski AJ, Lonigan CJ, Hart SA, Taylor J.** Examining the etiological associations among higher-order temperament dimensions. J Res Personal. 2014 Feb;**48**:51–60.
28. **Bartels M, Boomsma DI.** Born to be happy? The etiology of subjective well-being. Behav Genet. 2009 Nov;**39**(6):605–15.
29. **De Neve J-E, Christakis NA, Fowler JH, Frey BS.** Genes, economics, and happiness. J Neurosci Psychol Econ. 2012 Nov;**5**(4):193–211.
30. **Lykken D, Tellegen A.** Happiness is a stochastic phenomenon. Psychol Sci. 1996 May;**7**(3):186–9.
31. **Nes RB, Røysamb E, Tambs K, Harris JR, Reichborn-Kjennerud T.** Subjective well-being: Genetic and environmental contributions to stability and change. Psychol Med. 2006 Jul;**36**(7):1033–42.
32. **Stubbe JH, Posthuma D, Boomsma DI, De Geus EJC.** Heritability of life satisfaction in adults: A twin-family study. Psychol Med. 2005 Nov;**35**(11):1581–8.
33. **Jacobs N, Menne-Lothmann C, Derom C, Thiery E, van Os J, Wichers M.** Deconstructing the familiality of variability in momentary negative and positive affect. Acta Psychiatr Scand. 2013 Apr;**127**(4):318–27.
34. **Menne-Lothmann C, Jacobs N, Derom C, Thiery E, van Os J, Wichers M.** Genetic and environmental causes of individual differences in daily life positive affect and reward experience and its overlap with stress-sensitivity. Behav Genet. 2012 Sep;**42**(5):778–86.
35. **Strelau J, Zawadzki B, Oniszczenko W, Angleitner A, Riemann R.** Genetic and environmental determinants of emotions: Data based on cross-country twin studies on temperament. Pol Psychol Bull. 2002;**33**(1):5–13.
36. **Lemery-Chalfant K, Kao K, Swann G, Goldsmith HH.** Childhood temperament: Passive gene–environment correlation, gene–environment interaction, and the hidden importance of the family environment. Dev Psychopathol. 2013 Feb;**25**(01):51–63.
37. **Plomin R, Emde RN, Braungart JM, Campos J, Corley R, Fulker DW**, et al. Genetic change and continuity from fourteen to twenty months: The MacArthur Longitudinal Twin Study. Child Dev. 1993 Oct;**64**(5):1354–76.
38. **Hopwood CJ, Donnellan MB, Blonigen DM, Krueger RF, McGue M, Iacono WG**, et al. Genetic and environmental influences on personality trait stability and growth during the transition to adulthood: A three-wave longitudinal study. J Pers Soc Psychol. 2011 Mar;**100**(3):545–56.

39. **Buss AH, Plomin R.** Theory and measurement of EAS. Temperam Early Dev Personal Traits. 1984;98–130.

40. **Watson D, Clark LA, Carey G.** Positive and negative affectivity and their relation to anxiety and depressive disorders. J Abnorm Psychol. 1988 Aug;**97**(3):346–53.

41. **Eysenck HJ.** Student Selection by Means of Psychological Tests—a Critical Survey. Br J Educ Psychol. 1947 Feb 1;**17**(1):20–39.

42. **Lake RIE, Eaves LJ, Maes HHM, Heath AC, Martin NG.** Further evidence against the environmental transmission of individual differences in neuroticism from a collaborative study of 45,850 twins and relatives on two continents. Behav Genet. 2000 May;**30**(3):223–33.

43. **Rettew DC, Vink JM, Willemsen G, Doyle A, Hudziak JJ, Boomsma DI.** The Genetic Architecture of Neuroticism in 3301 Dutch Adolescent Twins as a Function of Age and Sex: A Study from the Dutch Twin Register. Twin Res Hum Genet. 2006 Feb;**9**(1):24–9.

44. **Saudino KJ, Gagne JR, Grant J, Ibatoulina A, Marytuina T, Ravich-Scherbo I,** et al. Genetic and environmental influences on personality in adult Russian twins. Int J Behav Dev. 1999 Jun;**23**(2):375–89.

45. **Yamagata S, Suzuki A, Ando J, Ono Y, Kijima N, Yoshimura K,** et al. Is the genetic structure of human personality universal? A cross-cultural twin study from North America, Europe, and Asia. J Pers Soc Psychol. 2006 Jun;**90**(6):987–98.

46. **Kandler C, Bleidorn W, Riemann R, Spinath FM, Thiel W, Angleitner A.** Sources of cumulative continuity in personality: A longitudinal multiple-rater twin study. J Pers Soc Psychol. 2010 Jun;**98**(6):995–1008.

47. **Laceulle OM, Ormel J, Aggen SH, Neale MC, Kendler KS.** Genetic and Environmental Influences on the Longitudinal Structure of Neuroticism A Trait-State Approach. Psychol Sci. 2013 Sep 1;**24**(9):1780–90.

48. **Jacobs N, van Os J, Derom C, Thiery E, Delespaul P, Wichers M.** Neuroticism explained? From a non-informative vulnerability marker to informative person–context interactions in the realm of daily life. Br J Clin Psychol. 2011 Mar;**50**(1):19–32.

49. **Eisenberg N, Fabes RA.** Emotion regulation and children's socioemotional competence. In: Balter L, Tamis-LeMonda CS, Balter L, Tamis-LeMonda CS, (eds). Child Psychology: A Handbook of Contemporary Issues, 2nd edn. New York, NY: Psychology Press; 2006, pp. 357–81.

50. **Gross JJ, Thompson RA.** Emotion regulation: Conceptual foundations. In: Gross JJ, Gross JJ, (eds). Handbook of Emotion Regulation. New York, NY: Guilford Press; 2007, pp. 3–24.

51. **Wang M, Saudino KJ.** Genetic and environmental influences on individual differences in emotion regulation and its relation to working memory in toddlerhood. Emotion. 2013 Dec;**13**(6):1055–67.

52. **Soussignan R, Boivin M, Girard A, Pérusse D, Liu X, Tremblay RE.** Genetic and environmental etiology of emotional and social behaviors in 5-month-old infant twins: Influence of the social context. Infant Behav Dev. 2009 Jan;**32**(1):1–9.

53. **Gagne JR, Saudino KJ.** Wait for it! A twin study of inhibitory control in early childhood. Behav Genet. 2010 May;**40**(3):327–37.

54. **Goldsmith HH, Pollak SD, Davidson RJ.** Developmental neuroscience perspectives on emotion regulation. Child Dev Perspect. 2008 Dec;**2**(3):132–40.

55. **Boomsma DI, Van Baal GC, Orlebeke JF.** Genetic influences on respiratory sinus arrhythmia across different task conditions. Acta Genet Medicae Gemellol Twin Res. 1990;**39**(2):181–91.

56. **Coccaro EF, Ong AD, Seroczynski AD, Bergeman CS.** Affective intensity and lability: Heritability in adult male twins. J Affect Disord. 2012 Feb;**136**(3):1011–6.

57. **Ouellet-Morin I, Boivin M, Dionne G, Lupien SJ, Arsenault L, Barr RG,** et al. Variations in heritability of cortisol reactivity to stress as a function of early familial adversity among 19-month-old twins. Arch Gen Psychiatry. 2008 Feb;**65**(2):211–8.

58. **Bogdan R, Pizzagalli DA.** The heritability of hedonic capacity and perceived stress: A twin study evaluation of candidate depressive phenotypes. Psychol Med. 2009 Feb;**39**(2):211–8.

59. **Racine SE, Keel PK, Burt SA, Sisk CL, Neale M, Boker S**, et al. Exploring the relationship between negative urgency and dysregulated eating: Etiologic associations and the role of negative affect. J Abnorm Psychol. 2013 May;**122**(2):433–44.
60. **Weinberg A, Venables NC, Proudfit GH, Patrick CJ.** Heritability of the neural response to emotional pictures: Evidence from ERPs in an adult twin sample. Soc Cogn Affect Neurosci. 2015 Mar;**10**(3):424–34.
61. **Anokhin AP, Golosheykin S, Heath AC.** Heritability of individual differences in cortical processing of facial affect. Behav Genet. 2010 Mar;**40**(2):178–85.
62. **Becker B, Mihov Y, Scheele D, Kendrick KM, Feinstein JS, Matusch A**, et al. Fear Processing and Social Networking in the Absence of a Functional Amygdala. Biol Psychiatry. 2012 Jul 1;**72**(1):70–7.
63. **Sullivan PF, Neale MC, Kendler KS.** Genetic epidemiology of major depression: Review and meta-analysis. Am J Psychiatry. 2000 Oct;**157**(10):1552–62.
64. **Shimada-Sugimoto M, Otowa T, Hettema JM.** Genetics of anxiety disorders: Genetic epidemiological and molecular studies in humans. Psychiatry Clin Neurosci. 2015 Jul;**69**(7):388–401.
65. **Van Houtem CMHH, Laine ML, Boomsma DI, Ligthart L, van Wijk AJ, De Jongh A.** A review and meta-analysis of the heritability of specific phobia subtypes and corresponding fears. J Anxiety Disord. 2013 May;**27**(4):379–88.
66. **Bartels M, van Beijsterveldt CEM, Derks EM, Stroet TM, Polderman TJC, Hudziak JJ**, et al. Young Netherlands Twin Register (Y-NTR): A longitudinal multiple informant study of problem behavior. Twin Res Hum Genet. 2007 Feb;**10**(1):3–11.
67. **Saudino KJ, Carter AS, Purper-Ouakil D, Gorwood P.** The etiology of behavioral problems and competencies in very young twins. J Abnorm Psychol. 2008 Feb;**117**(1):48–62.
68. **Mikolajewski AJ, Allan NP, Hart SA, Lonigan CJ, Taylor J.** Negative affect shares genetic and environmental influences with symptoms of childhood internalizing and externalizing disorders. J Abnorm Child Psychol. 2012 Sep 26;**41**(3):411–23.
69. **Schmitz S, Fulker DW, Plomin R, Zahn-Waxler C, Emde RN, DeFries JC.** Temperament and problem behavior during early childhood. Int J Behav Dev. 1999 Jun;**23**(2):333–55.
70. **Taylor J, Allan N, Mikolajewski AJ, Hart SA.** Common genetic and nonshared environmental factors contribute to the association between socioemotional dispositions and the externalizing factor in children. J Child Psychol Psychiatry. 2013 Jan;**54**(1):67–76.
71. **Gjone H, Stevenson J.** A longitudinal twin study of temperament and behavior problems: Common genetic or environmental influences? J Am Acad Child Adolesc Psychiatry. 1997 Oct;**36**(10):1448–56.
72. **Tackett JL, Lahey BB, van Hulle C, Waldman I, Krueger RF, Rathouz PJ.** Common genetic influences on negative emotionality and a general psychopathology factor in childhood and adolescence. J Abnorm Psychol. 2013 Nov;**122**(4):1142–53.
73. **Franz CE, York TP, Eaves LJ, Prom-Wormley E, Jacobson KC, Lyons MJ**, et al. Adult romantic attachment, negative emotionality, and depressive symptoms in middle aged men: A multivariate genetic analysis. Behav Genet. 2011 Jul;**41**(4):488–98.
74. **Hettema JM, Prescott CA, Kendler KS.** Genetic and environmental sources of covariation between generalized anxiety disorder and neuroticism. Am J Psychiatry. 2004 Sep;**161**(9):1581–7.
75. **Jardine R, Martin NG, Henderson AS, Rao DC.** Genetic covariation between neuroticism and the symptoms of anxiety and depression. Genet Epidemiol. 1984 Jan 1;**1**(2):89–107.
76. **Hansell NK, Wright MJ, Medland SE, Davenport TA, Wray NR, Martin NG**, et al. Genetic co-morbidity between neuroticism, anxiety/depression and somatic distress in a population sample of adolescent and young adult twins. Psychol Med. 2012 Jun;**42**(6):1249–60.
77. **Kendler KS, Neale MC, Kessler RC, Heath AC, Eaves LJ.** A longitudinal twin study of personality and major depression in women. Arch Gen Psychiatry. 1993 Nov;**50**(11):853–62.
78. **Rietschel L, Zhu G, Kirschbaum C, Strohmaier J, Wüst S, Rietschel M**, et al. Perceived stress has genetic influences distinct from neuroticism and depression. Behav Genet. 2014 Nov;**44**(6):639–45.

79. **Bergin J, Verhulst B, Aggen SH, Neale MC, Kendler KS, Bienvenu OJ**, et al. Obsessive compulsive symptom dimensions and neuroticism: An examination of shared genetic and environmental risk. Am J Med Genet B Neuropsychiatr Genet. 2014 Dec;**165**(8):647–53.

80. **Belmaker RH, Biederman J.** Genetic markers, temperament, and psychopathology. Biol Psychiatry. 1994 Jul;**36**(2):71–2.

81. **Wang M, Saudino KJ.** Positive affect: Phenotypic and etiologic associations with prosocial behaviors and internalizing problems in toddlers. Front Psychol [Internet]. 2015 Apr 9 [cited 2016 Apr 1];6. Available from: <http://www.ncbi.nlm.nih.gov/pmc/articles/PMC4391446/>.

82. **Bartels M, Cacioppo JT, van Beijsterveldt TCEM, Boomsma DI.** Exploring the association between well-being and psychopathology in adolescents. Behav Genet. 2013 May;**43**(3):177–90.

83. **Vaidyanathan U, Vrieze SI, Iacono WG.** The power of theory, research design, and transdisciplinary integration in moving psychopathology forward. Psychol Inq. 2015 Jul;**26**(3):209–30.

Chapter 3

Gene–environment interactions in humans across multiple units of analyses: A focus on psychopathology and imaging

Suzanne Vrshek-Schallhorn, Bradley M Avery, and Vaibhav Sapuram

3.1 Introduction

Imagine two hypothetical children—Casey and Sam—who both experience early emotional abuse. But as they develop, Casey shows resilience and goes on to thrive, while Sam struggles under this burden, and experiences depression in a downward spiral that may even plague her life course.

Understanding this variability in responding to circumstances across the lifespan has captured the attention of psychopathology researchers, the press, and the public alike. In an era of advancing molecular genetic sophistication, this attention has led to a singular question: Can we identify common genetic differences that aid in predicting how people will respond to their circumstances?

Such is the question at the heart of gene–environment interaction (G×E) research in psychopathology. G×Es characterize how genetic differences influence how people respond to environmental conditions, or vice versa. Measured G×E research capitalizes on behavioral genetic research showing that genetic factors contribute at least moderately to psychopathology, and lends specificity by identifying the individual common genetic variations involved.

The present chapter examines four overarching topics: theoretical models undergirding G×E research, a brief history of G×E and its controversies, the current state of G×E research and emerging approaches, and future directions. We give particular attention to how G×E research examines an array of outcomes in order to inform not only associations of genetic differences with diagnoses, but also their mechanisms. Within the field of psychopathology, G×E research examines proximal outcomes (physiological variables) and distal outcomes (affect, symptoms, and diagnoses) in an effort to inform a nomological network across multiple units of analysis. Among the most compelling of these intermediate outcomes have been physiological variables that cannot readily be biased by research participants—neuroendocrine responses to lab-induced stress and patterns of brain activity in functional magnetic resonance imaging.

Within this framework, we focus on the candidate-gene perspective (testing a priori hypotheses about specific genetic differences) as opposed to the genome-wide association perspective (examining genetic differences throughout the genome in case-control comparisons without specific hypotheses). Similarly, we focus on G×E interactions in most cases, except when research on the main effects of genetic variables is particularly informative.

Ultimately, the hope of G×E research in psychopathology is to use identified genetic risk factors to ameliorate suffering in a number of ways. We may one day be able to screen individuals and identify circumstances placing them at elevated risk, shunting individuals such as "Sam" in our opening example toward preventive interventions before pathology takes hold. Similarly, we may also improve our understanding of how genetic factors enable individuals to benefit more robustly than their peers from *positive* environments. But perhaps the greatest hope is that learning about the genetic risk factors and what environments trigger their actions will shed light on the full etiology of both disorders and positive developmental outcomes and lead to markedly enhanced interventions.

3.2 Gene–environment interaction: Theoretical models

Complex behavioral and health phenomena rarely boil down to the direct influence of individual genes (1), hindering the discovery of genes "responsible" for pathology. In light of this, two primary theoretical models are key for contemplating the role of genes in mental health and beyond.

Vulnerability-stress or *diathesis-stress models* suggest a latent biological vulnerability to psychopathology that is activated by a particular trigger or "stressor" (for a review, see 2). In this view, biological vulnerability is necessary but not sufficient to develop psychopathology. For example, although depression is often precipitated by a stressful life event (e.g. 3), if biological vulnerability to depression (e.g. genetic risk) is not present, depression is unlikely to develop despite that stress. Similarly, one may possess biological vulnerability but experience minimal stress, making it unlikely that one would develop depression.

An offshoot of the vulnerability-stress model is the *differential susceptibility model*, which proposes that one's genetic makeup may heighten sensitivity to both good and bad aspects of environmental influences, leading to both heightened positive and heightened negative outcomes (4). Accordingly, if an individual possesses genetic "risk" for psychopathology in the face of adverse life circumstances, in an enriching, positive environment, the same individual may flourish more than their counterparts. Although both diathesis-stress and differential susceptibility models predict interactive G×E effects (i.e. moderation), they differ in the precise form the interaction takes, as well as their implications for measurement of the environment and phenotypes (Figure 3.1).

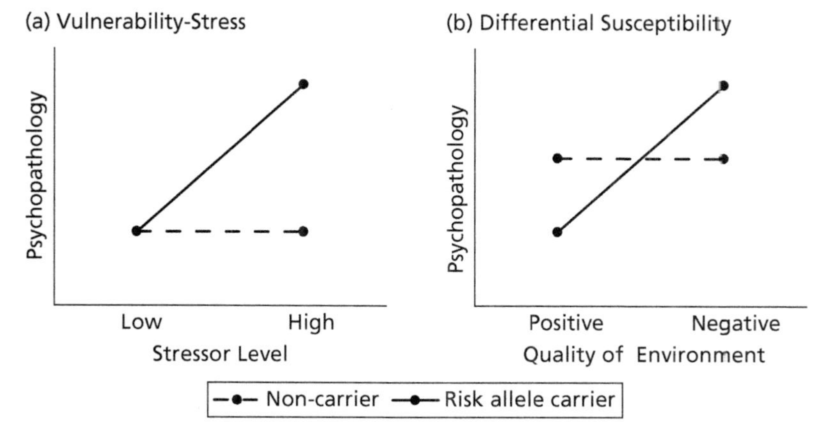

Figure 3.1 (a) Shows vulnerability-stress models in which a risk allele moderates the effect of stressful life events on risk for psychopathology. (b) Shows differential susceptibility models in which genotype differentially affects the outcome dependent on type of environment.

3.3 Brief history of gene–environment interaction and controversies

A handful of essential studies launched the field of measured G×E research. Caspi and colleagues (5) found that a polymorphism in gene encoding the monoamine oxidase A enzyme (MAOA)—relevant in metabolizing monoamines such as norepinephrine, serotonin, and dopamine—moderated the relationship between early childhood maltreatment and antisocial behavior in males. Also, in a report exploring the association between cannabis use and later development of schizophrenia, Caspi and colleagues (6) found that valine (Val) allele carriers of a polymorphism located in the catechol-*O*-methyltransferase (COMT) gene—critical in catabolizing catecholamines—were more susceptible to symptoms of psychosis as a function of their cannabis use than those homozygous for the methionine (Met) allele.

But perhaps the most influential, if controversial, of these early studies, was a report that carriers of the less transcriptionally efficient short ("S") allele for a functional polymorphism, 5-HTTLPR, in the serotonin transporter gene promoter were more susceptible to depression symptoms, diagnoses, and suicidal ideation after stressful life experiences than were long allele ("L") homozygotes (67). An earlier study on the 5-HTTLPR polymorphism also revealed that S-carriers also exhibited greater amygdala reactivity to fearful stimuli versus fixation in an fMRI paradigm, as compared to their L/L counterparts, providing an intermediate pathway by which the S-allele may heighten risk for psychopathology (7). Research into the 5-HTTLPR polymorphism, while captivating researchers and lay-people alike, has been deeply controversial. Conflicting meta-analyses add to the controversy (Table 3.1). Taken together, meta-analyses with fewer papers have generally drawn negative conclusions, while more inclusive meta-analyses have drawn positive conclusions. Regarding the status of G×E research broadly, we provide a comprehensive table of the findings of meta-analyses of G×E investigations in mental and physical health outcomes (Table 3.1).

Table 3.1. Extant G×E meta-analyses in mental and physical health

Citation	Genetic Target	# Studies Examined	Environmental Target and Key Moderators	Summary of Conclusions
Risch et al., 2009 (58)	Serotonin Transporter Gene (5HTTLPR)	14	Ordinal scale of stressful life events	No main effect or G×E effect of 5HTTLPR on depression
Munafò et al., 2009 (59)	Serotonin Transporter Gene (5HTTLPR)	5	Stressful life events dichotomized into absence or presence	No evidence that 5HTTLPR is associated with depression independent of or in interaction with life stress
Karg et al., 2011 (60)	Serotonin Transporter Gene (5HTTLPR)	54	Stressors by type and assessment method by type	Support for a 5HTTLPR G×E effect on depression. More robust associations for interview & objective stress assessments
Sharpley et al., 2014 (61)	Serotonin Transporter Gene (5HTTLPR)	81	Stressors by type and assessment method	Support for a 5HTTLPR G×E effect on depression. More robust associations for interview and objective stress assessments

Table 3.1. Continued

Citation	Genetic Target	# Studies Examined	Environmental Target and Key Moderators	Summary of Conclusions
van Ijzendoorn et al., 2012 (62)	Serotonin Transporter Gene (5HTTLPR)	30	Negative vs positive environments	Partial support for differential susceptibilityhypothesis. S-carriers showed more negative outcomes in adverse environments, and in primarily Caucasian samples, S-carriers also showed *better* outcomes in positive environments.
Kim-Cohen et al., 2006 (63)	Monoamine Oxidase-A (MAOA)	5	Physical abuse	In males, low activity MAOA associated with antisocial behaviors when preceded by early adversity.
Byrd & Manuck, 2014 (64)	Monamine Oxidase-A (MAOA)	27	Maltreatment or other childhood adversities	In males, ow activity genotype associated with antisocial behaviors following early adversity. In females, no overall relationship but opposite effect with high activity genotype inked with antisocial behavior following early adversity
Marcus et al., 2000 (65)	*N*-acetyltransferase 2 (NAT2)	16	Tobacco Use	Slow acetylators show a stronger relationship between c garette smoking and bladder cancer risk than fast acetylators
Zeiger et al., 2005 (66)	*Taq1*, Transforming Growth Factor α	5	Maternal cigarette smoking	Association of maternal smoking with cleft palate stronger in C2-carriers. No effect for cleft lip.

Note: Meta-analyses focused on main effects of genetic variants rather than interactions, such as main effects of variants found in catechol-*O*-methyl transferase and FKPB5, are not included here.

3.4 Current approaches in psychopathology G×E research

In the roughly 15 years since the first publication of a measured G×E effect in humans, there have been significant advancements in G×E research. Among these advancements are giving greater theoretical consideration to the selection of environment, to development, to bidirectional relationships between stress and depression (i.e. stress generation), and to additive genetic risk.

3.4.1 Interaction with what environment?

In a scathing review of 5-HTTLPR G×E depression research, life stress experts Monroe and Reid (8) decried the relative lack of care devoted to measuring life stress in studies of the

5-HTTLPR G×E interaction. In particular, they note the well-known poor validity of commonly used life event checklists, as well as the confounding of the predicted outcome with the participant's perception of life stress. As in other papers focused on methodological recommendations for the measurement of recent life stress (e.g. 9), they recommend using objective life stress interviews. They advocate the view that it is important to conceptualize stress as in the definition derived from physics—of pressure pushing down on an organism—rather than stress as defined in the Selye tradition, where stress represents the organism's response to that pressure. Defining stress as the organism's response to adversity confounds stress with its outcome, risk for psychopathology.

Beyond improving the quality of measurement, we take the perspective that G×E research can benefit from devoting as much care to selecting the "candidate environment" as most give to selecting candidate genes. For example, Vrshek-Schallhorn and colleagues (10) showed in two samples of emerging adults studied with repeated diagnostic and life stress interviews for five years, that interpersonal chronic and major episodic stress consistently uniquely predicted depressive episode onsets over and above non-interpersonal chronic and episodic stress. Moreover, the particular stressor mattered in a related G×E investigation: interpersonal major stressful life events and one form of chronic interpersonal stress (in family relationships) interacted with 5-HTTLPR genotype to predict major depressive episode onsets as hypothesized, while other forms of stress did not (11). We would never expect the same environment that elicits depression necessarily also to elicit schizophrenia, diabetes, or cancer, but we must apply the same scrutiny to our assumptions within disorders to boost effect sizes and power for G×E effects, as well as to clarify explanatory models.

3.4.2 Importance of developmental considerations.

Recent work also highlights the need for developmentally sensitive research designs in G×E endeavors. Evidence supports changes across development in all aspects of G×E effects: the amount and nature of genetic contributions to risk, the salience of particular stressors, and the manifestation of the outcome of interest.

First, a meta-analysis focused on behavioral genetic studies of adolescence and young adulthood demonstrated that levels of heritability (i.e. studied as main effects without considering G×E interactions) significantly increase between approximately age 10 and age 25 for externalizing behaviors (acting out) and internalizing (anxious, depressive) behaviors, but stay roughly static for attention-deficit hyperactive disorder (12). Furthermore, not only does level of heritability vary, but the genetic variants that contribute to these aggregate estimations also change. Using a developmental twin design and latent variable modeling, Kendler, Gardner, and Lichtenstein (13) provided evidence for a common genetic factor that contributes to symptoms of anxiety and depression, in an attenuating fashion from pre-adolescence through early adulthood. Further, during early adolescence, late adolescence, and early adulthood, new sets of risk genes began contributing at each stage—although *which* genes come online or attenuate in their contribution to risk is as yet untested. Few studies consider how specific molecular genetic risk factors vary in their contributions to psychological functioning across development, an area to which we ought to devote increased attention as we cultivate larger samples spanning developmentally sensitive periods.

Second, there is also evidence that the salience of potential stressors changes over development. For example, in a G×E investigation, romantic involvement at age 15 predicted concurrent depression symptoms, but the same was not true at age 20, when romantic involvement is more developmentally normative (14). In this investigation, romantic involvement at age 15 interacted with 5-HTTLPR genotype to predict age-20 depression symptoms. Future efforts ought to probe the developmental salience of stressors further in order to tailor the "candidate environment" in a better way.

Third, appropriate outcomes to measure vary over the course of development. For example, one G×E study predicted diagnoses of depression in older adolescents but focused on prediction of the severity of peak *symptom* onsets in early adolescent girls when full onsets are too rare to detect with sufficient power (15). Thus as predicted, Hankin and colleagues (16) showed that *late* adolescents demonstrated a significantly stronger G×E effect 5-HTTLPR with peer stress (lack of a supportive social circle) on depression diagnoses (broadly defined to include minor depressive episodes) than did *early* adolescents.

Finally, evidence of G×E effects across the developmental spectrum strengthens arguments for the influence of particular genes on a given outcome. In an example outside of psychopathology, additive risk from a collection of genetic variants identified through genome-wide association studies for obesity predicted greater weight gain in infancy, faster and higher adiposity rebound in childhood, and chronic obesity during adulthood (17). Similarly, an additive genetic risk variable developed from meta-analyses of nicotine dependence did not predict trying nicotine, but did predict converting to smoking daily in the teen years, more rapid progression to and longer duration of heavy smoking, and greater failure rates when attempting to quit, among other outcomes (18). Ultimately, if G×E research is to realize its full potential, it must be mindful of developmental considerations.

3.4.3 **Polygenic approaches to genetic and G×E research**

An exciting new direction for G×E research in psychopathology grew from the evidence supporting behavioral genetic assumptions: many common genetic variants each contribute small amounts of risk for pathology, and act together in an additive rather than multiplicative fashion to increase risk (19, 20).[1] Interest in additive genetic risk has led to the development of polygenic, additive risk variables. These are known by several names including polygenic risk scores and multilocus genetic risk profile scores (including biologically informed multilocus profile scores). All comprise either weighted or unweighted total numbers or proportions of "risk" alleles across multiple polymorphisms.

Genetic main effects research has examined these with respect to a wide array of outcomes spanning physical and mental health including obesity (17), nicotine dependence (18), as discussed at section 3.4.2, ventral-striatal neural reactivity to reward (21), high cholesterol levels (22), childhood intelligence (23), and intellectual and economic outcomes across development (24), among others. Many such reports draw on empirically indicated candidates from genome-wide association studies for their additive genetic risk variables, but others draw on theoretically or biologically indicated candidate genetic variants. For example, Nikolova and colleagues (21) developed their biologically informed dopaminergic multilocus risk profile score using five putatively functional polymorphisms from the DA system, which collectively account for almost 11% of variance in between-person ventral striatal fMRI reward-related activity.

[1] It merits noting that beyond additive effects of common variants, there is evidence for minor multiplicative effects (52, 53) leading to investigations in gene-gene interactions (54, 55). Further, for some mental health conditions with especially large genetic contributions, such as schizophrenia and autism spectrum disorder, there is also evidence for supporting contributions from rare genetic variants idiosyncratic to particular families (56), as well as contributions from de novo mutations (57).

Polygenic approaches are also being applied in G×E investigations. In the first application of polygenic risk in a G×E, Pagliaccio and colleagues (25) showed that additive risk from ten SNPs located in four stress-relevant genes (*CRHR1, NR3C2, NR3C1*, and *FKBP5*) predicted greater cortisol levels under lab stress, and interacted with early-life adversity to predict left hippocampal and amygdala volumes. In the first to apply this approach to diagnoses of psychopathology in a G×E test, Salvatore and colleagues (50) used a genome-wide polygenic score for externalizing disorders derived from an adult sample. This polygenic score significantly interacted with two risk factors for adolescent externalizing disorders (high peer substance use and low parental monitoring) to predict externalizing disorders in a separate sample of adolescents. Further, a novel serotonergic multilocus profile score from four functional SNPs and one meta-analytically indicated SNP, interacted with recent major interpersonal stressful events to predict major depressive episode onsets in emerging adults, and in a replication sample, interacted with recent interpersonal stress severity to predict peak past-year onsets of depressive symptoms in adolescent girls (15). Of note, in this report, higher numbers of so-called risk alleles both significantly increased risk for elevated depression symptoms under more stressful conditions, but protected *against* symptoms under less stressful conditions, consistent with differential susceptibility theoretical models (4).

3.4.4 Stress generation in G×E research

While diathesis-stress models in which stress precedes negative outcomes have received much attention, *stress generation* is also key to understanding and testing G×E effects. The stress generation model posits that individuals with depression (or in some studies, risk for depression) actively generate stress for themselves through their cognitive styles, attachment styles, traits, and behaviors (26, 27). According to this conceptualization, these individuals experience a greater number of dependent (i.e. self-induced) life stressors than their counterparts (26). Importantly, diathesis-stress and stress generation models are not mutually exclusive, and can be conceptualized in a bidirectional or transactional model (see Figure 3.2) whereby stress influences the development of depression, and depressive symptoms or characteristics generate greater number of stressors, perpetuating the cycle of depression.

The genetic underpinnings of stress generation have been explored in two studies of the 5-HTTLPR polymorphism, finding that S-carriers had a stronger relationship between depression at age 15 and dependent and interpersonal stressful life events at age 20 (28). Additionally, S-carriers experiencing low relational security at age 15 experienced more stressful life events at age 20. However, S-carriers experiencing high relational security at age 15 experienced *fewer* stressful life events at age 20 than their L/L counterparts (29), consistent with a differential susceptibility model.

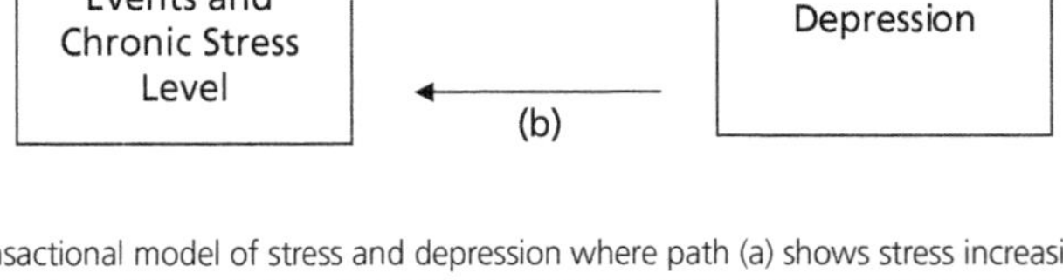

Figure 3.2

Note: Bidirectional/transactional model of stress and depression where path (a) shows stress increasing risk for depression, and path (b) shows stress generation whereby depression or depressogenic characteristics increases number of stressful life events or level of chronic stress.

3.5 Historical and current fMRI genetics

An emerging subfield examines G×E effects on neurobiological processes implicated in psychiatric disorders. Although this subfield encompasses diverse methodological paradigms, all seek to further our understanding of the transdiagnostic mechanisms that contribute to maladaptive behavior, from gene to cell to brain to behavior. Two particularly exciting paradigms within this subfield are imaging genetics (30), and imaging G×E (iGxE; 31), which use neuroimaging techniques to examine the influence of genetic main effects and G×E interactions respectively on neural features. Here we briefly review imaging genetics and iG×E studies involving 5-HTTLPR, as well as some methodological concerns specific to iG×E research. This section is not a comprehensive review of imaging genetics or iG×E studies broadly (for such a review, see 32), or even neuroimaging studies involving 5-HTTLPR specifically. Rather, it seeks succinctly to provide examples of how these methodological paradigms have been used to explore biological mechanisms linking genetic variants to maladaptive behaviors.

3.5.1 5-HTTLPR and imaging genetics

Given the plethora of research investigating behavioral effects of the 5-HTTLPR polymorphism, it is not surprising that this polymorphism has also been the focus of numerous imaging genetics and iG×E studies. Much research on 5-HTTLPR has focused on its effect on the amygdala (33), which exhibits markedly increased activity in response to threatening stimuli (34). For example, as noted in section 3.3, one of the earliest imaging genetics study found that S-carriers exhibited increased amygdala reactivity to angry and fearful faces relative to L/L individuals, and proposed that this mechanism may mediate the association between the 5-HTTLPR polymorphism and symptoms of psychopathology (7). Since the publication of this study, several other groups have replicated this finding (e.g. 35), and a meta-analysis indicated that 5-HTTLPR genotype accounts for up to 10% of the variance in amygdala reactivity to threatening stimuli (36). The 5-HTTLPR S-allele has also been associated with decreased grey matter volume in the amygdala and the perigenual anterior cingulate cortex (pACC), as well as decreased functional connectivity of these two regions, which was significantly associated with self-reported anxiety (37). Thus, it appears that an amygdala–pACC circuit may play an important role in internalizing symptoms, and may be affected by differences in serotonergic function conferred by 5-HTTLPR genotype (33).

3.5.2 5-HTTLPR in iG×E

Several iG×E studies have examined whether the 5-HTTLPR polymorphism interacts with life stress to differentially predict neural reactivity. For example, the first study to utilize the iG×E paradigm found that S-carriers exhibited increased resting amygdala and hippocampal activity in response to increasing levels of life stress, while L/L individuals exhibited decreased resting amygdala activity in response to increasing life stress (38). These researchers similarly found that S-carriers exhibited a significant positive association between rumination and life-stress, while L/L individuals exhibited the opposite pattern. These findings suggest that amygdala and hippocampal function may play an important role in linking 5-HTTLPR genotype and life stress to rumination, a critical risk factor for internalizing disorders (39). Additional studies have found that the short (S) allele is associated with greater increases in amygdala reactivity to fearful faces and amygdala–hypothalamus functional connectivity (40), as well as decreased hippocampal volume (41), in the context of greater life stress. Given evidence that stimulation of the amygdala elicits increased HPA activity, one interpretation of these findings is that the 5-HTTLPR S-allele might be associated with increased cortisol reactivity to threat (42) via its effects, in interaction with life stress, on the amygdala and amygdala–hypothalamus functional connectivity (40).

3.5.3 Methodological challenges for iG×E

Although iG×E paradigms have many unique benefits, they also present unique methodological challenges (for a review, see 31). First, although effect sizes in imaging genetics and imaging G×E studies are generally larger than those observed in G×E studies using diagnostic outcomes, they are still likely to be quite small (32, 43). Thus, the sample sizes required to obtain adequate statistical power can be very large: researchers have previously estimated that the analyses necessary to test mechanistic hypotheses in iG×E research require at least 500–1000 participants (31). Second, neuroimaging is expensive, compounding challenges caused by large sample size requirements. Third, participant recruitment for iG×E studies can be difficult. Participants must pass safety screenings for neuroimaging (e.g. many medical devices are barred), consent to both DNA analyses and neuroimaging, commit hours and often multiple visits to the lab, and disclose personal details to research staff during stress assessments. Fourth, as discussed in section 3.4.1), accumulating research suggests that interview measures more accurately assess life stress than checklist measures (8); however, these interviews are time and personnel intensive.

Because of these factors, iG×E research requires an enormous investment of resources on the part of research teams. However, these challenges are not insurmountable; several independent research teams have implemented iG×E paradigms in the past decade. Furthermore, an emerging zeitgeist of inter-institution collaboration and data pooling has resulted from acknowledgement of these challenges (32), and researchers who conduct iG×E research should strongly consider participating in such collaborations to obtain larger sample sizes and increase replicability. Last, recent theoretical advances such as the use of biologically informed multilocus profile scores in imaging genetics (e.g. 21) have resulted in increased statistical power (44), which may ameliorate these challenges.

3.6 Future directions

In addition to the emerging directions noted, we highlight four additional directions for future focus: transdiagnostic considerations, accounting for the role of culture, characterizing main effects of risky environments for less studied forms of psychopathology, and bolstering quality research practices.

3.6.1 Transdiagnostic G×E

Recently, funding agency priorities have fostered increased support for examining transdiagnostic dimensions (e.g. negative valence systems) across multiple units of analysis, ranging from genetics to neural activation to behavior (45). Such an approach acknowledges the need to "carve nature at its joints," and echoes many genetic researchers' longstanding interest in intermediate outcomes or "endophenotypes" (e.g. 46) that appear more proximal to biological (and thus genetic) functioning than more distal diagnoses of psychopathology. Future G×E research ought to increasingly apply such an approach to examining the impact of genetic variation not only on complex phenotypes such as neural activation, but also more mundane ones, such as affect in naturalistic settings, measured through experience sampling and daily diary methodologies. Evidence across a full spectrum of outcomes bolsters the nomological network for G×E effects.

3.6.2 Culturally-informed G×E

To date, G×E research has largely been conducted in high-income countries (68), with focus on individual variation at the expense of population-level characteristics including culture. We advocate a broader approach, accounting for the role of culture and its influence on psychopathological processes. The relativist viewpoint suggests that cultural context shapes both biological and psychological development, and that this culturally driven variation in development influences experience of psychopathology; in accord, evidence suggests that large variation exists in risk and protective factors for psychopathology across cultures (47). For example, in an analysis of cross-national samples in 29 countries from 124 prior publications, 5-HTTLPR S-allele frequency was higher in countries with a collectivist rather than individualistic culture. S-carrier frequency predicted lower mood and anxiety disorder prevalence, and this association was mediated by collectivist tendencies (48). This discrepancy in risk for psychopathology, explained by cultural differences, highlights the importance of cross-cultural consideration when interpreting G×E research. Thus, we suggest G×E interaction research moves forward with particular attention to replication across various ethnic and cultural groups, while attending to population stratification concerns, with the goal of identifying cultural variation in risk and protective factors.

3.6.3 Future directions for environmental conceptualization

An important precursor to G×E research for many forms of psychopathology will be initially defining the environmental precipitant. Many forms of psychopathology lack research on specific types of environmental hazards contribute to disorder onsets. We know most about environmental contributors to depression (e.g. major stressful life events, chronic stress; 10), schizophrenia (e.g. perinatal complications, 49), and externalizing disorders (e.g. peer models and insufficient supervision, 50), but have only hints about provocateurs of the anxiety disorders, despite widely held diathesis-stress models for these conditions. Defining environmental risk factors is an essential step.

3.6.4 Future directions for research practices

G×E research must also heed criticism of tenuous research practices, including underpowered sample sizes (51) and analytic flexibility. As other research areas move toward more transparent research practices, G×E research ought to lead the way. Finally, we urge greater collaboration that may make large sample sizes possible while preserving use of rigorous environmental and outcome measures.

3.7 Conclusion

Taken together, although gene–environment interaction research in psychopathology is in its relative infancy as a scientific discipline (just over 15 years from the first reported molecular genetic G×E effects in humans) and faces numerous hurdles, it is also making exciting and rapid progress. Strengths of this area include cross-cutting findings which show the influence of various genetic factors across multiple levels of analysis, demonstrating a continuum of effects beginning with intermediate phenotypes or endophenotypes and ranging to full-blown psychopathology, characterizing a nomological network that builds confidence in findings. The challenges ahead for G×E psychopathology research will be to shift toward polygenic models, and to adopt cutting-edge measures and conceptualizations of the environment.

References

1. **Kendler K.** "A gene for . . . ": The nature of gene action in psychiatric disorders. Am J Psychiatry. 2005;**162**(7):1243–52.
2. **Monroe S, Simons A.** Diathesis-stress theories in the context of life stress research: Implications for the depressive disorders. Psychol Bull. 1991;**110**(3):406–25.

3. **Kendler K, Karkowski L, Prescott C.** Causal relationship between stressful life events and the onset of major depression. Am J Psychiatry. 1999;**156**(6):837–41.
4. **Belsky J, Pluess M.** Beyond diathesis stress: Differential susceptibility to environmental influences. Psychol Bull. 2009;**135**(6):885–908.
5. **Caspi A, McClay J, Moffitt TE, Mill J, Martin J, Craig IW**, et al. Role of genotype in the cycle of violence in maltreated children. Science. 2002;**297**(5582):851–4.
6. **Caspi A, Moffitt TE, Cannon M, McClay J, Murray R, Harrington H**, et al. Moderation of the effect of adolescent-onset cannabis use on adult psychosis by a functional polymorphism in the catechol-O-methyltransferase gene: Longitudinal evidence of a gene x environment interaction. Biol Psychiatry. 2005;**57**(10):1117–27.
7. **Hariri A, Mattay VS, Tessitore A, Kolachana B, Fera F, Goldman D**, et al. Serotonin transporter genetic variation and the response of the human amygdala. Science. 2002;**297**(5580):400–3.
8. **Monroe S, Reid M.** Gene–environment interactions in depression research: Genetic polymorphisms and life-stress polyprocedures. Psychol Sci. 2008;**19**(10):947–56.
9. **Monroe S.** Modern approaches to conceptualizing and measuring human life stress. Annu Rev Clin Psychol. 2008;**4**:33–52.
10. **Vrshek-Schallhorn S, Stroud CB, Mineka S, Hammen C, Zinbarg RE, Wolitzky-Taylor K**, et al. Chronic and episodic interpersonal stress as statistically unique predictors of depression in two samples of emerging adults. J Abnorm Psychol. 2015;**124**(4):776–90.
11. **Vrshek-Schallhorn S, Mineka S, Zinbarg R, Craske M, Griffith J, Sutton J**, et al. Refining the candidate environment: Interpersonal stress, the serotonin transporter polymorphism, and gene–environment interactions in major depression. Clin Psychol Sci. 2014;**2**(3):235–48.
12. **Bergen SE, Gardner CO, Kendler KS.** Age-related changes in heritability of behavioral phenotypes over adolescence and young adulthood: A meta-analysis. Twin Res Hum Gen. 2007;**10**(03):423–33.
13. **Kendler K, Gardner C, Lichtenstein P.** A developmental twin study of symptoms of anxiety and depression: Evidence for genetic innovation and attenuation. Psychol Med. 2008;**38**(11):1567–75.
14. **Starr LR, Hammen C.** Genetic moderation of the association between adolescent romantic involvement and depression: Contributions of serotonin transporter gene polymorphism, chronic stress, and family discord. Dev Psychopathol. 2016;**28**(02):447–57.
15. **Vrshek-Schallhorn S, Stroud CB, Mineka S, Zinbarg RE, Adam EK, Redei E**, et al. Additive genetic risk from five serotonin system polymorphisms interacts with interpersonal stress to predict depression. J Abnorm Psychol. 2015;**124**(4):776–90.
16. **Hankin BL, Young JF, Abela JR, Smolen A, Jenness JL, Gulley LD**, et al. Depression from childhood into late adolescence: Influence of gender, development, genetic susceptibility, and peer stress. J Abnorm Psychol. 2015;**124**(4):803–16.
17. **Belsky D, Moffitt TE, Houts R, Bennett GG, Biddle AK, Blumenthal JA**, et al. Polygenic risk, rapid childhood growth, and the development of obesity: Evidence from a 4-decade longitudinal study. Arch Pediatr Adolesc Med. 2012;**166**(6):515–21.
18. **Belsky D, Moffitt TE, Baker TB, Biddle AK, Evans JP, Harrington H**, et al. Polygenic risk and the developmental progression to heavy, persistent smoking and nicotine dependence: Evidence from a 4-decade longitudinal study. JAMA Psychiatry. 2013;**70**(5):534–42.
19. **Fisher R.** The correlation of relatives on the assumption of Mendelian inheritance. Trans Royal Soc Edinburgh. 1918;**52**:399–433.
20. **Wright S.** Systems of mating. Genetics. 1921;**6**(2):111–78.
21. **Nikolova YS, Ferrell RE, Manuck SB, Hariri AR.** Multilocus genetic profile for dopamine signaling predicts ventral striatum reactivity. Neuropsychopharmacology. 2011;**36**(9):1940–7.
22. **Talmud PJ, Shah S, Whittall R, Futema M, Howard P, Cooper JA**, et al. Use of low-density lipoprotein cholesterol gene score to distinguish patients with polygenic and monogenic familial hypercholesterolaemia: A case-control study. Lancet. 2013;**381**(9874):1293–301.

23. **Benyamin B, Pourcain B, Davis OS, Davies G, Hansell NK, Brion M-J**, et al. Childhood intelligence is heritable, highly polygenic and associated with FNBP1L. Mol Psychiatry. 2014;**19**(2):253–8.
24. **Belsky D, Moffitt TE, Corcoran DL, Domingue B, Harrington H, Hogan S**, et al. The genetics of success: How single-nucleotide polymorphisms associated with educational attainment relate to life-course development. Psychol Sci. 2016;**27**(7):957–72.
25. **Pagliaccio D, Luby JL, Bogdan R, Agrawal A, Gaffrey MS, Belden AC**, et al. Stress-system genes and life stress predict cortisol levels and amygdala and hippocampal volumes in children. Neuropsychopharmacology. 2014;**39**(5):1245–53.
26. **Hammen C.** Generation of stress in the course of unipolar depression. J Abnorm Psychol. 1991;**100**(4):555–61.
27. **Liu RT, Alloy LB.** Stress generation in depression: A systematic review of the empirical literature and recommendations for future study. Clin Psychol Rev. 2010;**30**(5):582–93.
28. **Starr LR, Hammen C, Brennan PA, Najman JM.** Serotonin transporter gene as a predictor of stress generation in depression. J Abnorm Psychol. 2012;**121**(4):810.
29. **Starr LR, Hammen C, Brennan PA, Najman JM.** Relational security moderates the effect of serotonin transporter gene polymorphism (5-HTTLPR) on stress generation and depression among adolescents. J Abnorm Child Psychol. 2013;**41**(3):379–88.
30. **Hariri A, Weinberger DR.** Imaging genomics. Br Med Bull. 2003;**65**(1):259–70.
31. **Hyde LW, Bogdan R, Hariri AR.** Understanding risk for psychopathology through imaging gene–environment interactions. Trends Cogni Sci. 2011;**15**(9):417–27.
32. **Bogdan R, Pagliaccio D, Baranger DA, Hariri AR.** Genetic moderation of stress effects on corticolimbic circuitry. Neuropsychopharmacology. 2016;**41**(1):275–96.
33. **Hariri A, Drabant EM, Weinberger DR.** Imaging genetics: Perspectives from studies of genetically driven variation in serotonin function and corticolimbic affective processing. Biol Psychiatry. 2006;**59**(10):888–97.
34. **Whalen PJ.** Fear, vigilance, and ambiguity: Initial neuroimaging studies of the human amygdala. Curr Dir Psychol Sci. 1998;**7**(6):177–88.
35. **Bertolino A, Arciero G, Rubino V, Latorre V, De Candia M, Mazzola V**, et al. Variation of human amygdala response during threatening stimuli as a function of 5'HTTLPR genotype and personality style. Biol Psychiatry. 2005;**57**(12):1517–25.
36. **Munafò M, Brown SM, Hariri AR.** Serotonin transporter (5-HTTLPR) genotype and amygdala activation: A meta-analysis. Biol Psychiatry. 2008;**63**(9):852–7.
37. **Pezawas L, Meyer-Lindenberg A, Drabant EM, Verchinski BA, Munoz KE, Kolachana BS**, et al. 5-HTTLPR polymorphism impacts human cingulate-amygdala interactions: a genetic susceptibility mechanism for depression. Nat Neurosci. 2005;**8**(6):828–34.
38. **Canli T, Qiu M, Omura K, Congdon E, Haas BW, Amin Z**, et al. Neural correlates of epigenesis. PNAS. 2006;**103**(43):16033–8.
39. **Nolen-Hoeksema S.** The role of rumination in depressive disorders and mixed anxiety/depressive symptoms. J Abnorm Psychol. 2000;**109**(3):504.
40. **Alexander N, Klucken T, Koppe G, Osinsky R, Walter B, Vaitl D**, et al. Interaction of the serotonin transporter-linked polymorphic region and environmental adversity: increased amygdala-hypothalamus connectivity as a potential mechanism linking neural and endocrine hyperreactivity. Biol Psychiatry. 2012;**72**(1):49–56.
41. **Rabl U, Meyer BM, Diers K, Bartova L, Berger A, Mandorfer** D, et al. Additive gene–environment effects on hippocampal structure in healthy humans. J Neurosci. 2014;**34**(30):9917–26.
42. **Miller R, Wankerl M, Stalder T, Kirschbaum C, Alexander N.** The serotonin transporter gene-linked polymorphic region (5-HTTLPR) and cortisol stress reactivity: A meta-analysis. Mol Psychiatry. 2013;**18**:1018–24.
43. **Hibar DP, Stein JL, Renteria ME, Arias-Vasquez A, Desrivières S, Jahanshad N**, et al. Common genetic variants influence human subcortical brain structures. Nature. 2015;**520**(7546):224–9.

44. **Bogdan R, Hyde L, Hariri A.** A neurogenetics approach to understanding individual differences in brain, behavior, and risk for psychopathology. Mol Psychiatry. 2013;**18**(3):288–99.
45. **Cuthbert BN, Insel TR.** Toward the future of psychiatric diagnosis: The seven pillars of RDoC. BMC Medicine. 2013;**11**(1):126.
46. **Gottesman I, Gould T.** The endophenotype concept in psychiatry: Etymology and strategic intentions. Am J Psychiatry. 2003;**160**(4):636–45.
47. **Canino G, Alegría M.** Psychiatric diagnosis–is it universal or relative to culture? J Child Psychol Psychiatry. 2008;**49**(3):237–50.
48. **Chiao JY, Blizinsky KD.** Culture–gene coevolution of individualism–collectivism and the serotonin transporter gene. Proc Royal Soc B. 2010;**277**(1681):529–37.
49. **Mittal VA, Ellman LM, Cannon TD.** Gene–environment interaction and covariation in schizophrenia: The role of obstetric complications. Schizophr Bull. 2008;**34**(6):1083–94.
50. **Salvatore JE, Aliev F, Bucholz K, Agrawal A, Hesselbrock V, Hesselbrock M,** et al. Polygenic risk for externalizing disorders gene-by-development and gene-by-environment effects in adolescents and young adults. Clin Psychol Sci. 2015;**3**(2):189–201.
51. **Duncan LE, Keller MC.** A critical review of the first 10 years of candidate gene-by-environment interaction research in psychiatry. Am J Psychiatry. 2011;**168**(10):1041–9.
52. **Chipuer HM, Rovine MJ, Plomin R.** LISREL modeling: Genetic and environmental influences on IQ revisited. Intelligence. 1990;**14**(1):11–29.
53. **Lykken D.** The mechanism of emergenesis. Genes Brain Behav. 2006;**5**(4):306–10.
54. **Conway CC, Hammen C, Brennan PA, Lind PA, Najman JM.** Interaction of chronic stress with serotonin transporter and catechol-O-methyltransferase polymorphisms in predicting youth depression. Depress Anxiety. 2010;**27**(8):737–45.
55. **Kaufman J, Yang B, Douglas-Palumberi H, Grasso D, Lipschitz D, Houshyar S,** et al. Brain-derived neurotrophic factor-5-HTTLPR gene interactions and environmental modifiers of depression in children. Biol Psychiatry. 2006;**59**(8):673–80.
56. **Purcell SM, Moran JL, Fromer M, Ruderfer D, Solovieff N, Roussos P,** et al. A polygenic burden of rare disruptive mutations in schizophrenia. Nature. 2014;**506**(7487):185–90.
57. **Sanders SJ, Murtha MT, Gupta AR, Murdoch JD, Raubeson MJ, Willsey AJ,** et al. De novo mutations revealed by whole-exome sequencing are strongly associated with autism. Nature. 2012;**485**:237–41.
58. **Risch N, Herrell R, Lehner T, Liang K, Eaves L, Hoh J,** et al. Interaction between the serotonin transporter gene (5-HTTLPR), stressful life events, and risk of depression: A meta-analysis. JAMA. 2009;**301**(23):2462–71.
59. **Munafò M, Durrant C, Lewis G, Flint J.** Gene x environment interactions at the serotonin transporter locus. Biol Psychiatry. 2009;**65**(3):211–9.
60. **Karg K, Burmeister M, Shedden K, Sen S.** The serotonin transporter promoter variant (5-HTTLPR), stress, and depression meta-analysis revisited: Evidence of genetic moderation. Arch Gen Psychiatry. 2011;**68**(5):444–54.
61. **Sharpley CF, Palanisamy SK, Glyde NS, Dillingham PW, Agnew LL.** An update on the interaction between the serotonin transporter promoter variant (5-HTTLPR), stress and depression, plus an exploration of non-confirming findings. Behav Brain Res. 2014;**273**:89–105.
62. **van IJzendoorn M, Belsky J, Bakermans-Kranenburg M.** Serotonin transporter genotype 5HTTLPR as a marker of differential susceptibility? A meta-analysis of child and adolescent gene-by-environment studies. Transl Psychiatry. 2012;**2**(8):e147.
63. **Kim-Cohen J, Caspi A, Taylor A, Williams B, Newcombe R, Craig IW,** et al. MAOA, maltreatment, and gene–environment interaction predicting children s mental health: New evidence and a meta-analysis. Mol Psychiatry. 2006;**11**(10):903–13.
64. **Byrd AL, Manuck SB.** MAOA, childhood maltreatment, and antisocial behavior: Meta-analysis of a gene–environment interaction. Biol Psychiatry. 2014;**75**(1):9–17.

65. **Marcus PM, Hayes RB, Vineis P, Garcia-Closas M, Caporaso NE, Autrup H,** et al. Cigarette smoking, *N*-acetyltransferase 2 acetylation status, and bladder cancer risk: A case-series meta-analysis of a gene–environment interaction. Cancer Epidemiol Biomark Prev. 2000;**9**(5):461–7.

66. **Zeiger JS, Beaty TH, Liang K-Y.** Oral clefts, maternal smoking, and TGFA: A meta-analysis of gene–environment interaction. The Cleft Palate-Craniofacial Journal. 2005;**42**(1):58–63.

67. **Caspi A, Sugden K, Moffitt TE, Taylor A, Craig IW, Harrington H, McClay J, Mill J, Martin J, Braithwaite A, Poulton R.** Influence of life stress on depression: Moderation by a polymorphism in the 5-HTT gene. Science. 2003;**301**:386–9.

68. **Kohrt BA, Worthman CM, Ressler KJ, Mercer KB, Upadhaya N, Koirala S, Nepal MK, Sharma VD, Binder EB** (2015). Cross-cultural gene: Environment interactions in depression, post-traumatic stress disorder, and the cortisol awakening response: FKBP5 polymorphisms and childhood trauma in South Asia: GxE interactions in South Asia. Int Rev Psychiatry. 2015;**27**:180–96.

Chapter 4

Epigenetics and twin studies: A review and applications in human aggressive behavior

Jenny van Dongen and Dorret I Boomsma

4.1 Introduction

Twins arise under one of two special circumstances: when two mature oocytes are produced during ovulation and fertilized by two sperm cells (dizygotic twins) or when a single fertilized egg cell (zygote) splits into two or more (monozygotic twins or multiples). Dizygotic (DZ) twins are more common to older mothers; the prevalence of having DZ twins differs around the globe and being a mother of DZ twins has a genetic component (1). DZ twins each have a partially distinct DNA code and are genetically no different from non-twin siblings. Monozygotic (MZ) twins are also called "identical twins" because they usually look very similar and have identical DNA sequences. Their prevalence tends to be very similar in different parts of the world, and there are at present very few, if any, known causal factors to having MZ twins.

On average, dizygotic twins share 50% of all segregating variants in their DNA, while monozygotic twins share 100%. This information forms the basis for the classical twin design (2), which allows estimating the contribution of environmental and genetic influences to phenotypic variation in human populations, without the need to know the exact environmental influences responsible or to know the genomic location of any of the variants in the DNA that influence the trait. In fact, twin studies were already performed long before the discovery of DNA; early twin researchers were aware of the existence of heritable information but did not yet know that this heritable information lies encoded in DNA molecules. While certain epigenetic mechanisms, namely DNA methylation, were discovered around the same time as the DNA code, they have only been studied by twin researchers for 10 years or so when it became technically and financially feasible to measure DNA methylation on a genome-wide and epidemiological scale.

Epigenetics literally means "above genetics" (the word *epi* is Greek for "above") and it is currently defined by most as the study of molecular mechanisms—other than the DNA sequence itself—that influence gene expression and that are transmitted during somatic cell division (mitosis) from one cell generation to the next (3) (section 4.2). Epigenetic mechanisms are key regulators of cell differentiation and development, but only recently did it become clear that they are also involved in regulating modifiable gene expression programs that may respond to changes in internal and external conditions in differentiated (mature) cells. In the central nervous system, epigenetic mechanisms mediate crucial processes that rely on transient activation of gene expression such as synaptic plasticity, memory formation and consolidation, and learning (4, 5). Animal studies have shown that early-life conditions such as (a lack of) maternal care trigger long-lasting changes in epigenetic mechanisms that regulate the offspring's DNA in brain cells, thereby programming life-long stress response (6). Thus, epigenetic mechanisms can explain how environmental influences that do not directly induce any anatomical damage can still have long-lasting effects on physiology and behavior. It has been hypothesized that many environmental risk factors for psychiatric conditions, including prenatal exposures and early-life adversity, as well as genetic risk factors (e.g. SNPs) (7, 8), may act through epigenetic mechanisms.

MZ twins start out with a (nearly) identical DNA sequence, which explains why they are often so remarkably similar. During life, somatic mutations may arise is one twin, and not in the co-twin, but most current empirical evidence points to near identity in very old twins as well (9). Yet the epigenomes of MZ twins can acquire differences that may cause different usage of the same code. This can sometimes lead to remarkable phenotypic differences (10). The study of such phenotypically discordant persons with identical DNA sequence is regarded as a particularly strong research design to identify epigenetic mechanisms involved in complex traits (section 4.3.1). The study of monozygotic and dizygotic twins can also give insight into the causes of variance in epigenetic marks between people (section 4.4). In this chapter, we first describe the various epigenetic mechanisms employed in human cells. Next, the advantages of twin studies for epigenetic research are discussed and illustrated with examples of studies that have applied twin designs to study the connections between epigenetic mechanisms and complex (disease) traits, with a focus on traits related to cognitive and mental health.

4.2 Epigenetic mechanisms of gene expression regulation

Nearly every cell in the human body contains the same DNA sequence, but not all DNA is expressed in each cell at all times (expression refers to the synthesis of RNA, transcription, and subsequent translation of RNA into protein). A brain cell expresses different genes than, for example, a white blood cell, which gives rise to their distinct functions and morphology. In mature cells, some genes need to be permanently repressed (e.g. imprinted genes and pluripotency-associated genes; 11), while the expression of other genes must be flexible, allowing cells to respond to varying internal and external conditions by up-regulating genes of which more product is needed and down-regulating genes of which less product is needed. For example, contextual fear conditioning studies of rats have demonstrated that experience-driven behavioral changes are induced and maintained by changes in the expression level of certain genes in hippocampal cells (12).

Transcription of DNA requires that transcription factors bind to specific target sequences in the DNA (i.e. promoters). Whether a transcription factor can bind to its target depends on whether interaction partners such as co-activators and repressors are present, and on whether the local chromatin structure (chromatin refers to the complex of the DNA molecule and all attached histone proteins) permits such interactions to take place; in other words, whether the DNA molecule is accessible (13). There are numerous molecular mechanisms that collectively control chromatin structure within cell nuclei or regulate gene expression in other ways. They include chemical tags that are attached to the DNA molecule itself (DNA methylation) or to the histone proteins around which DNA is wrapped (histone modifications), and molecules that interact with DNA or RNA transcripts (e.g. non-coding RNAs, transcription factors, and methyl-CpG-binding proteins) (14). Each time a somatic cell divides, an identical copy of its DNA sequence is transferred to the daughter cells and some but not all of the DNA's regulatory molecules are copied as well, allowing mother cells to pass their identity to daughter cells. Epigenetics is strictly defined as the study of those molecular mechanisms—other than the DNA sequence itself—that influence gene expression and that are transmitted from one cell generation to the next through cell division (mitosis and possibly meiosis) (3). In practice, epigenetics is often used in a less strict sense to refer to DNA methylation (section 4.2.1) and all histone modifications (Box 4.1), although the extent to which histone modifications are retained during cell division is still unclear, but it is expected that transmission is not perfect (15).

4.2.1 DNA methylation

DNA methylation is the covalent attachment of a methyl-group (CH_3) to the DNA molecule. It is one of the best studied epigenetic mechanisms in humans, and is currently the only one that is suited for assessment in large-scale human epidemiological studies. In vertebrates, DNA methylation occurs mostly at cytosines located next to guanines (CpG sites). Non-CpG methylation is prevalent in human embryonic stem cells, but is thought to be very rare or absent in most differentiated somatic cells (16–18), except for cells in the adult brain where it was only recently discovered that methylation at CpA, CpT, and CpC is widespread (19). While (largely) absent at birth, the amount of non-CpG methylation increases between birth and adolescence in brain cells and it has been hypothesized that this may give rise to distinct functions in the adult compared to the child brain (20). Because the function of non-CpG methylation is still poorly understood, we will focus only on CpG methylation in the rest of this chapter.

DNA methylation is established and maintained by enzymes from the DNA methyltransferase (DNMT) family, including DNMT1, DNMT3a, and DNMT3b (21). Some studies have suggested that *de novo* DNA methylation represents a passive process that targets regulatory sequences in the DNA that are not occupied by transcription factors (22, 23). In total, an estimated 70–80% of CpG sites in the genome are methylated in mammalian cells (24). Many CpGs occur in clusters called CpG islands (CGIs). The promoters of approximately 70% of human genes overlap with a CpG island. The classic view is that methylation of CpGs in the promoter area of genes is associated with repression of gene transcription. It has now become clear that the effect of methylation on expression may vary depending on the exact location that is methylated (25), and there are indications that methylation at enhancers is more strongly related to the expression level of genes than methylation at promoters (25).

DNA methylation may impact on regulation of transcription through two main mechanisms. First, methylation of specific sequences may prevent the binding of regulator proteins to those sequences (e.g. transcription factors (26) and insulators (27)). Second, methylated CpGs may attract methyl-CpG-binding domain proteins, which are regulatory proteins that bind to methylated cytosine and recruit chromatin regulators such as histone deacetylases and chromatin remodeling complexes to the site (28, 29). Thus, rather than acting as independent mechanisms, it is thought that the different layers of epigenetic information (e.g. DNA methylation, histone modifications; Box 4.1) generally work together to regulate transcription. Methylation of promoter CGIs may contribute to long-term repression, for example at inactive X-chromosome genes in females (30) and imprinted genes (31). In comparison to other epigenetic marks, it has been suggested that CpG methylation may be best described as a "memory signal for the long-term maintenance of gene silencing" (32). For example, drugs that target histone modifications (histone deacetylase inhibitors, HDACi) in colon cancer cells can lead to transient reactivation of loci silenced by DNA methylation, but permanent reactivation can only be induced by DNA-demethylating drugs (32).

Multiple techniques exist to measure DNA methylation. In epidemiological studies, the Infinium HumanMethylation450 array (450K array) was a popular platform for assessing DNA methylation at a genome-wide scale in large cohorts (33, 34) and has now been succeeded by the MethylationEPIC BeadChip (850K array) (35). These arrays assess methylation level at over 485,000/850,000 methylation sites (mostly CpG) across a variety of regions in the human genome, including gene promoters, gene bodies, 3' regions of genes, and intergenic locations. The MethylationEPIC BeadChip covers > 90% of the probes from the Illumina 450k, and additionally includes CpGs in enhancer regions. The methylation status of an entire genome (methylome) can be obtained through bisulfite sequencing (BS-seq) (36) or methylated DNA immunoprecipitation followed by sequencing (MeDIP-seq) (37). Sequencing-based techniques are being used for assessing whole-genome methylation or fine mapping of regions detected in array-based experiments but are currently too expensive to use in large (epidemiological-scale) studies. (Box 4.1)

Box 4.1 Chromatin state and histone modifications

Within the nuclei of cells, DNA molecules are folded into smaller packages (chromosomes), with the help of histone proteins. Because histone proteins are positively charged while DNA is negatively charged, histones help to fold the DNA into a smaller volume through electrostatic interactions. The chromatin complex is visible under an electron microscope and reveals where transcription occurs. In such regions, the chromatin looks like "a string of beads"; a DNA string with so-called nucleosomes ("beads"), which are units of about 146 base pairs of DNA that are wrapped 1.65 times around a complex of histone proteins (38–40). Between each nucleosome there are little pieces of DNA (stretching on average 20 base pairs), called linker DNA. Nucleosome-free DNA contains regulatory elements including promoters, enhancers, silencer, insulators, and locus control regions (23).

The structure of chromatin ("chromatin state") varies along the genome of each cell. In genomic regions that are not transcribed, chromatin is folded further into a highly condensed conformation called "inactive chromatin" or "heterochromatin," while other regions are characterized by an open chromatin state ("active chromatin" or "euchromatin"), where the distance between nucleosomes is larger, thereby permitting transcription to occur (41). There are also various intermediate states. For example, in permissive or repressed states, DNA is accessible to become de-repressed or activated through interaction with, for example, transcription factors (42). The exact structure of chromatin of a genomic region is to an important extent controlled by covalent post-translational modifications of histone proteins.

The histone protein complex within nucleosomes is called "histone octamer" and consists of eight histone proteins: two copies of each of the histone types H2A, H2B, H3, and H4. A fifth type of histone (H1) can bind to the linker DNA between nucleosomes. Each histone protein is subject to numerous modifications at specific amino acids, including methylation, acetylation, phosphorylation, ubiquitination, and many more. Most of these histone modifications occur at the "tails" (amino termini) of histones that stick out from the nucleosome complex. In total, more than 100 different histone modifications are currently known (41), most not well understood, and it is thought that the exact chromatin state of a region lies written in the specific combination of histone modifications within that region ("the histone code") (43). For example, acetylation of the amino acid lysine characterizes transcriptionally active, accessible chromatin. Other modifications are associated with specific DNA elements. For example, histone H3 lysine 4 trimethylation (H3K4me3) is associated with promoters of actively transcribed genes (44). Histone modifications may exert their influence on chromatin structure by changing the electric charge of the histone (thereby modifying the strength of the interaction between the histone protein and the DNA molecule), and by recruiting other molecules to the DNA. Histone modifications and chromatin structure are mediated by a number of proteins that fall in one or more of the following categories: writers (responsible for depositing histone modifications, for example histone acetyl transferases; HATs; 45), erasers (which can remove modifications, for example histone deacetylases; HDACs), and readers (proteins that "interpret" the histone code by recruiting additional molecules that can stabilize or remodel specific chromatin states, upon recognition of specific histone modifications).

4.2.2 The role of epigenetic mechanisms in human traits: Establishing long-term effects of environmental influences and genetic variants on gene expression

A person's DNA sequence changes very little during the entire lifetime (*de novo* mutations may occur in some cells and lead to mosaicism among somatic cells). DNA methylation and other epigenetic marks are dynamic by nature; they change during the lifetime as part of developmental programs (11) and aging (46), and may also change in response to specific environmental exposures (47). The crucial role of epigenetic mechanisms in human development and learning is illustrated by the devastating effects of rare genetic mutations that disrupt key components of the epigenetic machinery. For example, mutations in the gene encoding methyl-CpG binding protein 2 (MeCP2) cause Rett syndrome, which is characterized at the cellular level by deficits in synaptic plasticity in the hippocampus, causing impairments in memory formation (48). Children with this neurodevelopmental syndrome seem to develop normally until 6–18 months, but after that age they start to lose thus far acquired motor skills and speech and develop a range of problems.

The responsiveness of epigenetic mechanisms to the external environment is exemplified by multiple human studies of specific exposures. Methylation levels of the *AHRR* gene and thousands of others are altered in individuals who smoke (49), and in babies of mothers who smoked during pregnancy (50), and correlate with gene expression levels (50). The loci affected are important to many traits including cardiovascular disease and cancer (49). Other chemical exposures linked to DNA methylation changes at specific loci in humans include toxins such as lead, air pollution, and endocrine-disrupting chemicals (51, 52). Besides the influence of chemical exposures, epigenetic mechanisms may also respond to the psychosocial environment. The strongest evidence for such effects comes from animal studies, and indicates that there may be particular developmental time windows during which epigenetic regulation in the brain may respond to such cues. Weaver and colleagues found that the quality of maternal care in rats, indexed by the frequency of pup licking and grooming and arched-back nursing, influences long-term epigenetic regulation of the glucocorticoid receptor in the hippocampus, which affects the offspring's lifelong stress response; but this effect was only inducible in the first week of the rat pup's life (6). In humans, associations have been reported between DNA methylation and factors including childhood trauma (53), abuse (54), parental stress during childhood (55), and socio-economic status (56). It is important to realize that while some environmentally induced epigenetic changes may be merely harmful, it is believed that modifiable epigenetic mechanisms exist to make it possible to adjust gene expression depending on the conditions encountered in life ("epigenetic programming") thereby allowing for a certain flexibility in development ("phenotypic plasticity").

The potential of DNA methylation to act as a long-term molecular memory of past environmental exposures makes it an extremely interesting target to study in relationship to complex traits such as psychiatric disorders. First, studying this molecular layer may give insight into the genes involved in a disorder. Second, it may identify genomic regions whose methylation level may act as a biomarker for the collective effects of disease-relevant (but often unknown) environmental exposures. Importantly, epigenetic regulation may also be a key mode of action of complex trait-associated genetic variants, as indicated by (i) observations that SNPs identified by GWAS of complex traits are enriched in regulatory regions of the genome, suggesting that many of these SNPs exert their effect on the phenotype by influencing gene regulation (rather than, e.g. changing the structure of proteins) and (ii) methylation QTL studies (57, 58), and classical twin studies (section 4.4) suggest that the levels of DNA methylation across the genome are to an important extent driven by information in the DNA sequence itself.

4.2.3 Identifying epigenetic mechanisms involved in human traits: epigenome-wide association studies

The epigenome-wide association study (EWAS) is a "hypothesis-free" research design that is increasingly employed to identify epigenetic mechanisms involved in a trait of interest. EWA studies test, for a large number of genomic locations, if an epigenetic mark (usually the level of DNA methylation) is associated with a trait. This is done by comparing DNA methylation levels between cases and controls or by testing whether a (linear) relationship exists between DNA methylation level and a continuous phenotype. For practical and ethical reasons, these studies are usually conducted with DNA extracted from easily accessible biological samples, such as peripheral blood, buccal swabs, or saliva, although some studies have also examined DNA extracted from skin biopsies, adipose tissue, and post-mortem brain tissue (59–61), and an empirical question of large interest and importance is the extent to which epigenetic marks in one tissue also reflect the epigenetic status in other tissues (section 4.5). A particularly strong design for EWAS is the monozygotic discordant twin design, in which genome-wide methylation levels are compared between affected twins and their unaffected, but genetically identical, co-twin (section 4.3).

4.3 Epigenetic differences in monozygotic twins

MZ twins may be less identical at the level of the DNA and its regulation than initially thought. First, *de novo* mutations events can cause small differences in the DNA sequence between monozygotic twins, although these events are rare (9, 62). Epigenetic differences are likely to arise between monozygotic twins, which may cause different expression of the DNA in each twin even if the DNA sequences are identical. Monozygotic twins already display differences in DNA methylation and gene expression in various tissues at birth (63–65). These differences may be caused by the prenatal environments experienced by each twin, for example in terms of nutrient supply or exposure to other substances passing the placenta (MZ twins may have different placentas) (66). After birth, epigenetic differences may arise between twins due to postnatal environmental exposures, although twins usually experience similar childhood environments (unless they are separated at birth and adopted by different families). It has been observed that the genome-wide methylation patterns of some twins diverge whereas other twins become more similar in the first 18 months after birth (67), suggesting that the prenatal environment may be an important initial driver of epigenetic variation and that the similar postnatal environment of twins or genetic influences on epigenetic mechanisms may outweigh some of the initial epigenetic differences established *in utero*.

Even if their environments would be entirely the same, monozygotic twins may still acquire epigenetic differences due to stochastic (random) effects, referring to the phenomenon that molecular control of gene regulation is not perfect. For example, the maintenance of DNA methylation in dividing cells by maintenance DNA methyltransferases is not 100% accurate (68). X-chromosome inactivation through epigenetic mechanisms in women is also thought to be controlled by a stochastic process, resulting in some cells inactivating the paternal X-chromosome and others the maternal X-chromosome. Female MZ twin have been described who are discordant for monogenic X-linked conditions due to differences in X-chromosome inactivation (69) (e.g. Duchenne muscular dystrophy; 70), where one twin "escapes" the disorder because the X-chromosome with the disease mutation is inactivated in a majority of cells ("skewed X-inactivation"). It has been reported that female monozygotic twin pairs discordant for bipolar disorder show greater differences in methylation of maternal and paternal X-chromosomes compared to concordant twin pairs, suggesting that skewed X-chromosome inactivation may contribute to discordance for bipolar disorder and that X-linked loci may be involved in the disorder (71).

4.3.1 The value of discordant monozygotic twins in epigenetics

The discordant monozygotic twin design can be seen as a case-control study featuring twin pairs of which one twin is the case and the co-twin is the control. Depending on the research question, monozygotic twins discordant for a particular disease, discordant for exposures such as smoking or a traumatic experience, or with a large difference in a continuous trait such as body weight, may be studied. Such studies may lead to identification of locations in the genome where epigenetic marks are connected to the disease, or the particular exposure.

The advantage of studying such questions in monozygotic twins is that many other factors that may differ between unrelated disease cases and controls are controlled for in discordant MZ twin studies by design, including age, sex, birth year (monozygotic twins always have the same age, sex, and even birth date), DNA sequence (*de novo* mutations excepted), many prenatal circumstances because twins inhabit the same womb at the same time (e.g. maternal genotype, maternal age, maternal health and lifestyle during pregnancy), and many postnatal environmental influences (both twins are raised at the same time by the same parents in the same house, are given the same food, etc.). By contrast, all of these factors may differ between unrelated cases and controls, where they may dilute, bias, or confound associations between disease status and DNA methylation levels. MZ twins also tend to have more similar white blood cell counts (because the levels of white blood cells are heritable; 72). This is an advantage when whole blood is used to study DNA methylation (section 4.5). An advantage specific to studies using methylation arrays is that the measurement of methylation differences between MZ co-twins is unlikely to be affected by SNPs within the target sequence of probes, which may potentially cause differential probe binding efficiency in individuals with different genotypes (73).

4.3.2 EWAS in discordant monozygotic twins: Aggressive behavior

We will illustrate the discordant MZ twin EWAS design by discussing a study of MZ twin pairs discordant for aggressive behavior conducted by ourselves and colleagues (74). DNA was extracted from whole blood of 20 discordant adult twin pairs (Figure 4.1) and DNA methylation was measured with the Illumina 450k array, which returns the methylation level (proportion of methylated alleles) at sites across the genome. The aggression scores of the twins are plotted in Figure 4.1a. In DNA that is derived from a mixture of cells (such as whole blood), the methylation level represents a continuous variable with values that may range between 0 and 1. For example, a methylation level of 1 means that all DNA strands had a methyl group attached at this position and a value of 0.5 means that 50% of all DNA strands had a methyl group attached at this position. This may occur when the location is methylated on one chromosome, as happens at many positions on one of the two copies of each X-chromosome in women, or when the position is methylated in a subset of all cells from which the DNA was extracted. After measuring methylation levels across the genome, the methylation data were adjusted for covariates, including technical batch information and white blood cell counts (section 4.5). Next, we compared, for each twin pair, the methylation level at each autosomal location in the aggression high-scoring twin compared to the low-scoring twin. As an example, methylation differences across the genome are plotted for two aggression discordant pairs in Figure 4.1b. The CpGs showing a large difference in methylation (e.g. greater than 0.3, Figure 4.1b) were generally twin pair-specific (only observed in one twin pair). This has also been observed in autism-discordant MZ twins (75). Next, we tested for consistent differences in DNA methylation levels between aggression-discordant MZ twins, by applying paired t-tests to all methylation sites (> 400,000). While studies that test a single hypothesis commonly use an alpha level of 0.05 in statistical tests, this alpha level is unsuited when a large number of statistical tests is performed. Therefore, in EWAS, it is common to apply Bonferroni correction for the number of tests performed or to control the false discovery rate (FDR), usually applying a threshold of 5%. In studies of Illumina 450k array data, Bonferroni correction comes down to applying an alpha of 0.05/~450.000 = 1×10^{-7} to evaluate whether a test is "genome-wide significant." Because methylation levels are to some extent correlated across locations, the tests performed for different locations are not fully independent. Therefore, Bonferroni correction for the total number of tests performed may be too conservative.

In our comparison of 20 aggression discordant twin pairs, no genome-wide significant differences were identified, but methylation sites with a lower p-value were located more often than expected by chance in or near genes related to various central nervous system processes. Figure 4.1c shows the methylation level in aggression-discordant twins at three CpG sites with the lowest p-value. The top site (cg21557159, $p = 5.7 \times 10^{-6}$) is located upstream of *RAB39*. In twins scoring high on aggressive behavior, the methylation level of this CpG was on average 1.4% lower ("hypomethylated") in whole blood compared to their low-scoring co-twin. *RAB39* is involved in cellular endocytosis, is thought to be ubiquitously expressed in human tissues, and plays an important role in neuronal cell differentiation (76). The results from this study indicate that aggression discordance is accompanied by small DNA methylation differences in blood at loci that may provide insight into the etiology of aggressive behavior. Yet, larger studies are needed to detect individual CpGs with these effect sizes at genome-wide significance (section 4.5).

4.4 The classical twin design

The aim of the classical twin design is to estimate how much of the variation of a trait between individuals in human populations (V_P, phenotypic variance) can be attributed to genetic variation (V_G, genetic variance) and how much to environmental and stochastic variation (V_E, environmental variance) (77). An outcome of interest, for example, methylation level at a particular site in the genome, is measured in MZ and DZ twin pairs, and the average resemblance of MZ and DZ pairs is quantified, typically by calculating the covariance or correlation between the trait value of twin 1 and the trait value of twin 2. If the correlation between MZ twins is larger than the correlation between DZ twins, this indicates that genetic variants in the DNA sequence contribute to variation of the trait (Figure 4.2).

The phenotypic covariance in MZ and DZ pairs can be used to estimate the variance due to multiple (unmeasured) sources: genetic variance, including variance due to additive genetic effects (V_A) and variance due to non-additive genetic effects (dominance variance; V_D), and variance due to environmental influences (V_E), including those that are shared by family members ("the common environment"; V_C) and influences that are unique to each individual ("the unique environment" (V_U), which includes stochastic effects). Because MZ twins are genetically (nearly) identical and DZ twins share on average 50% of genetic effects, the phenotypic covariance of MZ twins is expected to be equal to $V_A + V_D + V_C$, and the phenotypic covariance of DZ twins is expected to equal $0.5V_A + 0.25V_D + V_C$. Various software packages can be used to obtain estimates of V_A, V_D, or V_C and V_E based on the observed phenotypic covariance in twins. V_C and V_D cannot be estimated simultaneously in classical twin models (this is possible in extended twin-family designs). V_D is estimated if there is stronger evidence for non-additive effects (if the MZ correlation is more than twice as large as the DZ correlation), and V_C is estimated if there is stronger evidence for common environmental effects (if the MZ correlation is less than twice as large as the DZ correlation). Heritability is defined as the proportion of variation of a trait that is due to genetic variance (V_G/V_P). When only additive genetic effects are considered (V_A/V_P), we call this the narrow-sense heritability (denoted by the symbol h^2).

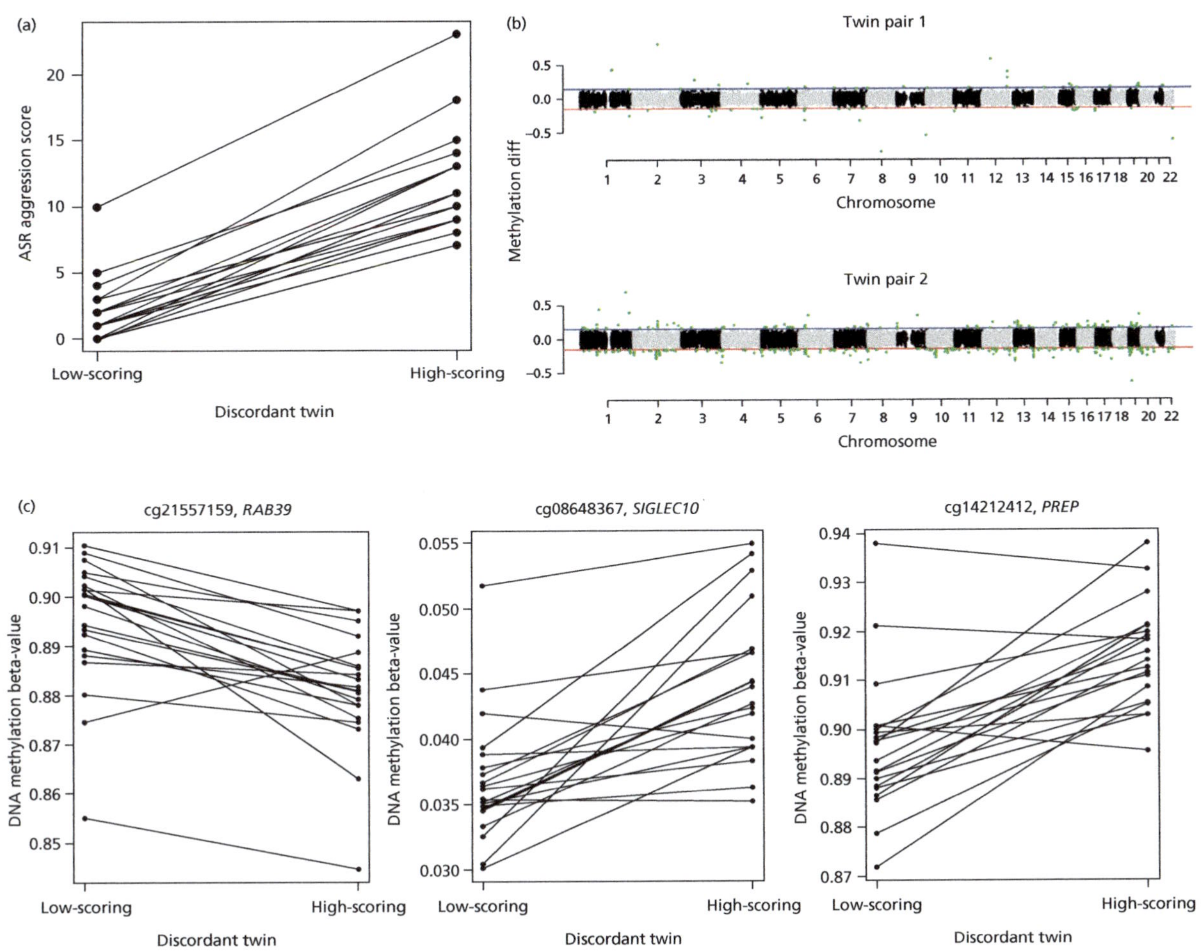

(a)
ASR aggression score
0
5
10
15
20
Low-scoring
High-scoring
Discordant twin
(b)
Twin pair 1
Methylation diff
0.5
0.0
−0.5
1 2 3 4 5 6 7 8 9 10 11 12 13 14 15 16 17 18 20 22
Chromosome
Twin pair 2
0.5
0.0
−0.5
1 2 3 4 5 6 7 8 9 10 11 12 13 14 15 16 17 18 20 22
Chromosome
(c)
cg21557159, RAB39
DNA methylation beta-value
0.91
0.90
0.89
0.88
0.87
0.86
0.85
Low-scoring
High-scoring
Discordant twin
cg08648367, SIGLEC10
DNA methylation beta-value
0.055
0.050
0.045
0.040
0.035
0.030
Low-scoring
High-scoring
Discordant twin
cg14212412, PREP
DNA methylation beta-value
0.94
0.93
0.92
0.91
0.90
0.89
0.88
0.87
Low-scoring
High-scoring
Discordant twin

Figure 4.1 Genome-wide methylation differences in monozygotic twins discordant for aggressive behavior. (a) Aggression scores of discordant MZ twins. The adult self-reported aggression score at the survey closest to the moment of blood draw is plotted for low-scoring and high-scoring twins of 20 discordant MZ twin pairs. The scores of co-twins are connected by lines. (b) Difference in methylation level (proportion, y-axis) across the genome (x-axis) in two twin pairs discordant for aggressive behavior. The numbers on the x-axis denote the number of each chromosome. The green dots denote methylation differences larger than 0.3 (30%). The figure illustrates that each twin pair shows a unique pattern of methylation differences, and that twin pair 2 shows more methylation differences greater than 0.3 compared to twin pair 1. (c) Methylation levels at the three most significant differentially methylated sites (lowest *p*-value) in aggression-discordant MZ twins. The residual DNA methylation values (adjusted for covariates) are plotted for low- and high-scoring twins of 20 discordant MZ twin pairs. The DNA methylation levels of co-twins are connected by lines.

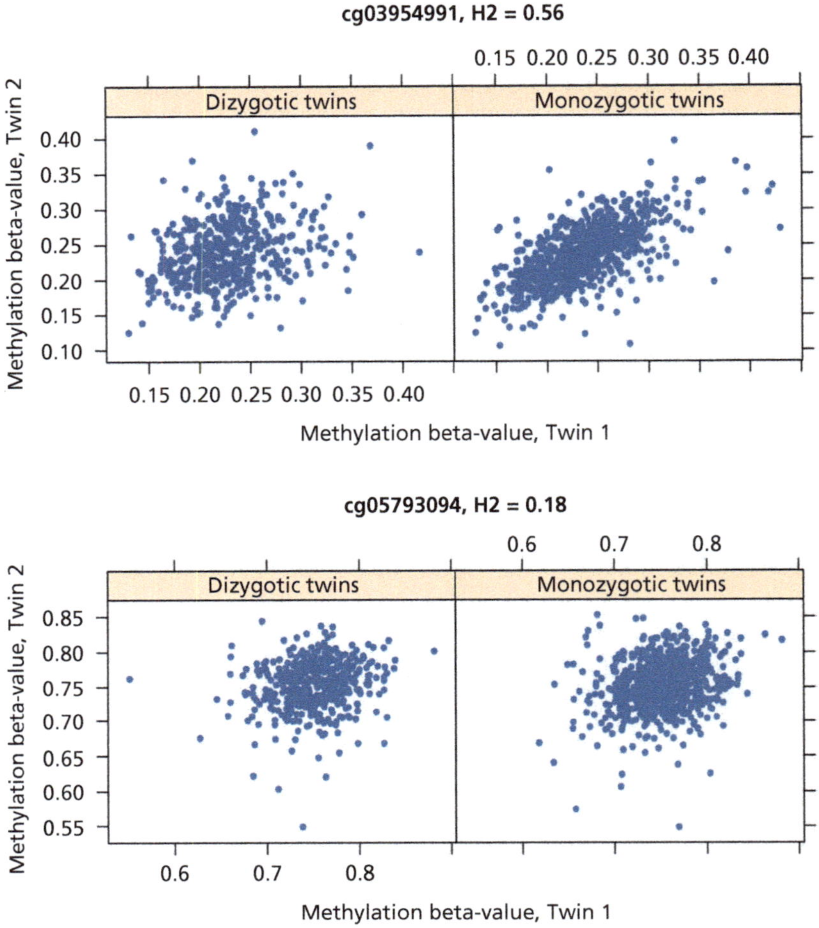

Figure 4.2 Resemblance of monozygotic and dizygotic twin pairs for DNA methylation level. This figure shows scatterplots of DNA methylation levels in blood from 769 MZ and 424 DZ twin pairs at two exemplary CpG sites: cg03954991, located in the promoter of *OSCP1* on chromosome 1 (upper panel), and cg05793094, located in the gene body of *TRIM10* on chromosome 6 (lower panel). The methylation level in twin 2 (y-axis) is plotted against the methylation level in twin 1 (x-axis) for all MZ twin pairs and all DZ twin pairs (each dot represents one twin pair). Methylation β-value stands for methylation proportion. The first CpG site (upper panel) is an example of a site where the pattern of resemblance of MZ and DZ twin pairs indicates that a large part (56%) of the total variation in methylation level is explained by genetic variation (heritability = 0.56). The data points are clearly more scattered, thus the correlation is smaller, in dizygotic twins compared to monozygotic twins. If we compare the scatter plots of MZ twins in the upper versus lower panel, we clearly see more scattering of data points in the lower figure, which means that MZ twins resemble each other less for methylation level at the second CpG site than they do for methylation level at the first CpG site. The lower panel is an example of a site where the pattern of resemblance of MZ and DZ twin pairs indicates that a large part of the total variation in methylation level between people is related to environmental and stochastic variation (82%) and only a small part (18%) of the variation is explained to genetic variation (heritability = 0.18).

4.4.1 Insights from classical twin studies of DNA methylation

Thus far, several classical twin studies have been applied to DNA methylation in accessible peripheral tissues at candidate genes or genome-wide (63, 78–89). These studies led to several interesting observations, which can be summarized in the following main conclusions: (i) genetic and environmental variation between individuals contributes to variation in the methylome; (ii) total variance and heritability of methylation levels vary across the genome: many sites show little variation between people and it is especially the sites with large individual differences that tend to show most heritable variation; (iii) the heritability of methylation levels may vary across tissues; (iv) the unique environmental variance in methylation levels across the genome in blood increases with age in adults; (v) part of the heritability of DNA methylation levels in blood is captured by common genome-wide SNPs.

4.4.2 Epigenetic effects in the classical twin design

Classical twin models (and extensions of it, such as multivariate and longitudinal models) have been and continue to be applied to a wide range of human traits. The classical twin model does not explicitly consider epigenetic effects. Having illustrated the importance of epigenetic mechanisms in this chapter, we will spend a few words on the implications of epigenetic sources of variations for classical twin modeling of phenotypic outcomes (such as depressive symptoms or height).

In the classical twin design, the (additive and non-additive) genetic effects (polygenic factors) represent the total effects of DNA sequence variants at all causal loci that contribute to variation in the trait of interest (e.g. depressive symptoms). Because epigenetic mechanisms such as DNA methylation regulate the expression of DNA throughout life, the impact of genetic variants carried by an individual on its phenotypic value will depend on the epigenetic regulation of the individual's DNA. We have described elsewhere (90) how we conceptualize epigenetic regulation within the classical twin design: the impact of the genetic component (polygenic factor) on the trait, rather than being fixed, will vary between individuals as a function of epigenetic individual differences (of genetic, environmental, and stochastic origin). In the classical twin design, this is ignored (polygenic effects are assumed to be fixed). We have shown with simulations that the classical twin design provides a good estimate of the average heritability in the population, though it does not allow estimation of the variation of the heritability due to epigenetic effects. Measured information on epigenetic variation at specific loci could be used to test for example if the heritability of a trait differs between individuals with high versus low expression or methylation of a gene of interest, using existing gene–environment models (91, 92).

4.5 Challenges

Several challenges apply to epigenetic studies described in this chapter and the broader field of human epigenetic epidemiology in general. A first important issue is the vulnerability of epigenetic studies to confounding. For example, methylation differences between disease discordant twins may be caused by medication that is used to treat the disease, or by differences in lifestyle. This may be taken into account, for example by adjusting for medication use or by checking whether the observed methylation difference remains after excluding twins using medication. A potential caveat is that information on potential confounders is not always available, although some exposures may be predicted based on methylation data. This approach has been applied to predict and adjust for smoking status in an EWAS of schizophrenia (93).

To establish whether methylation of a locus may contribute to causing the disease (rather than vice versa: the disease process itself causes the methylation change), longitudinal monozygotic twin designs are very valuable because such studies allow to distinguish epigenetic differences that were already present before disease onset from epigenetic differences that arose later. Another approach that may be used to investigate the direction of causation that has been applied to DNA methylation is Mendelian randomization (94).

A potential confounder in epigenetic studies of DNA extracted from samples of mixed cell types is cellular heterogeneity. DNA extracted from whole blood samples, for instance, comes from various types of white blood cells, the proportions of which may vary between individuals and within a person over time. DNA methylation patterns across the genome vary between different tissues and cell types (95). In fact, this information is being used to predict, based on genome-wide DNA methylation data, the proportions of various white blood cell types that were present in the sample from which DNA was extracted (96). If the cellular composition of whole blood samples differs between cases and controls (e.g. if one type of white blood cells is elevated in affected individuals), many loci will be associated with disease status that are merely markers of the difference in white blood cell proportions. It is therefore nowadays common practice to adjust for measured or predicted white blood cell count proportions (96) in EWAS to avoid this type of confounding

Epigenetic regulation is largely tissue-specific, and it is often impossible to obtain DNA from the tissue that is involved in the disease of interest, such as the brain. However, epigenetic variation in easily accessible tissues such as blood and buccal cells may give some insight about processes in other tissues, including the brain, and may give insight into systemic disease effects (97). Although most variation in DNA methylation appears to be unique to each tissue, a small proportion is correlated across tissues.

MZ discordant twin studies often have more power than case-control studies of unrelated participants. For example, it has been estimated for Illumina 450k array data that 178 pairs of discordant twins or 211 unrelated case-control pairs are needed to achieve 80% power to detect a mean methylation difference of 7% at genome-wide significance (98). Given that many discordant MZ twin studies of DNA methylation have reported smaller effect sizes (generally in the range of a 1–5% mean methylation difference), even larger samples may be needed to detect such sites at genome-wide significance. Studies that have been published thus far frequently were underpowered to detect the effect sizes that they report. Importantly, the effect size obviously is not known in advance. Particularly valuable is replication in an independent sample. If a finding is replicated in an independent sample in a tissue relevant to disease (e.g. brain tissue in psychiatric disorders), this especially adds confidence in the finding. This has been done for instance in a study of methylation differences in buccal cells from depression discordant twin pairs, which followed up and replicated some of the top differentially methylated loci in twins in post-mortem cerebellum tissue from depression cases and controls (99).

4.6 **Conclusions and future directions**

It is clear that the development of many human traits (especially those relating to behavior and brain functions) involves continuous interactions between the DNA sequence, gene expression, the developing trait, and the individual's external environment. The classical twin model, when applied to cross-sectional data, may be viewed as one snapshot of the sources that turn out to explain most of the individual differences in the outcome at the end of this cascade of events up until the age where the outcome is assessed. More in-depth questions about the causes of variation across life can be examined with longitudinal epigenetic data from twins. For example, longitudinal DNA samples collected in MZ and DZ twins across important periods of life (e.g. birth, early childhood, puberty, adolescence, middle-age, and advanced age) would allow to investigate when epigenetic variation at loci that contribute to the development of diseases arises, and what the role of genetic and environmental influence is in the emergence of disease-relevant epigenetic changes over time. Such studies may for instance identify life periods where the impact of family environment or unique environmental influences on epigenetic mechanisms is most pronounced, providing insights into the origin and timing of (potentially modifiable) epigenetic pathways that regulate the expression of heritable diseases. Studies of discordant monozygotic twins with DNA and phenotype information collected at multiple time points will be valuable to investigate dynamic relationships between disease and epigenetic signatures to allow insight into for example the role of epigenetic mechanisms in disease patterns of remission and relapse (which are often observed in psychiatric diseases such as depression, addiction and psychosis).

Epigenetic mechanisms are an attractive target for disease treatment. Many potential epigenetic drugs are being investigated, especially in the area of cancer treatment, that are targeting various components of the epigenetic machinery, including DNA methylation and histone modifications (100). One epigenetic drug that is already applied in treatment of psychiatric disorders is valproic acid, which is a non-specific histone deacetylase (HDAC) inhibitor (Box 4.1). Current epigenetic drugs act on general epigenetic processes, and a challenge will be to target specific genomic regions or gene networks within specific tissues/cells. It also remains to be seen to what extent modification of gene expression programs in adulthood will be able to reverse neurodevelopmental disorders. Encouragingly, it has been reported that the neurological deficits of Rett syndrome in mice carrying a mutation in *Mecp2* can be rescued even in adult animals by restoring *Mecp2* gene expression (101).

Perhaps the most important value of monozygotic twins to epigenetic research lies in the potential to unravel to what extent the expression of (heritable) human diseases is modifiable by stochastic or environmentally induced epigenetic changes. Clearly, the best way to see how much an epigenome may change due to stochastic causes or under the influence of environment is by studying the epigenomes of individuals who have identical DNA-sequences but who may become exposed to different environments. In this sense, the monozygotic twin design may be viewed as a natural experiment. We anticipate that epigenetic studies will increasingly use meta-analysis designs and replication strategies similar to GWAS and that an important role for MZ twin studies in the future will lie in follow-up analysis of loci identified in ordinary case-control EWAS to investigate whether epigenetic variation at these loci is associated with the trait or disease independent of genetic variation, and to investigate whether epigenetic variation at these loci may be subject to environmental influences. Obviously, twin studies, in contrast to experimental animal studies, do not involve an experimental intervention that is under the control of the researcher. Rather, twin researchers rely on naturally occurring exposures in human populations and on the natural occurrence of discordance for disease or behavioral phenotypes in monozygotic twins, although it is conceivable that the epigenetic effects of, for example, lifestyle interventions such as running therapy or diet, can be studied in MZ twin pairs, where one twin receives the treatment and the other twin is put on a waiting list or receives a placebo.

Acknowledgements

This work was supported by: Biobanking and Biomolecular Resources Research Infrastructure (BBMRI –NL, 184.021.007); Royal Netherlands Academy of Science Professor Award (PAH/6635) to DIB. JvD is supported by ACTION. ACTION receives funding from the European Union Seventh Framework Program (FP7/2007-2013) under grant agreement no 602768.

References

1. **Mbarek H, Steinberg S, Nyholt DR, Gordon SD, Miller MB, McRae AF**, et al. Identification of common genetic variants influencing spontaneous dizygotic twinning and female fertility. Am J Hum Genet. 2016;**98**(5):898–908.
2. **Boomsma D, Busjahn A, Peltonen L.** Classical twin studies and beyond. Nat Rev Genet. 2002;**3**(11):872–82.
3. **Russo VEA, Martienssen RA, Riggs AD.** Epigenetic mechanisms of gene regulation. Cold Spring Harb Lab Press. 1996;692.
4. **Miller CA, Sweatt JD.** Covalent modification of DNA regulates memory formation. Neuron. 2007;**53**(6):857–69.
5. **Miller CA, Campbell SL, Sweatt JD.** DNA methylation and histone acetylation work in concert to regulate memory formation and synaptic plasticity. Neurobiol Learn Mem. 2008;**89**(4):599–603.
6. **Weaver ICG, Cervoni N, Champagne F a, D'Alessio AC, Sharma S, Seckl JR**, et al. Epigenetic programming by maternal behavior. Nat Neurosci. 2004;**7**(8):847–54.
7. **Maurano MT, Humbert R, Rynes E, Thurman RE, Haugen E, Wang H**, et al. Systematic localization of common disease-associated variation in regulatory DNA. Science (80). 2012;**337**(6099):1190–5.
8. **Gamazon ER, Badner J a, Cheng L, Zhang C, Zhang D, Cox NJ**, et al. Enrichment of cis-regulatory gene expression SNPs and methylation quantitative trait loci among bipolar disorder susceptibility variants. Mol Psychiatry. 2012;(18):340–6.
9. **Ye K, Beekman M, Lameijer E-W, Zhang Y, Moed MH, van den Akker EB**, et al. Aging as accelerated accumulation of somatic variants: Whole-genome sequencing of centenarian and middle-aged monozygotic twin pairs. Twin Res Hum Genet. 2013;**16**(6):1026–32.
10. **Zwijnenburg PJG, Meijers-Heijboer H, Boomsma DI.** Identical but not the same: The value of discordant monozygotic twins in genetic research. Am J Medl Genet B Neuropsychiatr Genet. 2010;**153**:1134–49.
11. **Reik W.** Stability and flexibility of epigenetic gene regulation in mammalian development. Nature. 2007;**447**(7143):425–32.
12. **Lubin FD, Roth TL, Sweatt JD.** Epigenetic regulation of BDNF gene transcription in the consolidation of fear memory. J Neurosci. 2008;**28**(42):10576–86.
13. **Li B, Carey M, Workman JL.** The role of chromatin during transcription. Cell. 2007;**128**:707–19.
14. **Goldberg AD, Allis CD, Bernstein E.** Epigenetics: A landscape takes shape. Cell. 2007;**128**: 635–8.
15. **Bernstein BE, Meissner A, Lander ES.** The mammalian epigenome. Cell. 2007;**128**:669–81.
16. **Laurent L, Wong E, Li G, Huynh T, Tsirigos A, Ong CT**, et al. Dynamic changes in the human methylome during differentiation. Genome Res. 2010;**20**(3):320–31.
17. **Lister R, Pelizzola M, Dowen RH, Hawkins RD, Hon G, Tonti-Filippini J**, et al. Human DNA methylomes at base resolution show widespread epigenomic differences. Nature. 2009;**462**(7271):315–22.
18. **Ramsahoye BH, Biniszkiewicz D, Lyko F, Clark V, Bird a P, Jaenisch R.** Non-CpG methylation is prevalent in embryonic stem cells and may be mediated by DNA methyltransferase 3a. PNAS. 2000;**97**(10):5237–42.
19. **Lister R, Mukamel EA, Nery JR, Urich M, Puddifoot CA, Johnson ND**, et al. Global epigenomic reconfiguration during mammalian brain development. Science. 2013;**341**(6146):1237905.
20. **Luo C, Ecker JR.** Exceptional epigenetics in the brain. Science (80-). 2015;**348**(6239):1094–5.
21. **Goll MG, Bestor TH.** Eukaryotic cytosine methyltransferases. Annu Rev Biochem. 2005;**74**:481–514.
22. **Stadler MB, Murr R, Burger L, Ivanek R, Lienert F, Schöler A**, et al. DNA-binding factors shape the mouse methylome at distal regulatory regions. Vol. 480, Nature. 2011. p. 490–5.
23. **Thurman RE, Rynes E, Humbert R, Vierstra J, Maurano MT, Haugen E**, et al. The accessible chromatin landscape of the human genome. Nature. 2012;**489**(7414):75–82.
24. **Bird A, Taggart M, Frommer M, Miller OJ, Macleod D.** A fraction of the mouse genome that is derived from islands of nonmethylated, CpG-rich DNA. Cell. 1985;**40**(1):91–9.

25. **Jones PA.** Functions of DNA methylation: islands, start sites, gene bodies and beyond. Nat Rev Genet. 2012;**13**(7):484–92.

26. **Watt F, Molloy PL.** Cytosine methylation prevents binding to DNA of a HeLa cell transcription factor required for optimal expression of the adenovirus major late promoter. Genes Dev. 1988;**2**(9):1136–43.

27. **Hark AT, Schoenherr CJ, Katz DJ, Ingram RS, Levorse JM, Tilghman SM.** CTCF mediates methylation-sensitive enhancer-blocking activity at the H19/Igf2 locus. Nature. 2000;**405**(6785):486–9.

28. **Nan X, Ng HH, Johnson CA, Laherty CD, Turner BM, Eisenman RN,** et al. Transcriptional repression by the methyl-CpG-binding protein MeCP2 involves a histone deacetylase complex. Nature. 1998;**393**(6683):386–9.

29. **Wade PA, Gegonne A, Jones PL, Ballestar E, Aubry F, Wolffe AP.** Mi-2 complex couples DNA methylation to chromatin remodelling and histone deacetylation. Nat Genet. 1999;**23**:1–5.

30. **Yasukochi Y, Maruyama O, Mahajan MC, Padden C, Euskirchen GM, Schulz V** et al. X chromosome-wide analyses of genomic DNA methylation states and gene expression in male and female neutrophils. PNAS. 2010;**107**(8):3704–9.

31. **Choufani S, Shapiro JS, Susiarjo M, Butcher DT, Grafodatskaya D, Lou Y,** et al. A novel approach identifies new differentially methylated regions (DMRs) associated with imprinted genes. Genome Res. 2011;**21**(3):465–76.

32. **Raynal NJM, Si J, Taby RF, Gharibyan V, Ahmed S, Jelinek J,** et al. DNA methylation does not stably lock gene expression but instead serves as a molecular mark for gene silencing memory. Cancer Res. 2012;**72**(5):1170–81.

33. **Bibikova M, Barnes B, Tsan C, Ho V, Klotzle B, Le JM,** et al. High density DNA methylation array with single CpG site resolution. Genomics. 2011;**98**(4):288–95.

34. **Sandoval J, Heyn HA, Moran S, Serra-Musach J, Pujana MA, Bibikova M,** et al. Validation of a DNA methylation microarray for 450,000 CpG sites in the human genome. Epigenetics. 2011;**6**(6):692–702.

35. **Moran S, Arribas C, Esteller M.** Validation of a DNA methylation microarray for 850,000 CpG sites of the human genome enriched in enhancer sequences. Epigenomics. 2016;**8**(3):389–99.

36. **Krueger F, Kreck B, Franke A, Andrews SR.** DNA methylome analysis using short bisulfite sequencing data. Nat Methods. 2012;**9**(2):145–51.

37. **Lister R, Ecker JR.** Finding the fifth base: Genome-wide sequencing of cytosine methylation. Vol. **19**, Genome Research. 2009. p. 959–66.

38. **Luger K, Mäder a W, Richmond RK, Sargent DF, Richmond TJ.** Crystal structure of the nucleosome core particle at 2.8 A resolution. Nature. 1997;**389**(6648):251–60.

39. **Olins AL, Olins DE.** Spheroid chromatin units (v bodies). Science. 1974;**183**(4122):330–2.

40. **Olins DE, Olins AL.** Chromatin history: Our view from the bridge. Nat Rev Mol Cell Biol. 2003;**4**(10):809–14.

41. **Kouzarides T.** Chromatin modifications and their function. Cell. 2007;**128**:693–705.

42. **Ernst J, Kheradpour P, Mikkelsen TS, Shoresh N, Ward LD, Epstein CB,** et al. Mapping and analysis of chromatin state dynamics in nine human cell types. Nature. 2011;**473**(7345):43–9.

43. **Jenuwein T, Allis CD.** Translating the histone code. Science (80-). 2001;**293**(5532):1074–80.

44. **Santos-Rosa H, Schneider R, Bannister AJ, Sherriff J, Bernstein BE, Emre NCT,** et al. Active genes are tri-methylated at K4 of histone H3. Nature. 2002;**419**(6905):407–11.

45. **Brown CE, Lechner T, Howe L, Workman JL.** The many HATs of transcription coactivators. Trends Biochem Sci. 2000;**25**:15–9.

46. **Horvath S.** DNA methylation age of human tissues and cell types. Genome Biol. 2013;**14**(10):R115.

47. **Feil R, Fraga MF.** Epigenetics and the environment: emerging patterns and implications. Nat Rev Genet. 2012;**13**(2):97–109.

48. **Amir RE, Van den Veyver I. B, Wan M, Tran CQ, Francke U, Zoghbi HY.** Rett syndrome is caused by mutations in X-linked MECP2, encoding methyl-CpG-binding protein 2. Nat Genet. 1999;**23**:185–8.

49. **Joehanes R, Just AC, Marioni RE, Pilling LC, Reynolds LM, Mandaviya PR,** et al. Epigenetic signatures of cigarette smoking. Circ Cardiovasc Genet. 2016;**9**(5):436–47.
50. **Joubert BR, Felix JF, Yousefi P, Bakulski KM, Just AC, Breton C,** et al. DNA Methylation in newborns and maternal smoking in pregnancy: Genome-wide Consortium Meta-analysis. Am J Hum Genet. 2016;**98**(4):680–96.
51. **Panni T, Mehta AJ, Schwartz JD, Baccarelli AA, Just AC, Wolf K,** et al. Genome-wide analysis of DNA methylation and fine particulate matter air pollution in three study populations: KORA F3, KORA F4, and the normative aging study. Environ Health Perspect. 2016;**124**(7):983–90.
52. **Hanna CW, Bloom MS, Robinson WP, Kim D, Parsons PJ, vom Saal FS,** et al. DNA methylation changes in whole blood is associated with exposure to the environmental contaminants, mercury, lead, cadmium and bisphenol A, in women undergoing ovarian stimulation for IVF. Hum Reprod. 2012;**27**(5):1401–10.
53. **Houtepen LC, Vinkers CH, Carrillo-Roa T, Hiemstra M, van Lier PA, Meeus W,** et al. Genome-wide DNA methylation levels and altered cortisol stress reactivity following childhood trauma in humans. Nat Commun. 2016;7:10967.
54. **Suderman M, Borghol N, Pappas JJ, Pinto Pereira SM, Pembrey M, Hertzman C,** et al. Childhood abuse is associated with methylation of multiple loci in adult DNA. BMC Med Genomics. 2014;**7**:1–12.
55. **Essex MJ, Thomas Boyce W, Hertzman C, Lam LL, Armstrong JM, Neumann SMA,** et al. Epigenetic vestiges of early developmental adversity: Childhood stress exposure and DNA methylation in adolescence. Child Dev. 2013;**84**(1):58–75.
56. **Needham BL, Smith JA, Zhao W, Wang X, Mukherjee B, Kardia SLR,** et al. Life course socioeconomic status and DNA methylation in genes related to stress reactivity and inflammation: The multi-ethnic study of atherosclerosis. Epigenetics. 2015;**10**(10):958–69.
57. **Bonder MJ, Luijk R, Zhernakova D V, Moed M, Deelen P, Vermaat M,** et al. Disease variants alter transcription factor levels and methylation of their binding sites. Nat Genet. 2016;**49**:131–8.
58. **Hannon E, Spiers H, Viana J, Pidsley R, Burrage J, Murphy TM,** et al. Methylation QTLs in the developing brain and their enrichment in schizophrenia risk loci. Nat Neurosci. 2015;**19**:1–9.
59. **Murphy TM, Crawford B, Dempster EL, Hannon E, Burrage J, Turecki G,** et al. Methylomic profiling of cortex samples from completed suicide cases implicates a role for PSORS1C3 in major depression and suicide. Transl Psychiatry. 2017;7(1):e989.
60. **Dick KJ, Nelson CP, Tsaprouni L, Sandling JK, Aïssi D, Wahl S,** et al. DNA methylation and body-mass index: A genome-wide analysis. Lancet. 2014;**383**(9933):1990–8.
61. **Wahl S, Drong A, Lehne B, Loh M, Scott WR, Kunze S,** et al. Epigenome-wide association study of body mass index, and the adverse outcomes of adiposity. Nature. 2016;**541**:81–6.
62. **Li R, Montpetit A, Rousseau M, Wu SYM, Greenwood CMT, Spector TD,** et al. Somatic point mutations occurring early in development: A monozygotic twin study. J Med Genet. 2014;**51**(1):28–34.
63. **Ollikainen M, Smith KR, Joo EJH, Ng HK, Andronikos R, Novakovic B,** et al. DNA methylation analysis of multiple tissues from newborn twins reveals both genetic and intrauterine components to variation in the human neonatal epigenome. Hum Mol Genet. 2010;**19**(21):4176–88.
64. **Gordon L, Joo JE, Powell JE, Ollikainen M, Novakovic B, Li X,** et al. Neonatal DNA methylation profile in human twins is specified by a complex interplay between intrauterine environmental and genetic factors, subject to tissue-specific influence. Genome Res. 2012;**22**(8):1395–406.
65. **Gordon L, Joo JHE, Andronikos R, Ollikainen M, Wallace EM, Umstad MP,** et al. Expression discordance of monozygotic twins at birth: Effect of intrauterine environment and a possible mechanism for fetal programming. Epigenetics. 2011;**6**(5):579–92.
66. **Marceau K, McMaster MTB, Smith TF, Daams JG, van Beijsterveldt CEM, Boomsma DI,** et al. The prenatal environment in twin studies: A review on chorionicity. Behav Genet. 2016;**46**(3):286–303.
67. **Martino D, Loke YJ, Gordon L, Ollikainen M, Cruickshank MN, Saffery R,** et al. Longitudinal, genome-scale analysis of DNA methylation in twins from birth to 18 months of age reveals rapid epigenetic change in early life and pair-specific effects of discordance. Genome Biol. 2013;**14**(5):R42.

68. **Goyal R, Reinhardt R, Jeltsch A.** Accuracy of DNA methylation pattern preservation by the Dnmt1 methyltransferase. Nucleic Acids Res. 2006;**34**(4):1182–8.
69. **Tiberio G.** MZ female twins discordant for X-linked diseases: A review. Acta Genet Med Gemellol (Roma). 1994;**43**(3–4):207–14.
70. **Richards CS, Watkins SC, Hoffman EP, Schneider NR, Milsark IW, Katz KS,** et al. Skewed X inactivation in a female MZ twin results in Duchenne muscular dystrophy. Am J Hum Genet. 1990;**46**(4):672–81.
71. **Rosa A, Picchioni MM, Kalidindi S, Loat CS, Knight J, Toulopoulou T,** et al. Differential methylation of the X-chromosome is a possible source of discordance for bipolar disorder female monozygotic twins. Am J Med Genet Part B Neuropsychiatr Genet. 2008;**147**(4):459–62.
72. **Lin BD, Hottenga J-J, Abdellaoui A, Dolan C V, de Geus EJC, Kluft C,** et al. Causes of variation in the neutrophil-lymphocyte and platelet-lymphocyte ratios: A twin-family study. Biomark Med. 2016;**10**(10):1061–72.
73. **Daca-Roszak P, Pfeifer A, Żebracka-Gala J, Rusinek D, Szybińska A, Jarząb B,** et al. Impact of SNPs on methylation readouts by Illumina Infinium HumanMethylation450 BeadChip Array: Implications for comparative population studies. BMC Genomics. 2015;**16**(1):1003.
74. **van Dongen J, Nivard MG, Baselmans BML, Zilhão NR, Ligthart L, Heijmans BT,** et al. Epigenome-wide association study of aggressive behavior. Twin Res Hum Genet. 2015;**18**(6):686–98.
75. **Wong CCY, Meaburn EL, Ronald A, Price TS, Jeffries AR, Schalkwyk LC,** et al. Methylomic analysis of monozygotic twins discordant for autism spectrum disorder and related behavioural traits. Mol Psychiatry. 2014;**19**(4):495–503.
76. **Mori Y, Matsui T, Omote D, Fukuda M.** Small GTPase Rab39A interacts with UACA and regulates the retinoic acid-induced neurite morphology of Neuro2A cells. Biochem Biophys Res Commun. 2013;**435**(1):113–9.
77. **van Dongen J, Slagboom PE, Draisma HHM, Martin NG, Boomsma DI.** The continuing value of twin studies in the omics era. Nat Rev Genet. 2012;**13**(9):640–53.
78. **van Dongen J, Nivard MG, Willemsen G, Hottenga J-J, Helmer Q, Dolan C V,** et al. Genetic and environmental influences interact with age and sex in shaping the human methylome. Nat Commun. 2016;7:11115.
79. **McRae AF, Powell JE, Henders AK, Bowdler L, Hemani G, Shah S,** et al. Contribution of genetic variation to transgenerational inheritance of DNA methylation. Genome Biol. 2014;**15**(5):R73.
80. **van Dongen J, Ehli EA, Slieker RC, Bartels M, Weber ZM, Davies GE,** et al. Epigenetic variation in monozygotic twins: A genome-wide analysis of DNA methylation in buccal cells. Genes (Basel). 2014;**5**(2):347–65.
81. **Boks MP, Derks EM, Weisenberger DJ, Strengman E, Janson E, Sommer IE,** et al. The relationship of DNA methylation with age, gender and genotype in twins and healthy controls. PLoS One. 2009;**4**(8).
82. **Coolen MW, Statham AL, Qu W, Campbell MJ, Henders AK, Montgomery GW,** et al. Impact of the genome on the epigenome is manifested in DNA methylation patterns of imprinted regions in monozygotic and dizygotic twins. PLoS One. 2011;**6**(10).
83. **Gervin K, Hammerø M, Akselsen HE, Moe R, Nygard H, Brandt I,** et al. Extensive variation and low heritability of DNA methylation identified in a twin study. Genome Res. 2011;**21**(11):1813–21.
84. **Heijmans BT, Kremer D, Tobi EW, Boomsma DI, Slagboom PE.** Heritable rather than age-related environmental and stochastic factors dominate variation in DNA methylation of the human IGF2/H19 locus. Hum Mol Genet. 2007;**16**(5):547–54.
85. **Wong CCY, Caspi A, Williams B, Craig IW, Houts R, Ambler A,** et al. A longitudinal study of epigenetic variation in twins. Epigenetics. 2010;**5**(6):516–26.
86. **Kaminsky Z, Petronis A, Wang S-C, Levine B, Ghaffar O, Floden D,** et al. Epigenetics of personality traits: An illustrative study of identical twins discordant for risk-taking behavior. Twin Res Hum Genet. 2008;**11**(1):1–11.

87. **Bell JT, Tsai PC, Yang TP, Pidsley R, Nisbet J, Glass D,** et al. Epigenome-wide scans identify differentially methylated regions for age and age-related phenotypes in a healthy ageing population. PLoS Genet. 2012;**8**(4).
88. **Grundberg E, Meduri E, Sandling JK, Hedman ÅK, Keildson S, Buil A,** et al. Global analysis of dna methylation variation in adipose tissue from twins reveals links to disease-associated variants in distal regulatory elements. Am J Hum Genet. 2013;**93**(5):876–90.
89. **Livshits G, Gao F, Malkin I, Needhamsen M, Xia Y, Yuan W,** et al. Contribution of heritability and epigenetic factors to skeletal muscle mass variation in United Kingdom twins. J Clin Endocrinol Metab. 2016;**101**(6):2450–9.
90. **Dolan C V, Nivard MG, Dongen J Van, Sluis, Sophie van der, Boomsma DI.** Methylation as an epigenetic source of random genetic effects in the classical twin design. Adv Genomics Genet. 2015;(5):305–15.
91. **Nivard MG, Middeldorp CM, Lubke G, Hottenga J-J, Abdellaoui A, Boomsma DI,** et al. Detection of gene–environment interaction in pedigree data using genome-wide genotypes. Eur J Hum Genet. 2016;**24**(July 2015):1–7.
92. **Purcell S.** Variance components models for gene-environment interaction in twin analysis. Twin Res. 2002;**5**(6):554–71.
93. **Hannon E, Dempster E, Viana J, Burrage J, Smith AR, Macdonald R,** et al. An integrated genetic-epigenetic analysis of schizophrenia: Evidence for co-localization of genetic associations and differential DNA methylation. Genome Biol. 2016;**17**(1):176.
94. **Dekkers KF, Slagboom PE, Jukema JW.** The multifaceted interplay between lipids and epigenetics. Curr Opin Lipidol. 2016;**27**(3):288–94.
95. **Reinius LE, Acevedo N, Joerink M, Pershagen G, Dahlen SE, Greco D,** et al. Differential DNA methylation in purified human blood cells: Implications for cell lineage and studies on disease susceptibility. PLoS One. 2012;7(1932–6203 (Electronic)):e41361.
96. **Houseman E, Accomando WP, Koestler DC, Christensen BC, Marsit CJ, Nelson HH,** et al. DNA methylation arrays as surrogate measures of cell mixture distribution. BMC Bioinformatics. 2012;**13**(1):86.
97. **Bakulski KM, Halladay A, Hu VW, Mill J, Fallin MD.** Epigenetic Research in Neuropsychiatric Disorders: The "Tissue Issue." Curr Behav Neurosci Reports. 2016;3(3):264–74.
98. **Tsai PC, Bell JT.** Power and sample size estimation for epigenome-wide association scans to detect differential DNA methylation. Int J Epidemiol. 2015;**44**(4):1429–41.
99. **Dempster EL, Wong CCY, Lester KJ, Burrage J, Gregory AM, Mill J,** et al. Genome-wide methylomic analysis of monozygotic twins discordant for adolescent depression. Biol Psychiatry. 2014;**76**(12):977–83.
100. **Park M, Keung AJ, Khalil AS.** The epigenome: the next substrate for engineering. Genome Biol. 2016;**17**(1):183.
101. **Guy J, Gan J, Selfridge J, Cobb S, Bird A.** Reversal of neurological defects in a mouse model of Rett syndrome. Science. 2007;**315**(5815):1143–7.

Chapter 5

Genome-wide association studies

Thomas W Mühleisen and Sven Cichon

5.1 The evolution of genetic analysis of behavioral traits and disorders

Human behavior, with its many phenotypic facets, shows considerable interindividual variability. Over a long period of time, scientific discussions tried to answer the question of whether nature or nurture make the most important contributions to individual differences in behavioral phenotypes or traits. Twin and adoption studies provide strong evidence that they are in fact substantially influenced by genetic factors, but these studies explain only a portion of the phenotypic variance. It is now widely acknowledged that behavioral traits are genetically complex which means that genetic and environmental factors contribute to them. Identifying molecular mechanisms through which genetic factors lead to the observed individual differences is an important step to understanding the biology of these traits in general and their influence on neuropsychiatric disorders. The question is: how can genes or genetic variants involved be identified? In principle, there are different genetic strategies that can be applied, but their success is critically influenced by the genetic architecture of the trait under study, that is, the nature and effect size of the contributing genetic variants.

Twenty to thirty years ago, the genetic architecture of complex behavioral traits and disorders that are common in the general population was unclear. There was hope that monogenic forms exist in at least a fraction of the individuals or patients which were caused by a single mutation somewhere in the genome. Such forms would best be identified by systematic linkage analysis in pedigrees of families showing the trait. The principle is straightforward: using a few hundred evenly spaced genetic markers across the whole genome (either microsatellite markers or single-nucleotide polymorphisms (SNPs)) that allow researchers to trace back the inheritance of each homologous chromosome pair in each family linkage has the potential to locate the trait-causing genes by the virtue of chromosomal position alone, without any prior knowledge of the molecular basis of the trait. Hopes of such monogenic forms of behavioral traits and disorders have not materialized since then.

During the 1990s and early 2000s, another major research strategy to identify genes involved in genetically complex phenotypes was represented by candidate gene association studies. The principle of these studies is to detect allelic variants of a particular gene that are significantly more common (overrepresented) or less common (underrepresented) in individuals showing the phenotype of interest than they are in the comparison population. Candidate gene studies are statistically more powerful than linkage studies in detecting allelic variants of small effect (due to reduced penetrance), provided that these variants are relatively common in the population (> 5% allele frequency). The candidate genes were usually selected based on molecular hypotheses of an involvement in the etiology of the trait, for example, a biological pathway. A disadvantage of these focused candidate gene studies is that they leave a large proportion of the genome uninvestigated. A systematic, comprehensive investigation of the genetic variability present in the candidate genes could not be performed due to limited knowledge on their variability in the population.

Retrospectively, the majority of these studies were heavily underpowered, using sample sizes of several hundred individuals at most. As a consequence, a confusing picture resulted in the literature, with numerous association findings that did not replicate in independent studies (1). One possible explanation for these inconclusive findings was that the chosen candidate genes were not the right ones or that the studied genetic variation at each gene locus was incomplete. Researchers began to conceptualize more systematic ways to perform studies of associations and the idea to do these in a genome-wide fashion was born. Still, many prerequisites were missing at that point, such as comprehensive information on the number and location of SNPs in the population as well as technological solutions to undertake SNP genotyping in a rapid and cost-effective way. However, an amazing development took place at the turn of the millennium that eventually paved the way for the first genome-wide association studies (GWAS).

5.2 The GWAS era

The completion of the Human Genome Project (1990–2006) marked the beginning of a new era in human genetics. For the first time, a complete "reference sequence" was available for the 3 billion bases of DNA that make up our nuclear genome (2). The published sequence was invariable, but it was clear at that time that genetic differences between individuals exist. Some data on variability at selected chromosomal loci (preferably coding regions) were already available, but a systematic overview resulting from the investigation of large numbers of individuals was still missing. In addition to that, an overview of population differences was needed. To address these questions, the International HapMap Project was formed. In the following, strong efforts were made to describe the extent of interindividual genetic variation systematically. The genomic variability observed was in fact quite high: two randomly chosen genomes from a population differ on average at every 1000th base position which makes approximately 3 million base differences across the genome (3). These variable positions, SNPs, are the most abundant type of genetic variation in humans. The majority of SNPs have two alleles in the population: a more frequent (major) allele and a less frequent (minor) allele. By definition, the minor allele frequency (MAF) of an SNP is at least 1%, single-base changes with a lower MAF are called rare variants. Each SNP has a unique identifier called reference sequence identifier (rsID), for example, rs1064395, and is stored in a public database named dbSNP (4).

Another important feature of the genome that was discovered at that time and that had a strong influence on the conception of GWAS was the haplotype block structure. The biological explanation behind this structure is that the human genome is made up of recombination hotspots, where most of the recombinations in meiosis take place. The genomic stretches between these recombination hotspots are the haplotype blocks. SNPs within a block tend to be inherited together. The phenomenon that addresses the strong correlation of SNPs is designated as linkage disequilibrium (LD). Genetically, they carry the same information. This has important consequences for the selection of SNPs for systematic genetic association studies: to do this in the most efficient way, it is sufficient to know the haplotype block structure of the genome and select the minimum set of SNPs in each block for genotyping that describes all common haplotypes of a block in a population. These SNPs are the so-called haplotype tagging SNPs or tag SNPs. Due to different population histories across the world, there are some differences in haplotype block structures between populations. Most of the common haplotypes are present in all populations, however at different haplotype frequencies. It is the merit of the International HapMap Project (2002–2009) that soon after the completion of the Human Genome Project it provided a comprehensive catalog of SNPs and information on the haplotype block structure and tagged SNPs in different populations. This information was subsequently used by companies such as Illumina and Affymetrix to develop

SNP microarrays containing several hundred thousand SNPs that allow rapid and cost-effective genotyping of SNPs across the genome (5,6).

With the advent of next-generation sequencing and the availability of whole-genome sequences from thousands of individuals, the 1000 Genomes Project replaced the HapMap Project and added more SNPs, rare single-base variation as well as small structural variants in further reference populations (7). Driven by these data, the latest SNP microarrays investigate millions of common and low-frequency SNPs (down to 1% MAF). To date, between 9 and 13 Mio SNPs are deposited in dbSNP, depending on the population origin (African, American, Asian, European).

These academic and technological efforts have enabled GWAS in tens of thousands of individuals, at affordable costs, to shed light on the genetic architecture of complex traits. In the following, we describe how a GWAS is conducted.

5.3 The concept of GWAS

A GWAS allows a systematic, genome-wide search for genetic variants that are significantly more frequent or less frequent in individuals with a particular trait compared to individuals without the trait. The goal is to identify causal variants (genes and regulatory elements of genes) underlying the biology of the trait. A major advantage of this design is that there is no requirement for an a priori hypothesis about a biological mechanism of the investigated trait. There is a common misconception on GWAS that it is a hypothesis-free research strategy. This is incorrect since it is based on a very clear hypothesis that SNPs exist in the genome, possibly falling into distinct biological pathways, that influence the susceptibility for the complex trait under study. GWAS therefore have the potential to identify both expected as well as novel biological mechanisms involved in the trait.

At the time when GWAS were initiated, there was a vivid discussion in the scientific community about the nature of the genetic factors underlying complex traits. The "Common Disease, Common Variant" (CDCV) hypothesis argued that genetic variations with appreciable frequency in the population at large, but relatively low penetrance (or the probability that a carrier of the relevant variants will express the disease), are the major contributors to genetic susceptibility to common diseases. The "Common Disease, Rare Variant" (CDRV) hypothesis, on the other hand, argued that multiple rare DNA sequence variations, each with relatively high penetrance, are the major contributors to genetic susceptibility to common diseases. Today, it is widely accepted that phenotypic variation in complex traits is mediated by both common and rare variants. Both hypotheses have their place in current research efforts (8). For the detection of common variants contributing to a complex trait, GWAS are the strategy of choice to identify them. For rare variants, next-generation sequencing-based studies in either samples of unrelated individuals (for a disease, patients, and healthy controls) or families with the trait will be more successful.

In the following, we describe the concept of a GWAS. The workflow and example focus on the case-control design. Over the years, protocols of quality control (QC) have been developed aiming to identify and remove DNA samples and markers that introduce a bias to the study (false-positive and false-negative associations). It is beyond the scope of this chapter to describe an analysis plan in detail. However, we would like to discuss some important aspects of QC during the workflow.

Before starting a GWAS, samples are recruited from a population (Figure 5.1a). The case group comprises individuals with a clinical diagnosis, according to standardized diagnostic instruments. Care should be taken when selecting a control group, it should be taken from the same population as the case group to avoid population-specific genotype differences. Furthermore, there should be no relatedness between individuals of the case group, control group, and between both groups (no first- and second-degree relatives).

Once venous blood samples have been taken from all the individuals who will be included in the study, lab work is necessary to generate the genotype data needed for the GWAS. First, genomic DNA is extracted from venous blood of each individual and introduced to high-throughput genotyping (Figure 5.1b). Using commercially available SNP microarrays, genotyping can be performed in parallel for a large number of SNPs. A microarray scanner reads out the genotype information on the microarray which is represented by a flourescent signal for each SNP genotype that is placed at a defined position on the microarray. The SNPs included in the SNP microarrays have two alleles, A and B. Homozygosity for allele A (AA) produces a red fluorescent signal, homozygosity for allele B (BB) a green signal, and heterozygosity (AB) is indicated by a yellow signal (superimposed red and green). The read-out (also named "genotype calling") is done in an automated fashion by the microarray reader, using a mathematical clustering algorithm. All signals are translated into the actual genotypes of the respective SNPs (e.g. red signal at position XY represents genotype G/G for SNP rs12345). Depending on the quality of each fluorescent signal, the calling algorithm also provides a confidence value (probability) for each called genotype. If a threshold is not reached, the genotype is considered "un-called." High failure rates per SNP and per individual (> 5% missingness) reduce the genomic coverage and power of a GWAS. Therefore, imputation algorithms have been developed to infer missing SNPs and genotypes using phased haplotypes from reference populations like HapMap or 1000 Genomes (9). Imputation is regularly used for meta-analysis of GWAS to harmonize SNP sets from individual GWAS that have been genotyped by different types of microarrays. Further sample QC comprises removal of individuals with discordant sex, outlying rate of heterozygosity, duplicated or related samples, or divergent ancestry. Remaining fine-scale population substructures can be corrected during the association analysis (covariates in test model). Additional marker QC includes removal of SNPs showing deviation from Hardy–Weinberg equilibrium, different missingness between cases and controls, or low MAF (depending on the sample size).

For association analysis, phenotype and genotype data are grouped using a contingency table of disease status by SNP genotype, for example, patients versus controls by AA, AB, and BB. Evidence for association is sought by a statistical analysis, for example, the chi-square test, using genotype counts (2×3 table, two degrees of freedom) or allele counts (2×2 table, one degree of freedom). Statistical power to detect an association depends on the sample size, the distribution of effect sizes of causal variants that are segregating in the population, the frequencies of those variants, and the LD between tag SNPs and the unknown causal variants (10). For an overview of association results, P-values on the Y-axis (red and blue dots) are plotted against chromosomal positions of the SNPs on the X-axis, generating the so-called Manhattan plot (Figure 5.1c). A genome-wide significant SNP (green dot) is indicated by a P value smaller than 5×10^{-8}. This threshold is considered to be a *gold standard* for a significant SNP in a systematic GWAS, taking into account the enormous multiple testing burden. The introduction of this threshold has effectively helped to minimize false-positive findings. Genome-wide significance is based on the assumption that the number of independent single-marker tests is about 1 million at a type I error rate of 0.05 ($0.05/10^6 = 5 \times 10^{-8}$). This definition follows evidence from the HapMap and 1000 Genomes Projects: genomic variability is reflected by 1 million representative SNPs that are not correlated (not in LD) with each other. Despite the strong statistical evidence that a SNP with a genome-wide significant p-value is a "true" association finding, there always remains a certain likelihood that the result is a chance finding. Therefore, replication of the top SNPs in an independent sample of cases and controls has become an important step to confirm a genome-wide significant association finding from a GWAS. In an example from the Molecular Causes of Major Mood Disorders and Schizophrenia (MooDS) consortium, the Manhattan plot shows five top loci associated with bipolar disorder (BD). One of them is located on chromosome 5p15.31 (encircled green dot) where SNP rs17826816 maps to gene *ADCY2*, a key enzyme in cAMP signaling (11). In fact, the minor allele G exhibit a risk effect by showing an odds ratio of 1.19 (Figure 5.1d). This association result indicates that individuals who carry two copies of this variant (GG homozygotes, black) are more susceptible to this neuropsychiatric disorder than GA heterozygotes (black and white) or AA homozygotes (white).

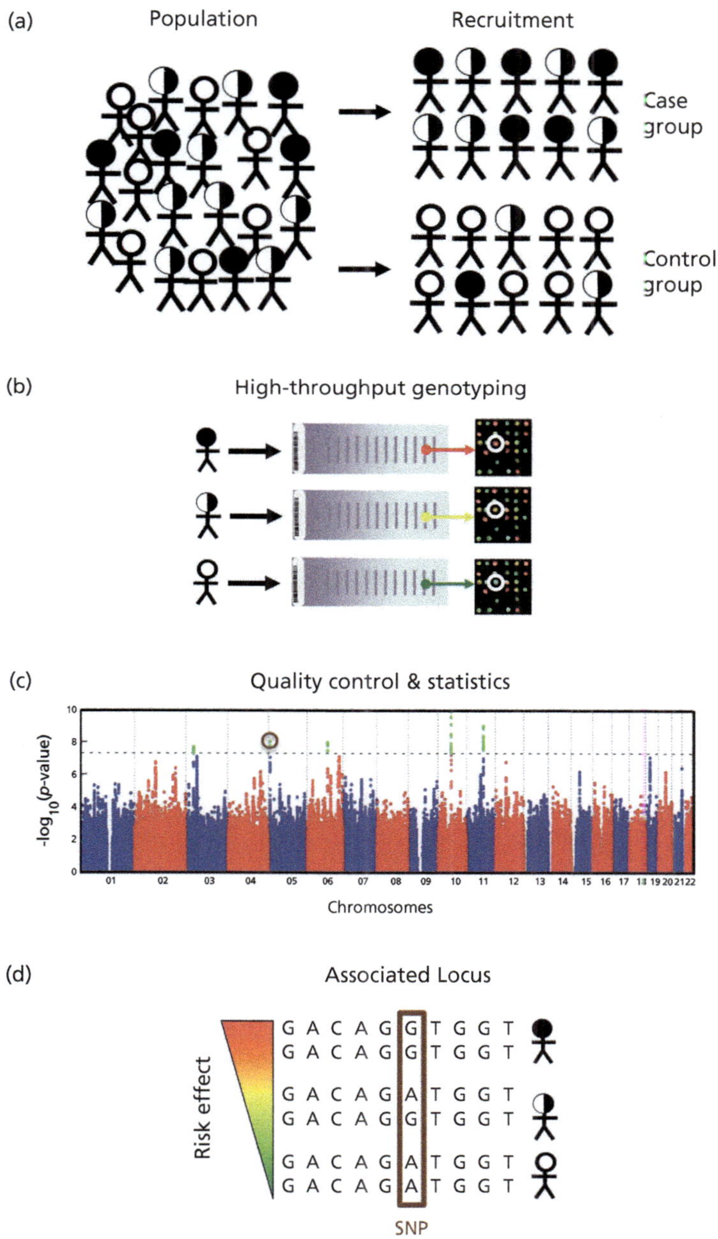

Figure 5.1 A four-step workflow of a case-control GWAS: Recruitment of samples (a); genotyping of samples (b); quality control and statistical analysis of data (c); identification of an associated SNP at a top locus (d).

Comprehensive guidelines for the design, QC, and analysis of GWAS have been reviewed elsewhere (12,13,14,15).

5.4 What we have learned from GWAS

After frustrating experiences with inconsistent and non-replicable findings from linkage and candidate gene studies, GWAS have provided novel insights into genes and their biological pathways involved in complex traits. GWAS have also demonstrated their potential to guide discovery of drug targets. In schizophrenia, for instance, a GWAS provided genome-wide significant evidence for the *DRD2* gene, validating the long-standing hypothesis of disturbed dopamine neurotransmission in disease etiology as well as the role that this receptor plays for the clinical application of antipsychotics (16). Another example refers to the association between BD and L-type voltage-gated calcium channel genes, especially *CACNA1C*, highlighting an involvement of ion channelopathies and the therapeutic potential of channel antagonists in BD (17). Very recently, a GWAS of insomnia with a sample size of about 1.3 million individuals was considered to be the largest genetic study ever and implicated 956 genes awaiting functional characterization (18).

Since the first GWAS was published by Klein and colleagues (19) who investigated age-related macular degeneration, more than 3,000 have been performed reporting more than 10,000 genome-wide significant SNP-trait associations until now. Top findings and sample characteristics from these studies are easily accessible through a public database entitled NHGRI-EBI Catalog of published GWAS (20). To be included, a finding from GWAS must meet eligibility criteria; for example, the data must provide a genome-wide coverage of more than 100,000 SNPs and the most significant association must show a p-value of $> 1 \times 10^{-5}$ based on the whole study sample (initial plus replication). To enable a structured query and visualization of data, SNP-trait associations are extracted from the literature and mapped to hierarchical terms of an ontology (Figure 5.2).

One limitation of GWAS is the burden of multiple comparisons and thereby the need to apply rigid statistical correction procedure to reduce the number of false-positive findings. There is general agreement in the scientific community that a credible GWAS finding needs a genome-wide significant p-value in the discovery (GWAS) step and statistically significant replication in at least one independent study sample. Results that fulfill these criteria are generally highly trustable.

Another limitation comes from the fact that many SNPs in a region showing association are in LD and that it is statistically very difficult to identify the responsible functional or causal SNP. However, methods like Mendelian randomization have been developed to test causal relationships between SNPs and traits in large samples (10). Apart from this, functional studies are clearly needed to follow up the initial GWAS findings and clarify the functional effects and pathophysiology.

Given the fact that effect sizes of SNPs in a polygenic architecture are smaller than previously hypothesized (genotype relative risk of 1.2 or less), single GWAS including hundreds or a few thousand subjects are often too small to provide sufficient power to reveal significance when accounting for the large number of independent test. Hence, combining association data from multiple GWAS into a well-powered single meta-analysis is utilized by consortia. Most GWAS consortia collect aggregated association statistics (summary statistics) from collaborators: marker name, marker call rate or imputation quality, effect allele with size (beta or odds ratio) and frequency, and P-value. The Psychiatric Genomic Consortium (PGC), a world-leading consortium of neuropsychiatric and behavioral disorders, follows another approach. The PGC collects individual-level genotype and phenotype data (21). The advantage is that complex QC and analysis protocols can be done centrally without burdening a large number of local analysts and

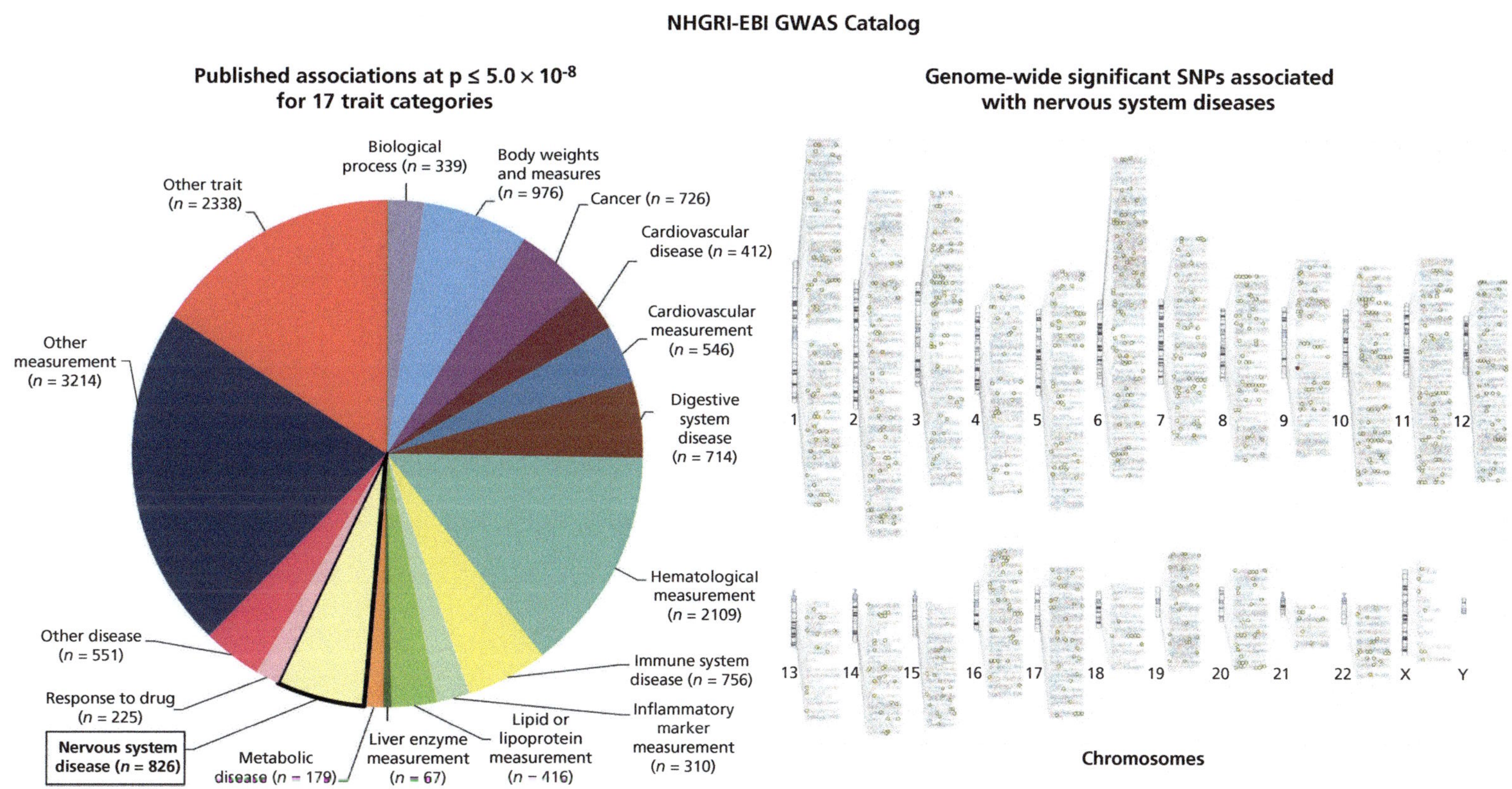

Figure 5.2 A lookup from the GWAS Catalog (as of Feb 12th 2018). The pie chart (left) summarizes the published genome-wide significant SNP-trait associations (in brackets) that have been mapped to 17 trait categories (in colors). The ideogram (right) displays the distribution of top loci (yellow dots) from "nervous system diseases" across the genome.

without the need of subsequent troubleshooting. However, data protection rules referring to the risk of participant identification may limit the sharing of individual-level data.

Another important merit of GWAS is that it has substantially contributed to reveal the polygenicity of complex traits and to estimate the contribution of common variants to its heritability (SNP-based heritability) (22). Both discoveries have facilitated the development of polygenic risk scores (PRS) to estimate disease risk and etiological overlaps across different diseases, for example, between BD, major depressive disorder (MDD), and schizophrenia (23). Until now, PRS have not been applied in a clinical study for the prediction of personalized risk of complex disease (10). However, PRS could be applied to distinguish etiological subgroups of a disorder that have different prognosis and may require different treatment. GWAS data enable further follow-up studies of which we present just a few examples. One approach is the gene set enrichment analysis (GSEA). To perform a GSEA, multiple SNPs that have surpassed a certain *P*-value threshold in a GWAS, for example, nominal significance ($P < 0.05$), are selected and mapped to genes. Then, an algorithm seeks for a significant enrichment of GWAS-supported genes in biological pathways (24). Other approaches run network algorithms on tissue- and cell type-specific gene expression to prioritize candidate genes from GWAS (25). Within a gene, it is relatively straightforward to prioritize likely functional or causal variants by annotating their predicted effects in the coding (missense, nonsense, synonymous, splice site) or intronic and 5' or 3' untranslated regions. Since many loci identified through GWAS are located outside coding regions, SNPs in non-coding and regulatory regions seem to significantly contribute to complex traits (26). Hence, tools for the analysis of allelic SNP variants on transcription factor binding sites, histone modifications, enhancers/silencers, and expressed quantitative trait loci (eQTL) are widely used. A short list of tools and databases for analysis and interpretation of SNP and GWAS data is provided at the end of the chapter.

Compared to BD, MDD, and schizophrenia, the outcome from GWAS of behavioral traits and disorders have lagged behind until recently. A major reason was that the initial sample sizes have been too small and underpowered. If we look at genome-wide significant findings in schizophrenia as a function of sample size and compare them to more successful GWAS of human traits, the schizophrenia results are well comparable to those of adult height. For schizophrenia, there is an inflection point at about 15,000 cases. Beyond this point, there is an almost linear relationship between sample size and the number of novel discoveries: approximately four new genome-wide significant SNPs per 1,000 additional cases. Hence, future GWAS of behavioral phenotypes should aim to further increase sample sizes to get a better yield of susceptibility loci. In doing so, a GWAS of neuroticism with 329,821 individuals has identified 116 genome-wide significant loci (27). For attention deficit hyperactivity disorder, 12 loci reaching genome-wide significance have been detected in 20,183 cases, while one genome-wide significant hit has been found for anorexia nervosa in 3,495 cases, as well as one for Tourette's syndrome in 4,232 cases (21).

Without doubt, GWAS is a success story for genetic analysis of complex traits and diseases, and it will also provide comprehensive insights into the biology of emotional traits and disorders. Recently, very large population-based studies including various traits/phenotypes have been launched and will also shed more light on their underlying biology. For instance, the UK Biobank Project has performed GWAS of various body traits and diseases using about 500,000 participants and provide these data to the scientific community (28). Results from the Enhancing Neuro Imaging Genetics through Meta Analysis (ENIGMA) consortium are an impressive example that the GWAS concept works on structural brain phenotypes, for example, hippocampus volume (29). Recently, the PGC has started the "Brainstorm inititative" to further exploit GWAS data by analyzing genetic correlations among brain disorders with all obtainable quantitative traits, for example, neuroimaging, eQTL, personality, IQ, etc. (21).

5.5 Future directions

Additional data to disentangle polygenic architectures will come from next-generation sequencing studies which may add further susceptibility genes (where no SNP associations have been described) and provide further support for already known genes (22). Whole-genome sequencing will also add information on the impact of rare structural variation (at the moment, microarray data have allowed to identify large structural variants (copy number variants, CNVs); not much is currently known about the impact of smaller structural variants, such as insertion/deletions (InDels) or microdeletions/microduplications in the size range between 1 kb and 50 kb.

With GWAS and also next-generation sequencing based studies, we now have very potent tools at hands to unravel the biological underpinnings of emotional traits/disorders. Clinicians, molecular geneticists, and data analysts are equally important and need to work hand in hand to make GWAS a success and identify common susceptibility factors for the traits/disorders under study. The critical task in the future will be to continue the clinical efforts and assemble study sample sizes that we now know are needed to get comprehensive information on the genetic architecture of these traits. Sample collections are expensive and time-consuming and should receive funding from the national funding agencies. Today, many researchers believe that sample sizes of several hundred thousand up to a million individuals (either individuals with a continuous trait/disorder or patients and controls) will eventually be needed to identify the majority of the biological pathways involved in multifactorial, polygenic traits/disorders. While it may take another five to ten years until this important milestone will be reached for the majority of brain disorders, it will clearly be to the benefit of the patients.

5.6 Tools and databases

Basic portals for SNPs, genes, and GWAS

- **1000 Genomes Project**[1] (7) provides news and a browser about the project data; the resource of the **HapMap Project**[2] has been retired due to a data security issue and users are asked to use the 1000 Genome's homepage.
- **dbSNP**[3] (4) is the database for SNP-specific information.
- **Ensembl Genome Browser**[4] (30) and **UCSC Genome Browser**[5] (31) are the most popular browsers for genomic data.
- **Epigenome Gateway**[6] leads to **Roadmap EpiGenome Browser** and **WashU Epigenome Browser**, both annotate SNPs and genes for epigenetic data from the ENCODE (32) and ROADMAP (33) projects.
- **GWAS Catalog**[7] (20) is the database for top findings and summary statistics from published GWAS.

---

1 <http://www.internationalgenome.org/home>

2 <https://www.ncbi.nlm.nih.gov/variation/news/NCBI_retiring_HapMap/>

3 <https://www.ncbi.nlm.nih.gov/SNP/>

4 <http://www.ensembl.org/index.html>

5 <http://genome.ucsc.edu/>

6 <http://epigenomegateway.wustl.edu>

7 <http://www.ebi.ac.uk/gwas/>

- **MSigDB**[8] (34) is one of the most comprehensive databases of gene sets for performing a GSEA.

Tools to prepare and conduct a GWAS

- **Michigan Imputation Server**[9] and **Sanger Imputation Server**[10] are free services for whole-genome imputation of variants on the basis of 1000 Genomes and other reference panels.
- **GWAtoolbox** (35) and **EasyQC** (12) are R software packages for QC of GWAS data.
- **GWAMA** (36), **METAL** (37), and **PLINK** (38) are toolsets for meta-analysis of multiple GWAS.
- **GenomeSpace**[11] (39) is a cloud-based framework to conduct integrative analyses of genomic data using various tools and recipes for common tasks.

Prioritizing of candidate SNPs and genes through prediction of function

- **BRAINEAC**[12] (40) analyses SNPs and genes for expressed quantitative trait loci (eQTL) in ten regions of the adult human brain; **GTEx**[13] (41) provides eQTL data from multiple tissues of the human body including the brain.
- **Ensembl Variant Effect Predictor**[14] (42) offers analysis, annotation, and prioritization of known and unknown variants in coding and non-coding regions.
- **FUMA GWAS**[15] (43) is a web-based toolset for functional mapping of SNPs and genes from summary statistics of GWAS.
- **MAGENTA**[16] (44) performs a GSEA on GWAS data to test if sets of functionally related genes are enriched for associations with a polygenic trait or disease.
- **HaploReg**[17] (45) and **RegulomeDB**[18] (46) evaluate SNPs for potential regulatory function in non-coding regions such as transcription factor binding, histone modifications, and other features.
- **HaploView**[19] (47) analyzes and visualizes SNPs in LD and haplotype maps.

Acknowledgements

The authors would like to thank Ms Dominique Hilger (Institute of Neuroscience and Medicine, INM-1, Research Center Jülich, Jülich, Germany) for excellent assistance during the writing of the manuscript and, in particular, with the conception and preparation of the figures.

[8] <http://software.broadinstitute.org/gsea/msigdb>

[9] <https://imputationserver.sph.umich.edu/index.html>

[10] <https://imputation.sanger.ac.uk>

[11] <http://www.genomespace.org/>

[12] <http://www.braineac.org>

[13] <https://www.gtexportal.org/home/>

[14] <http://www.ensembl.org/info/docs/tools/vep/index.html>

[15] <http://fuma.ctglab.nl/>

[16] <https://software.broadinstitute.org/mpg/magenta/>

[17] <http://compbio.mit.edu/HaploReg>

[18] <http://www.regulomedb.org/index>

[19] <https://www.broadinstitute.org/haploview/haploview>

References

1. **Sullivan PF, Daly MJ, O'Donovan M.** Genetic architectures of psychiatric disorders: the emerging picture and its implications. Nat Rev Genet. 2012;**13**(8):537–51.
2. **International Human Genome Sequencing Consortium.** Initial sequencing and analysis of the human genome. Nature. 2001;**409**:860–921.
3. **The International HapMap Consortium.** A haplotype map of the human genome. Nature. 2005;**437**(7063):1299–320.
4. **Sherry ST, Ward MH, Kholodov M, Baker J, Phan L, Smilgielski EM** et al. dbSNP: the NCBI database of genetic variation. Nucleic Acids Res. 2001;**29**(1):308–11.
5. **Bumgarner R.** DNA microarrays: Types, Applications and their future. Curr Protoc Mol Biol. 2013;Chapter 22:Unit 22.1.
6. **Lamy P, Grove J, Wiuf C.** A review of software for microarray genotyping. Hum Genomics. 2011;**5**:304–9.
7. **The 1000 Genomes Project Consortium.** An integrated map of genetic variation from 1,092 human genomes. Nature. 2012;**491**(7422):56–65.
8. **Schork NJ, Murray SS, Frazer KA, Topol EJ.** Common vs. rare allele hypotheses for complex diseases. Curr Opin Genet Dev. 2009;June **19**(3).212–9.
9. **Marchini J, Howie B, Myers SR, McVean G, Donnely P.** (2007) A new multipoint method for genome-wide association studies by imputation of genotypes. Nat Genet. 2007;**39**:906–13.
10. **Visscher PM, Wray NR, Zhang Q, Sklar P, McCarthy MI, Brown MA,** et al. 10 Years of GWAS Discovery: Biology, Function, and Translation. Am J Hum Genet. 2017;**101**(1):5–22.
11. **Mühleisen TW, Leber M, Schulze TG, Strohmaier J, Degenhardt F, Treutlein J,** et al. Genome-wide association study reveals two new risk loci for bipolar disorder. Nat Commun. 2014;**5**:3339.
12. **Winkler TW, Day FR, Croteau-Chonka DC, Wood AR, Locke AE, Mägi R,** et al. Quality control and conduct of genome-wide association meta-analyses. Nat Protoc. 2014;**9**(5):1192–212.
13. **Zeggini E, Morris A.** Analysis of Complex Disease Association Studies—A Practical Guide. London: Academic Press; 2011.
14. **Anderson CA, Pettersson FH, Clarke GM, Cardon LR, Morris AP, Zondervan KT.** Data quality control in genetic case-control association studies. Nat Protoc. 2010;**5**:1564–73.
15. **Zondervan KT, Cardon LR.** Designing candidate gene and genome-wide case-control association studies. Nat Protoc. 2007;**2**(10):2492–501.
16. **Schizophrenia Working Group of the Psychiatric Genomics Consortium.** Biological insights from 108 schizophrenia-associated genetic loci. Nature. 2014;**511**:421–7.
17. **Cipriani A, Saunders K, Attenburrow MJ, Stefaniak J, Panchal P, Stockton S,** et al. A systematic review of calcium channel antagonists in bipolar disorder and some considerations for their future development. Mol Psychiatry. 2016;**21**(10):1324–32.
18. **Jansen PR, Watanabe K, Stringer S, Skene N, Bryois J, Hammerschlag AR,** et al. Genome-wide analysis of insomnia (N=1,331,010) identifies novel loci and functional pathways. bioRxiv. 2018:S214973.
19. **Klein RJ, Zeiss C, Chew EY, Tsai J-Y, Sackler RS, Haynes C,** et al. Complement factor H polymorphism in age-related macular degeneration. Science. 2005;**308**(5720):385–9.
20. **MacArthur J, Bowler E, Cerezo M, Gil L, Hall P, Hastings E,** et al. The new NHGRI-EBI Catalog of published genome-wide association studies (GWAS Catalog). Nucleic Acids Res. 2017;**45**(Issue D1):D896–D901.
21. **Sullivan PF, Agrawal A, Bulik CM, Andreassen OA, Børglum AD, Breen G,** et al Psychiatric genomics: An update and an agenda. Am J Psychiatry. 2017;**175**(1):15–27.
22. **Timpson NJ, Greenwood CMT, Soranzo N, Lawson DL, Richards JB.** Genetic architecture: the shape of the genetic contribution to human traits and disease. Nat Rev Genet. 2018;**19**(2):110–24.
23. **Cross-Disorder Group of the Psychiatric Genomics Consortium.** Identification of risk loci with shared effects on five major psychiatric disorders: A genome-wide analysis. Lancet. 2013;**381**(9875):1371–9.

24. **Mühleisen TW, Reinbold CS, Forstner AJ, Abramova LI, Alda M, Babadjanova G,** et al. Gene set enrichment analysis and expression pattern exploration implicate an involvement of neurodevelopmental processes in bipolar disorder. J Affect Disord. 2018;**228**:20–25.
25. **Eising E, Huisman SMH, Mahfouz A, Vijfhuizen LS, Anttila V, Winsvold BS,** et al. Gene co-expression analysis identifies brain regions and cell types involved in migraine pathophysiology: a GWAS-based study using the Allen Human Brain Atlas. Hum Genet. 2016;**135**(4):425–439.
26. **Schork AJ, Thompson WK, Pham P, Torkamani A, Roddey JC, Sullivan PF, Dale AM.** All SNPs are not created equal: genome-wide association studies reveal a consistent pattern of enrichment among functionally annotated SNPs. PLoS Genetics. 2013;**9**:e1003449.
27. **Luciano M, Hagenaars SP, Davies G, Hill WD, Clarke T-K, Shirali M,** et al. Association analysis in over 329,000 individuals identifies 116 independent variants influencing neuroticism. Nat Genet. 2018;**50**:6–11.
28. **Bycroft C, Freeman C, Petkova D, Band G, Elliott LT, Sharp K,** et al. Genome-wide genetic data on ~500,000 UK Biobank participants. bioRxiv. 2017;166298.
29. **Thompson PM, Stein JL, Medland SE, Hibar DP, Vasquez AA, Renteria ME,** et al. The ENIGMA Consortium: large-scale collaborative analyses of neuroimaging and genetic data. Brain Imaging Behav. 2014;**8**:153–82.
30. **Zerbino DR, Achuthan P, Akanni W, Amode MR, Barrell D, Bhai J,** et al. Ensembl 2018. Nucleic Acids Res. 2018;**46**:D754–61.
31. **Casper J, Zweig AS, Villareal C, Tyner C, Speir ML, Rosenbloom KR,** et al. The UCSC Genome Browser database: 2018 update. Nucleic Acids Res. 2018;**46**(D1):D762–9.
32. **The ENCODE Project Consortium.** A user's guide to the encyclopedia of DNA elements (ENCODE). PLoS Biology. 2011;**9**:e1001046.
33. **Zhou X, Li D, Zhang B, Lowdon RF, Rockweiler NB, Sears RL, Madden PA,** et al. Epigenomic annotation of genetic variants using the Roadmap Epigenome Browser. Nature Biotechnol. 2015;**33**(4):345–6.
34. **Liberzon A, Birger C, Thorvaldsdóttir H, Ghandi M, Mesirov J P & Tamayo P.** The Molecular Signatures Database (MSigDB) hallmark gene set collection. Cell Systems. 2015;**1**(6):417–25.
35. **Fuchsberger C, Taliun D, Pramstaller PP, Pattaro C.** GWAtoolbox: an R package for fast quality control and handling of genome-wide association studies meta-analysis data. Bioinformatics. 2012;**28**(3):444–5.
36. **Mägi R & Morris AP.** GWAMA: software for genome-wide association meta-analysis. BMC Bioinformatics. 2010;**11**(1):288.
37. **Willer CJ, Li Y, Abecasis GR.** METAL: fast and efficient meta-analysis of genomewide association scans. Bioinformatics. 2010;**26**(17):2190–1.
38. **Purcell S, Neale B, Todd-Brown K, Thomas L, Ferreira MA, Bender D,** et al. PLINK: a tool set for whole-genome association and population-based linkage analyses. Am J Hum Genet. 2007;**81**(3):559–75.
39. **Qu K, Garamszegi S, Wu F, Thorvaldsdottir H, Liefeld T, Ocana M,** et al. Integrative genomic analysis by interoperation of bioinformatics tools in GenomeSpace. Nat Methods. 2016;**13**:245–247.
40. **Ramasamy A, Trabzuni D, Guelfi S, Varghese V, Smith C, Walker R,** et al. Genetic variability in the regulation of gene expression in ten regions of the human brain. Nat Neurosci. 2014;**17**:1418–28.
41. **The GTEx Consortium.** The Genotype-Tissue Expression (GTEx) project. Nat Genet. 2013;**45**:580–5.
42. **McLaren W, Gil L, Hunt SE, Singh Riat H, Ritchie GRS, Thormann A,** et al. The Ensembl Variant Effect Predictor. Genome Biol. 2016;**17**:122.
43. **Watanabe K, Taskesen E, van Bochoven A & Posthuma D.** Functional mapping and annotation of genetic associations with FUMA. Nature Commun. 2017;**8**:1826.
44. **Segrè AV, Groop L, Mootha VK, Daly MJ & Altshuler D.** Common Inherited Variation in Mitochondrial Genes Is Not Enriched for Associations with Type 2 Diabetes or Related Glycemic Traits. PLoS Genetics. 2010;**6**(8):e1001058.
45. **Ward LD & Kellis M.** HaploReg v4: systematic mining of putative causal variants, cell types, regulators and target genes for human complex traits and disease. Nucleic Acids Res. 2016;**44**:D877–81.
46. **Boyle AP, Hong EL, Hariharan M, Cheng Y, Schaub MA, Kasowski M,** et al. Annotation of functional variation in personal genomes using RegulomeDB. Genome Res. 2012;**22**(9):1790–7.
47. **Barrett JC, Fry B, Maller J, Daly MJ.** Haploview: analysis and visualization of LD and haplotype maps. Bioinformatics. 2005;**21**(2):263–5.

Chapter 6

Gene–environment interactions in animal models of depression and anxiety

Daniela Felice, Anand Gururajan, Olivia F O'Leary, and John F Cryan

6.1 Introduction

"Genes load the gun, and environment pulls the trigger." Cynthia M. Bulik, 2005

Depression and anxiety disorders account for half of all mental disorders, globally (1). Indeed, the World Health Organization predicts that by the year 2030, depression will be the second leading cause of disease burden worldwide (2). The etiology of these disorders involves a complex interaction of genetics and environmental risk factors mediated by various epigenetic mechanisms (3). While stress is a major risk factor for developing these disorders, differences in genetic factors cause individuals to respond differently to the similar stressors. Thus, some individuals are more vulnerable than others to developing these stress-related psychiatric disorders (4) and this is supported by clinical evidence showing interactions between the environment, genotype, and the increased risk of depression and anxiety disorders (5).

Animal models are powerful tools which we can use to study depression and anxiety as they allow us to control for a range of factors (e.g. genetic, environmental, social, developmental) which is not possible in a clinical setting (6–8). Moreover, the development of advanced techniques in molecular biology has facilitated the generation of genetically modified animal models, allowing the investigation of underlying cellular processes and the identification of novel therapeutic targets. However, it is important to recognize the limitations of using these models, particularly as some symptoms of depression, such as low mood or feelings of worthlessness, cannot be sufficiently modelled (9).

In this chapter, we will briefly review the most studied candidate genes for gene–environment G×E interaction in depression and anxiety disorders followed by an outline of candidate genes for future G×E studies

6.2 Preclinical evidence of G×E environment interactions associated with depression and anxiety disorders

Candidate genes for gene by environment interactions include those involved in the neuroendocrine stress response (i.e. the hypothalamic–pituitary–adrenal axis, HPA), the serotonergic, GABAergic, and brain-derived neurotrophic factor (BDNF) systems. In the following paragraphs we review the preclinical studies that have explored the interactions between these genes and stress in the context of depression and anxiety disorders.

6.2.1 HPA axis G×E interactions

The HPA axis is a neuroendocrine axis of the stress response. In response to stress, neurons in the paraventricular nucleus of the hypothalamus (PVN) release corticotropin-releasing hormone (CRH), which induces the formation and release of adrenocorticotropin hormone (ACTH) from the anterior pituitary gland. Once released, ACTH reaches the adrenal glands via the bloodstream and promotes the synthesis and secretion of glucocorticoids (stress hormones), such as cortisol in humans and corticosterone in rodents. In the brain, glucocorticoids bind to high-affinity mineralocorticoid receptors (MR) or lower-affinity glucocorticoid receptors (GRs). The high-affinity MR are bound at low circulating levels of glucocorticoids and are thought to be important in ambient glucocorticoid signaling processes such as regulation of basal secretion during the circadian cycle (10). By contrast, the low affinity GR are bound only when levels of circulating glucocorticoids levels are high, such as during exposure to stress. These GR receptors located in the PVN are important for mediating negative feedback mechanisms controlling glucocorticoid release (11).

Several preclinical studies have investigated the role of HPA axis in depression and anxiety disorders. Mice chronically treated with corticosterone display depression and anxiety-related behaviors (12–14). CRH knockout mice show no behavioral abnormalities (15). By contrast, knockout of CRH and CRHR1 in specific brain regions in mice decreases anxiety-like behavior (16, 17). On the other hand, mice transiently overexpressing CRH in the forebrain display anxiogenic- and despair-like behavior (18).

During development there is a critical time window, the stress hypo-responsive period, in which basal levels of glucocorticoid and stress-induced glucocorticoid responses are low (19). In mice, this critical period is postnatal day 1–10, in rats, postnatal day 3–14, and in humans is postnatal months 6–12 (20). Early-life stress especially during this period can alter HPA axis activity and induce permanent changes in sensitivity to stress in adulthood (21). Specifically, stressful events alter the expression of stress-responsive genes including GR and CRF via epigenetic mechanisms (22). Indeed, Meaney and colleagues provided evidence that higher maternal care during the first postnatal days reduced the stress response, ACTH plasma levels, CRH mRNA hypothalamus levels, and increased hippocampal mRNA GR levels in the offspring (23, 24). Interestingly, Francis and colleagues showed an intergenerational transmission of HPA axis modulation which is an epigenetic phenomenon (24). Weaner and colleagues showed that maternal care behavior in rats (low/high arch back nursing (ABN) and licking grooming (LG) behavior) induced higher DNA methylation of the hippocampal GR gene promoter in offspring (25). These epigenetic changes were reversed by cross-fostering and associated with changes in histone acetylation and transcription factor (NGFI-A) binding to the exon 1_7 GR promoter (25). In addition, intracerebroventricular infusion of a histone deacetylase inhibitor was able to reverse histone acetylation, DNA methylation, NGFI-A binding, GR expression, and HPA responses to stress (25). Notably, NGFI-A is required for epigenetic programming of GR expression (26).

Similarly to maternal care, early-life stress has been shown to induce epigenetic changes in the offspring that are transmitted across generations (27, 28). For example, early-life stress induced increased arginine vasopressin (AVP) expression in hypothalamic neurons and this was associated with hypomethylation of an AVP enhancer region that serves as DNA-binding sites for the methyl CpG–binding protein 2 (MeCP2) (27). Maternal stress during early gestation (exposure to mild stressors) induced depressive-like behaviors and exaggerated stress responses in male but not female offspring. These behavioral changes were associated with changes in GR and MR gene methylation (29). Similarly, Grundwald and Brunton observed that prenatal stress (repeated social stress) affects behavior of the offspring (anxiety-like behavior) and HPA axis activity (hyperactive HPA axis) in male but not female rats (30). The effects of the prenatal stress were transmitted to the second generation in male rats (30). In addition, early prenatal stress differentially affected the expression of critical genes for growth and development in male and female in the placentas, suggesting sex differences in the placental epigenetic machinery (29). Paternal stress has also been shown to modulate HPA axis activity in the offspring and this is associated with altered content of sperm microRNAs (31). Paternal stress may also affect offspring HPA axis by intervening in the epigenetic regulation of oocyte development or even earlier during spermatogenesis (31).

Taken together, environmental factors (maternal care, early-life stress) can induce genetic changes in the offspring in HPA axis-associated genes and activity. Those changes are transmitted across generations and may be sex specific.

6.2.2 Serotonin (5-HT) G×E interactions

One of the most explored hypotheses for the pathophysiology of depression and anxiety disorders is the monoamine hypothesis which suggests that depression and anxiety disorders may be a result of an imbalance of monoamine levels such as serotonin (5-HT), dopamine, and noradrenaline (32). This hypothesis was conceived based on the observation that antidepressants, such as monoamine oxidase inhibitors and tricyclic antidepressants, increased monoamine levels. The efficacy of selective serotonin inhibitors (SSRIs) which increase extracellular serotonin levels, and are effective in treating depression and some anxiety disorders, further supported the monoamine hypothesis (33). Moreover, clinical data also support this hypothesis. Indeed, tryptophan/5-HT depletion induces a greater reduction in mood in individuals with a family history of depression (34, 35), depressed suicide victims display decreased midbrain (36) and CSF (37) 5-HT levels.

Several preclinical studies on depression and anxiety-like behavior have explored the impact of environmental changes particularly in early life in rodents with mutations in the serotonin transporter (SERT), which is the protein responsible for serotonin reuptake from the synaptic cleft and is the target for selective serotonin reuptake inhibitor (SSRI) antidepressant drugs (38). In one of the first studies, SERT knockout, heterozygous SERT mice, and wildtype mice were exposed to foot shock during early postnatal life (PND 7–13). The authors observed that SERT knockout but not heterozygous or wildtype mice displayed increased anxiety-like and depression-like behaviors in adulthood and also decreased exploratory activity. Indeed, postnatal stress alone did not have any significant impact on behavior in wildtype mice (38).

Similarly, SERT knockout mice raised in a threatening environment (dams exposed to soiled bedding of unfamiliar adult males) exhibited pronounced anxiety-like behavior when compared to wildtype and heterozygous SERT mice in mice when tested at postnatal day 50–59 days (39). However, in another study by the same authors, exposure in early life to either unfamiliar male bedding or communal nesting did not affect anxiety-like behavior in SERT knockout mice at postnatal day 22 (40). The authors mentioned that one cause for this discrepancy could be that in the first study, they observed a higher level of mortality among the dams suggesting that may be exposed to additional stress of unknown nature (40). Another explanation could be the different age of the mice when tested, since unfamiliar male bedding nesting may have different effects at different life stages with possibly more pronounced effects in late adolescence (40). Incidentally, Kloke and colleagues observed no effects of exposure to unfamiliar male bedding in early life on anxiety-like behavior at postnatal day 51 in SERT knockout and wildtype mice (41). Another study assessed the effects of early-life stress (exposure to high or low maternal care) in heterozygous SERT mice (42). The SERT mutation enhanced early-life stress-induced depression and anxiety-like behavior in mice in adulthood (42). Furthermore, the study showed that SERT heterozygous mice experiencing low maternal care displayed decreased serotonin turnover in hippocampus and striatum (42). In another study, prenatal stress (maternal restraint) differentially affected SERT heterozygous female and male mice. Indeed, female but not male mice exposed to prenatal stress displayed increased depression-like behavior as observed in the forced swim test (43). However, the increased immobility in the forced swim test may be due to a decreased locomotor activity (43).

All the studies discussed above assessed the interaction of SERT genetic mutation with early-life stress in mice. Few studies have explored the impact of stress in adulthood in SERT genetically modified mice. A study from the group of Etienne and colleagues assessed the effects of unpredictable chronic mild stress (UCMS) in adult male and female SERT knockout mice (44). While UCMS induced depression/anxiety-like behavior regardless of genotype (44), intriguingly, female knockout mice were more susceptible to the effects of UCMS. Specifically, females exposed to unpredictable mild stress displayed higher emotionality than males as assessed in the open field (OF), elevated plus maze (EPM), and novelty suppressed feeding (NSF) tests, with a more anxious and depressive-like phenotype with increasing age (3, 6, and 12 months) (44). Nevertheless, this is in line with clinical evidence demonstrating that depression is more frequent in women (45). In another study, Jansen and colleagues assessed the effects of repeated social experience as "winners" (interaction with a non-aggressive mouse strain) or "losers" (interaction with an aggressive mouse strain) in wildtype, SERT heterozygous, or knockout mice (46). While repeated social stress increased anxiety-like levels (assessed in the dark light, elevated plus maze, and open field) in all genotypes, SERT knockout mice exposed to the negative stimuli (loser experience) displayed enhanced anxiety-like phenotype and increased HPA axis activity (46). Interestingly, stressed SERT heterozygous mice exhibit decreased serotonin turnover in the frontal cortex, a brain area implicated in the control of fear and avoidance responses and in susceptibility to depression, when compared to non-stressed SERT heterozygous mice and wildtype mice, suggesting that loss of SERT may increase susceptibility to emotional responses to a social threat (47).

Recently the effects of adverse (dangerous environment) or beneficial (safe environment) life conditions were investigated across the lifespan in SERT knockout mice (48). While the authors did not find G×E interactions, both genetic modifications and environmental challenges independently affected behavior. Specifically, SERT knockout mice displayed increased anxiety-like behavior and decreased exploratory behavior when compared with wildtype and heterozygous SERT mice on PND 75–77 independently from the environmental challenges (adverse or beneficial) (48). Mice exposed to early beneficial conditions followed by adverse conditions in adulthood displayed a less anxious-like phenotype and increased exploratory activity when analysis was performed without genotype as between-subject factor (48). To date, this has been the only study to assess the effects of environmental changes in different phases of life (early life, adolescence, and adulthood).

Overall, the studies suggest that with deletion of SERT increased vulnerability to the negative effects of stressors applied at various stages of life (early-life stress, postnatal, adolescence, or adulthood) (39, 42, 43, 46, 47, 48), although those effects are rather subtle and are not consistent (39, 41, 44). Interestingly, however, it seems that that female mice with impaired SERT functioning are more vulnerable to the negative impact of stress (43, 44), and this effect is consistent with clinical evidence reporting that women are more susceptible to stress (49).

6.2.3 γ-Aminobutyric acid (GABA) G×E interactions

γ-Aminobutyric acid (GABA), the main inhibitory neurotransmitter in the mammalian nervous system, is thought to play a key role in the pathophysiology of depression and anxiety disorders. Indeed, clinical, preclinical, and post-mortem studies strongly implicate GABAergic dysfunction in depression (50, 51). GABA acts in the central nervous system (CNS) on two types of receptors: ionotropic $GABA_A$ and $GABA_C$ receptors, and metabotropic $GABA_B$ receptors.

Several knockout mice for $GABA_A$ and $GABA_B$ receptors or their subunits have been generated to study their precise role in depression and anxiety-related behaviors. $GABA_A$ heterozygous γ 2 knockout mice and γ 2 knockdown mice exhibit increased anxiety-like behaviors (52, 53). In addition, heterozygous γ 2 knockout mice exhibited decreased adult hippocampal neurogenesis as well as increased anxiety- and depressive-like behavior (54). Interestingly, $GABA_A$ heterozygous γ 2 knockout mice displayed increased basal corticosterone levels which correlated with the emergence of an anxiety/depressive-like behavior during postnatal development (55).

The development of mice with altered $GABA_B$ receptor function has confirmed a clear role for this metabotropic receptor in modulating depression and anxiety-like behavior. Mice lacking either the $GABA_{B1}$ or $GABA_{B2}$ receptor subunits displayed an anxious phenotype and antidepressant-like phenotype in the forced swim test (56, 57). Interestingly, chronic treatment with a $GABA_B$ receptor antagonist has been shown to induce antidepressant-like effects in mice (58) and increase adult hippocampal neurogenesis *in vivo* (58) and *in vitro* (59). Furthermore, $GABA_{B1b}$ knockout mice displayed increased adult hippocampal neurogenesis (60). The development of conditional knockout mice for $GABA_B$ receptor could be very important to assess its role in neurogenesis and how this links back to the development of anxiety/depressive-like behavior.

A study from our group assessed how early-life stress interacts with the $GABA_B$ receptors and found that $GABA_{B1a}$ and $GABA_{B1b}$ receptor subunits differentially regulate stress resilience. Specifically, $GABA_{B1a}$ receptor knockout mice were more susceptible than $GABA_{B1b}$ receptor knockout mice to stress-induced (maternal separation, early-life stress) anhedonia and psychosocial stress (social defeat, adulthood) -induced social withdrawal, two core symptoms of depression (60). In addition, $GABA_{B1b}$ receptor knockout mice displayed increased adult hippocampal neurogenesis and were more resilient to stress-induced decreases in the survival of newly born cell in the adult hippocampus (60). Moreover, $GABA_B$ receptor pharmacological activation but not blockade during postnatal life increased anxiety-like behavior in mice (61).

A proteomic analysis revealed the existence of the KCTD protein family which consists of four sequence-related cytosolic proteins (KCTD8, KCTD12, KCTD12b, and KCTD16) that bind as tetramers to the C-terminal domain of the $GABA_{B2}$ subunit, influencing the pharmacology and kinetics of the receptor response (62). One study showed that KCTD proteins display differential temporal and spatial distribution in the mouse brain (63). KCTD12 downregulation has been associated with psychiatric disorders including bipolar disorder (64, 65). KTCD12 knockout mice displayed increased fear learning of a discrete auditory-conditioned stimulus and altered neuronal excitability (64). More studies are required to further investigate the interactions between $GABA_B$ and KTCD loss of function and stress in the context of depression and anxiety-like behavior.

Overall, these results suggest a role for GABA system in modulating sensitivity in response to stressful stimuli during early life or adulthood.

6.2.4 Brain-derived neurotrophic factor (BDNF)

The neurotrophic theory of depression states that a deficiency in neurotrophic factors such as brain-derived neurotrophic factor (BDNF) is involved in the pathophysiology of depression and in the mechanism of antidepressant action (66–69). BDNF promotes adult hippocampal neurogenesis (birth of new neurons), which is a putative antidepressant mechanism of action (70), and antidepressant treatment increases BDNF expression in the hippocampus (71, 72). Repeated (midbrain) or a single (dentate gyrus of the hippocampus) BDNF infusions are able to induce antidepressant-like effects in rat models of depression (learned helplessness and forced swim test)

(73, 74). By contrast, BDNF infusion in the ventral tegmental area (VTA) induced depression-like behavior in rats (75). Accordingly, depressed naive patients show lower BDNF serum levels than antidepressant treated or healthy patients (76).

In a study by Seo and colleagues, maternal separation and chronic restraint stress induced several epigenetic changes of BDNF associated with depression-like behavior in rats, and escitalopram treatment was able to reverse the phenotype (77). Importantly, early-life stress has been shown to induce epigenetic changes in the rat hippocampus which affect BDNF expression (77). Maternal separation (early-life stress) and restraint stress (adulthood stress) reduced BDNF exon IV mRNA levels and the acetylation of histone H3 and H4 at BDNF promoter IV in the adult rat hippocampus (77). The latter was exacerbated in early-life-stressed rats by chronic restraint stress in adulthood. In addition, maternal separation and restraint stress enhanced MeCP2 levels at BDNF promoter IV and histone deacetylase 5 (HDAC5) mRNA levels (77). Hypoacetylation of histone H3 and H4 and methylation of CpG islands are thought to decrease/silence transcriptional activity at the promoter. MeCP2 is thought to mediate the gene silencing effects of DNA methylation interacting with the histone deacetylase (HDAC)/Sin3 complex. Thus, early-life stress induced epigenetic modification in the BDNF promoter, affecting its expression. Notably, those effects were reversed by chronic escitalopram treatment (77).

Further supporting antidepressant regulation of stress-induced epigenetic modifications, Nestler and colleagues investigated the effects social defeat stress following chronic imipramine treatment on hippocampal BDNF gene expression and chromatin remodeling (78). As expected, chronic social defeat induced social avoidance in mice and this was reversed by chronic antidepressant treatment (imipramine, fluoxetine). Stress decreased BDNF expression in the mouse hippocampus (*Bdnf* transcripts III and IV) and increased repressive histone methylation at their corresponding promoters. Chronic treatment with imipramine increased BDNF levels and increased histone acetylation at these promoters. Moreover, treatment with imipramine induced downregulation of histone deacetylase (HDAC) 5. In addition, the authors observed that viral-mediated HDAC5 overexpression in the hippocampus prevented imipramine effects highlighting the key role of histone remodeling in antidepressant-induced regulation of the epigenetic effects of stress on BDNF (78).

Important interactions between early-life stress and BDNF heterozygous mice have also been shown (79). In this study, high/low maternal care induced subtle changes in anxiety-like behavior (decreased time grooming and sniffing in the elevated plus maze) in wildtype and BDNF heterozygous mice. However, those effects were greater in BDNF heterozygous mice (open field: decreased time in the center, decreased time of center visits, increased time in the corners, increased latency; elevated plus maze: decreased visits to the open arms, increased time sniffing, decreased unprotected head dipping) (79). On the other hand, chronic administration of the stress hormone, corticosterone, during the adolescent period did not induce depressive or anxiety-like behaviors in BDNF heterozygous rats but did disrupt extinction learning (80).

In two studies, BDNF heterozygous mice exhibited depressive-like and anxiety-like behaviors when exposed to stress (isolation stress, three weeks) (81) or poor housing conditions, respectively (82). The effects of poor housing were reversed by environmental enrichment (82). Another study showed sex-dependent effects in response to UCMS. Female mice with loss of BDNF function specifically in the forebrain (conditional knockout mice) were more vulnerable to the effects of UCMS when compared to wildtypes (83). In another study, BDNF heterozygous mice were unaffected by the UCMS paradigm compared to wildtype mice but displayed a blunted response to antidepressant treatment (imipramine) (84). Interestingly, male $BDNF^{+/Met}$ heterozygous mice were more sensitive to seven days of restraint stress, displaying higher HPA axis activity, increased depressive and anxiety-like behaviors, and deficits in working memory compared to the wildtype stressed group (85). Depressive-like behavior was reversed by acute treatment with desipramine but not fluoxetine (85), suggesting that this allele also plays a role in the antidepressant response.

Overall, some (79, 82, 85) but not all (83, 84) studies suggest that loss of BDNF function may increase susceptibility to stress-induced psychopathologies and female biological sex may increase this vulnerability (83)

6.3 **Novel candidate target genes for the treatment of depression and anxiety disorders**

In the last few decades, researchers have been searching for new genes targets for the treatment of depression and anxiety disorders. Among those, glutamate receptors may represent an ideal candidate (86). Glutamate is the major excitatory neurotransmitter in the mammalian central nervous system and acts via two classes of receptors, ligand gated ion channels (ionotropic receptors) and G-protein coupled (metabotropic) receptors. There are three types of ionotropic receptors, α-amino-3-hydroxy-5-methyl-4-isoxazolepropionic acid (AMPA), *N*-methyl-D-aspartate (NMD), and kainate receptors, while the metabotropic receptors exhibit different subtypes, including group I (mGlul and mGlu 5), group II (mGlu 2 and mGlu3), and group III (mGlu4, mGlu7, and mGlu8) receptors. Evidence suggests crucial roles for the mGlu2/3 and mGlu5 receptor in the treatment of anxiety, stress disorders, and depression (87, 88). Recently, a meta-analysis study revealed an association of GRIK4 (Glutamate Ionotropic Receptor Kainate Type Subunit 4) polymorphism (rs1954787) and responsiveness to antidepressant treatment in depressed patients (89). In addition, the glutamatergic genes GRIA3 (Glutamate Ionotropic Receptor AMPA Type Subunit 3) and GRIK2 (glutamate ionotropic receptor kainate type subunit 2) were associated with treatment-emergent suicidal ideation (90). Moreover, the polymorphisms within the metabotropic glutamate receptor 1 gene (*GRM1*) have been associated with depression (91).

Other candidate target genes for G×E studies are related to the immune system. In particular, the interleukin genes have been largely investigated in light of the emerging role of the immune system in depression and anxiety disorders (92, 93). For instance, IL-1β single nucleotide polymorphisms have been associated with depressive and anxiety symptoms in the context of life stress (94, 95). In addition, polymorphisms in genes which are involved in T-cell function have been associated with depressive symptoms and antidepressant response (96).

Ghrelin is an orexigenic hormone that increases appetite and food intake (97). Recent preclinical evidence suggests that ghrelin may be linked to stress-induced psychopathologies (98–101). Similarly, clinical studies have reported a role for a ghrelin polymorphism in depression (102, 103).

Galanin (GAL) is an estrogen-inducible neuropeptide, highly expressed in brain regions, reported to play a possible role in mood disorders. Some preclinical studies (104–106) and human studies (107, 108) suggest that galanin modulates depression and anxiety-related behavior as well as antidepressant response. Bagdy and colleagues reported that variants in genes for galanin (GAL) and its receptors (GALR1, GALR2, GALR3) may be risk factors for depression and anxiety disorders in individuals exposed to early-life stress (109). Intriguingly, the effects of galanin polymorphisms may be gender specific and depend on estrogens levels (108).

Overall, glutamate, ghrelin, and galanin, and immune system-related genes may represent the future targets for developing therapeutic approaches for depression and anxiety disorders. However, future clinical and preclinical studies assessing the effects of these candidate G×E interactions in psychiatric disorders are mandatory. Finally, in light of the brain–gut axis theory of psychiatric disorders (110), changes in diet or microbiota composition should be evaluated in order to further investigate the interactions between gut microbiota and host genetics in depression and

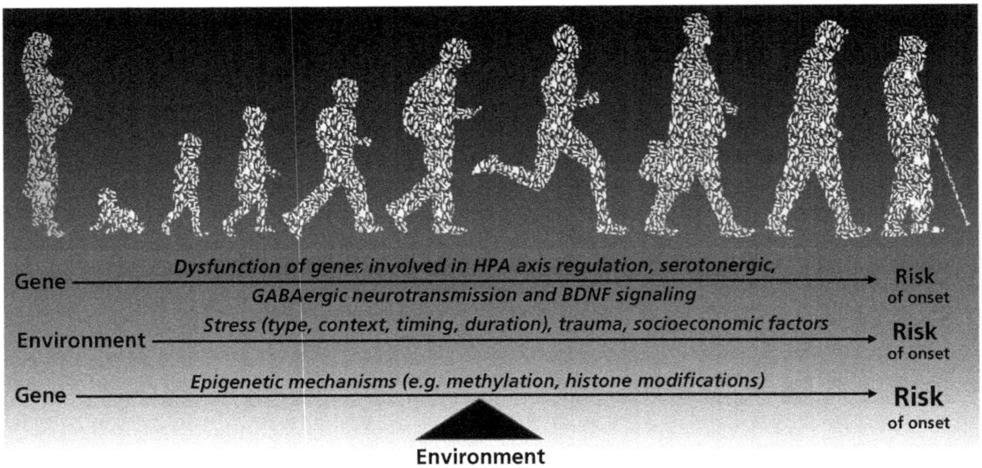

Figure 6.1 Genetic and environmental risk factors can independently influence risk of onset of affective disorders. However, interactions between factors can have a greater effect that is mediated by putative epigenetic mechanisms. These interactions can occur at specific stages of development or throughout the lifespan of an individual to modify and increase overall risk of onset.

anxiety. Indeed, the ancient concept "we are what we eat" is recruiting increasing support and gut microbiota modifications hold promise as a therapeutic strategy.

6.4 Conclusions and future perspectives

Overall, it is clear that the genotype interacts with the environment in a complex way in determining susceptibility or resilience to psychiatric disorders. However, there are few robust and replicated studies in the literature. One reason for this could be that psychiatric disorders are complex diseases with multifactorial genetic and environmental causes, not single genes or life events. Nonetheless, as we have outlined above, there are some clear candidate genes (Figure 6.1). Future studies should expand on the knowledgebase by using high-resolution techniques such as transcription activator-like effector nucleases (TALENS), clustered regularly interspaced short palindromic repeats (CRISPR), optogenetics, and designer receptors exclusively activated by designer drugs (DREADDs).

Over the last few years, there has been much hype over the benefits to be gained from personalizing therapeutic approaches to the treatment of psychiatric disorders based on genetic makeup. However, it may be argued that population-wide interventions may be on the whole more effective than targeting the single individual to decrease the overall incidence of these disorders in today's society. These interventions could range from promoting healthy lifestyles (diet and exercise) to creating community safe spaces where affected individuals can go to seek support. Nevertheless, the prospect of having a comprehensive genetic profile together with a detailed history could be of great value for treatment-resistant patients.

References

1. **Vos T, Mathers C, Herrman H, Harvey C, Gureje O, Bui D**, et al. The burden of mental disorders in Victoria, 1996. Soc Psychiatry Psychiatr Epidemiol. 2001 Feb;**36**(2):53–62.
2. **Mathers CD, Loncar D.** Projections of global mortality and burden of disease from 2002 to 2030. PLoS Med. 2006;**3**(11):e442.1

3. **Klengel T, Binder EB.** Epigenetics of stress-related psychiatric disorders and gene x environment interactions. Neuron. 2015 Jun 17;**86**(6):1343–57.
4. **McEwen BS, Bowles NP, Gray JD, Hill MN, Hunter RG, Karatsoreos IN**, et al. Mechanisms of stress in the brain. Nat Neurosci. 2015 Oct;**18**(10):1353–63.
5. **Sharma S, Powers A, Bradley B, Ressler KJ.** Gene × environment determinants of stress- and anxiety-related disorders. Annu Rev Psychol. 2016;**67**:239–61.
6. **Cryan JF, Holmes A.** The ascent of mouse: Advances in modelling human depression and anxiety. Nat Rev Drug Discov. 2005 Sep;**4**(9):775–90.
7. **Czeh B, Fuchs E, Wiborg O, Simon M.** Animal models of major depression and their clinical implications. Prog Neuropsychopharmacol Biol Psychiatry. 2016 Jan 4;**64**:293–310.
8. **Berton O, Hahn CG, Thase ME.** Are we getting closer to valid translational models for major depression? Science. 2012 Oct 05;**338**(6103):75–9.
9. **Cryan JF, Mombereau C.** In search of a depressed mouse: Utility of models for studying depression-related behavior in genetically modified mice. Mol Psychiatry. 2004 Apr;**9**(4):326–57.
10. **Dallman MF, Levin N, Cascio CS, Akana SF, Jacobson L, Kuhn RW.** Pharmacological evidence that the inhibition of diurnal adrenocorticotropin secretion by corticosteroids is mediated via type I corticosterone-preferring receptors. Endocrinology. 1989 Jun;**124**(6):2844–50.
11. **De Kloet ER, Vreugdenhil E, Oitzl MS, Joels M.** Brain corticosteroid receptor balance in health and disease. Endocr Rev. 1998 Jun;**19**(3):269–301.
12. **David DJ, Samuels BA, Rainer Q, Wang JW, Marsteller D, Mendez I**, et al. Neurogenesis-dependent and -independent effects of fluoxetine in an animal model of anxiety/depression. Neuron. 2009 May 28;**62**(4):479–93.
13. **Zhao Y, Ma R, Shen J, Su H, Xing D, Du L.** A mouse model of depression induced by repeated corticosterone injections. Eur J Pharmacol. 2008 Feb 26;**581**(1–2):113–20.
14. **Gregus A, Wintink AJ, Davis AC, Kalynchuk LE.** Effect of repeated corticosterone injections and restraint stress on anxiety and depression-like behavior in male rats. Behav Brain Res. 2005 Jan 06;**156**(1):105–14.
15. **Muglia L, Jacobson L, Dikkes P, Majzoub JA.** Corticotropin-releasing hormone deficiency reveals major fetal but not adult glucocorticoid need. Nature. 1995 Feb 2;**373**(6513):427–32.
16. **Zhang R, Asai M, Mahoney CE, Joachim M, Shen Y, Gunner G**, et al. Loss of hypothalamic corticotropin-releasing hormone markedly reduces anxiety behaviors in mice. Mol Psychiatry. 2016 Sep 6. PubMed PMID: 27595593.
17. **Muller MB, Zimmermann S, Sillaber I, Hagemeyer TP, Deussing JM, Timpl P**, et al. Limbic corticotropin-releasing hormone receptor 1 mediates anxiety-related behavior and hormonal adaptation to stress. Nat Neurosci. 2003 Oct;**6**(10):1100–7.
18. **Kolber BJ, Boyle MP, Wieczorek L, Kelley CL, Onwuzurike CC, Nettles SA**, et al. Transient early-life forebrain corticotropin-releasing hormone elevation causes long-lasting anxiogenic and despair-like changes in mice. J Neurosci. 2010 Feb 17;**30**(7):2571–81.
19. **Schapiro S, Geller E, Eiduson S.** Neonatal adrenal cortical response to stress and vasopressin. Proc Soc Exp Biol Med. 1962 Apr;**109**:937–41.
20. **Daskalakis NP, De Kloet ER, Yehuda R, Malaspina D, Kranz TM.** Early life stress effects on glucocorticoid-BDNF interplay in the hippocampus. Front Mol Neurosci. 2015;**8**:68.
21. **de Kloet ER, Sibug RM, Helmerhorst FM, Schmidt MV.** Stress, genes and the mechanism of programming the brain for later life. Neurosci Biobehav Rev. 2005 Apr;**29**(2):271–81.
22. **Hunter RG.** Epigenetic effects of stress and corticosteroids in the brain. Front Cell Neurosci. 2012;**6**:18.
23. **Liu D, Diorio J, Tannenbaum B, Caldji C, Francis D, Freedman A**, et al. Maternal care, hippocampal glucocorticoid receptors, and hypothalamic–pituitary–adrenal responses to stress. Science. 1997 Sep 12;**277**(5332):1659–62.

24. **Francis D, Diorio J, Liu D, Meaney MJ.** Nongenomic transmission across generations of maternal behavior and stress responses in the rat. Science. 1999 Nov 05;**286**(5442):1155–8.
25. **Weaver IC, Cervoni N, Champagne FA, D'Alessio AC, Sharma S, Seckl JR**, et al. Epigenetic programming by maternal behavior. Nat Neurosci. 2004 Aug;**7**(8):847–54.
26. **Weaver IC, D'Alessio AC, Brown SE, Hellstrom IC, Dymov S, Sharma S**, et al. The transcription factor nerve growth factor-inducible protein a mediates epigenetic programming: Altering epigenetic marks by immediate-early genes. J Neurosci. 2007 Feb 14;**27**(7):1756–68.
27. **Murgatroyd C, Patchev AV, Wu Y, Micale V, Bockmuhl Y, Fischer D**, et al. Dynamic DNA methylation programs persistent adverse effects of early-life stress. Nat Neurosci. 2009 Dec;**12**(12):1559–66.
28. **Franklin TB, Russig H, Weiss IC, Graff J, Linder N, Michalon A**, et al. Epigenetic transmission of the impact of early stress across generations. Biol Psychiatry. 2010 Sep 01;**68**(5):408–15.
29. **Mueller BR, Bale TL.** Sex-specific programming of offspring emotionality after stress early in pregnancy. J Neurosci. 2008 Sep 03;**28**(36):9055–65.
30. **Grundwald NJ, Brunton PJ.** Prenatal stress programs neuroendocrine stress responses and affective behaviors in second generation rats in a sex-dependent manner. Psychoneuroendocrinology. 2015 Dec;**62**:204–16.
31. **Rodgers AB, Morgan CP, Bronson SL, Revello S, Bale TL.** Paternal stress exposure alters sperm microRNA content and reprograms offspring HPA stress axis regulation. J Neurosci. 2013 May 22;**33**(21):9003–12.
32. **O'Leary OF, Cryan JF.** The behavioral genetics of serotonin: Relevance to anxiety and depression. In: Christian PM, Barry LJ (eds). Handbook of Behavioral Neuroscience, Vol. **21**: Amsterdam: Elsevier; 2010, pp. 749–89.
33. **Morilak DA, Frazer A.** Antidepressants and brain monoaminergic systems: A dimensional approach to understanding their behavioural effects in depression and anxiety disorders. Int J Neuropsychopharmacol. 2004 Jun;7(2):193–218.
34. **Benkelfat C, Ellenbogen MA, Dean P, Palmour RM, Young SN.** Mood-lowering effect of tryptophan depletion. Enhanced susceptibility in young men at genetic risk for major affective disorders. Arch Gen Psychiatry. 1994 Sep;**51**(9):687–97.
35. **van der Veen FM, Evers EA, Deutz NE, Schmitt JA.** Effects of acute tryptophan depletion on mood and facial emotion perception related brain activation and performance in healthy women with and without a family history of depression. Neuropsychopharmacology. 2007 Jan;**32**(1):216–24..
36. **Shaw DM, Camps FE, Eccleston EG.** 5-Hydroxytryptamine in the hind-brain of depressive suicides. Br J Psychiatry. 1967 Dec;**113**(505):1407–11.
37. **Asberg M.** Neurotransmitters and suicidal behavior. The evidence from cerebrospinal fluid studies. Ann N Y Acad Sci. 1997 Dec 29;**836**:158–81.
38. **Carroll JC, Boyce-Rustay JM, Millstein R, Yang R, Wiedholz LM, Murphy DL**, et al. Effects of mild early life stress on abnormal emotion-related behaviors in 5-HTT knockout mice. Behav Genet. 2007 Jan;**37**(1):214–22.
39. **Heiming RS, Jansen F, Lewejohann L, Kaiser S, Schmitt A, Lesch KP**, et al. Living in a dangerous world: The shaping of behavioral profile by early environment and 5-HTT genotype. Front Behav Neurosci. 2009;**3**:26.
40. **Heiming RS, Bodden C, Jansen F, Lewejohann L, Kaiser S, Lesch KP**, et al. Living in a dangerous world decreases maternal care: A study in serotonin transporter knockout mice. Horm Behav. 2011 Sep;**60**(4):397–407.
41. **Kloke V, Heiming RS, Bolting S, Kaiser S, Lewejohann L, Lesch KP**, et al. Unexpected effects of early-life adversity and social enrichment on the anxiety profile of mice varying in serotonin transporter genotype. Behav Brain Res. 2013 Jun 15;**247**:248–58.
42. **Carola V, Frazzetto G, Pascucci T, Audero E, Puglisi-Allegra S, Cabib S**, et al. Identifying molecular substrates in a mouse model of the serotonin transporter × environment risk factor for anxiety and depression. Biol Psychiatry. 2008 May 1;**63**(9):840–6.

43. **van den Hove DL, Jakob SB, Schraut KG, Kenis G, Schmitt AG, Kneitz S**, et al. Differential effects of prenatal stress in 5-Htt deficient mice: Towards molecular mechanisms of gene × environment interactions. PLoS One. 2011;**6**(8):e22715.
44. **Joeyen-Waldorf J, Edgar N, Sibille E.** The roles of sex and serotonin transporter levels in age- and stress-related emotionality in mice. Brain Res. 2009 Aug 25;**1286**:84–93.
45. **Albert PR.** Why is depression more prevalent in women? J Psychiatry Neurosci. 2015 Jul;**40**(4):219–21.
46. **Jansen F, Heiming RS, Lewejohann L, Touma C, Palme R, Schmitt A**, et al. Modulation of behavioural profile and stress response by 5-HTT genotype and social experience in adulthood. Behav Brain Res. 2010 Feb 11;**207**(1):21–9.
47. **Bartolomucci A, Carola V, Pascucci T, Puglisi-Allegra S, Cabib S, Lesch KP**, et al. Increased vulnerability to psychosocial stress in heterozygous serotonin transporter knockout mice. Dis Model Mech. 2010 Jul–Aug;**3**(7–8):459–70.
48. **Bodden C, Richter SH, Schreiber RS, Kloke V, Gerss J, Palme R**, et al. Benefits of adversity?! How life history affects the behavioral profile of mice varying in serotonin transporter genotype. Front Behav Neurosci. 2015;**9**:47.
49. **Sandanger I, Nygard JF, Sorensen T, Moum T.** Is women's mental health more susceptible than men's to the influence of surrounding stress? Soc Psychiatry Psychiatr Epidemiol. 2004 Mar;**39**(3):177–84.
50. **Cryan JF, Slattery DA.** GABAB receptors and depression. Current status. Adv Pharmacol. 2010;**58**:427–51.
51. **Mohler H.** The GABA system in anxiety and depression and its therapeutic potential. Neuropharmacology. 2012 Jan;**62**(1):42–53.
52. **Chandra D, Korpi ER, Miralles CP, De Blas AL, Homanics GE.** GABAA receptor gamma 2 subunit knockdown mice have enhanced anxiety-like behavior but unaltered hypnotic response to benzodiazepines. BMC Neurosci. 2005 Apr 25;**6**:30.
53. **Crestani F, Lorez M, Baer K, Essrich C, Benke D, Laurent JP**, et al. Decreased GABAA-receptor clustering results in enhanced anxiety and a bias for threat cues. Nat Neurosci. 1999 Sep;**2**(9):833–9.
54. **Earnheart JC, Schweizer C, Crestani F, Iwasato T, Itohara S, Mohler H**, et al. GABAergic control of adult hippocampal neurogenesis in relation to behavior indicative of trait anxiety and depression states. J Neurosci. 2007 Apr 04;**27**(14):3845–54.
55. **Shen Q, Lal R, Luellen BA, Earnheart JC, Andrews AM, Luscher B.** gamma-Aminobutyric acid-type A receptor deficits cause hypothalamic–pituitary–adrenal axis hyperactivity and antidepressant drug sensitivity reminiscent of melancholic forms of depression. Biol Psychiatry. 2010 Sep 15;**68**(6):512–20.
56. **Mombereau C, Kaupmann K, Froestl W, Sansig G, van der Putten H, Cryan JF.** Genetic and pharmacological evidence of a role for GABA(B) receptors in the modulation of anxiety- and antidepressant-like behavior. Neuropsychopharmacology. 2004 Jun;**29**(6):1050–62.
57. **Mombereau C, Kaupmann K, Gassmann M, Bettler B, van der Putten H, Cryan JF.** Altered anxiety and depression-related behaviour in mice lacking GABAB(2) receptor subunits. Neuroreport. 2005 Feb 28;**16**(3):307–10.
58. **Felice D, O'Leary OF, Pizzo RC, Cryan JF.** Blockade of the GABA(B) receptor increases neurogenesis in the ventral but not dorsal adult hippocampus: Relevance to antidepressant action. Neuropharmacology. 2012 Dec;**63**(8):1380–8.
59. **Giachino C, Barz M, Tchorz JS, Tome M, Gassmann M, Bischofberger J**, et al. GABA suppresses neurogenesis in the adult hippocampus through GABAB receptors. Development. 2014 Jan;**141**(1):83–90.
60. **O'Leary OF, Felice D, Galimberti S, Savignac HM, Bravo JA, Crowley T**, et al. GABAB(1) receptor subunit isoforms differentially regulate stress resilience. PNAS. 2014 Oct 21;**111**(42):15232–7.
61. **Sweeney FF, O'Leary OF, Cryan JF.** Activation but not blockade of GABAB receptors during early-life alters anxiety in adulthood in BALB/c mice. Neuropharmacology. 2014 Jun;**81**:303–10.
62. **Pinard A, Seddik R, Bettler B.** GABAB receptors: Physiological functions and mechanisms of diversity. Adv Pharmacol. 2010;**58**:231–55.

63. **Metz M, Gassmann M, Fakler B, Schaeren-Wiemers N, Bettler B.** Distribution of the auxiliary GABAB receptor subunits KCTD8, 12, 12b, and 16 in the mouse brain. J Comp Neurol. 2011 Jun 1;**519**(8):1435–54.
64. **Cathomas F, Stegen M, Sigrist H, Schmid L, Seifritz E, Gassmann M,** et al. Altered emotionality and neuronal excitability in mice lacking KCTD12, an auxiliary subunit of GABAB receptors associated with mood disorders. Transl Psychiatry. 2015;5:e510.
65. **Lee MT, Chen CH, Lee CS, Chen CC, Chong MY, Ouyang WC,** et al. Genome-wide association study of bipolar I disorder in the Han Chinese population. Mol Psychiatry. 2011 May;**16**(5):548–56.
66. **O'Leary OF, Castren E.** Neurotrophic factors and antidepressant action: Recent advances. In: Cryan JF, Leonard BE (eds). Depression: From Psychopathology to Pharmacotherapy. Mod Trends Pharmacopsychiatry.. Basel: Karger; 2010, pp. 199–223.
67. **Duman RS, Heninger GR, Nestler EJ.** A molecular and cellular theory of depression. Arch Gen Psychiatry. 1997 Jul;**54**(7):597–606.
68. **Nestler EJ, Barrot M, DiLeone RJ, Eisch AJ, Gold SJ, Monteggia LM.** Neurobiology of depression. Neuron. 2002 Mar 28;**34**(1):13–25.
69. **Autry AE, Monteggia LM.** Brain-derived neurotrophic factor and neuropsychiatric disorders. Pharmacol Rev. 2012;**64**(2):238–58.
70. **Waterhouse EG, An JJ, Orefice LL, Baydyuk M, Liao GY, Zheng K,** et al. BDNF promotes differentiation and maturation of adult-born neurons through GABAergic transmission. J Neurosci. 2012 Oct 10;**32**(41):14318–30.
71. **Duman RS, Vaidya VA.** Molecular and cellular actions of chronic electroconvulsive seizures. J ECT. 1998 Sep;**14**(3):181–93.
72. **Castren E, Rantamaki T.** Role of brain-derived neurotrophic factor in the aetiology of depression: Implications for pharmacological treatment. CNS Drugs. 2010 Jan;**24**(1):1–7.
73. **Shirayama Y, Chen AC, Nakagawa S, Russell DS, Duman RS.** Brain-derived neurotrophic factor produces antidepressant effects in behavioral models of depression. J Neurosci. 2002 Apr 15;**22**(8):3251–61.
74. **Siuciak JA, Lewis DR, Wiegand SJ, Lindsay RM.** Antidepressant-like effect of brain-derived neurotrophic factor (BDNF). Pharmacol Biochem Behav. 1997 Jan;**56**(1):131–7.
75. **Eisch AJ, Bolanos CA, de Wit J, Simonak RD, Pudiak CM, Barrot M,** et al. Brain-derived neurotrophic factor in the ventral midbrain–nucleus accumbens pathway: A role in depression. Biol Psychiatry. 2003 Nov 15;**54**(10):994–1005.
76. **Shimizu E, Hashimoto K, Okamura N, Koike K, Komatsu N, Kumakiri C,** et al. Alterations of serum levels of brain-derived neurotrophic factor (BDNF) in depressed patients with or without antidepressants. Biol Psychiatry. 2003 Jul 1;**54**(1):70–5.
77. **Seo MK, Ly NN, Lee CH, Cho HY, Choi CM, Nhu le H,** et al. Early life stress increases stress vulnerability through BDNF gene epigenetic changes in the rat hippocampus. Neuropharmacology. 2016 Jun;**105**:388–97.
78. **Tsankova NM, Berton O, Renthal W, Kumar A, Neve RL, Nestler EJ.** Sustained hippocampal chromatin regulation in a mouse model of depression and antidepressant action. Nat Neurosci. 2006 Apr;**9**(4):519–25.
79. **Carola V, Gross C.** BDNF moderates early environmental risk factors for anxiety in mouse. Genes Brain Behav. 2010 Jun 1;**9**(4):379–89.
80. **Gururajan A, Hill RA, van den Buuse M.** Brain-derived neurotrophic factor heterozygous mutant rats show selective cognitive changes and vulnerability to chronic corticosterone treatment. Neuroscience. 2015 Jan 22;**284**:297–310.
81. **Duman CH, Schlesinger L, Kodama M, Russell DS, Duman RS.** A role for MAP kinase signaling in behavioral models of depression and antidepressant treatment. Biol Psychiatry. 2007 Mar 01;**61**(5):661–70.

82. **Chourbaji S, Brandwein C, Vogt MA, Dormann C, Hellweg R, Gass P.** Nature vs. nurture: Can enrichment rescue the behavioural phenotype of BDNF heterozygous mice? Behav Brain Res. 2008 Oct 10;**192**(2):254–8.
83. **Autry AE, Adachi M, Cheng P, Monteggia LM.** Gender-specific impact of brain-derived neurotrophic factor signaling on stress-induced depression-like behavior. Biol Psychiatry. 2009 Jul 1;**66**(1):84–90.
84. **Ibarguen-Vargas Y, Surget A, Vourc'h P, Leman S, Andres CR, Gardier AM,** et al. Deficit in BDNF does not increase vulnerability to stress but dampens antidepressant-like effects in the unpredictable chronic mild stress. Behav Brain Res. 2009 Sep 14;**202**(2):245–51.
85. **Yu H, Wang DD, Wang Y, Liu T, Lee FS, Chen ZY.** Variant brain-derived neurotrophic factor Val66Met polymorphism alters vulnerability to stress and response to antidepressants. J Neurosci. 2012 Mar 21;**32**(12):4092–101.
86. **Sanacora G, Treccani G, Popoli M.** Towards a glutamate hypothesis of depression: An emerging frontier of neuropsychopharmacology for mood disorders. Neuropharmacology. 2012 1;**62**(1):63–77.
87. **Peterlik D, Flor PJ, Uschold-Schmidt N.** The emerging role of metabotropic glutamate receptors in the pathophysiology of chronic stress-related disorders. Curr Neuropharmacol 2016;**14**(5):514–39.
88. **Golubeva AV, Moloney RD, O'Connor RM, Dinan TG, Cryan JF.** Metabotropic glutamate receptors in central nervous system diseases. Current Drug Targets. 2016;**17**(5):538–616.
89. **Kawaguchi DM, Glatt SJ.** GRIK4 polymorphism and its association with antidepressant response in depressed patients: A meta-analysis. Pharmacogenomics. 2014 Aug;**15**(11):1451–9.
90. **Laje G, Paddock S, Manji H, Rush AJ, Wilson AF, Charney D,** et al. Genetic markers of suicidal ideation emerging during citalopram treatment of major depression. Am J Psychiatry. 2007 Oct;**164**(10):1530–8.
91. **Menke A, Samann P, Kloiber S, Czamara D, Lucae S, Hennings J,** et al. Polymorphisms within the metabotropic glutamate receptor 1 gene are associated with depression phenotypes. Psychoneuroendocrinology. 2012 Apr;**37**(4):565–75.
92. **Miller AH, Raison CL.** The role of inflammation in depression: From evolutionary imperative to modern treatment target. Nat Rev Immunol. 2016 Jan;**16**(1):22–34.
93. **Dantzer R, O'Connor JC, Freund GG, Johnson RW, Kelley KW.** From inflammation to sickness and depression: When the immune system subjugates the brain. Nat Rev Neurosci. 2008 Jan;**9**(1):46–56.
94. **Kovacs D, Eszlari N, Petschner P, Pap D, Vas S, Kovacs P,** et al. Effects of IL1B single nucleotide polymorphisms on depressive and anxiety symptoms are determined by severity and type of life stress. Brain Behav Immun. 2016 Aug;**56**:96–104.
95. **Baune BT, Dannlowski U, Domschke K, Janssen DG, Jordan MA, Ohrmann P,** et al. The interleukin 1 beta (IL1B) gene is associated with failure to achieve remission and impaired emotion processing in major depression. Biol Psychiatry. 2010 Mar 15;**67**(6):543–9.
96. **Wong ML, Dong C, Maestre-Mesa J, Licinio J.** Polymorphisms in inflammation-related genes are associated with susceptibility to major depression and antidepressant response. Mol Psychiatry. 2008 Aug;**13**(8):800–12.
97. **Howick K, Griffin BT, Cryan JF, Schellekens H.** From belly to brain: Targeting the ghrelin receptor in appetite and food intake regulation. Int J Mol Sci. 2017 Jan 27;**18**(2).
98. **Lutter M, Sakata I, Osborne-Lawrence S, Rovinsky SA, Anderson JG, Jung S,** et al. The orexigenic hormone ghrelin defends against depressive symptoms of chronic stress. Nat Neurosci. 2008 Jul;**11**(7):752–3.
99. **Kanehisa M, Akiyoshi J, Kitaichi T, Matsushita H, Tanaka E, Kodama K,** et al. Administration of antisense DNA for ghrelin causes an antidepressant and anxiolytic response in rats. Prog Neuropsychopharmacol Biol Psychiatry. 2006 Dec 30;**30**(8):1403–7.
100. **Chuang JC, Zigman JM.** Ghrelin's roles in stress, mood, and anxiety regulation. Int J Pept. 2010;2010.
101. **Meyer RM, Burgos-Robles A, Liu E, Correia SS, Goosens KA.** A ghrelin-growth hormone axis drives stress-induced vulnerability to enhanced fear. Mol Psychiatry. 2014 Dec;**19**(12):1284–94.

102. **Nakashima K, Akiyoshi J, Hatano K, Hanada H, Tanaka Y, Tsuru J**, et al. Ghrelin gene polymorphism is associated with depression, but not panic disorder. Psychiatr Genet. 2008 Oct;**18**(5):257.

103. **Hansson C, Annerbrink K, Nilsson S, Bah J, Olsson M, Allgulander** C, et al. A possible association between panic disorder and a polymorphism in the preproghrelingene. Psychiatry Res. 2013 Mar 30;**206**(1):22–5.

104. **Kozlovsky N, Matar MA, Kaplan Z, Zohar J, Cohen H.** The role of the galaninergic system in modulating stress-related responses in an animal model of posttraumatic stress disorder. Biol Psychiatry. 2009 Mar 01;**65**(5):383–91.

105. **Holmes PV, Blanchard DC, Blanchard RJ, Brady LS, Crawley JN.** Chronic social stress increases levels of preprogalanin mRNA in the rat locus coeruleus. Pharmacol Biochem Behav. 1995 Apr;**50**(4):655–60.

106. **Kuteeva E, Hokfelt T, Wardi T, Ogren SO.** Galanin, galanin receptor subtypes and depression-like behaviour. Cell Mol Life Sci. 2008 Jun;**65**(12):1854–63.

107. **Murck H, Held K, Ziegenbein M, Kunzel H, Holsboer F, Steiger A.** Intravenous administration of the neuropeptide galanin has fast antidepressant efficacy and affects the sleep EEG. Psychoneuroendocrinology. 2004 Oct;**29**(9):1205–11..

108. **Unschuld PG, Ising M, Roeske D, Erhardt A, Specht M, Kloiber S**, et al. Gender-specific association of galanin polymorphisms with HPA-axis dysregulation, symptom severity, and antidepressant treatment response. Neuropsychopharmacology. 2010 Jun;**35**(7):1583–92.

109. **Juhasz G, Hullam G, Eszlari N, Gonda X, Antal P, Anderson IM**, et al. Brain galanin system genes interact with life stresses in depression-related phenotypes. PNAS. 2014 Apr 22;**111**(16):E1666–73.

110. **Cryan JF, Dinan TG.** Mind-altering microorganisms: The impact of the gut microbiota on brain and behaviour. Nat Rev Neurosci. 2012 Oct;**13**(10):701–12.

Chapter 7

Methods and theoretical approaches: Genetic animal models of emotional disorders and convergence with human data

Celine L St Pierre, Kayvon Sharif, Emily Funsten, Abraham A Palmer, and Clarissa C Parker

7.1 Advantages of rodent models

Human studies have clearly demonstrated that diseases like anxiety, depression, and stress-related disorders are influenced by genetics. However, conducting research on humans is limited by numerous practical, financial, and ethical concerns (1). One complementary approach relies on the use of rodent models to clarify the genetic determinants of psychiatric disorders.

Rats and mice have advantages for the understanding of human disease above and beyond what is possible when studying humans directly. First, they allow for environmental factors to be held constant or systematically varied in order to explore interactions between genotype and environment. Since external factors partially regulate how depression, stress, and anxiety symptoms manifest, utilizing rodents exploits this important source of variability (2). Second, almost all human genes known to be associated with disease have orthologs in the rat and mouse (3–5) and multiple disease genes identified in rodents have also been shown to play a role in human diseases (6). Another advantage is that brain tissue can be obtained under ideal laboratory conditions, potentially following invasive procedures, some of which are not possible in humans (7). For example, obtaining analogous brain tissue from humans is especially challenging because humans have vastly different life histories; this confounds the analysis of gene expression and makes obtaining matched control tissue virtually impossible. Furthermore, pharmacological manipulations can easily be applied in rodent models as additional evidence for the validity of a specific phenotype and can help distinguish between different aspects of a disorder (8, 9). Finally, rodent genomes can be easily and directly manipulated through a wide range of techniques, making them an incredibly powerful and almost limitless genetic tool (10–12).

7.2 Defining traits: Intermediate phenotypes

Nonetheless, attempts to mimic the clinical constructs of depression, stress, and anxiety in rodents are not without caveats and require a reductionist approach. Obvious cognitive differences exist between species such that creating a rodent model that fully recapitulates every aspect of a human disorder is unfeasible (2). One reason is that these disorders involve emotions and experiences that may be unique to humans. Because we cannot ask a rat about its feelings, we instead make inferences about its mental state by observing its behavior and physiological responses (13,

14). Thus, the endpoint of rodent studies is behaviorally or physiologically based, rather than centered on self-reported emotional states.

Another consideration is that clinical definitions of human psychiatric disorders are heterogeneous and encompass many symptoms, not all of which are required for a diagnosis. Two individuals could be diagnosed with the same disorder but share no common symptoms. Different symptoms most likely have separate genetic or environmental causes, so attempting to study them all at once in rodents is a misguided approach (15). This view is supported by an increasing number of studies showing that genes associated with psychiatric disorders do not align themselves along specific DSM diagnoses (16, 17). Finally, the diagnostic criteria for most psychiatric disorders are qualitative, rely on subjective reports, and do not currently include any measurable behaviors or biological traits (18, 19).

The concept of an intermediate phenotype has been helpful in addressing these issues. Intermediate phenotypes were originally defined as measurable components, not externally apparent, that lay along the pathway between genotype and phenotype (20). The term was later adapted to bridge the gap between the overt behavioral symptoms of psychiatric disorders and their elusive genetic underpinnings (21, 22). It has long been clear that clinical phenotypes are not optimal for mapping studies since they are affected by a large number of genetic and environmental influences. Intermediate phenotypes attempt to deconstruct the clinical outcome into related biological variables that are hypothetically regulated by fewer genes (22). A genetic locus should therefore be easier to identify because it should have a larger effect on the intermediate phenotype than on the total psychiatric disorder (23). Despite this logical advantage, the construct was not widely utilized until more recently. Growing frustration with unsuccessful linkage and association studies, together with more sophisticated animal models and genomic analysis tools, has resulted in the reemergence of use of intermediate phenotypes (22).

The study of psychiatric disorders through the investigation of intermediate phenotypes has also gained traction following the National Institute of Mental Health's (NIMH) recent development of the Research Domain Criteria (RDoC) initiative; a research framework for studying mental disorders (24). Due to the complexity of many psychiatric disorders, the RDoC emphasizes the mapping of disorders and illnesses to specific biological systems (25). To achieve this goal, the priority is shifted away from employing DSM-style diagnoses involving symptom clusters and towards neurological, physiological, and behavioral constructs that are specific, observable, and quantifiable (26, 27). These psychophysiological intermediate phenotypes are more objective, more quantifiable, can be measured across degrees of severity, may represent more discrete symptom categories (and thus lead to more discrete neurobiological causes), and are often useful across multiple species (28).

7.3 Reverse genetic approaches in rodent models

Genotype-to-phenotype (also known as reverse genetic) approaches refer to a method of discovering the function of a gene by examining the phenotypic effects that result from a targeted mutation (most commonly deletion). They are most useful for examining the influence of a single gene on a phenotype of interest and can be used to test statistical associations identified in human genetic studies. Thus, reverse genetic approaches in rodents are commonly used to elucidate the underlying mechanism by which genes influence intermediate phenotypes in humans.

Reverse genetic approaches include gene ablation (knockouts), changes or replacement of genes (knockins), addition of extra copies of genes or addition of nonnative sequences (transgenics), or the use of viral vectors or short interfering RNA segments (miRNA knockdowns) (29). Additional levels of complexity and precision can be added by using various conditional and inducible systems that allow for tissue- or temporally specific changes in gene expression or even control by exogenous triggers (tetracycline, tamoxifen, or light via various optogenetic systems; 30–33). Historically, the mouse was the most commonly used model organism for reverse genetic approaches due to the success of embryonic stem (ES) cell-based targeting technology (34), whereas rats lacked convenient tools for genome modification. However, some progress has been made with rat ES cell-based gene targeting technology (35–38) and the application of nuclease-mediated targeting by clustered regularly interspaced short palindromic repeats (CRISPR/Cas), zinc-finger nucleases (ZFNs), and transcription activator-like effector (TALE) domains for efficient TALE nucleases (TALENs) have been used to modify endogenous genes quickly and efficiently across a variety of cell types and model organisms (10, 11, 39–41). These approaches enable precise genome engineering to generate modifications such as point mutations, accurate insertions and deletions, and conditional knockout and knockin rats. Because of these advances in genome editing technology, the decision to use mice or rats now depends on the behavioral endpoint being studied, rather than technical advantages as a genetic organism (42).

7.4 Forward genetic approaches in rodent models

Phenotype-to-genotype (also known as forward genetic) approaches begin with the measurement of the trait of interest in order to uncover the underlying genetic architecture in a population. Forward genetic strategies offer an unbiased approach to identifying the genes and alleles that give rise to variability in a given phenotype. Three of the most commonly used approaches are selective breeding, mutagenesis, and association mapping.

Selective breeding entails creating rodent lines that are highly divergent for the phenotype of study. It is particularly valuable for determining the heritability of a trait as well as pleiotropy. One of the most valuable aspects of selectively bred lines is that they allow researchers to explore precursors of traits of interest. Most psychiatric disorders are the product of gene by environment interactions; therefore it is critical to understand the variables that precede them or that render certain individuals particularly vulnerable, in order to best prevent them. Selective breeding has been performed in rodents for a wide range of traits including conditioned fear, anxiety-like behavior, depression-like behavior, stress reactivity, and coping styles (43–53). However, while selective breeding can provide strong evidence of genetic contributions to phenotypes of interest, it rarely allows for a single gene to be identified.

Mutagenesis-based screens offer an unbiased discovery platform to identify novel genes associated with traits of interest. In this approach, mutagens such as *N*-ethyl-*N*-nitrosourea (ENU) are used to induce single nucleotide variants randomly in a population. This is followed by screening wildtypes and mutants for extreme phenotypes. Mutagenesis is analogous to studies of highly penetrant but rare familial disorders and *de novo* mutations in humans. ENU can produce either loss-of-function or gain-of-function mutations (54). This technique has been used in large-scale ENU-mutagenized rodent populations to find mutants that display abnormalities across a broad range of phenotypes, including fear learning, antidepressant activity, stress responsiveness, and anxiety-related traits (55–58). In some cases, these neurobehavioral mutants have led to identifying candidate genes (59). Causal mutations are relatively easy to find because they occur on an otherwise isogenic background. The advent of next-generation sequencing has vastly accelerated the task of identifying such mutations (60).

Association mapping is another forward genetic approach for identifying alleles that contribute to natural variation in behavioral or physiological phenotypes. Here, numerous subjects are phenotyped and genotyped across the entire genome in order to scan for single-nucleotide polymorphisms (SNPs) that are significantly correlated to variation in a trait of interest. The most associated SNP is not necessarily the causal mutation; rather it "tags" a region (haplotype) that contains the causal variant (61). Such genetic regions are hypothesized to affect the given trait; when the trait is quantitative (e.g. height) such regions are sometimes referred to as a quantitative trait loci (QTL).

Traditionally, association studies in rodents utilized relatively simple populations such as F_2 crosses between two inbred strains or recombinant inbred (RI) lines to determine the location of QTLs. Because these populations have limited numbers of recombinations, the identified QTLs were large, often measuring tens of mega-bases and containing hundreds of genes (62, 63). More recently, association studies related to emotional disorders are being conducted using highly recombinant populations such as advanced intercross lines (64,65), heterogeneous stocks (66–72), the Hybrid Mouse Diversity Panel (73), and commercially available outbred stocks (74–77). These populations involve complex breeding schemes, increased genetic diversity, and show a rapid breakdown of linkage disequilibrium that allows for increasingly high-resolution mapping (62, 63, 78). With advances in genotyping technology, and the development of statistical software to address issues of familial relatedness, it has become inexpensive and relatively straightforward to perform genome-wide association studies in hundreds or thousands of rodents. This approach promises to streamline what has historically been a very slow and expensive process: the definitive identification of the genes that underlie QTLs.

Thus, forward genetics provide an unbiased approach to identify genetic variations associated with trait variations in animal models. These genes of interest can then be studied in human populations to evaluate whether they contribute to variability in corresponding intermediate phenotypes.

7.5 Rodent models of anxiety, depression, and stress

Three sets of criteria are commonly used to assess rodent models of human psychiatric disorders: predictive validity, face validity, and construct validity (see 79 for an excellent review on validity in animal models of anxiety and depression). In order to have predictive validity, the model should be sensitive to clinically effective treatments, both pharmacological and otherwise (80). In order to possess face validity, responses observed in the rodent model should be similar to behavioral, neuroanatomical, and physiological responses in humans (81). Construct validity requires that the etiological processes underlying the disorder be similar in rodents and humans (82). We describe some of the most common behavioral tests that are used in rodents to study anxiety, depression, and stress-related disorders in humans. We evaluate their validity across the three aforementioned domains, highlight their strengths and weaknesses, and then discuss promising intermediate phenotypes that can be used to evaluate the physiological underpinnings that contribute to risk of development of psychiatric disorders. In particular, we focus on intermediate phenotypes with translational utility and provide examples in which using one or some of these intermediate phenotypes has led to an enhanced understanding of the molecular pathways underlying a disorder, promising candidate genes, or novel treatment strategies.

7.6 Anxiety

Like humans, rodents react to perceived or actual threats with characteristic autonomic nervous system responses and defensive behaviors. Anxiety-like traits can thus be studied in rodents using behavioral paradigms that model these aspects of human anxiety (13). These behavioral paradigms are easily quantified and their predictive validity has been established by showing sensitivity to drugs known classically to be anxiogenic and anxiolytic in humans. In addition, they have some construct validity in that they are also associated with the same brain regions and neurotransmitter systems dysregulated in human anxiety. Most are ethological tests, which assess a rodent's spontaneous or natural behavior (e.g. its preferences, aversions, and responses to stressful stimuli) and do not involve conditioned-based procedures (83). These behavioral tests have many overlapping features but still vary in their design and interpretation.

In rodents, tests of anxiety-like behaviors are generally based on approach-avoidance conflicts, and most paradigms are analogous to acute anxiety episodes in humans rather than chronic anxiety. These behavioral assays exploit the opposing motivations of rats and mice to avoid bright, open, and potentially dangerous areas (avoidance) and to explore new spaces to assess the availability of food, escape routes, and mating partners (approach) (84). Common tests of anxiety-like behavior include the light/dark box test, elevated plus/zero mazes, defensive burying, and the open field test. Higher anxiety-like behavior is signified by a shift towards avoidance tactics, which model the characteristic fear of public or open spaces in human agoraphobia (83). Despite the wide array of tests measuring anxiety-like behavior in rodents, they share many of the same limitations. First, these tests may be difficult to interpret because they may reflect altered motivation to explore a novel apparatus (approach, novelty-seeking) rather than differences in avoidance. They can also be confounded by differences in locomotor activity (85) as well as by pre-exposure to the testing apparatus (86, 87). One final issue that should be considered when using these tests is the difference between state and trait anxiety. State anxiety refers to anxiety at the time of assessment, and may be more likely to represent an adaptive and evolutionarily advantageous response to an unfamiliar testing environment. In contrast, trait anxiety refers to a general tendency to experience anxiety across situations, and may reflect a pathological anxiety state. Currently, few trait anxiety rodent models are used. Different neurocircuitry may underlie state versus trait anxiety and the interaction of a rodent's trait anxiety with experimentally induced state anxiety may interfere with responses to pharmacological interventions and complicate interpretation of results (88).

An alternative measure of unconditioned anxiety is the startle response, which is elicited by the presentation of a high-intensity stimulus (usually a noise or flash of light) The startle response is a short latency reflex mediated by a well-characterized brainstem and spinal cord pathway that is conserved across species, including humans. In humans, eye-blink is commonly measured as the startle response, while in rodents the whole-body jump is defined as the magnitude of startle (89). Several experimental paradigms exist for distinguishing various forms of startle reactivity that involve different, yet overlapping neurocircuitry (90). These include general startle response, fear-potentiated startle response, and context-potentiated startle response. Importantly, the startle response has good predictive validity as it is increased during aversive motivational states (e.g. fear, anxiety, withdrawal from drugs of abuse) in both humans and rodents, as well as has good face validity for studying anxiety disorders (90–92). For example, alterations in the startle response are observed in panic disorder, post-traumatic stress disorder (PTSD), obsessive compulsive disorder, and generalized anxiety disorder (93–96). Thus, rodent models of anxiety based on the startle response have good face, predictive, and construct validity, and are especially promising for translational studies of anxiety disorders in humans.

Genes associated with anxiety-like behaviors in rodents have been identified that may also influence anxiety disorders in humans. For example, accumulating evidence from both forward and reverse genetic approaches in mice point to the g protein-coupled regulator *Rgs2* (regulator of G-protein signaling 2) as a likely candidate gene for anxiety in humans. Over 15 years ago, Oliveira-Dos-Santos and colleagues (97) reported that *Rgs2* deficient mice displayed increased anxiety-like behavior in the light/dark box test as well as an increased acoustic startle response. Just a few years later, a forward genetic study in commercially available outbred mice demonstrated that *Rgs2* modulated anxiety-like behavior in the open field (98, 99). More recently, Lifschytz and colleagues

(100) suggested that the relationship between *Rgs2* and anxiety-like behavior may involve changes in serotonergic receptor expression. Mice carrying a mutation resulting in lower *RGS2* gene expression had significantly lower 5-HT1A and 5-HT1B receptor expression in the raphe nucleus, as well as increased anxiety-like behavior in the elevated plus maze. In humans, variants in *RGS2* have been associated with anxiety-related temperament (101), generalized anxiety disorder (102), post-traumatic stress symptoms (103), behavioral avoidance (104), and panic disorder with and without comorbid agoraphobia (104–106); but see (107–109). Yet despite the parallel correlations between *RGS2* and anxiety disorders and between dysregulated startle responsiveness and anxiety disorders, no research has been published to date examining the relationship between *RGS2* and the startle response in humans. Therefore, further investigation into *RGS2* and the startle response may provide useful information regarding the neural processes that contribute to the pathophysiology of anxiety disorders. Rodent studies have not only the potential to explore the molecular events driving these observed correlations, but also the capacity to identify the specific conditions in which polymorphisms in *RGS2* influence the startle response and anxiety-like traits.

7.7 **Depression**

Major depressive disorder (MDD) is complex disease with substantial clinical and etiological heterogeneity. One of the primary predisposing factors to MDD in humans is exposure to trauma or stress (110, 111). As a result, tests for assessing depression-related behavior in rodents often involve exposure to stressful stimuli. Two of the most common depression-related tests are the forced swim test (FST) and the tail suspension test (TST). In the FST, as originally described by Porsolt (112, 113), rodents are forced to swim in an inescapable cylinder filled with water. The time a rodent spends without making any movements beyond those required to keep its head above water is measured. The development of immobility is thought to reflect either a failure of persistence in escape-directed behavior (i.e. behavioral despair) or the development of passive behavior that disengages the animal from active forms of coping with stressful stimuli (114). In the TST, immobility is measured in rodents hung upside-down by their tail (115). Both tests have good predictive validity in that administration of various classes of antidepressants before either test causes rodents to engage in escape-directed behaviors for longer periods of time than vehicle-treated controls (115, 116).

Despite their widespread use, the FST and TST have poor face validity and mixed results regarding construct validity. The neuropsychological mechanisms underlying the development of immobility are unclear and may be confounded by factors that influence locomotor activity (112, 117–119). Some even argue that the development of immobility is not a depression-like phenotype at all but instead represents an adaptive learning process that enhances likelihood of survival by conserving energy (120). There is evidence suggesting both tests have good construct validity in that they are sensitive to factors that influence depression in humans, such as genetic predisposition, prior exposure to stress, sleep deprivation, and drug withdrawal (115, 116). However, both tests involve exposing rodents to short-term stress, which is quite different from the long-lasting pathology observed in depressed humans (80). Similarly, many antidepressants that decrease immobility acutely in rodents take weeks before they are effective in alleviating depression in humans (121), suggesting that antidepressant action in the FST and TST may not provide insight into the etiology of depression.

While it is broadly accepted that genetic factors and stressful life events contribute to the development of the disorder, the underlying pathophysiology of MDD remains poorly understood. As a result, many researchers have turned to an intermediate phenotype-based approach, in which individual symptoms of MDD are modeled independently using quantifiable and objective endpoints. One promising endpoint involves the measurement of gene expression levels in animal models in order to identify the molecular mechanisms associated with pathology in humans. Brain-derived neurotrophic factor (BDNF) is the most abundant neurotrophin in the central nervous system. It promotes neurogenesis, neuronal survival, axonal and dendritic growth, and synapse formation (122–126). Importantly, BDNF levels in brain and serum are reduced by stress and MDD (127–131), and are increased by antidepressants and successful recovery (132–135). These findings have implicated BDNF in the pathophysiology of depression as well as in the mechanism of action of antidepressant treatments. Recent findings suggest that post-translational processing of BDNF gene product produces different protein products that alter signaling through neurotrophic tyrosine receptor kinase type 2 (TrkB) and the pan-neurotrophin receptor $p75^{NTR}$ (136). As a result, many researchers have focused on BDNF levels as both an intermediate phenotype and potential treatment target for MDD (137).

For example, it has been shown that a common single nucleotide polymorphism (SNP) in the human *BDNF* gene (Val66Met) results in variable BDNF protein function (138) and other depression-relevant intermediate phenotypes such as smaller hippocampal volume (139, 140) and disruptions in hippocampal functioning (141, 142) during hippocampal-dependent memory tasks (but see 143, 144). *BDNF* Val66Met has been modeled in mice by introducing the variant into the endogenous mouse *BDNF* gene. The Met-allele in mice leads to diminished activity-dependent BDNF secretion, smaller hippocampal volume, reduced dendritic arborization and complexity of hippocampal neurons, and impaired hippocampal-dependent learning and memory (141, 145). Furthermore, Met/Met mice display diminished behavioral responsiveness to antidepressants and fail to increase BDNF levels and hippocampal synaptic plasticity after treatment with the antidepressant fluoxetine (146, 147). This reverse genetic approach enables a mechanistic examination of the observed phenotypes in a way that would not be possible in humans.

More recently, Malki and colleagues (148) used a forward genetic approach to study the molecular mechanisms underlying MDD. They examined gene expression differences in the prefrontal cortex between Flinders Sensitive and Resistant Lines (FSL/FRL) of rat. Flinders rats represent one of the most robust genetic models of MDD and display many characteristics resembling the human depression phenotype including depression-related behavior, anhedonia, sleep disturbances, and decreased appetite (49). Genes that were differentially expressed in the prefrontal cortex (PFC) of FSL and FRL rats were then used to inform probe set selection in a comparable case-control human PFC mRNA data set. Using this approach, researchers observed an upregulation of the *NTRK2* gene in MDD patients. *NTRK2* encodes TrkB; binding of BDNF to its high-affinity TrkB receptor is the primary mechanism by which BDNF signaling occurs. These findings demonstrate the advantage of using rodent models of disease to inform human studies by providing a hypothesis-free candidate gene selection approach to identify those genes that have a higher probability of being involved in human psychopathology.

7.8 **Stress**

Post-traumatic stress disorder (PTSD) is a psychiatric disorder that develops after exposure to stressful or traumatic events. Although risk for developing PTSD is demonstrably heritable (149, 150), only 20–30% of individuals exposed to severe trauma develop PTSD (151). This indicates the presence of protective factors conveying potential resilience but can also make gene identification difficult. PTSD symptoms manifest in a number of ways following the event(s) and can include hyperarousal, re-experiencing of the traumatic event, avoidant behavior, anxiety, negative cognition, distress, and neuroendocrine dysregulation. While these responses can be objectively assessed, other symptoms such as intrusive memories and dreams, dissociation, and affective numbing cannot be modeled in rodents. Finally, PTSD is often comorbid with affective disorders, such as depression and bipolar disorder (152). As a result of this complexity, rodent models do not replicate the human condition in its entirety but instead attempt to mimic specific intermediate phenotypes associated with PTSD. As such, the predictive validity of models based on only a subset of symptoms requires regular re-evaluation to maintain and improve their translational relevance (153, 154).

Traditionally, rodent models of PTSD focused on exposure to trauma/stress and abnormal fear regulation in conditioned fear paradigms. Trauma/stress models are largely based on face validity, and entail exposure to stimuli believed to elicit an extreme sense of threat or helplessness (153, 155). The stressors can be physical (footshock, immobilization stress) or psychosocial (early-life stress, social defeat). A limitation of some of these models is that the stressful event affects all animals similarly or that individual differences are not reported. Rather, comparisons are only made between naïve versus trauma-exposed subjects (for notable exceptions, see 156–158). This contrasts with human PTSD in which only a subset of trauma-exposed individuals goes on to develop the disorder. Therefore, to improve face validity, trauma/stress models should aim to study individual differences in vulnerability in order to provide a greater understanding of the factors influencing resilience and susceptibility to trauma.

Exposure to stress alone does not sufficiently explain the behavioral and physiological responses that persist long after trauma exposure (159). This has led some to claim that abnormal fear regulation may underlie the development of PTSD (160, 161). Following exposure to traumatic or stressful events, the normal and adaptive fear response involves consolidation and expression of fear memories in dangerous situations *and* the extinction of fear behaviors in safe situations. When this process becomes dysregulated, fear sensitization and fear generalization can occur. These abnormal fear responses characterize PTSD in humans (162–166) and make rodent models of fear learning useful in understanding the neurogenetic mechanisms underlying PTSD. The typical conditioned fear paradigm involves a rodent being placed in a neutral context (chamber) and simultaneously experiencing an aversive unconditioned stimulus (electrical shock) and neutral conditioned stimulus (tone). These cues become paired to the point where the rodent will display a natural defensive "freezing" response when either tested in the originally neutral context or presented with the originally neutral stimulus in a different context (167, 168). The conditioned fear responses can be extinguished by repeatedly presenting the conditioned stimulus without the unconditioned stimulus (169). However, fear conditioning alone is a normal phenomenon in both rodents and humans, and is not sufficient to produce the PTSD phenotype (159).

A better understanding of the pathophysiology of PTSD has led to improved rodent models with superior construct validity. Models include exposure to predator scent (170–173), serial exposure to multiple stressors (174–177), and stress-enhanced fear learning (178–181). They encompass a variety of behavioral and physiological end points, such as abnormal fear extinction, enhanced acoustic startle, abnormal hippocampal structure/functioning, and alterations in neuroendocrine signaling. Importantly, PTSD in humans is also characterized by changes in neuroendocrine responses to stress. These responses, both acute and chronic, are heavily mediated and regulated by the hypothalamic–pituitary–adrenal (HPA) axis (182). For example, altered cortisol levels are observed in individuals with PTSD (183) and are believed to be associated with dysfunction of the HPA axis (184). As such, HPA-axis dysregulation may prove to be a promising intermediate phenotype for PTSD. In rats, the chronic plus acute prolonged stress paradigm is used to model the stressful events that can lead to HPA-axis dysregulation and PTSD in humans (176, 177). The paradigm consists of chronic unpredictable cold stress for 14 days, followed by a series of acute stressors such as social defeat, immobilization, and cold swim (176). This model enhances acute fear responses, impairs extinction of conditioned fear, increases anxiety-like behavior in the defensive burying test, reduces expression of glucocorticoid receptors in the prefrontal cortex, and blunts the HPA response to subsequent stressors (176). These effects resemble many aspects of human PTSD, including impaired fear extinction, a shift from active to passive coping strategies, enhanced anxiety, and altered HPA-axis reactivity (177).

Investigating the pathophysiology of HPA-axis regulation has uncovered a particular gene of interest, *FKBP5*, which encodes for widespread glucocorticoid receptor (GR) sensitivity and HPA-axis feedback (185, 186). Binder and colleagues (187) initially discovered that specific alleles in *FKBP5* were associated with decreased cortisol-binding affinity to GR and increased negative feedback in patients with affective disorders. Further studies reported that *FKBP5* is similarly associated in clinical patients with PTSD and other stress-related disorders (185). Patients who experienced childhood trauma (188) and healthy patients with altered stress activity (189) also exhibited genetic variations in *FKBP5*. Certain SNPs increased *FKBP5* expression, decreased GR binding affinity, and increased negative feedback in the HPA axis (190). These mechanistic effects result in altered stress responses and functional changes in the HPA axis which can give rise to both affective disorders and PTSD.

Human studies provide strong evidence that PTSD is correlated with specific *FKBP5* variants that modify how the HPA axis functions (185). Unfortunately, age, duration, and type of trauma as well as age of symptom onset are all examples of environmental factors that vary across individuals and are difficult to control. Most likely as a result, other groups have not supported a main effect of *FKBP5*-alleles on stress-related phenotypes and disorders across populations with differing degrees and temporalities of stress exposure (191–194). These conflicting data make interpreting their true relationship difficult.

Rodent behavioral paradigms have provided major insight into the neurobiological mechanisms underlying stress response and shed more light on the functional role of *FKBP5* in HPA-axis dysregulation and PTSD. Conditioned fear models have shown that administering dexamethasone before testing enhances the extinction of conditioned fear memories in male Sprague-Dawley rats (195). A putative PTSD model that combines the conditioned fear and immobilization stress tests has also confirmed that dexamethasone dose-dependently enhances fear extinction and dynamically influences *FKBP5* expression in wildtype C57BL/6J mice (196). Considering that dexamethasone is a systemic GR agonist and specific *FKBP5* variants are thought to decrease GR binding affinity (190), then normalizing GR function may be one way to eradicate fear memories associated with PTSD in humans.

Reverse genetic approaches have also proved advantageous in this realm. Mice with *Fkbp5* gene deletions were found to demonstrate robust neuroendocrine, physiological, and behavior changes induced by either a chronic social defeat paradigm (197) or an immobilization stress test (198). After undergoing these stress exposure models, both cohorts of *FKBP5* knockout mice displayed better coping mechanisms to aversive situations, as measured by increased swimming behavior in the FST. These mice also showed higher GR sensitivity, lower adrenal gland weights, increased HPA-axis regulation, lower corticosterone levels, and normalized neuroendocrine responses (197–199). In human studies, there is evidence linking these same changes to the psychopathology of PTSD (200, 201).

Finally, using intermediate phenotypes in rodent models has led to novel putative treatment strategies. One example of how these models can be useful for drug development comes from the triple reuptake inhibitor antidepressant RO-05, which has been shown to regulate both GR and *FKBP5* expression in rats (202). A chronic mild stress paradigm was used to induce a depressive-like state that gradually developed in response to a series of unpredictable stressors, such as temporary food/water deprivation, continuous illumination, 45° cage tilt, and damp bedding (203, 204). The model elicited an array of neurobiological changes in these rats, including a hyperactive

HPA axis, reduced exploratory behaviors, anhedonia, and increased *FKBP5* expression and GR translocation (202). Administering RO-05 significantly reversed these physiological and behavioral effects of HPA-axis dysregulation as well as reinstated a state of stable neuroendocrine activity (202). By using rodent models to improve understanding of the relationship between *FKBP5* and HPA-axis dysregulation, novel drugs can be tested that show promise in treating PTSD symptoms in humans.

7.9 Future directions

Utilizing intermediate phenotypes to break down complex psychiatric traits in animal models has proven a fruitful strategy. Highly recombinant rodent populations now permit fine-mapping of genes associated with anxiety, mood, and stress-related phenotypes that may also be present in humans; and the advent of CRISPR/Cas technology allows for rapid functional validation of the candidate genes that are finally beginning to emerge from large-scale human GWAS approaches. We believe that the most promising future research will combine these advancements in forward and reverse genetic techniques with the inclusion of carefully selected intermediate phenotypes relevant to the disorder under investigation.

Acknowledgements

A.A.P. was supported by P50DA037844 and C.C.P was supported by NIH P20GM103449, NIH R15DA041618, and a NARSAD Young Investigator Grant from the Brain & Behavior Research Foundation.

References

1. **Kapp MB.** Ethical and legal issues in research involving human subjects: Do you want a piece of me? J Clin Pathol. 2006 Apr;**59**(4):335–9.
2. **Hovatta I, Barlow C.** Molecular genetics of anxiety in mice and men. Ann Med. 2008;**40**(2):92–109.
3. **Adams DJ, Doran AG, Lilue J, Keane TM.** The Mouse Genomes Project: A repository of inbred laboratory mouse strain genomes. Mamm Genome. 2015 Oct;**26**(9–10):403–12.
4. **Gibbs RA, Weinstock GM, Metzker ML, Muzny DM, Sodergren EJ, Scherer S**, et al. Genome sequence of the Brown Norway rat yields insights into mammalian evolution. Nature. 2004 Apr 1;**428**(6982):493–521.
5. **Mouse Genome Sequencing Consortium, Waterston RH, Lindblad-Toh K, Birney E, Rogers J, Abril JF**, et al. Initial sequencing and comparative analysis of the mouse genome. Nature. 2002 Dec 5;**420**(6915):520–62.
6. **Aitman TJ, Critser JK, Cuppen E, Dominiczak A, Fernandez-Suarez XM, Flint J**, et al. Progress and prospects in rat genetics: A community view. Nat Genet. 2008 May;**40**(5):516–22.
7. **Huang G-J, Shifman S, Valdar W, Johannesson M, Yalcin B, Taylor MS**, et al. High resolution mapping of expression QTLs in heterogeneous stock mice in multiple tissues. Genome Res. 2009 Jun;**19**(6):1133–40.
8. **Sarkis EH.** "Model" behavior. Science. 2000 Mar 24;**287**(5461):2160–2.
9. **Seong E, Seasholtz AF, Burmeister M.** Mouse models for psychiatric disorders. Trends Genet. 2002 Dec;**18**(12):643–50.
10. **Geurts AM, Cost GJ, Freyvert Y, Zeitler B, Miller JC, Choi VM**, et al. Knockout Rats via embryo microinjection of zinc-finger nucleases. Science. 2009 Jul 24;**325**(5939):433.
11. **Sander JD, Joung JK.** CRISPR-Cas systems for editing, regulating and targeting genomes. Nat Biotechnol. 2014 Apr;**32**(4):347–55.

12. **Thomas KR, Capecchi MR.** Targeted disruption of the murine int-1 proto-oncogene resulting in severe abnormalities in midbrain and cerebellar development. Nature. 1990 Aug 30;**346**(6287):847–50.
13. **Leonardo ED, Hen R.** Genetics of affective and anxiety disorders. Annu Rev Psychol. 2006;**57**:117–37.
14. **Steimer T.** Animal models of anxiety disorders in rats and mice: Some conceptual issues. Dialogues Clin Neurosci. 2011;**13**(4):495–506.
15. **Lander ES, Schork NJ.** Genetic dissection of complex traits. Science. 1994 Sep 30;**265**(5181):2037–48.
16. **Sebat J, Levy DL, McCarthy SE.** Rare structural variants in schizophrenia: One disorder, multiple mutations; one mutation, multiple disorders. Trends Genet. 2009 Dec;**25**(12):528–35.
17. **State MW.** Another piece of the autism puzzle. Nat Genet. 2010 Jun;**42**(6):478–9.
18. **Demyttenaere K, Bruffaerts R, Posada-Villa J, Gasquet I, Kovess V, Lepine JP** et al. Prevalence, severity, and unmet need for treatment of mental disorders in the World Health Organization World Mental Health Surveys. JAMA. 2004 Jun 2;**291**(21):2581–90.
19. **Lang PJ, McTeague LM, Bradley MM.** RDoC, DSM, and the reflex physiology of fear: A biodimensional analysis of the anxiety disorders spectrum. Psychophysiology. 2016 Mar 1;**53**(3):336–47.
20. **John B, Lewis KR.** Chromosome variability and geographic distribution in insects. Science. 1966 May 6;**152**(3723):711–21.
21. **Gottesman II, Shields J.** Schizophrenia and Genetics: A Twin Study Vantage Point. Vol. **xviii**. Oxford: Academic Press; 1972.
22. **Gottesman II, Gould TD.** The endophenotype concept in psychiatry: Etymology and strategic intentions. Am J Psychiatry. 2003 Apr;**160**(4):636–45.
23. **Goldman D, Ducci F.** Deconstruction of vulnerability to complex diseases: Enhanced effect sizes and power of intermediate phenotypes. Scientific World Journal. 2007;**7**:124–30.
24. **Montalvo-Ortiz JL, Gelernter J, Hudziak J, Kaufman J.** RDoC and translational perspectives on the genetics of trauma-related psychiatric disorders. Am J Med Genet Part B Neuropsychiatr Genet. 2016 Jan;**171**(1):81–91.
25. **Young G.** PTSD, Endophenotypes, the RDoC, and the DSM-5. Psychol Inj Law. 2014 Mar 5;**7**(1):75–91.
26. **Glatt CE, Lee FS.** Common polymorphisms in the age of Research Domain Criteria (RDoC): Integration and translation. Biol Psychiatry. 2016 Jan 1;**79**(1):25–31.
27. **Kraemer HC.** Research Domain Criteria (RDoC) and the DSM-Two methodological approaches to mental health diagnosis. JAMA Psychiatry. 2015 Dec 1;**72**(12):1163–4.
28. **Ray LA, Bujarski S, Roche DJO.** Subjective response to alcohol as a research domain criterion. Alcohol Clin Exp Res. 2016 Jan;**40**(1):6–17.
29. **Crawley JN.** What's Wrong With My Mouse: Behavioral Phenotyping of Transgenic and Knockout Mice. Chichester: John Wiley & Sons; 2007.
30. **Deisseroth K.** Optogenetics: 10 years of microbial opsins in neuroscience. Nat Neurosci. 2015 Sep;**18**(9):1213–25.
31. **Fenno L, Yizhar O, Deisseroth K.** The development and application of optogenetics. Annu Rev Neurosci. 2011;**34**:389–412.
32. **Mallo M.** Controlled gene activation and inactivation in the mouse. Front Biosci J Virtual Libr. 2006;**11**:313–27.
33. **Stieger K, Belbellaa B, Le Guiner C, Moullier P, Rolling F.** In vivo gene regulation using tetracycline-regulatable systems. Adv Drug Deliv Rev. 2009 Jul 2;**61**(7–8):527–41.
34. **Capecchi MR.** Gene targeting in mice: Functional analysis of the mammalian genome for the twenty-first century. Nat Rev Genet. 2005 Jun;**6**(6):507–12.
35. **Buehr M, Meek S, Blair K, Yang J, Ure J, Silva J,** et al. Capture of authentic embryonic stem cells from rat blastocysts. Cell. 2008 Dec 26;**135**(7):1287–98.
36. **Kawamata M, Ochiya T.** Generation of genetically modified rats from embryonic stem cells. PNAS. 2010 Aug 10;**107**(32):14223–8.

37. **Li P, Tong C, Mehrian-Shai R, Jia L, Wu N, Yan Y**, et al. Germline competent embryonic stem cells derived from rat blastocysts. Cell. 2008 Dec 26;**135**(7):1299–310.
38. **Tong C, Li P, Wu NL, Yan Y, Ying Q-L.** Production of p53 gene knockout rats by homologous recombination in embryonic stem cells. Nature. 2010 Sep 9;**467**(7312):211–3.
39. **Cui X, Ji D, Fisher DA, Wu Y, Briner DM, Weinstein EJ.** Targeted integration in rat and mouse embryos with zinc-finger nucleases. Nat Biotechnol. 2011 Jan;**29**(1):64–7.
40. **Wood AJ, Lo T-W, Zeitler B, Pickle CS, Ralston EJ, Lee AH**, et al. Targeted genome editing across species using ZFNs and TALENs. Science. 2011 Jul 15;**333**(6040):307.
41. **Richardson CD, Ray GJ, DeWitt MA, Curie GL, Corn JE.** Enhancing homology-directed genome editing by catalytically active and inactive CRISPR-Cas9 using asymmetric donor DNA. Nat Biotechnol. 2016 Jan 20;
42. **Parker CC, Chen H, Flagel SB, Geurts AM, Richards JB, Robinson TE**, et al. Rats are the smart choice: Rationale for a renewed focus on rats in behavioral genetics. Neuropharmacology. 2014 Jan;**76 Pt B**:250–8.
43. **Choi K, Le T, McGuire J, Xing G, Zhang L, Li H**, et al. Expression pattern of the cannabinoid receptor genes in the frontal cortex of mood disorder patients and mice selectively bred for high and low fear. J Psychiatr Res. 2012 Jul;**46**(7):882–9.
44. **Coyner J, McGuire JL, Parker CC, Ursano RJ, Palmer AA, Johnson LR.** Mice selectively bred for high and low fear behavior show differences in the number of pMAPK (p44/42 ERK) expressing neurons in lateral amygdala following Pavlovian fear conditioning. Neurobiol Learn Mem. 2014 Jul;**112**:195–203.
45. **Blizard DA, Adams N.** The Maudsley reactive and nonreactive strains: A new perspective. Behav Genet. 2002 Sep;**32**(5):277–99.
46. **Landgraf R, Kessler MS, Bunck M, Murgatroyd C, Spengler D, Zimbelmann M**, et al. Candidate genes of anxiety-related behavior in HAB/LAB rats and mice: Focus on vasopressin and glyoxalase-I. Neurosci Biobehav Rev. 2007;**31**(1):89–102.
47. **Landgraf R, Wigger A.** High vs low anxiety-related behavior rats: An animal model of extremes in trait anxiety. Behav Genet. 2002 Sep;**32**(5):301–14.
48. **Muigg P, Scheiber S, Salchner P, Bunck M, Landgraf R, Singewald N.** Differential stress-induced neuronal activation patterns in mouse lines selectively bred for high, normal or low anxiety. PloS One. 2009;**4**(4):e5346.
49. **Neumann ID, Wegener G, Homberg JR, Cohen H, Slattery DA, Zohar J**, et al. Animal models of depression and anxiety: What do they tell us about human condition? Prog Neuropsychopharmacol Biol Psychiatry. 2011 Aug 1;**35**(6):1357–75.
50. **Henn FA, Vollmayr B.** Stress models of depression: Forming genetically vulnerable strains. Neurosci Biobehav Rev. 2005;**29**(4–5):799–804.
51. **Wegener G, Mathe AA, Neumann ID.** Selectively bred rodents as models of depression and anxiety. Curr Top Behav Neurosci. 2012;**12**:139–87.
52. **Touma C, Bunck M, Glasl L, Nussbaumer M, Palme R, Stein H**, et al. Mice selected for high versus low stress reactivity: A new animal model for affective disorders. Psychoneuroendocrinology. 2008 Jul;**33**(6):839–62.
53. **Piras G, Giorgi O, Corda MG.** Effects of antidepressants on the performance in the forced swim test of two psychogenetically selected lines of rats that differ in coping strategies to aversive conditions. Psychopharmacology (Berl). 2010 Sep;**211**(4):403–14.
54. **Keays DA, Nolan PM.** N-ethyl-N-nitrosourea mouse mutants in the dissection of behavioural and psychiatric disorders. Eur J Pharmacol. 2003 Nov 7;**480**(1–3):205–17.
55. **Cook MN, Dunning JP, Wiley RG, Chesler EJ, Johnson DK, Miller DR**, et al. Neurobehavioral mutants identified in an ENU-mutagenesis project. Mamm Genome. 2007 Aug;**18**(8):559–72.
56. **Eisener-Dorman AF, Bailey JS, Grabowski-Boase L, Huitron-Resendiz S, Roberts AJ, Wiltshire T**, et al. Characterization of Highper, an ENU-induced mouse mutant with abnormal psychostimulant and stress responses. Psychopharmacology (Berl). 2013 Jan;**225**(2):407–19.

57. **Reijmers LG, Coats JK, Pletcher MT, Wiltshire T, Tarantino LM, Mayford M.** A mutant mouse with a highly specific contextual fear-conditioning deficit found in an *N*-ethyl-*N*-nitrosourea (ENU) mutagenesis screen. Learn Mem Cold Spring Harb N. 2006 Apr;**13**(2):143–9.
58. **Umemori J, Takao K, Koshimizu H, Hattori S, Furuse T, Wakana S**, et al. ENU-mutagenesis mice with a non-synonymous mutation in Grin1 exhibit abnormal anxiety-like behaviors, impaired fear memory, and decreased acoustic startle response. BMC Res Notes. 2013;**6**:203.
59. **Vitaterna MH, King DP, Chang AM, Kornhauser JM, Lowrey PL, McDonald JD**, et al. Mutagenesis and mapping of a mouse gene, Clock, essential for circadian behavior. Science. 1994 Apr 29;**264**(5159):719–25.
60. **Simon MM, Moresco EMY, Bull KR, Kumar S, Mallon A-M, Beutler B**, et al. Current strategies for mutation detection in phenotype-driven screens utilising next generation sequencing. Mamm Genome. 2015;**26**(9–10):486–500.
61. **Stranger BE, Stahl EA, Raj T.** Progress and promise of genome-wide association studies for human complex trait genetics. Genetics. 2011 Feb;**187**(2):367–83.
62. **Parker CC, Palmer AA.** Dark matter: Are mice the solution to missing heritability? Front Genet. 2011;**2**:32.
63. **Solberg Woods LC.** QTL mapping in outbred populations: Successes and challenges. Physiol Genomics. 2014 Feb 1;**46**(3):81–90.
64. **Parker CC, Carbonetto P, Sokoloff G, Park YJ, Abney M, Palmer AA.** High-resolution genetic mapping of complex traits from a combined analysis of F2 and advanced intercross mice. Genetics. 2014 Sep;**198**(1):103–16.
65. **Parker CC, Sokoloff G, Cheng R, Palmer AA.** Genome-wide association for fear conditioning in an advanced intercross mouse line. Behav Genet. 2012 May;**42**(3):437–48.
66. **Baud A, Guryev V, Hummel O, Johannesson M, Rat Genome Sequencing and Mapping Consortium, Flint J.** Genomes and phenomes of a population of outbred rats and its progenitors. Sci Data. 2014;**1**:140011.
67. **Díaz-Morán S, Palència M, Mont-Cardona C, Cañete T, Blázquez G, Martínez-Membrives E**, et al. Gene expression in hippocampus as a function of differential trait anxiety levels in genetically heterogeneous NIH-HS rats. Behav Brain Res. 2013 Nov 15;**257**:129–39.
68. **Goodson M, Rust MB, Witke W, Bannerman D, Mott R, Ponting CP**, et al. Cofilin-1: A modulator of anxiety in mice. PLoS Genet. 2012;**8**(10):e1002970.
69. **Hettema JM, Webb BT, Guo A-Y, Zhao Z, Maher BS, Chen X**, et al. Prioritization and association analysis of murine-derived candidate genes in anxiety-spectrum disorders. Biol Psychiatry. 2011 Nov 1;**70**(9):888–96.
70. **Logan RW, Robledo RF, Recla JM, Philip VM, Bubier JA, Jay JJ**, et al. High-precision genetic mapping of behavioral traits in the diversity outbred mouse population. Genes Brain Behav. 2013 Jun;**12**(4):424–37.
71. **Valdar W, Solberg LC, Gauguier D, Burnett S, Klenerman P, Cookson WO**, et al. Genome-wide genetic association of complex traits in heterogeneous stock mice. Nat Genet. 2006 Aug;**38**(8):879–87.
72. **Yalcin B, Flint J, Mott R.** Using progenitor strain information to identify quantitative trait nucleotides in outbred mice. Genetics. 2005 Oct;**171**(2):673–81.
73. **Miller BH, Schultz LE, Gulati A, Su AI, Pletcher MT.** Phenotypic characterization of a genetically diverse panel of mice for behavioral despair and anxiety. PloS One. 2010;**5**(12):e14458.
74. **Ghazalpour A, Doss S, Kang H, Farber C, Wen P-Z, Brozell A**, et al. High-resolution mapping of gene expression using association in an outbred mouse stock. PLoS Genet. 2008;**4**(8):e1000149.
75. **Yalcin B, Nicod J, Bhomra A, Davidson S, Cleak J, Farinelli L**, et al. Commercially available outbred mice for genome-wide association studies. PLoS Genet. 2010 Sep;**6**(9):e1001085.
76. **Nicod J, Davies RW, Cai N, Hassett C, Goodstadt L, Cosgrove C**, et al. Genome-wide association of multiple complex traits in outbred mice by ultra-low-coverage sequencing. Nat Genet. 2016 Aug;**48**(8):912–18.

77. **Parker CC, Gopalakrishnan S, Carbonetto P, Gonzales NM, Leung E, Park YJ**, et al. Genome-wide association study of behavioral, physiological and gene expression traits in outbred CFW mice. Nat Genet. 2016 Aug;**48**(8):919–26.
78. **Mott R, Flint J.** Dissecting quantitative traits in mice. Annu Rev Genomics Hum Genet. 2013;**14**:421–39.
79. **Belzung C, Lemoine M.** Criteria of validity for animal models of psychiatric disorders: Focus on anxiety disorders and depression. Biol Mood Anxiety Disord. 2011 Nov 7;**1**:9.
80. **Nestler EJ, Hyman SE.** Animal models of neuropsychiatric disorders. Nat Neurosci. 2010 Oct;**13**(10):1161–9.
81. **Fernando ABP, Robbins TW.** Animal models of neuropsychiatric disorders. Annu Rev Clin Psychol. 2011;**7**(1):39–61.
82. **Chadman KK, Yang M, Crawley JN.** Criteria for validating mouse models of psychiatric diseases. Am J Med Genet B Neuropsychiatr Genet. 2009 Jan 5;**150B**(1):1–11.
83. **Hohoff C.** Anxiety in mice and men: A comparison. J Neural Transm Vienna Austria 1996. 2009 Jun;**116**(6):679–87.
84. **Sousa N, Almeida OFX, Wotjak CT.** A hitchhiker's guide to behavioral analysis in laboratory rodents. Genes Brain Behav. 2006;**5** Suppl 2:5–24.
85. **Kliethermes CL.** Anxiety-like behaviors following chronic ethanol exposure. Neurosci Biobehav Rev. 2005 Jan;**28**(8):837–50.
86. **File SE.** The interplay of learning and anxiety in the elevated plus-maze. Behav Brain Res. 1993 Dec 20;**58**(1–2):199–202.
87. **File SE, Zangrossi H, Viana M, Graeff FG.** Trial 2 in the elevated plus-maze: A different form of fear? Psychopharmacology (Berl). 1993;**111**(4):491–4.
88. **Steckler T, Stein M, Holmes A.** Developing novel anxiolytics: Improving preclinical detection and clinical assessment. In: McArthur R, Borsini F (eds). Animal and Translational Models in Psychiatric Disorders, Vol 1: Psychiatric Disorders. London: Academic Press; 2008, pp. 117–32.
89. **Davis M.** The role of the amygdala in fear-potentiated startle: Implications for animal models of anxiety. Trends Pharmacol Sci. 1992 Jan;**13**(1):35–41.
90. **Davis M, Walker DL, Miles L, Grillon C.** Phasic vs sustained fear in rats and humans: Role of the extended amygdala in fear vs anxiety. Neuropsychopharmacology. 2010 Jan;**35**(1):105–35.
91. **Grillon C.** Models and mechanisms of anxiety: Evidence from startle studies. Psychopharmacology (Berl). 2008 Aug;**199**(3):421–37.
92. **Fendt M, Koch M.** Translational value of startle modulations. Cell Tissue Res. 2013 Oct;**354**(1):287–95.
93. **Davis M, Falls WA, Campeau S, Kim M.** Fear-potentiated startle: A neural and pharmacological analysis. Behav Brain Res. 1993 Dec 20;**58**(1–2):175–98.
94. **Grillon C.** Startle reactivity and anxiety disorders: Aversive conditioning, context, and neurobiology. Biol Psychiatry. 2002 Nov 15;**52**(10):958–75.
95. **McTeague LM, Lang PJ.** The anxiety spectrum and the reflex physiology of defense: From circumscribed fear to broad distress. Depress Anxiety. 2012 Apr;**29**(4):264–81.
96. **Hoffman KL.** New dimensions in the use of rodent behavioral tests for novel drug discovery and development. Expert Opin Drug Discov. 2016 Apr;**11**(4):343–53.
97. **Oliveira-Dos-Santos AJ, Matsumoto G, Snow BE, Bai D, Houston FP, Whishaw IQ**, et al. Regulation of T cell activation, anxiety, and male aggression by RGS2. PNAS. 2000 Oct 24;**97**(22):12272–7.
98. **Talbot CJ, Nicod A, Cherny SS, Fulker DW, Collins AC, Flint J.** High-resolution mapping of quantitative trait loci in outbred mice. Nat Genet. 1999 Mar;**21**(3):305–8.
99. **Yalcin B, Willis-Owen SAG, Fullerton J, Meesaq A, Deacon RM, Rawlins JNP**, et al. Genetic dissection of a behavioral quantitative trait locus shows that Rgs2 modulates anxiety in mice. Nat Genet. 2004 Nov;**36**(11):1197–202.

100. **Lifschytz T, Broner EC, Zozulinsky P, Slonimsky A, Eitan R, Greenbaum L,** et al. Relationship between Rgs2 gene expression level and anxiety and depression-like behaviour in a mutant mouse model: Serotonergic involvement. Int J Neuropsychopharmacol. 2012 Oct;**15**(9):1307–18.

101. **Smoller JW, Paulus MP, Fagerness JA, Purcell S, Yamaki LH, Hirshfeld-Becker D,** et al. Influence of RGS2 on anxiety-related temperament, personality, and brain function. Arch Gen Psychiatry. 2008 Mar;**65**(3):298–308.

102. **Koenen KC, Amstadter AB, Ruggiero KJ, Acierno R, Galea S, Kilpatrick DG,** et al. RGS2 and generalized anxiety disorder in an epidemiologic sample of hurricane-exposed adults. Depress Anxiety. 2009;**26**(4):309–15.

103. **Amstadter AB, Koenen KC, Ruggiero KJ, Acierno R, Galea S, Kilpatrick DG** et al. Variant in RGS2 moderates posttraumatic stress symptoms following potentially traumatic event exposure. J Anxiety Disord. 2009 Apr;**23**(3):369–73.

104. **Hohoff C, Weber H, Richter J, Domschke K, Zwanzger PM, Ohrmann P,** et al. RGS2 genetic variation: Association analysis with panic disorder and dimensional as well as intermediate phenotypes of anxiety. Am J Med Genet Part B Neuropsychiatr Genet Off. 2015 Apr;**168B**(3):211–22.

105. **Hommers L, Raab A, Bohl A, Weber H, Scholz C-J, Erhardt A,** et al. MicroRNA hsa-miR-4717-5p regulates RGS2 and may be a risk factor for anxiety-related traits. Am J Med Genet Part B Neuropsychiatr. 2015 Jun;**168B**(4):296–306.

106. **Otowa T, Shimada T, Kawamura Y, Sugaya N, Yoshida E, Inoue K,** et al. Association of RGS2 variants with panic disorder in a Japanese population. Am J Med Genet Part B Neuropsychiatr Genet. 2011 Jun;**156B**(4):430–4.

107. **Hettema JM, Sun C, Chen X, Kendler KS.** Genetic association study between RGS2 and anxiety-related phenotypes. Psychiatr Genet. 2013 Apr;**23**(2):92.

108. **Mouri K, Hishimoto A, Fukutake M, Nishiguchi N, Shirakawa O, Maeda K.** Association study of *RGS2* gene polymorphisms with panic disorder in Japanese. Kobe J Med Sci. 2010;**55**(5):E116–21.

109. **Strug LJ, Suresh R, Fyer AJ, Talati A, Adams PB, Li W,** et al. Panic disorder is associated with the serotonin transporter gene (*SLC6A4*) but not the promoter region (5-HTTLPR). Mol Psychiatry. 2010 Feb;**15**(2):166–76.

110. **Kessler RC.** The effects of stressful life events on depression. Annu Rev Psychol. 1997;**48**:191–214.

111. **Sullivan PF, Neale MC, Kendler KS.** Genetic epidemiology of major depression: Review and meta-analysis. Am J Psychiatry. 2000 Oct;**157**(10):1552–62.

112. **Porsolt RD, Le Pichon M, Jalfre M.** Depression: A new animal model sensitive to antidepressant treatments. Nature. 1977 Apr 21;**266**(5604):730–2.

113. **Porsolt RD, Anton G, Blavet N, Jalfre M.** Behavioural despair in rats: A new model sensitive to antidepressant treatments. Eur J Pharmacol. 1978 Feb 15;**47**(4):379–91.

114. **Lucki I.** The forced swimming test as a model for core and component behavioral effects of antidepressant drugs. Behav Pharmacol. 1997 Nov;**8**(6–7):523–32.

115. **Cryan JF, Mombereau C, Vassout A.** The tail suspension test as a model for assessing antidepressant activity: Review of pharmacological and genetic studies in mice. Neurosci Biobehav Rev. 2005;**29**(4–5):571–625.

116. **Cryan JF, Valentino RJ, Lucki I.** Assessing substrates underlying the behavioral effects of antidepressants using the modified rat forced swimming test. Neurosci Biobehav Rev. 2005;**29**(4–5):547–69.

117. **Panconi E, Roux J, Altenbaumer M, Hampe S, Porsolt RD.** MK-801 and enantiomers: Potential antidepressants or false positives in classical screening models? Pharmacol Biochem Behav. 1993 Sep;**46**(1):15–20.

118. **Pryce CR, Klaus F.** Translating the evidence for gene association with depression into mouse models of depression-relevant behaviour: Current limitations and future potential. Neurosci Biobehav Rev. 2013 Sep;**37**(8):1380–402.

119. **Slattery DA, Cryan JF.** The ups and downs of modelling mood disorders in rodents. ILAR J Natl Res Counc Inst Lab Anim Resour. 2014;**55**(2):297–309.

120. **de Kloet ER, Molendijk ML.** Coping with the forced swim stressor: Towards understanding an adaptive mechanism. Neural Plast. 2016;2016:6503162.

121. **Mann JJ.** The medical management of depression. N Engl J Med. 2005 Oct 27;**353**(17):1819–34.

122. **Teng HK, Teng KK, Lee R, Wright S, Tevar S, Almeida RD**, et al. ProBDNF induces neuronal apoptosis via activation of a receptor complex of p75NTR and sortilin. J Neurosci. 2005 Jun 1;**25**(22):5455–63.

123. **Woo NH, Teng HK, Siao C-J, Chiaruttini C, Pang PT, Milner TA**, et al. Activation of p75NTR by proBDNF facilitates hippocampal long-term depression. Nat Neurosci. 2005 Aug;**8**(8):1069–77.

124. **Koshimizu H, Kiyosue K, Hara T, Hazama S, Suzuki S, Uegaki K**, et al. Multiple functions of precursor BDNF to CNS neurons: Negative regulation of neurite growth, spine formation and cell survival. Mol Brain. 2009;**2**:27.

125. **Deinhardt K, Kim T, Spellman DS, Mains RE, Eipper BA, Neubert TA**, et al. Neuronal growth cone retraction relies on proneurotrophin receptor signaling through Rac. Sci Signal. 2011;**4**(202):ra82.

126. **Zagrebelsky M, Korte M.** Form follows function: BDNF and its involvement in sculpting the function and structure of synapses. Neuropharmacology. 2014 Jan;**76 Pt** C:628–38.

127. **Duman RS, Monteggia LM.** A neurotrophic model for stress-related mood disorders. Biol Psychiatry. 2006 Jun 15;**59**(12):1116–27.

128. **Bocchio-Chiavetto L, Bagnardi V, Zanardini R, Molteni R, Nielsen MG, Placentino A**, et al. Serum and plasma BDNF levels in major depression: A replication study and meta-analyses. World J Biol Psychiatry. 2010 Sep;**11**(6):763–73.

129. **Guilloux J-P, Douillard-Guilloux G, Kota R, Wang X, Gardier AM, Martinowich K**, et al. Molecular evidence for BDNF- and GABA-related dysfunctions in the amygdala of female subjects with major depression. Mol Psychiatry. 2012 Nov;**17**(11):1130–42.

130. **Tripp A, Oh H, Guilloux J-P, Martinowich K, Lewis DA, Sibille E.** Brain-derived neurotrophic factor signaling and subgenual anterior cingulate cortex dysfunction in major depressive disorder. Am J Psychiatry. 2012 Nov;**169**(11):1194–202.

131. **Molendijk ML, Spinhoven P, Polak M, Bus B a. A, Penninx BWJH, Elzinga BM.** Serum BDNF concentrations as peripheral manifestations of depression: Evidence from a systematic review and meta-analyses on 179 associations (*N*=9484). Mol Psychiatry. 2014 Jul;**19**(7):791–800.

132. **Shimizu E, Hashimoto K, Okamura N, Koike K, Komatsu N, Kumakiri C**, et al. Alterations of serum levels of brain-derived neurotrophic factor (BDNF) in depressed patients with or without antidepressants. Biol Psychiatry. 2003 Jul 1;**54**(1):70–5.

133. **Sen S, Duman R, Sanacora G.** Serum brain-derived neurotrophic factor, depression, and antidepressant medications: Meta-analyses and implications. Biol Psychiatry. 2008 Sep 15;**64**(6):527–32.

134. **Castrén E, Rantamäki T.** The role of BDNF and its receptors in depression and antidepressant drug action: Reactivation of developmental plasticity. Dev Neurobiol. 2010 Apr;**70**(5):289–97.

135. **McMurray KMJ, Ramaker MJ, Barkley-Levenson AM, Sidhu PS, Elkin PK, Reddy MK**, et al. Identification of a novel, fast-acting GABAergic antidepressant. Mol Psychiatry. 2017 Mar 21;

136. **Castrén E, Kojima M.** Brain-derived neurotrophic factor in mood disorders and antidepressant treatments. Neurobiol Dis. 2016 Jul 15;

137. **Donaldson ZR, Hen R.** From psychiatric disorders to animal models: A bidirectional and dimensional approach. Biol Psychiatry. 2015 Jan 1;**77**(1):15–21.

138. **Egan MF, Kojima M, Callicott JH, Goldberg TE, Kolachana BS, Bertolino A**, et al. The BDNF val66met polymorphism affects activity-dependent secretion of BDNF and human memory and hippocampal function. Cell. 2003 Jan 24;**112**(2):257–69.

139. **Molendijk ML, Bus BAA, Spinhoven P, Kaimatzoglou A, Oude Voshaar RC, Penninx BWJH**, et al. A systematic review and meta-analysis on the association between BDNF val(66)met and hippocampal

volume—a genuine effect or a winners curse? Am J Med Genet Part B Neuropsychiatr Genet. 2012 Sep;**159B**(6):731–40.

140. **Pezawas L, Verchinski BA, Mattay VS, Callicott JH, Kolachana BS, Straub RE**, et al. The brain-derived neurotrophic factor val66met polymorphism and variation in human cortical morphology. J Neurosci. 2004 Nov 10;**24**(45):10099–102.
141. **Chen Z-Y, Jing D, Bath KG, Ieraci A, Khan T, Siao C-J**, et al. Genetic variant BDNF (Val66Met) polymorphism alters anxiety-related behavior. Science. 2006 Oct 6;**314**(5796):140–3.
142. **Hariri AR, Goldberg TE, Mattay VS, Kolachana BS, Callicott JH, Egan MF**, et al. Brain-derived neurotrophic factor val66met polymorphism affects human memory-related hippocampal activity and predicts memory performance. J Neurosci. 2003 Jul 30;**23**(17):6690–4.
143. **Gyekis JP, Yu W, Dong S, Wang H, Qian J, Kota P**, et al. No association of genetic variants in BDNF with major depression: A meta- and gene-based analysis. Am J Med Genet Part B Neuropsychiatr Genet. 2013 Jan;**162B**(1):61–70.
144. **Harrisberger F, Smieskova R, Schmidt A, Lenz C, Walter A, Wittfeld K**, et al. BDNF Val66Met polymorphism and hippocampal volume in neuropsychiatric disorders: A systematic review and meta-analysis. Neurosci Biobehav Rev. 2015 Aug;**55**:107–18.
145. **Martinowich K, Manji H, Lu B.** New insights into BDNF function in depression and anxiety. Nat Neurosci. 2007 Sep;**10**(9):1089–93.
146. **Bath KG, Jing DQ, Dincheva I, Neeb CC, Pattwell SS, Chao MV**, et al. BDNF Val66Met impairs fluoxetine-induced enhancement of adult hippocampus plasticity. Neuropsychopharmacology. 2012 Apr;**37**(5):1297–304.
147. **Pattwell SS, Bath KG, Perez-Castro R, Lee FS, Chao MV, Ninan I.** The BDNF Val66Met polymorphism impairs synaptic transmission and plasticity in the infralimbic medial prefrontal cortex. J Neurosci. 2012 Feb 15;**32**(7):2410–21.
148. **Malki K, Pain O, Tosto MG, Du Rietz E, Carboni L, Schalkwyk LC.** Identification of genes and gene pathways associated with major depressive disorder by integrative brain analysis of rat and human prefrontal cortex transcriptomes. Transl Psychiatry. 2015;**5**:e519.
149. **Afifi TO, Asmundson GJG, Taylor S, Jang KL.** The role of genes and environment on trauma exposure and posttraumatic stress disorder symptoms: A review of twin studies. Clin Psychol Rev. 2010 Feb;**30**(1):101–12.
150. **Almli LM, Fani N, Smith AK, Ressler KJ.** Genetic approaches to understanding post-traumatic stress disorder. Int J Neuropsychopharmacol. 2014 Feb;**17**(2):355–70.
151. **Breslau N, Davis GC, Andreski P, Peterson E.** Traumatic events and posttraumatic stress disorder in an urban population of young adults. Arch Gen Psychiatry. 1991 Mar;**48**(3):216–22.
152. **American Psychiatric Association**. Diagnostic and Statistical Manual of Mental Disorders, 5th edn. Arlington, VA: American Psychiatric Association; 2013.
153. **Matar MA, Zohar J, Cohen H.** Translationally relevant modeling of PTSD in rodents. Cell Tissue Res. 2013 Oct;**354**(1):127–39.
154. **Cohen H, Matar MA, Zohar J.** Maintaining the clinical relevance of animal models in translational studies of post-traumatic stress disorder. ILAR J Natl Res Counc Inst Lab Anim Resour. 2014;**55**(2):233–45.
155. **Whitaker AM, Gilpin NW, Edwards S.** Animal models of post-traumatic stress disorder and recent neurobiological insights. Behav Pharmacol. 2014 Sep;**25**(5–6):398–409.
156. **Krishnan V, Han M-H, Graham DL, Berton O, Renthal W, Russo SJ**, et al. Molecular adaptations underlying susceptibility and resistance to social defeat in brain reward regions. Cell. 2007 Oct 19;**131**(2):391–404.
157. **Siegmund A, Kaltwasser SF, Holsboer F, Czisch M, Wotjak CT.** Hippocampal N-acetylaspartate levels before trauma predict the development of long-lasting posttraumatic stress disorder-like symptoms in mice. Biol Psychiatry. 2009 Feb 1;**65**(3):258–62.

158. **Holmes A, Singewald N.** Individual differences in recovery from traumatic fear. Trends Neurosci. 2013 Jan;**36**(1):23–31.
159. **Pitman RK, Rasmusson AM, Koenen KC, Shin LM, Orr SP, Gilbertson MW**, et al. Biological studies of post-traumatic stress disorder. Nat Rev Neurosci. 2012 Nov;**13**(11):769–87.
160. **Yehuda R, LeDoux J.** Response variation following trauma: A translational neuroscience approach to understanding PTSD. Neuron. 2007 Oct 4;**56**(1):19–32.
161. **Risbrough VB, Glenn DE, Baker DG.** On the road to translation for PTSD treatment: Theoretical and practical considerations of the use of human models of conditioned fear for drug development. Curr Top Behav Neurosci. 2016 Jun 17;
162. **Orr SP, Metzger LJ, Lasko NB, Macklin ML, Peri T, Pitman RK.** De novo conditioning in trauma-exposed individuals with and without posttraumatic stress disorder. J Abnorm Psychol. 2000 May;**109**(2):290–8.
163. **Peri T, Ben-Shakhar G, Orr SP, Shalev AY.** Psychophysiologic assessment of aversive conditioning in posttraumatic stress disorder. Biol Psychiatry. 2000 Mar 15;**47**(6):512–9.
164. **Guthrie RM, Bryant RA.** Extinction learning before trauma and subsequent posttraumatic stress. Psychosom Med. 2006 Apr;**68**(2):307–11.
165. **Milad MR, Orr SP, Lasko NB, Chang Y, Rauch SL, Pitman RK.** Presence and acquired origin of reduced recall for fear extinction in PTSD: Results of a twin study. J Psychiatr Res. 2008 Jun;**42**(7):515–20.
166. **Glover EM, Phifer JE, Crain DF, Norrholm SD, Davis M, Bradley B**, et al. Tools for translational neuroscience: PTSD is associated with heightened fear responses using acoustic startle but not skin conductance measures. Depress Anxiety. 2011 Dec 21;**28**(12):1058–66.
167. **Rudy JW, Huff NC, Matus-Amat P.** Understanding contextual fear conditioning: Insights from a two-process model. Neurosci Biobehav Rev. 2004 Nov;**28**(7):675–85.
168. **VanElzakker MB, Dahlgren MK, Davis FC, Dubois S, Shin LM.** From Pavlov to PTSD: The extinction of conditioned fear in rodents, humans, and in anxiety disorders. Neurobiol Learn Mem. 2014 Sep;**113**:3–18.
169. **Milad MR, Rauch SL, Pitman RK, Quirk GJ.** Fear extinction in rats: Implications for human brain imaging and anxiety disorders. Biol Psychol. 2006 Jul;**73**(1):61–71.
170. **Blanchard RJ, Nikulina JN, Sakai RR, McKittrick C, McEwen B, Blanchard DC.** Behavioral and endocrine change following chronic predatory stress. Physiol Behav. 1998 Feb 15;**63**(4):561–9.
171. **Adamec R, Head D, Blundell J, Burton P, Berton O.** Lasting anxiogenic effects of feline predator stress in mice: Sex differences in vulnerability to stress and predicting severity of anxiogenic response from the stress experience. Physiol Behav. 2006 Jun 15;**88**(1–2):12–29.
172. **Adamec R, Strasser K, Blundell J, Burton P, McKay DW.** Protein synthesis and the mechanisms of lasting change in anxiety induced by severe stress. Behav Brain Res. 2006 Feb 28;**167**(2):270–86.
173. **Adamec R, Muir C, Grimes M, Pearcey K.** Involvement of noradrenergic and corticoid receptors in the consolidation of the lasting anxiogenic effects of predator stress. Behav Brain Res. 2007 May 16;**179**(2):192–207.
174. **Liberzon I, López JF, Flagel SB, Vázquez DM, Young EA.** Differential regulation of hippocampal glucocorticoid receptors mRNA and fast feedback: Relevance to post-traumatic stress disorder. J Neuroendocrinol. 1999 Jan;**11**(1):11–7.
175. **Kohda K, Harada K, Kato K, Hoshino A, Motohashi J, Yamaji T**, et al. Glucocorticoid receptor activation is involved in producing abnormal phenotypes of single-prolonged stress rats: A putative post-traumatic stress disorder model. Neuroscience. 2007 Aug 10;**148**(1):22–33.
176. **Green MK, Rani CSS, Joshi A, Soto-Piña AE, Martinez PA, Frazer A**, et al. Prenatal stress induces long term stress vulnerability, compromising stress response systems in the brain and impairing extinction of conditioned fear after adult stress. Neuroscience. 2011 Sep 29;**192**:438–51.
177. **Roth MK, Bingham B, Shah A, Joshi A, Frazer A, Strong R**, et al. Effects of chronic plus acute prolonged stress on measures of coping style, anxiety, and evoked HPA-axis reactivity. Neuropharmacology. 2012 Nov;**63**(6):1118–26.

178. **Servatius RJ, Ottenweller JE, Bergen MT, Soldan S, Natelson BH.** Persistent stress-induced sensitization of adrenocortical and startle responses. Physiol Behav. 1994 Nov;**56**(5):945–54.

179. **Pynoos RS, Ritzmann RF, Steinberg AM, Goenjian A, Prisecaru I.** A behavioral animal model of posttraumatic stress disorder featuring repeated exposure to situational reminders. Biol Psychiatry. 1996 Jan 15;**39**(2):129–34.

180. **Rau V, DeCola JP, Fanselow MS.** Stress-induced enhancement of fear learning: An animal model of posttraumatic stress disorder. Neurosci Biobehav Rev. 2005;**29**(8):1207–23.

181. **Rau V, Fanselow MS.** Exposure to a stressor produces a long lasting enhancement of fear learning in rats. Stress Amst Neth. 2009 Mar;**12**(2):125–33.

182. **Juruena MF.** Early-life stress and HPA axis trigger recurrent adulthood depression. Epilepsy Behav. 2014 Sep;**38**:148–59.

183. **Morris MC, Compas BE, Garber J.** Relations among posttraumatic stress disorder, comorbid major depression, and HPA function: A systematic review and meta-analysis. Clin Psychol Rev. 2012 Jun;**32**(4):301–15.

184. **Castro-Vale I, van Rossum EFC, Machado JC, Mota-Cardoso R, Carvalho D.** Genetics of glucocorticoid regulation and posttraumatic stress disorder: What do we know? Neurosci Biobehav Rev. 2016 Apr;**63**:143–57.

185. **Binder EB.** The role of FKBP5, a co-chaperone of the glucocorticoid receptor in the pathogenesis and therapy of affective and anxiety disorders. Psychoneuroendocrinology. 2009 Dec;**34** Suppl 1:S186–95.

186. **Yehuda R, Flory JD, Pratchett LC, Buxbaum J, Ising M, Holsboer F.** Putative biological mechanisms for the association between early life adversity and the subsequent development of PTSD. Psychopharmacology (Berl). 2010 Oct;**212**(3):405–17.

187. **Binder EB, Salyakina D, Lichtner P, Wochnik GM, Ising M, Pütz B,** et al. Polymorphisms in FKBP5 are associated with increased recurrence of depressive episodes and rapid response to antidepressant treatment. Nat Genet. 2004 Dec;**36**(12):1319–25.

188. **Xie P, Kranzler HR, Poling J, Stein MB, Anton RF, Farrer LA,** et al. Interaction of FKBP5 with childhood adversity on risk for post-traumatic stress disorder. Neuropsychopharmacology. 2010 Jul;**35**(8):1684–92.

189. **Ising M, Depping A-M, Siebertz A, Lucae S, Unschuld PG, Kloiber S,** et al. Polymorphisms in the FKBP5 gene region modulate recovery from psychosocial stress in healthy controls. Eur J Neurosci. 2008 Jul;**28**(2):389–98.

190. **Zannas AS, Binder EB.** Gene-environment interactions at the FKBP5 locus: Sensitive periods, mechanisms and pleiotropism. Genes Brain Behav. 2014 Jan;**13**(1):25–37.

191. **Binder EB, Bradley RG, Liu W, Epstein MP, Deveau TC, Mercer KB,** et al. Association of FKBP5 polymorphisms and childhood abuse with risk of posttraumatic stress disorder symptoms in adults. JAMA. 2008 Mar 19;**299**(11):1291–305.

192. **Logue MW, Baldwin C, Guffanti G, Melista E, Wolf EJ, Reardon AF,** et al. A genome-wide association study of post-traumatic stress disorder identifies the retinoid-related orphan receptor alpha (RORA) gene as a significant risk locus. Mol Psychiatry. 2013 Aug;**18**(8):937–42.

193. **Solovieff N, Roberts AL, Ratanatharathorn A, Haloosim M, De Vivo I, King AP,** et al. Genetic Association analysis of 300 genes identifies a risk haplotype in SLC18A2 for post-traumatic stress disorder in two independent samples. Neuropsychopharmacology. 2014 Jul;**39**(8):1872–9.

194. **Nievergelt CM, Maihofer AX, Mustapic M, Yurgil KA, Schork NJ, Miller MW,** et al. Genomic predictors of combat stress vulnerability and resilience in U.S. Marines: A genome-wide association study across multiple ancestries implicates PRTFDC1 as a potential PTSD gene. Psychoneuroendocrinology. 2015 Jan;**51**:459–71.

195. **Yang Y-L, Chao P-K, Lu K-T.** Systemic and intra-amygdala administration of glucocorticoid agonist and antagonist modulate extinction of conditioned fear. Neuropsychopharmacology. 2006 May;**31**(5):912–24.

196. **Sawamura T, Klengel T, Armario A, Jovanovic T, Norrholm SD, Ressler KJ**, et al. Dexamethasone treatment leads to enhanced fear extinction and dynamic Fkbp5 regulation in amygdala. Neuropsychopharmacology. 2016 Feb;**41**(3):832–46.
197. **Hartmann J, Wagner KV, Liebl C, Scharf SH, Wang X-D, Wolf M**, et al. The involvement of FK506-binding protein 51 (FKBP5) in the behavioral and neuroendocrine effects of chronic social defeat stress. Neuropharmacology. 2012 Jan;**62**(1):332–9.
198. **Touma C, Gassen NC, Herrmann L, Cheung-Flynn J, Büll DR, Ionescu IA**, et al. FK506 binding protein 5 shapes stress responsiveness: Modulation of neuroendocrine reactivity and coping behavior. Biol Psychiatry. 2011 Nov 15;**70**(10):928–36.
199. **Schmidt U, Buell DR, Ionescu IA, Gassen NC, Holsboer F, Cox MB**, et al. A role for synapsin in FKBP51 modulation of stress responsiveness: Convergent evidence from animal and human studies. Psychoneuroendocrinology. 2015 Feb;**52**:43–58.
200. **Webster MJ, Knable MB, O'Grady J, Orthmann J, Weickert CS.** Regional specificity of brain glucocorticoid receptor mRNA alterations in subjects with schizophrenia and mood disorders. Mol Psychiatry. 2002;7(9):985–94, 924.
201. **de Kloet ER, Joëls M, Holsboer F.** Stress and the brain: From adaptation to disease. Nat Rev Neurosci. 2005 Jun;**6**(6):463–75.
202. **Xing Y, Hou J, Meng Q, Yang M, Kurihara H, Tian J.** Novel antidepressant candidate RO-05 modulated glucocorticoid receptors activation and FKBP5 expression in chronic mild stress model in rats. Neuroscience. 2015 Apr 2;**290**:255–65.
203. **Willner P.** The chronic mild stress procedure as an animal model of depression: Valid, reasonably reliable, and useful. Psychopharmacology (Berl). 1997;**134**(4):371–7.
204. **Willner P.** Chronic mild stress (CMS) revisited: Consistency and behavioural-neurobiological concordance in the effects of CMS. Neuropsychobiology. 2005;**52**(2):90–110.

Chapter 8

Optogenetic and chemogenetic technologies for advanced functional investigations of the neural correlates of emotions

Alexandre Surget and Catherine Belzung

8.1 Introduction

About 150 years ago, Charles Darwin (1) highlighted the adaptive value of emotions which he considered crucial for the species' survival. For example, fear and anxiety enable optimization of the subject's behavior when facing a danger, and thus survival when, for example, an animal is faced with a predator. Aggression enables individuals to compete for resources, while disgust enables one to avoid spoiled food, for example. Consequently, the function of emotions has been highly conserved across the phylum and emotional responses are present in all species. The presence of emotions can be described according to two different approaches: (i) a categorical approach that focuses on which emotions are present in a given species, claiming, for example, that primary or basic emotions, such as fear, are present in all vertebrates, while secondary or self-conscious emotions, such as guilt, are present only in humans; or (ii) a dimensional approach that describes the emotions according to different components (such as intensity, valence, arousal, peripheral manifestations, etc.) and analyzes which of them are present in a given organism. Regarding this view, one can mention that the components crucial for human emotions have been well described in the psychological literature. According to this theoretical framework, emotions might be based either on simple automatic cognitive processes enabling the individual to assess the environmental challenges and to cope with them (this is the view offered by appraisal theories such as the one proposed by Klaus Scherer; see, e.g. 2) or include more sophisticated components such as found, for example, in the multilevel cognitive model proposed by Pierre Philippot (see, e.g. 3), which includes emotional regulation and perception of bodily changes. In fact, during evolution, emotional responses have progressively increased in complexity: in humans, the emotional phenomenon requires some high-order cognitive processes such as autonoetic consciousness, or some efficient emotional regulation strategies that appeared very late during evolution. Other, less sophisticated cognitive processes necessary for the generation of emotions such as detection of novelty, of pleasantness (valence), of intensity, and of goal conduciveness of a stimulus, ability to select coping strategies, or to express cognitive bias, are present in most species, and particularly in non-primate mammals such as rodents (4). Further, emotional behaviors are accompanied by bodily changes including modification of physiological variables such as respiration and heart rate (4).

The cerebral underpinnings underlying emotions have been much investigated in human subjects using neurofunctional imaging methodologies, which enable the investigation of neural correlates of emotions while a normal subject is processing various emotional-relevant stimuli. With this experimental strategy, the neural networks underlying emotions have been investigated according to either the dimensional or the categorical approach. The first consists of exploring the brain circuits underlying processing of specific dimensions of emotions, whatever their category, and therefore include investigations focusing on the brain regions processing the intensity of a stimulus, its valence, or activated while a subject is representing his bodily state or when regulating emotions, in a manner independent from any specific emotion. Regarding this approach, one can, for example, mention research investigating the brain circuits underlying valence: it found that valence (positive and negative) is related to activity in brain networks including areas like the medial prefrontal cortex (mPFC), the anterior cingulate (ACC), the visual cortex, the cerebellum, the supplementary motor area (SMA), the supramarginal gyrus, and the somatosensory cortex (5). Another example concerns investigation of the circuits underlying emotional regulation that highlighted a network including the dorsolateral and the ventrolateral prefrontal cortex, the SMA, the parietal cortex, and the ventral ACC (6). The second approach consists in investigating the network that is recruited during various categorical emotional states. Here, findings generally highlight a very distributed network for processing emotions. According to a meta-analysis, specific neural activations are shared between multiple emotions: for example, activation of the amygdala is seen during experience and perception of fear, but also during experience and perception of other emotions including anger, disgust, happiness, and sadness (7). According to another meta-analysis (see, e.g. 8), each specific emotion is systematically associated with a specific pattern of brain activation: according to this view, fear is associated with the amygdala; anger with the orbitofrontal cortex; disgust with the insula, ventral prefrontal cortex, and amygdala; sadness with the mPFC; and happiness with the rostral ACC. Furthermore, from this point of view, fear will be related to the amygdala as well as to structures connected to the amygdala: the network will thus consist of a set of subcortical structures processing fear (amygdala, bed nucleus of stria terminalis (BNST), and ventral hippocampus): (i) modulated by cortical areas such as the orbital and medial prefrontal cortex, including cingulate cortex, that enable an efficient regulation of fear; and (ii) projecting to a set of brain regions implementing emotional responses such as the hypothalamus, the periaqueductal gray (PAG), brainstem nuclei, as well as the intermediolateral cell column of the spinal cord and peripheral components of the autonomic nervous system (7). Therefore, dysfunctions of this system can be the consequence of either (i) a dysfunction of the subcortical areas processing fear (amygdala nuclei, BNST, or hippocampus); or (ii) from a modification of the inputs to or outputs of this system.

This categorical approach investigating neural correlates of emotions has also been undertaken in rodents using, for example, immediate early gene (IEG) mapping. Using this approach, activation of several brain areas including the mPFC, BNST, lateral septum, hypothalamic regions, locus coeruleus, and parts of the amygdala and PAG has been reported after confrontation of rats or mice with anxiogenic situations such as confrontation with a novel environment, conditioned fear, air puff to eyes, or injection of anxiogenic compounds (9). This indicates that similar brain circuits are activated by anxiogenic stimuli in humans and rodents, highlighting the fact that these networks are conserved in evolution. This network similarity is very crucial, as animal studies enable to investigate the contribution of these structures in a more accurate way and with higher resolutions via (i) observational approaches, to investigate association between neuronal activity and emotions, and (ii) interventional approaches, using manipulations increasing or decreasing the activity of these circuits such as lesions, for example, in order to establish causal relationships between neuronal activities and functions.

8.2 Novel methods to investigate the neural substrates of emotions

A major challenge for neuroscience is to provide a mechanistic explanation of brain's emergent properties, that is, how brain functions such as consciousness, reasoning, feelings, or emotions arise out of chemical and electrical operations in neuronal circuits. Therefore, the development of novel technologies that can accurately record or manipulate neuronal activity within defined neuronal subpopulations in intact organisms is essential for advancing our understanding. In this regard, the recent emergence of optogenetics and chemogenetics represents a great improvement in our understanding of the computational mechanisms and neuronal networks that underlie brain functions in general and emotions in particular. These novel technologies are both based on the expression of transfected genetically engineered proteins, preferably in a specified cell population, allowing monitoring or controlling neuronal activity at cellular or subcellular levels. However, these two methods differ according to the effectors/actuators used: light (opto-) for optogenetics and small diffusible molecules (chemo-) for chemogenetics.

8.2.1 Optogenetics

Optogenetics refers to the utilization of molecular tools to induce photosensitivity in a genetically and/or a spatially defined cell population (10–12). The aim of this method is to provide transfected cells with (i) either a protein able to generate activity-dependent light emission to monitor cell activity during observational approach (optogenetic indicators); or (ii) a protein whose conformation may shift under a certain light wavelength, causing changes in cell properties and enabling investigators to manipulate cell activity for interventional approach (optogenetic actuators).

For both approaches, optogenetic tools provide great benefits for investigations of neural circuits. An important part of these advantages is related to light properties (10, 13). First, light minimally interacts with the endogenous cellular machinery (i.e. the cells that do not express the transfected protein). Indeed, light does not serve as a natural agent of interaction or communication between brain cells (unlike electrical and chemical agents) and has no apparent noxious effects on tissue at the light power applied in standard protocols (unlike electrical, magnetic, or mechanical stimulations, radiations, or temperature). Accordingly, light delivery and emission have the advantage of having a minimal influence on biological processes in brain tissues (if any). Second, optogenetic methods can achieve excellent spatial resolutions. This is particularly the case for an observational approach where light emission from indicators can be detected at cellular or subcellular levels (μm range), even in large fields of view (up to mm^2 range) (14, 15). Moreover, a similar range of resolution can be achievable in cell cultures and brain slices for optogenetic interventions, while light delivery in deep brain regions in *in vivo* experiments remains possible via stereotaxic methods but at the cost of the spatial resolution which runs down to a tenth of a micrometer precision. Third, another strength offered by optogenetics is its excellent temporal resolution. Light can be delivered at a range of a submillisecond precision for interventional approach, and fast neuronal events can be monitored by various devices in observational approaches, although the temporal scale of observable or controllable events depends more in practice on the reactivity and kinetic properties of the optogenetic indicators or actuators (usually at the millisecond range). Hence, with a direct control over light stimulation frequency, it becomes possible to dictate accurately the activity frequency of neurons expressing optogenetic actuators. Finally, a decisive advantage offered by optogenetic methods is the possibility to target a molecularly defined subset of cells within a region of interest (ROI). Indeed, because optogenetic indicators and actuators are genetically engineered proteins, they enable investigators to create genetic constructs or transgenic strategies for cell type-specific expression, providing an unprecedented specificity for observational or interventional approaches.

Altogether, these properties offer a unique and prominent place for optogenetics in the neuroscience toolbox, allowing precise timescale experiments, accurate real-time activity monitoring of hundreds/thousands of neurons, as well as sophisticated functional investigations of circuits and microcircuits through gain- or loss-of-function of precise events in neuronal subpopulations or projections. Traditional methods hardly bear comparison for most of these characteristics. Indeed, these methods do not provide such a combination of high spatiotemporal resolution, cell type-dependent effects, and strong selectivity. Instead, they display a trade-off between these parameters, sometimes performing well in one aspect but displaying poor performance in others. For instance, most of the methods used to manipulate neuronal activity are based either on electrical and magnetic stimulations, which are significantly limited by the lack of cell-type specificity and lower spatial resolution, or on pharmacological interventions which are limited by poor spatial and temporal resolutions.

Optogenetics displayed some limitations too, which are mainly caused by light scattering and absorption in biological tissues. Light is indeed estimated to lose more than 90% of its initial power density after only 0.5–1.5 mm penetration in the brain, the exact distance being dependent on the light wavelength (13). As a result, light must be emitted or recorded tightly to the ROI and the control or monitoring of neuronal activity will be restricted to this area. This causes two major practical inconveniences for *in vivo* experiments: (i) invasiveness (implantation of optic fiber or light source/sensor is required, notably for deep or non-superficial brain regions), and (ii) the necessity to head-restrain or tether the animal to the light source or to the recording device which may limit extensively the behavioral experiments that can be done (NB: wireless implantable systems for light delivery are now commercially available, but they are not providing similar levels of performances yet). On the other hand, it is noteworthy that this limitation is also behind one of the most beneficial contributions of optogenetics, which offers the possibility of implementing a projection-targeting strategy. Indeed, through the placement of the optic fiber it becomes possible to investigate the functional role of a particular projection of a specific neuronal population.

8.2.2 Chemogenetics

Chemogenetics is another experimental strategy for the control of activity of molecular-defined subsets of cells. The chemogenetic actuators include engineered G-protein-coupled receptors (GPCR) and ligand-gated ion channels that can be activated by diffusible drug-like compounds (16, 17). A great asset of chemogenetics is that actuators are genetically engineered proteins, as in optogenetics, and therefore precise cell-type targeting can be achieved using specific cell-type promoters or transgenic strategies based on recombinase systems. Hence, chemogenetic actuators allow manipulating membrane potential, neuronal activity, or intracellular signaling in target neurons, too. However, because this method is not based on photosensitivity, it does not benefit from precise millisecond-range temporal resolutions as well as from the relative innocuity and specificity offered by light properties in optogenetics. Hence, three essential features will be required for an efficient chemogenetic approach: (i) high affinity of the actuators for its exogenous "naturally-inert" ligand (ii) low affinity for all endogenous ligands; and (iii) minimal constitutive activity. When these three features are met, chemogenetics can display high specificity and can sidestep most of adverse, unwanted effects. However, millisecond-range temporal resolution remains unreachable using chemical approaches. Temporal properties of chemogenetic actuators depend mainly on the ligand-receptor affinity and the pharmacokinetic properties of the ligand (as it occurs with standard pharmacological approaches). Hence, various factors influence the time between ligand administration and actuator activation: route of administration, solubility, blood–brain barrier permeability, metabolism, etc. Commonly, effects are expected to peak after 30 minutes following a peripheral administration and may progressively decline in the subsequent 2 hours (18). Hence, temporal resolution will represent the major characteristic that differentiates optogenetic and chemogenetic actuators, with an obvious advantage of optogenetics for the precise temporal control of neuronal activity, while chemogenetics can represent a valuable alternative in studies where neuronal activity must be modulated constantly over long periods of times (from hours to weeks).

Regarding spatial resolution, chemogenetic approaches rely mostly on cell-type specificity, rather than on spatially confined actions. Indeed, standard procedures in chemogenetics are generally based on a peripheral administration of the exogenous ligand, allowing the diffusion of the compound within the whole organism and resulting in a low spatial resolution. Because actuator expression is normally aimed to be restricted to a molecular-defined subset of cell type, effects occur specifically within this target neuronal population and low spatial resolution may not be a problem. However, molecular-defined neuronal populations are commonly not confined to a single restricted brain area and may be expressed in various brain areas related to distinct functions. In order to circumvent such potential confounding effects, it is possible to carry out localized intracranial microinjections into target brain areas, leading to higher spatial resolution, similar to optogenetic approaches, but at the cost of more invasiveness. Indeed, an authentic and outstanding asset of standard chemogenetic approaches over optogenetics is the simplicity, high flexibility, and minimal invasiveness these approaches offer: standard procedures are commonly based on peripheral administration, with a large choice of route (intraperitoneal, intravenous, subcutaneous, oral) or preparations (in food or drink), and no specialized tools or material (as optic fibers, light source generators, or light sensors in optogenetics).

Accordingly, optogenetic and chemogenetic approaches share several benefits (i.e. cell-type specificity, accurate control of neuronal activity) but also display some different pros and cons (regarding spatiotemporal resolution, invasiveness, unspecific effects, etc.). The choice of the approach used in an experiment will depend mainly on the degree of spatiotemporal resolutions required for each study, optogenetics being better suited for high spatiotemporal precision, while chemogenetics is appropriate for prolonged experiments. Accordingly, rather than two competitive approaches, they are complementary technologies with distinct advantages and disadvantages. It is even possible to combine various optogenetic and/or chemogenetic tools in a single experiment: multiplexing in the same animal neuronal activity monitoring, gain- and/or loss-of-functions, different target cell types, and different spatiotemporal resolutions, for example.

8.3 Optogenetic- and chemogenetic-associated components

Optogenetic and chemogenetic technologies arose out at the confluence of progress in molecular biology, biotechnology, engineering, and neuroscience. Various components are therefore put together allowing effective monitoring and manipulation of neuronal activity. We can distinguish four principal components involved in both approaches: (i) indicator or actuator genes; (ii) transduction or transgenic strategy; (iii) targeting strategy; (iv) effector delivery.

8.3.1 Indicator and actuator genes

8.3.1.1 Optogenetic indicators

Optogenetic indicators are genetically encoded light sensors that provide a method for the optical reading of cellular activity with high spatiotemporal resolution and cell-type specificity.

The development of optogenetic indicators originated from the discovery, characterization, and cloning of the green fluorescent protein (GFP) of a jellyfish (*Aequorea Victoria*) in the 1990s (19, 20). From that starting point, small GFP-like compounds have been fashioned from the original protein to improve its fluorescent properties, to broaden emission spectrum, and to promote stability in mammal cells. Novel fluorophores started to be discovered in other species too, such as the DsRed from *Discosoma Sp.* (21, 22). The transformation of such compounds into operative indicators was made possible following another step of optimization aiming at (i) making them sensitive to physicochemical changes in target cells, or (ii) merging them with proteins sensitive to neuronal activity (sensors). As a consequence, physicochemical conditions or intracellular events became able to influence the conformational state of the modified/merged proteins whose change directly governs the light emission from the fluorophore domain.

In this category, we can find compounds that account for synaptic activity, like the synaptopHluorine (23). These genetically engineered proteins were made from the merging of VAMP2 (a protein associated to synaptic vesicles) with a GFP-like compound sensitive to pH (exhibiting reduced fluorescence in acidic environment). The resulting chimeric proteins are then able to elevate their fluorescence emission strongly when synaptic vesicles release their content into an extracellular medium, providing an optical correlate of synaptic activity.

Other optogenetic indicators have been constructed to monitor neuronal activity. Within this type of indicators, genetically encoded calcium indicators are used as indirect sensors of action potentials (APs). They are built from the merge of calcium chelators with modified GFP-like domains (24, 25). The GCaMP protein is an important member of this family whose usage has spread widely since its first report (26). When cytosolic calcium concentration is reduced, the conformational state of the GCaMP protein prevents GFP-like domain to emit light. When neuronal activity increases, AP firing is associated with intracellular calcium rise, leading to a conformational change and promoting fluorescence emission. The most recent variants of GCaMP have been shown to reliably detect cytosolic calcium fluctuations from AP train and even a single spike (25).

Another example of this category is the FRET-based indicators. FRET (Forster Resonance Energy Transfer) is a mechanism of energy transfer between two fluorophores (27). Accordingly, FRET-based indicators contain two fluorophore domains: a first, the donor, is able to produce a fluorescent emission spectrum that overlaps with the fluorescent excitation spectrum of the second fluorophore domain, the acceptor. However, the ability of the donor to excite the acceptor depends not only on the wavelength of fluorescent emission/excitation but also on the distance and the relative orientation of the two fluorophore domains. Hence, FRET-based indicators commonly possess another domain, the sensor, that can change the conformational state depending on physicochemical parameters or neuronal activity. As a result, conformational changes impact the relative distance and orientation between the two fluorophore domains, affecting energy transfer and the light emitted by the acceptor. This category includes a large variety of sensors, including FRET-based calcium indicators (e.g. cameleon) (28), chloride indicators (e.g. clomeleon) (29), glutamate indicators (iGluSnFR) (30), and transmembrane potential indicators (e.g. GEVIs, VSFP) (31).

8.3.1.2 **Optogenetic actuators**

The capacity to monitor neuronal activity in living organisms is an important step toward the identification of the neural correlates of brain functions. However, to obtain a more complete picture it is important to investigate the causal relationship between neuronal activities and functions: is the activity of a targeted neuronal population necessary or sufficient for the investigated functions? In the last decade, optogenetic actuators have emerged as the finest method within the modern neuroscience toolbox to answer such questions.

It is important to emphasize that optogenetics actuators are not simply tools for photoexcitation or photoinhibition of targeted cells or projections. They are expected to deliver gain- or loss-of-function of precise events, which means that millisecond precision must be considered as an essential feature of true optogenetic approaches (10). With this in mind, efforts in the development of optogenetic actuators have mainly focused on single-component microbial opsins (type I) where light sensor and ion conductance are gathered into a single genetically encoded protein, contrary to multi-component systems. Indeed, single-component optogenetic actuators are easier to implement (reduced genome size), facilitate viral transfection, and display shorter-timescale temporal properties. Our examples and descriptions will therefore focus on single-component systems. In an effort to implement such single-component optogenetic actuators into mammalian organisms, diverse molecular modifications of the original microbial opsin genes have been incorporated. More precisely, substitutions of algal codons by mammalian codons contributed to optimize actuator expression and properties: higher expression, stronger photocurrent, reduced desensitization, changes in kinetic decay and wavelength excitation, improved channel selectivity, subcellular/neurite addressing, and membrane trafficking.

Optogenetic actuators can be divided into four major categories. The first category represents the actuator for fast neuronal excitation. The typical microbial opsin used in this category is the channelrhodopsins (ChRs). These opsins were originally isolated from the green algae *Chlamydomonas reinhardtii* (32). The ChRs are membrane channels that display light-activated cation-flux properties. Accordingly, light at a certain wavelength (usually blue light) can open the channel, which produces high Na^+ currents and small Ca^{2+} currents favoring neuronal depolarization with submillisecond-range temporal resolution. This property provides the possibility to produce spike trains that mimic "natural" activities of the targeted neuronal population. Moreover, many ChR variants with distinct properties have been recently developed and are now included in the optogenetic actuator palette. This offers the possibility of choosing an actuator for fast excitation that best suits each experimental condition.

The second category of optogenetic actuators groups the opsins for fast neuronal inhibition. This category includes the halorhodopsins (NpHRs), a light-activated chloride pump, the bacteriorhodopsins (BRs), and archeorhodopsins (Arch, ArchT), and light-activated proton pumps (33, 34). Under an appropriate light wavelength, these pumps induce a hyperpolarization and inhibit the activity of the targeted cells until the light is switched off.

A third category includes the bistable or step-function opsins (SFOs). These opsins enable step-like control of membrane potential. They usually display weak photocurrents but significantly increased off-kinetics (from 2 s to 30 min) which can produce an enduring neuronal activity following an appropriate light pulse (e.g. blue light) and until the closure of the channel. Termination can also be pre-empted by applying a light pulse of a different wavelength (e.g. yellow light).

The fourth category includes actuators aimed to control intracellular signaling optically: the OptoXRs. This class of opsin groups together chimera-based proteins fashioned by the merge between vertebrate rhodopsin (light sensor) and conventional GPCR (35). The single-component OptoXR actuators allow controlling a large variety of intracellular signaling under appropriate light wavelength. More recently new classes of OptoXRs have been developed aimed, for example, at light-controlling intracellular compounds such as small GTPase (36) and adenylate cyclase (37).

8.3.1.3 Chemogenetic actuators

The chemogenetic actuators group together exogenous chemically activated receptors and ligand-gated ion channels that are expressed in targeted neuronal populations and that can be activated by diffusible drug-like compounds. Typical chemogenetic actuators include the Designer Receptors Exclusively Activated by Designer Drugs (DREADDs) (16, 38). These receptors were originally fashioned from a modified muscarinic, GPCR, like the human M4 muscarinic DREADD (hM4Di), which is coupled to Gi signaling, or like the human M3 muscarinic DREADD (hM3Dq), which is coupled to Gq signaling. Muscarinic DREADDs have been designed to be activated by a specific exogenous diffusible ligand: the clozapine-*N*-oxide (CNO), which is an inert metabolite of the atypical antipsychotic clozapine. Other DREADDs have been developed more recently which use other exogenous ligands and/or signaling pathways. For instance, the kappa-opioid receptor DREADD (KORD) that is coupled to Gi is insensitive to CNO or natural ligand of opioid receptors but can be activated by the inert ligand salvinorin B. Among other chemogenetic actuators, we can mention the RASSL (39) or the allatostatin receptor (40) for the GPCR family, and the Trpv1, $P2X_2$, or LGCICs for the exogenous ligand-gated ion channels (41-43).

8.3.2 Gene expression

We can distinguish three main methods to deliver the genetic construct that codes for the indicator/actuator into the target cells: transfection by electroporation, viral transduction, or transgenic animals. Electroporation strategy may be easily implemented in *in vitro* experiments but are less simple and less advantageous than viral transduction in *in vivo* experiments. The viral transduction is certainly the method most used in laboratories as it has numerous advantages: rapidity, flexibility, potency (high gene copy numbers/cell), and capability of multiplexing genetic and anatomical specificity. Finally, the generation of transgenic mouse lines is another strategy that can directly express indicator/actuator within molecular-defined cell types. Although this method provides a reliable cell-type specific expression, the efforts and costs associated with the creation and maintenance of transgenic lines make this strategy prohibitive for many labs.

8.3.3 Targeting strategy

Cell-type specificity can be obtained by different (non-exclusive) means:

(1) *The virus tropism and properties*. Virus tropism and properties can bias the expression of indicator/actuator gene toward a cell type. For example, it has been shown that expression of transgenes under the same promoters with two different viruses, an adeno-associated virus 2 (AAV2) and a lentivirus (LV) in the somatosensory cortex were biased to a preferential expression into different cell types: the LV showing a bias toward excitatory neurons, the AAV2 showing a preferential transgene expression in inhibitory neurons (44). However, the tropism strategy has a limited reliability and, most importantly, a precise cell-type specificity is almost unattainable with this single strategy while it can be easily achieved through other strategies.

(2) *The transgene promoter*. This is a widely used and reliable method for cell-type specific expression of the indicator/actuator. In this case, the transgene is placed directly under the control of a promoter specific of the targeted cell type. However, two concerns may sometimes be associated with this strategy: (i) the genome size that can be packaged in the virus is limited (particularly in AAV ~4.7 kb) and many cell-type promoters have sizes that extend this limitation making it impossible to choose this targeting strategy; (ii) promoters specific to certain cell types may sometime lack strength and result in a specific but very weak expression of the indicator/actuator, thus affecting the capacity to monitor or control neuronal activity reliably.

(c) *Viral vector and Cre-driver lines*. In order to circumvent the limitations imposed by virus packaging capacity, it is possible to combine virus injection with Cre-driver mouse lines. These mouse lines express a Cre recombinase under the control of a promoter specific of molecularly-defined cell populations. The Cre recombinase is an enzyme capable to delete or recombine DNA associated to specific sequences (LoxP or floxed sequences). With this strategy, the indicator/actuator transgene inserted in the virus is placed under the control of a small, ubiquitous, strongly active promoter. In addition, either a floxed stop codon is placed upstream to the transgene, or the transgene is double-floxed and its reading frame is inverted. The virus is then injected into the brain region of the Cre-driver mouse line. Accordingly, the virus will transduce most of neurons within the site of injection. However, only the cells that express the Cre recombinase will be able to either delete the stop codon or flip the inverted transgene, leading in both cases to the specific and strong expression of the indicator/actuator. With the growing number of different Cre driver mouse lines currently available, this strategy is very powerful and versatile (45). Other inducible/conditional expression systems have been extensively developed in the field (e.g. Flp recombinase, Tet-On/Tet-off) and can display the same benefits as the Cre system for optogenetic and chemogenetic approaches.

In addition to cell-type specificity, targeting can even be improved by strategies based on spatially confined microinjections of the virus bearing the indicator/actuator gene as well as on spatially confined light delivery/recording (for optogenetic actuator only). Such a targeting strategy is therefore based on the ability selectively to drive or inhibit cell populations defined by their location, wiring, or projections. Indeed, indicator/actuator gene products are transported along dendrites and axons. This property allows transduction of cell bodies in one brain region and illumination of projections in another (and vice-versa).

8.4 Examples of research using optogenetic actuators to decipher anxiety-related neural circuits

In this section, we aim to provide several examples of research that has been undertaken to specify neural circuits underlying emotions using optogenetics. In order to render this easier to understand, we illustrate it by focusing on one particular emotion: anxiety. Indeed, this emotion is highly conserved in rodents and many investigations have focused on it. Therefore, using this particular example it will be easier to show how optogenetics has enabled a deeper understanding of the fear network, highlighting the contribution of region-specific cell types, the function of particular neural projections to the different cognitive/physiological components of anxiety, and enabling an advanced understanding of the way cells process the selection of an appropriate response.

8.4.1 Assessment of anxiety behavior in rodents

Anxiety behavior is usually triggered by confronting rodents with an unknown place that usually comprises anxiogenic parts (e.g. the unprotected arms of an elevated plus maze or the center of a big arena). In such a situation, rodents generally display a preference for the most protected areas (closed arm of an elevated plus maze or most parts of the open field that are close to the walls): these are viewed as anxiety-like behaviors.

8.4.2 Contribution of region-specific cell types to anxiety-like behaviors

Optogenetics enables us to investigate the contribution of specific subareas of a given structure to anxiety-related behaviors, because in some cases these sub-areas express specific proteins that can be selectively targeted by optogenetic tools. This is, for example, the case of the BNST, a brain area that has been shown to be involved in anxiety-related behaviors. Indeed, using the inhibitory

Natronomonas pharaonis halorhodopsin (eNpHR3.0) and yellow light administration to the BNST, Kim and colleagues (46) showed anxiolysis in mice subjected to the elevated plus maze and to the open field. Conversely, stimulation of the somata of the BNST neurons with the excitatory channelrhodopsin-2 (ChR2) elicited the opposite pattern, that is, anxiety-like behaviors. Taken together, these experiments established a causal role of the BNST in anxiety-like behaviors. However, the BNST is a very heterogeneous conglomerate of subnuclei including anterior parts, posterior sections as well as the oval nucleus that may support different role in anxiety-related behaviors.

Hence other experiments focused on the functions of these subregions. Using the same behavioral situations, Henckens and colleagues (47) focused on the posterior BNST (pBNST) and, more specifically, on the cells expressing corticotropin-releasing factor receptor type 2 (CRFR2). Therefore, they crossbred CRFR2-Cre mice with channelrhodopsin- (ChR2) conditional mice, implanted these mice with fiber-optic cannulas into the pBNST bilaterally and delivered light into this region. They showed that activation of pBNST CRFR2 neurons reduced anxiety-like behaviors while inhibition of pBNST CRFR2 neurons elicited the opposite effects.

Other experiments focused on the function of the oval nucleus of the BNST (ovBNST). Kim and colleagues (46) targeted cells expressing dopamine D1a receptors, whose expression is restricted to the ovBNST. They injected a viral vector containing a Cre-dependent eNpHR3.0 transgene into the BNST of dopamine receptor 1a (Drd1a::Cre) mutated mice showing restricted Cre expression in the ovBNST and delivered yellow light. They found that this was sufficient to elicit anxiolytic-like behaviors in the elevated plus maze. Conversely, stimulation of the ovBNST using ChR2 had the opposite effect. This indicates a specific role of the ovBNST in anxiety-related behaviors. Interestingly, the ovBNST sends inhibitory projections to the antero-dorsal BNST (adBNST), a region receiving afferents from the basolateral amygdala (BLA) and sending outputs to the ventral tegmental area (VTA), the lateral hypothalamus (LH), and the parabrachial nucleus (PB); adBNST activity is associated to anxiolytic effects.

8.4.3 Contribution of specific inputs/outputs to anxiety-like behaviors or to components of the anxiety response

Even if optogenetic manipulation of the BLA cells had no effect per se on anxiety-related behaviors, activation of specific outputs of the BLA impacted strongly on anxiety behaviors or altered specifically some features of the anxiety response. For example, the stimulation of BLA terminals in the central amygdala (CeA) elicited anxiolytic-like effects while inhibition of the same projection induced the opposite effects (48). This was achieved by injecting, in the BLA, AAVs containing a sequence coding for ChR2 or eNpHR3.0 (AAV5-CamKIIα-hChR2-eYFP or AAV5-CamKIIα-eNpHR3.0-eYFP respectively) under the control of a CamKIIα promoter in order to target the BLA glutamatergic projections to the CeA.

As already mentioned, emotions such as anxiety are conceptualized through their various features/dimensions, including the appraisal of valence, as reflected in behavioral and physiological responses. In an impressive set of experiments, Kim and colleagues (46) were able to demonstrate a dissociation of anxiety-related changes in valence, in respiration, and in avoidance behaviors, depending upon the projection from the BLA to the BNST, and from distinct adBNST outputs. First, using AAV-CamKIIα-eNpHR3.0-eYFP injection in the BLA and light-induced inactivation of the BLA terminals in the adBNST, they showed that anxiogenic-like effects were elicited in the elevated plus maze and in the open-field test, while the light-induced activation of the BLA terminals in the adBNST using AAV-CamKIIα-ChR2-eYFP elicited anxiolytic-like effects. This suggests an involvement of the BLA-BNST projections in anxiety. The anxiety-related response included avoidance of aversive spaces (such as reduced exploration of open arms of the elevated plus maze or of the central part of the open field), increase in respiration rate and negative valence (determined by avoidance in a real-time place preference paradigm). Most interestingly, using the optogenetic toolbox, Kim and colleagues (46) were able to dissect these different components as they revealed that the different outputs of the adBNST were selectively involved in these different aspects. Indeed, they showed that photostimulation of the adBNST-LH projections decreased anxiety-related behaviors in the elevated plus maze and in the open field, but had no effect on the anxiety-related change in respiration rate and valence indicating that this pathway selectively modulates the behavioral but not the physiological or the appetitive features of anxiety. Similarly, they showed that photoactivation of the adBNST-PB output selectively affected the respiration rate, but not the other components of the anxiety response, while the adBNST-VTA pathway was involved selectively in the appetitive aspects. Thus, optogenetics enables us to decipher the different features associated with spontaneous anxiety. This functional segregation would not have been possible using other methods.

8.4.4 Contribution of specific cells in the selection of the fear response

Finally, when confronted with a threat, rodents select the most appropriate behavioral response, which can be either a passive response (freezing) or an active response (flight). In fact, according to the situation and particularly according to the distance from the threat, it will be more adaptive to escape or to remain undetected. In a recent set of experiments, Fadok and colleagues (49) have tried to explore the neural substrate of this response selection, focusing on the CeA. Interestingly, the CeA comprises several kinds of neurons, including neurons expressing CRF and somatostatin-positive (SOM+) neurons. To determine the role of defined CeA neuronal subpopulations for expression of "flight" responses, authors used the inhibitory opsin archaerhodopsin (Arch), which was Cre-dependently expressed in CRF^+ or SOM^+ neurons. They found that inhibition of CRF^+, but not SOM^+ neurons, completely abolished conditioned flight behavior. Conversely, photostimulation of ChR2-expressing CRF^+ neurons decreased freezing and led to more active behaviors, while activation of SOM^+ neurons decreased flight and caused an increase in freezing behaviors. Interestingly, electrophysiological experiments revealed that the CRF^+ cells, which are crucial for flight, provide direct inhibition to the SOM^+ neurons, which are involved in freezing behavior, and vice versa. This might allow for rapid switching between the two behavioral strategies, according to the context.

Using optogenetics, many other neuronal circuits have been described as underlying anxiety, highlighting the contribution of the septo-hippocampal system and inputs from different parts of the prefrontal cortex on subcortical limbic structures such as the amygdala. However, an exhaustive description of this circuit is beyond the reach of this chapter which intends instead simply to illustrate the impact of the optogenetic toolbox on anxiety research, by using some specific examples. However, further reviews are available (50, 51).

8.5 Conclusion

Optogenetics and chemogenetics enabled a revolution in research in the field of the neurobiology of emotions. It provided a more complex view of the different cells and pathways involved in typical emotions, and enabled us to start dissecting the contribution of specific outputs to different emotional components. These methods have allowed unprecedented progress in this field in the last decade.

References

1. **Darwin C.** On the Origin of Species. New York, NY: Appleton and Co.; 1871.
2. **Leventhal H, Scherer K.** The relationship of emotion to cognition: A functional approach to a semantic controversy. Cognition and Emotion. **1** (1987) 3–28.
3. **Philippot P, Schaefer A.** Emotion: Current issues and future directions. In: Mayne TJ, Bonano GA (eds). Emotion and Memory. New York, NY: Guilford Press; 2001, pp. 82–122.
4. **Belzung C, Philippot P.** Anxiety from a phylogenetic perspective: Is there a qualitative difference between human and animal anxiety? Neural Plasticity. 2007;59676.
5. **Rohr CS, Okon-Singer H, Craddock RC, Villringer A, Margulies, DS.** Affect and the brain's functional organization: A resting-state connectivity approach. PLoS One. 2013;**8**(7):e68015.
6. **Etkin A, Buchel C, Gross JJ.** The neural bases of emotion regulation. Nat Rev Neurosci. 2015;**16**(11):693–700.
7. **Adolphs R.** The biology of fear. Curr Biol. 2013;**23**(2):R79–93.
8. **Hamann S.** Mapping discrete and dimensional emotions onto the brain: Controversies and consensus. Trends Cogn Sci. 2012;**16**(9):458–66.
9. **Singewald N.** Altered brain activity processing in high-anxiety rodents revealed by challenge paradigms and functional mapping. Neurosci Biobehav Rev. 2007;**31**(1):8–40.
10. **Yizhar O, Fenno LE, Davidson TJ, Mogri M, Deisseroth K.** Optogenetics in neural systems. Neuron 2011;**71**(1):9–34.
11. **Dugue GP, Tricoire L.** Principles and applications of optogenetics in neuroscience. Med Sci (Paris). 2015;**31**(3):291–303.
12. **Deisseroth K.** Optogenetics: 10 years of microbial opsins in neuroscience. Nature Neurosci. 2015;**18**(9):1213–25.
13. **Aravanis AM, Wang LP, Zhang F, Meltzer LA, Mogri MZ, Schneider MB,** et al. An optical neural interface: In vivo control of rodent motor cortex with integrated fiberoptic and optogenetic technology. J Neural Eng. 2007;**4**(3):S143–56.
14. **Peron S, Chen TW, Svoboda K.** Comprehensive imaging of cortical networks. Curr Opin Neurobiol. 2015;**32**:115–23.
15. **Birkner A, Tischbirek CH, Konnerth A.** Improved deep two-photon calcium imaging in vivo. Cell Calcium. 2017;**64**:29–35.
16. **Whissell PD, Tohyama S, Martin LJ.** The use of DREADDs to deconstruct behavior. Front Genet. 2016;7:70.
17. **Wulff P, Arenkiel BR.** Chemical genetics: Receptor-ligand pairs for rapid manipulation of neuronal activity. Curr Opin Neurobiol. 2012;**22**(1):54–60.
18. **Guettier JM, Gautam D, Scarselli M, Ruiz de Azua I, Li JH, Rosemond E,** et al. A chemical-genetic approach to study G protein regulation of beta cell function in vivo. PNAS. 2009;**106**(45):19197–202.
19. **Ormo M, Cubitt AB, Kallio K, Gross LA, Tsien RY, Remington SJ.** Crystal structure of the Aequorea victoria green fluorescent protein. Science. 1996;**273**(5280):1392–5.
20. **Prasher DC, Eckenrode VK, Ward WW, Prendergast FG, Cormier MJ.** Primary structure of the Aequorea victoria green-fluorescent protein. Gene. 1992;**111**(2):229–33.
21. **Matz MV, Fradkov AF, Labas YA, Savitsky AP, Zaraisky AG, Markelov ML,** et al. Fluorescent proteins from nonbioluminescent Anthozoa species. Nat Biotechnol. 1999;**17**(10):969–73.
22. **Shaner NC, Patterson GH, Davidson MW.** Advances in fluorescent protein technology, J Cell Sci. 2007;**120**(Pt 24):4247–60.
23. **Miesenbock G, De Angelis DA, J.E. Rothman JE.** Visualizing secretion and synaptic transmission with pH-sensitive green fluorescent proteins. Nature, 1998;**394**(6689):192–5.
24. **Akerboom J, Rivera JD, Guilbe MM, Malave EC, Hernandez HH, L. Tian L,** et al. Crystal structures of the GCaMP calcium sensor reveal the mechanism of fluorescence signal change and aid rational design. J Biol Chem. 2009;**284**(10):6455–64.

25. **Chen TW, Wardill TJ, Sun Y, Pulver SR, Renninger SL, Baohan A**, et al. Ultrasensitive fluorescent proteins for imaging neuronal activity. Nature. 2013;**499**(7458):295–300.
26. **Nakai J, Ohkura M, Imoto K.** A high signal-to-noise Ca(2+) probe composed of a single green fluorescent protein. Nat Biotechnol. 2001;**19**(2):137–41.
27. **Tsien RY.** Indicators based on fluorescence resonance energy transfer (FRET). Cold Spring Harb Protoc. 2009;(7): pdb top57.
28. **Horikawa K, Yamada Y, Matsuda T, Kobayashi K, Hashimoto M, Matsu-ura T**, et al. Spontaneous network activity visualized by ultrasensitive Ca(2+) indicators, yellow Cameleon-Nano. Nat Methods. 2010;**7**(9):729–32.
29. **Berglund K, Schleich W, Krieger P, Loo LS, Wang D, Cant NB**, et al. Imaging synaptic inhibition in transgenic mice expressing the chloride indicator, Clomeleon. Brain Cell Biol. 2006;**35**(4–6):207–28.
30. **Marvin JS, Borghuis BG, Tian L, Cichon J, Harnett MT, Akerboom J**, et al. An optimized fluorescent probe for visualizing glutamate neurotransmission. Nat Methods. 2013;**10**(2):162–70.
31. **Mutoh H, Knopfel T.** Probing neuronal activities with genetically encoded optical indicators: From a historical to a forward-looking perspective. Pflugers Arch. 2013;**465**(3):361–71.
32. **Nagel G, Ollig D, Fuhrmann M, Kateriya S, Musti AM, Bamberg E**, et al. Channelrhodopsin-1: A light-gated proton channel in green algae. Science. 2002;**296**(5577):2395–8.
33. **Oesterhelt D, Stoeckenius W.** Functions of a new photoreceptor membrane. PNAS. 1973;**70**(10):2853–7.
34. **Matsuno-Yagi A, Mukohata Y.** Two possible roles of bacteriorhodopsin; a comparative study of strains of Halobacterium halobium differing in pigmentation. Biochem Biophys Res Commun. 1977;**78**(1):237–43.
35. **Kim JM, Hwa J, Garriga P, Reeves PJ, RajBhandary UL, Khorana HG.** Light-driven activation of beta 2-adrenergic receptor signaling by a chimeric rhodopsin containing the beta 2-adrenergic receptor cytoplasmic loops. Biochemistry. 2005;**44**(7):2284–92.
36. **Yazawa M, Sadaghiani AM, Hsueh B, Dolmetsch RE.** Induction of protein-protein interactions in live cells using light. Nat Biotechnol. 2009;**27**(10):941–5.
37. **Stierl M, Stumpf P, Udwari D, Gueta R, Hagedorn R, Losi A**, et al. Light modulation of cellular cAMP by a small bacterial photoactivated adenylyl cyclase, bPAC, of the soil bacterium Beggiatoa. J Biol Chem. 2011;**286**(2):1181–8.
38. **Armbruster BN, Li X, Pausch MH, Herlitze S, Roth BL.** Evolving the lock to fit the key to create a family of G protein-coupled receptors potently activated by an inert ligand. PNAS. 2007;**104**(12):5163–8.
39. **Coward P, Wada HG, Falk MS, Chan SD, Meng F, Akil H**, et al. Controlling signaling with a specifically designed Gi-coupled receptor. PNAS. 1998;**95**(1):352–7.
40. **Lechner HA, Lein ES, Callaway EM.** A genetic method for selective and quickly reversible silencing of Mammalian neurons. J Neurosci. 2002;**22**(13):5287–90.
41. **BZemelman BV, Nesnas N, Lee GA, Miesenbock G.** Photochemical gating of heterologous ion channels: Remote control over genetically designated populations of neurons. PNAS. 2003;**100**(3):1352–7.
42. **Guler AD, Rainwater A, Parker JG, Jones GL, Argilli E, Arenkiel BR**, et al. Transient activation of specific neurons in mice by selective expression of the capsaicin receptor. Nat Commun. 2012;**3**:746.
43. **Magnus CJ, Lee PH, Atasoy D, Su HH, Looger LL, Sternson SM.** Chemical and genetic engineering of selective ion channel-ligand interactions. Science. 2011;**333**(6047):1292–6.
44. **Nathanson JL, Yanagawa Y, Obata K, Callaway EM.** Preferential labeling of inhibitory and excitatory cortical neurons by endogenous tropism of adeno-associated virus and lentivirus vectors. Neuroscience. 2009;**161**(2):441–50.
45. **Tsien JZ.** Cre-Lox Neurogenetics: 20 years of versatile applications in brain research and counting. Front Genet. 2016;**7**:19.

46. **Kim SY, Adhikari A, Lee SY, Marshel JH, Kim CK, Mallory CS**, et al. Diverging neural pathways assemble a behavioural state from separable features in anxiety. Nature. 2013;**496**(7444):219–23.
47. **Henckens MJ, Printz Y, Shamgar U, Dine J, Lebow M, Drori Y**, et al. CRF receptor type 2 neurons in the posterior bed nucleus of the stria terminalis critically contribute to stress recovery. Mol Psychiatry. 2017;**22**(12):1691–700.
48. **Tye KM, Prakash R, Kim SY, Fenno LE, Grosenick L, Zarabi H**, et al. Amygdala circuitry mediating reversible and bidirectional control of anxiety. Nature. 2011;**471**(7338):358–62.
49. **Fadok JP, Krabbe S, Markovic M, Courtin J, Xu C, Massi L**, et al. A competitive inhibitory circuit for selection of active and passive fear responses. Nature. 2017;**542**(7639):96–100.
50. **Tye KM, Deisseroth K.** Optogenetic investigation of neural circuits underlying brain disease in animal models. Nat Rev Neurosci. 2012;**13**(4):251–66.
51. **Tovote P, Fadok JP, Luthi A.** Neuronal circuits for fear and anxiety. Nat Rev Neurosci. 2015;**16**(6):317–31.

Part 2

Cognitive mechanisms

Chapter 9

Fear learning and extinction

Tina B Lonsdorf

9.1 Fear conditioning, extinction, and return of fear as translational models for psychiatry

Fear-, anxiety-, and stress-related disorders are highly prevalent and place a heavy burden on affected individuals as well as a major economic burden on society (1–4).

The development, treatment, and relapse of these disorders can be modeled experimentally by employing fear conditioning, extinction, and return of fear paradigms. Understanding the molecular pathways that mediate experimental fear acquisition and extinction might therefore make an important contribution to the study of anxiety pathophysiology and resilience as well as offering potential avenues for prevention and intervention.

This chapter summarizes the current literature on the impact of genetic factors on experimental fear acquisition, extinction, and the return of fear, starting with a brief historical synopsis.

9.1.1 A historical perspective focusing on individual differences and genetic factors

This year, we celebrate the 100th anniversary of the publication of the first experimental study on fear conditioned reflexes in humans (5) (for a comprehensive historical overview, see (6)). During the past century, fear acquisition and extinction, both as experimental procedures and as theoretical mechanisms, have attracted substantial interest. This can not only be attributed to these procedures providing a connection link between emotional and memory process (7) but also because of their strong translational value (8). Results derived from fear acquisition and extinction research have indeed cross-fertilized neurobiological research and basic research on learning principles (8,9) and they still serve as prime experimental models for the study of the pathogenesis and treatment of anxiety disorders, respectively.

The advent of fear conditioning research was marked by John Watson and Rosalie Rayner's renowned experiment with "little Albert," which represents a convincing case of a "conditioned emotional reaction" after pairing an initially neutral stimulus (white rat) with a potent aversive event (noise) (10). Subsequently, this work was extended by Mary Cover Jones in describing methods for "the elimination of fear in children" (11) and the famous case report of "little Peter" (12) which has heavily inspired modern behavioral treatment of anxiety disorders (13).

Fear conditioning, which will be used in the following as an umbrella term subsuming fear acquisition, extinction, and return of fear procedures, subsequently became an epitome of learning and environmentalism. Hence, historically, Pavlovian learning processes have been assumed to be universal whereas variability in behavior and deviations from invariant learning patterns has generally been regarded as "noise" (14). Until the 1980s, research focused on "average responding," yet, there is strong evidence since Pavlov (15,16) that individuals respond differently to identical conditioning procedures.

This approach, focusing on individual differences, regained interest towards the end of the twentieth century, with accumulating evidence from epidemiological studies showing strong familial aggregation of anxiety disorders (for comprehensive reviews and meta-analysis, see (17–19)). Thereby, twin studies attribute an important source of familial risk (i.e. 30–40%) to additive genetic factors, which ultimately sparked the "search for anxiety genes" (17).

The twenty-first century brought strengthened interest in individual differences in fear acquisition and extinction processes. Ironically, despite their once iconic status in behaviorism, it is now generally accepted that individual differences exert substantial impact on fear acquisition and extinction processes. Concurrent major technological developments that allowed for faster, cheaper, and more accessible genotyping methods supported and advanced the search for genetic pathways contributing to individual differences in fear and anxiety in humans as well as their neurobiological underpinnings (reviewed in 20–22).

The shifts in beliefs about whether human behavior is determined by genes (i.e. nature) or by environmental factors (i.e. nurture) is hence reflected in research on fear conditioning with an explosion of interest in "gene–environment interaction" (G×E) in the last decade.

Before discussing the genetic contributions to fear conditioning, the basic experimental procedures for fear acquisition, extinction, and return of fear, as well as their clinical implications, will be introduced.

9.1.2 Fear acquisition

Learning to predict danger from previous experience is critical to an organism's survival and well-being. Fear acquisition through Pavlovian conditioning imbues a relatively neutral stimulus (the to-be conditioned stimulus, or CS+) with fear-evoking properties as the result of its co-occurrences with an aversive event (the unconditioned stimulus, US) that threatens the well-being of the organism. In cognitive terms, the organism learns that the CS is a reliable predictor of the dangerous US, evokes anticipatory (fear) reactions, and mobilizes defensive reaction mechanisms (i.e. conditioned reactions, CRs). Note that the term *fear acquisition* thereby refers to the underlying theoretical process, whereas the term *fear acquisition training* is used to refer to the experimental procedure (23). As the term "fear" may imply the presence of a conscious feeling, the term "*threat conditioning*" has recently been suggested to be more appropriate than fear conditioning (24).

Fear acquisition plays a major role in psychological theories of anxiety disorders such as phobias (25,26), panic disorder (27), as well as post-traumatic stress disorder (PTSD) (28). Importantly, however, a clear distinction exists between fear- and anxiety-related processes. Thereby, *fear* represents the response to a specific, stimulus-driven threat ('phasic'), whereas *anxiety* represents a sustained defensive state which is not linked to a specific stimulus (*cf.* 29,30). These theoretical constructs are thought to have their parallels in human psychopathology with some anxiety and stress-related disorders linked to phasic fear (e.g. phobias, PTSD) while others are linked to sustained anxiety (e.g. generalized anxiety disorder, panic disorder). Corresponding procedural variations (cue versus context conditioning) have been developed (31–33).

Fear acquisition training protocols in rodents usually employ *single-cue* designs (CS+ only) and hence compare conditioned responding in freezing, defensive, or avoidance behavior to this cue with either the absence of the cue or a control group. The majority of human studies in turn employ *differential* protocols in which one CS (CS+) is predictive of the US, while a second one is not (CS–). Differential conditioning involves "excitatory" learning to

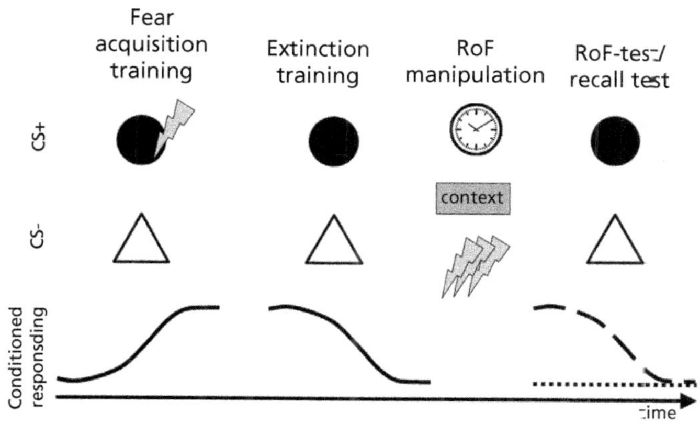

Figure 9.1 Schematic illustration of the phases of a fear conditioning experiment, conditioned stimuli (CS+, CS−) as well as the development of the conditioned response (black line) over the different phases.

Note that the clock indicates passing of time leading to *spontaneous recovery*, the "context" icon indicates contextual change leading to *renewal*, and the bolds indicate *reinstatement* US to induced return of fear (RoF). Also note that *extinction recall* and *RoF test*, in particular with respect to spontaneous recovery, do not differ *procedurally* but can only be differentiated conceptually by the prediction of the dominant memory trace at test (i.e. fear or extinction memory dominance leading to expression of conditioned responding or not, respectively).

The upper dashed line indicates return of fear (i.e. recall of the fear memory trace) and the lower dotted line indicates the absence of return of fear (i.e. recall of the extinction memory trace).

the CS+ but may also involve "inhibitory" learning to the CS−, which signals the absence of danger ("safety stimulus"). Conditioned responding in autonomic, verbal, behavioral, and/or defensive reactions is thereby quantified as the difference in response amplitude/strength to the CS+ as compared to the CS−. Importantly, patients suffering from anxiety disorders have been shown to display deficits of fear acquisition, in particular safety signal processing (34,35).

In sum, fear acquisition has been established as an outstanding, valid, and widely used translational model for the experimental investigation of mechanisms underlying pathological anxiety (8,36; for a critical discussion, see (37)).

9.1.3 **Fear extinction**

In changing environments, responding has to be adapted constantly. For instance, when a previously threatening stimulus has lost its predictive power, defensive responding will cease. Hence, the presentation of unreinforced CS+ presentations (in the procedure of *extinction training*) leads to a gradual waning of (differential; i.e. CS+ > CS−) conditioned responding, a (theoretical) process referred to as *extinction*. Importantly, patients suffering from anxiety disorders have shown deficits in extinction learning (34,35).

Generally, extinction has obvious implications for the treatment of anxiety disorders (8,38) and has inspired highly efficient exposure treatments of anxiety disorders (13).

Importantly, extinction does not, under most circumstances, lead to erasure of the "excitatory" fear-conditioning trace (i.e. CS+/US association). In most cases, extinction generates a competing fear-inhibitory extinction memory trace (CS−/no US association) which co-exists with and inhibits the fear memory trace after successful extinction (39,40). Procedurally, one must discriminate between *immediate extinction* training (common in human research), which directly follows upon fear acquisition, and *delayed extinction* training (common in rodent research), which occurs after a certain time interval following fear acquisition. Despite similar procedural implementation and terminology, immediate and delayed extinction differ in extinction speed and subsequently occuring return of fear (see section 9.1.4; 23,41).

In sum, fear extinction has been established as a prime model for the experimental study of treatment of anxiety (for a review, see (8); for a critical consideration, see (37)). Critically, maladaptive responding can not only arise from excessive fear learning but also from the failure to extinguish previously acquired fears (34,35), or a combination thereof.

9.1.4 Return of fear and extinction recall

High relapse rates after initial treatment success represent a major limitation to long-term remission of anxiety disorders despite effective psychological and pharmacological interventions (42). Advancing the understanding of clinical relapse may hence provide a first step towards individually tailored prevention programs to enhance long-term remission.

Relapse or the resurgence of fear can be modelled in experimental fear conditioning through so-called *return of fear* procedures following (successful) extinction training. Return of fear can be experimentally induced by the mere passage of time (*spontaneous recovery*), induction of contextual change (*renewal*), or by exposure to unsignaled US (*reinstatement*) (43; for an overview, see, e.g. (44)). Different return of fear phenomena are thereby thought to model different (clinical) mechanisms underlying clinical relapse, such as insufficient generalization of safety learning from the safe therapeutic situation to everyday situations (renewal; 45), or relapse following exposure to life adversity (46).

In human research, the return of fear phenomena has only been the focus of research for a decade since they have been more widely employed as outcome measures of pharmacological or behavioral manipulations of fear or extinction memory consolidation (47–50).

On a mechanistic level, insufficient expression of extinction memories (i.e. *extinction recall*) is thought to result in re-emergence of conditioned responding (*return of fear*; RoF) due to dominance of the fear over the extinction memory trace (51).

9.2 Fear conditioning as a prime model for genetic studies

Studying the genetic underpinnings of fear conditioning requires some theoretical and procedural considerations. Genetic studies are reasonable only if there is evidence that the to-be-studied phenotype is clearly defined, heritable, and can be reliably quantified. Furthermore, it can be highly advantageous when the neural network underlying phenotypic expression is well delineated and when data on the biological underpinnings exist (20,21). In this respect, fear conditioning represents a particularly promising model for genetic studies.

9.2.1 Precision of phenotypic definition

Both the accurate *definition* and, relatedly, the *quantification* of the phenotype are critical to the success of molecular genetic studies. More precisely, replication failures may partly result from ambiguous definitions of the phenotype (52–54).

In addition to the *definition* of the phenotype, the relation of the phenotype to genetic factors is crucial. As genetic influences on psychiatric disorders transcend the boundaries of binary diagnostic categories (55), the study of dimensional phenotypes (17) that map symptoms with a potential biological and genetic underpinning is currently the focus of research. This approach is employed by the Research Domain Criteria (RDoC) initiated by National Institute of Mental Health (NIMH) (56,57), which suggests investigations on multiple levels of analyses (gene, molecules, neural circuits, physiology, behavior). Fear acquisition and extinction therefore represent prime experimental procedures for the study of "acute threat," one of the constructs included in RDoC (22,38).

The operationalization and quantification of conditioned responding, however, warrants some methodological considerations to put potentially divergent findings into perspective. More precisely, results depend in part on subtle procedural choices, the nature of the outcome measure (verbal, behavioral, physiological), and the number of trials considered, as well as the statistical quantification (e.g. CS+ > CS−) and corrections of responding, which vary widely across studies (for a discussion, see (23)).

9.2.2 Heritability of the phenotype

Established heritability of the to-be-studied phenotype is a major prerequisite for considering its specific genetic underpinnings. With respect to fear conditioning, both human (58,59) and animal studies (60) have shown that genetic factors represent a significant source of individual variation in the habituation, acquisition, and extinction of fear.

In rodents, heritability has been convincingly demonstrated by selective breeding (61–64) and robust strain differences (60,65,66) in conditionability, extinction learning (67), and extinction recall (68). In humans, about one-third of the variance in both human fear acquisition (59) and the risk of developing anxiety disorders (69) has been attributed to genetic factors.

Importantly, however, in contrast to the genotype, the phenotypic expression (i.e. fear and extinction learning) is inherently dynamic, unstable, and reactive to environmental and developmental demands and processes (70,71), which results in a certain degree of variance in heritability estimates.

9.2.3 Heritability measures of phenotypic outcome measures

Heritability of the phenotype is, however, critically contingent upon heritability of the outcome measures employed for the quantification of conditioned responding. In humans, conditioned responding is most commonly assessed through physiological indicators such as skin conductance responses (SCRs), defensive reflex activation such as fear potentiated startle (FPS), verbal self-reports of US expectancy or fear, as well as neural activation as assessed by functional magnetic resonance imaging (fMRI).

Importantly, twin studies have demonstrated heritability of SCRs during fear acquisition (35–45%; 59). Furthermore, human eye-blink conditioning (72) and the overall startle magnitude in humans are heritable (37–67%; 73–75). Evidence for the role of genetic factors on the affect-modulated startle in turn is absent (73,75) or preliminary (76). Particularly, heritability of *fear*-potentiated startle in humans is limited to instructed threat studies in patients suffering from anxiety and depression and their non-affected family members (77,78), whereas there is clear evidence for inheritance in rodents (79,80).

Studies on the heritability of task-related brain reactivity are, however, scarce with respect to regions involved during fear and threat processing. Initial evidence suggests amygdala activation to faces to be a temporally stable and trait-like measure (81). Furthermore, brain activation as assessed by fMRI has been shown to be generally moderately heritable (82,83).

Importantly, despite of evidence of heritability across read-out measures, the underlying genetic factors do not necessarily converge (84), which has implications for the selection of read-out measures and advocates a multi-level approach.

9.2.4 Reliability of phenotype quantification

In addition to a precise definition and a heritable base, phenotypes for genetic studies must be quantified reliably and reproducibly over time (i.e. *test-retest reliability*).

With respect to fear conditioning, within-subject reproducibility and test-retest reliability has been established for conditioned SCRs across time intervals ranging from weeks (85,86) to eight months (87). This has been taken to conclude that conditioned SCRs represents "a genetically determined trait factor" which is appropriate "to assess individual differences in conditionability" (*cf.*, 85), even though situational factors also contribute to variability. Furthermore, temporal stability has also been reported for differential conditioned startle responding as well as US expectancy ratings during fear acquisition (87).

Importantly, significant test-retest reliability was observed for maximum CS+ responding (SCRs: 85,86), CS– responding (SCRs: 85), as well as CS+/CS– discrimination during fear acquisition (SCRs: 86; SCRs, FPS, and US expectancy: 87), extinction recall and renewal (first two trials each in SCRs: 86), as well as fear generalization (SCRs, FPS, US expectancy: 87).

Responding during extinction (last two trials) in turn did not display significant test-retest reliability (SCRs: 86), possibly due to floor level responding.

Importantly, these findings are based on multiple fear-conditioning sessions employing identical (85–87) or different experimental (87) stimuli across session, with the former showing somewhat compromised stability.

With respect to genetic studies of fear conditioning, the different read-out measures of conditioned responding may display different levels of penetrance. As between-subject variation in physiological measures is substantial, it is somewhat smaller for verbal measures of conditioned fear (87), and this has implications for selection of outcome measures for genetic studies of fear conditioning.

9.2.5 Well-delineated neural network

The neural network underlying fear conditioning and extinction has been delineated in detail with striking similarities in rodents and humans (7,88–92) and involves the amygdala, the insula, the hippocampus, as well as the medial and prefrontal cortex and dorsal anterior cingulate areas. Importantly, neural correlates of CS+/CS– discrimination during fear acquisition (i.e. amygdala activation) correlate with individual differences in CS+/CS– discrimination in SCRs (93,94).

Generally, it has been argued that neuroimaging phenotypes may be more closely related to and hence exert a "stronger penetrance" on the neurobiological level (e.g. brain activation) than behavioral phenotypes because brain structure and function are considered to be more closely related to the genes function than behavior (95). Hence, phenotypes known to originate from well-described neural circuits may facilitate the selection of candidate genes.

9.3 Searching for "anxiety genes"

Human studies have initially focused on establishing a heritable base of anxiety and anxiety disorders. This provided a rationale for more mechanistically oriented genetic studies. Rodent work and more recent work in humans have investigated and advanced the specific genetic determinants underlying fear and anxiety processing in general, and fear conditioning processes in particular. After initial optimism that the mapping of the human genome would bring genetic markers for psychopathology within reach, it has now been established that a plethora of single genes with small effects each contribute to the genetic underpinnings of complex disorders.

In the following, evidence from genetic epidemiological studies will be briefly summarized. Genetic studies on conditionability in humans have mainly focused on genetic association studies in biologically plausible candidate genes, which will then be described in more detail.

9.3.1 Genetic epidemiological studies

Evidence from epidemiological studies, which investigate the role of genetic factors in families and populations, has clearly shown strong familial aggregation of all major subtypes of anxiety disorders (for comprehensive reviews and meta-analysis, see (17–19,96)).

Thereby, twin studies (see Chapter 6 of this volume) have attributed the major source of familial risk (i.e. 20–43%) to additive genetic factors, (17,97,98) which established a moderate genetic predisposition.

Similarly, strains of genetically diverse mice and rats behave differently in laboratory tests of anxiety and fear and to anxiolytic treatment (reviewed in (69)). Remarkably, a strain can be most anxious in one test but the least anxious in another, which might be attributed to different aspects of anxiety and an impact of different genetic factors (reviewed in (69)).

9.3.2 Genetic association studies

Association analyses investigate the relation between a phenotype and specific genetic variants and have become the predominant method in the past decades. Two approaches are employed: candidate gene studies and whole-genome association studies.

9.3.2.1 Candidate gene studies

Candidate gene studies have been at the forefront of genetic association studies in human fear conditioning. Studies have focused primarily on genes that are *a priori* implicated in fear and anxiety processing based on *biological plausibility* (21,99,100). Biological plausibility, for instance, is based on the genes' function and its supposed relevance to fear and anxiety processing, an approach which has been coined the "*candidate gene approach*" (see (101) for a critique). Once a target gene is identified, (functional) polymorphisms in this gene are selected and tested for a statistical association with fear-conditioning measures. In interpreting findings, it must be considered that neurotransmitter systems, such as the serotonin (5-HT) system, have widespread pleiotropic effects on biological processes. Hence, subtle changes in the functioning of this system, as induced by a single genetic polymorphism, can generally not be expected to be more specific than the systems general function.

In the following, two prime examples of the most reliable and prominent genetic association findings in the field are outlined. A comprehensive in-depth review of the experimental (see 21,22) and clinical (102) literature is beyond the scope of this chapter.

9.3.2.1.1 The serotonin transporter-linked polymorphic region Strong evidence implicates the serotonin (5-HT) system in emotional learning and memory (for a review, see (103)) in general and the acquisition and expression of fear in particular (for a review, see (104,105)). Major drugs used to treat patients suffering from pathological anxiety and depression target the 5-HT system, such as selective serotonin reuptake inhibitors (SSRIs), which block the functioning of the serotonin transporter (5-HTT). A functional polymorphism in the *5-HTT* promoter region (*5-HTT*LPR, where LPR refers to "linked polymorphic region") has been identified that comprises a 43bp ins/del polymorphism, resulting in a short (s) and a long (l) version that differ in transcriptional efficacy (106).

The low-efficacy *5-HTT*LPR S-allele has consistently been associated with facilitated fear conditioning in uninstructed (Pavlovian) designs (107–109), instructed fear learning (110,111), as well as observational fear learning (112), as assessed by SCRs and FPS in humans. Thereby, genotype-related activation of the dorsomedial prefrontal cortex during fear expression has been shown to mediate genotype-related individual differences in physiological responding (SCRs and FPS; 110). Furthermore, homozygous S-carriers displayed stronger differential (i.e. CS+ > CS–) activation in in a number of other areas of the neural network linked to fear processing such as the amygdala and/or the insula during fear conditioning (113,114), as well as the threat of shock experiments (115). Importantly, similar findings have been reported in rodents (116–118).

9.3.2.1.2 Brain-derived neurotrophic factor (*BDNF* val66met) Another prominent example is a functional single-nucleotide polymorphism in the pro-domain of the gene coding for brain-derived neurotrophic factor (*BDNF* val66met). There is a rich preclinical literature on BDNF effects and hippocampus, as well as amygdala-dependent learning in general (119–122), and on anxiety-like behavior in the open field test, elevated plus maze as well as fear acquisition and extinction particularly (123,124; e.g. 125–128)

In humans, a polymorphism in the *BDNF* gene, which codes for the replacement of valine by methionine at codon 66 in the *BDNF* gene transcript (*BDNF* Val66Met), has been identified which leads to reduced activity-dependent secretion of the BDNF protein (125,129). Thereby, the Met allele has been linked to attenuated fear acquisition and its retention, as assessed by FPS, as well as stronger CS discrimination in the amygdala and the subgenual anterior cingulate cortex in fMRI (130–132) (but see (133)). Furthermore, the Met allele has been linked to stronger fear generalization (134) and to deficits in extinction learning in rodents and humans (128), even though results derived from work in humans in this study are ambiguous (for a discussion, see (21)), and they have not been replicated (131–133). The *BDNF* Val66Met polymorphism has been linked to response to CBT, which heavily relies on the principles of extinction learning (47,133,135; see Chapter 23 of this volume).

Hence, there is ample evidence for a general impact of *BDNF* genetic variation on (fear) learning and memory derived from experimental work in rodents and humans.

9.3.2.2 Genome-wide association studies

The recent availability of high-throughput genotyping methods has nourished hopes for the identification of novel biological pathways and candidate genes through genome-wide association studies (GWAS; see Chapter 7 of this volume) in the past decade. GWAS investigate the association of a phenotype with a large number (> 500.000) of markers across the entire genome in a very large number (i.e. thousands) of individuals.

To date, several GWAS on anxiety-related traits or psychopathological conditions (136–139), however not on experimental fear conditioning, have yielded promising but also sometimes disappointing (i.e. limited genome-wide significance) and partly inconsistent results. The fact that heritability estimates outnumber the percentage of variance explained by already-identified genetic factors is thereby generally referred to as "missing heritability" (140,141). As no GWAS on fear conditioning is available to date, we refer to other sources for an overview on results derived from other related traits (for an overview, see (102)).

9.3.2.3 Gene–environment interaction

Gene–environment interaction (G×E) refers to an association between environmental experience and a phenotype that is contingent on a specific genotype (see Chapters 3 and 4 of this volume). In human fear conditioning research, for instance, the impact of the *5-HTT* LPR on neural activation of the insula and the occipital cortex during fear conditioning (142), and the amygdala during immediate extinction (113), has been shown to be modulated by life adversity, with similar findings in rodents (117,143,144).

Environmental factors become biologically embedded in part through epigenetic mechanisms (see Chapter 5). Recently, epigenetic processes contributing to the formation, storage, and updating of fear and extinction memories, as well as their clinical implications (145–147) and possible heritability (148), have been discussed. Corresponding data recently demonstrated epigenetic correlates of cognitive behavioral treatment outcome (149,150), which together represent a promising avenue for the development of new behavioral and pharmacological prevention and intervention programs.

9.4 Conclusions and future perspectives

In closing, studying the genetic basis of experimental fear acquisition, extinction, and return of fear has potential to deepen our mechanistic neurobiological understanding of risk and resilience trajectories to stress- and anxiety-related pathology.

Whereas initial work in humans followed hypothesis-driven candidate gene approaches, recent methodological advances such as the feasibility of large-scale sequencing- and array-based techniques, as well as new statistical developments and machine-learning algorithms, hold promise of advancing the field further. The interplay between these methodological advances and analysis of "big data" originating from translational collaborative research centers (151–153) or consortia (e.g. 55), all hold promise to aid unraveling the neurobiology of fear acquisition, extinction, and return of fear.

Synergistic cooperation between the fields of molecular genetics, neuroimaging, psychophysiology, and psychopharmacology will not only improve our mechanistic understanding but can also be expected to contribute to the development of specifically tailored (individualized) behavioral and pharmacological intervention and prevention programs.

References

1. **Fineberg NA, Haddad PM, Carpenter L, Gannon B, Sharpe R, Young AH**, et al. The size, burden and cost of disorders of the brain in the UK. J Psychopharmacol. 2013 Sep;**27**(9):761–70.
2. **Greenberg PE, Sisitsky T, Kessler RC, Finkelstein SN, Berndt ER, Davidson JR**, et al. The economic burden of anxiety disorders in the 1990s. J Clin Psychiatry. 1999 Jul;**60**(7):427–35.
3. **Kessler RC, Aguilar-Gaxiola S, Alonso J, Chatterji S, Lee S, Ormel J**, et al. The global burden of mental disorders: An update from the WHO World Mental Health (WMH) Surveys. Epidemiol Psichiatr Soc. 2009;**18**(1):23–33.
4. **Wittchen HU, Jacobi F, Rehm J, Gustavsson A, Svensson M, Jönsson B**, et al. The size and burden of mental disorders and other disorders of the brain in Europe 2010. Eur Neuropsychopharmacol. 2011 Sep;**21**(9):655–79.
5. **Watson JB.** The place of the conditioned-reflex in psychology. Psychological Review. 1916;**23**(2):89–116.
6. **Eelen P, Vervliet B.** Fear conditioning and clinical implications: what can we learn from the past. In: *Fear and Learning—From Basic Processes to Clinical Implications*. Washington, DC: American Psychological Association; 2006, pp. 17–35.
7. **LeDoux JE.** Emotion circuits in the brain. Annu Rev Neurosci. 2000;**23**:155–84.
8. **Milad MR, Quirk GJ.** Fear extinction as a model for translational neuroscience: Ten years of progress. Annu Rev Psychol. 2012;**63**:129–51.
9. **Maren S.** Neurobiology of Pavlovian fear conditioning. Annu Rev Neurosci. 2001;**24**:897–931.
10. **Watson JB, Rayner R.** Conditioned emotional reactions. J Exp Psychol. 1920;**3**(1):1–14.
11. **Cover Jones M.** Anticipatory and inhibitory characteristics of delayed conditioned reactions. J Exp Psychol. 1934;**17**(5):603–20.

12. **Jones MC.** A laboratory study of fear: The case of Peter. 1924. J Genet Psychol. 1991 Dec;**152**(4): 462–9.
13. **Barlow DH.** *Anxiety and Its Disorders: The Nature and Treatment of Anxiety and Panic*. London: Guilford Press; 2002.
14. **Kosslyn SM, Cacioppo JT, Davidson RJ, Hugdahl K, Lovallo WR, Spiegel D**, et al. Bridging psychology and biology. The analysis of individuals in groups. Am Psychol. 2002 May;**57**(5):341–51.
15. **Pavlov I.** *Conditioned Reflexes*. London: Oxford University Press; 1927.
16. **Pavlov IP.** Croonian Lecture: Certain problems in the physiology of the cerebral hemispheres. Proc R Soc Lond B. 1928 Jun 1;**103**(722):97–110.
17. **Hettema JM, Neale MC, Kendler KS.** A review and meta-analysis of the genetic epidemiology of anxiety disorders. Am J Psychiatry. 2001 Oct 1;**158**(10):1568–78.
18. **Weissman MM.** Family genetic studies of panic disorder. J Psychiatr Res. 1993;**27** Suppl 1:69–78.
19. **Lonsdorf TB, Baas JMP.** Genetics in experimental psychopathology: From laboratory models to therapygenetics. Where do we go from here? Psychopathology Review. 2015;**a4**(2):169–188.
20. **Lonsdorf TB, Kalisch R.** A review on experimental and clinical genetic associations studies on fear conditioning, extinction and cognitive-behavioral treatment. Transl Psychiatry. 2011;**1**:e41.
21. **Sumner JA, Powers A, Jovanovic T, Koenen KC.** Genetic influences on the neural and physiological bases of acute threat: A research domain criteria (RDoC) perspective. Am J Med Genet B Neuropsychiatr Genet. 2015 Sep 17;
22. **Lonsdorf TB, Menz MM, Andreatta M, Fullana MA, Golkar A, Haaker J**, et al. Don't fear "fear conditioning": Methodological considerations for the design and analysis of studies on human fear acquisition, extinction, and return of fear. Neuroscience & Biobehavioral Reviews [Internet]. [cited 2017 Mar 14]; Available from: <http://www.sciencedirect.com/science/article/pii/S0149763416308466>.
23. **LeDoux JE.** Coming to terms with fear. PNAS. 2014 Feb 25;**111**(8):2871–8.
24. **Ohman A, Mineka S.** Fears, phobias, and preparedness: Toward an evolved module of fear and fear learning. Psychol Rev. 2001 Jul;**108**(3):483–522.
25. **Seligman MEP.** Phobias and preparedness. Behavior Therapy. 1971 Jul;**2**(3):307–20.
26. **Bouton ME, Mineka S, Barlow DH.** A modern learning theory perspective on the etiology of panic disorder. Psychol Rev. 2001 Jan;**108**(1):4–32.
27. **Orr SP, Metzger LJ, Lasko NB, Macklin ML, Peri T, Pitman RK.** De novo conditioning in trauma-exposed individuals with and without posttraumatic stress disorder. J Abnorm Psychol. 2000 May;**109**(2):290–8.
28. **Davis M.** Are different parts of the extended amygdala involved in fear versus anxiety? Biol Psychiatry. 1998 Dec 15;**44**(12):1239–47.
29. **Davis M, Walker DL, Miles L, Grillon C.** Phasic vs sustained fear in rats and humans: Role of the extended amygdala in fear vs anxiety. Neuropsychopharmacology. 2010 Jan;**35**(1):105–35.
30. **Baas JM, Nugent M, Lissek S, Pine DS, Grillon C.** Fear conditioning in virtual reality contexts: A new tool for the study of anxiety. Biol Psychiatry. 2004 Jun 1;**55**(11):1056–60.
31. **Grillon C.** Startle reactivity and anxiety disorders: Aversive conditioning, context, and neurobiology. Biol Psychiatry. 2002;**52**:958–75.
32. **Grillon C, Baas JM, Cornwell B, Johnson L.** Context conditioning and behavioral avoidance in a virtual reality environment: Effect of predictability. Biol Psychiatry. 2006;**60**(7):752–9.
33. **Duits P, Cath DC, Lissek S, Hox JJ, Hamm AO, Engelhard IM**, et al. Updated meta-analysis of classical fear conditioning in the anxiety disorders. Depress Anxiety. 2015 Feb 1;**32**(4):239–53.
34. **Lissek S, Powers A, McClure E, Phelps E, Woldehawariat G, Grillon C**, et al. Classical fear conditioning in the anxiety disorders: A meta-analysis. Behav Res Ther. 2005 Nov;**43**(11):1391–424.
35. **Vervliet B, Raes F.** Criteria of validity in experimental psychopathology: Application to models of anxiety and depression. Psychol Med. 2013 Nov;**43**(11):2241–4.

36. **Scheveneels S, Boddez Y, Vervliet B, Hermans D.** The validity of laboratory-based treatment research: Bridging the gap between fear extinction and exposure treatment. Behav Res Ther. 2016 Nov;**86**:87–94.
37. **Anderson KC, Insel TR.** The promise of extinction research for the prevention and treatment of anxiety disorders. Biol Psychiatry. 2006 Aug 15;**60**(4):319–21.
38. **Bouton ME.** Context and behavioral processes in extinction. LearnMem. 2004;**11**(5):485–94.
39. **Myers KM, Davis M.** Mechanisms of fear extinction. Molecular Psychiatry. 2007;**12**(2):120–50.
40. **Maren S.** Nature and causes of the immediate extinction deficit: A brief review. Neurobiol Learn Mem. 2014 Sep;**113**:19–24.
41. **Yonkers KA, Bruce SE, Dyck IR, Keller MB.** Chronicity, relapse, and illness—course of panic disorder, social phobia, and generalized anxiety disorder: Findings in men and women from 8 years of follow-up. Depress Anxiety. 2003;**17**(3):173–9.
42. **Bouton ME.** Context, ambiguity, and unlearning: Sources of relapse after behavioral extinction. Biol Psychiatry. 2002 Nov 15;**52**(10):976–86.
43. **Vervliet B, Baeyens F, Van den Bergh O, Hermans D.** Extinction, generalization, and return of fear: A critical review of renewal research in humans. Biol Psychol. 2013 Jan;**92**(1):51–8.
44. **Vervliet B, Craske MG, Hermans D.** Fear Extinction and relapse: State of the art [Internet]. Rochester, NY: Social Science Research Network; 2013 Mar [cited 2013 Apr 24]. Report No.: ID 2242342. Available from: <http://papers.ssrn.com/abstract=2242342>.
45. **Scharfenort R, Menz M, Lonsdorf TB.** Adversity-induced relapse of fear: Neural mechanisms and implications for relapse prevention from a study on experimentally induced return-of-fear following fear conditioning and extinction. Transl Psychiatry. 2016 Jul 19;**6**(7):e858.
46. **Fitzgerald PJ, Seemann JR, Maren S.** Can fear extinction be enhanced? A review of pharmacological and behavioral findings. Brain Res Bull. 2014;**105**:46–60.
47. **Haaker J, Gaburro S, Sah A, Gartmann N, Lonsdorf TB, Meier K**, et al. Single dose of L-dopa makes extinction memories context-independent and prevents the return of fear. PNAS. 2013;**110**:E2428–36.
48. **Kindt M, Soeter M, Vervliet B.** Beyond extinction: erasing human fear responses and preventing the return of fear. Nature Neurosci. 2009;**12**:256–8.
49. **Schiller D, Monfils M-H, Raio CM, Johnson DC, LeDoux JE, Phelps EA.** Preventing the return of fear in humans using reconsolidation update mechanisms. Nature. 2010;**463**:49–53
50. **Bouton ME.** A learning theory perspective on lapse, relapse, and the maintenance of behavior change. Health Psychol. 2000 Jan;**19**(1 Suppl):57–63.
51. **Cross-Disorder Phenotype Group of the Psychiatric GWAS Consortium, Craddock N, Kendler K, Neale M, Nurnberger J, Purcell S**, et al. Dissecting the phenotype in genome-wide association studies of psychiatric illness. Br J Psychiatry. 2009 Aug;**195**(2):97–9.
52. **Green AE, Munafò MR, DeYoung CG, Fossella JA, Fan J, Gray JR.** Using genetic data in cognitive neuroscience: From growing pains to genuine insights. Nat Rev Neurosci. 2008 Sep;**9**(9):710–20.
53. **Noble EP.** The D2 dopamine receptor gene: A review of association studies in alcoholism and phenotypes. Alcohol. 1998 Jul;**16**(1):33–45.
54. **Cross-Disorder Group of the Psychiatric Genomics Consortium.** Identification of risk loci with shared effects on five major psychiatric disorders: A genome-wide analysis. Lancet. 2013 Apr 20;**381**(9875):1371–9.
55. **Glatt CE, Lee FS.** Common polymorphisms in the age of Research Domain Criteria (RDoC): Integration and rranslation. Biol Psychiatry. 2016 Jan 1;**79**(1):25–31.
56. **Insel T, Cuthbert B, Garvey M, Heinssen R, Pine DS, Quinn K**, et al. Research Domain Criteria (RDoC): Toward a new classification framework for research on mental disorders. Am J Psychiatry. 2010 Jul 1;**167**(7):748–51.
57. **Fredrikson M, Annas P, Hettema JM.** Different genetic factors underlie fear conditioning and episodic memory. Psychiatr Genet. 2015 Aug;**25**(4):155–62.

58. **Hettema JM, Annas P, Neale MC, Kendler KS, Fredrikson M.** A twin study of the genetics of fear conditioning. Arch Gen Psychiatry. 2003 Jul;**60**(7):702–8.
59. **Royce JR.** Avoidance conditioning in nine strains of inbred mice using optimal stimulus parameters. Behav Genet. 1972 Mar;**2**(1):107–10.
60. **Choi KH, Le T, McGuire J, Coyner J, Higgs BW, Diglisic S**, et al. Expression profiles of mitochondrial genes in the frontal cortex and the caudate nucleus of developing humans and mice selectively bred for high and low fear. PLoS ONE. 2012;**7**(11):e49183.
61. **de Castro Gomes V, Landeira-Fernandez J.** Amygdaloid lesions produced similar contextual fear conditioning disruption in the Carioca high- and low-conditioned freezing rats. Brain Res. 2008 Oct 3;**1233**:137–45.
62. **Ponder CA, Huded CP, Munoz MB, Gulden FO, Gilliam TC, Palmer AA.** Rapid selection response for contextual fear conditioning in a cross between C57BL/6J and A/J: behavioral, QTL and gene expression analysis. Behav Genet. 2008 May;**38**(3):277–91.
63. **Radcliffe RA, Lowe MV, Wehner JM.** Confirmation of contextual fear conditioning QTLs by short-term selection. Behav Genet. 2000 May;**30**(3):183–91.
64. **Imada H.** Emotional reactivity and conditionability in four strains of rats. J Comp Physiol Psychol. 1972 Jun;**79**(3):474–80.
65. **Smith DR, Gallagher M, Stanton ME.** Genetic background differences and nonassociative effects in mouse trace fear conditioning. Learn Mem. 2007 Jan 9;**14**(9):597–605.
66. **Chang C, Maren S.** Strain difference in the effect of infralimbic cortex lesions on fear extinction in rats. Behav Neurosci. 2010 Jun;**124**(3):391–7.
67. **Shumake J, Furgeson-Moreira S, Monfils MH.** Predictability and heritability of individual differences in fear learning. Anim Cogn. 2014 Sep;**17**(5):1207–21.
68. **Gordon JA, Hen R.** Genetic approaches to the study of anxiety. Annu Rev Neurosci. 2004;**27**:193–222.
69. **Schulze TG, McMahon FJ.** Defining the phenotype in human genetic studies: forward genetics and reverse phenotyping. Hum Hered. 2004;**58**(3–4):131–8.
70. **Shechner T, Hong M, Britton JC, Pine DS, Fox NA.** Fear conditioning and extinction across development: Evidence from human studies and animal models. Biol Psychol. 2014 Jul;**100**:1–12.
71. **Merrill KA, Steinmetz JE, Viken RJ, Rose RJ.** Genetic influences on human conditionability: A twin study of the conditioned eyeblink response. Behav Genet. 1999 Mar;**29**(2):95–102.
72. **Anokhin AP, Golosheykin S, Heath AC.** Genetic and environmental influences on emotion-modulated startle reflex: A twin study. Psychophysiology. 2007 Jan;**44**(1):106–12.
73. **Hasenkamp W, Epstein MP, Green A, Wilcox L, Boshoven W, Lewison B**, et al. Heritability of acoustic startle magnitude, prepulse inhibition, and startle latency in schizophrenia and control families. Psychiatry Res. 2010 Jul 30;**178**(2):236–43.
74. **Vaidyanathan U, Malone SM, Miller MB, McGue M, Iacono WG.** Heritability and molecular genetic basis of acoustic startle eye blink and affectively modulated startle response: A genome-wide association study. Psychophysiology. 2014 Dec;**51**(12):1285–99.
75. **Carlson SR, Katsanis J, Iacono WG, McGue M.** Emotional modulation of the startle reflex in twins: Preliminary findings. Biol Psychol. 1997 Oct 10;**46**(3):235–46.
76. **Grillon C, Dierker L, Merikangas KR.** Fear-potentiated startle in adolescent offspring of parents with anxiety disorders. Biol Psychiatry. 1998 Nov 15;**44**(10):990–7.
77. **Grillon C, Warner V, Hille J, Merikangas KR, Bruder GE, Tenke CE**, et al. Families at high and low risk for depression: A three-generation startle study. Biol Psychiatry. 2005 May 1;**57**(9):953–60.
78. **Falls WA, Carlson S, Turner JG, Willott JF.** Fear-potentiated startle in two strains of inbred mice. Behav Neurosci. 1997 Aug;**111**(4):855–61.
79. **Mccaughran JR.** JA, **Bell III J, Hitzemann RJ.** Fear-potentiated startle response in mice: Genetic analysis of the C57BL/6J and DBA/2J intercross. Pharmacol Biochem Behav. 2000 Feb;**65**(2):301–12.

80. **Manuck SB, Brown SM, Forbes EE, Hariri AR.** Temporal stability of individual differences in amygdala reactivity. Am J Psychiatry. 2007 Oct;**164**(10):1613–4.

81. **Blokland GAM, de Zubicaray GI, McMahon KL, Wright MJ.** Genetic and environmental influences on neuroimaging phenotypes: A meta-analytical perspective on twin imaging studies. Twin Res Hum Genet. 2012 Jun;**15**(3):351–71.

82. **Koten JW, Wood G, Hagoort P, Goebel R, Propping P, Willmes K,** et al. Genetic contribution to variation in cognitive function: An FMRI study in twins. Science. 2009 Mar 27;**323**(5922):1737–40.

83. **Hettema JM, Annas P, Neale MC, Fredrikson M, Kendler KS.** The genetic covariation between fear conditioning and self-report fears. Biol Psychiatry. 2008 Mar 15;**63**(6):587–93.

84. **Fredrikson M, Annas P, Georgiades A, Hursti T, Tersman Z.** Internal consistency and temporal stability of classically conditioned skin conductance responses. Biol Psychol. 1993 Apr;**35**(2):153–63.

85. **Zeidan MA, Lebron-Milad K, Thompson-Hollands J, Im JJY, Dougherty DD, Holt DJ,** et al. Test-retest reliability during fear acquisition and fear extinction in humans. CNS Neurosci Ther. 2012 Apr;**18**(4):313–7.

86. **Torrents-Rodas D, Fullana MA, Bonillo A, Andión O, Molinuevo B, Caseras X,** et al. Testing the temporal stability of individual differences in the acquisition and generalization of fear. Psychophysiology. 2014 Jul;**51**(7):697–705.

87. **Fanselow MS, LeDoux JE.** Why we think plasticity underlying Pavlovian fear conditioning occurs in the basolateral amygdala. Neuron. 1999 Jun;**23**(2):229–32.

88. **Fullana MA, Harrison BJ, Soriano-Mas C, Vervliet B, Cardoner N, Àvila-Parcet A,** et al. Neural signatures of human fear conditioning: An updated and extended meta-analysis of fMRI studies. Mol Psychiatry [Internet]. 2015 Jun 30 [cited 2015 Jul 22]; Available from: <http://www.nature.com/mp/journal/vaop/ncurrent/abs/mp201588a.html>.

89. **Herry C, Ferraguti F, Singewald N, Letzkus JJ, Ehrlich I, Lüthi A.** Neuronal circuits of fear extinction. Eur J Neurosci. 2010 Feb;**31**(4):599–612.

90. **Milad MR, Rauch SL.** The role of the orbitofrontal cortex in anxiety disorders. Ann N Y Acad Sci. 2007 Dec;**1121**:546–61.

91. **Sotres-Bayon F, Quirk GJ.** Prefrontal control of fear: More than just extinction. Curr Opin Neurobiol. 2010 Apr;**20**(2):231–5.

92. **Furmark T, Fischer H, Wik G, Larsson M, Fredrikson M.** The amygdala and individual differences in human fear conditioning. Neuroreport. 1997 Dec 22;**8**(18):3957–60.

93. **MacNamara A, Rabinak CA, Fitzgerald DA, Zhou XJ, Shankman SA, Milad MR,** et al. Neural correlates of individual differences in fear learning. Behav Brain Res. 2015;**287**:34–41.

94. **Meyer-Lindenberg A.** The future of fMRI and genetics research. Neuroimage. 2012 Aug 15;**62**(2):1286–92.

95. **Merikangas KR, Low NCP.** Genetic epidemiology of anxiety disorders. Handb Exp Pharmacol. 2005;(169):163–79.

96. **Hettema JM, Prescott CA, Myers JM, Neale MC, Kendler KS.** The structure of genetic and environmental risk factors for anxiety disorders in men and women. Arch Gen Psychiatry. 2005 Feb 1;**62**(2):182–9.

97. **Kendler KS.** Twin studies of psychiatric illness: An update. Arch Gen Psychiatry. 2001 Nov;**58**(11):1005–14.

98. **Kwon JM, Goate AM.** The candidate gene approach. Alcohol Res Health. 2000;**24**(3):164–8.

99. **Patnala R, Clements J, Batra J.** Candidate gene association studies: A comprehensive guide to useful in silico tools. BMC Genetics. 2013 May 9;**14**(1):39.

100. **Tabor HK, Risch NJ, Myers RM.** Candidate-gene approaches for studying complex genetic traits: Practical considerations. Nat Rev Genet. 2002 May;**3**(5):391–7.

101. **Erhardt A, Spoormaker VI.** Translational approaches to anxiety: Focus on genetics, fear extinction and brain imaging. Curr Psychiatry Rep. 2013 Dec;**15**(12):417.

102. **Meneses A, Liy-Salmeron G.** Serotonin and emotion, learning and memory. Rev Neurosci. 2012;**23**(5–6):543–53.
103. **Bauer EP.** Serotonin in fear conditioning processes. Behav Brain Res. 2015 Jan 15;**277C**:68–77.
104. **Homberg JR.** Serotonergic modulation of conditioned fear. Scientifica. 2012 Oct 24;**2012**:e821549.
105. **Heils A, Teufel A, Petri S, Seemann M, Bengel D, Balling U,** et al. Functional promoter and polyadenylation site mapping of the human serotonin (5-HT) transporter gene. J Neural Transm, General Section. 1995;**102**(3):247–54.
106. **Garpenstrand H, Annas P, Ekblom J, Oreland L, Fredrikson M.** Human fear conditioning is related to dopaminergic and serotonergic biological markers. Behav Neurosci. 2001 Apr;**115**(2):358–64.
107. **Julia Wendt JN.** Genetic influences on the acquisition and inhibition of fear. Int J Psychophysiol. 2014;
108. **Lonsdorf T, Weike A, Nikamo P, Schalling M, Hamm A, Ohman A.** Genetic Gating of human fear learning and extinction: Possible implications for gene–environment interaction in anxiety disorder. Psychological Science. 2009 Feb;**20**(2):198–206.
109. **Klumpers F, Kroes MC, Heitland I, Everaerd D, Akkermans SEA, Oosting RS,** et al. Dorsomedial prefrontal cortex mediates the impact of serotonin transporter linked polymorphic region genotype on anticipatory threat reactions. Biol Psychiatry. 2015 Oct 15;**78**(8):582–9.
110. **Klumpers F, Heitland I, Oosting RS, Kenemans JL, Baas JMP.** Genetic variation in serotonin transporter function affects human fear expression indexed by fear-potentiated startle. Biol Psychol. 2012 Feb;**89**(2):277–82.
111. **Crisan L, Pana S, Vulturar R, Heilman R, Szekely R, Druga B,** et al. Genetic contributions of the serotonin transporter to social learning of fear and economic decision making. Soc Cogn Affect Neurosci. 2009 Dec;**4**(4):399–408.
112. **Hermann A, Küpper Y, Schmitz A, Walter B, Vaitl D, Hennig J,** et al. Functional gene polymorphisms in the serotonin system and traumatic life events modulate the neural basis of fear acquisition and extinction. PLoS ONE. 2012;**7**(9):e44352.
113. **Klucken T, Schweckendiek J, Blecker C, Walter B, Kuepper Y, Hennig J,** et al. The association between the 5-HTTLPR and neural correlates of fear conditioning and connectivity. Soc Cogn Affect Neurosci. 2014 Aug 19;
114. **Drabant EM, Ramel W, Edge MD, Hyde LW, Kuo JR, Goldin PR,** et al. Neural mechanisms underlying 5-HTTLPR-related sensitivity to acute stress. Am J Psychiatry. 2012 Apr;**169**(4):397–405.
115. **Holmes A, Yang RJ, Lesch K-P, Crawley JN, Murphy DL.** Mice lacking the serotonin transporter exhibit 5-HT1A receptor-mediated abnormalities in tests for anxiety-like behavior. Neuropsychopharmacology. 2003 Sep 10;**28**(12):2077–88.
116. **Jansen F, Heiming RS, Lewejohann L, Touma C, Palme R, Schmitt A,** et al. Modulation of behavioural profile and stress response by 5-HTT genotype and social experience in adulthood. Behav Brain Res. 2010 Feb 11;**207**(1):21–9.
117. **Murphy DL, Lerner A, Rudnick G, Lesch K-P.** Serotonin transporter: gene, genetic disorders, and pharmacogenetics. Mol Interv. 2004 Apr;**4**(2):109–23.
118. **Ou L-C, Gean P-W.** Regulation of amygdala-dependent learning by brain-derived neurotrophic factor is mediated by extracellular signal-regulated kinase and phosphatidylinositol-3-kinase. Neuropsychopharmacology. 2006 Feb;**31**(2):287–96.
119. **Rattiner LM, Davis M, Ressler KJ.** Differential regulation of brain-derived neurotrophic factor transcripts during the consolidation of fear learning. Learn Mem. 2004 Dec;**11**(6):727–31.
120. **Rattiner LM, Davis M, Ressler KJ.** Brain-derived neurotrophic factor in amygdala-dependent learning. Neuroscientist. 2005 Aug;**11**(4):323–33.
121. **Tyler WJ, Alonso M, Bramham CR, Pozzo-Miller LD.** From acquisition to consolidation: on the role of brain-derived neurotrophic factor signaling in hippocampal-dependent learning. Learn Mem. 2002 Oct;**9**(5):224–37.

122. **Andero R, Ressler KJ.** Fear extinction and BDNF: Translating animal models of PTSD to the clinic. Genes Brain Behav. 2012 Jul;**11**(5):503–12.

123. **Autry AE, Monteggia LM.** Brain-derived neurotrophic factor and neuropsychiatric disorders. Pharmacol Rev. 2012 Jan 4;**64**(2):238–58.

124. **Chen Z-Y, Patel PD, Sant G, Meng C-X, Teng KK, Hempstead BL,** et al. Variant brain-derived neurotrophic factor (BDNF) (Met66) alters the intracellular trafficking and activity-dependent secretion of wild-type BDNF in neurosecretory cells and cortical neurons. J Neurosci. 2004 May 5;**24**(18):4401–11.

125. **Cunha C, Brambilla R, Thomas KL, Cunha C, Brambilla R, Thomas KL.** A simple role for BDNF in learning and memory? Front Mol Neurosci. 2010;3:1.

126. **Martinowich K, Manji H, Lu B.** New insights into BDNF function in depression and anxiety. Nat Neurosci. 2007 Sep;**10**(9):1089–93.

127. **Soliman F, Glatt CE, Bath KG, Levita L, Jones RM, Pattwell SS,** et al. A genetic variant BDNF polymorphism alters extinction learning in both mouse and human. Science. 2010 Feb 12;**327**(5967):863–6.

128. **Egan MF, Kojima M, Callicott JH, Goldberg TE, Kolachana BS, Bertolino A,** et al. The BDNF val66met polymorphism affects activity-dependent secretion of BDNF and human memory and hippocampal function. Cell. 2003 Jan 24;**112**(2):257–69.

129. **Hajcak G, Castille C, Olvet DM, Dunning JP, Roohi J, Hatchwell E.** Genetic variation in brain-derived neurotrophic factor and human fear conditioning. Genes Brain Behav. 2009 Feb;**8**(1):80–5.

130. **Lonsdorf T, Weike A, Golkar A, Schalling M, Hamm A, Ohman A.** Amygdala-dependent fear conditioning in humans is modulated by the BDNFval66met polymorphism. Behav Neurosci. 2010 Feb;**124**(1):9–15.

131. **Lonsdorf TB, Golkar A, Lindström KM, Haaker J, Öhman A, Schalling M,** et al. BDNFval66met affects neural activation pattern during fear conditioning and 24 h delayed fear recall. Soc Cogn Affect Neurosci. 2014 Aug 7;nsu102.

132. **Fullana MA, Alonso P, Gratacòs M, Jaurrieta N, Jiménez-Murcia S, Segalàs C,** et al. Variation in the BDNF Val66Met polymorphism and response to cognitive-behavior therapy in obsessive-compulsive disorder. Eur Psychiatry. 2012 Jul;**27**(5):386–90.

133. **Mühlberger A, Andreatta M, Ewald H, Glotzbach-Schoon E, Tröger C, Baumann C,** et al. The BDNF Val66Met polymorphism modulates the generalization of cued fear responses to a novel context. Neuropsychopharmacology. 2013 Nov 19;**39**(5):1187–95.

134. **Felmingham KL, Dobson-Stone C, Schofield PR, Quirk GJ, Bryant RA.** The brain-derived neurotrophic factor Val66Met polymorphism predicts response to exposure therapy in posttraumatic stress disorder. Biol Psychiatry. 2013 Jun;**73**(11):1059–63.

135. **Erhardt A, Czibere L, Roeske D, Lucae S, Unschuld PG, Ripke S,** et al. TMEM132D, a new candidate for anxiety phenotypes: evidence from human and mouse studies. Mol Psychiatry. 2011 Jun;**16**(6):647–63.

136. **Erhardt A, Akula N, Schumacher J, Czamara D, Karbalai N, Müller-Myhsok B,** et al. Replication and meta-analysis of TMEM132D gene variants in panic disorder. Transl Psychiatry. 2012;**2**:e156.

137. **Otowa T, Maher BS, Aggen SH, McClay JL, van den Oord EJ, Hettema JM.** Genome-wide and gene-based association studies of anxiety disorders in European and African American samples. PLoS One [Internet]. 2014 Nov 12 [cited 2016 Jun 22];**9**(11). Available from: <http://www.ncbi.nlm.nih.gov/pmc/articles/PMC4229211/>.

138. **Deckert J, Weber H, Villmann C, Lonsdorf TB, Richter J, Andreatta M,** et al. GLRB allelic variation associated with agoraphobic cognitions, increased startle response and fear network activation: A potential neurogenetic pathway to panic disorder. Mol Psychiatry. 2017 Feb 7;

139. **Eichler EE, Flint J, Gibson G, Kong A, Leal SM, Moore JH,** et al. Missing heritability and strategies for finding the underlying causes of complex disease. Nat Rev Genet. 2010 Jun;**11**(6):446–50.

140. **Manolio TA, Collins FS, Cox NJ, Goldstein DB, Hindorff LA, Hunter DJ**, et al. Finding the missing heritability of complex diseases. Nature. 2009 Oct 8;**461**(7265):747–53.

141. **Klucken T, Alexander N, Schweckendiek J, Merz CJ, Kagerer S, Osinsky R**, et al. Individual differences in neural correlates of fear conditioning as a function of 5-HTTLPR and stressful life events. Soc Cogn Affect Neurosci. 2013 Mar;**8**(3):318–25.

142. **Carola V, Frazzetto G, Pascucci T, Audero E, Puglisi-Allegra S, Cabib S**, et al. Identifying molecular substrates in a mouse model of the serotonin transporter x environment risk factor for anxiety and depression. Biol Psychiatry. 2008 May 1;**63**(9):840–6.

143. **Lewejohann L, Kloke V, Heiming RS, Jansen F, Kaiser S, Schmitt A**, et al. Social status and day-to-day behaviour of male serotonin transporter knockout mice. Behav Brain Res. 2010 Aug 25;**211**(2):220–8.

144. **Kwapis JL, Wood MA.** Epigenetic mechanisms in fear conditioning: Implications for treating post-traumatic stress disorder. Trends Neurosci. 2014 Dec;**37**(12):706–20.

145. **Stafford JM, Maughan DK, Ilioi EC, Lattal KM.** Exposure to a fearful context during periods of memory plasticity impairs extinction via hyperactivation of frontal-amygdalar circuits. Learn Mem. 2013;**20**(3):156–63.

146. **Zovkic IB, Sweatt JD.** Epigenetic mechanisms in learned fear: Implications for PTSD. Neuropsychopharmacology. 2013 Jan;**38**(1):77–93.

147. **Dias BG, Ressler KJ.** Parental olfactory experience influences behavior and neural structure in subsequent generations. Nat Neurosci. 2014 Jan;**17**(1):89–96.

148. **Roberts S, Lester KJ, Hudson JL, Rapee RM, Creswell C, Cooper PJ**, et al. Serotonin tranporter methylation and response to cognitive behaviour therapy in children with anxiety disorders. Transl Psychiatry. 2014 Oct 14;**4**(10):e467.

149. **Ziegler C, Richter J, Mahr M, Gajewska A, Schiele MA, Gehrmann A**, et al. MAOA gene hypomethylation in panic disorder-reversibility of an epigenetic risk pattern by psychotherapy. Transl Psychiatry. 2016;6:e773.

150. **A Reif JR.** MAOA and mechanisms of panic disorder revisited: from bench to molecular psychotherapy. Mol Psychiatry. 2014;**19**(1):122–8.

151. **Kuhn M, Haaker J, Glotzbach-Schoon E, Schümann D, Andreatta M, Mechias M-L**, et al. Converging evidence for an impact of a functional NOS gene variation on anxiety-related processes. Soc Cogn Affect Neurosc [Internet]. 2016 Jan 8; Available from: <http://scan.oxfordjournals.org/content/early/2016/01/08/scan.nsv151.abstract>.

152. **Straube B, Reif A, Richter J, Lueken U, Weber H, Arolt V**, et al. The functional-1019C/G HTR1A polymorphism and mechanisms of fear. Transl Psychiatry. 2014;**4**:e490.

Chapter 10

Emotional action control: The role of serotonin in health and disease

Inge Volman, Hanneke Den Ouden, and Karin Roelofs

10.1 Introduction

Emotionally salient situations elicit strong biases for approach or avoidance behavior (1–3). For example, seeing your child may elicit a strong urge to approach him/her, whereas an aggressive person makes you want to avoid the scene. Emotional biases likely rely on computational shortcuts, potentially hard-wired evolutionary old response mechanisms (4). It is suggested that they have evolved from early mechanisms facilitating exploration of resources as well as harm avoidance (5).

Crucially, sometimes these biases are inappropriate, for example when a parent needs to let their child explore the world and become independent, or an individual must approach a dangerous situation when the threat it poses affects a loved one. Importantly, emotional biases prime an individual's behavior but they do not always help an individual in reaching a given goal; one must not always follow them blindly. The latter, also known as instrumental action plans, consider the costs and potential consequences of the situation (6). When emotional biases and instrumental actions pull in opposite directions, that is, one elicits approach and the other avoidance, then the two systems are in conflict (7–9). Emotion control plays a crucial role in coordinating approach–avoidance responses in these situations, often mediating reduced influence of emotional predispositions and enhancement of the instrumental action strength.

This chapter focuses on the impact of emotions on instrumental, goal-directed behavior as well as the ability to overcome or control the influence of emotional predispositions. We will first discuss the behavioral effects of emotion on action and control thereof, then address its genetic basis, and relate this to associated neuropsychiatric disorders with known heritability. We will focus on the neurotransmitter serotonin (5-HT), which appears to play a key role in increased risk for psychopathologies related to emotion processing (10–14). We will end with a research agenda to advance this emerging field of emotional action control.

10.2 Emotion and action biases

Influential theories have suggested that emotions are reactions to salient events that prepare action biases and action readiness (15, 16). More recent componential emotion theories have argued that when exposed to an emotional event, the individual evaluates this on several dimensions, including intrinsic pleasantness. This component is regarded as most relevant to the initiation of fast action biases (17, 18). Intrinsic pleasantness relates to whether a stimulus itself is in general (sub)consciously considered to result in a positive or negative outcome, independent of the individual's goal (19). The underlying evaluative process of emotions is called appraisal, a process that triggers physiological reactions and action biases, which in turn feeds back into the appraisal process (17, 20). According to these appraisal theories, emotional approach and avoidance behaviors exemplify action biases following the intrinsic pleasantness evaluation.

A positive evaluation of an object or situation typically elicits approach, while negative evaluations trigger avoidance or withdrawal (1–3, 20). A striking example was described by Darwin (1872), who consciously decided not to make a withdrawal movement if a viper snake struck at him. The snake was behind a thick glass plate in a zoo therefore the situation did not create any real danger, but Darwin wrote: "As soon as the blow was struck, my resolution went for nothing, and I jumped a yard or two backwards with astonishing rapidity" (21). Fast affective approach and avoidance actions occur following an automatic and possibly innate response program (15, 17, 21–23). These behavioral biases crucially predispose most, if not all, emotional actions to either approach or avoidance.

Emotional response biases can be experimentally studied in humans using so-called affective "approach–avoidance tasks" (AA tasks) (1–3, 24). Typically, participants are instructed to respond to emotional stimuli by making approach and avoidance movements with their arm, leading to a facilitation of so-called congruent stimulus–action mappings (with medium effect sizes (25)).

Another well-known behavioral dimension elicited by emotionally valenced stimuli is the increased invigoration of action elicited by a rewarding stimulus and behavioral inhibition under threat of punishment (so-called Pavlovian biases). Emotional approach and avoidance biases as well as emotional invigoration and inhibition are often considered to be similar systems (26). Importantly though, avoidance reflects action and can thereby exclusively be considered not to be similar to inhibition of action; avoidance might even reflect an additional behavioral dimension. This difference might seem trivial in current experiments of emotional action biases and result in similar observable behavioral outcomes, but it is likely to involve different underlying mechanisms. Deakin (27) proposes that this might be a potential difference between fight (for proximal threat) and freeze (for distal threat), most likely subserved by different serotonergic systems. Freeze/fight/flight studies (28) or tasks that separate approach–avoidance and invigoration–inhibition on a trial-by-trial basis might separate these effects.

10.3 **Emotion-driven action tendencies affect instrumental behavior**

While action tendencies elicited by emotional cues allow us to respond rapidly and usually adequately, they may also interfere with ongoing, instrumental behavior (Figure 10.1). This can be observed in situations where the action tendency runs counter to the instrumental goal at hand. For example, in a series of studies across species, Breland and Breland (23) conditioned animals to pick up and deposit an object in order to obtain food. As the object became associated with the food outcome, the animals developed problems with completing this task, displaying their typical food consumption behaviors on the object itself instead of food (e.g. "washing" the object in raccoons, "rooting" the object in pigs). The animals, ranging from cats to dogs to cows to porpoises, were "trapped" by strong instinctive behaviors, interfering with the goal of actually obtaining the food reward.

The influence of these response tendencies on instrumental behavior was quantified using so-called Pavlovian-to-Instrumental transfer (PIT) paradigms. PIT paradigms combine Pavlovian and instrumentally conditioned cues to measure the degree to which instrumental decisions are affected by the presentation of emotionally valenced, but instrumentally irrelevant, Pavlovian stimuli (Figure 10.1b) (29–31). The effect of the emotional response bias on goal-directed action is observed when positively/negatively conditioned Pavlovian stimuli invigorate/inhibit lever-pressing on an instrumental task. Stimuli with potentially innate values, such as emotional faces,

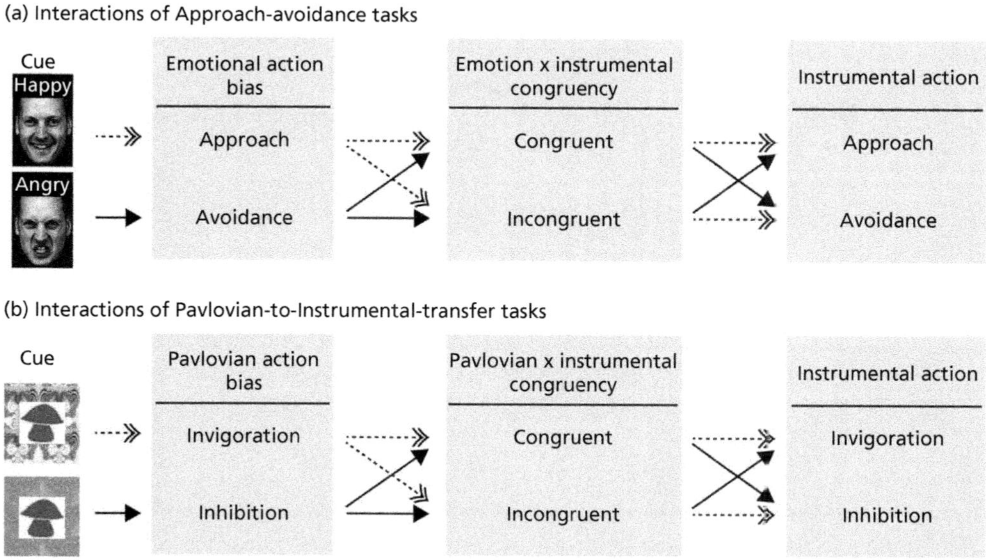

Figure 10.1 Examples of emotional action control tasks. (A) Approach-avoidance tasks: positive (here: happy faces) and negative (here: angry faces) elicit emotional action tendencies to approach or avoid, respectively. These biases can be congruent or incongruent with the instrumental instruction. Instrumental action that are incongruent with the emotional action bias require more cognitive control and take longer to execute (7). (B) Pavlovian-to-instrumental-transfer tasks: instrumental stimuli (here: mushroom) are presented simultaneously with positive or negative background (arbitrary stimuli that have acquired valence through conditioning). This background elicits a certain Pavlovian action bias. This can be congruent or incongruent with the pre-learned instrumental response (based on mushroom type), eventually leading to an invigoration (Go) or inhibition (NoGo) response for a correct answer.

can influence instrumental behavior in much the same way (32, 33). For example, during a probabilistic learning task, participants approached and avoided targets to obtain monetary reward versus punishment. Ly and colleagues (33) found that angry (versus happy) face primes preceding an instrumental cue induced an avoidance (versus approach) bias on the instrumental response. In addition, the magnitude of the freezing reaction (putatively reflecting a form of Pavlovian inhibition) induced by the faces was significantly correlated to the approach–avoidance bias. This study combines features of AA tasks (Figure 10.1a) and PIT tasks (Figure 10.1b) allowing tests of approach–avoidance biases as well as action inhibition.

10.4 **Emotional control**

The ability to detect when emotional biases are interfering with our goals, and consequently inhibit or modify them, enables flexible control of our behavior (35, 36). The control of emotional behavior regulates the influence of emotional biases in order to implement specific instrumental goals. Emotional action biases elicit bottom-up emotional influences on behavior, whereas emotional control triggers top-down processes to override the emotional action biases.

The ability to exert emotional control can be observed during incongruent conditions of tasks in which instrumental demands are interfering with the emotion/Pavlovian-elicited action biases (Figure 10.1). For example, during incongruent conditions of the approach–avoidance task,

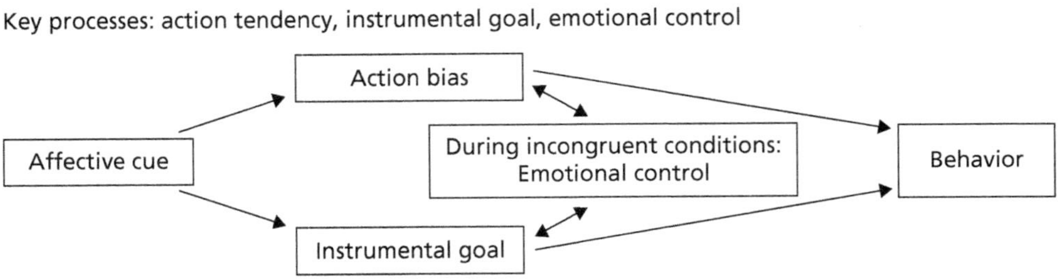

Figure 10.2 Overview of three main processes underlying emotional action control: (I) response biases, such as tendencies to avoid negative and approach positive situations; (II) instrumental goal to approach, avoid, invigorate, or inhibit; and (III) during incongruent conditions, emotional control. These processes together lead to the observed behavioral outcome, which would be the instrumental action on correct trials of AA-tasks and PIT tasks (7, 8, 34).

participants are instructed to avoid positive and approach negative situations. In order to overcome the emotional action biases for positive-approach and negative-avoid and to enable swift behavior, the biases need to be suppressed and instrumental instructions facilitated (7). The importance of this top-down coordination mechanism has been illustrated by neurostimulation. Our study showed that reduction of prefrontal-driven emotion action control led to a stronger influence of emotional biases and more errors during incongruent trials (8).

Importantly, emotional control is a relatively unacknowledged aspect of fast emotional action implementation. This is most likely to be due to the fact that action biases, instrumental goals, and emotional control (Figure 10.2) temporarily overlap, making it difficult to disentangle the influence of each component on the observed behavior (Figure 10.2). Studies investigating the neural mechanisms underlying emotional actions have begun to look into the specifics of emotional control (Box 10.1).

10.5 **Serotonin and emotional biasing of action**

5-HT plays an important role during the regulation of social and emotional processes and behaviors (27, 37, 38). 5-HT most likely has a critical function of eliciting avoidance behavior as well as active inhibition (39). It has been suggested that it specifically affects negative avoidance biases, as well as reduces behavioral adaptations after punishment (38, 40), although there is also evidence for an influence of positive reinforcement (39, 41). 5-HT reuptake inhibitors are used for treatment of a range of emotional disorders, although not always effectively (14). Importantly, a genetic variation of 5-HT transporter availability (*5-HTTLPR*) predisposes to the development of social psychopathologies, such as anxiety-, depression-, and aggression-related disorders, especially after a strong emotional event (10–13). Together, this suggests that the 5-HT pathway is a prime candidate to influence emotional action control.

Acute tryptophan depletion (ATD) is a dietary intervention to deplete the 5-HT precursor tryptophan, reducing precursor availability and, as a result, lowering central levels of 5-HT (42). Several ATD studies have shown that it affects aspects of emotional action control (38). For example, ATD leads to a disinhibition of responses to aversive stimuli, as it abolished aversive PIT, that is, the inhibition of ongoing instrumental behavior by negatively conditioned stimuli (29). We also showed recently that ATD released action inhibition during high motivational situations irrespective of motivational direction (43), whereas other studies have shown a specific punishment-induced disinhibition (44–45, although see 46). Taken together, these ATD studies suggest that lowering central levels of 5-HT leads to a release of motivated (aversive) inhibition (29).

10.6 Genetic variation of the serotonergic pathway

The most widely reported genetic modulation of the serotonergic pathway in humans is the functional polymorphism *5-HTTLPR*. This variation on the 5-HT transporter gene (*SLC6A4*) is located on chromosome 17. The transcriptional activity of *SLC6A4* is modulated by a repetitive sequence in the promoter region, the *5-HTTLPR* polymorphism. This sequence consists of two alleles, which can be a long (L) or short (S) version. The length of each allele specifies the amount of transcribed messenger (m)RNA and 5-HT transporter proteins (47). The 5-HT transporter facilitates the reuptake of extracellular and synaptic 5-HT, thereby affecting 5-HT interaction with postsynaptic receptors. A repetitive sequence containing one or two short alleles is associated with a reduced availability and functioning of the 5-HT transporter, as compared to those homozygous for the L-allele (LL- or non-carriers). Furthermore, these S-allele carriers (heterozygous or homozygous) putatively show significantly reduced 5-HT reuptake (48).

There has been little behavioral research into emotion action control in relation to *5-HTTLPR* variability. Interesting though, manipulating positive and negative framing of value-based decisions (eliciting a framing bias) affected the choice made by individuals homozygous for the S-allele. Framing biases are observed in tasks where people make choices between receiving a definite amount of money or taking a gamble. When the option is phrased in terms of gains, people are biased toward choosing the former, yet when the decision is phrased in terms of losses, they choose the latter (the frame effect). Responses in line with this framing bias can be conceptualized as the "congruent" response, in contrast to responses going against the framing bias (i.e. the certain option when playing to avoid a loss, and the uncertain option for the gain frame). In this study, S-allele carriers showed a stronger framing bias than L-homozygotes (49).

Moreover, this enhanced framing bias was associated with greater amygdala activation when following, rather than going counter to, the framing bias. In addition, genotype modulated frontal (ACC)-amygdala coupling, with L-homozygotes showing larger coupling than S-carriers when overcoming the framing bias (49). This conforms to other findings: that S-carriers of the *5-HTTLPR* polymorphism are characterized by enhanced amygdala processing during threat, as well as altered functional and anatomical connectivity between the amygdala and PFC (10, 50–54). Similarly, we showed that when emotional control is needed during the AA task, S-allele carriers show enhanced amygdala activity and reduced inhibitory regulation by the aPFC to the amygdala (Box 10.1). Moreover, the dorsal medial PFC has been shown to mediate the relation between stronger threat anticipation and endogenous action biases (psychophysiological responses to threat; i.e. skin conductance, startle responses) in S-allele carriers (55), suggesting a potential mechanism for enhanced anxiety responses as a function of *5-HTTLPR*.

Alterations in emotional action control could potentially underlie the increased vulnerability of developing emotional disorders (see also Box 10.1). Studying the mechanisms underlying emotion action control linked with this polymorphism might therefore be important to disentangle the precise mechanisms behind genetic effects (54). For example, in a prospective longitudinal study, alterations in freezing reactions in 5-HTTLPR S-allele carrier infants (too long or absence of freezing) were associated with increased internalizing symptoms in early and late adolescence, respectively (122).

A recent dynamic causal modeling study confirmed this mechanism to account for group differences in neural control over emotional action in relation to the *5-HTTLPR* polymorphism. We tested whether increased amygdala reactivity to emotional faces previously observed in S-allele carriers (50) might be associated with a reduced inhibitory control of the aPFC (54). Twenty-four S-allele carriers and 24 L-homozygotes completed the AA task (Figure 10.1) in the magnetic resonance (MR) scanner, after which directionality of emotional processing was modeled.

Box 10.1 Neural mechanisms underlying emotional action control and 5-HTTLPR

Emotional action control studied using approach–avoidance tasks provides evidence for a consistent and important function of the lateral anterior prefrontal cortex (aPFC). In a line of studies, we indicate that, after receiving information on the context, the aPFC plays a crucial role during emotional action control to bias processing towards rule-driven (instrumental) areas in the parietal cortex and away from areas generating emotional biases, such as the amygdala (7, 8, 34, 54). Inhibition of the amygdala by the aPFC seemed especially relevant during emotional action control, most likely diminishing the influence of emotional action biases (54).

Importantly, when emotional biases need to be overridden by another action (e.g. approach threat instead of avoid threat), S-allele carriers show enhanced amygdala activity and reduced inhibitory regulation by the aPFC to the amygdala (Figure 10.3). This suggests that prefrontal emotional control acts specifically on emotional biases driven by the amygdala. Interestingly, this effect is similar to findings in psychopathic criminals during the same task, namely reduced communication between the aPFC and amygdala during emotional action control (54). The striking similarity of networks involved in the S-allele carriers and the psychopathic criminals confirm that prefrontal inhibition of the amygdala is an important mechanism during emotional control and a reduction thereof is a risk factor for the development of social psychopathologies. This can be corroborated by multiple studies testing the neural mechanisms of a diverse range of psychopathologies during emotional processing/behavior (56–59).

Other genetic variants affecting 5-HT function might well influence emotional action control. However, to date research on this is very limited. The STTP/rs3813034 variant, a polymorphism of 5-HT transporter polyadenylation, is such a genetic variation affecting 5-HT. So-called G-allele

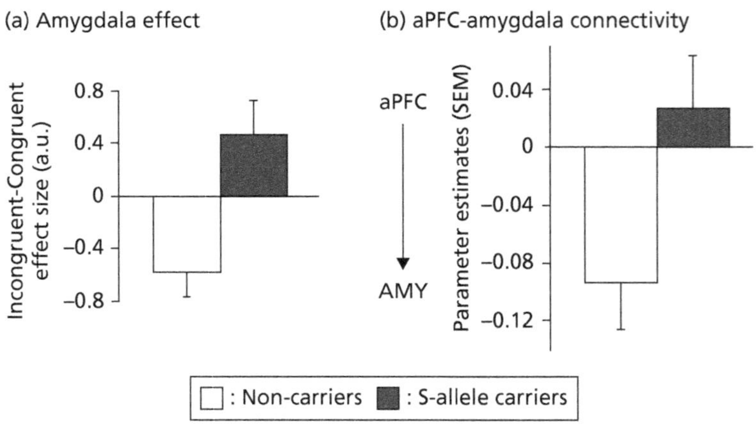

Figure 10.3 (A) Increased amygdala activity and (B) reduced inhibition from aPFC to amygdala in S-allele carriers compared to homozygote *ll*-carriers (non-carriers) during emotional action control (incongruent—congruent conditions of the AA-task). aPFC-amygdala connectivity was assessed using dynamic causal modeling (60). Figure from (54).

carriers have reduced 5-HT transporter levels and show an association with heightened anxiety and depressive symptoms, as well as impaired fear-related processing (61). Also, initial animal studies have tested effects of 5-HT variability using genetic knockout models. These studies provide additional indications of genetic 5-HT effects on emotional action control. For example, mice with genetic variants that cause lower 5-HT brain function show more persistent pre-learned Pavlovian bias behavior counter to their goals compared to those with more normal 5-HT function (62). This behavioral pattern has been supported by neurotoxic lesions causing 5-HT depletion and has been linked to poor inhibitory control of behavior (63).

10.7 Genetic vulnerability for psychiatric disorders

The 5-*HTTLPR* has been suggested to predispose to the development of social psychopathologies, such as anxiety-, depression-, and aggression-related disorders, especially after a strong emotional event (10, 11, 13), although see (64, 65). For example, in a seminal paper, Caspi and colleagues showed that the amount of stressful life events influences the occurrence and severity of depression in *5-HTTLPR* S-allele carriers (66). Alterations in the control of emotions could potentially underlie this increased vulnerability (37). Studying (the neural correlates underlying) emotional action control in this polymorphism has therefore also been suggested to be important to disentangle the precise mechanisms behind these gene–environment interactions (37).

Patients with a social anxiety disorder (SAD) experience extreme fear of negative evaluations by others in social situations and show strong tendencies to avoid social situations (67). Even if their instrumental goal is set to approach a social situation, they have difficulties overriding their emotional avoidance biases (68, 69). The AA task has been considered as a measure of in-situation avoidance biases in socially anxious individuals, showing increased reaction times when approaching emotional stimuli (68, 69). A new promising intervention for SAD treatment seems to be retraining of emotional action biases, a method focused on increasing social approach (70).

Depressed patients show a reduced influence of Pavlovian stimuli on instrumental actions or even an absence of normal positive-approach and negative-avoidance biases. Critically, depressive patients that showed enhanced Pavlovian influences on avoidance behavior specifically showed better recovery (71). Genetic factors including serotonergic variants make a strong contribution to the comorbidity between affective and anxiety disorders (72). Interestingly, up to 60% of anxiety or depressive patients suffer from depressive/anxiety comorbidity (73, 74). A currently common treatment of anxiety and depression is the administration of serotonergic-based medication, like 5-HT-reuptake inhibitors. However, they still have limited efficacy, for reasons not sufficiently known.

Patients with aggression-related disorders also have problems with emotional action control (75), being characterized by increased social approach behavior (76), leading to an increased chance of violent encounters. In humans, the S-allele of the *5-HTTLPR* polymorphism has been repeatedly shown to affect risk for aggression and violence (77–79), although one study showed an association between the L-allele and aggression (80). Environmental factors seem to contribute to these effects (81–83). Epigenetic effects of this gene are beginning to be explored, suggesting enhancing effects of childhood aggression on 5-HT transporter gene methylation (84). However, the relation between 5-HT and aggression is complex. Genetic rodent models of reduced as well as excessive 5-HT synthesis can both lead to aggressive phenotypes, which might be due to different developmental pathways (85, 86). Developmental pharmacologic 5-HT transporter manipulation studies show that prenatal or postnatal inhibition of 5-HT results in either increased or decreased aggression displayed when the animal is mature (with decreased versus increased anxiety behavior), respectively (87, 88). Different 5-HT receptors may play different roles in aggression, however the exact relations remain unclear (89, 90).

Similarly, the relation between 5-HT polymorphisms, aggression, and emotional action control remains unexplored. Emotional action control tasks have shown that individuals with psychopathy exhibit reduced avoidance tendencies corroborative of severe abnormalities during every day interactions, such as increased aggressive behavior in social situations (76, 91). Furthermore, we found that the neural network underlying emotional action control was different in psychopathic criminals compared to healthy controls (92). In this study, 15 psychopathic criminals and 19 matched healthy controls performed an AA task in the MR scanner. While the healthy participants displayed the normal pattern of increased aPFC response and negative coupling between the aPFC and amygdala during emotional control, this effect was reduced in psychopathic individuals, a pattern highly similar to that previously observed in S-allele carriers (54). Especially in those psychopathic individuals with high testosterone values, an important modulator of social emotional behavior (93–96), showed reduced aPFC and aPFC-amygdala coupling during emotional control (92).

10.8 **Other neurotransmitter/hormonal systems and emotional action control**

5-HT interactions with hormones and other neurotransmitters, such as testosterone and dopamine, have an influence on emotional action control as well. Especially low levels of serotonin and high testosterone levels are associated with increased aggression in humans (97, 98). Testosterone and 5-HT have been found to interact, leading to altered aggressive behavior (97–100), and testosterone administration facilitated amygdala reactivity during emotional control of actions in response to social threat (101). The triple imbalance hypothesis of reactive aggression might have additional explanatory power here as well. This hypothesis suggests that individuals who show reactive aggression in social situations have altered levels of testosterone and cortisol. High levels of endogenous testosterone and low levels of endogenous cortisol together with low levels of cortical 5-HT predispose individuals to reactive aggression (113). S-allele carriers of the *5-HTTLPR* polymorphism might be at particular risk of showing aggressive behavior if accompanied by a high testosterone/cortisol ratio, and anxiety if accompanied by low testosterone/cortisol ratio. However, this remains hypothetical and would need to be tested in experimental settings.

The neurotransmitter dopamine shows involvement in affective motivational processes leading to behavioral invigoration as well as general PIT influences (102). For example, enhancing dopamine levels using levodopa enhances Pavlovian approach behavior in a risky choice task (103). Similarly, genetic variability of dopamine transporter (DAT gene) affects the strength of emotional approach–avoidance biases on action outcome (104).

10.9 **Outlook**

In sum, it is of theoretical as well as clinical relevance to gain more insight into the serotonergic and genetic mechanisms underlying emotional action control. Several lines of research will be important to extend this goal in the future, including translation from subtle experimental paradigms to stronger emotional provocations, establishing the interaction details between emotional biases, instrumental planning, and emotional control as well as testing the alterations in patient populations. Emotional action control might be of crucial influence to understand the interactions between 5-HT and emotional psychopathologies. This review extends animal research, showing that also in humans the serotonergic system is critically involved in behavioral flexibility by facilitating inhibition of suboptimal responses.

Tasks studying emotional action control generally use a subtle affective provocation (29, 43, 44, 92, 105), and how these translate to more intense emotional interactions that may occur in everyday life remains to be seen. Animal models and studies in this chapter suggest that when prefrontal inhibition of the amygdala turns out to be insufficient, a stronger influence of emotional action biases on behavior is likely (54, 92, 106). Also, it is highly expected that more imminent situations will elicit stronger emotional action biases. In these situations, emotional predispositions on actions will potentially increase in strength and "healthy" emotional control might not be sufficient to implement contradictory instrumental goals successfully. Future research might indicate whether highly arousing emotional stimuli indeed enhance emotional biases and PIT effects, and whether this makes it more difficult to apply emotional action control. An indication that this might indeed be true is provided by a study that provoked social stress in SAD patients, enhancing arousal and increasing avoidance tendencies during the AA task (107).

This chapter focused on different systems influencing emotional behavior, namely emotional/Pavlovian action biases, instrumental actions, and emotional action control (Figure 10.2). However, these interact strongly and together lead to the observed behavioral outcome. Crucially, all current explanations of emotional health and disease frequently focus on one or several of these components in isolation (or regard them as exclusive antagonists) and neglect putative interactions (6, 108, 109). For example, social anxiety disorders are often related to either strong avoidance biases or enhanced action inhibition (67, 110). Aggressive behavior, on the other hand, is commonly viewed as either increased emotional action bias to approach or loss of action inhibition (111–113). These characterizations most likely underestimate our emotional behavior's complexity and diversity (114). To advance understanding of the combined effect of emotional action biases, instrumental planning, and emotional control on behavior it will be important to investigate a complementary account: in humans, optimal emotional behavior arises not from emotional biases and instrumental action selection systems individually but from their interaction, with emotional control acting though prefrontal systems when in conflict.

One aspect that has not received much attention in this chapter is the upside of carrying an S-allele of the *5-HTTLPR* polymorphism (115). S-allele carriers have indeed been associated with being more vulnerable to negative environments, but they have also been shown to profit more from positive environments (116). This includes the display of more positive emotional expressions, better stress reactivity, and reduced externalizing behavior (117–119). A possible explanation of these effects might be given by gene–environment interactions (G×E), such as effects of *5-HTT* gene variations and early life stress on the neuropeptide corticotrophin-releasing factor (CRF; important for stress coping) (120). However, the effects in relation to fast behavioral responses and emotional action control are unknown to date.

Finally, emotional action control might be crucial in explaining relations between the serotonergic system and (vulnerability to) psychopathology. Influences of the serotonergic system on psychopathology, as well as potential environmental modulations, have been suggested as well as disputed in the literature (10, 121). Similar alterations of the neural mechanism underlying emotional action control can be observed for *5-HTTLPR* polymorphic variability as well as aggressive disorders. Future studies might show whether this extends to affective disorders such as anxiety and depression. For now, we would like to hypothesize from the evidence presented in this chapter that the potential to employ emotional action control when needed, might mediate adaptive behavioral responses and, in relation to the serotonergic system, affects vulnerability versus resilience to psychopathologies.

10.10 **Conclusion**

Initial studies suggest a significant relation between 5-HT and associated genes on emotional action control. Stronger affective biases, reduced behavioral flexibility and emotional control have been found by pharmacological studies lowering central 5-HT levels and genetic studies of reduced 5-HT reuptake. Moreover, there seems to be a profound relation between emotional action control and serotonin to modulate susceptibility for emotional disorders. Together, this suggests that studying emotional action control in relation to 5-HT as well as emotional disorders might provide fruitful therapeutic target points, potentially aimed at training emotional action control together with serotonergic-based medication.

References

1. **Chen M, Bargh JA.** Consequences of automatic evaluation: Immediate behavioral predispositions to approach or avoid the stimulus. Personal Soc Psychol Bull. 1999;**25**:215–24.
2. **Solarz AK.** Latency of instrumental responses as a function of compatibility with the meaning of eliciting verbal signs. J Exp Psychol. 1960;**59**:239–45.
3. **Rotteveel M, Phaf RH.** Automatic affective evaluation does not automatically predispose for arm flexion and extension. Emotion. 2004;**4**:156–72.
4. **LeDoux J.** Rethinking the emotional brain. Neuron. 2012;**73**:653–76.
5. **Panksepp J.** Toward a general psychobiological theory of emotions. Behav Brain Sci. 1982;**5**:407–67.
6. **Dolan RJ, Dayan P.** Goals and habits in the brain. Neuron. 2013;**80**:312–25.
7. **Roelofs K, Minelli A, Mars RB, van Peer J, Toni I.** On the neural control of social emotional behavior. Soc Cogn Affect Neurosci. 2009;**4**:50–8.
8. **Volman I, Roelofs K, Koch S, Verhagen L, Toni I.** Anterior prefrontal cortex inhibition impairs control over social emotional actions. Curr Biol. 2011;**21**:1766–70.
9. **Cavanagh JF, Eisenberg I, Guitart-Masip M, Huys Q, Frank MJ.** Frontal theta overrides pavlovian learning biases. J Neurosci. 2013;**33**:8541–8.
10. **Caspi A, Hariri AR, Holmes A, Uher R, Moffitt TE.** Genetic sensitivity to the environment: The case of the serotonin transporter gene and its implications for studying complex diseases and traits. Am J Psychiatry. 2010;**167**:509–27.
11. **Garcia LF, Aluja A, Fibla J, Cuevas L, García O.** Incremental effect for antisocial personality disorder genetic risk combining 5-HTTLPR and 5-HTTVNTR polymorphisms. Psychiatry Res. 2010;**177**:161–6.
12. **Juhasz G, Gonda X, Hullam G, Eszlari N, Kovacs D, Lazary J,** et al. Variability in the effect of 5-HTTLPR on depression in a large European population: the role of age, symptom profile, type and intensity of life stressors. PLoS One. 2015;**10**:e0116316.
13. **Karg K, Burmeister M, Shedden K, Sen S.** The serotonin transporter promoter variant (5-HTTLPR), stress, and depression meta-analysis revisited: Evidence of genetic moderation. Arch Gen Psychiatry. 2011;**68**:444–54.
14. **Deakin JF.** The role of serotonin in panic, anxiety and depression. Int Clin Psychopharmacol. 1998;**13** Suppl 4:S1–5.
15. **Frijda NH.** The laws of emotion. Am Psychol. 1988;**43**:349–58.
16. **James W.** What is an emotion? Mind. 1884;**9**:188–205.
17. **Grandjean D, Scherer KR.** Unpacking the cognitive architecture of emotion processes. Emotion. 2008;**8**:341–51.
18. **Scherer KR.** Emotions are emergent processes: They require a dynamic computational architecture. Philos Trans R Soc Lond B. 2009;**364**:3459–74.
19. **Lanctôt N, Hess U.** The timing of appraisals. Emotion. 2007;**7**:207–12.

20. **Sander D, Grandjean D, Scherer KR.** A systems approach to appraisal mechanisms in emotion. Neural Networks. 2005;**18**:317–52.
21. **Darwin C.** *The Expression of the Emotions in Man and Animals.* 2nd edn. New York, NY: Penguin Group, 1872.
22. **Hershberger WA.** An approach through the looking-glass. Anim Learn Behav. 1986;**14**:443–51.
23. **Breland K, Breland M.** The misbehavior of organisms. Am Psychol. 1961;**16**:681–4.
24. **Rinck M, Becker ES.** Approach and avoidance in fear of spiders. J Behav Ther Exp Psychiatry. 2007;**38**:105–20.
25. **Phaf RH, Mohr SE, Rotteveel M, Wicherts JM.** Approach, avoidance, and affect: a meta-analysis of approach-avoidance tendencies in manual reaction time tasks. Emot Sci. 2014;**5**:378.
26. **Huys QJM, Cools R, Gölzer M, Friedel E, Heinz A, Dolan RJ**, et al. Disentangling the roles of approach, activation and valence in instrumental and pavlovian responding. PLoS Comput Biol. 2011;**7**:e1002028.
27. **Deakin J.** The origins of "5-HT and mechanisms of defence" by Deakin and Graeff: a personal perspective. J Psychopharmacol. 2013;**27**:1084–9.
28. **Hagenaars MA, Oitzl M, Roelofs K.** Updating freeze: Aligning animal and human research. Neurosci Biobehav Rev. 2014;**47**:165–76.
29. **Geurts DEM, Huys QJM, den Ouden HEM, Cools R.** Serotonin and aversive Pavlovian control of instrumental behavior in humans. J Neurosci. 2013;**33**:18932–9.
30. **Guitart-Masip M, Duzel E, Dolan RJ, Dayan P.** Action versus valence in decision making. Trends Cogn Sci. 2014;**18**:194–202.
31. **Holmes NM, Marchand AR, Coutureau E.** Pavlovian to instrumental transfer: A neurobehavioural perspective. Neurosci Biobehav Rev. 2010;**34**:1277–95.
32. **Ly V, Bergmann TO, Gladwin TE**, Volman I, **Usberti N, Cools R**, et al. Reduced affective biasing of instrumental action with tDCS over the prefrontal cortex. Brain Stimul. 2016;**9**:380–7.
33. **Ly V, Huys QJ, Stins JF, Roelofs K, Cools R.** Individual differences in bodily freezing predict emotional biases in decision making. Front Behav Neurosci. 2014;**8**:1–9.
34. **Volman I, Toni I, Verhagen L, Roelofs K.** Endogenous testosterone modulates prefrontal-amygdala connectivity during social emotional behavior. Cereb Cortex. 2011;**21**:2282–90.
35. **Gross JJ.** Emotion regulation: Affective, cognitive, and social consequences. Psychophysiology. 2002;**39**:281–91.
36. **Lee TW, Dolan RJ, Critchley HD.** Controlling emotional expression: Behavioral and neural correlates of nonimitative emotional responses. Cereb Cortex. 2008;**18**:104–13.
37. **Canli T, Lesch K-P.** Long story short: the serotonin transporter in emotion regulation and social cognition. Nat Neurosci. 2007;**10**:1103–9.
38. **Faulkner P, Deakin JFW.** The role of serotonin in reward, punishment and behavioural inhibition in humans: Insights from studies with acute tryptophan depletion. Neurosci Biobehav Rev. 2014;**46**:365–78.
39. **Dayan P, Huys Q.** Serotonin's many meanings elude simple theories. Elife. 2015;**4**:e07390.
40. **den Ouden HE, Daw ND, Fernandez G, Elshout JA, Rijpkema M, Hoogman M**, et al. Dissociable effects of dopamine and serotonin on reversal learning. Neuron. 2013;**80**:1090–100.
41. **Cohen JY, Amoroso MW, Uchida N.** Serotonergic neurons signal reward and punishment on multiple timescales. Elife. 2015;**4**:e06346.
42. **Crockett MJ, Clark L, Roiser JP, Robinson OJ, Cools R, Chase HW**, et al. Converging evidence for central 5-HT effects in acute tryptophan depletion. Mol Psychiatry. 2012;**17**:121–3.
43. **den Ouden HE, Swart JC, Schmidt K, Fekkes D, Geurts DE, Cools R.** Acute serotonin depletion releases motivated inhibition of response vigour. Psychopharmacology. 2015;**232**:1303–12.
44. **Crockett MJ, Clark L, Robbins TW.** Reconciling the role of serotonin in behavioral inhibition and aversion: Acute tryptophan depletion abolishes punishment-induced inhibition in humans. J Neurosci. 2009;**29**:11993–9.

45. **Crockett MJ, Clark L, Apergis-Schoute AM, Morein-Zamir S, Robbins TW.** Serotonin modulates the effects of pavlovian aversive predictions on response vigor. Neuropsychopharmacology. 2012;**37**:2244–52.
46. **Hebart MN, Gläscher J.** Serotonin and dopamine differentially affect appetitive and aversive general Pavlovian-to-instrumental transfer. Psychopharmacology. 2015;**232**:437–51.
47. **Heils A, Teufel A, Petri S, Stöber G, Riederer P, Bengel D**, et al. Allelic variation of human serotonin transporter gene expression. J Neurochem. 1996;**66**:2621–4.
48. **Lesch KP, Bengel D, Heils A, Sabol SZ, Greenberg BD, Petri S**, et al. Association of anxiety-related traits with a polymorphism in the serotonin transporter gene regulatory region. Science. 1996;**274**:1527–31.
49. **Roiser JP, de Martino B, Tan GCY, Kumaran D, Seymour B, Wood NW**, . A genetically mediated bias in decision making driven by failure of amygdala control. J Neurosci. 2009;**29**:5985–91.
50. **Hariri AR, Mattay VS, Tessitore A, Kolachana B, Fera F, Goldman D**, et al. Serotonin transporter genetic variation and the response of the human amygdala. Science. 2002;**297**:400–3.
51. **Heinz A, Braus DF, Smolka MN, Wrase J, Puls I, Hermann D**, et al. Amygdala-prefrontal coupling depends on a genetic variation of the serotonin transporter. Nat Neurosci. 2005;**8**:20–1.
52. **Pacheco J, Beevers C, Benavides C, McGeary J, Stice E, Schnyer DM.** Frontal-limbic white matter pathway associations with the serotonin transporter gene promoter region (5-HTTLPR) polymorphism. J Neurosci. 2009;**29**:6229–33.
53. **Pezawas L, Meyer-Lindenberg A, Drabant EM, Verchinski BA, Munoz KE, Kolachana BS**, et al. 5-HTTLPR polymorphism impacts human cingulate-amygdala interactions: A genetic susceptibility mechanism for depression. Nat Neurosci. 2005;**8**:828–34.
54. **Volman I, Verhagen L, den Ouden HE, Fernández G, Rijpkema M, Franke B**, et al. Reduced serotonin transporter availability decreases prefrontal control of the amygdala. J Neurosci. 2013;**33**:8974–9.
55. **Klumpers F, Kroes MC, Heitland I, Everaerd D, Akkermans SE, Oosting RS**, et al. Dorsomedial prefrontal cortex mediates the impact of serotonin transporter linked polymorphic region genotype on anticipatory threat reactions. Biol Psychiatry. 2015;**78**:582–9.
56. **Guyer AE, Lau JY, McClure-Tone EB, Parrish J, Shiffrin ND, Reynolds RC**, et al. Amygdala and ventrolateral prefrontal cortex function during anticipated peer evaluation in pediatric social anxiety. Arch Gen Psychiatry. 2008;**65**:1303–12.
57. **Hoptman MJ, D'Angelo D, Catalano D, Mauro CJ, Shehzad ZE, Kelly AM**, et al. Amygdalofrontal functional disconnectivity and aggression in schizophrenia. Schizophr Bull. 2010;**36**:1020–8.
58. **Kim MJ, Loucks RA, Palmer AL, Brown AC, Solomon KM, Marchante AN**, et al. The structural and functional connectivity of the amygdala: from normal emotion to pathological anxiety. Behav Brain Res. 2011;**223**:403–10.
59. **Rudie JD, Shehzad Z, Hernandez LM, Colich NL, Bookheimer SY, Iacoboni M**, et al. Reduced functional integration and segregation of distributed neural systems underlying social and emotional information processing in autism spectrum disorders. Cereb Cortex. 2012;**22**:1025–37.
60. **Stephan KE, Penny WD, Moran RJ, den Ouden HE, Daunizeau J, Friston KJ.** Ten simple rules for dynamic causal modeling. Neuroimage. 2010;**49**:3099–109.
61. **Hartley CA, McKenna MC, Salman R, Holmes A, Casey BJ, Phelps EA**, et al. Serotonin transporter polyadenylation polymorphism modulates the retention of fear extinction memory. PNAS. 2012;**109**:5493–8.
62. **Campus P, Accoto A, Maiolati M, Latagliata C, Orsini C.** Role of prefrontal 5-HT in the strain-dependent variation in sign-tracking behavior of C57BL/6 and DBA/2 mice. Psychopharmacology. 2016;**233**:1157–69.
63. **Harrison AA, Everitt BJ, Robbins TW.** Central 5-HT depletion enhances impulsive responding without affecting the accuracy of attentional performance: Interactions with dopaminergic mechanisms. Psychopharmacology. 1997;**133**:329–42.

64. **Lopizzo N, Bocchio Chiavetto L, Cattane N, Plazzotta G, Tarazi FI, Pariante CM**, et al. Gene-environment interaction in major depression: Focus on experience-dependent biological systems. Front Psychiatry. 2015;6.

65. **Munafò MR, Zammit S, Flint J.** Practitioner review: A critical perspective on gene-environment interaction models—What impact should they have on clinical perceptions and practice? J Child Psychol Psychiatry. 2014;**55**:1092–101.

66. **Caspi A, Sugden K, Moffitt TE, Taylor A, Craig IW, Harrington H**, et al. Influence of Life Stress on Depression: Moderation by a Polymorphism in the 5-HTT Gene. Science. 2003;**301**:386–9.

67. **Clark D, Wells A.** A cognitive model of social phobia. In: Heimberg R (ed.). *Social Phobia: Diagnosis, Assessment, and Treatment* (pp. 69–93). London: Guilford Press, 1995.

68. **Heuer K, Rinck M, Becker ES.** Avoidance of emotional facial expressions in social anxiety: The approach-avoidance task. Behav Res Ther. 2007;**45**:2990–3001.

69. **Roelofs K, Putman P, Schouten S, Lange WG, Volman I, Rinck M.** Gaze direction differentially affects avoidance tendencies to happy and angry faces in socially anxious individuals. Behav Res Ther. 2010;**48**:290–4.

70. **Taylor CT, Amir N.** Modifying automatic approach action tendencies in individuals with elevated social anxiety symptoms. Behav Res Ther. 2012;**50**:529–36.

71. **Huys QJM, Gölzer M, Friedel E, Heinz A, Cools R, Dayan P**, et al. The specificity of Pavlovian regulation is associated with recovery from depression. Psychol Med. 2016;**46**:1027–35.

72. **Cerdá M, Sagdeo A, Johnson J, Galea S.** Genetic and environmental influences on psychiatric comorbidity: A systematic review. J Affect Disord. 2010;**126**:14–38.

73. **Brady EU, Kendall PC.** Comorbidity of anxiety and depression in children and adolescents. Psychol Bull. 1992;**111**:244–55.

74. **Hirschfeld RMA.** The comorbidity of major depression and anxiety disorders: recognition and management in primary care. Prim Care Companion J Clin Psychiatry. 2001;**3**:244–54.

75. **Gray J.** A critique of Eysenck's theory of personality. In: **Eysenck HJ** (ed.). A Model for Personality. New York, NY: Springer, 1981, pp. 246–76.

76. **von Borries AKL, Volman I, de Bruijn ER, Bulten BH, Verkes RJ, Roelofs K.** Psychopaths lack the automatic avoidance of social threat: Relation to instrumental aggression. Psychiatry Res. 2012;**200**:761–6.

77. **Hallikainen T, Saito T, Lachman HM, Volavka J, Pohjalainen T, Ryynänen OP**, et al. Association between low activity serotonin transporter promoter genotype and early onset alcoholism with habitual impulsive violent behavior. Mol Psychiatry. 1999;**4**:385–8.

78. **Haberstick BC, Smolen A, Hewitt JK.** Family-based association test of the 5HTTLPR and aggressive behavior in a general population sample of children. Biol Psychiatry. 2006;**59**:836–43.

77. **Hallikainen T, Saito T, Lachman HM, Volavka J, Pohjalainen T, Ryynänen OP**, et al. Association between low activity serotonin transporter promoter genotype and early onset alcoholism with habitual impulsive violent behavior. Mol Psychiatry. 1999;**4**:385–8.

79. **Beitchman JH, Baldassarra L, Mik H, De Luca V, King N, Bender D**, et al. Serotonin transporter polymorphisms and persistent, pervasive childhood aggression. Am J Psychiatry. 2006;**163**:1103–5.

80. **May ME, Lightfoot DA, Srour A, Kowalchuk RK, Kennedy CH.** Association between serotonin transporter polymorphisms and problem behavior in adult males with intellectual disabilities. Brain Res. 2010;**1357**:97–103.

81. **Verona E, Joiner TE, Johnson F, Bender TW.** Gender specific gene–environment interactions on laboratory-assessed aggression. Biol Psychol. 2006;**71**:33–41.

82. **Conway CC, Keenan-Miller D, Hammen C, Lind PA, Najman JM, Brennan PA.** Coaction of stress and serotonin transporter genotype in predicting aggression at the transition to adulthood. J Clin Child Adolesc Psychol. 2012;**41**:53–63.

83. **Reif A, Rösler M, Freitag CM, Schneider M, Eujen A, Kissling C**, et al. Nature and nurture predispose to violent behavior: serotonergic genes and adverse childhood environment. Neuropsychopharmacology. 2007;**32**:2375–83.

84. **Wang D, Szyf M, Benkelfat C, Provençal N, Turecki G, Caramaschi D**, et al. Peripheral SLC6A4 DNA methylation is associated with in vivo measures of human brain serotonin synthesis and childhood physical aggression. PLoS One. 2012;7.
85. **Osipova DV, Kulikov AV, Popova NK.** C1473G polymorphism in mouse tph2 gene is linked to tryptophan hydroxylase-2 activity in the brain, intermale aggression, and depressive-like behavior in the forced swim test. J Neurosci Res. 2009;**87**:1168–74.
86. **Mosienko V, Bert B, Beis D, Matthes S, Fink H, Bader M**, et al. Exaggerated aggression and decreased anxiety in mice deficient in brain serotonin. Transl Psychiatry. 2012;**2**:e122.
87. **Yu Q, Teixeira CM, Mahadevia D, Huang Y, Balsam D, Mann JJ**, et al. Dopamine and serotonin signaling during two sensitive developmental periods differentially impact adult aggressive and affective behaviors in mice. Mol Psychiatry. 2014;**19**:688–98.
88. **Kiryanova V, Dyck RH.** Increased aggression, improved spatial memory, and reduced anxiety-like behaviour in adult male mice exposed to fluoxetine early in life. Dev Neurosci. 2014;**36**:396–408.
89. **Takahashi A, Quadros IM, de Almeida RM, Miczek KA.** Behavioral and pharmacogenetics of aggressive behavior. Curr Top Behav Neurosci. 2012;**12**:73–138.
90. **Bortolato M, Pivac N, Muck Seler D, Nikolac Perkovic M, Pessia M, Di Giovanni G.** The role of the serotonergic system at the interface of aggression and suicide. Neuroscience. 2013;**236**:160–85.
91. **Glenn AL, Raine A.** Psychopathy and instrumental aggression: Evolutionary, neurobiological, and legal perspectives. Int J Law Psychiatry. 2009;**32**:253–8.
92. **Volman I, von Borries AKL, Bulten BH, Verkes RJ, Toni I, Roelofs K.** Testosterone modulates altered prefrontal control of emotional actions in psychopathic offenders. eNeuro. 2016;3.
93. **Archer J.** Testosterone and human aggression: An evaluation of the challenge hypothesis. Neurosci Biobehav Rev. 2006;**30**:319–45.
94. **Bos PA, Panksepp J, Bluthé RM, Van Honk J.** Acute effects of steroid hormones and neuropeptides on human social-emotional behavior: A review of single administration studies. Front Neuroendocrinol. 2012;**33**:17–35.
95. **Eisenegger C, Haushofer J, Fehr E.** The role of testosterone in social interaction. Trends Cogn Sci. 2011;**15**:263–71.
96. **van Honk J, Tuiten A, Verbaten R, van den Hout M, Koppeschaar H, Thijssen J**, et al. Correlations among salivary testosterone, mood, and selective attention to threat in humans. Horm Behav. 1999;**36**:17–24.
97. **Birger M, Swartz M, Cohen D, Alesh Y, Grishpan C, Kotelr M.** Aggression: The testosterone-serotonin link. Isr Med Assoc J. 2003;**5**:653–8.
98. **Kuepper Y, Alexander N, Osinsky R, Mueller E, Schmitz A, Netter P**, et al. Aggression—interactions of serotonin and testosterone in healthy men and women. Behav Brain Res. 2010;**206**:93–100.
99. **Higley JD, Mehlman PT, Poland RE, Taub DM, Vickers J, Suomi SJ**, et al. CSF testosterone and 5-HIAA correlate with different types of aggressive behaviors. Biol Psychiatry. 1996;**40**:1067–82.
100. **Kubala KH, McGinnis MY, Anderson GM, Lumia AR.** The effects of an anabolic androgenic steroid and low serotonin on social and non-social behaviors in male rats. Brain Res. 2008;**1232**:21–9.
101. **Radke S, Volman I, Mehta P, van Son V, Enter D, Sanfey A**, et al. Testosterone biases the amygdala toward social threat approach. Sci Adv. 2015;**1**:e1400074–e1400074.
102. **Salamone JD, Pardo M, Yohn SE, López-Cruz L, SanMiguel N, Correa M.** Mesolimbic dopamine and the regulation of motivated behavior. Curr Top Behav Neurosci 2016;**27**:231–57.
103. **Rutledge RB, Skandali N, Dayan P, Dolan RJ.** Dopaminergic modulation of decision making and subjective well-being. J Neurosci. 2015;**35**:9811–22.
104. **Enter D, Colzato LS, Roelofs K.** Dopamine transporter polymorphisms affect social approach-avoidance tendencies. Genes, Brain Behav. 2012;**11**:671–6.
105. **Guitart-Masip M, Huys QJ, Fuentemilla L, Dayan P, Duzel E, Dolan RJ.** Go and no-go learning in reward and punishment: Interactions between affect and effect. Neuroimage. 2012;**62**:154–66.

106. **Quirk GJ, Gehlert DR.** Inhibition of the amygdala: Key to pathological states? Ann N Y Acad Sci. 2003;**985**:263–72.

107. **Roelofs K, van Peer J, Berretty E, Jong PD, Spinhoven P, Elzinga BM.** Hypothalamus–pituitary–adrenal axis hyperresponsiveness is associated with increased social avoidance behavior in social phobia. Biol Psychiatry. 2009;**65**:336–43.

108. **Cromwell HC, Panksepp J.** Rethinking the cognitive revolution from a neural perspective: How overuse/misuse of the term "cognition" and the neglect of affective controls in behavioral neuroscience could be delaying progress in understanding the BrainMind. Neurosci Biobehav Rev. 2011;**35**:2026–35.

109. **Crone EA, Dahl RE.** Understanding adolescence as a period of social-affective engagement and goal flexibility. Nat Rev Neurosci. 2012;**13**:636–50.

110. **Hofmann SG.** Cognitive factors that maintain social anxiety disorder: A comprehensive model and its treatment implications. Cogn Behav Ther. 2007;**36**:193–209.

111. **Carré JM, McCormick CM, Hariri AR.** The social neuroendocrinology of human aggression. Psychoneuroendocrinology. 2011;**36**:935–44.

112. **Siever LJ.** Neurobiology of aggression and violence. Am J Psychiatry. 2008;**165**:429–42.

113. **Van Honk J, Harmon-Jones E, Morgan BE, Schutter DJ.** Socially explosive minds: The triple imbalance hypothesis of reactive aggression. J Pers. 2010;**78**:67–94.

114. **Iordan a D, Dolcos S, Dolcos F.** Neural signatures of the response to emotional distraction: A review of evidence from brain imaging investigations. Front Hum Neurosci. 2013;**7**:200.

115. **Homberg JR, Lesch KP.** Looking on the bright side of serotonin transporter gene variation. Biol Psychiatry. 2011;**69**:513–9.

116. **van Ijzendoorn MH, Belsky J, Bakermans-Kranenburg MJ.** Serotonin transporter genotype 5HTTLPR as a marker of differential susceptibility? A meta-analysis of child and adolescent gene-by-environment studies. Transl Psychiatry. 2012;**2**:e147.

117. **Haase CM, Beermann U, Saslow LR, Shiota MN, Saturn SR, Lwi SJ**, et al. Short alleles, bigger smiles? The effect of 5-HTTLPR on positive emotional expressions. Emotion. 2015;**15**:438–48.

118. **Sumner JA, McLaughlin KA, Walsh K, Sheridan MA, Koenen KC.** Caregiving and 5-HTTLPR genotype predict adolescent physiological stress reactivity: Confirmatory tests of gene × environment interactions. Child Dev. 2015;**0**:1–10.

119. **Brett ZH, Humphreys KL, Smyke AT, Gleason MM, Nelson CA, Zeanah CH**, et al. Serotonin transporter linked polymorphic region (5-HTTLPR) genotype moderates the longitudinal impact of early caregiving on externalizing behavior. Dev Psychopathol. 2015;**27**:7–18.

120. **van der Doelen RH, Arnoldussen IA, Ghareh H, van Och L, Homberg JR, Kozicz T**, et al. Early life adversity and serotonin transporter gene variation interact to affect DNA methylation of the corticotropin-releasing factor gene promoter region in the adult rat brain. Dev Psychopathol. 2015;**27**:123–35.

121. **Munafò MR, Durrant C, Lewis G, Flint J.** Gene X environment interactions at the serotonin transporter locus. Biol Psychiatry. 2009;**65**:211–9.

122. **Niermann HCM, Tyborowska A, Cillessen AHN, van Donkelaar MM, Lammertink F, Gunnar MR**, et al. The relation between infant freezing and the development of internalizing symptoms in adolescence: A prospective longitudinal study. Dev Sci. 2018. doi.org/10.1111/desc.12763.

Chapter 11

Genetics of emotion regulation: A systematic review

Andrei C Miu and Mirela I Bîlc

11.1 Introduction

In recent decades, there has been an upsurge of interest in the study of emotions, largely fueled by methodological advances in inducing and assessing affective responses and their biological mechanisms (1,2). To date, extensive evidence has accumulated on the widespread influence of emotions on cognition (3–5), behavior (6,7), and health (8–10). The study of emotions pervades virtually all research areas in psychology and cognitive neuroscience.

One of the general findings, which applies to both emotion–cognition interactions and the relations between emotions and mental health, is that the effects of emotions can be adaptive or maladaptive, depending on their frequency, intensity, duration, and type (11). This apparent heterogeneity in the effects of emotions highlights the essential contribution of the processes by which people can modulate their emotional responses, coined "emotion regulation" (12,13). People are frequently trying to down- or up-regulate their emotions in order to achieve hedonic (i.e. feeling good) or instrumental (e.g. using emotions to their advantage) goals (14). There is evidence, for instance, that the stage of the emotional response in which people initiate regulation efforts (e.g. 15), the regulation strategy that they choose (e.g. 16), and their ability to implement that strategy (e.g. 17), all play a major role in determining the effects of emotions. Research on emotion regulation has rapidly taken center stage in affective science and has recently advanced to the genetic and neural level.

Work on the biological underpinnings of emotion regulation has been stimulated by a recent shift of focus in clinical psychology and psychiatry, placing emphasis on transdiagnostic mechanisms in psychopathology. The Research Domain Criteria (RDoC) is a framework put forward by the National Institute of Mental Health, which describes several mechanisms (i.e. negative and positive valence systems, cognitive systems, social processes, arousal, and regulatory systems) that are thought to contribute to the development and persistence of multiple mental disorders (18). The aim of the RDoC is to characterize these transdiagnostic mechanisms at several levels of analysis, based on genetic, physiological, behavioral, and subjective approaches. It has been recently argued that emotion regulation could be added to the major transdiagnostic mechanisms of psychopathology (19), considering the fact that emotion regulation difficulties have been associated with many mental disorders (11,20), and they cannot be reduced to any of the systems that were originally included in the RDoC. In this context, genetic studies could contribute to understanding the contribution of emotion regulation to comorbidity of mental disorders, and to explaining the significant genetic overlap between these disorders (e.g. 21).

The present chapter aims to describe the emerging literature on genetic contributions to emotion regulation. After defining emotion and introducing the process model of emotion regulation (12,13), which has guided much emotion regulation research in recent decades, this chapter will systematically review twin and candidate gene studies on this topic. In comparison to previous reviews, which have focused on specific candidate genes (22,23) or have emphasized the importance and agenda of this line of work (24,25), the present chapter will provide a broad and systematic overview. All available genetic studies, identified through a systematic search, will be reviewed, provided that they included assessments of specific emotion regulation strategies. That is, in order to avoid methodological and conceptual heterogeneity and confusion, we chose to leave out research on coping and emotional intelligence, two related, but distinct approaches to the regulation of emotion (for review, see 26,27). A systematic review of genetic studies on coping is available for interested readers (28), and two chapters in the present volume offer complementary views focused on emotional resilience (Chapter 20) and emotional action control (see Chapter 10). The final section of this chapter will outline several important issues that could advance research on the genetics of emotion regulation in the future.

11.2 **Emotion and emotion regulation**

Early attempts to define emotions (e.g. 29,30–32) have played a major role in the return of emotion in mainstream psychology and neuroscience. Following the decline of psychoanalysis and the rise of cognitive psychology, emotions were initially viewed as too subjective and irregular to lend themselves to scientific scrutiny. However, critical approaches to defining emotions, progress in inducing and assessing emotions in animals and humans, and success in uncovering some of the neural underpinnings of emotions have gradually re-established interest in studying emotions (for review, see 33). A major contribution has been made by cognitive theories of emotion, which built on the work of Schachter and Singer (34) and have compellingly demonstrated that emotions are driven by cognitive appraisals (for review, see 35).

To date, there is general agreement that emotions are generated through the appraisal of events in relation to current goals (13,36,37). In other words, the way external (e.g. an approaching job interview) or internal (e.g. an accelerating heart beat) events are interpreted in relation to oneself will determine whether and which emotion will arise. If someone believes that a situation may threaten or, to the contrary, offer an opportunity for achieving personal goals, this appraisal will generate negative or positive emotions, respectively. Appraisal is multidimensional and includes judgments of pleasantness, anticipated effort, certainty, and control or responsibility, for instance, and there is evidence that each emotion is characterized by a unique pattern of appraisals (38). For example, both fear and anger are driven by appraisals of high unpleasantness and anticipated effort, but they are differentiated by appraisals of control and certainty, with the former involving a lower degree of control and higher uncertainty (38).

Drawing on this literature, Gross (12) has systematically described the generation of affect as a sequence of four stages, focused on: (i) the onset of an event; (ii) its attentional processing; (iii) its interpretation or appraisal relative to current goals; and (iv) the development and expression of the emotional response. This so-called modal model of emotion has offered an heuristic framework for a process approach to emotion regulation (12,13), which has laid the foundation for and guided much of the recent research on this topic. In comparison to previous approaches (e.g. coping), the process model has stimulated a much wider interest in the study of emotion regulation (39) due to at least three fundamental theoretical contributions: (i) it provided a clear-cut definition of emotion regulation, directly derived from the widely accepted modal model of emotion; (ii) it highlighted the categorization of emotion regulation strategies according to the stage of emotion generation that they target; and (iii) it advanced the novel hypothesis that the earlier the stage of emotion generation in which a regulation strategy intervenes, the better its efficiency (for review, see 13,40,41).

According to the process model (12,13), emotion regulation is initiated when emotions themselves are appraised in relation to current goals. If emotions are judged to help or hinder the individual's objectives, the goal of up- or down-regulating them will be activated and the individual will try to employ one or more strategies in order to modulate his or her affective responses in the desired direction.

In addition to defining emotion regulation, the process model proposed that strategies are categorized according to the stage of the emotional response that they target (12,13). Strategies such as situation selection or situation modification attempt to change emotions by intervening on the event that was appraised and is expected to generate an emotion. Other strategies target the cognitive processes that contribute to the generation of emotion, mainly attention and appraisal. By driving attention away or toward the emotional event, or indeed, by changing the way it was initially appraised, cognitive strategies such as distraction and rumination, and reappraisal, respectively, are able to alter emotional responses. Finally, there are strategies by which individuals attempt to regulate emotions by modulating their behavioral expression. A commonly used strategy, for example, is expressive suppression, in which someone tries to hide emotion by inhibiting its manifestations in voice, facial expression, gestures, and body posture.

One of the first and most important predictions drawn from the process model has suggested that the efficiency of emotion regulation strategies differs depending on their timing: the earlier the stage of emotion generation in which they act, the better their efficiency. From this temporal perspective, two categories of emotion regulation strategies have been outlined: antecedent-focused, which include situation-related and cognitive strategies; and response-focused, which include behavioral strategies (42). The prediction was that, by acting in the early stages of emotion generation, before emotional responses have fully developed, antecedent-focused strategies may be more efficient than response-focused strategies. Pioneering studies in which cognitive reappraisal (i.e. a cognitive, antecedent-focused strategy) was compared to expressive suppression (i.e. a behavioral, response-focused strategy) have supported this hypothesis, and found that the former is more efficient in terms of modulating subjective experience and autonomic activity (15), and reducing the cognitive costs of emotions (43). Subsequent studies have extended the evidence by showing, for instance, that unlike suppression, reappraisal does not disrupt communication in social interaction (44) and it is associated with greater well-being in the long-term (45).

While reappraisal has shown many benefits at the subjective, cognitive, and physiological level, there are situations in which its efficiency is limited. For instance, when it is initiated later in the course of emotion, after affective responses have developed, reappraisal is less efficient in reducing negative affect (46) and autonomic activity (47). In these studies, reappraisal was compared to distraction, another cognitive emotion regulation strategy by which one tries to divert his or her attention away from the emotional event. Late distraction is relatively more efficient in reducing negative affect and autonomic activity, but, given that it involves disengaging from the target situation, it is also associated with impaired memory for that situation (48). Reappraisal leaves memory intact (43,48), but it impairs cognitive control when it is initiated late in the course of emotion (48). These findings have suggested that emotion regulation strategies (e.g. distraction compared to reappraisal) that target earlier stages in the emotion generation process (e.g. attention compared to appraisal) require less effort, and this may explain why they are more efficient when initiated after emotion has escalated (42). They have also emphasized that in describing the efficiency of emotion regulation strategies, one needs to consider aspects such as the intensity of emotion and situational demands (e.g. the need to subsequently recall the emotional event) (49).

Rumination is another cognitive emotion regulation strategy that has been extensively studied (for review, see 50,51). Rumination involves focusing attention on the emotional event and its implications for oneself (52), and it has been contrasted to distraction and reappraisal in several studies. By asking participants to focus on their feelings or on a neutral topic such as the layout of the local shopping center, Nolen-Hoeksema and Morrow (53) have induced rumination and distraction, respectively, and assessed their effects on mood. They found that rumination increased and distraction decreased negative affect in healthy volunteers with higher levels of depressive symptoms (53). In contrast to distraction, rumination induced cognitive biases such as increased negative interpretation of hypothetical situations, pessimistic predictions about the future, and reduced ability to find solutions to interpersonal problems (54).

More recent studies (e.g. 55) have described two types of rumination, brooding and reflective pondering, with different effects on affect. Both involve focusing on the emotional event, but the goals are different: brooding involves dwelling on the negative consequences of the event, while reflective pondering involves attempts to understand the reasons for one's emotion. There is evidence that brooding is more specifically associated with emotional problems (55).

In summary, emotion regulation is initiated when individuals believe that their current emotion might impede or serve their goals. Multiple emotion regulation strategies have been described (e.g. cognitive reappraisal, distraction, rumination, expressive suppression) and they are commonly used in attempts to down- or up-regulate emotions. While antecedent-focused strategies are generally more efficient than response-focused strategies, due to their intervention at an early stage of the emotional response, their efficiency depends on other aspects such as the intensity of emotion and the need to attend to the situation that generated the emotional response. In the following section, we will briefly describe the links between emotion regulation and mental health.

11.2.1 Emotion regulation and psychopathology

Emotional problems are pervasive in psychopathology and they may stem from altered emotional reactivity, emotion regulation, or both (11). Problems with emotion and emotion regulation are hard to disentangle (56), especially considering that regulation efforts may be automatic and emotion regulation measures are limited to consciously controlled processes. Nonetheless, emotion regulation difficulties have been described in multiple mental disorders (for review, see 11,19,57) and while they do not exclude concurrent emotional reactivity (e.g. intensity, duration, frequency, type) problems, they clearly show that emotion regulation plays specific and important roles in psychopathology.

Based on predictions from the process model of emotion regulation, extensive studies have shown that reappraisal is associated with positive, and expressive suppression with negative consequences on mental health (for review, see 11). For instance, these relations were found using self-reported measures of habitual reappraisal and suppression, which indicated opposite links between these two emotion regulation strategies and depressive symptoms (45). Other evidence has indicated that the habitual use of reappraisal increases and that of suppression decreases during cognitive-behavioral psychotherapy for social anxiety, and variation in the use of both emotion regulation strategies correlates with the severity of social anxiety symptoms on a weekly basis (58). Furthermore, persistent reductions in social anxiety symptoms are predicted by increases in the self-reported success (i.e. perceived efficiency of regulation attempts) (58) and self-efficacy (i.e. confidence that emotion regulation will succeed) (59) of reappraisal during psychotherapy.

Individual differences in rumination have also been extensively linked to psychopathology. Rumination is increased in depression and anxiety (e.g. 60,61,62), and the level of rumination predicts the onset and duration of depressive episodes (e.g. 63). A recent meta-analysis (20) has indicated that the increased use of rumination is one of the best characterized emotion regulation difficulties in psychopathology.

While these studies focused on self-reported measures of various aspects (e.g. habitual use, efficiency, self-efficacy) of emotion regulation, other studies have used performance-based measures of the success of emotion regulation. For example, reappraisal ability has been assessed in cognitive tasks in which individuals are exposed to emotional stimuli such as images or films, with instructions to naturally respond to or decrease the emotional impact of these stimuli using reappraisal (64). The differences between the intensity of emotions in the two experimental conditions (i.e. look and decrease) has been used as an index of reappraisal ability. Using this experimental measure, evidence has emerged that reappraisal is associated with decreased depressive symptoms in individuals who report high levels of recent stress (65). In a similar fashion, the ability to suppress the behavioral expression of emotion has also been assessed in laboratory conditions which involved instructions to use this strategy while viewing emotional material, and both self-reported and independent evaluations of its success (66). This experimental measure of suppression ability was linked to distress following traumatic events (66).

This literature has suggested that psychopathology may be characterized by multiple emotion regulation difficulties. A recent extension of the process model (67) has provided a framework for categorizing and assessing emotion regulation difficulties associated with psychopathology. This model argues that there are three stages in the process of emotion regulation in which failure may occur: (i) identifying emotions that need regulation; (ii) selecting an emotion regulation strategy; and (iii) implementing the selected strategy. Emotion dysregulation may involve problems in one or more of these stages, such as not engaging in emotion regulation when it would be helpful to do so (i.e. emotion regulation failure), or using a form of emotion regulation that is poorly matched to the situation (i.e. emotion misregulation) (19,57). For instance, over- or underrepresentation of emotions that need to be regulated may characterize panic disorder (68) and manic states in bipolar I disorder (69), respectively. Selection of maladaptive emotion regulation strategies may be involved, for example, in depression, in which rumination is often preferred to reappraisal (20). Implementation of distraction may also be altered in depression considering that patients with major depression have difficulty recalling pleasant memories and show reduced emotional benefits from this process (70). In addition, processing dynamics difficulties, involving problems with interrupting the implementation of a regulation strategy and switching to another strategy, have also been described in affective and anxiety disorders (for review, see 71,72).

11.3 Systematic review

11.3.1 Search procedure

Articles were identified by searching the PubMed, PsycINFO, and Web of Science electronic databases, from their inception to September 2017, crossing keywords, including truncated terms, related to genetics (i.e. "genetic" or "polymorphism") and emotion regulation (i.e. "emotion regulation" or "expressive suppression" or "reappraisal" or "rumination" or "acceptance" or "distraction"). We also checked the reference list of eligible articles and related reviews.

Research reports were included if they involved a genetic measure such as heritability in twin studies or genotype in candidate gene studies, as well as a self-reported or experimental assessment of at least one emotion regulation strategy. Studies on coping and emotional intelligence were excluded because the measures derived from these concepts focus on more general domains (e.g. emotion- versus problem-focused coping), rather than specific regulation strategies. Furthermore, studies in which emotion regulation was not distinguished from emotional reactivity (e.g. emotional intelligence) or was inferred from its consequences on behavior (e.g. emotional resilience) rather than directly assessed were also excluded.

The relevance of every article was first assessed based on the title and abstract and then the full text. Two assessors independently examined full texts and selected eligible studies.

11.3.2 Search results and review

The systematic search returned 553 references. After screening the titles and abstracts, 468 articles were excluded because they did not meet the inclusion criteria or they were conference abstracts or duplicates. However, two extended abstracts, which reported quantitative data in more detail, were included. The full text of the remaining 85 articles was read and another 70 articles were excluded because although they used the concept of emotion regulation, they relied on conceptually distinct assessments of coping and emotional intelligence, or they did not directly assess emotion regulation. An additional 39 references were identified in the reference list of eligible research reports and related reviews, out of which 27 were also excluded because, after reading them in full, it became clear that they did not meet the inclusion criteria. Another report, recently submitted for publication by our group, was also added. Therefore, the systematic review included a total of 28 articles: 4 twin studies (Table 11.1), and 24 candidate gene studies (Table 11.2).

11.3.2.1 Reappraisal

The heritability of habitual reappraisal has been investigated in a recent twin study (73). Using a self-report assessment of the frequency of using reappraisal in daily life, this study indicated that a model including additive genetic and non-shared environmental influences provided the best

Table 11.1 Characteristics of twin studies on emotion regulation

Reference	Sample size (*N*)	Emotion regulation strategy (assessment method)	Other phenotypic measures (assessment instrument)
McRae et al. (73)	743 twin pairs (*N* = 1486)	Reappraisal (Emotion Regulation Questionnaire) Suppression (Emotion Regulation Questionnaire)	Trait neuroticism (NEO-Five Factor Inventory) Adaptive emotional functioning (BRISC Brief Risk–Resilience Index for Screening)
Chen & Li (89)	674 twin pairs (*N* = 1348)	Rumination (Children's Response Styles Questionnaire)	Depressive symptoms (Children's Depression Inventory: self-report and parent report versions)
Moore et al. (88)	378 twin pairs (*N* = 756)	Rumination (Ruminative Responses Scale) Distraction (Response Style Questionnaire)	Depressive symptoms (composite score including: Children's Depression Inventory; Early Adolescent Temperament Questionnaire—Revised; Health and Behavior Questionnaire; major depression disorder symptoms from the Diagnostic Interview Schedule for Children—IV)
Johnson et al. (90)	316 twin pairs (*N* = 632)	Rumination (Ruminative Responses Scale—10 items; Rumination–Reflection Questionnaire)	Depressive symptoms (composite score including: The Center for Epidemiological Studies Depression Scale; major depression disorder symptoms from the Diagnostic Interview Schedule—IV)

Table 11.2 Characteristics of candidate gene studies on emotion regulation

Reference	Sample size (*N*)	Gene (polymorphism)	Environmental assessments	Emotion regulation strategy (assessment method)	Other phenotypic measures
Schardt et al. (74)	*N* = 37	*SLC6A4* (43 bp indel or *5-HTTLPR*)	NA	Habitual reappraisal (Emotion Regulation Questionnaire) Reappraisal ability (cognitive task) Habitual suppression (Emotion Regulation Questionnaire)	Neuroticism (NEO Five Factor Inventory) Harm avoidance (Temperament and Character Inventory)
Miu et al. (76)	*N* = 182	*SLC6A4* (43 bp indel or *5-HTTLPR*; SNP rs25531)	NA	Habitual Reappraisal (Emotion Regulation Questionnaire; Cognitive Emotion Regulation Questionnaire)	Social anxiety symptoms (Liebowitz Social Anxiety Scale—Self-Report)
Ford et al. (79)	*N* = 205	*SLC6A4* (43 bp indel or *5-HTTLPR*; SNP rs25531)	Stress (Adolescent Life Events Questionnaire)	Habitual Reappraisal (Emotion Regulation Questionnaire adapted for children)	Depressive symptoms (Children's Depression Inventory)
Firk et al. (80)	*N* = 30	*SLC6A4* (43 bp indel or *5-HTTLPR*; SNP rs25531)	NA	Reappraisal ability (cognitive task)	NA
Miu et al. (81)	*N* = 254	*BDNF* (SNP rs6265 or Val66Met)	Childhood maltreatment (Childhood Experience of Care and Abuse Questionnaire)	Reappraisal ability (cognitive task)	Depressive symptoms (Beck Depression Inventory)
Bîlc et al. (82)	*N* = 266	*BDNF* (SNP rs6265 or Val66Met)	Childhood trauma (Childhood Traumatic Events Scale)	Reappraisal ability (cognitive task)	NA
Hilt et al. (91)	*N* = 200 (100 mother-daughter pairs)	*BDNF* (SNP rs6265 or Val66Met)	NA	Rumination (Response Style Questionnaire; Children's Response Style Questionnaire)	Depressive symptoms (Beck Depression Inventory; Structured Clinical Interview for DSM-IV; Computerized Diagnostic Interview Schedule for Children)

Stone et al. (93)	*N* = 228 (Study I) *N* = 87 (Study II)	*BDNF* (SNP rs6265 or Val66Met) *SLC6A4* (43 bp indel or *5-HTTLPR*; SNP rs25531)	History of maternal depression (Schedule for Affective Disorder and Schizophrenia-Lifetime Version)	Rumination (Study I: Children's Response Styles Scales; Study II: Ruminative Responses Scale)	Co-rumination frequency (Co-Rumination Questionnaire) Depressive symptoms (Children's Depression Inventory)
Van Hulle et al. (94)	*N* = 441	*BDNF* (SNP rs6265 or Val66Met) *CRHR1* gene (SNPs rs242924; rs720936)	History of maternal depression (Composite International Diagnostic Index)	Rumination (Ruminative Responses Scale)	Depressive Symptoms (Children's Depression Inventory) Pubertal Status (Pubertal Developmental Scale)
Beevers et al. (95)	*N* = 71	*BDNF* (SNP rs6265 or Val66Met) *SLC6A4* gene (43 bp indel or *5-HTTLPR*; SNP rs25531)	NA	Rumination (Ruminative Responses Scale)	Depressive Symptoms (Beck Depression Inventory; Inventory for Diagnosing Depression -Lifetime version)
Clasen et al. (96)	*N* = 273	*BDNF* (SNP rs6265 or Val66Met) *SLC6A4* gene (43 bp indel or *5-HTTLPR*; SNP rs25531)	Recent stressful events (Adverse Events Questionnaire)	Rumination (Ruminative Responses Scale)	Depressive Symptoms (Beck Depression Inventory; Inventory for Diagnosing Depression -Lifetime version)
Antypa et al. (97)	*N* = 250	*SLC6A4* gene (43 bp indel or *5-HTTLPR*; SNP rs25531)	Childhood emotional abuse (Childhood Trauma Questionnaire)	Rumination (Leiden Index of Depression Sensitivity)	Depressive and anxiety symptoms (Major Depression Questionnaire; Hospital Anxiety and Depression Scale) Neuroticism (Revised NEO Personality Inventory)
Canli et al. (98)	*N* = 21	*SLC6A4* gene (43 bp indel or *5-HTTLPR*; SNP rs25531)	Life stress history (Self-report version based on items from the Life History Calendar)	Rumination (Ruminative Responses Scale)	Extraversion and neuroticism (Revised NEO Personality Inventory) Positive and negative affect (Profile of Mood States)

(*continued*)

Table 11.2 Continued

Reference	Sample size (*M*)	Gene (polymorphism)	Environmental assessments	Emotion regulation strategy (assessment method)	Other phenotypic measures
Knyazev et al. (99)	*N* = 156	*SLC6A4* gene (43 bp indel or *5-HTTLPR*; SNP rs25531)	Stress (Holmes-Rahe stress scale)	Habitual suppression (Emotion Regulation Questionnaire; White Bear Suppression Inventory) Habitual reappraisal (Emotion Regulation Questionnaire) Rumination (Ruminative Responses Scale)	Depressive Symptoms (Beck Depression Inventory)
Eszlari et al. (100)	*N* = 1498	*HTR2A* (SNP rs3125)	Childhood adversity (Childhood Trauma Questionnaire)	Rumination (Ruminative Responses Scale)	Depressive symptoms (Brief Symptom Inventory)
Juhasz et al. (101)	*N* = 1269	*BDNF* (SNPs rs12273363; rs962369; rs988748; rs7127507; rs6265; rs1519480) *CREB1* (SNP rs2253206) *NTRK2* gene (SNPs rs1187323; rs1187326)	Stressful events (List of Life-Threatening Experiences) Childhood adversity (Childhood Trauma Questionnaire)	Rumination (Ruminative Responses Scale)	Depressive symptoms (Brief Symptom Inventory; Background Questionnaire)
Lazary et al. (102)	*N* = 651 *N* = 1174	*CREB1* (SNP rs2253206) *KCNJ6* gene (SNP rs2070995)	NA	Rumination (Ruminative Responses Scale)	Personality traits (Revised NEO Personality Inventory; Tridimensional Personality Questionnaire)
Pap et al. (105)	*N* = 939	*COMT* (SNPs rs933271; rs740603; rs4680; rs4646316)	NA	Rumination (Ruminative Responses Scale)	NA
Whitmer & Gotlib (108)	*N* = 634	*DRD2* (SNP rs6277)	NA	Rumination (Ruminative Responses Scale)	Depressive symptoms (Beck Depression Inventory)

Woody et al. (109)	*N* = 255	*CRHR1* (SNPs rs7209436; rs110402; rs242924)	History of maternal depression (Structured Clinical Interview for DSM-IV Axis I Disorders)	Rumination (Children's Response Style Scale)	Depressive symptoms (Schedule for Affective Disorders and Schizophrenia for School-Age Children—Present and Lifetime Version; Children's Depression Inventory)
Klok et al. (114)	*N* = 150	*NR3C2* (SNPs rs2070951; rs5522; six other new SNPs)	NA	Rumination (Leiden Index of Depression Sensitivity, LEIDS-R)	Neuroticism (Revised NEO Personality Inventory) Depressive Symptoms (Major Depression Questionnaire; Hospital Anxiety and Depression Scale)
Borelli et al. (118)	*N* = 99	*FKBP5* (SNP rs3800373)	Maternal overcontrol (UCS Parental Overcontrol Scale)	Habitual Suppression (Emotion Regulation for Children and Adolescents) Rumination (Children's Response Style Questionnaire)	Attachment (Security Scale) Depressive symptoms (Children's Depression Inventory) Cognitive ability task (Challenge Task) Respiratory Sinus Arrhythmia
Eszlari et al. (123)	*N* = 895 *N* = 1309	*MTHFR* (SNP rs1801133) *MTHFD1L* (SNP rs11754661)	NA	Rumination (Ruminative Responses Scale)	Depressive symptoms (New Mood Questionnaire; Brief Symptom Inventory)
Kim et al. (124)	*N* = 99 (Koreans) *N* = 152 (Americans)	*OXTR* (SNP rs53576)	NA	Habitual reappraisal (Emotion Regulation Questionnaire) Habitual suppression (Emotion Regulation Questionnaire)	NA

Abbreviations: NA, not applicable; SNP, single nucleotide polymorphism.

fit to the data. Habitual reappraisal showed low heritability (0.20) and the rest of its variance was explained by non-shared environmental contributions. There were common genetic and non-shared environmental influences between habitual reappraisal, on the one hand, and a self-report measure of adaptive emotional functioning including negative interpretation biases, emotional resilience, and social skills, on the other (73). In contrast, negative emotionality, as indicated by trait neuroticism, showed no significant genetic and environmental factors in common with habitual reappraisal. Therefore, this study suggested that reappraisal habits are mostly influenced by individual experience, and confirmed that some of these factors also contribute to emotional well-being. Notably, the low heritability of reappraisal has not been interpreted as discouraging in the search for candidate genes considering that, for instance, it could involve a simpler genetic background with fewer but larger genetic influences and gene–environment (G×E) interactions. In addition, reappraisal ability, which is not necessarily linked to the habitual use of this strategy, may show a distinct genetic background (73).

Several candidate gene studies examined habitual reappraisal. The first study (74) focused on the 43-bp insertion/deletion polymorphism in the serotonin transporter gene (*SLC6A4*) promoter (*5-HTTLPR*) (75) and found that carriers of the short (S) low-expressing allele reported lower levels of habitual reappraisal. This association was replicated in another study (76), in which genotyping included the rs25531 single-nucleotide polymorphism (SNP) in the long (L) high-expressing allele of *5-HTTLPR* (77,78), and habitual reappraisal was concurrently assessed by two self-report measures. The results converged on a gene–dosage relation between the low-expressing alleles (S or L_G, grouped as S') and habitual reappraisal, with the lowest levels of reappraisal in participants with two such alleles (S'S'), intermediary levels in those with one allele (S'L'), and the highest levels in those with no such allele (i.e. L_A homozygotes or L'L'). Moreover, this study showed that individual differences in habitual reappraisal were a significant mediator between the *5-HTTLPR* genotype and social anxiety symptoms (76). To our knowledge, only one other study (79) examined the association between *5-HTTLPR* and habitual reappraisal, in a sample of children. The results did not support the genetic association found in adults, but habitual reappraisal was involved in a three-way interaction with stress and the *5-HTTLPR* genotype, which was related to depressive symptoms. That is, higher levels of stress were associated with more depressive symptoms in S-allele carriers compared to L-allele homozygotes, but only at lower levels of habitual reappraisal (79).

Several candidate gene studies have also assessed reappraisal ability, using cognitive tasks in which reappraisal is instructed and the success of altering emotional responses to unpleasant images using this strategy is quantified (see 64). This task has been used in two genetic neuroimaging studies (74,80), which investigated the association between *5-HTTLPR* and reappraisal ability assessed at the subjective (i.e. based on affect ratings) and neural level. Both studies involved small samples and, in addition, comparison between results is limited by differences in affect rating considering that participants were asked to assess emotional valence (i.e. the unpleasantness of images) in the former study (74) and emotional arousal in the latter (80). Moreover, there were differences in the *5-HTTLPR* genotyping, with SNP rs25531 genotyped only in the latter study. The results indicated that reappraisal ability, as indicated by variation in emotional arousal during reappraisal compared to passive viewing, is reduced in carriers of two low-expressing *5-HTTLPR*/rs25531 alleles (80). Emotional valence ratings did not capture this effect (74), but this may have been due to the more limited variance of responses on this dimension of affect. The effects of *5-HTTLPR* genotype extended to the neural level, with increased prefrontal activity (74,80) and larger decreases of amygdala responses to fear-related images (74) during reappraisal in S-allele carriers. However, differences in the analysis of brain activity also preclude direct comparison between these studies.

Two recent candidate gene studies (81,82) have investigated reappraisal ability in larger samples. These studies were informed by a previous meta-analysis (83), which found consistent evidence that depression is influenced by a G×E interaction involving the rs6265 SNP in the brain-derived neurotrophic factor (*BDNF*) gene, also known as *BDNF* Val66Met, and the history of childhood maltreatment (see Chapter 21 this volume). Previous studies at the cellular level have indicated that neurons carrying the Met allele display reduced activity-dependent release of the BDNF protein (84), and this effect is echoed by multiple differences in neuroplasticity (85) and learning (86) at the systems level. As expected, reappraisal ability was reduced in people with a history of childhood maltreatment, but only in those who carried the *BDNF* Met allele (81). This finding parallels similar results on depression (83) and supports the view that reappraisal ability may represent an intermediate phenotype of psychopathology. Our study also suggested that the *BDNF* × childhood maltreatment interaction should be interpreted from a differential susceptibility perspective (87), considering the fact that Met carriers showed both the lowest and the highest level of reappraisal ability, depending on whether they did or did not have a history of childhood maltreatment (81). This G×E interaction was recently replicated in another similarly large sample (82), in which *BDNF* Met carriers with a history of childhood maltreatment, but also other types of trauma (i.e. parental conflicts, death of close others, and severe illness), displayed altered reappraisal ability.

In summary, available evidence from a twin study indicates that reappraisal has low heritability and non-shared environment explains most of the variance in the habitual use of this strategy. There are common genetic and environmental factors between habitual reappraisal and emotional functioning, which argues for the former as a potential intermediate phenotype of psychopathology. Evidence from candidate genes studies is also limited, but available evidence suggests that *5-HTTLPR* is associated with reappraisal frequency and ability, and *BDNF* Val66Met moderates the impact of childhood maltreatment on reappraisal ability.

11.3.2.2 Distraction

Genetic influences on distraction were investigated only in one twin study (88). Heritability explained 30% of the variance of distraction, and the remaining 70% of the variance was attributed to non-shared environment. The relation between distraction and depressive symptoms was also partially explained by common genetic and non-shared environmental factors. Considering that this relation was negative, these results suggest that the same genetic and non-shared environmental factors that contribute to the increased use of distraction may protect against depression (88).

11.3.2.3 Rumination

Three twin studies have investigated the heritability of rumination and its covariance with depressive symptoms. Two of these studies were in children and adolescents and found that genetic influences accounted for around 20% (88,89) of individual differences in rumination. The remaining variance was either exclusively (88) or mostly (89) explained by non-shared environment, with some of the variance (i.e. 20%) also attributed to shared environment in the latter study. This study included younger (11–14 years) and older (14–17 years) adolescents, and follow-up analyses suggested that the shared environment contribution to rumination is specific to young ages. Both twin studies (88,89) found that common genetic factors accounted for around 60% of the correlation between rumination and depressive symptoms, and the remaining variance was explained by non-shared environment. In a similar investigation carried out in adults, the heritability of rumination was estimated at 37% and 41% in women and men, respectively (90). The remaining variance was accounted for by non-shared environment and over 70% of the overlap between rumination and depressive symptoms was explained by common genetic influences (90).

Multiple candidate genes have been examined in relation to rumination. Interestingly, this line of research preceded twin studies and was driven by the hypothesis that rumination, which had already been extensively linked to depression, may represent a cognitive endophenotype of this mental disorder and may thus more readily capture genetic influences. One of the first candidate genes was *BDNF*, for which there had been mixed evidence on its association with depression. Hilt and colleagues (91) investigated the relation between *BDNF* Val66Met and rumination in a sample of adolescent girls and their mothers. Over half of the mothers had a history of depression, with early or adult onset. *BDNF* Met homozygotes were very few, which is typical of Caucasian samples (92), and they were excluded from analyses. Results indicated that adolescents who were homozygous for the *BDNF* Val allele reported higher levels of rumination and depressive symptoms. In contrast, a subgroup of mothers who carried the *BDNF* Met allele and had a history of adult-onset depression showed the same characteristics (91).

Several studies followed up the association between *BDNF* and rumination, and its possible moderation by developmental stage. Given the reduced frequency of the *BDNF* Met allele, Val homozygotes were compared to Met carriers in these studies. In a sample of children and adolescents, with ages between 8 and 14 years, Stone and colleagues (93) found no relation between *BDNF* Val66Met and rumination in the whole sample. However, a follow-up analysis that compared participants under and over the age of 11 (labelled "children" versus "adolescents") suggested that an effect may be found in the latter group, with higher rumination in Val homozygotes compared to Met carriers (93). It is noteworthy that the beginning of adolescence may be more reliably determined based on pubertal development, that is, the stage of physical changes that are driven by the rise of sexual hormones at puberty. Therefore, another study (94), which assessed pubertal development, may have been in a better position to study the interaction of *BDNF* and development in rumination. *BDNF* Val66Met was associated with rumination in the more mature, but not the less mature adolescents (94). However, in this study, the presence of the *BDNF* Met allele was associated with higher levels of rumination. Higher rumination in Met carriers was also reported in two other studies in adults (95,96).

The relation between *BDNF* Val66Met and rumination may also involve environmental factors and other genetic differences. Stressful events have been implicated in the etiology and pathogenesis of depression, and their interaction with genetic differences may increase the risk of depression by modulating the susceptibility to rumination. One study found that, indeed, the number of stressful events was positively related to the level of rumination and this association was enhanced in *BDNF* Met homozygotes compared to Val homozygotes (96). Moreover, this study also suggested that the effects of stressful events on rumination may be modulated by other genetic differences. While results in a small sample suggested no association between *5-HTTLPR* and rumination (95), two studies (97,98) reported opposing relations between negative life events and rumination depending on this genotype, with a positive association in carriers of low-expressing S' alleles and a negative association in L' homozygotes. Furthermore, Antypa and colleagues (97) found that individuals with the former *5-HTTLPR* genotype showed both the lowest and highest level of rumination, depending on whether their exposure to childhood stressful events was low or high, respectively. The *5-HTTLPR* genotype may therefore be interpreted as a differential susceptibility factor, which, depending on the quality of environment, may be associated with both benefits (i.e. low levels of rumination) and costs (i.e. high levels of rumination). Clasen and colleagues (96) found a similar interaction between the number of life-stressful events and the *5-HTTLPR* genotype, with a more positive slope in carriers of two S' alleles compared to L' homozygotes. They also indicated that the contributions of *BDNF* Val66Met and *5-HTTLPR* to individual differences in rumination following exposure to adverse environment may be additive. The total number of *BDNF* Met and *5-HTTLPR* S' alleles predicted a stronger association between the number of stressful events and the level of rumination (96). The *5-HTTLPR* × recent stressful events in rumination was not replicated in another study (99).

There is evidence that variation in the serotonin receptor 2A (*HTR2A*) gene may also contribute to individual differences in rumination. In two independent samples, Eszlari and colleagues (100) found that a SNP in the *HTR2A* gene moderates the impact of childhood trauma on brooding rumination, but not reflection.

Haplotypes and interactions between multiple genes involved in the same molecular pathway have also been investigated in relation to rumination. Juhasz and colleagues (101) genotyped polymorphisms in the *BDNF* gene, the neurotrophic tyrosine kinase receptor type 2 (*NTRK2*) gene, which encodes the BDNF receptor, and the cyclic adenosine monophosphate response element-binding protein 1 (*CREB1*) gene, a transcription factor which modulates the expression of the former two genes. They found that carriers of the minor, functionally less active alleles of both *BDNF* Val66Met and *CREB1* rs2553206 polymorphisms (i.e. the Met and A allele, respectively) reported reduced rumination (101). Individual differences in rumination were a mediator in the relation between these genotypes and depressive symptoms. In addition, a *BDNF* haplotype based on 6 SNPs, including Val66Met (rs6265), was also associated with rumination. Individuals with the TACTAT haplotype, which includes the Met or A allele of Val66Met in the fifth position, reported reduced levels of rumination (101). By contrast, carriers of the *BDNF* Met and *CREB1* A alleles, as well as those with the *NTRK2* T allele at the rs1187326 locus, showed increased risk for lifetime depression if they had a history of childhood maltreatment. However, this G×E interaction was independent of rumination. These results generally support the relation between *BDNF* Val66Met and rumination, and they expand these relations to other functionally linked genes such as *CREB1*. They also indicate that the *BDNF* Met allele at the rs6265 locus may be associated with both adaptive effects (i.e. reduced rumination) in the absence of childhood maltreatment, and maladaptive effects following exposure to this type of adverse event. Moreover, these results suggest that rumination is not the only mechanism involved in the relation between *BDNF* Val66Met genotype and depression, which warrants future studies on other emotion regulation strategies.

Another study (102) found that the same polymorphism (rs2553206) in the *CREB1* gene interacts with another polymorphism (rs2070995) in the *KCNJ6* gene, encoding a G-protein-activated potassium channel involved in neuronal excitability and synaptic signaling, and this interaction is related to rumination. While neither polymorphism was associated with individual differences in rumination, their interaction was significant in two independent samples. Moreover, the same epistatic or gene–gene (G×G) interaction was linked to Cluster C personality disorders characterized by anxious, fearful, and obsessive–compulsive traits, which may reflect increased rumination (102). This further emphasizes that the genetic underpinnings of rumination may involve complex interactions between multiple genes.

Other studies on rumination focused on genes related to dopamine signaling, including the catechol-O-methyltransferase (*COMT*) gene. COMT is one of the enzymes which degrade catecholamines such as dopamine, and may play a particularly important role in dopamine inactivation and efficient synaptic signaling in areas of the brain (e.g. prefrontal cortex) in which dopamine transporters are scarce (103). One of the first functional polymorphisms identified in this gene is *COMT* Val158Met (rs4680), and there is evidence that the Met allele is associated with a four-times more reduced activity of the COMT enzyme (104). This effect may become apparent during stress, when there is increased dopamine release (103), as well as in conditions that increasingly recruit prefrontal circuits, such as emotion regulation. Indeed, one study (105) reported an association between several polymorphisms in the *COMT* gene and rumination. *COMT* Val158Met was not independently related to rumination, but it was part of a four SNP haplotype which showed a significant relation to the habitual use of this strategy. Depending on the combination with alleles at the other three loci, both the Met and the Val alleles of the *COMT* Val158Met were associated with increased rumination (105).

The C957T (rs6277) polymorphism in the dopamine receptor D2 (*DRD2*) gene has also been associated with rumination. This SNP alters the translation and stability of DRD2 and results in a reduced availability of these receptors in individuals with the C- compared to the T-allele. Considering previous evidence that linked C957T to difficulties in learning from punishment (106), which are characteristic of depression (107), the association between this polymorphism and rumination was investigated in depressed and non-depressed individuals (108). The level of rumination was higher in C-allele homozygotes compared to the other genotype groups, but this effect was qualified by a genotype–depression interaction. The genetic association was only supported in depressed individuals, a finding which may be explained by the higher frequency of rumination in these individuals and the possibility that learning from punishment, which is altered in depression, is one of the underlying cognitive mechanisms (108).

Rumination has also been linked with polymorphisms in several genes related to the activity of the hypothalamic–pituitary–adrenal (HPA) axis. For instance, a recent study (109) followed up the association between a haplotype comprised of several SNPs in the corticotropin-releasing hormone receptor 1 (*CRHR1*) gene and depression, and focused on rumination. Previous studies (110,111) have indicated that individuals with the TAT haplotype at the locus of three SNPs in the *CRHR1* (i.e. rs7209436, rs110402, and rs242924), and a history of child maltreatment, show reduced depressive symptoms. In light of this evidence, Woody and colleagues (109) investigated whether the protective effect of this haplotype extended to rumination. They assessed rumination in a sample of children and focused on the interaction between the *CRHR1* TAT haplotype and maternal depression. This interaction was significant and maternal depression was associated with increased rumination only in children with no copies of the TAT haplotype (109).

Polymorphisms in the mineralocorticoid receptor (*NR3C2*) gene, which also plays an important role in cortisol response, have also been linked to individual differences in rumination. A haplotype comprising two SNPs in this gene (i.e. -2G/C or rs2070951, and I180V or rs5522) has been associated with enhanced translation and capacity to transactivate target genes in cell lines (112), and increased cortisol response to social stress (113). Given the involvement of cortisol response in psychopathology, Klok and colleagues (114) investigated the association between an extended *NR3C2* haplotype, including the two previously characterized and another six novel SNPs, and rumination in a sample of students. They found that one of the haplotypes was indeed related to reduced levels of rumination, as well as decreased aggression and risk aversion.

Studies on rumination have also focused on *FKBP5*, another gene related to the HPA axis. This gene encodes a co-chaperone molecule for the glucocorticoid receptor, which, together with another chaperone, hsp90, participate in the translocation of the glucocorticoid receptor complex in the nucleus and the modulation of glucocorticoid-responsive genes (115). An *FKBP5* haplotype comprising four SNPs was associated with increased risk for post-traumatic stress disorder in adults with a history of child abuse (116). Other studies have replicated and extended the evidence linking this gene to affective and anxiety disorders (for review, see 117) and a recent study has also approached rumination. In a sample of children, Borelli and colleagues (118) found a significant interaction between one of the known SNPs in the *FKBP5* gene and child attachment. Specifically, they reported that attachment security was related to decreased rumination only in children who carried the A-allele of SNP rs3800373, but not in those homozygous for the C-allele.

Other genetic studies in rumination focused on genes related to folate metabolism, which has been linked to psychopathology. Folate deficiency and increased homocysteine have been found in depression and other mental disorders (e.g. 119), and their role as etiologic factors or side effects of poor diet has been debated (120). In favor of the former view, these metabolic differences have also been related to altered monoamine metabolism (i.e. reduced cerebrospinal fluid levels of serotonin, dopamine, and norepinephrine catabolytes) in depression (121). Moreover, recent evidence from genome-wide association studies (GWAS) (122) suggests that genetic differences related to the folate metabolism are linked to affective disorders. In light of this evidence, Eszlari and colleagues (123) investigated the association between two SNPs in the methylenetetrahydrofolate reductase (*MTHFR*) gene and the human mitochondrial monofunctional 10-formyl-tetrahydrofolate synthetase (*MTHFD1L*) gene, and rumination. They found that only SNP rs11754661 in the *MTHFD1L* gene was significantly associated with the level of rumination, with higher levels in carriers of the A-allele compared to G-allele homozygotes. Moreover, individual differences in rumination explained the association between the *MTHFD1L* genotype and depressive symptoms (123).

In summary, individual differences in rumination show low or moderate heritability. The remaining variance in rumination is explained by non-shared environment. The candidate gene literature on rumination is more extensive compared to the other emotion regulation strategies and has focused on genes related to several molecular pathways: (i) neurotrophins (*BDNF, NTRK2, CREB1*); (ii) monoamine (i.e. dopamine, serotonin) metabolism (*COMT*) and signaling (*5-HTTLPR, DRD2*); (iii) neuronal excitability (*KCNJ6*); (iv) HPA axis activity (*CRHR1, NR3C2, FKBP5*); and (v) folate metabolism (*MTHFD1L*). These studies have suggested potential gene × development (G×D) (e.g. *BDNF* Val66Met and pubertal stage interaction; *FKBP5* and childhood attachment interaction), G×E (e.g. *BDNF* Val66Met and stressful events), and epistatic or G×G (e.g. *BDNF* Val66Met and *5-HTTLPR; BDNF* Val66Met and *CREB1; CREB1* and *KCNJ6*) interactions. There is also evidence that rumination is a mediator in the link between some of these genotypes and depression.

11.3.2.4 Suppression

To date, only one twin study (73) examined the heritability of expressive suppression. Using a self-report measure of the habitual use of this strategy, it was found that genetic influences and non-shared environment explained 35% and 65% of the variance of suppression. Moreover, some of the genetic and non-shared environmental influences were common between suppression and neuroticism (73).

Evidence from candidate gene studies on suppression is also very limited. One of the studies described in the section on rumination (118) also assessed habitual suppression in children genotyped for rs3800373 in the *FKBP5* gene. Results indicated that attachment security was negatively related to habitual suppression in children, but the effect was limited to those who carried the C-allele of the *FKBP5* polymorphism (118). Another study (99) reported that the interaction between *5-HTTLPR* and recent stressful events was not significantly associated with differences in habitual suppression.

Given that there are cultural differences in the habitual use of expressive suppression, with higher levels in Asian compared to European Americans (45), a candidate gene study (124) has examined the hypothesis that genetic factors moderate these cultural differences (i.e. gene–culture interaction or G×C). Korean and American individuals were genotyped for SNP rs53576 in the oxytocin receptor (*OXTR*) gene and reported habitual suppression. Koreans reported higher levels of habitual suppression compared to Americans, but this effect was qualified by a G×C interaction. Compared to A-allele homozygotes, G-allele carriers showed higher levels of suppression in the Korean sample and lower levels of suppression in the American sample. These results suggested that the *OXTR* G allele was associated with increased sensitivity to cultural norms regarding suppression. This interpretation was also supported by the increased similarity of Asian Americans to European Americans, rather than Koreans (124).

In summary, habitual suppression has displayed moderate heritability and most of its variance has been explained by non-shared environment. While the relation between *5-HTTLPR* and suppression was not supported in one study, two other studies reported positive findings related to the G×E interaction between *FKBP5* and childhood attachment, and the G×C interaction between *OXTR* and cultural norms related to emotion expression.

11.4 **Conclusions and future directions**

Research on the genetics of emotion regulation is a recent development, but several lines of evidence have already emerged. Four emotion regulation strategies (i.e. reappraisal, distraction, rumination, suppression) have been investigated in twin studies and they all showed low or moderate heritability, and predominant contributions from non-shared environment. Through this combination of underlying factors, emotion regulation resembles most other psychological characteristics such as personality and intelligence (125,126) (for review, see Chapter 2 this volume). The results from twin studies were corroborated by significant genetic associations with functional polymorphisms in multiple genes, as well as G×E interactions. Moreover, this literature has uncovered G×D interactions suggesting that genetic influences on emotion regulation are moderated by developmental changes in adolescence, and G×C interactions which have taken environmental differences to a higher level.

The future potential of this research area could draw from recent advances in both emotion regulation research and genetics. First, as illustrated by the studies reviewed in this chapter, there is potential in the availability of multiple methods to assess emotion regulation, using self-report and experimental measures. For instance, to date, candidate gene studies on rumination have exclusively relied on questionnaires, and the accessibility of this method has probably contributed to the success of collecting data from larger samples and afforded more efforts being invested in identifying new gene candidates. Evidence on reappraisal is more limited in terms of the average sample size and the number of genes investigated, but its complementary strength is that it has employed both self-report and performance-based measures. Using multiple methods to assess emotion regulation will be important in future research, considering that these assessments do not completely overlap (e.g. 17,127) and they are differentially linked to psychopathology (e.g. 128). In light of this evidence, a recent twin study (73) in which self-reported habitual reappraisal showed low heritability emphasized the need to follow up this finding using an experimental measure of reappraisal ability. Indeed, the two measures show different aspects of reappraisal (i.e. frequency and efficiency), and may involve distinct combinations of genetic and environmental influences. Future studies could also take advantage of established experimental methods to induce and assess other emotion regulation strategies such as rumination and distraction (e.g. 53), as well as suppression (e.g. 44,66).

Another development which could inform future genetic studies is related to the increased focus on difficulties in specific stages of emotion regulation. The extended model of emotion regulation (67) has emphasized that psychopathology may involve one or more failures in identifying the emotions that need regulation, choosing the best regulation strategies, and successfully implementing them, for example. Recent research has started to uncover differences in regulation effort between emotions (e.g. 129) and the moderator role of emotional awareness in emotion regulation (e.g. 130). Moreover, there is increasing evidence that emotion regulation choice or the preference for certain regulation strategies over others is not only influenced by individual differences such as habits, but also by characteristics of the emotional response (e.g. intensity) (16) and the situation (e.g. the expectancy to encounter a similar event in the future) (131). Finally, difficulties in implementing specific emotion regulation strategies, including problems in interrupting a strategy and switching to another one, have started to be examined in psychopathology (for review, see 19,57). Some of these emotion regulation difficulties could be transdiagnostic, while others may be specific to certain mental disorders. Considering that interest in the genetics of emotion regulation stems from the hypothesis that emotion regulation is a mechanism of psychopathology and may therefore represent a valuable intermediate phenotype, future twin and candidate studies could increasingly focus on stage-specific difficulties in emotion regulation.

Recent progress in understanding emotion regulation has also come from field studies, which have employed ecological momentary assessment. This approach, also known as experience sampling, involves administering short questionnaires over the phone using applications that deliver prompts at random times throughout the study interval. In addition to minimizing recall biases, this approach also allows analyzing within-individual variation and its situational and psychological correlates (for review, see 132,133). Recent studies on emotion regulation have indicated that multiple regulation strategies are typically employed in daily life (e.g. 134,135). Therefore, this work highlights the importance of differences in emotion regulation repertoire (i.e. the number of available strategies), and flexibility in adapting emotion regulation to context, which may contribute to risk for psychopathology (for review, see 71,136). To date, only one study focused on the influence of *5-HTTLPR* on emotion dynamics in daily life (137), and future studies on emotion regulation are warranted.

Another avenue for future genetic research is related to using neural measures of emotion regulation. Functional magnetic resonance imaging (fMRI) studies have examined brain activity during the implementation of emotion regulation strategies such as reappraisal and suppression (138), and distraction (139). Extensive evidence has correlated reappraisal ability with the functional connectivity between prefrontal areas (e.g. ventromedial and ventral anterior cingulate areas) (140), as well as dorsomedial, dorsolateral, and ventrolateral prefrontal areas and posterior parietal areas (141), which have been involved in cognitive control, and the amygdala, which has been involved in emotion generation. A pioneering neuroimaging genetics study has found that cingulate–amygdala functional coupling may be reduced in carriers of the S-allele of *5-HTTLPR*, and this neural phenotype explained a large portion of the variance in self-reported anxiety (142) (see Chapter 16 this volume). Similar studies have followed, some of which were reviewed in this chapter (e.g. 74,80). For instance, functional connectivity of the amygdala with frontal cortical areas and other brain regions was recently related to G×E interactions between several polymorphisms in genes related to HPA activity including *CRHR1*, *NR3C1*, and *FKBP2*, and stressful events in children (143). However, emotion regulation has not been directly assessed in most of these studies and therefore, future research could help bridge the gap between neural activity and emotional experience, and describe the potentially different impact of genetic differences on these two levels. Available evidence (e.g. 74) suggests that neural measures may capture genetic influences better than self-report measures of reappraisal, for example, which is in line with the endophenotype theory. However, this hypothesis needs to be instantiated further in larger samples and in relation to other emotion regulation strategies.

Future studies could also investigate genetic and environmental influences on emotion regulation change following psychotherapy. Both self-reported changes in habitual emotion regulation (58) and self-efficacy (59), for instance, and neural changes in emotion regulation (58,144) support the view that emotion regulation is a mechanism of change in cognitive-behavioral therapy. Therefore, genetic influences on emotion regulation could extend to the susceptibility to emotion regulation change following psychotherapy. This hypothesis has been supported in recent therapygenetics research, which showed that the response to cognitive-behavioral therapy may be associated with genotypes such as *5-HTTLPR* and *BDNF* Val66Met (for review, see Chapter 26 this volume). However, these studies exclusively focused on outcomes related to emotional symptoms.

This evidence could be extended by investigating genetic influences on emotion regulation change throughout and following psychotherapy. From an endophenotype perspective, it is possible that assessing this mechanism of change instead of emotional symptoms may more readily capture genetic effects.

Finally, in light of the burgeoning evidence on the involvement of emotion regulation in psychopathology and the genetic underpinnings of emotion regulation, GWAS are in line (see Chapters 5 and 14 this volume). To our knowledge, there is only one published GWAS on emotion regulation difficulties (145) assessed using a questionnaire that focuses on general domains of emotion and emotion regulation (e.g. emotion intensity and the influence of emotion on cognition and behavior), which cut across emotion regulation strategies (and which was therefore, not reviewed in this chapter). Moreover, recent GWASs on neuroticism—a dimension of personality which reflects emotional instability (see Chapter 19 this volume), and shares genetic and environmental influences with habitual suppression (73)—have been successful in identifying multiple candidate genes (e.g. 146,147). Until recently, GWAS have mostly focused on SNPs and their marginal effects on phenotypes, but methods to investigate larger structural variants (e.g. copy number variations) (e.g. 148) and genome-wide G×E interactions (e.g. 149) are underway and will be extremely useful in characterizing the complex genetics of emotion regulation.

Acknowledgements

This work was supported by grant PN-III-P4-ID-PCE-2016-0840 from the Romanian Ministry of Research and Innovation, through CNCS–UEFISCDI.

References

1. **Davidson RJ, Scherer KR, Goldsmith HH.** *Handbook of Affective Sciences.* New York, NY: Oxford University Press; 2009.
2. **Gross JJ, Barrett LF.** The emerging field of affective science. Emotion. 2013;**13**(6):997–8.
3. **Dolan RJ.** Emotion, cognition, and behavior. Science. 2002;**298**(5596):1191–4.
4. **Dolcos F, Denkova E.** Current emotion research in cognitive neuroscience: Linking enhancing and impairing effects of emotion on cognition. Emot Rev. 2014;**6**(4):362–75.
5. **Power M, Dalgleish T.** *Cognition and Emotion: From Order to Disorder*, 2nd edn. New York, NY: Psychology Press; 2008.
6. **Bagozzi RP, Baumgartner H, Pieters R, Zeelenberg M.** The role of emotions in goal-directed behavior. In: Ratneshwar S, Mick DG, Huffman C (eds). *The Why of Consumption.* New York, NY: Routledge; 2000, pp. 36–58.
7. **Baumeister RF, Vohs KD, DeWall CN, Zhang LQ.** How emotion shapes behavior: Feedback, anticipation, and reflection, rather than direct causation. Personality and Social Psychology Review. 2007;**11**(2):167–203.
8. **DeSteno D, Gross JJ, Kubzansky L.** Affective science and health: the importance of emotion and emotion regulation. Health Psychol. 2013;**32**(5):474–86.
9. **McEwen BS.** Stress, adaptation, and disease: Allostasis and allostatic load. Ann N Y Acad Sci. 1998;**840**:33–44.
10. **Ryff CD, Singer BH.** The role of emotion on pathways to positive health. In: Davidson RJ, Scherer KR, Goldsmith HH (eds). *Handbook of Affective Sciences.* New York, NY: Oxford University Press; 2003, pp. 1083–104.
11. **Gross JJ, Jazaieri H.** Emotion, emotion regulation, and psychopathology: An affective science perspective. Clin Psychol Sci 2014;**2**(4):387–401.
12. **Gross JJ.** The emerging field of emotion regulation: an integrative review. Rev Gen Psychol. 1998;**2**(3):271–99.
13. **Gross JJ.** Emotion regulation: Conceptual and empirical foundations. In: Gross JJ, editor. *Handbook of Emotion Regulation.* New York, NY: Guilford Press; 2014, pp. 3–20.

14. **Tamir M.** Why do people regulate their emotions? A taxonomy of motives in emotion regulation. Pers Soc Psychol Rev. 2016;**20**(3):199–222.
15. **Gross JJ.** Antecedent- and response-focused emotion regulation: Divergent consequences for experience, expression, and physiology. J Pers Soc Psychol. 1998;**74**(1):224–37.
16. **Sheppes G, Scheibe S, Suri G, Gross JJ.** Emotion-regulation choice. Psychol Sci. 2011;**22**(11):1391–6.
17. **McRae K, Jacobs SE, Ray RD, John OP, Gross JJ.** Individual differences in reappraisal ability: Links to reappraisal frequency, well-being, and cognitive control. J Res Pers. 2012;**46**(1):2–7.
18. **Insel T, Cuthbert B, Garvey M, Heinssen R, Pine DS, Quinn K,** et al. Research domain criteria (RDoC): Toward a new classification framework for research on mental disorders. Am J Psychiatry. 2010;**167**(7):748–51.
19. **Fernandez KC, Jazaieri H, Gross JJ.** Emotion regulation: A transdiagnostic perspective on a new RDoC domain. Cognit Ther Res. 2016;**40**(3):426–40.
20. **Aldao A, Nolen-Hoeksema S, Schweizer S.** Emotion-regulation strategies across psychopathology: A meta-analytic review. Clin Psychol Rev. 2010;**30**(2):217–37.
21. **Doherty JL, Owen MJ.** Genomic insights into the overlap between psychiatric disorders: implications for research and clinical practice. Genome Med. 2014;**6**(4):29.
22. **Canli T, Lesch KP.** Long story short: the serotonin transporter in emotion regulation and social cognition. Nat Neurosci. 2007;**10**(9):1103–9.
23. **Hariri AR, Holmes A.** Genetics of emotional regulation: The role of the serotonin transporter in neural function. Trends Cogn Sci. 2006;**10**(4):182-91.
24. **Canli T, Ferri J, Duman EA.** Genetics of emotion regulation. Neuroscience. 2009;**164**(1):43–54.
25. **Hawn SE, Overstreet C, Stewart KE, Amstadter AB.** Recent advances in the genetics of emotion regulation: A review. Curr Opin Psychol. 2015;**3**:108–16.
26. **Compas BE, Jaser SS, Dunbar JP, Watson KH, Bettis AH, Gruhn MA,** et al. Coping and emotion regulation from childhood to early adulthood: Points of convergence and divergence. Aust J Psychol. 2014;**66**(2):71–81.
27. **Pena-Sarrionandia A, Mikolajczak M, Gross JJ.** Integrating emotion regulation and emotional intelligence traditions: A meta-analysis. Front Psychol. 2015;**6**:160.
28. **Dunn SH, Conley YP.** A systematic review of genetic influences on coping. Biol Res Nurs. 2015;**17**(1):87–93.
29. **Frijda NH.** Emotion, cognitive structure, and action tendency. Cogn Emot. 1987;**1**(2):115–43.
30. **Lang PJ.** The varieties of emotional experience: A meditation on James–Lange theory. Psychol Rev. 1994;**101**(2):211–21.
31. **Lazarus RS.** Cognition and motivation in emotion. Am Psychol. 1991;**46**(4):352-67.
32. **LeDoux JE.** Cognitive–emotional interactions in the brain. Cogn Emot. 1989;**3**(4):267–89.
33. **Dalgleish T, Dunn BD, Mobbs D.** Affective neuroscience: Past, present, and future. Emot Rev. 2009;**1**(4):355–68.
34. **Schachter S, Singer JE.** Cognitive, social, and physiological determinants of emotional state. Psychol Rev. 1962;**69**:379–99.
35. **Ackermann KH, Adams N, Adler C, Ahammed Z, Ahmad S, Allgower C,** et al. Elliptic flow in Au+Au collisions at square root(S)NN = 130 GeV. Phys Rev Lett. 2001;**86**(3):402–7.
36. **Moors A, Ellsworth PC, Scherer KR, Frijda NH.** Appraisal theories of emotion: State of the art and future development. Emot Rev. 2013;**5**(2):119–24.
37. **Oatley K, Johnson-Laird PN.** Cognitive approaches to emotions. Trends Cogn Sci. 2014;**18**(3):134–40.
38. **Smith CA, Ellsworth PC.** Patterns of cognitive appraisal in emotion. J Pers Soc Psychol. 1985;**48**(4):813–38.
39. **Webb TL, Miles E, Sheeran P.** Dealing with feeling: A meta-analysis of the effectiveness of strategies derived from the process model of emotion regulation. Psychol Bull. 2012;**138**(4):775–808.

40. **Gross JJ.** Emotion regulation: Affective, cognitive, and social consequences. Psychophysiology. 2002;**39**(3):281–91.
41. **Gross JJ.** Emotion regulation: Taking stock and moving forward. Emotion. 2013;**13**(3):359–65.
42. **Sheppes G, Gross JJ.** Is timing everything? Temporal considerations in emotion regulation. Pers Soc Psychol Rev. 2011;**15**(4):319–31.
43. **Richards JM, Gross JJ.** Emotion regulation and memory: The cognitive costs of keeping one's cool. J Pers Soc Psychol. 2000;**79**(3):410–24.
44. **Butler EA, Wilhelm FH, Gross JJ.** Respiratory sinus arrhythmia, emotion, and emotion regulation during social interaction. Psychophysiology. 2006;**43**(6):612–22.
45. **Gross JJ, John OP.** Individual differences in two emotion regulation processes: Implications for affect, relationships, and well-being. J Pers Soc Psychol. 2003;**85**(2):348–62.
46. **Sheppes G, Meiran N.** Better late than never? On the dynamics of online regulation of sadness using distraction and cognitive reappraisal. Pers Soc Psychol Bull. 2007;**33**(11):1518–32.
47. **Sheppes G, Catran E, Meiran N.** Reappraisal (but not distraction) is going to make you sweat: Physiological evidence for self-control effort. Int J Psychophysiol. 2009;**71**(2):91–6.
48. **Sheppes G, Meiran N.** Divergent cognitive costs for online forms of reappraisal and distraction. Emotion. 2008;**8**(6):870–4.
49. **Sheppes G, Gross JJ.** Emotion generation and emotion regulation: Moving beyond traditional dual-process accounts. In: Gawronski B, Sherman JW, Trope Y (eds). *Dual-Process Theories of the Social Mind*. New York, NY: Guilford Press; 2014, pp. 483–93.
50. **Nolen-Hoeksema S, Wisco BE, Lyubomirsky S.** Rethinking rumination. Perspect Psychol Sci. 2008;**3**(5):400–24.
51. **Thomsen DK.** The association between rumination and negative affect: A review. Cogn Emot. 2006;**20**(8):1216–35.
52. **Nolen-Hoeksema S.** Responses to depression and their effects on the duration of depressive episodes. J Abnorm Psychol. 1991;**100**(4):569–82.
53. **Nolen-Hoeksema S, Morrow J.** Effects of rumination and distraction on naturally occurring depressed mood. Cogn Emot. 1993;7(6):561–70.
54. **Lyubomirsky S, Nolen-Hoeksema S.** Effects of self-focused rumination on negative thinking and interpersonal problem solving. J Pers Soc Psychol. 1995;**69**(1):176–90.
55. **Treynor W, Gonzalez R, Nolen-Hoeksema S.** Rumination reconsidered: A psychometric analysis. Cognit Ther Res. 2003;**27**(3):247–59.
56. **Gross JJ, Sheppes G, Urry HL.** Emotion generation and emotion regulation: A distinction we should make (carefully). Cogn Emot. 2011;**25**(5):765–81.
57. **Sheppes G, Suri G, Gross JJ.** Emotion regulation and psychopathology. Annu Rev Clin Psychol. 2015;**11**:379–405.
58. **Goldin PR, Ziv M, Jazaieri H, Weeks J, Heimberg RG, Gross JJ.** Impact of cognitive-behavioral therapy for social anxiety disorder on the neural bases of emotional reactivity to and regulation of social evaluation. Behav Res Ther. 2014;**62**:97–106.
59. **Goldin PR, Ziv M, Jazaieri H, Werner K, Kraemer H, Heimberg RG,** et al. Cognitive reappraisal self-efficacy mediates the effects of individual cognitive-behavioral therapy for social anxiety disorder. J Consult Clin Psychol. 2012;**80**(6):1034–40.
60. **Aldao A, Nolen-Hoeksema S.** Specificity of cognitive emotion regulation strategies: a transdiagnostic examination. Behav Res Ther. 2010;**48**(10):974–83.
61. **D'Avanzato C, Joormann J, Siemer M, Gotlib IH.** Emotion regulation in depression and anxiety: Examining diagnostic specificity and stability of strategy use. Cognit Ther Res. 2013;**37**(5):968–80.
62. **McLaughlin KA, Hatzenbuehler ML, Mennin DS, Nolen-Hoeksema S.** Emotion dysregulation and adolescent psychopathology: A prospective study. Behav Res Ther. 2011;**49**(9):544–54.

63. **Nolen-Hoeksema S.** The role of rumination in depressive disorders and mixed anxiety/depressive symptoms. J Abnorm Psychol. 2000;**109**(3):504–11.

64. **Ochsner KN, Ray RD, Cooper JC, Robertson ER, Chopra S, Gabrieli JD**, et al. For better or for worse: Neural systems supporting the cognitive down- and up-regulation of negative emotion. Neuroimage. 2004;**23**(2):483–99.

65. **Troy AS, Wilhelm FH, Shallcross AJ, Mauss IB.** Seeing the silver lining: Cognitive reappraisal ability moderates the relationship between stress and depressive symptoms. Emotion. 2010;**10**(6):783–95.

66. **Bonanno GA, Papa A, Lalande K, Westphal M, Coifman K.** The importance of being flexible—The ability to both enhance and suppress emotional expression predicts long-term adjustment. Psychol Sci. 2004;**15**(7):482–7.

67. **Gross JJ.** Emotion regulation: Current status and future prospects. Psychol Inq. 2015;**26**(1):1–26.

68. **McNally RJ.** Anxiety sensitivity and panic disorder. Biol Psychiatry. 2002;**52**(10):938–46.

69. **Gruber J.** A review and synthesis of positive emotion and reward disturbance in bipolar disorder. Clin Psychol Psychother. 2011;**18**(5):356–65.

70. **Joormann J, Siemer M, Gotlib IH.** Mood regulation in depression: Differential effects of distraction and recall of happy memories on sad mood. J Abnorm Psychol. 2007;**116**(3):484–90.

71. **Bonanno GA, Burton CL.** Regulatory flexibility: An individual differences perspective on coping and emotion regulation. Perspect Psychol Sci. 2013;**8**(6):591–612.

72. **Mancebo MC, Eisen JL, Grant JE, Rasmussen SA.** Obsessive compulsive personality disorder and obsessive compulsive disorder: Clinical characteristics, diagnostic difficulties, and treatment. Ann Clin Psychiatry. 2005;**17**(4):197–204.

73. **McRae K, Rhee SH, Gatt JM, Godinez D, Williams LM, Gross JJ.** Genetic and environmental influences on emotion regulation: A twin study of cognitive reappraisal and expressive suppression. Emotion. 2017;**17**(5):772–7.

74. **Schardt DM, Erk S, Nusser C, Nothen MM, Cichon S, Rietschel M**, et al. Volition diminishes genetically mediated amygdala hyperreactivity. Neuroimage. 2010;**53**(3):943–51.

75. **Lesch KP, Bengel D, Heils A, Sabol SZ, Greenberg BD, Petri S**, et al. Association of anxiety-related traits with a polymorphism in the serotonin transporter gene regulatory region. Science. 1996;**274**(5292):1527–31.

76. **Miu AC, Vulturar R, Chis A, Ungureanu L, Gross JJ.** Reappraisal as a mediator in the link between 5-HTTLPR and social anxiety symptoms. Emotion. 2013;**13**(6):1012–22.

77. **Hu XZ, Lipsky RH, Zhu G, Akhtar LA, Taubman J, Greenberg BD**, et al. Serotonin transporter promoter gain-of-function genotypes are linked to obsessive-compulsive disorder. Am J Hum Genet. 2006;**78**(5):815–26.

78. **Wendland JR, Martin BJ, Kruse MR, Lesch KP, Murphy DL.** Simultaneous genotyping of four functional loci of human SLC6A4, with a reappraisal of 5-HTTLPR and rs25531. Mol Psychiatry. 2006;**11**(3):224–6.

79. **Ford BQ, Mauss IB, Troy AS, Smolen A, Hankin B.** Emotion regulation moderates the risk associated with the 5-HTT gene and stress in children. Emotion. 2014;**14**(5):930–9.

80. **Firk C, Siep N, Markus CR.** Serotonin transporter genotype modulates cognitive reappraisal of negative emotions: a functional magnetic resonance imaging study. Soc Cogn Affect Neurosci. 2013;**8**(3):247–58.

81. **Miu AC, Carnuta M, Vulturar R, Szekely-Copindean RD, Bilc MI, Chis A**, et al. BDNF Val66Met polymorphism moderates the link between child maltreatment and reappraisal ability. Genes Brain Behav. 2017;**16**(4):419–26.

82. **Bilc MI, Vulturar R, Chiş A, Buciuman M, Nutu D, Bunea I**, et al. Childhood trauma and emotion regulation: The moderator role of BDNF Val66Met. Neurosci Lett. 2018;**685**:7–11

83. **Hosang GM, Shiles C, Tansey KE, McGuffin P, Uher R.** Interaction between stress and the BDNF Val66Met polymorphism in depression: A systematic review and meta-analysis. BMC Med. 2014;**12**:7.

84. **Egan MF, Kojima M, Callicott JH, Goldberg TE, Kolachana BS, Bertolino A**, et al. The BDNF val66met polymorphism affects activity-dependent secretion of BDNF and human memory and hippocampal function. Cell. 2003;**112**(2):257–69.
85. **Bath KG, Lee FS.** Variant BDNF (Val66Met) impact on brain structure and function. Cogn Affect Behav Neurosci. 2006;**6**(1):79–85.
86. **Dincheva I, Glatt CE, Lee FS.** Impact of the BDNF Val66Met polymorphism on cognition: implications for behavioral genetics. Neuroscientist. 2012;**18**(5):439–51.
87. **Belsky J, Pluess M.** Beyond diathesis stress: differential susceptibility to environmental influences. Psychol Bull. 2009;**135**(6):885–908.
88. **Moore MN, Salk RH, Van Hulle CA, Abramson LY, Hyde JS, Lemery-Chalfant K**, et al. Genetic and environmental influences on rumination, distraction, and depressed mood in adolescence. Clin Psychol Sci. 2013;**1**(3):316–22.
89. **Chen J, Li X.** Genetic and environmental influences on adolescent rumination and its association with depressive symptoms. J Abnorm Child Psychol. 2013;**41**(8):1289–98.
90. **Johnson DP, Whisman MA, Corley RP, Hewitt JK, Friedman NP.** Genetic and environmental influences on rumination and its covariation with depression. Cogn Emot. 2014;**28**(7):1270–86.
91. **Hilt LM, Sander LC, Nolen-Hoeksema S, Simen AA.** The BDNF Val66Met polymorphism predicts rumination and depression differently in young adolescent girls and their mothers. Neurosci Lett. 2007;**429**(1):12–6.
92. **Petryshen TL, Sabeti PC, Aldinger KA, Fry B, Fan JB, Schaffner SF**, et al. Population genetic study of the brain-derived neurotrophic factor (BDNF) gene. Mol Psychiatry. 2010;**15**(8):810–5.
93. **Stone LB, McGeary JE, Palmer RH, Gibb BE.** Identifying genetic predictors of depression risk: 5-HTTLPR and BDNF Val66Met polymorphisms are associated with rumination and co-rumination in adolescents. Front Genet. 2013;**4**:246.
94. **Van Hulle CA, Clifford S, Moore MN, Lemery-Chalfant K, Goldsmith HH.** Partial replication of two rumination-related candidate gene studies. Cogn Emot. 2017;**31**(5):963–71.
95. **Beevers CG, Wells TT, McGeary JE.** The BDNF Val66Met polymorphism is associated with rumination in healthy adults. Emotion. 2009;**9**(4):579–84.
96. **Clasen PC, Wells TT, Knopik VS, McGeary JE, Beevers CG.** 5-HTTLPR and BDNF Val66Met polymorphisms moderate effects of stress on rumination. Genes Brain Behav. 2011;**10**(7):740–6.
97. **Antypa N, Van der Does AJ.** Serotonin transporter gene, childhood emotional abuse and cognitive vulnerability to depression. Genes Brain Behav. 2010;**9**(6):615–20.
98. **Canli T, Qiu M, Omura K, Congdon E, Haas BW, Amin Z**, et al. Neural correlates of epigenesis. PNAS. 2006;**103**(43):16033–8.
99. **Knyazev GG, Bazovkina DV, Savostyanov AN, Naumenko VS, Kuznetsova VB, Proshina EA.** Suppression mediates the effect of 5-HTTLPR by stress interaction on depression. Scand J Psychol. 2017;**58**(5):373–8.
100. **Eszlari N, Kovacs D, Petschner P, Gonda X, Elliott R, Anderson I**, et al. Brooding subtype of rumination is modulated by the interplay between serotonin receptor 2A gene and childhood adversity. Eur Neuropsychopharmacol. 2016;**26**:S77–S8.
101. **Juhasz G, Dunham JS, McKie S, Thomas E, Downey D, Chase D**, et al. The CREB1–BDNF–NTRK2 pathway in depression: Multiple gene–cognition–environment interactions. Biol Psychiatry. 2011;**69**(8):762–71.
102. **Lazary J, Juhasz G, Anderson IM, Jacob CP, Nguyen TT, Lesch KP**, et al. Epistatic interaction of CREB1 and KCNJ6 on rumination and negative emotionality. Eur Neuropsychopharmacol. 2011;**21**(1):63–70.
103. **Stein DJ, Newman TK, Savitz J, Ramesar R.** Warriors versus worriers: The role of COMT gene variants. CNS Spectr. 2006;**11**(10):745–8.
104. **Lachman HM, Papolos DF, Saito T, Yu YM, Szumlanski CL, Weinshilboum RM.** Human catechol-O-methyltransferase pharmacogenetics: Description of a functional polymorphism and its potential application to neuropsychiatric disorders. Pharmacogenetics. 1996;**6**(3):243–50.

105. **Pap D, Juhasz G, Bagdy G.** Association between the COMT gene and rumination in a Hungarian sample. Neuropsychopharmacol Hung. 2012;**14**(4):285–92.
106. **Frank MJ, Hutchison K.** Genetic contributions to avoidance-based decisions: Striatal D2 receptor polymorphisms. Neuroscience. 2009;**164**(1):131–40.
107. **Whitmer AJ, Frank MJ, Gotlib IH.** Sensitivity to reward and punishment in major depressive disorder: Effects of rumination and of single versus multiple experiences. Cogn Emot. 2012;**26**(8):1475–85.
108. **Whitmer AJ, Gotlib IH.** Depressive rumination and the C957T polymorphism of the DRD2 gene. Cogn Affect Behav Neurosci. 2012;**12**(4):741–7.
109. **Woody ML, Kudinova AY, McGeary JE, Knopik VS, Palmer RH, Gibb BE.** Influence of maternal depression on children's brooding rumination: Moderation by CRHR1 TAT haplotype. Cogn Emot. 2016;**30**(2):302–14.
110. **Bradley RG, Binder EB, Epstein MP, Tang Y, Nair HP, Liu W**, et al. Influence of child abuse on adult depression: moderation by the corticotropin-releasing hormone receptor gene. Arch Gen Psychiatry. 2008;**65**(2):190–200.
111. **Polanczyk G, Caspi A, Williams B, Price TS, Danese A, Sugden K**, et al. Protective effect of CRHR1 gene variants on the development of adult depression following childhood maltreatment: Replication and extension. Arch Gen Psychiatry. 2009;**66**(9):978–85.
112. **van Leeuwen N, Bellingrath S, de Kloet ER, Zitman FG, DeRijk RH, Kudielka BM**, et al. Human mineralocorticoid receptor (MR) gene haplotypes modulate MR expression and transactivation: Implication for the stress response. Psychoneuroendocrinology. 2011;**36**(5):699–709.
113. **DeRijk RH, Wust S, Meijer OC, Zennaro MC, Federenko IS, Hellhammer DH** et al. A common polymorphism in the mineralocorticoid receptor modulates stress responsiveness. J Clin Endocrinol Metab. 2006;**91**(12):5083–9.
114. **Klok MD, Giltay EJ, Van der Does AJ, Geleijnse JM, Antypa N, Penninx BW**, et al. A common and functional mineralocorticoid receptor haplotype enhances optimism and protects against depression in females. Transl Psychiatry. 2011;**1**:e62.
115. **Davies TH, Ning YM, Sanchez ER.** A new first step in activation of steroid receptors: hormone-induced switching of FKBP51 and FKBP52 immunophilins. J Biol Chem. 2002;**277**(7):4597–600.
116. **Binder EB, Bradley RG, Liu W, Epstein MP, Deveau TC, Mercer KB**, et al. Association of FKBP5 polymorphisms and childhood abuse with risk of posttraumatic stress disorder symptoms in adults. JAMA. 2008;**299**(11):1291–305.
117. **Binder EB.** The role of FKBP5, a co-chaperone of the glucocorticoid receptor in the pathogenesis and therapy of affective and anxiety disorders. Psychoneuroendocrinology. 2009;**34** Suppl 1:S186–95.
118. **Borelli JL, Smiley PA, Rasmussen HF, Gomez A, Seaman LC, Nurmi EL.** Interactive effects of attachment and FKBP5 genotype on school-aged children's emotion regulation and depressive symptoms. Behav Brain Res. 2017;**325**(Pt B):278–89.
119. **Assies J, Mocking RJ, Lok A, Ruhe HG, Pouwer F, Schene AH.** Effects of oxidative stress on fatty acid- and one-carbon-metabolism in psychiatric and cardiovascular disease comorbidity. Acta Psychiatr Scand. 2014;**130**(3):163–80.
120. **Reynolds EH.** Folic acid, ageing, depression, and dementia. BMJ. 2002;**324**(7352):1512–5.
121. **Bottiglieri T, Laundy M, Crellin R, Toone BK, Carney MW, Reynolds EH.** Homocysteine, folate, methylation, and monoamine metabolism in depression. J Neurol Neurosurg Psychiatry. 2000;**69**(2):228–32.
122. **Network and Pathway Analysis Subgroup of Psychiatric Genomics Consortium.** Psychiatric genome-wide association study analyses implicate neuronal, immune and histone pathways. Nat Neurosci. 2015;**18**(2):199–209.
123. **Eszlari N, Kovacs D, Petschner P, Pap D, Gonda X, Elliott R**, et al. Distinct effects of folate pathway genes MTHFR and MTHFD1L on ruminative response style: A potential risk mechanism for depression. Transl Psychiatry. 2016;**6**:e745.

124. **Kim HS, Sherman DK, Mojaverian T, Sasaki JY, Park J, Suh EM,** et al. Gene-Culture Interaction: Oxytocin Receptor Polymorphism (OXTR) and Emotion Regulation. Soc Psychol Pers Sci. 2011;**2**(6):665–72.
125. **Polderman TJ, Benyamin B, de Leeuw CA, Sullivan PF, van Bochoven A, Visscher PM,** et al. Meta-analysis of the heritability of human traits based on fifty years of twin studies. Nat Genet. 2015;**47**(7):702–9.
126. **Bouchard TJ.** Genetic influence on human psychological traits—A survey. Curr Dir Psychol Sci. 2004;**13**(4):148–51.
127. **Gutentag T, Halperin E, Porat R, Bigman YE, Tamir M.** Successful emotion regulation requires both conviction and skill: Beliefs about the controllability of emotions, reappraisal, and regulation success. Cogn Emot. 2017;**31**(6):1225–33.
128. **Ford BQ, Karnilowicz HR, Mauss IB.** Understanding reappraisal as a multicomponent process: The psychological health benefits of attempting to use reappraisal depend on reappraisal success. Emotion. 2017;**17**(6):905–11.
129. **Dixon-Gordon KL, Aldao A, De Los Reyes A.** Emotion regulation in context: Examining the spontaneous use of strategies across emotional intensity and type of emotion. Pers Individ Dif. 2015;**86**:271–6.
130. **Szczygieł D, Buczny J, Bazińska R.** Emotion regulation and emotional information processing: The moderating effect of emotional awareness. Pers Individ Dif. 2012;**52**(3):433–7.
131. **Sheppes G, Scheibe S, Suri G, Radu P, Blechert J, Gross JJ.** Emotion regulation choice: A conceptual framework and supporting evidence. J Exp Psychol Gen. 2014;**143**(1):163–81.
132. **Ebner-Priemer UW, Trull TJ.** Ecological momentary assessment of mood disorders and mood dysregulation. Psychol Assess. 2009;**21**(4):463–75.
133. **Myin-Germeys I, Oorschot M, Collip D, Lataster J, Delespaul P, van Os J.** Experience sampling research in psychopathology: Opening the black box of daily life. Psychol Med. 2009;**39**(9):1533–47.
134. **Brans K, Koval P, Verduyn P, Lim YL, Kuppens P.** The regulation of negative and positive affect in daily life. Emotion. 2013;**13**(5):926–39.
135. **Heiy JE, Cheavens JS.** Back to basics: A naturalistic assessment of the experience and regulation of emotion. Emotion. 2014;**14**(5):878–91.
136. **Aldao A.** The future of emotion regulation research: Capturing context. Perspect Psychol Sci. 2013;**8**(2):155–72.
137. **van Roekel E, Verhagen M, Engels R, Kuppens P.** Variation in the serotonin transporter polymorphism (5-HTTLPR) and inertia of negative and positive emotions in daily life. Emotion. 2018;**18**(2):229–36.
138. **Goldin PR, McRae K, Ramel W, Gross JJ.** The neural bases of emotion regulation: reappraisal and suppression of negative emotion. Biol Psychiatry. 2008;**63**(6):577–86.
139. **McRae K, Hughes B, Chopra S, Gabrieli JD, Gross JJ, Ochsner KN.** The neural bases of distraction and reappraisal. J Cogn Neurosci. 2010;**22**(2):248–62.
140. **Etkin A, Buchel C, Gross JJ.** The neural bases of emotion regulation. Nat Rev Neurosci. 2015;**16**(11):693–700.
141. **Buhle JT, Silvers JA, Wager TD, Lopez R, Onyemekwu C, Kober H,** et al. Cognitive reappraisal of emotion: A meta-analysis of human neuroimaging studies. Cereb Cortex. 2014;**24**(11):2981–90.
142. **Pezawas L, Meyer-Lindenberg A, Drabant EM, Verchinski BA, Munoz KE, Kolachana BS,** et al. 5-HTTLPR polymorphism impacts human cingulate-amygdala interactions: A genetic susceptibility mechanism for depression. Nat Neurosci. 2005;**8**(6):828–34.
143. **Pagliaccio D, Luby JL, Bogdan R, Agrawal A, Gaffrey MS, Belden AC,** et al. Amygdala functional connectivity, HPA axis genetic variation, and life stress in children and relations to anxiety and emotion regulation. J Abnorm Psychol. 2015;**124**(4):817–33.

144. **Goldin PR, Ziv M, Jazaieri H, Hahn K, Heimberg R, Gross JJ.** Impact of cognitive behavioral therapy for social anxiety disorder on the neural dynamics of cognitive reappraisal of negative self-beliefs: randomized clinical trial. JAMA Psychiatry. 2013;**70**(10):1048–56.

145. **Powers A, Almli L, Smith A, Lori A, Leveille J, Ressler KJ**, et al. A genome-wide association study of emotion dysregulation: Evidence for interleukin 2 receptor alpha. J Psychiatr Res. 2016;**83**:195–202.

146. **Lo MT, Hinds DA, Tung JY, Franz C, Fan CC, Wang Y**, et al. Genome-wide analyses for personality traits identify six genomic loci and show correlations with psychiatric disorders. Nat Genet. 2017;**49**(1):152–6.

147. **Smith DJ, Escott-Price V, Davies G, Bailey ME, Colodro-Conde L, Ward J**, et al. Genome-wide analysis of over 106 000 individuals identifies 9 neuroticism-associated loci. Mol Psychiatry. 2016;**21**(6):749–57.

148. **Manolio TA, Collins FS, Cox NJ, Goldstein DB, Hindorff LA, Hunter DJ**, et al. Finding the missing heritability of complex diseases. Nature. 2009;**461**(7265):747–53.

149. **Murcray CE, Lewinger JP, Gauderman WJ.** Gene–environment interaction in genome-wide association studies. Am J Epidemiol. 2009;**169**(2):219–26.

Chapter 12

Emotional memory

Mana R Ehlers and Rebecca M Todd

12.1 Introduction

Do you remember the moment you learned that two planes had flown into the towers of the World Trade Center in New York? Most of us who are old enough to have experienced the event retain a vivid memory of where we were and what we were doing that day in 2001. Indeed, a good deal of research supports the intuitive observation that we typically remember emotionally arousing events more clearly (although not more accurately) than mundane ones, reliving the birth of a child or a teenage humiliation with a high degree of vividness decades later (1). The exceptionally vivid qualities of emotionally laden memories have been observed for hundreds of years, raising many theories about possible mechanisms of recall. For example, popular theories in the sixteenth century held that moisture and temperature in the brain are important mediators of emotional memory function (2). In contrast, in the early nineteenth century, under the influence of Franz Josef Gall's doctrine of phrenology, it was thought that emotion and memory were entirely separate processes (2). Over recent decades, our understanding of emotionally enhanced memory has been strongly informed by research in non-human animals, which has provided empirical evidence for underlying neural mechanisms. Still more recently, cognitive neuroscientists have endeavored to investigate patterns of human brain activation underlying the memory-enhancing effect of emotions. This program of research extends to examination of individual differences in the emotional memory system due to both genetic predisposition and environmental factors.

This chapter will provide an overview of emotional memory from a cognitive neuroscience perspective, integrating knowledge about key neurotransmitter systems and genes. We will focus on *emotional arousal* (as opposed to *valence*, or positivity–negativity) as a key driving property of enhanced encoding and consolidation of memories for emotionally salient events. In order to elucidate this idea, we will discuss the relation to stress and genetic polymorphisms related to the arousal/norepinephrine system. The review will cover both explicit and implicit emotional memory processes.

12.2 Emotional memory

Typically, when we describe the kind of emotional memories that transfer us back to another time—the scent of freshly baked cookies deeply connected to a feeling of safety and home in childhood, or a song that may remind us of a loved one who has passed away—we refer to long-term, explicit, or episodic memory of emotionally salient events. However, emotional memory is an umbrella term that covers distinct memory systems. Emotional memory can refer to both *explicit* (conscious) episodic memories in which we are able to time travel to relive emotionally salient events (3) as well as *implicit* (unconscious) memories of associations between actions or neutral cues and emotionally arousing events. In the latter case, we can find ourselves responding emotionally to a cue that is associated with pain or pleasure without being able to recall the specific event in which the association was formed.

When we characterize the kind of emotional event that is remembered, we typically describe it by its valence (how pleasant or unpleasant it is) and by its arousal (how exciting it is, often reflected in effects of physiological arousal). In the current chapter, when we talk about emotional enhancement of memory or learning, we will primarily discuss it irrespective of its valence and refer to all memory content that is emotionally arousing. However, it is important to emphasize that the relation between emotion and memory is also defined by the emotional state of the learner at the time the memory is acquired. Beyond memories for major historical events, such as 9/11, or personal milestones such as births, deaths, and weddings, a state of high arousal or stress will typically result in enhanced memory for a situation that is otherwise mundane (4).

Not all of us experience these emotional enhancement effects equally. The field is just beginning to appreciate the range of individual differences in the degree of emotional enhancement of memory in healthy populations (5). Moreover, at the extreme ends of the range, susceptibility to emotional enhancement of memory can be maladaptive (6). Even when exposed to the same event or when learning the same emotional association, only some individuals develop traumatic memories or develop long-lasting or life-crippling associations or behaviors (7). This raises the question of which factors account for such individual differences. Previous research has shown that common genetic variations are among the factors that can influence capacity and vividness of emotional memory.

This chapter will review what is currently known about the role of emotional arousal of memory content and internal state of the learner in influencing both explicit and implicit memory performance. We will then focus on evidence on the role played by common genetic polymorphisms influencing arousal and neurotransmitter systems in modulating brain and behavior related to emotional memory performance.

12.3 Explicit emotional memory

Effects of enhanced episodic memory for emotional relative to non-emotional material are now well established by a large body of literature (for review see 8–10). Certainly, this phenomenon makes intuitive sense; from an evolutionary perspective it is highly adaptive to deploy one's limited cognitive resources to remember what is motivationally relevant. For example, remembering how and when to avoid life-threatening danger or to secure food and reproductive resources is crucial for survival. Thus, investigation of the neurocognitive mechanisms underlying this evolutionarily conserved capacity has been an important focus in emotional memory research.

Researchers characterize general episodic memory processes in three stages. *Encoding* refers to the initial encounter with information as well as its conversion into storable material (11). Encoding is followed by *consolidation*, the phase in which storage of information is stabilized and transferred into long-term memory. Short-term consolidation harnesses molecular processes necessary for the formation of new synaptic connections and structural change of existing ones that takes place over the course of hours (12). Finally, *retrieval* refers to the process of accessing the remembered content. Across all stages of memory, research has shown that emotional enhancement of memory is subserved by the engagement of additional brain regions, which modulate neural systems that mediate episodic memory formation (13). Moreover, emotional events recruit a unique cascade of hormones and neurotransmitters that facilitate multiple stages of memory formation (14).

More precisely, specific cognitive and neurochemical factors have been found to facilitate emotional content at each stage of episodic memory formation and recall (9). At the time of encoding, at the neurobiological level, enhanced physiological arousal linked to greater amygdala activation enhances memory (15, 16). At the psychological level, attentional processes facilitate prioritization of emotionally salient aspects of the environment (17), and this attentional enhancement is in turn associated with heightened emotional memory (18). Such biases towards the motivationally salient are particularly potent in situations where attentional resources are limited (17, 19). Similarly, emotional memories have an advantage during consolidation, as hormones and neurotransmitters acting on the amygdala and other brain regions in response to emotionally arousing stimuli promote long-term memory consolidation (20). Such neurochemical advantages may be complemented by differences in psychological functioning following salient events, as enhanced consolidation has also been associated with more frequent rehearsal of emotionally salient memories (21). Finally, emotional events also show an advantage during memory retrieval, and this effect too is associated with greater amygdala activation (22). At the psychological level, at least in some populations (older adults), effects of emotion on retrieval have been associated with greater elaboration and cognitive control (23). We will elaborate further on the evidence for neurobiological processes subserving emotional enhancement of episodic memory at each stage.

12.3.1 **Neuroanatomical substrates of emotionally enhanced memory**

Building on non-human animal research, studies of human lesion patients provided the first evidence of human brain systems critical for emotional enhancement of memory. In particular, early findings indicated that the amygdala is crucial for emotional memory formation at every stage (24, 25). Neurological conditions that selectively affect the amygdala are rare (26), and patients who suffer from them have been intensely studied (27, 28). One of those rare conditions is Urbach–Wiethe syndrome, which in individual cases leads to targeted pathology of bilateral amygdala (28). Post-surgical studies of patients who underwent unilateral medial temporal lobectomy due to epilepsy have further enhanced our understanding of the role of the amygdala in emotional episodic memory (8). While there are conflicting findings related to specific processes impaired by amygdala lesions, possibly due to differences in the extent of damage and when in life the damage occurred, injury to the medial temporal lobe or amygdala has been consistently linked to impairment of emotional memory (8, 28, 29). Following up on these seminal lesion studies, neuroimaging studies have helped elaborate extended brain networks implicated in emotional enhancement of memory at each stage.

12.3.1.1 Memory phases

Encoding. Recent research has provided more detailed mapping of the brain circuitry subserving each stage of emotional memory. Enhancement of attentional processes during encoding are mediated by networks in which the amygdala and ventral prefrontal cortices (orbitofrontal/ventromedial prefrontal) serve as hubs for prioritization of emotionally relevant sensory processing (Figure 12.1) (30, 31). During encoding, the amygdala is sensitive to signals representing emotionally relevant content from both cortical and subcortical sources, and plays a role rerouting resources to other brain systems to enhance visual processing, attention, and memory (32, 33). Specifically, it enhances visual cortex activation for enhanced "seeing" of emotionally salient aspects of the world (34, 35). Co-activation between the amygdala, visual cortex, and ventromedial prefrontal cortex (VMPFC) have further been associated with prioritization of emotional stimuli (19), as well as emotional enhancement of perceived perceptual vividness, which in turn predicts subsequent memory vividness (36, 37). Crucially, some patients with amygdala lesions have been found to lack the capacity for emotional prioritization at encoding (38), suggesting impaired capacity to segregate the significant from the mundane may partly underlie their deficits in emotional memory (39).

In addition, emotional arousal at encoding—at least under some circumstances—has been found to narrow the focus of attention to encode central information preferentially, enhancing memory for negative information at the expense of peripheral or contextual information (40). However, there is conflicting evidence about whether negative information enhances encoding of gist or detail (41–43). Critically, characteristic effects of emotional information at encoding are impaired in patients with amygdala lesions (44). In turn, such amygdala activity during encoding has been shown to directly predict subsequent memory performance, including memory vividness (45, 46), and interact with activation in high-level visual cortex to support subsequent recollection detail (47). Hippocampus activation is also important for successful encoding of emotional episodic memories. Studies in patient populations have revealed that pathology of both the amygdala and hippocampus predict memory performance and that their activity during encoding is mutually dependent (44). Additional neuroimaging studies have firmly supported the importance of the hippocampus in encoding of emotionally enhanced memory (21, 48). Thus, convergent evidence suggests that the amygdala and hippocampus, as well as VMPFC, function as key nodes in extended networks mediating enhanced encoding of emotional material.

Consolidation. Human and non-human animal studies have demonstrated that increasing arousal/amygdala activation after encoding of an event typically boosts subsequent memory in interaction with arousal at encoding (49, 50). According to the modulation hypothesis (51), the role of the amygdala during memory consolidation is to modulate activity in other brain regions critical for memory formation such as the hippocampus, the caudate nucleus, insula, entorhinal cortex, and prefrontal regions (52, 53). Memory maintenance in the cortex is further promoted through amygdala activation of the stress hormone system (14). Nonetheless, post-encoding arousal can also have negative effects on subsequent memory, and the direction of effects has been found to depend in part on how arousal interacts with task-related goals (54). Yet, whatever the direction of effects, the amygdala has been found to function as a central hub mediating effects of emotional arousal on memory consolidation (55).

Retrieval. While the modulation hypothesis emphasizes amygdala involvement in encoding and consolidation of emotional memories, fMRI studies have demonstrated an important role for activation of the amygdala, alongside other brain systems, during memory retrieval (22, 56). For example, one study found enhanced activity in several structures related to episodic memory during recognition of stimuli that were learned in an emotional relative to neutral context. Among those regions were the amygdala, other regions sensitive to emotional salience such as the OFC and anterior cingulate cortex, as well as hippocampus and prefrontal regions associated with episodic memory systems (57). Another fMRI study demonstrated enhanced activity in amygdala and other temporal lobe regions for successful retrieval of emotions relative to neutral pictures one year after encoding (56). Thus, as in encoding, the amygdala plays an important role in emotional enhancement of memory alongside nodes of episodic memory networks.

12.3.2 Neuromodulators

The large number of studies demonstrating an enhancement of emotional memory—particularly for events that elicit arousal—suggests that the release of stress hormones plays an important role in modulating influences on memory processes (51). The release of norepinephrine (NE) from the locus coeruleus (LC) of the brainstem is central to our physical response to arousing stimuli. In concert with glucocorticoids (GCs), these stress hormones act on different brain regions, and a major target is the amygdala, which is particularly rich in NE (58, 59) and GC receptors (60). It is through the action of neuromodulators released in response to arousal, that the amygdala exerts its influence on other regions, such as the hippocampus, to promote consolidation of emotionally arousing information.

Norepinephrine. The amygdala is thought to tag emotionally relevant stimuli by noradrenergic action that in turn provides the basis of an emotional memory trace (61). Our own BANE model (31) has emphasized the role of the LC in modulating nodes of affective salience networks in the emotional enhancement of encoding processes (Figure 12.1). Early pharmacological studies demonstrated that the adrenergic system in the amygdala is essential for emotional memory performance. In humans, one seminal study demonstrated that a blockade of beta-adrenergic receptors impairs emotional memory performance (62). In this study, the beta-adrenergic antagonist propranolol was used to block noradrenergic action before participants were told an emotional or neutral story. The study revealed that the application of propranolol impaired memory for the emotional but not for the neutral story.

Neuroimaging studies further demonstrated that the memory impairment associated with noradrenergic beta receptor blockade is associated with reduced amygdala activity (63, 64). Similarly, stimulation of the noradrenergic system leads to enhanced memory for emotional events (65, 66). Thus, a series of studies provide evidence for a noradrenergic modulation of amygdala activation as an essential component of emotional memory formation.

Glucocorticoids. Noradrenergic activation of the amygdala has been further shown to be a prerequisite for the effects of glucocorticoids (GCs) on emotional memory as demonstrated by an abolished effect of GC manipulation after amygdala lesion or beta receptor blockade (61, 67). GCs are released upon hypothalamic–pituitary–adrenal (HPA) axis activation, which constitutes the slower stress response relative to the fast release of epinephrine and norepinephrine (61). Many animal studies have demonstrated that enhanced GC levels during learning facilitate memory consolidation (68, 69).

Several human studies also report beneficial effects of GCs or acute stress exposure on emotional memory consolidation. GC treatment or stress exposure before or immediately after learning have been associated with enhanced memory retrieval, especially for emotionally arousing material (50, 70). Consistent with the role of amygdala and GC activity in memory consolidation, a recent study demonstrated that individual variations in cortisol levels in response to emotional stimuli correlated with amygdala activation (71). In contrast, both animal and human research have shown that GCs have detrimental effects on memory retrieval (72–74), an effect that is once again dependent on noradrenergic action in the basolateral amygdala (75, 76).

Taken together, animal and human research demonstrate the importance of the interplay of NE and GCs for emotional memory consolidation, consistent with a key role for arousal—in the content of the encoded information or the emotional state of the learner (e.g. acute stress)—in emotional memory formation.

12.4 **Implicit emotional memory**

As mentioned earlier, emotional memory is not confined to declarative or conscious episodic memory. Non-declarative or implicit emotional memory is most widely studied in the form of conditioning, which can be either appetitive (reward learning) or aversive (fear conditioning), and fear-conditioning studies are very prevalent in the literature. In fear conditioning, an animal or human rapidly (often unconsciously) learns to associate an initially neutral stimulus (CS+) with an aversive stimulus or event (US) that is associated with an automatic fear response (77). After learning, the presentation of the CS+ alone leads to a fear response until a continuous presentation of the CS+ without its associated US leads to extinction of fear. Fear conditioning has been widely studied in non-human animals as well as in humans.

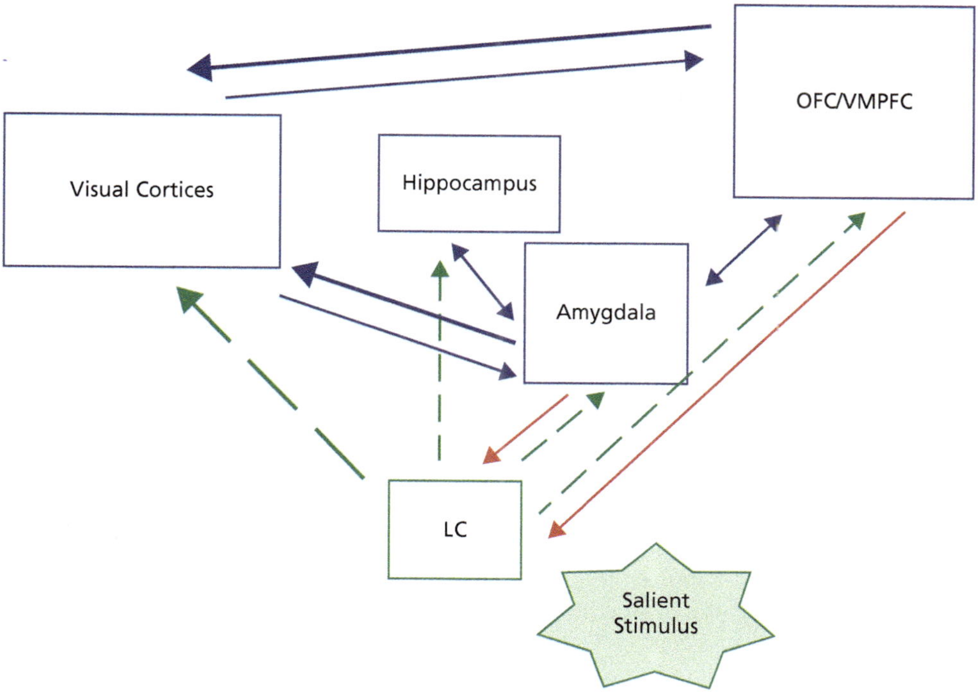

Figure 12.1 Key pathways emphasized by BANE model. Green dashed lines indicate norepinephrine (NE) pathways. Red lines indicate projections to the locus coeruleus (LC). Thicker lines indicate direct modulation of visual cortex activity in affect-biased attention. NE activity is implicated in both stimulus encoding and selective attention (149). A salient stimulus activates LC neurons, which project widely to cortical and subcortical regions. LC neurons are highly sensitive to previously acquired associations between a stimulus and punishment or reward. Descending influences from amygdala (central nucleus) and ventral prefrontal cortices (OFC/VMPFC) provide information about contextually determined relevance, which can then modulate the pattern of LC firing accordingly (150). Activity from LC can modulate activity in visual cortex directly, facilitating gating and tuning of neuronal activity and enhancing perceptual acuity directly as well as via the amygdala and prefrontal cortices. The amygdala receives contextual information from the hippocampus as well as prefrontal regions and in turn modulates hippocampus activity related to memory encoding and consolidation.

Reprinted from *Behavioural and Brain Research*, 259(1), Markovic J, Anderson AK, and Todd RM, 'Tuning to the significant: Neural and genetic processes underlying affective enhancement of visual perception and memory', pp. 229–41,

12.4.1 Animal literature

Rodents are most often the experimental animal of choice in fear-conditioning studies. Tones, lights, or experimental chambers (for context conditioning) are typical CS+ stimuli, whereas mild electric shocks are often used as a US. Robust fear conditioning can be achieved after only a few continuous pairings of the CS with the US. The contingency between CS and US markedly influences the degree of fear learning (77). The typical fear response in rats includes a range of behaviors such as freezing (78) or potentiated startle (79). Fear can also be learned quickly through instrumental or operant conditioning. In this active form of conditioning, the animal learns that a certain action is followed by an aversive event. A typical paradigm is the passive avoidance task.

In this design, an animal learns that an automatic behavior such as avoiding the light by entering a dark compartment of a box is associated with, for example, a footshock (80).

Because of the reliability and simplicity, as well as the high degree of control the experimenter has over all aspects of the learning process, fear conditioning has been used as a tool to understand the emergence of fear and the cellular and molecular mechanisms of emotional learning (81).

12.4.2 Human literature

Similar to animal research, there is a wide range of possible CS and US, an experimenter can choose from when studying fear conditioning in humans (82). While auditory or even olfactory CS can be found in the literature, visual conditioned stimuli are by far the most common type. A majority of researchers employs photographs specifically of human faces or geometrical figures. Similarly, US differ in stimulus modality as well as in salience and unpleasantness. Among the most common US, are electric shocks, air blasts, and auditory cues. Fear learning can be assessed by skin conductance response (SCR), eye blink reflex, and subjective stimulus ratings (8).

12.4.3 Neuroanatomical substrates of implicit fear learning

An extensive body of literature based on research from several laboratories performed in the 1980s provides us with a relatively clear picture of the neuroanatomy underlying fear conditioning in rodents. The amygdala plays a central role in fear conditioning, as it integrates information about the CS and US and controls fear responses via projections to autonomic and endocrine control systems in the brainstem (81).

In humans, the essential role of the amygdala in fear conditioning was initially demonstrated by lesion studies. Lesions of the central nucleus of the amygdala result in attenuation of heart rate conditioning (83). Lesions of the amygdala are further associated with impairments in both cue and context conditioning. In contrast, targeted lesions of the hippocampus lead to impaired context conditioning but not simple cue conditioning (84), indicating a dissociation between the roles of these two structures. These findings support the proposal that the amygdala but not the hippocampus exerts an associative role during all forms of fear conditioning.

While it is well established that the VMPFC is involved in extinction of learned fear association by suppressing amygdala function through interneurons (85, 86), recent findings suggest that the prefrontal cortex also modulates fear response in the amygdala and is essential for fear expression (87). In conclusion, the non-human animal literature supports the notion that the amygdala in interplay with hippocampus and ventral prefrontal cortex is a key player in the neural circuit underlying fear conditioning.

In line with the animal literature, human research has revealed that fear conditioning and fear responses depend on an intact amygdala and hippocampus (88, 89). However, unlike rodents, humans can acquire conscious knowledge of stimulus associations (90). Previous research has shown that the explicit knowledge of stimulus relationships depends on the hippocampus, while damage to the amygdala impairs the formation on implicit emotional associations (91). Thus, there is evidence that in humans the amygdala is key to the evolutionarily conserved capacity to associate stimuli with punishment (and reward), whereas a hippocampal contribution is more important for declarative memory.

12.4.4 Neuromodulators

Recent studies further suggest that neuromodulators such as norepinephrine (NE) and dopamine (DA) regulate plasticity processes during fear learning (92). We review the specific role of each of these neuromodulators in the following.

Norepinephrine. Previous research has produced mixed results regarding the involvement of the noradrenergic system in fear conditioning. While some studies demonstrate that pharmacological blockade of norepinephrine in the lateral nucleus of the amygdala abolishes fear acquisition in rats (93, 94), another study suggested that these effects are specific to contextual fear conditioning, at least in humans (95). Consistent with the latter finding is another animal study showing that infusions of norepinephrine into the basolateral amygdala after training enhances consolidation and hence contextual fear conditioning (96). A recent review paper suggests that the inconsistent findings are likely to stem from effects specific to tasks, the specific noradrenergic receptor involved, and, importantly, from interactions with other neurotransmitter systems (92). Current evidence supports the view that NE affects the acquisition but not consolidation of implicit fear memory by promoting plasticity (92), while NE affects all stages of explicit memory formation as reviewed above.

Dopamine. Several studies also suggest a role of dopaminergic action in the acquisition of fear memories (97–99). An influence of dopamine on fear expression was further demonstrated by a study showing that injection of an indirect dopaminergic agonist can potentiate fear responses (100). However, much of the previous literature revealed inconsistent results regarding the effect of dopaminergic transmission on fear learning and expression. The mixed results can be partially explained by differences in pharmacological manipulation and the influence of different dopaminergic receptor subtypes. A detailed review of dopaminergic pathways involved in fear conditioning can be found elsewhere (101).

Glucocorticoids. The effects of GC treatment or acute stress on fear consolidation are similar to those described for declarative memory. The literature shows general agreement that GCs facilitate contextual fear conditioning (102, 103), while other studies also report a potentiation of cue conditioning (104, 105). The discrepancy might be explained by differences in GC dosage or stress intensity (102). Some studies suggest that GC effects on fear conditioning are dependent on noradrenergic activity in the basolateral amygdala (BLA) as reported for explicit memory (104).

12.5 Genetic influences on emotional memory

Genetic variability is a major source of individual differences in memory performance. Estimates from twin studies suggest that 30–60 % of variability in episodic memory can be accounted for by genetic variation (106, 107). However, as described earlier, emotional memory is a complex construct involving different memory systems and brain circuits that are modulated by neurotransmitters and hormones in unique ways. As a result, single genetic variations may contribute to only certain aspects of memory formation or recall (108).

Genetic research is not only of interest to researchers who want to understand genetic sources of individual differences in memory performance. Well-studied candidate genes for which there is sufficient knowledge about biological mechanisms can also be used as a tool to investigate effects of neuromodulators and hormones on emotional memory and learning processes in more detail (109, 110). The advantage of examining genetic variations that affect neurotransmitter systems, for example, is that we can directly and non-invasively observe the effect of naturally occurring differences in modulator action in humans (111). This provides an ecologically more valid manipulation than strong pharmacological or anatomical manipulations may make possible. The following section will discuss recent findings from genome-wide association studies (GWAS) and selected candidate gene approaches.

There are few GWAS that have directly searched for genes related to emotional memory. However, a couple of studies have identified gene loci that are related to episodic memory performance (112–114). These studies demonstrate the great advantage of GWAS, which is to discover novel genes and subsequently molecular pathways without a severe bias towards those mechanisms that are well studied. On the downside, a direct interpretation of the findings is often difficult and it needs much further investigation to identify the biological consequences of the discovered genes. Thus, while GWAS can greatly help the field to discover novel variations and learn more about the underlying molecular mechanisms of emotional memory, candidate genes with pre-existing biological consequences should be employed simultaneously to improve our understanding of the mechanisms identified to date (108, 110).

12.5.1 Noradrenergic polymorphism

The *ADRA2b* gene codes for noradrenergic alpha2b receptors that are primarily expressed as presynaptically located autoreceptors. A common deletion variant in this gene is characterized by an inframe deletion of three glutamic acid residues in the third extracellular loop of the receptor (115) and is thought to result in greater NE availability (116). More specifically, previous findings suggest that reduced activity of the adrenergic autoreceptor results in greater extracellular NE availability (36).

In a seminal study, de Quervain and colleagues (116) tested 435 participants for the *ADRA2b* deletion genotype after they completed an emotional memory task. Participants were presented with neutral and emotionally arousing pictures, which had to be recalled freely after a short delay period. While on average all participants showed enhanced memory for emotional relative to neutral material, carriers of the deletion variant of the *ADRA2b* polymorphism showed significantly greater enhancement for emotionally arousing material relative to non-carriers.

While the authors speculated that the memory enhancement was likely to be due to noradrenergic influences on consolidation, another study has demonstrated that deletion carriers perceive emotional pictures more vividly than non-carriers (36), consistent with the idea that arousal-related influences on perception and attention are in effect during encoding of emotional memories and are potentiated in deletion carriers (117). This idea is further supported by a study revealing that *ADRA2b* deletion carriers show greater perceptual awareness of emotional stimuli relative to non-carriers under situations of limited attentional resources, which may explain the stronger emotional bias observed in memory (118). There is also evidence directly linking the subjective, affective experience at encoding to subsequent memory retrieval, an effect that is only observed in deletion carriers (119).

Neuroimaging findings are in line with the notion that genetically influenced noradrenergic activity modulates memory-related processes during encoding. Rasch and colleagues (120) established that deletion carriers exhibit greater amygdala activation during encoding of emotional compared to neutral pictures. Another study suggested that some of these effects may be situation- or context-dependent and be potentiated under acute stress or arousal (121). Finally, other research suggests that emotional memory retrieval is also affected by the *ADRA2b* genotype (122, 123).

Taken together, there is substantial evidence supporting the emotional enhancement of attention and memory by genetically based differences in NE availability, as indicated by *ADRA2b* genotype. These findings are consistent with the extensive literature reviewed earlier suggesting that increased NE availability in response to emotionally salient events aids explicit memory formation. It remains to be investigated how this common genetic variation affects implicit forms of emotional memory.

12.5.2 Dopaminergic polymorphism

The Val158met variation of the gene regulating catechol-O-methyltransferase (COMT), which is an enzyme that degrades dopamine and NE, has been associated with emotional enhancement of perception and memory. This variation is characterized by a valine (Val) to methionine (Met) mutation at position 158. Carriers of two Met-alleles (Met/Met) show up to 75% reduced COMT enzyme activity, which is associated with higher extracellular DA levels, in the prefrontal cortex (PFC) (124). In contrast, Val/Val carriers show enhanced DA uptake in the midbrain (125). As a result, Met/Met carriers show increased performance in tasks that depend on working memory (126, 127).

As described earlier, dopaminergic acitvity has been associated with fear conditioning (97–99). In addition, there is some evidence that DA modulates the amygdala's response to emotional stimuli (128, 129). Carriers of the *COMT* Met/Met-allele have been found to show greater amygdala activation reactivity to emotionally negative but not positive scenes (130). Behaviorally, Met-allele carriers also show a selective effect for aversive or threatening stimuli as demonstrated by increased startle responses (131, 132).

While an influence of the *COMT* polymorphism on amygdala reactivity has been demonstrated in several studies, direct effects on subjective arousal or attentional processes related to encoding processes have not been reported (133). However, consistent with the role of DA in fear conditioning and the link between the Met/Met-allele and enhanced tonic levels of prefrontal dopamine (134), it has been reported that Met carriers show impaired extinction after fear learning (135) as well as reconsolidation (136). There is less evidence linking the *COMT* polymorphims to explicit emotional memory.

In conclusion, genetic variations in the prefrontal dopaminergic system have been shown to affect responses to aversive stimuli. It should be noted that effects of *COMT* may be due to its effects on NE activity and cannot be cleanly tied to DA, and some studies' reliability may be hampered by small sample sizes. One meta-analysis concluded that *COMT* genotype had little if any reliable effect on cognitive function (137). Nonetheless, as reviewed above, several studies suggest a role for DA in implicit fear learning. It is likely that the interplay of hundreds of genetic polymorphisms affecting different dopamingeric pathways together would result in marked differences in fear learning.

12.5.3 Serotonergic polymorphism

Serotonin has also been related to emotional memory performance. Non-human animal studies have implicated the serotonergic system in fear conditioning (138). Pharmacological manipulations in humans have provided evidence that increased extracellular serotonin levels are associated with reduced startle response and increased episodic memory for positive relative to negative material (139). Consistently, depletion of the serotonin precursor tryptophan impaired recall for positive stimuli (140). Thus, available data suggest that serotonergic influences on emotional memory may be specific to positively arousing content.

A variable number of tandem repeats (VNTRs) in the promoter region (5HTTLPR) of the human serotonin transporter gene has been identified (141). The short (S-) allele has been linked to neuroticism and harm avoidance (142, 143). Furthermore, some studies have found that carriers of the S-allele who have been exposed to life stress are more vulnerable to clinical anxiety and depression (144, 145). Several studies found that the S-allele is associated with enhanced amygdala response to negative facial expressions (143, 146). However, it should be noted that here again meta-analyses have failed to confirm previously reported associations with harm avoidance, although associations with trait neuroticism have remained reliable (147). To date, there is no direct evidence showing altered emotional memory performance in S-allele carriers; however, the current findings show an association with affective biases that may underlie emotional memory formation and learning (148).

12.6. Conclusion

In the current chapter, after giving an overview of what is known about emotional enhancement of memory, we presented evidence that supports the idea that both explicit and implicit emotional memory depend to a large extent on noradrenergic influences on the amygdala.

The amygdala is an anatomically central brain region essential for integrating and conveying information about the (emotional) salience of stimuli and events. This is accomplished by tagging information as emotionally/motivationally relevant, a process largely dependent on the activation of the LC/NE system in interaction with stress hormones and dopamine. Noradrenergic activity in the amygdala is necessary for all stages of memory formation spanning different memory systems. One area remaining to be investigated is the link between the amygdala and NE in appetitive conditioning. Much of the research in the past decades has focused on dopaminergic influences on reward-based learning, but since the LC/NE system is activated in response to emotional stimuli regardless of valence, there is reason to believe that NE may also be involved in appetitive learning.

We further summarized research on selected genetic polymorphisms known to affect involved neurotransmitter systems and have discussed what is known about their role in emotional memory performance. Future research on genetic differences should use several approaches to improve understanding of the complex heritability of such a complex cognitive function. GWAS can help to identify new genetic variations as well as gene clusters; candidate gene approaches with well mapped out biological consequences can be used as tools to improve understanding of how naturally occurring variations affect emotional memory on a molecular level. Finally, multicenter studies should invest in integrating behavioral and brain imaging findings with life experience, psychopathology, and large-scale genetic analyses in large sample sizes to better understand individual differences in such a complex but important function as emotional memory.

References

1. **Reisberg D, Heuer F.** Memory for emotional events. In: Reisberg D, Hertel P (eds). Memory and Emotion. London: Oxford University Press; 2004, pp. 3–41.
2. **Holzapfel W.** Emotions and memory intertwined—A historical perspective. Revista de historia de la psicología. 2001;**22**:383–8.
3. **Tulving E.** Episodic memory and common sense: How far apart? Philos T Roy Soc B. 2001;**356**(1413):1505–15.
4. **Joels M, Pu Z, Wiegert O, Oitzl MS, Krugers HJ.** Learning under stress: How does it work? Trends Cogn Sci. 2006;**10**(4):152–8.
5. **Hamann S, Canli T.** Individual differences in emotion processing. Curr Opin Neurobiol. 2004;**14**(2):233–8.
6. **de Quervain DJ, Aerni A, Schelling G, Roozendaal B.** Glucocorticoids and the regulation of memory in health and disease. Front Neuroendocrinol. 2009;**30**(3):358–70.
7. **Holmes A, Singewald N.** Individual differences in recovery from traumatic fear. Trends Neurosci. 2013;**36**(1):23–31.
8. **LaBar KS, Cabeza R.** Cognitive neuroscience of emotional memory. Nat Rev Neurosci. 2006;**7**(1):54–64.
9. **Hamann S.** Cognitive and neural mechanisms of emotional memory. Trends Cogn Sci. 2001;**5**(9):394–400.
10. **Phelps EA, Sharot T.** How (and why) emotion enhances the subjective sense of recollection. Curr Dir Psychol Sci. 2008;**17**(2):147–52.
11. **Baddeley A, Eysenck MW, Anderson MC.** Memory. London: Psychology Press; 2009.

12. **Frankland PW, Bontempi B.** The organization of recent and remote memories. Nat Rev Neurosci. 2005;**6**(2):119–30.
13. **Tulving E.** Episodic memory: From mind to brain. Annu Rev Psychol. 2002;**53**:1–25.
14. **Roozendaal B, McEwen BS, Chattarji S.** Stress, memory and the amygdala. Nat Rev Neurosci. 2009;**10**(6):423–33.
15. **Cahill L, Haier RJ, Fallon J, Alkire MT, Tang C, Keator D**, et al. Amygdala activity at encoding correlated with long-term, free recall of emotional information. PNAS. 1996;**93**(15):8016–21.
16. **Canli T, Zhao Z, Brewer J, Gabrieli JD, Cahill L.** Event-related activation in the human amygdala associates with later memory for individual emotional experience. J Neurosci. 2000;**20**(19):RC99.
17. **Anderson AK.** Affective influences on the attentional dynamics supporting awareness. J Exp Psychol Gen. 2005;**134**(2):258–81.
18. **Talmi D, Anderson AK, Riggs L, Caplan JB, Moscovitch M.** Immediate memory consequences of the effect of emotion on attention to pictures. Learn Mem. 2008;**15**(3):172–82.
19. **Lim SL, Padmala S, Pessoa L.** Segregating the significant from the mundane on a moment-to-moment basis via direct and indirect amygdala contributions. PNAS. 2009;**106**(39):16841–6.
20. **McGaugh JL.** Memory—a century of consolidation. Science. 2000;**287**(5451):248–51.
21. **Kensinger EA, Corkin S.** Two routes to emotional memory: Distinct neural processes for valence and arousal. PNAS. 2004;**101**(9):3310–5.
22. **Buchanan TW.** Retrieval of emotional memories. Psychol Bull. 2007;**133**(5):761–79.
23. **Mather M, Knight M.** Goal-directed memory: The role of cognitive control in older adults' emotional memory. Psychol Aging. 2005;**20**(4):554–70.
24. **Cahill L, Babinsky R, Markowitsch HJ, McGaugh JL.** The amygdala and emotional memory. Nature. 1995;**377**(6547):295–6.
25. **LeDoux JE.** Emotional memory systems in the brain. Behav Brain Res. 1993;**58**(1-2):69–79.
26. **Angrilli A, Mauri A, Palomba D, Flor H, Birbaumer N, Sartori G**, et al. Startle reflex and emotion modulation impairment after a right amygdala lesion. Brain. 1996;**119** (Pt 6):1991–2000.
27. **Adolphs R, Tranel D, Damasio H, Damasio A.** Impaired recognition of emotion in facial expressions following bilateral damage to the human amygdala. Nature. 1994;**372**(6507):669–72.
28. **Markowitsch HJ, Calabrese P, Wurker M, Durwen HF, Kessler J, Babinsky R**, et al. The amygdala's contribution to memory—a study on two patients with Urbach–Wiethe disease. Neuroreport. 1994;**5**(11):1349–52.
29. **Adolphs R, Cahill L, Schul R, Babinsky R.** Impaired declarative memory for emotional material following bilateral amygdala damage in humans. Learn Mem. 1997;**4**(3):291–300.
30. **Pourtois G, Schettino A, Vuilleumier P.** Brain mechanisms for emotional influences on perception and attention: What is magic and what is not. Biol Psychol. 2013;**92**(3):492–512.
31. **Markovic J, Anderson AK, Todd RM.** Tuning to the significant: Neural and genetic processes underlying affective enhancement of visual perception and memory. Behav Brain Res. 2014;**259**:229–41.
32. **Pessoa L.** Emotion and cognition and the amygdala: From "what is it?" to "what's to be done?." Neuropsychologia. 2010;**48**(12):3416–29.
33. **Pessoa L, Adolphs R.** Emotion processing and the amygdala: From a "low road" to "many roads" of evaluating biological significance. Nat Rev Neurosci. 2010;**11**(11):773–83.
34. **Pessoa L, Kastner S, Ungerleider LG.** Attentional control of the processing of neural and emotional stimuli. Brain Res Cogn Brain Res. 2002;**15**(1):31–45.
35. **Sabatinelli D, Bradley MM, Fitzsimmons JR, Lang PJ.** Parallel amygdala and inferotemporal activation reflect emotional intensity and fear relevance. Neuroimage. 2005;**24**(4):1265–70.
36. **Todd RM, Ehlers MR, Muller DJ, Robertson A, Palombo DJ, Freeman N**, et al. Neurogenetic variations in norepinephrine availability enhance perceptual vividness. J Neurosci 2015;**35**(16):6506–16.

37. **Todd RM, Talmi D, Schmitz TW, Susskind J, Anderson AK.** Psychophysical and neural evidence for emotion-enhanced perceptual vividness. J Neurosci. 2012;**32**(33):11201–12.
38. **Anderson AK, Phelps EA.** Lesions of the human amygdala impair enhanced perception of emotionally salient events. Nature. 2001;**411**(6835):305–9.
39. **Adolphs R, Tranel D.** In: Aggleton JP (ed.). The Amygdala: A Functional Analysis. New York, NY: Oxford University Press; 2000, pp. 587–630.
40. **Kensinger EA.** Remembering the details: Effects of emotion. Emot Rev. 2009;**1**(2):99–113.
41. **Adolphs R, Denburg NL, Tranel D.** The amygdala's role in long-term declarative memory for gist and detail. Behav Neurosci. 2001;**115**(5):983–92.
42. **Bookbinder SH, Brainerd CJ.** Emotionally negative pictures enhance gist memory. Emotion. 2016;**17**(1):102–119.
43. **Levine LJ, Edelstein RS.** Emotion and memory narrowing: A review and goal-relevance approach. Cognition Emotion. 2009;**23**(5):833–75.
44. **Richardson MP, Strange BA, Dolan RJ.** Encoding of emotional memories depends on amygdala and hippocampus and their interactions. Nat Neurosci. 2004;**7**(3):278–85.
45. **Cahill L, Haier RJ, Fallon J, Alkire MT, Tang C, Keator D**, et al. Amygdala activity at encoding correlated with long-term, free recall of emotional information. P Natl Acad Sci USA. 1996;**93**(15):8016–21.
46. **Kensinger EA, Addis DR, Atapattu RK.** Amygdala activity at encoding corresponds with memory vividness and with memory for select episodic details. Neuropsychologia. 2011;**49**(4):663–73.
47. **Kensinger EA, Garoff-Eaton RJ, Schacter DL.** How negative emotion enhances the visual specificity of a memory. J Cognitive Neurosci. 2007;**19**(11):1872–87.
48. **Todd RM, Schmitz TW, Susskind J, Anderson AK.** Shared neural substrates of emotionally enhanced perceptual and mnemonic vividness. Front Behav Neurosci. 2013;**7**:40.
49. **Anderson AK, Wais PE, Gabrieli JDE.** Emotion enhances remembrance of neutral events past. PNAS. 2006;**103**(5):1599–604.
50. **Cahill L, Gorski L, Le K.** Enhanced human memory consolidation with post-learning stress: Interaction with the degree of arousal at encoding. Learn Mem. 2003;**10**(4):270–4.
51. **McGaugh JL, McIntyre CK, Power AE.** Amygdala modulation of memory consolidation: Interaction with other brain systems. Neurobiol Learn Mem. 2002;**78**(3):539–52.
52. **McGaugh JL.** The amygdala modulates the consolidation of memories of emotionally arousing experiences. Annu Rev Neurosci. 2004;**27**:1–28.
53. **Packard MG, Cahill L, McGaugh JL.** Amygdala modulation of hippocampal-dependent and caudate nucleus-dependent memory processes. PNAS. 1994;**91**(18):8477–81.
54. **Sakaki M, Fryer K, Mather M.** Emotion strengthens high-priority memory traces but weakens low-priority memory traces. Psychol Sci. 2014;**25**(2):387–95.
55. **McGaugh JL.** Making lasting memories: Remembering the significant. PNAS. 2013;**110** Suppl 2:10402–7.
56. **Dolcos F, LaBar KS, Cabeza R.** Remembering one year later: Role of the amygdala and the medial temporal lobe memory system in retrieving emotional memories. PNAS. 2005;**102**(7):2626–31.
57. **Smith AP, Henson RN, Dolan RJ, Rugg MD.** fMRI correlates of the episodic retrieval of emotional contexts. Neuroimage. 2004;**22**(2):868–78.
58. **Scheinin M, Lomasney JW, Hayden-Hixson DM, Schambra UB, Caron MG, Lefkowitz RJ**, et al. Distribution of alpha 2-adrenergic receptor subtype gene expression in rat brain. Brain Res Mol Brain Res. 1994;**21**(1-2):133–49.
59. **Young WS, 3rd, Kuhar MJ.** Noradrenergic alpha 1 and alpha 2 receptors: Light microscopic autoradiographic localization. PNAS. 1980;**77**(3):1696–700.
60. **Han F, Ding J, Shi Y.** Expression of amygdala mineralocorticoid receptor and glucocorticoid receptor in the single-prolonged stress rats. BMC Neurosci. 2014;**15**:77.

61. **Wolf OT.** The influence of stress hormones on emotional memory: Relevance for psychopathology. Acta Psychol (Amst). 2008;**127**(3):513–31.
62. **Cahill L, Prins B, Weber M, McGaugh JL.** Beta-adrenergic activation and memory for emotional events. Nature. 1994;**371**(6499):702–4.
63. **Strange BA, Dolan RJ.** Beta-adrenergic modulation of emotional memory-evoked human amygdala and hippocampal responses. PNAS. 2004;**101**(31):11454–8.
64. **van Stegeren AH, Goekoop R, Everaerd W, Scheltens P, Barkhof F, Kuijer JP,** et al. Noradrenaline mediates amygdala activation in men and women during encoding of emotional material. Neuroimage. 2005;**24**(3):898–909.
65. **Clark KB, Naritoku DK, Smith DC, Browning RA, Jensen RA.** Enhanced recognition memory following vagus nerve stimulation in human subjects. Nat Neurosci. 1999;**2**(1):94–8.
66. **Southwick SM, Davis M, Horner B, Cahill L, Morgan CA, 3rd, Gold PE,** et al. Relationship of enhanced norepinephrine activity during memory consolidation to enhanced long-term memory in humans. Am J Psychiatry. 2002;**159**(8):1420–2.
67. **Roozendaal B, Okuda S, de Quervain DJ, McGaugh JL.** Glucocorticoids interact with emotion-induced noradrenergic activation in influencing different memory functions. Neuroscience. 2006;**138**(3):901–10.
68. **de Kloet ER, Oitzl MS, Joels M.** Stress and cognition: Are corticosteroids good or bad guys? Trends Neurosci. 1999;**22**(10):422–6.
69. **Roozendaal B.** 1999 Curt P. Richter award. Glucocorticoids and the regulation of memory consolidation. Psychoneuroendocrinology. 2000;**25**(3):213–38.
70. **Buchanan TW, Lovallo WR.** Enhanced memory for emotional material following stress-level cortisol treatment in humans. Psychoneuroendocrinology. 2001;**26**(3):307–17.
71. **van Stegeren AH, Wolf OT, Everaerd W, Scheltens P, Barkhof F, Rombouts SARB.** Endogenous cortisol level interacts with noradrenergic activation in the human amygdala. Neurobiol Learn Mem. 2007;**87**(1):57–66.
72. **de Quervain DJF, Roozendaal B, McGaugh JL.** Stress and glucocorticoids impair retrieval of long-term spatial memory. Nature. 1998;**394**(6695):787–90.
73. **Kuhlmann S, Kirschbaum C, Wolf OT.** Effects of oral cortisol treatment in healthy young women on memory retrieval of negative and neutral words. Neurobiol Learn Mem. 2005;**83**(2):158–62.
74. **Buchanan TW, Tranel D, Adolphs R.** Impaired memory retrieval correlates with individual differences in cortisol response but not autonomic response. Learn Mem. 2006;**13**(3):382–7.
75. **Roozendaal B, Hahn EL, Nathan SV, de Quervain DJ, McGaugh JL.** Glucocorticoid effects on memory retrieval require concurrent noradrenergic activity in the hippocampus and basolateral amygdala. J Neurosci. 2004;**24**(37):8161–9.
76. **Roozendaal B, de Quervain DJF, Schelling G, McGaugh JL.** A systemically administered beta-adrenoceptor antagonist blocks corticosterone-induced impairment of contextual memory retrieval in rats. Neurobiol Learn Mem. 2004;**81**(2):150–4.
77. **Rescorla RA.** Probability of shock in the presence and absence of CS in fear conditioning. J Comp Physiol Psychol. 1968;**66**(1):1–5.
78. **Blanchard DC, Blanchard RJ.** Innate and conditioned reactions to threat in rats with amygdaloid lesions. J Comp Physiol Psychol. 1972;**81**(2):281–90.
79. **Davis M.** Neurobiology of fear responses: The role of the amygdala. J Neuropsychiatry Clin Neurosci. 1997;**9**(3):382–402.
80. **Nagel JA, Kemble ED.** Effects of amygdaloid lesions on the performance of rats in four passive avoidance tasks. Physiol Behav. 1976;**17**(2):245–50.
81. **LeDoux JE.** Emotion circuits in the brain. Annu Rev Neurosci. 2000;**23**:155–84.
82. **Sehlmeyer C, Schoning S, Zwitserlood P, Pfleiderer B, Kircher T, Arolt V,** et al. Human fear conditioning and extinction in neuroimaging: A systematic review. PLoS One. 2009;**4**(6):e5865.

83. **Kapp BS, Frysinger RC, Gallagher M, Haselton JR.** Amygdala central nucleus lesions: Effect on heart rate conditioning in the rabbit. Physiol Behav. 1979;**23**(6):1109–17.
84. **Phillips RG, LeDoux JE.** Differential contribution of amygdala and hippocampus to cued and contextual fear conditioning. Behav Neurosci. 1992;**106**(2):274–85.
85. **Maren S, Quirk GJ.** Neuronal signalling of fear memory. Nat Rev Neurosci. 2004;**5**(11):844–52.
86. **Sotres-Bayon F, Bush DE, LeDoux JE.** Emotional perseveration: An update on prefrontal–amygdala interactions in fear extinction. Learn Mem. 2004;**11**(5):525–35.
87. **Sotres-Bayon F, Quirk GJ.** Prefrontal control of fear: More than just extinction. Curr Opin Neurobiol. 2010;**20**(2):231–5.
88. **Peper M, Karcher S, Wohlfarth R, Reinshagen G, LeDoux JE.** Aversive learning in patients with unilateral lesions of the amygdala and hippocampus. Biol Psychol. 2001;**58**(1):1–23.
89. **LaBar KS, LeDoux JE, Spencer DD, Phelps EA.** Impaired fear conditioning following unilateral temporal lobectomy in humans. J Neurosci. 1995;**15**(10):6846–55.
90. **LaBar KS, Disterhoft JF.** Conditioning, awareness, and the hippocampus. Hippocampus. 1998;**8**(6):620–6.
91. **Bechara A, Tranel D, Damasio H, Adolphs R, Rockland C, Damasio AR.** Double dissociation of conditioning and declarative knowledge relative to the amygdala and hippocampus in humans. Science. 1995;**269**(5227):1115–8.
92. **Johansen JP, Cain CK, Ostroff LE, LeDoux JE.** Molecular mechanisms of fear learning and memory. Cell. 2011;**147**(3):509–24.
93. **Schulz B, Fendt M, Schnitzler HU.** Clonidine injections into the lateral nucleus of the amygdala block acquisition and expression of fear-potentiated startle. Eur J Neurosci. 2002;**15**(1):151–7.
94. **Bush DE, Caparosa EM, Gekker A, Ledoux J.** Beta-adrenergic receptors in the lateral nucleus of the amygdala contribute to the acquisition but not the consolidation of auditory fear conditioning. Front Behav Neurosci. 2010;**4**:154.
95. **Grillon C, Cordova J, Morgan CA, Charney DS, Davis M.** Effects of the beta-blocker propranolol on cued and contextual fear conditioning in humans. Psychopharmacology (Berl). 2004;**175**(3):342–52.
96. **LaLumiere RT, Buen TV, McGaugh JL.** Post-training intra-basolateral amygdala infusions of norepinephrine enhance consolidation of memory for contextual fear conditioning. J Neurosci. 2003;**23**(17):6754–8.
97. **Greba Q, Gifkins A, Kokkinidis L.** Inhibition of amygdaloid dopamine D2 receptors impairs emotional learning measured with fear-potentiated startle. Brain Res. 2001;**899**(1-2):218–26.
98. **Guarraci FA, Frohardt RJ, Falls WA, Kapp BS.** The effects of intra-amygdaloid infusions of a D2 dopamine receptor antagonist on Pavlovian fear conditioning. Behav Neurosci. 2000;**114**(3):647–51.
99. **Guarraci FA, Frohardt RJ, Kapp BS.** Amygdaloid D1 dopamine receptor involvement in Pavlovian fear conditioning. Brain Res. 1999;**827**(1-2):28–40.
100. **Miczek KA, Luttinger D.** Differential attenuation of two kinds of conditioned suppression by d-amphetamine and pentobarbital. J Pharmacol Exp Ther. 1978;**205**(2):282–90.
101. **Pezze MA, Feldon J.** Mesolimbic dopaminergic pathways in fear conditioning. Prog Neurobiol. 2004;**74**(5):301–20.
102. **Cordero MI, Sandi C.** A role for brain glucocorticoid receptors in contextual fear conditioning: Dependence upon training intensity. Brain Res. 1998;**786**(1-2):11–7.
103. **Pugh CR, Tremblay D, Fleshner M, Rudy JW.** A selective role for corticosterone in contextual-fear conditioning. Behav Neurosci. 1997;**111**(3):503–11.
104. **Roozendaal B, Hui GK, Hui IR, Berlau DJ, McGaugh JL, Weinberger NM.** Basolateral amygdala noradrenergic activity mediates corticosterone-induced enhancement of auditory fear conditioning. Neurobiol Learn Mem. 2006;**86**(3):249–55.
105. **Zorawski M, Killcross S.** Posttraining glucocorticoid receptor agonist enhances memory in appetitive and aversive Pavlovian discrete-cue conditioning paradigms. Neurobiol Learn Mem. 2002;**78**(2):458–64.

106. **McClearn GE, Johansson B, Berg S, Pedersen NL, Ahern F, Petrill SA**, et al. Substantial genetic influence on cognitive abilities in twins 80 or more years old. Science. 1997;**276**(5318):1560–3.
107. **Volk HE, McDermott KB, Roediger HL, 3rd, Todd RD.** Genetic influences on free and cued recall in long-term memory tasks. Twin Res Hum Genet. 2006;**9**(5):623–31.
108. **Papassotiropoulos A, de Quervain DJ.** Genetics of human episodic memory: Dealing with complexity. Trends Cogn Sci. 2011;**15**(9):381–7.
109. **Zhu M, Zhao S.** Candidate gene identification approach: Progress and challenges. Int J Biol Sci. 2007;**3**(7):420–7.
110. **Tabor HK, Risch NJ, Myers RM.** Candidate-gene approaches for studying complex genetic traits: Practical considerations. Nat Rev Genet. 2002;**3**(5):391–7.
111. **Todd RM, Ehlers MR, Anderson AK.** What BANE can offer GANE: Individual differences in function of hotspot mechanisms. Behav Brain Sci. 2016;**39**:e226.
112. **Luksys G, Fastenrath M, Coynel D, Freytag V, Gschwind L, Heck A**, et al. Computational dissection of human episodic memory reveals mental process-specific genetic profiles. PNAS. 2015;**112**(35):E4939–48.
113. **Papassotiropoulos A, Stefanova E, Vogler C, Gschwind L, Ackermann S, Spalek K**, et al. A genome-wide survey and functional brain imaging study identify CTNNBL1 as a memory-related gene. Mol Psychiatry. 2013;**18**(2):255–63.
114. **Papassotiropoulos A, Stephan DA, Huentelman MJ, Hoerndli FJ, Craig DW, Pearson JV**, et al. Common Kibra alleles are associated with human memory performance. Science. 2006;**314**(5798):475–8.
115. **Small KM, Brown KM, Forbes SL, Liggett SB.** Polymorphic deletion of three intracellular acidic residues of the alpha 2B-adrenergic receptor decreases G protein-coupled receptor kinase-mediated phosphorylation and desensitization. J Biol Chem. 2001;**276**(7):4917–22.
116. **de Quervain DJ, Kolassa IT, Ertl V, Onyut PL, Neuner F, Elbert T**, et al. A deletion variant of the alpha2b-adrenoceptor is related to emotional memory in Europeans and Africans. Nat Neurosci. 2007;**10**(9):1137–9.
117. **Todd RM, Palombo DJ, Levine B, Anderson AK.** Genetic differences in emotionally enhanced memory. Neuropsychologia. 2011;**49**(4):734–44.
118. **Todd RM, Muller DJ, Lee DH, Robertson A, Eaton T, Freeman N**, et al. Genes for emotion-enhanced remembering are linked to enhanced perceiving. Psychol Sci. 2013;**24**(11):2244–53.
119. **Todd RM, Muller DJ, Palombo DJ, Robertson A, Eaton T, Freeman N**, et al. Deletion variant in the ADRA2B gene increases coupling between emotional responses at encoding and later retrieval of emotional memories. Neurobiol Learn Mem. 2014;**112**:222–9.
120. **Rasch B, Spalek K, Buholzer S, Luechinger R, Boesiger P, Papassotiropoulos A**, et al. A genetic variation of the noradrenergic system is related to differential amygdala activation during encoding of emotional memories. PNAS. 2009;**106**(45):19191–6.
121. **Cousijn H, Rijpkema M, Qin S, van Marle HJ, Franke B, Hermans EJ**, et al. Acute stress modulates genotype effects on amygdala processing in humans. PNAS. 2010;**107**(21):9867–72.
122. **Li S, Weerda R, Guenzel F, Wolf OT, Thiel CM.** ADRA2B genotype modulates effects of acute psychosocial stress on emotional memory retrieval in healthy young men. Neurobiol Learn Mem. 2013;**103**:11–8.
123. **Li S, Weerda R, Milde C, Wolf OT, Thiel CM.** ADRA2B genotype differentially modulates stress-induced neural activity in the amygdala and hippocampus during emotional memory retrieval. Psychopharmacology (Berl). 2015;**232**(4):755–64.
124. **Chen J, Lipska BK, Halim N, Ma QD, Matsumoto M, Melhem S**, et al. Functional analysis of genetic variation in catechol-*O*-methyltransferase (COMT): Effects on mRNA, protein, and enzyme activity in postmortem human brain. Am J Hum Genet. 2004;**75**(5):807–21.
125. **Meyer-Lindenberg A, Kohn PD, Kolachana B, Kippenhan S, McInerney-Leo A, Nussbaum R**, et al. Midbrain dopamine and prefrontal function in humans: Interaction and modulation by COMT genotype. Nat Neurosci. 2005;**8**(5):594–6.

126. **Weinberger DR, Egan MF, Bertolino A, Callicott JH, Mattay VS, Lipska BK**, et al. Prefrontal neurons and the genetics of schizophrenia. Biol Psychiatry. 2001;**50**(11):825–44.

127. **Raz N, Dahle CL, Rodrigue KM, Kennedy KM, Land S.** Effects of age, genes, and pulse pressure on executive functions in healthy adults. Neurobiol Aging. 2011;**32**(6):1124–37.

128. **Hariri AR, Mattay VS, Tessitore A, Fera F, Smith WG, Weinberger DR.** Dextroamphetamine modulates the response of the human amygdala. Neuropsychopharmacology. 2002;**27**(6):1036–40.

129. **Tessitore A, Hariri AR, Fera F, Smith WG, Chase TN, Hyde TM**, et al. Dopamine modulates the response of the human amygdala: A study in Parkinson's disease. J Neurosci. 2002;**22**(20):9099–103.

130. **Smolka MN, Schumann G, Wrase J, Grusser SM, Flor H, Mann K**, et al. Catechol-*O*-methyltransferase val158met genotype affects processing of emotional stimuli in the amygdala and prefrontal cortex. J Neurosci. 2005;**25**(4):836–42.

131. **Montag C, Buckholtz JW, Hartmann P, Merz M, Burk C, Hennig J**, et al. COMT genetic variation affects fear processing: Psychophysiological evidence. Behav Neurosci. 2008;**122**(4):901–9.

132. **Smolka MN, Buhler M, Schumann G, Klein S, Hu XZ, Moayer M**, et al. Gene–gene effects on central processing of aversive stimuli. Mol Psychiatry. 2007;**12**(3):307–17.

133. **Gibbs AA, Naudts KH, Spencer EP, David AS.** The role of dopamine in attentional and memory biases for emotional information. Am J Psychiatry. 2007;**164**(10):1603–9; quiz 24.

134. **Bilder RM, Volavka J, Lachman HM, Grace AA.** The catechol-*O*-methyltransferase polymorphism: Relations to the tonic-phasic dopamine hypothesis and neuropsychiatric phenotypes. Neuropsychopharmacology. 2004;**29**(11):1943–61.

135. **Lonsdorf TB, Weike AI, Nikamo P, Schalling M, Hamm AO, Ohman A.** Genetic gating of human fear learning and extinction: Possible implications for gene–environment interaction in anxiety disorder. Psychol Sci. 2009;**20**(2):198–206.

136. **Agren T, Furmark T, Eriksson E, Fredrikson M.** Human fear reconsolidation and allelic differences in serotonergic and dopaminergic genes. Transl Psychiatry. 2012;**2**:e76.

137. **Barnett JH, Scories L, Munafo MR.** Meta-analysis of the cognitive effects of the catechol-*O*-methyltransferase gene Val158/108Met Polymorphism. Biol Psychiat. 2008;**64**(2):137–44.

138. **Buhot MC, Martin S, Segu L.** Role of serotonin in memory impairment. Ann Med. 2000;**32**(3):210–21.

139. **Harmer CJ, Shelley NC, Cowen PJ, Goodwin GM.** Increased positive versus negative affective perception and memory in healthy volunteers following selective serotonin and norepinephrine reuptake inhibition. Am J Psychiatry. 2004;**161**(7):1256–63.

140. **Klaassen T, Riedel WJ, Deutz NE, Van Praag HM.** Mood congruent memory bias induced by tryptophan depletion. Psychol Med. 2002;**32**(1):167–72.

141. **Heils A, Teufel A, Petri S, Stober G, Riederer P, Bengel D**, et al. Allelic variation of human serotonin transporter gene expression. J Neurochem. 1996;**66**(6):2621–4.

142. **Lesch KP, Bengel D, Heils A, Sabol SZ, Greenberg BD, Petri S**, et al. Association of anxiety-related traits with a polymorphism in the serotonin transporter gene regulatory region. Science. 1996;**274**(5292):1527–31.

143. **Munafo MR, Brown SM, Hariri AR.** Serotonin transporter (5-HTTLPR) genotype and amygdala activation: A meta-analysis. Biol Psychiatry. 2008;**63**(9):852–7.

144. **Canli T.** Toward a neurogenetic theory of neuroticism. Ann N Y Acad Sci. 2008;**1129**:153–74.

145. **Fox NA, Nichols KE, Henderson HA, Rubin K, Schmidt L, Hamer D**, et al. Evidence for a gene–environment interaction in predicting behavioral inhibition in middle childhood. Psychol Sci. 2005;**16**(12):921–6.

146. **Dannlowski U, Konrad C, Kugel H, Zwitserlood P, Domschke K, Schoning S**, et al. Emotion specific modulation of automatic amygdala responses by 5-HTTLPR genotype. Neuroimage. 2010;**53**(3):893–8.

147. **Munafo MR, Freimer NB, Ng W, Ophoff R, Veijola J, Miettunen J**, et al. 5-HTTLPR genotype and anxiety-related personality traits: A meta-analysis and new data. Am J Med Genet B. 2009;**150b**(2):271–81.

148. **Beevers CG, Wells TT, Ellis AJ, McGeary JE.** Association of the serotonin transporter gene promoter region (5-HTTLPR) polymorphism with biased attention for emotional stimuli. J Abnorm Psychol. 2009;**118**(3):670–81.

149. **Sara SJ.** The locus coeruleus and noradrenergic modulation of cognition. Nat Rev Neurosci. 2009;**10**(3):211–23.

150. **Aston-Jones G, Cohen JD.** An integrative theory of locus coeruleus-norepinephrine function: Adaptive gain and optimal performance. Annu Rev Neurosci. 2005;**28**:403–50.

Chapter 13

Genetics of decision-making

Joshua C Gray, Sandra Sanchez-Roige,
Abraham A Palmer, Harriet de Wit,
and James MacKillop

13.1 Introduction

Decision-making is a fundamental component of human behavior. Effectively considering trade-offs and selecting beneficial choices is critical for healthy normative behavior, and, conversely, persistently maladaptive decision-making is a cardinal feature of several psychiatric conditions, most notably substance use disorders (1–3). This chapter reviews evidence of the heritability of specific decision-making phenotypes and their association with addiction, taking the perspective that measures of decision-making may serve as promising intermediate phenotypes (i.e. intervening mechanisms that connect genetic variation with vulnerability for addictive disorders). The chapter focuses on the genetic correlates of the three most widely studied experimental measures of decision-making: impulsivity, as measured by delayed reward discounting, a behavioral economic index of capacity to delay gratification; disadvantageous decision-making, as measured using the Iowa Gambling Task; and risk sensitivity, as measured by the Balloon Analogue Risk Task. Each of the three phenotypes will be reviewed in terms of the nature of the phenotype; the phenotype's relationship to addiction; and studies linking genetic variation to systematic differences in decision-making. The chapter concludes with considerations and future priorities for exploring this nascent area.

13.1.1 Delayed reward discounting

One of the most widely studied measures of decision-making is delayed reward discounting (DRD), a behavioral economic measure of how much an individual discounts future rewards relative to immediate rewards (4, 5). The most common version of a DRD task presents a participant with a series of binary choices between a small immediate reward or a larger reward delivered at a later date (e.g. "Would you rather have $11 today or $30 in 7 days?") (6). A similar task has been developed for animal models, providing a series of choices between a lever providing a small number of food pellets after no delay and a lever providing a larger number of food pellets after a delay (e.g. 7). All of the responses from the DRD task are used to generate a function that summarizes the overall devaluation of long-term rewards as a function of the delay to receipt (for a discussion of different analysis approaches, see 8, 9). Higher rates of DRD reflect lower capacity to delay gratification and are considered a form of impulsivity. High DRD rates have been robustly associated with numerous neuropsychiatric traits, including addiction across multiple substances (10, 11), obesity and eating disorders (12, 13), and attention deficit hyperactivity disorder (ADHD) (14).

One approach used to investigate the genetic factors contributing to DRD is to compare phenotypic differences across inbred rodent strains in a controlled environment. In mice and rats, 16–66% of variability in DRD rates has been accounted for by strain differences (15–19). In humans, similar heritability estimates have been found in twin studies, with estimates showing increased heritability with age (i.e. 12 [30%] and 14 [51%], (20); 16 [35–46%], 17 [47–51%], and 18 [55–62%] (21–23)). This trend of increasing heritability with age is intriguing, possibly reflecting developmental differences that occur during adolescent brain maturation of prefrontal regions implicated in delayed gratification (24–27).

DRD has also been associated with having a family history of addiction in some studies (28–30), but not others (31–33). However, it is important to note that the presence of a family history of addiction reflects confounded genetic and environmental risk factors. Similarly, pre-existing differences in impulsivity have been correlated with differences in alcohol-related behavior across rodent strains (e.g. 34). These findings suggest that DRD has phenotypic overlap with addiction, is itself heritable, and shares heritability with addiction. It is thus plausible that DRD and addiction are co-heritable traits and have overlapping risk genes.

Candidate gene studies that focus on a single locus or a small number of loci in a putatively relevant gene have concentrated primarily on systems related to monoamine neurotransmitters, particularly dopamine (see Table 13.1). The most studied locus to date is rs4680, a common variant that falls in the *COMT* (catechol-*O*-methyl transferase) gene which codes for the eponymous enzyme that breaks down dopamine. Adult carriers of the Val-allele have been reported to exhibit increased DRD (35–38; but see 39), whereas the Met-allele has been reported to be the risk allele in young adults (38, 40). The Met/Met genotype has been associated with reduced levels of enzymatic activity of catechol-*O*-methyl transferase and higher levels of dopamine primarily in the prefrontal cortex (41–43). Supporting this hypothesis, Val/Val carriers show reduced activity in the left dorsal prefrontal cortex (dPFC) during a resting state paradigm (36). However, another group found that Val/Val carriers exhibited increased dPFC activity while performing the DRD task (35). Overall, these studies present preliminary and sometimes contradictory evidence regarding the role of rs4680 in DRD; larger sample sizes will be needed to more directly resolve these conflicting findings.

The exon 3 VNTR polymorphism in the dopamine D_4 receptor gene, *DRD4* VNTR has also been explored in several studies. *DRD4* VNTR is typically divided into a long form (L; ≥ 7 repeats) variation, associated with blunted response to dopamine, and a short form (S; < 7 repeats) variation, associated with increased response to dopamine (44). Four different association studies have not supported a direct relationship between this VNTR and DRD (40, 45–47), although one reported an association between the L-allele and decreased DRD rates (39). *DRD4* VNTR has been reported to exhibit some context dependent relationships, such as the L-allele interacting with the T-allele of rs1800497 in one investigation (46) and the L-allele interacting with low socioeconomic status (SES) in another (47). Given these varying findings, it will be important for future studies to explore the potential of *DRD4* VNTR as a differential susceptibility gene (see 48) to clarify if the relationship between DRD and genetic variation is contingent upon other genes or environmental exposures.

Another locus, rs1800497, within the *ANKK1* gene associated with dopamine D_2 receptor density and proximal to the D_2 receptor gene (*DRD2*) (49–51), has been associated with DRD (37, 46); however, this association has not been replicated in several other independent studies (39, 52). Relatedly, a distinct variant within the *DRD2* gene, the T-allele of rs6277, has been associated with higher DRD rates in individuals of Asian ancestry (52).

Unrelated to dopamine genes, higher DRD rates have been associated with the C-allele of rs1360780, within the FK506 binding protein 5 gene (*FKBP5*); this gene is also implicated in cortisol response (53). Furthermore, three loci, all in high linkage disequilibrium and within the α_{2A} adrenergic receptor gene (*ADRA2A*), have been associated with increased DRD in cocaine users but not healthy controls (54). Relatedly, higher *ADRA2A* mRNA expression levels were associated

Table 13.1 Decision-making candidate gene studies

Task	Study	Sample	Sex	Polymorphism(s)	Primary findings
DRD	Boettiger et al., 2007 (35)	10 HA 9 AD	42%	*COMT* Val158Met (rs4680)	Val ↑
	Smith & Boettiger, 2012 (38)	70 HA 72 LA	53% F	*COMT* Val158Met (rs4680)	Val/Val ↑ HA only Met ↑ LA only
	Gianotti et al., 2012 (36)	73 HA	100% M	*COMT* Val158Met (rs4680)	Val ↑
	MacKillop et al., 2015 (37)	713 DS	40% F	*COMT* Val158Met (rs4680) *ANKK1* (rs1800497)	Val ↑ T ↑
	Gray & MacKillop 2014[1] (39)	175 WG	14% F	*COMT* Val158Met (rs4680) *ANKK1* (rs1800497) *DRD4* VNTR	= = 7+-VNTR ↓ *DRD4* x ANKK1 =
	Paloyelis et al., 2010 (40)	36 AADHD 32 CON	100% M	*COMT* Val158Met (rs4680) *DRD4* VNTR *DAT1 10/6* VNTR	Met/Met ↑ = 10/6 HZ ↑ CON
	Kawamura et al., 2013a (52)	91 HA	27% F	*DRD2* (rs6277) *ANKK1* (rs1800497)	T ↑ =
	Eisenberg et al., 2007 (46)	195 LA	42% F	*ANKK1* (rs1800497) *DRD4* VNTR	T ↑ *DRD4* (7+-VNTR) x ANKK1 (T) ↑
	Sweitzer et al., 2013 (47)	546 HA	50% F	*DRD4* VNTR	= 7-VNTR + low SES ↑
	Garcia et al., 2010 (45)	181 LA	65% F	*DRD4* VNTR	=
	Kawamura et al., 2013b (53)	91 HA	27% F	*FKBP5* (rs1360780)	C ↑
	Havranek et al., 2015 (54)	94 HA 129 CU	27% F	*ADRA2A* (rs1800544[2])	G ↑ CU only

IGT	Roussos et al., 2008 (87)	107 HA	100% M	*COMT* (rs4818)	C ↑
	de Souza et al., 2016 (88)	192 HA	56% F	*COMT* (rs4680)	Val/Val ↓ (T 41-100) F only
	da Rocha et al., 2011 (74)	122 OCD	47% F	*BDNF* val66met (rs6265)	Met ↑
	Ness et al., 2011 (126)	245 HA	52% F	15 loci *GRIN2B*	=
	Malloy-Diniz et al., 2013 (86)	195 HA	56% F	*COMT* Val158Met (rs4680)	Met/Met ↑
	He et al., 2012 (89)	556 HA	55% F	*COMT* Val158Met (rs4680)	Met + stress ↑ Val/Val + warmth ↓
	van den Bos et al., 2009[3] (83)	70 HA	100% F	*COMT* Val158Met (rs4680)	Met/Met ↑ (vs. Val/Val) *COMT* (Met/Met) x 5-HTTLPR (S/S) ↑
	O'Brien et al., 2014 (67)	55 FHA offspring 31 CON (ages 16-34)	48% F	5-HTTLPR *COMT* Val158Met (rs4680)	S/S ↑ =
	Gu et al., 2013 (80)	465 SCZ 448 CON	36% F	5-HTTLPR; *HTR1A* (rs6295)	L/L ↑ (T 1-40) G/G ↑ (T 1-40)
	Kang et al., 2010 (85)	168 HA	45% F	*BDNF* Val66Met (rs6265); *COMT* Val158Met (rs4680)	Met ↑ (T 41-100) = No epistatic interactions
	Ha et al., 2009 (76)	159 HA	47% F	5-HTTLPR DRD4 VNTR	= = DRD4 (2R+) x 5-HTTLPR (S/S) ↓ In the absence of the S/S genotype: 2R- ↓ 2R+
	Verdejo-García et al., 2013[4] (82)	86 MU 58 CON	31% F	5-HTTLPR COMT Val158Met (rs4680)	S/S MU ↑ S/S CON =
	Miu et al., 2012 (73)	135 HA	87%	5-HTTLPR	S ↓
	Lovallo et al., 2014 (81)	314 HA	57% F	5-HTTLPR	= FH+ L_a ↑ FH- L_a

(*continued*)

Table 13.1 Continued

Task	Study	Sample	Sex	Polymorphism(s)	Primary findings
	da Rocha et al., 2008 (70)	49 OCD	53% F	5-HTTLPR	La/La ↓ (T 41–100)
	Homberg et al., 2008 (71)	88 HA	100% F	5-HTTLPR	S/S ↑ (T 41–100)
	Stoltenberg et al., 2011 (78)	391 HA	65% F	5-HTTLPR	L/L ↑ (T 1- 20)
	Must et al., 2007 (72)	124 MDD	67% F	5-HTTLPR	S ↑
	Lage et al., 2011 (127)	127 HA	59% F	5-HTTLPR	=
	He et al., 2010 (79)	572 HA	55% F	5-HTTLPR	S/S ↑ (T 1–40)
	Tenconi et al., 2016 (75)	98 HA 91 AN	100%	5-HTTLPR	=
	Stoltenberg et al., 2010 (77)	188 HA	62% F	5-HTTLPR	= S/S and S/L 5-HTTLPR ↓ (T 1-20) vs. L/L M only
	Maurex et al., 2009 (90)	48 BPD 30 HA	100% F	TPH-1 haplotype	TPH-1 haplotype ↑ BPD only
BART	Crişan et al. 2009 (111)	32 HA	72% F	5-HTTLPR	L/L ↑
	Mata et al., 2012 (108)	322 HA	73% F	*DAT1* VNTR	9-VNTR ↑
	Lovallo et al., 2014 (81)	314 HA	57% F	5-HTTLPR	=
	Lancaster et al., 2012 (105)	70 HA	61% F	*COMT* Val158Met (rs4680)	Met/Met ↑
	Lancaster et al., 2015 (106)	101 HA	53% F	*COMT* Val158Met (rs4680)	=
	Amstadter et al., 2012 (107)	223 EA	44% F	*COMT* Val158Met (rs4680)	Met ↑ F only
	Kohno et al., 2016 (109)	65 HA	43% F	Dopamine risk score (*DRD2, DRD3, DRD4, DAT1, COMT*)	↓↑ (curvilinear) reduced DLPFC function increased midbrain function

Note: Only loci that have at least one positive finding are included in the table.[1] Eleven additional exploratory loci not depicted were also identified in this study; [2] rs521674 and rs602618 were also tested, but were in near total linkage disequilibrium; [3] direct effects of 5-HTTLPR were tested with the larger sample in He et al., 2010; [4] only interaction effects were tested in this study; ↓: low impulsivity/better performance; ↑: high impulsivity/worse performance; =: no effect; HA: healthy adults; LA: late adolescents; EA: early adolescents; M: males; F: females; AD: alcohol dependence; DS: daily smokers, WG: weekly gamblers; AADHD: adolescents with ADHD; CU: cocaine users; HZ: homozygotes; DLPFC: dorsolateral prefrontal cortex; MDD: major depressive disorder; T: trials; B: block; FHA: family history of alcohol use disorder; FH+: family history of alcohol or substance use disorder; CON: control; SCZ: schizophrenia; MU: marijuana users; AN: lifetime anorexia; L_a: long "high activity" allele.

with reduced DRD, suggesting that the combined impact of cocaine use on adrenergic receptors and genetic risk lead to increased risk for impulsivity. This study indicates the importance of considering environmental (e.g. repeated cocaine exposure) and genetic effects simultaneously.

Collectively, the preceding lines of research suggest that, from a genetic standpoint, the preclinical studies, twin studies, and family history studies support the hypothesis of DRD as a heritable characteristic. Candidate gene studies remain at an early stage and have generated mixed findings. Furthermore, they are limited by small sample sizes and heterogeneous populations varying in age, ethnicity, and substance use history, increasing the risk for false positives.

13.1.2 **Iowa Gambling Task**

In the Iowa Gambling Task (IGT), participants receive $2,000 of hypothetical money and select from four decks of cards, each with reward–punishment ratios reflecting profiles that are advantageous (smaller, more frequent rewards with fewer punishments, leading to a net gain) or disadvantageous (larger, less frequent rewards, with more punishments, leading to a net loss). The task consists of 100 trials, and after an initial period of sampling each deck (i.e. the first 20 trials of the task), participants with healthy executive functioning increasingly select from the advantageous decks whereas individuals with addictive disorders (55–59), ADHD (60), eating disorders (61), or a damaged ventromedial prefrontal cortex (62, 63) tend to continue to favor net loss decks. An analog of the task has been adapted successfully for use in rodents (for a review, see 64).

Heritability estimates from twin studies suggest that 20–46% of the variability in IGT risk behavior can be explained by genetic factors (65), with age being an important modulator (11–13 [35%], 14–15 [20%], and 16–18 [46%]). Although it is also possible that the variability across age is in part due to varying subject numbers across each wave (n = 343, n = 831, and n = 451, respectively). Relatedly, poorer performance has been identified in individuals with family history of alcoholism (66, 67) and obsessive–compulsive disorder (68). Based on these findings, there is initial evidence for the role of genetic factors in IGT.

Candidate gene studies on the IGT have focused on serotoninergic systems and growth factor pathways (Table 13.1). The short (S) version of the 5-HTTLPR polymorphism, near the transcription initiation site of the serotonin transporter gene *SLC6A4*, is associated with reduced serotonin transporter protein availability, compared to the long (L) version (for a review, see 69). The associations of 5-HTTLPR with IGT performance are multifarious: some studies found the S or S/S genotypes were associated with worse performance (67, 70–72), another found S-carriers had higher scores (73), and many others found no association (74–76). Other 5-HTTLPR findings have been confined to main effects or gender specific findings in the first 20 trials (77, 78) or 40 trials of the IGT (79, 80), which is surprising given the beginning of the task is thought to be largely random sampling. Finally, several studies have identified interaction effects relating to poorer performance including 5-HTTLPR carriers of L "high activity" variants with a family history of alcoholism versus those without a family history of alcoholism (81), S/S cannabis users versus S/S controls (82), and various interactions with rs4680, rs6295, or *DRD4* VNTR variants (see Table 13.1; 76, 80, 83).

Other genes have been considered to a lesser degree. In studies of rs6265, located in the brain-derived neurotrophic factor (*BDNF*) gene implicated in proliferation of neurotransmitter systems (84), Met/Met and Val/Met carriers exhibited poorer performance on the IGT (74, 85). With regard to *COMT*, three studies identified a relationship between Met/Met of rs4680 (or the equivalent C genotype of *COMT* variant rs4818) and poor IGT performance (83, 86, 87), one study found that Val/Val carriers exhibited better IGT performance (in women only; 88), and three others found no relationship (67, 85, 89). An additional study identified an interaction such that

Met carriers who experienced stressful life events had worse IGT performance, whereas Val/Val individuals with high parental warmth exhibited better IGT performance (89). Finally, one study suggests that the frequency of a tryptophan hydroxylase 1 gene (*TPH1*) haplotype that is implicated in serotonin synthesis is associated with IGT performance in women with borderline personality disorder, but not healthy women (90).

Taking the aforementioned findings together, no consistent pattern emerges from the candidate gene studies. Given the heterogeneity of findings, differential task variables used, and variable samples (e.g. healthy, male versus female, individuals with schizophrenia, obsessive–compulsive disorder, borderline personality disorder), the reports of significant associations should be interpreted with caution. Clearly more work is needed to clarify genetic correlates of IGT performance.

13.1.3 Balloon Analogue Risk Task

The Balloon Analogue Risk Task (BART) is a computerized task that measures risk-taking by presenting a series of virtual balloons (91). Each time a button is pressed to inflate the balloon, a potential monetary reward increases in amount, but each inflation also increases the risk the balloon will pop. At any point in a trial an individual can cease pumping up the balloon and take their earnings. Therefore, performance reflects a person's willingness to trade a larger reward for greater risk of loss. BART risk-taking performance has been associated with various forms of addictive behavior (92–95; but see 96–99).

Evidence from the only twin study to date revealed a wide range in estimated heritability ranging from non-significant to 55%, with evidence of greater heritability in males (28% at age 12 and 55% at age 14) (100). Evidence for co-heritability of addiction and BART performance is present in studies that found that adolescents (101, 102) and adults with a family history of heavy alcohol use (102) exhibited greater risk-taking on the BART (but see 103). Furthermore, one study found that maternal risk-taking (as measured by the BART) predicted alcohol use escalation in their children from grade 8–10 (104).

Genes associated with dopaminergic transmission have been explored in relation to BART performance (Table 13.1). For rs4680 (*COMT* Val158Met), the Met/Met genotype was found to exhibit greater risk-taking (i.e. more pumps) than Met/Val or Val/Val genotypes in healthy adults (105; but see 106). Furthermore, Met carrier female (but not male) adolescents were reported to exhibit greater reward-seeking (107). An investigation of the 3' UTR of the dopamine transporter gene, *DAT1* VNTR, found that 9-repeat carriers exhibit greater reward-seeking than 10/10 and 11-repeat carriers (108). Finally, a recent study aggregated 5 dopamine related genes to compute an overall dopamine risk score (109). They identified a significant curvilinear relationship between this score and performance on the BART task (i.e. high/low dopamine levels being associated with higher BART risk-taking). This score was found to correlate with increased midbrain activity and reduced dorsolateral prefrontal cortex activity while completing the BART, suggesting that this genetic vulnerability may result in disrupted fronto-striatal activity necessary for adaptive decision-making (110). With regard to serotonin transmission, one study found 5-HTTLPR L/L carriers exhibited higher BART risk taking (111), while another found no relationship (81).

In sum, there is preliminary evidence that BART risk-taking is a heritable trait, but candidate gene research is inchoate at this point.

13.2 Future directions

This chapter reviewed the evidence for genetic associations with multiple forms of decision-making, particularly with an eye towards understanding the genetics of addiction. The extant literature is virtually entirely candidate gene investigations that focused on theoretically relevant neurotransmitter systems (e.g. dopaminergic and serotonergic). However, despite some evidence of heritability for all of the decision-making phenotypes, the candidate gene studies reveal a highly mixed patterns of findings. This is challenging to interpret because the existing studies are often underpowered (median n = 159 in Table 13.1); comprise samples varying considerably in characteristics, such as age and psychopathology (see Table 13.1); have a very low replication rate; and may well be susceptible to the "file drawer problem" (i.e. a bias towards only publishing positive results; 112, 113). As future studies emerge, it will be important to combine findings into meta-analyses both to examine effects in aggregate and by subpopulation (e.g. by age, gender, ethnicity, and presence of psychopathology) to clarify true signal versus noise.

It is also important to remember that it is increasingly understood that many loci shape variation in decision-making (114), like most human traits. As such, well-powered genome-wide association studies (GWAS) provide an agnostic approach by scanning the whole genome for novel loci associated with the traits herein described (115, 116). Furthermore, studies should incorporate the contribution of the environment, particularly early-life stress, as those most at risk in adverse environments have been found to fare best in favorable environments (117). An additional future direction is to better characterize overlapping heritability between decision-making and addiction, and across different decision-making phenotypes, by using polygenic approaches (e.g. 118). These approaches may provide insights into both the basic biology of decision-making and the associated psychiatric disorders.

While these decision-making phenotypes offer some promise, others may also join the ranks as candidate intermediate phenotypes if similar evidence accrues. One such candidate intermediate phenotype is probability discounting, a behavioral economic measure of risk sensitivity. Probability discounting measures how much a subject devalues a reward based on its level of uncertainty. Similar to DRD, it is typically assessed using a decision-making task that presents smaller guaranteed rewards and larger uncertain rewards (e.g. "$40 for sure" or "a 30-in-31 chance of winning $60 [97% chance]") (119). By varying the amounts and probabilities of the uncertain rewards, a probability discounting function can be generated, reflecting the extent to which a person devalues a certain reward across varying parameters. Although probability discounting and DRD reflect similar processes, they represent different forms of discounted utility. Probability discounting is associated with gambling disorder (119) and gambling treatment response (120), and varies between rodent strains (121), suggesting that it is a generally heritable trait. However, it has not been investigated in relation to genetic variation (to our knowledge).

13.3 **Conclusion**

Understanding genetic influences on decision-making remains at an early stage. Studies using quantitative genetics and animal models provide evidence that experimental decision-making performance is heritable. However, genetic studies in humans have generated a highly heterogeneous and mixed empirical literature. Two clear next steps are the need to expand sample sizes to increase power and, in turn, the reliability of the observed findings and to expand the genomic scope using GWAS. These directions should be pursued because the identification of genetic influences on decision-making is both a scientifically important question, but also may have the potential to inform the treatment for a variety of conditions. In the same way that neuroimaging research led to transcranial magnetic stimulation and transcranial direct current stimulation as treatment strategies (e.g. 122–125), ascertaining the genetic determinants of maladaptive decision-making could lead to pharmacogenetic strategies to alter the underlying biological decision-making circuits. For these applications to be realized, however, a sound empirical foundation will be essential and should be the near-term priority for this line of inquiry.

References

1. **Volkow ND, Baler RD.** NOW vs LATER brain circuits: Implications for obesity and addiction. Trends Neurosci. 2015;**38**(6):345–52.
2. **Tang Y-Y, Posner MI, Rothbart MK, Volkow ND.** Circuitry of self-control and its role in reducing addiction. Trends Cogn Sci. 2015;**19**(8):439–44.
3. **Ekhtiari H, Victor TA, Paulus MP.** Aberrant decision-making and drug addiction—how strong is the evidence? Curr Opin Behav Sci. 2017;**13**:25–33.
4. **Ainslie G.** Specious reward: A behavioral theory of impulsiveness and impulse control. Psychol Bull. 1975;**82**(4):463–96.
5. **Rachlin H, Green L.** Commitment, choice and self-control. J Exp Anal Behav. 1972;**17**(1):15–22.
6. **Kirby KN, Petry NM, Bickel WK.** Heroin addicts have higher discount rates for delayed rewards than non-drug-using controls. J Exp Psychol. 1999/04/01. 1999;**128**(1):78–87.
7. **Perry JL, Carroll ME.** The role of impulsive behavior in drug abuse. Psychopharmacol. 2008;**200**(1):1–26.
8. **Green L, Myerson J.** A discounting framework for choice with delayed and probabilistic rewards. Psychol Bull. 2004;**130**(5):769–92.
9. **Marzilli Ericson KM, White JM, Laibson D, Cohen JD.** Money earlier or later? Simple heuristics explain intertemporal choices better than delay discounting does. Psychol Sci. 2015;**26**(6):826–33.
10. **Amlung MT, Vedelago L, Acker J, Balodis I, MacKillop J.** Steep delay discounting and addictive behavior: A meta-analysis of continuous associations. Addiction. 2016;**112**(1):51–62.
11. **MacKillop J, Amlung MT, Few LR, Ray LA, Sweet LH, Munafò MR.** Delayed reward discounting and addictive behavior: A meta-analysis. Psychopharmacology (Berl). 2011;**216**(3):305–21.
12. **Amlung MT, Petker T, Jackson J, Balodis I, MacKillop J.** Steep discounting of delayed monetary and food rewards in obesity: A meta-analysis. Psychol Med. 2016;**46**(11):2423–34.
13. **McClelland J, Dalton B, Kekic M, Bartholdy S, Campbell IC, Schmidt U.** A systematic review of temporal discounting in eating disorders and obesity: Behavioural and neuroimaging findings. Neurosci Biobehav Rev. 2016;**71**:506–28.
14. **Jackson JNS, MacKillop J.** Attention deficit hyperactivity disorder and monetary delay discounting: A meta-analysis of case-control studies. Biol Psychiatry Cogn Neurosci Neuroimaging. 2016;**1**(4):316–25.
15. **Isles AR, Humby T, Walters E, Wilkinson LS.** Common genetic effects on variation in impulsivity and activity in mice. J Neurosci. 2004;**24**(30):6733–40.
16. **Anderson KG, Woolverton WL.** Effects of clomipramine on self-control choice in Lewis and Fischer 344 rats. Pharmacol Biochem Behav. 2005;**80**(3):387–93.
17. **Madden GJ, Smith NG, Brewer AT, Pinkston JW, Johnson PS.** Steady-state assessment of impulsive choice in Lewis and Fischer 344 rats: Between-condition delay manipulations. J Exp Anal Behav. 2008;**90**(3):333–44.
18. **Stein JS, Pinkston JW, Brewer AT, Francisco MT, Madden GJ.** Delay discounting in Lewis and Fischer 344 rats: Steady-state and rapid-determination adjusting-amount procedures. J Exp Anal Behav. 2012;**97**(3):305–21.
19. **Richards JB, Lloyd DR, Kuehlewind B, Militello L, Paredez M, Solberg Woods L**, et al. Strong genetic influences on measures of behavioral-regulation among inbred rat strains. Genes Brain Behav. 2013;**12**(5):490–502.
20. **Anokhin AP, Golosheykin S, Grant JD, Heath AC.** Heritability of delay discounting in adolescence: A longitudinal twin study. Behav Genet. 2011;**41**(2):175–83.
21. **Anokhin AP, Grant JD, Mulligan RC, Heath AC.** The genetics of impulsivity: Evidence for the heritability of delay discounting. Biol Psychiatry. 2015;**77**(10):887–94.
22. **Isen JD, Sparks JC, Iacono WG.** Predictive validity of delay discounting behavior in adolescence: A longitudinal twin study. Exp Clin Psychopharmacol. 2014;**22**(5):434–43.

23. **Sparks JC, Isen JD, Iacono WG.** Preference on cash-choice task predicts externalizing outcomes in 17-year-olds. Behav Genet. 2014;**44**(2):102–12.

24. **Carter RM, Meyer JR, Huettel SA.** Functional neuroimaging of intertemporal choice models: A review. J Neurosci Psychol Econ. 2010;**3**:27–45.

25. **Steinberg L.** A behavioral scientist looks at the science of adolescent brain development. Brain Cogn. 2010;**72**(1):160–4.

26. **Peters J, Büchel C.** The neural mechanisms of inter-temporal decision-making: Understanding variability. Trends Cogn Sci. 2011;**15**(5):227–39.

27. **Luo S, Ainslie G, Pollini D, Giragosian L, Monterosso JR.** Moderators of the association between brain activation and farsighted choice. Neuroimage. 2012;**59**(2):1469–77.

28. **Dougherty DM, Charles NE, Mathias CW, Ryan SR, Olvera RL, Liang Y,** et al. Delay discounting differentiates pre-adolescents at high and low risk for substance use disorders based on family history. Drug Alcohol Depend. 2014;**143**:105–11.

29. **Acheson A, Vincent AS, Sorocco KH, Lovallo WR.** Greater discounting of delayed rewards in young adults with family histories of alcohol and drug use disorders: Studies from the Oklahoma family health patterns project. Alcohol Clin Exp Res. 2011;**35**(9):1607–13.

30. **VanderBroek L, Acker J, Palmer AA, de Wit H, MacKillop J.** Interrelationships among parental family history of substance misuse, delay discounting, and personal substance use. Psychopharmacology (Berl). 2016;**233**(1):39–48.

31. **Herting MM, Schwartz D, Mitchell SH, Nagel BJ.** Delay discounting behavior and white matter microstructure abnormalities in youth with a family history of alcoholism. Alcohol Clin Exp Res. 2010;**34**(9):1590–602.

32. **Petry NM, Kirby KN, Kranzler HR.** Effects of gender and family history of alcohol dependence on a behavioral task of impulsivity in healthy subjects. J Stud Alcohol Drugs. 2002;**63**(1):83–90.

33. **Sanchez-Roige S, Stephens DN, Duka T.** Heightened impulsivity: Associated with family history of alcohol misuse, and a consequence of alcohol intake. Alcohol Clin Exp Res. 2016;**40**(10):2208–17.

34. **Linsenbardt DN, Smoker MP, Janetsian-Fritz SS, Lapish CC.** Impulsivity in rodents with a genetic predisposition for excessive alcohol consumption is associated with a lack of a prospective strategy. Cogn Affect Behav Neurosci. 2016;1–17.

35. **Boettiger CA, Mitchell JM, Tavares VC, Robertson M, Joslyn G, D'Esposito M,** et al. Immediate reward bias in humans: Fronto-parietal networks and a role for the catechol-O-methyltransferase 158(Val/Val) genotype. J Neurosci. 2007;**27**(52):14383–91.

36. **Gianotti LRR, Figner B, Ebstein RP, Knoch D.** Why some people discount more than others: Baseline activation in the dorsal PFC mediates the link between COMT genotype and impatient choice. Front Neurosci. 2012;**6**(54).

37. **MacKillop J, Gray JC, Bidwell LC, Bickel WK, Sheffer CE, McGeary JE.** Genetic influences on delay discounting in smokers: Examination of a priori candidates and exploration of dopamine-related haplotypes. Psychopharmacology (Berl). 2015;**232**(20):3731–9.

38. **Smith CT, Boettiger CA.** Age modulates the effect of COMT genotype on delay discounting behavior. Psychopharmacology (Berl). 2012;**222**(4):609–17.

39. **Gray JC, MacKillop J.** Genetic basis of delay discounting in frequent gamblers: Examination of a priori candidates and exploration of a panel of dopamine-related loci. Brain Behav. 2014;**4**(6):812–21.

40. **Paloyelis Y, Asherson P, Mehta MA, Faraone S V, Kuntsi J.** DAT1 and COMT effects on delay discounting and trait impulsivity in male adolescents with attention deficit/hyperactivity disorder and healthy controls. Neuropsychopharmacology. 2010;**35**(12):2414–26.

41. **Weinshilboum RM, Otterness DM, Szumlanski CL.** Methylation pharmacogenetics: Catechol-O-methyltransferase, thiopurine methyltransferase, and histamine N-methyltransferase. Annu Rev Pharmacol Toxicol. 1999;**39**(1):19–52.

42. **Chen J, Lipska BK, Halim N, Ma QD, Matsumoto M, Melhem S**, et al. Functional analysis of genetic variation in catechol-*O*-methyltransferase (COMT): Effects on mRNA, protein, and enzyme activity in postmortem human brain. Am J Hum Genet. 2004;**75**(5):807–21.

43. **Tunbridge EM, Bannerman DM, Sharp T, Harrison PJ.** Catechol-o-methyltransferase inhibition improves set-shifting performance and elevates stimulated dopamine release in the rat prefrontal cortex. J Neurosci. 2004;**24**(23):5331–5.

44. **Asghari V, Sanyal S, Buchwaldt S, Paterson A, Jovanovic V, Van Tol HHM.** Modulation of intracellular cyclic AMP levels by different human dopamine D4 receptor variants. J Neurochem. 1995;**65**(3):1157–65.

45. **Garcia JR, MacKillop J, Aller EL, Merriwether AM, Wilson DS, Lum JK.** Associations between dopamine D4 receptor gene variation with both infidelity and sexual promiscuity. PLoS One. 2010;**5**(11):e14162.

46. **Eisenberg DTA, MacKillop J, Modi M, Beauchemin J, Dang D, Lisman SA**, et al. Examining impulsivity as an endophenotype using a behavioral approach: A DRD2 Taql A and DRD4 48-bp VNTR association study. Behav Brain Funct. 2007;**3**(1).

47. **Sweitzer MM, Halder I, Flory JD, Craig AE, Gianaros PJ, Ferrell RE**, et al. Polymorphic variation in the dopamine D4 receptor predicts delay discounting as a function of childhood socioeconomic status: Evidence for differential susceptibility. Soc Cogn Affect Neurosci. 2013;**8**(5):499–508.

48. **Bakermans-Kranenburg MJ, van Ijzendoorn MH.** Differential susceptibility to rearing environment depending on dopamine-related genes: New evidence and a meta-analysis. Dev Psychopathol. 2011;**23**(1):39–52.

49. **Jönsson EG, Nöthen MM, Grünhage F, Farde L, Nakashima Y, Propping P**, et al. Polymorphisms in the dopamine D2 receptor gene and their relationships to striatal dopamine receptor density of healthy volunteers. Mol Psychiatry. 1999;**4**(3):290–6.

50. **Pohjalainen T, Rinne JO, Någren K, Lehikoinen P, Anttila K, Syvälahti EKG**, et al. The A1 allele of the human D2 dopamine receptor gene predicts low D2 receptor availability in healthy volunteers. Mol Psychiatry. 1998;**3**(3):256–60.

51. **Savitz J, Hodgkinson CA, Martin-Soelch C, Shen P-H, Szczepanik J, Nugent AC**, et al. DRD2/ANKK1 Taq1A polymorphism (rs1800497) has opposing effects on D2/3 receptor binding in healthy controls and patients with major depressive disorder. Int J Neuropsychopharmacol. 2013;**16**(9):2095–101.

52. **Kawamura Y, Takahashi T, Liu X, Nishida N, Noda Y, Yoshikawa A**, et al. Variation in the DRD2 gene affects impulsivity in intertemporal choice. Open J Psychiatry. 2013a;3(1):26–31.

53. **Kawamura Y, Takahashi T, Liu X, Nishida N, Tokunaga K, Ukawa K**, et al. DNA polymorphism in the FKBP5 gene affects impulsivity in intertemporal choice. Asia Pac Psychiatry. 2013b;**5**(1):31–8.

54. **Havranek MM, Hulka LM, Tasiudi E, Eisenegger C, Vonmoos M, Preller KH**, et al. α2A -Adrenergic receptor polymorphisms and mRNA expression levels are associated with delay discounting in cocaine users. Addict Biol. 2015;**22**(2):561–9.

55. **Bechara A, Dolan S, Denburg N, Hindes A, Anderson SW, Nathan PE.** Decision-making deficits, linked to a dysfunctional ventromedial prefrontal cortex, revealed in alcohol and stimulant abusers. Neuropsychologia. 2001;**39**(4):376–89.

56. **Bechara A, Dolan S, Hindes A.** Decision-making and addiction (part II): Myopia for the future or hypersensitivity to reward? Neuropsychologia. 2002;**40**(10):1690–705.

57. **Dolan SL, Bechara A, Nathan PE.** Executive dysfunction as a risk marker for substance abuse: The role of impulsive personality traits. Behav Sci Law. 2008;**26**(6):799–822.

58. **Bechara A, Damasio H.** Decision-making and addiction (part I): Impaired activation of somatic states in substance dependent individuals when pondering decisions with negative future consequences. Neuropsychologia. 2002;**40**(10):1675–89.

59. **Hammers DB, Suhr JA.** Neuropsychological, impulsive personality, and cerebral oxygenation correlates of undergraduate polysubstance use. J Clin Exp Neuropsychol. 2010;**32**(6):599–609.

60. **Groen Y, Gaastra GF, Lewis-Evans B, Tucha O.** Risky behavior in gambling tasks in individuals with ADHD—a systematic literature review. PLoS One. 2013;**8**(9):e74909.
61. **Guillaume SS, Gorwood P, Jollant F, Van den Eynde F, Courtet P, Richard-Devantoy S**, et al. Impaired decision-making in symptomatic anorexia and bulimia nervosa patients: A meta-analysis. Psychol Med. 2015;**45**(16):3377–91.
62. **Bechara A, Damasio AR, Damasio H, Anderson SW.** Insensitivity to future consequences following damage to human prefrontal cortex. Cognition. 1994;**50**(1–3):7–15.
63. **Bechara A, Damasio H, Tranel D, Anderson SW.** Dissociation Of working memory from decision making within the human prefrontal cortex. J Neurosci. 1998;**18**(1):428–37.
64. **Bos R van den, Koot S, Visser L de.** A rodent version of the Iowa Gambling Task: 7 years of progress. Front Psychol. 2014;**5**:203.
65. **Tuvblad C, Gao Y, Wang P, Raine A, Botwick T, Baker LA.** The genetic and environmental etiology of decision-making: A longitudinal twin study. J Adolesc. 2013;**36**(2):245–55.
66. **Lovallo WR, Yechiam E, Sorocco KH, Vincent AS, Collins FL.** Working memory and decision-making biases in young adults with a family history of alcoholism: Studies from the Oklahoma family health patterns project. Alcohol Clin Exp Res. 2006;**30**(5):763–73.
67. **O'Brien JW, Lichenstein SD, Hill SY.** Maladaptive decision making and substance use outcomes in high-risk individuals: Preliminary evidence for the role of 5-HTTLPR variation. J Stud Alcohol Drugs. 2014;**75**(4):643–52.
68. **Viswanath B, Janardhan Reddy YC, Kumar KJ, Kandavel T, Chandrashekar CR.** Cognitive endophenotypes in OCD: A study of unaffected siblings of probands with familial OCD. Prog Neuropsychopharmacol Biol Psychiatry. 2009;**33**(4):610–15.
69. **Canli T, Lesch K-P.** Long story short: The serotonin transporter in emotion regulation and social cognition. Nat Neurosci. 2007;**10**(9):1103–9.
70. **da Rocha FF, Malloy-Diniz L, Lage NV, Romano-Silva MA, de Marco LA, Correa H.** Decision-making impairment is related to serotonin transporter promoter polymorphism in a sample of patients with obsessive-compulsive disorder. Behav Brain Res. 2008;**195**(1):159–63.
71. **Homberg JR, van den Bos R, den Heijer E, Suer R, Cuppen E.** Serotonin transporter dosage modulates long-term decision-making in rat and human. Neuropharmacology. 2008;**55**(1):80–4.
72. **Must A, Juhász A, Rimanóczy A, Szabó Z, Kéri S, Janka Z.** Major depressive disorder, serotonin transporter, and personality traits: Why patients use suboptimal decision-making strategies? J Affect Disord. 2007;**103**(1–3):273–6.
73. **Miu AC, Crişan LG, Chiş A, Ungureanu L, Drugă B, Vulturar R.** Somatic markers mediate the effect of serotonin transporter gene polymorphisms on Iowa Gambling Task. Genes, Brain Behav. 2012;**11**(4):398–403.
74. **da Rocha FF, Malloy-Diniz L, Lage NV, Corrêa H.** The relationship between the Met allele of the BDNF Val66Met polymorphism and impairments in decision making under ambiguity in patients with obsessive-compulsive disorder. Genes, Brain Behav. 2011;**10**(5):523–9.
75. **Tenconi E, Degortes D, Clementi M, Collantoni E, Pinato C, Forzan M**, et al. Clinical and genetic correlates of decision making in anorexia nervosa. J Clin Exp Neuropsychol. 2016;**38**(3):327–37.
76. **Ha RY, Namkoong K, Kang JI, Kim YT, Kim SJ.** Interaction between serotonin transporter promoter and dopamine receptor D4 polymorphisms on decision making. Prog Neuropsychopharmacology Biol Psychiatry. 2009;**33**(7):1217–22.
77. **Stoltenberg SF, Vandever JM.** Gender moderates the association between 5-HTTLPR and decision-making under ambiguity but not under risk. Neuropharmacology. 2010;**58**(2):423–8.
78. **Stoltenberg SF.** Serotonin transporter (5-HTTLPR) genotype and childhood trauma are associated with individual differences in decision making. Front Genet. 2011;**2**.
79. **He Q, Xue G, Chen C, Lu Z, Dong Q, Lei X**, et al. Serotonin transporter gene-linked polymorphic region (5-HTTLPR) influences decision making under ambiguity and risk in a large Chinese sample. Neuropharmacology. 2010;**59**(6):518–26.

80. **Gu H, Liu C, Chen M, Zhang Q, Zhai J, Wang K**, et al. The combined effects of the 5- HTTLPR and HTR1A rs6295 polymorphisms modulate decision making in schizophrenia patients. Genes, Brain Behav. 2013;**12**(1):133–9.

81. **Lovallo WR, Enoch M-A, Yechiam E, Glahn DC, Acheson A, Sorocco KH**, et al. Differential impact of serotonin transporter activity on temperament and behavior in persons with a family history of alcoholism in the Oklahoma Family Health Patterns Project. Alcohol Clin Exp Res. 2014;**38**(6):1575–81.

82. **Verdejo-García A, Fagundo AB, Cuenca A, Rodriguez J, Cuyás E, Langohr K**, et al. COMT val158met and 5-HTTLPR genetic polymorphisms moderate executive control in cannabis users. Neuropsychopharmacology. 2013;**38**(8):1598–606.

83. **van den Bos R, Homberg J, Gijsbers E, den Heijer E, Cuppen E.** The effect of COMT Val158 Met genotype on decision-making and preliminary findings on its interaction with the 5-HTTLPR in healthy females. Neuropharmacology. 2009;**56**(2):493–8.

84. **Bath K, Lee F.** Variant BDNF (Val66Met) impact on brain structure and function. Cogn Affect Behav Neurosci. 2006;**6**(1):79–85.

85. **Kang JI, Namkoong K, Ha RY, Jhung K, Kim YT, Kim SJ.** Influence of BDNF and COMT polymorphisms on emotional decision making. Neuropharmacology. 2010;**58**(7):1109–13.

86. **Malloy-Diniz LF, Lage GM, Campos SB, de Paula JJ, de Souza Costa D, Romano-Silva MA**, et al. Association between the Catechol O-methyltransferase (COMT) Val158met polymorphism and different dimensions of impulsivity. PLoS One. 2013;**8**(9):e73509.

87. **Roussos P, Giakoumaki SG, Pavlakis S, Bitsios P.** Planning, decision-making and the COMT rs4818 polymorphism in healthy males. Neuropsychologia. 2008;**46**(2):757–63.

88. **de Souza Costa D, Bechara A, de Paula JJ, Romano-Silva MA, Correa H, Lage GM**, et al. Influence of COMT Val158Met polymorphism on emotional decision-making: A sex-dependent relationship? Psychiatry Res. 2016;**246**:650–5.

89. **He Q, Xue G, Chen C, Lu Z-L, Chen C, Lei X**, et al. COMT Val158Met polymorphism interacts with stressful life events and parental warmth to influence decision making. Sci Rep. 2012;2.

90. **Maurex L, Zaboli G, Wiens S, Asberg M, Leopardi R, Ohman A.** Emotionally controlled decision-making and a gene variant related to serotonin synthesis in women with borderline personality disorder. Scand J Psychol. 2009;**50**(1):5–10.

91. **Lejuez CW, Read JP, Kahler CW, Richards JB, Ramsey SE, Stuart GL**, et al. Evaluation of a behavioral measure of risk taking: The Balloon Analogue Risk Task (BART). J Exp Psychol Appl. 2002;**8**(2):75–84.

92. **Bornovalova MA, Daughters SB, Hernandez GD, Richards JB, Lejuez CW.** Differences in impulsivity and risk-taking propensity between primary users of crack cocaine and primary users of heroin in a residential substance-use program. Exp Clin Psychopharmacol. 2005;**13**(4):311–18.

93. **Hanson KL, Thayer RE, Tapert SF.** Adolescent marijuana users have elevated risk-taking on the balloon analog risk task. J Psychopharmacol. 2014;**28**(11):1080–7.

94. **Lejuez CW, Aklin WM, Jones HA, Richards JB, Strong DR, Kahler CW**, et al. The Balloon Analogue Risk Task (BART) differentiates smokers and nonsmokers. Exp Clin Psychopharmacol. 2003;**11**(1):26–33.

95. **Lejuez CW, Aklin W, Bornovalova M, Moolchan ET.** Differences in risk-taking propensity across inner-city adolescent ever- and never-smokers. Nicotine Tob Res. 2005;**7**(1):71–9.

96. **Ashenhurst JR, Jentsch JD, Ray LA.** Risk-taking and alcohol use disorders symptomatology in a sample of problem drinkers. Exp Clin Psychopharmacol. 2011;**19**(5):361–70.

97. **Courtney KE, Arellano R, Barkley-Levenson E, Gálvan A, Poldrack RA, Mackillop J**, et al. The relationship between measures of impulsivity and alcohol misuse: An integrative structural equation modeling approach. Alcohol Clin Exp Res. 2012;**36**(6):923–31.

98. **Dean AC, Sugar CA, Hellemann G, London ED.** Is all risk bad? Young adult cigarette smokers fail to take adaptive risk in a laboratory decision-making test. Psychopharmacology (Berl). 2011;**215**(4):801–11.

99. **Ryan KK, MacKillop J, Carpenter MJ.** The relationship between impulsivity, risk-taking propensity and nicotine dependence among older adolescent smokers. Addict Behav. 2013;**38**(1):1431–4.

100. **Anokhin AP, Golosheykin S, Grant J, Heath AC.** Heritability of risk-taking in adolescence: A longitudinal twin study. Twin Res Hum Genet. 2009;**12**(4):366–71.

101. **Euser AS, Greaves-Lord K, Crowley MJ, Evans BE, Huizink AC, Franken IHA.** Blunted feedback processing during risky decision making in adolescents with a parental history of substance use disorders. Dev Psychopathol. 2013;**25**(4 Pt 1):1119–36.

102. **Gorka SM, Liu H, Klein D, Daughters SB, Shankman SA.** Is risk-taking propensity a familial vulnerability factor for alcohol use? An examination in two independent samples. J Psychiatr Res. 2015;**68**:54–60.

103. **Dougherty DM, Lake SL, Mathias CW, Ryan SR, Bray BC, Charles NE,** et al. Behavioral impulsivity and risk-taking trajectories across early adolescence in youths with and without family histories of alcohol and other drug use disorders. Alcohol Clin Exp Res. 2015;**39**(8):1501–9.

104. **Banducci AN, Felton JW, Dahne J, Ninnemann A, Lejuez CW.** Maternal risk taking on the balloon analogue risk task as a prospective predictor of youth alcohol use escalation. Addict Behav. 2015;**49**:40–5.

105. **Lancaster TM, Linden DE, Heerey EA.** COMT val158met predicts reward responsiveness in humans. Genes, Brain Behav. 2012;**11**(8):986–92.

106. **Lancaster TM, Heerey EA, Mantripragada K, Linden DEJ.** Replication study implicates COMT val158met polymorphism as a modulator of probabilistic reward learning. Genes, Brain Behav. 2015;**14**(6):486–92.

107. **Amstadter AB, Macpherson L, Wang F, Banducci AN, Reynolds EK, Potenza MN,** et al. The relationship between risk-taking propensity and the COMT Val(158)Met polymorphism among early adolescents as a function of sex. J Psychiatr Res. 2012;**46**(7):940–5.

108. **Mata R, Hau R, Papassotiropoulos A, Hertwig R.** DAT1 polymorphism is associated with risk taking in the Balloon Analogue Risk Task (BART). PLoS One. 2012;**7**(6):e39135.

109. **Kohno M, Nurmi EL, Laughlin CP, Morales AM, Gail EH, Hellemann GS,** et al. Functional genetic variation in dopamine signaling moderates prefrontal cortical activity during risky decision making. Neuropsychopharmacology. 2016;**41**(3):695–703.

110. **Dalley JW, Everitt BJ, Robbins T.** Impulsivity, compulsivity, and top-down cognitive control. Neuron. 2011;**69**(4):680–94.

111. **Crişan LG, Pană S, Vulturar R, Heilman RM, Szekely R, Drugă B,** et al. Genetic contributions of the serotonin transporter to social learning of fear and economic decision making. Soc Cogn Affect Neurosci. 2009;**4**(4):399–408.

112. **Munafò M, Clark T, Johnstone E, Murphy M, Walton R.** The genetic basis for smoking behavior: A systematic review and meta-analysis. Nicotine Tob Res. 2004;**6**(4):583–97.

113. **Rosenthal R.** The file drawer problem and tolerance for null results. Psychol Bull. 1979;**86**(3):638–41.

114. **Chabris CF, Lee JJ, Cesarini D, Benjamin DJ, Laibson DI.** The fourth law of behavior genetics. Curr Dir Psychol Sci. 2015;**24**(4):304–12.

115. **Chabris CF, Hebert BM, Benjamin DJ, Beauchamp J, Cesarini D, van der Loos M,** et al. Most reported genetic associations with general intelligence are probably false positives. Psychol Sci. 2012;**23**(11):1314–23.

116. **Samek DR, Bailey J, Hill KG, Wilson S, Lee S, Keyes MA,** et al. A test-replicate approach to candidate gene research on addiction and externalizing disorders: A collaboration across five longitudinal studies. Behav Genet. 2016;**46**(5):608–26.

117. **Bakermans-Kranenburg MJ, van IJzendoorn MH.** The Hidden Efficacy of Interventions: Gene×Environment Experiments from a Differential Susceptibility Perspective. Annu Rev Psychol. 2015;**66**(1):381–409.

118. **Ashenhurst JR, Harden KP, Corbin WR, Fromme K.** Alcohol-related genes show an enrichment of associations with a persistent externalizing factor. J Abnorm Psychol. 2016;**125**(7) 933–45.

119. **Madden GJ, Petry NM, Johnson PS.** Pathological gamblers discount probabilistic rewards less steeply than matched controls. Exp Clin Psychopharmacol. 2009/10/07. 2009;**17**(5):283–90.
120. **Petry NM.** Discounting of probabilistic rewards is associated with gambling abstinence in treatment-seeking pathological gamblers. J Abnorm Psychol. 2012;**121**(1):151–9.
121. **Wilhelm CJ, Mitchell SH.** Rats bred for high alcohol drinking are more sensitive to delayed and probabilistic outcomes. Genes, Brain Behav. 2008;7(7):705–13.
122. **Cho SS, Ko JH, Pellecchia G, Van Eimeren T, Cilia R, Strafella AP.** Continuous theta burst stimulation of right dorsolateral prefrontal cortex induces changes in impulsivity level. Brain Stimul. 2010;3(3):170–6.
123. **Cho SS, Koshimori Y, Aminian K, Obeso I, Rusjan P, Lang AE**, et al. Investing in the future: Stimulation of the medial prefrontal cortex reduces discounting of delayed rewards. Neuropsychopharmacology. 2015;**40**(3):546–53.
124. **Figner B, Knoch D, Johnson E.** Lateral prefrontal cortex and self-control in intertemporal choice. Nat Neurosci. 2010;**13**:538–9.
125. **He Q, Chen M, Chen C, Xue G, Feng T, Bechara A.** Anodal stimulation of the left DLPFC increases IGT scores and decreases delay discounting rate in healthy males. Front Psychol. 2016;7.
126. **Ness V, Arning L, Niesert HE, Stüttgen MC, Epplen JT, & Beste C.** (2011). Variations in the GRIN2B gene are associated with risky decision-making. Neuropharmacology, **61**(5–6), 950–956.
127. **Lage GM, Malloy-Diniz LF, Matos LO, Bastos MA, Abrantes SS, & Corrêa H.** (2011). Impulsivity and the 5-HTTLPR polymorphism in a non-clinical sample. PLoS One, **6**(2), e16927.

Part 3

Biological mechanisms

Chapter 14

Missing heritability in studies of trait anxiety and amygdala function: Is the solution in plain sight?

Turhan Canli

14.1 Heritability misunderstood

Fifty years of behavioral genetic studies have amassed a substantial empirical foundation on the heritability of traits. In the most comprehensive meta-analysis ever compiled, Polderman and colleagues included variance measures of 17,804 traits from 2,748 publications obtained from more than 14.5 million partly dependent twin pairs (1). They found a 49% heritability across all traits, and twin data that for 69% of all traits were consistent with a model in which the observed variance could be explained by non-shared environmental variance and additive genetic variance (which is represented by the sum of all alleles affecting the observed trait) alone. Among the top 20 most-investigated traits, 12 were consistent with such a model, and 8 traits were inconsistent with such a model, suggesting the presence of additional shared environmental and/or *non-additive* genetic variables (which is represented by *interactions* between alleles within a gene or across genes affecting the observed trait).

The most comprehensive meta-analysis on personality traits was conducted by Vukasovic and Bratko, who combed through six decades of behavioral genetics research pertaining to four influential personality models (2). These models included conceptualizations that were influential in the 1970s and 1980s, such as Eysenck's three-factor model of extraversion, neuroticism, and psychoticism; and Tellegen's model of positive emotionality, negative emotionality, and constraint. Included in the meta-analysis were also two more contemporary personality models, both based on a five-factor structure: Goldberg's extraversion, emotional stability, agreeableness, conscientiousness, and intellect; and Costa and McCrae's extraversion, neuroticism, agreeableness, conscientiousness, and openness. Despite the structural differences across these models, Vukasovic and Bratko's meta-analysis found that heritability estimates varied little, and assigned about 40% of the observed variance to genetic factors.

The main finding by Vukasovic and Bratko was that study design significantly affected heritability estimates, which were 47% and 22% for twin studies and for family/adoption studies, respectively. Years earlier, Plomin and colleagues (3) noticed the same pattern, which they speculated reflected the additional contribution of *non-additive* (on top of additive) genetic effects observed in twin studies. As stated earlier, the distinction between additive and non-additive genetics effects is that the former represent a simple sum of, whereas the latter represent an interaction between, all alleles that contribute to a trait. Whereas both twin and family/adoption studies capture additive genetic effects, only twin studies capture the additional non-additive effects, because only monozygotic twins share all additive and non-additive effects (they have identical genomes). The plausible presence of non-additive genetic effects vastly complicates identification of individual genes associated with any particular behavioral trait, a point made by Plomin and colleagues (3).

Finally, a high "heritability" estimate is not equivalent to a high "genetic" contribution, although it is commonly misunderstood to mean just that. Instead, heritability is calculated on the basis of measures of variance (the ratio of genetic to total (i.e. genetic plus environmental) variance), with the counterintuitive consequence that phenotypes that are fully genetically determined (i.e. not affected by any environmental factors) have no calculable heritability, because the underlying variance measures would be zero. Furthermore, because the heritability estimate is based in part on environmental variance, heritability can be high if environmental variance is low, and low if environmental variance is high, given the same genetic variance (4). This may explain why many traits show decreasing heritability with age (1). The lesson from these considerations is that a high heritability estimate does not necessarily equate to a high contribution of genetic factors, which may be one reason why molecular genetic studies have not detected a large number of trait-associated genes to date, as I will discuss in the next section.

14.2 The genetic basis of trait anxiety: A tale of two approaches

14.2.1 Candidate genes

Efforts to identify specific genes associated with anxiety began with a candidate gene approach, in which a structural variant (polymorphisms) within an *a priori* gene of interest is associated with measures of a phenotype of interest (5–19). In the first study of this kind, which was foundational to the field of "molecular psychology" (5,7), Lesch and colleagues (20) discovered that the short variant of the serotonin (5-HT) transporter-linked polymorphic region, 5-HTTLPR, was associated with elevated levels of neuroticism, a form of trait anxiety. This result has been replicated and confirmed by a number of meta-analyses (21–24), although these also showed that the strength of the association depends in part on the selection of trait measurement instruments, and others criticized the literature on statistical grounds (25,26). As the number of study extensions grew, later meta-analyses further confirmed associations between the 5-HTTLPR and endophenotypes of anxious traits, such as cortisol stress reactivity (27) and amygdala activity (26). More recently, Munafo and colleagues analyzed the literature on 5-HTTLPR and amygdala activation with regard to potential citation distortions (28). They found that for the network of 25 studies cited in one particular meta-analysis, 39% of replication studies but only 25% of non-replications studies were later cited. In a response, Kaufman (29) noted that a focus on 5-HTTLPR amygdala activation ignores the larger context in which such associations are investigated in the contemporary literature, which includes many other contributing variables or outcome measures (e.g. brain circuits, gene networks, gene–environment interactions, epigenetics).

Following up on my earlier discussion of additive and non-additive effects, there is evidence for both with respect to 5-HTTLPR function. In collaboration with Lesch and colleagues, we showed that for the 5-HTTLPR polymorphism and a SNP within the tryptophan hydroxylase-2 gene (*TPH2*), there was an additive effect on the response to negative stimuli as a function of the number of risk alleles carried by these two genes (30,31). Non-additive interactive effects were observed between the 5-HTTLPR and *TPH2* in a study using the five factor model (FFM), in which the presence of the 5-HTTLPR short allele moderated the effect of *TPH2* T/T on measures of Conscientiousness (32). In collaboration with Klein and colleagues, we observed another non-additive effect between 5-HTTLPR and the *BDNF* Val66Met polymorphisms, in which cortisol baseline and response to an acute stressor was moderated by the interaction of the 5-HTTLPR short and the Met-*BDNF* allele (33). In more recent studies, 5-HTTLPR was found to interact with *COMT* Val158Met in measures of gray matter volume in parahippocampal gyrus, amygdala, hippocampus, vermis of cerebellum, and the putamen/insula (34); with *BDNF* Val66Met in a putative marker of serotonin levels in neocortex, amygdala, caudate, hippocampus, and putamen (35), in measures of serotonin transporter binding (36), and in measures of cortical thickness (37); and with the 5-HT rs25531 SNP polymorphism correlating with resting state functional connectivity between the amygdala and the fusiform gyrus (38). These studies await further replication with larger samples, but they reinforce the view that genetic effects on trait anxiety, or behavioral phenotypes in general, are likely to include both additive and non-additive effects.

Another gene that has received much attention is FK506 binding protein 5 (*FKBP5*), a glucocorticoid receptor-regulating co-chaperone of hsp-90. Binder and colleagues identified polymorphisms within this gene that were associated with depression (39), which was also confirmed in a 10-year longitudinal prospective study in interaction with life adversity (40), and with post-traumatic stress disorder (PTSD) in interaction with a history of child abuse (41). These gene–environment interactions are mediated through epigenetic mechanisms, such as allele-specific DNA demethylation (42) through reducing the activity of DNA methyltransferase 1 (DNMT1) (43).

FKBP5 has been shown to have both additive and non-additive effects with other genes. Additive effects were reported in one study between *FKBP5* allelic variation and corticotropin releasing hormone high-affinity binding protein (*CRHBP*) in the context of childhood trauma and suicide risk (44). These investigators found that, among individuals with high levels of childhood trauma, the prevalence of suicide attempt was 0.24 in those who only carried the minor allele of both *FKBP5* rs3800373 and *CRHBP* rs7728378; it was 0.30–0.35 in carriers of either major homozygote, and 0.58 in carriers of both major homozygotes. In another study, PTSD risk was associated with the number of risk alleles in *FKBP5* (rs9470080), *COMT* (rs4680), *CHRNA5* (rs16969968), and *CRHR1* (rs110402) (45). One recent study conceptualized *FKBP5* as belonging to a set of genes regulating hypothalamic–pituitary–adrenal (HPA) axis function, whose collective contributions were captured in a “biologically informed multilocus profile score (BIMPS)” (46). The combined contributions of *FKBP5* rs1360780, *CRHR1* rs110402, and *NR3C2* rs5522/rs4635799 interacted with measures of early life stress correlating with amygdala activation during a face-matching task. Consistent with non-additive genetic effects, *FKBP5* rs1360780 has been shown to interact with the gene encoding the kainic acid-type glutamate receptor 1 (KA1) subunit (*GRIK4*) rs12800734 SNP in a study of antidepressant treatment response (47).

As 5-HTTLPR, *FKBP5* genetic variation has been associated with differential amygdala measures. The first imaging study found that six *FKBP5* SNP's risk alleles (rs7748266, rs1360780, rs9296158, rs3800373, rs9470080, and rs9394309) previously associated with either psychopathology and/or HPA axis function interacted with a history of emotional neglect to correlate with amygdala activation during a face-matching task (48). In conjunction with three other genes (*CRHR1, NR3C2, NR3C1*), *FKBP5* genetic variation interacted with early life stress to correlate with both amygdala and hippocampal volumes in pre-school-aged children (49), and this interaction was later expanded to include measures of amygdala activation and further moderation by sex and pubertal development (50). A history of childhood abuse in homozygous carriers of the *FKBP5* rs1360780 T-allele was associated with reduced gray matter volume not only in the amygdala but also in hippocampus/parahippocampus, anterior and middle cingulate cortex, and insula (51). However, in a Japanese sample, the *FKBP5* rs1360780 T-allele was associated with larger amygdala volume (52), suggesting potential cultural or ethnic moderators, and another study failed to observed any association between amygdala or hippocampal volume and *FKBP5* genotype (53). In a functional activation study using a face-matching task in young adults from a high-risk community sample, the *FKBP5* rs1360780 T-allele was associated with highest amygdala activation, as well as increased coupling with hippocampus and orbitofrontal cortex (54).

Additional support for a role of *FKBP5* in the amygdala comes from a post-mortem study of suicide victims (55). These investigators found that both FKBP5 gene expression and protein levels were reduced by 38% and 41%, respectively, in the amygdala of suicide victims ($N = 13$), relative to controls ($N = 13$).

Although these candidate-gene-based studies have produced a large literature of gene–brain–behavior studies, the variance explained by these genes is small and replication for many candidate genes not discussed here has been poor (56,57). An alternative to the candidate-gene approach is the genome-wide association study (GWAS) approach, which eschews the limitations of small-scale a priori gene assessments for theory-free, large-scale, data-driven discovery.

14.2.2 Genome-wide Association Studies

GWAS of personality began in 2008 with a study of neuroticism (58). Conducted on a discovery cohort of 2,000 individuals and a replication cohort of 1,500 individuals, this study reported a significant association between degree of neuroticism and a SNP (rs702543) located within the phosphodiesterase 4D gene (*PDE4D*), although replication efforts in additional smaller samples failed to confirm the association, but were confirmed in a later GWAS of more than 2,200 individuals (59). Curiously, no study has since been published (based on a PubMed search conducted in June 2017) on *PDE4D* and neuroticism. Clinicaltrials.gov lists two completed studies on PDE4D and cognition from 2013 and 2016, respectively, neither of which has produced any publications to date.

Non-replication of GWA studies of complex behavioral traits are common, and, given the evidence for very small contributions to the observed variance by individual SNPs, the trend in GWA studies is towards ever-larger samples. One approach to increase in sample size has been the use of meta-analytic techniques. For example, a meta-analysis on GWAS neuroticism data was conducted on a cohort of 6,200 and replication cohort of 6,000 individuals, which identified polymorphisms within the two genes, *LMAN1L* (Lectin, Mannose Binding 1 Like) and *SCAMP2* (Secretory Carrier Membrane Protein 2) as being associated with neuroticism (60). However, a later meta-analysis based on 17,000 individuals did not replicate this association, nor any other SNP association with neuroticism (61). A PubMed search (as of June 2017) did again not identify any other publications investigating the link between neuroticism and either *LMAN1L* or *SCAMP2*, and no clinical trials are listed including these genes as search terms.

One problem with comparing personality associations across GWA studies is that there is a great deal of heterogeneity in personality assessment tools. A meta-analysis of 79 personality studies found low-to-moderate convergent validity across instruments that purportedly measured the same personality construct (62): for example, correlations across personality measures ranged from 0.32–0.66 for "emotional stability." This problem has been addressed by the development of a technique to "harmonize" data across personality instruments (63). Applied to neuroticism data from 29 cohorts totaling more than 63,000 individuals and a replication cohort of over 10,000 individuals, a follow-up study observed a significant association between neuroticism and a SNP (rs35855737) located within *MAGI1* (Membrane Associated Guanylate Kinase, WW And PDZ Domain Containing 1 gene) in the large discovery cohort, but not in the smaller replication cohort (64). There is no record of additional follow-up studies on this association in PubMed (as of June 2017) or at <http://www.clinicaltrials.gov>.

The largest GWAS of neuroticism, as of June 2017, combined the previous study of 63,000 with an additional set of cohorts from the UK Biobank for a total sample of over 170,000 individuals (and almost 300,000 individuals for measures on well-being) (65). This study identified eleven SNPs, six of which were located within an inversion polymorphism detected in a prior study that also contributed to the current analysis. Yet, the total of all measured SNPs accounted for a very small proportion of variance, yielding only a heritability estimate of 0.9%. Combined with an estimated 0.01% contribution from the most significant SNP, the conclusion was that hundreds or thousands of SNPs were still missing (assuming an additive genetic model), suggesting the need for even larger samples to attain the needed statistical power.

14.2.3 Missing heritability hiding in plain sight

The preceding sections illustrate the contrast between the candidate gene and GWAS approaches: one focuses on single or small numbers of genes and is based on *a priori* knowledge about biological systems, whereas the other is theory-free, data-driven, and probes the entire genome. The former has produced replicable findings for only a very small number of gene polymorphisms, the latter has produced no discoveries that have consistently replicated. Either way, a large amount of genetic variance remains unaccounted.

I suggest that missing heritability may be hiding in plain sight: in the genetically encoded machinery of gene regulation, with differential activation of this machinery enabled through environmental influences and epigenetic inheritance. Indeed, I think a critical shortcoming of both approaches has been an over-reliance on DNA sequence variation as the critical determinant of genetic individual differences, rather than variation in the expression of these sequences at the mRNA or protein level. I suggest four hiding spots for missing heritability, three of which involve "non-additive genetic" effects and one which involves non-genetic epigenetic inheritance.

One hiding spot for non-additive genetic effects lies in those protein coding sequences that regulate DNA transcription and translation. Although the central dogma states that DNA is transcribed to mRNA which is translated to protein, the correlation between these levels of gene expression is only poor to moderate. For example, mRNA-protein correlation coefficients across 79 different types of human tissues ranged only from 0.12–0.31 (66). Similarly, there is a threefold mismatch between the abundance of mRNAs encoding ribosomal proteins (required for gene transcription) and their encoded proteins, suggesting translational regulation (67). Such non-linearity could contribute to significant individual differences.

Another hiding spot for non-additive genetic effects lies in regions of the genome that encode non-coding RNAs (which are typically not captured in GWAS). Examples include ribosomal and transfer RNAs, which play a role in gene expression; long non-coding (lnc) RNAs, which regulate developmental programs (68), allelic expression (69), and which are responsive to stress (70); and microRNAs, which are involved in mRNA degradation or translational repression (71). Indeed, one prior study has shown that *FKBP5* expression is regulated by microRNA-511 (72).

A third hiding spot for non-additive genetic effects could lie within DNA sequences that can carry epigenetic marks, that is, transgenerational *genetic* inheritance of epigenetic information. For example, DNA methylation occurs preferentially at DNA CG-repeat sequences (so-called CpG islands), so that a mutation or polymorphism altering a CpG locus could disrupt the transmission of existing DNA methylation marks, or prevent the acquisition of future DNA methylation marks in the offspring. Evidence for the genetic transmission of DNA methylation comes from twin studies showing higher concordance rates across MZ pairs than DZ pairs or unrelated subjects (73), with DNA variance estimated to account for about 20% of individual differences in DNA methylation variation (74).

DNA methylation is of particular interest in complex behavioral traits including trait anxiety because it can be moderated by life experiences, and thus translate environmental variables into gene regulation. DNA methylation has been shown to vary as a function of early life adversity: first in rats (75), and later in post-mortem brain tissue of suicide victims (76). Methylation of the serotonin transporter gene (*SLC6A4*) has also been associated with a history of childhood adversity (77,78), and is further moderated by 5-HTTLPR genotype (79, 80). Furthermore, *SLC6A4* methylation was associated with amygdala activation during a face-matching task and with reduced mRNA expression in postmortem amygdala (81).

An intriguing possibility is that individual differences may also be bestowed by transgenerational non-genetic, that is, *epigenetic* inheritance of epigenetic information. This would mean that epigenetic marks would be transmitted across generations in a manner that is independent of the underlying DNA sequence. A DNA methylation study in mono- and dizygotic twins observed significantly higher epigenetic differences among DZ then MZ twins (82). The authors favored the interpretation that these differences reflected epigenomic rather than DNA sequence differences, on the basis of *in silico* SNP analyses and animal studies.

Direct experimental evidence for epigenetic inheritance was provided in a mouse study of maternal immune activation (MIA) affecting later social and fear behavior and amygdala gene expression (83). These investigators induced prenatal immune activation by the viral mimetic poly(I:C) in pregnant dams, which produced reduced sociability and increased fear to a Pavlovian conditioned cue in both the F_1 and F_2 generations, and even in F_3 animals derived from the paternal lineage. F_1 and F_2 generations were further analyzed for genome-wide transcriptional changes in the amygdala, which revealed 2,217 differentially expressed genes (DEGs) in F_1 and 4,015 DEGs in F_2. Transmission through the paternal lineage implies transgenerational non-genetic inheritance through male germ cells. This is consistent with other work that implied transgenerational epigenetic programming through male germ cells in retained histone modification and DNA methylation through sperm RNAs (84,85). Finally, MIA induced by viral mimetic poly(I:C) was also shown to alter epigenetic regulators in the offspring by changing the global histone acetylation profile in the hippocampus, reducing hippocampal histone deacetylase (HDAC) 1 levels, and inducing changes in the serotonin transporter promoter and gene expression (86).

14.3 **Conclusions**

In conclusion, I have argued that both the candidate gene and GWAS approaches, although complementary, have failed to account fully for the observed heritability of trait anxiety and other complex behavioral traits. Given that heritability estimates are based on both (additive and non-additive) genetic and environmental variance, I have suggested that missing heritability may be hidden in plain sight: in DNA coding regions, non-coding regions, and/or in patterns of epigenetic inheritance, which itself can be manifested in either genetic or non-genetic transmission pathways.

These musings can serve as a framework for future work directed at capturing complete DNA sequences and their corresponding -omic (transcriptome, proteome, methylome, etc) expression profiles from brain regions that are associated with behaviorally relevant phenotypes, and across generations. Exploration of these spaces may ferret out the hiding spots of missing heritability of complex traits, after all.

References

1. **Polderman TJ, Benyamin B, de Leeuw CA, Sullivan PF, van Bochoven A, Visscher PM**, et al. Meta-analysis of the heritability of human traits based on fifty years of twin studies. Nat Genet. 2015;**47**(7):702–9.

2. **Vukasovic T, Bratko D.** Heritability of personality: A meta-analysis of behavior genetic studies. Psychol Bull. 2015;**141**(4):769–85.
3. **Plomin R, Corley R, Caspi A, Fulker DW, DeFries J.** Adoption results for self-reported personality: Evidence for nonadditive genetic effects? J Pers Soc Psychol. 1998;**75**(1):211–8.
4. **Johnson W, Penke L, Spinath FM.** Heritability in the era of molecular genetics: Some thoughts for understanding genetic influences on behavioural traits. Eur J Pers. 2011;**25**:254–66.
5. **Canli T.** Toward a "molecular psychology" of personality. In: John O, Robbins RW, Pervin LA (eds). *Handbook of Personality: Theory and Research*, 3rd edn. New York, NY: Guilford Press; 2008, pp. 311–27.
6. **Canli T.** Toward a neurogenetic theory of neuroticism. Ann N Y Acad Sci. 2008;**1129**:153–74.
7. **Canli T.** Molecular psychology: A brief history and introduction. In: Canli T (ed.). *The Oxford Handbook of Molecular Psychology*. New York, NY: Oxford University Press; 2015.
8. **Canli T, Ferri J, Duman EA.** Genetics of emotion regulation. Neuroscience. 2009;**164**(1):43–54.
9. **Canli T, Lesch K-P.** Long story short: The serotonin transporter in emotion regulation and social cognition. Nat Neurosci. 2007;**10**(9):1103–9.
10. **Congdon E, Canli T.** Genomic imaging of personality: Towards a molecular neurobiology of impulsivity. In: Boyle G, Matthews G, Saklowsky D (eds). *The Sage Handbook of Personality Theory and Assessment: Personality Measurement and Testing*, Vol **2**. Thousand Oaks, CA: Sage Publications; 2008, pp. 334–51.
11. **Congdon E, Canli T.** A neurogenetic approach to impulsivity. J Pers. 2008;**76**(6):1447–84.
12. **Congdon E, Canli T.** The endophenotype of impulsivity: Reaching consilience through behavioral, genetic, and neuroimaging approaches. Behav Cogn Neurosci Rev. 2005;**4**(4):262–81.
13. **Ebstein RP, Zhong S, Chark R, San Lai P, Chew SH.** Modeling the genetics of social cognition in the laboratory: Definition, measurement, and association. In: Canli T (ed.). *The Oxford Handbook of Molecular Psychology*. New York, NY: Oxford University Press; 2015; 167–209.
14. **Bertolino A, Rubino V, Sambataro F, Blasi G, Latorre V, Fazio L,** et al. Prefrontal-hippocampal coupling during memory processing is modulated by COMT Val158Met genotype. Biol Psychiatry. 2006.
15. **Buckholtz JW, Meyer-Lindenberg A.** MAOA and the neurogenetic architecture of human aggression. Trends Neurosci. 2008;**31**(3):120–9.
16. **Meyer-Lindenberg A, Buckholtz JW, Kolachana B, A RH, Pezawas L, Blasi G,** et al. Neural mechanisms of genetic risk for impulsivity and violence in humans. PNAS. 2006;**103**(16):6269–74.
17. **Meyer-Lindenberg A, Nichols T, Callicott JH, Ding J, Kolachana B, Buckholtz J,** et al. COMT haplotype variation affects human prefrontal function. Mol Psychiatry. 2006;**11**(9):797.
18. **Hariri AR, Holmes A.** Genetics of emotional regulation: The role of the serotonin transporter in neural function. Trends Cogn Sci. 2006;**10**(4):182–91.
19. **Hariri AR, Drabant EM, Weinberger DR.** Imaging genetics: Perspectives from studies of genetically driven variation in serotonin function and corticolimbic affective processing. Biol Psychiatry. 2006;**59**(10):888–97.
20. **Lesch KP, Bengel D, Heils A, Sabol SZ, Greenberg BD, Petri S,** et al. Association of anxiety-related traits with a polymorphism in the serotonin transporter gene regulatory region. Science. 1996;**274**(5292):1527–31.
21. **Munafo MR, Clark TG, Flint J.** Does measurement instrument moderate the association between the serotonin transporter gene and anxiety-related personality traits? A meta-analysis. Mol Psychiatry. 2005;**10**(4):415–9.
22. **Sen S, Burmeister M, Ghosh D.** Meta-analysis of the association between a serotonin transporter promoter polymorphism (5-HTTLPR) and anxiety-related personality traits. Am J Med Genet. 2004;**127B**(1):85–9.
23. **Schinka JA, Busch RM, Robichaux-Keene N.** A meta-analysis of the association between the serotonin transporter gene polymorphism (5-HTTLPR) and trait anxiety. Mol Psychiatry. 2004;**9**(2):197–202.

24. **Munafo MR, Freimer NB, Ng W, Ophoff R, Veijola J, Miettunen J**, et al. 5-HTTLPR genotype and anxiety-related personality traits: A meta-analysis and new data. Am J Med Genet B Neuropsychiatr Genet. 2009;**150B**(2):271–81.
25. **Ficks CA, Waldman ID.** Candidate genes for aggression and antisocial behavior: A meta-analysis of association studies of the 5HTTLPR and MAOA-uVNTR. Behav Genet. 2014;**44**(5):427–44.
26. **Murphy SE, Norbury R, Godlewska BR, Cowen PJ, Mannie ZM, Harmer CJ**, et al. The effect of the serotonin transporter polymorphism (5-HTTLPR) on amygdala function: A meta-analysis. Mol Psychiatry. 2013;**18**(4):512–20.
27. **Miller R, Wankerl M, Stalder T, Kirschbaum C, Alexander N.** The serotonin transporter gene-linked polymorphic region (5-HTTLPR) and cortisol stress reactivity: A meta-analysis. Mol Psychiatry. 2013;**18**(9):1018–24.
28. **Bastiaansen JA, de Vries YA, Munafo MR.** Citation distortions in the literature on the serotonin-transporter-linked polymorphic region and amygdala Activation. Biol Psychiatry. 2015;**78**(8):e35–6.
29. **Kaufman J.** Reply To: Citation distortions in the literature on the serotonin-transporter-linked polymorphic region and amygdala activation. Biol Psychiatry. 2015;**78**(8):e37–8.
30. **Herrmann MJ, Huter T, Muller F, Muhlberger A, Pauli P, Reif A**, et al. Additive effects of serotonin transporter and tryptophan hydroxylase-2 gene variation on emotional processing. Cereb Cortex. 2007;**17**(5):1160–3.
31. **Canli T, Congdon E, Constable TR, Lesch KP.** Additive effects of serotonin transporter and tryptophan hydroxylase-2 gene variation on neural correlates of affective processing. Biol Psychol. 2008;**79**(1):118–25.
32. **Lehto K, Vaht M, Maestu J, Veidebaum T, Harro J.** Effect of tryptophan hydroxylase-2 gene polymorphism G-703 T on personality in a population representative sample. Prog Neuro-psychopharmacolo Biol psychiatry. 2015;**57**:31–5.
33. **Dougherty LR, Klein DN, Congdon E, Canli T, Hayden EP.** Interaction between 5-HTTLPR and BDNF Val66Met polymorphisms on HPA axis reactivity in preschoolers. Biol Psychol. 2010;**83**(2):93–100.
34. **Radua J, El-Hage W, Monte GC, Gohier B, Tropeano M, Phillips ML**, et al. COMT Val158Met * SLC6A4 5-HTTLPR interaction impacts on gray matter volume of regions supporting emotion processing. Soc Cogn Affect Neurosci. 2014;**9**(8):1232–8.
35. **Fisher PM, Holst KK, Adamsen D, Klein AB, Frokjaer VG, Jensen PS**, et al. BDNF Val66met and 5-HTTLPR polymorphisms predict a human in vivo marker for brain serotonin levels. Hum Brain Mapp. 2015;**36**(1):313–23.
36. **Fisher PM, Ozenne B, Svarer C, Adamsen D, Lehel S, Baare WFC**, et al. BDNF val66met association with serotonin transporter binding in healthy humans. Transl Psychiat. 2017;**7**(2):e1029.
37. **Jaworska N, MacMaster FP, Foster J, Ramasubbu R.** The influence of 5-HTTLPR and Val66Met polymorphisms on cortical thickness and volume in limbic and paralimbic regions in depression: A preliminary study. BMC Psychiatry. 2016;**16**:61.
38. **Kruschwitz JD, Walter M, Varikuti D, Jensen J, Plichta MM, Haddad L**, et al. 5-HTTLPR/rs25531 polymorphism and neuroticism are linked by resting state functional connectivity of amygdala and fusiform gyrus. Brain Struct Func. 2015;**220**(4):2373–85.
39. **Binder EB, Salyakina D, Lichtner P, Wochnik GM, Ising M, Putz B**, et al. Polymorphisms in FKBP5 are associated with increased recurrence of depressive episodes and rapid response to antidepressant treatment. Nat Genet. 2004;**36**(12):1319–25.
40. **Zimmermann P, Bruckl T, Nocon A, Pfister H, Binder EB, Uhr M**, et al. Interaction of FKBP5 gene variants and adverse life events in predicting depression onset: Results from a 10-year prospective community study. Am J Psychiatry. 2011;**168**(10):1107–16.
41. **Binder EB, Bradley RG, Liu W, Epstein MP, Deveau TC, Mercer KB**, et al. Association of FKBP5 polymorphisms and childhood abuse with risk of posttraumatic stress disorder symptoms in adults. JAMA. 2008;**299**(11):1291–305.

42. **Klengel T, Mehta D, Anacker C, Rex-Haffner M, Pruessner JC, Pariante CM,** et al. Allele-specific FKBP5 DNA demethylation mediates gene-childhood trauma interactions. Nat Neurosci. 2013;**16**(1):33–41.

43. **Gassen NC, Fries GR, Zannas AS, Hartmann J, Zschocke J, Hafner K,** et al. Chaperoning epigenetics: FKBP51 decreases the activity of DNMT1 and mediates epigenetic effects of the antidepressant paroxetine. Sci Signal. 2015;**8**(404):ra119.

44. **Roy A, Hodgkinson CA, Deluca V, Goldman D, Enoch M-A.** Two HPA axis genes, CRHBP and FKBP5, interact with childhood trauma to increase the risk for suicidal behavior. J Psychiatr Res. 2012;**46**(1):72–9.

45. **Boscarino JA, Erlich PM, Hoffman SN, Zhang X.** Higher FKBP5, COMT, CHRNA5, and CRHR1 allele burdens are associated with PTSD and interact with trauma exposure: Implications for neuropsychiatric research and treatment. Neuropsychiatr Dis Treat. 2012;**8**:131–9.

46. **Iorio CRD, Carey CE, Michalski LJ, Corral-Frias NS, Conley ED, Hariri AR,** et al. Hypothalamic–pituitary–adrenal axis genetic variation and early stress moderates amygdala function. Psychoneuroendocrinology. 2017;**80**:170–8.

47. **Horstmann S, Lucae S, Menke A, Hennings JM, Ising M, Roeske D,** et al. Polymorphisms in GRIK4, HTR2A, and FKBP5 show interactive effects in predicting remission to antidepressant treatment. Neuropsychopharmacology. 2010;**35**(3):727–40.

48. **White MG, Bogdan R, Fisher PM, Munoz KE, Williamson DE, Hariri AR.** FKBP5 and emotional neglect interact to predict individual differences in amygdala reactivity. Genes, Brain, Behav. 2012;**11**(7):869–78.

49. **Pagliaccio D, Luby JL, Bogdan R, Agrawal A, Gaffrey MS, Belden AC,** et al. Stress-system genes and life stress predict cortisol levels and amygdala and hippocampal volumes in children. Neuropsychopharmacology. 2014;**39**(5):1245–53.

50. **Pagliaccio D, Luby JL, Bogdan R, Agrawal A, Gaffrey MS, Belden AC,** et al. HPA axis genetic variation, pubertal status, and sex interact to predict amygdala and hippocampus responses to negative emotional faces in school-age children. Neuroimage. 2015;**109**:1–11.

51. **Grabe HJ, Wittfeld K, Van der Auwera S, Janowitz D, Hegenscheid K, Habes M,** et al. Effect of the interaction between childhood abuse and rs1360780 of the FKBP5 gene on gray matter volume in a general population sample. Hum Brain Mapp. 2016;**37**(4):1602–13.

52. **Hirakawa H, Akiyoshi J, Muronaga M, Tanaka Y, Ishitobi Y, Inoue A,** et al. FKBP5 is associated with amygdala volume in the human brain and mood state: A voxel-based morphometry (VBM) study. Int J Psychiatry Clin Pract. 2016;**20**(2):106–15.

53. **Gerritsen L, Milaneschi Y, Vinkers C, van Hemert B, van Velzen L, Schmaal L,** et al. HPA Axis Genes, and their Interaction with Childhood Maltreatment, are Related to Cortisol Levels and Stress-Related Phenotypes. Neuropsychopharmacology. 2017.

54. **Holz NE, Buchmann AF, Boecker R, Blomeyer D, Baumeister S, Wolf I,** et al. Role of FKBP5 in emotion processing: Results on amygdala activity, connectivity and volume. Brain Struct Funct. 2015;**220**(3):1355–68.

55. **Perez-Ortiz JM, Garcia-Gutierrez MS, Navarrete F, Giner S, Manzanares J.** Gene and protein alterations of FKBP5 and glucocorticoid receptor in the amygdala of suicide victims. Psychoneuroendocrinology. 2013;**38**(8):1251–8.

56. **Arango C.** Candidate gene associations studies in psychiatry: Time to move forward. Eur Arch Psychiatry Clin Neurosci. 2017;**267**(1):1–2.

57. **Balestri M, Calati R, Serretti A, De Ronchi D.** Genetic modulation of personality traits: A systematic review of the literature. Int Clin Psychopharmacol. 2014;**29**(1):1–15.

58. **Shifman S, Bhomra A, Smiley S, Wray NR, James MR, Martin NG,** et al. A whole genome association study of neuroticism using DNA pooling. Mol Psychiatry. 2008;**13**(3):302–12.

59. **Calboli FC, Tozzi F, Galwey NW, Antoniades A, Mooser V, Preisig M,** et al. A genome-wide association study of neuroticism in a population-based sample. PLoS One. 2010;**5**(7):e11504.

60. **Luciano M, Huffman JE, Arias-Vasquez A, Vinkhuyzen AA, Middeldorp CM, Giegling I,** et al. Genome-wide association uncovers shared genetic effects among personality traits and mood states. Am J Med Genet B Neuropsychiatr Genet. 2012;**159B**(6):684–95.
61. **de Moor MH, Costa PT, Terracciano A, Krueger RF, de Geus EJ, Toshiko** T, et al. Meta-analysis of genome-wide association studies for personality. Mol Psychiatry. 2012;**17**(3):337–49.
62. **Pace VL, Brannick MT.** How similar are personality scales of the "same" construct? A meta-analytic investigation. Pers Indiv Differ. 2010;**49**(7):669–76.
63. **van den Berg SM, de Moor MH, McGue M, Pettersson E, Terracciano A, Verweij KJ,** et al. Harmonization of Neuroticism and Extraversion phenotypes across inventories and cohorts in the Genetics of Personality Consortium: An application of Item Response Theory. Behav Genet. 2014;**44**(4):295–313.
64. **Genetics of Personality C, de Moor MH, van den Berg SM, Verweij KJ, Krueger RF, Luciano M,** et al. Meta-analysis of genome-wide association studies for neuroticism, and the polygenic association with major depressive disorder. JAMA Psychiatry. 2015;**72**(7):642–50.
65. **Okbay A, Baselmans BM, De Neve JE, Turley P, Nivard MG, Fontana MA,** et al. Genetic variants associated with subjective well-being, depressive symptoms, and neuroticism identified through genome-wide analyses. Nat Genet. 2016;**48**(6):624–33.
66. **Ostlund G, Sonnhammer EL.** Quality criteria for finding genes with high mRNA-protein expression correlation and coexpression correlation. Gene. 2012;**497**(2):228–36.
67. **Gupta V, Warner JR.** Ribosome-omics of the human ribosome. RNA (New York, NY). 2014;**20**(7):1004–13.
68. **Zhang X-Q, Wang Z-L, Poon M-W, Yang J-H.** Spatial-temporal Transcriptional Dynamics of Long Non-coding RNAs in Human Brain. Hum Mol Genetics. 2017.
69. **Kung JTY, Colognori D, Lee JT.** Long noncoding RNAs: Past, present, and future. Genetics. 2013;**193**(3):651–69.
70. **Amaral PP, Dinger ME, Mattick JS.** Non-coding RNAs in homeostasis, disease and stress responses: an evolutionary perspective. Brief Funct Genomics. 2013;**12**(3):254–78.
71. **Wilczynska A, Bushell M.** The complexity of miRNA-mediated repression. Cell Death Differ. 2015;**22**(1):22–33.
72. **Zheng D, Sabbagh JJ, Blair LJ, Darling AL, Wen X, Dickey CA.** MicroRNA-511 Binds to FKBP5 mRNA, Which Encodes a Chaperone Protein, and Regulates Neuronal Differentiation. J Biol Chem. 2016;**291**(34):17897–906.
73. **Bell JT, Spector TD.** DNA methylation studies using twins: What are they telling us? Genome Biol. 2012;**13**(10):172.
74. **McRae AF, Powell JE, Henders AK, Bowdler L, Hemani G, Shah S,** et al. Contribution of genetic variation to transgenerational inheritance of DNA methylation. Genome Biol. 2014;**15**(5):R73.
75. **Weaver IC, Cervoni N, Champagne FA, D'Alessio AC, Sharma S, Seckl JR,** et al. Epigenetic programming by maternal behavior. Nat Neurosci. 2004;**7**(8):847–54. Epub 2004 Jun 27.
76. **McGowan PO, Sasaki A, D'Alessio AC, Dymov S, Labonte B, Szyf M,** et al. Epigenetic regulation of the glucocorticoid receptor in human brain associates with childhood abuse. Nat Neurosci. 2009;**12**(3):342–8.
77. **Beach SR, Brody GH, Todorov AA, Gunter TD, Philibert RA.** Methylation at 5HTT mediates the impact of child sex abuse on women's antisocial behavior: An examination of the Iowa adoptee sample. Psychosom Med. 2011;**73**(1):83–7.
78. **Beach SR, Brody GH, Todorov AA, Gunter TD, Philibert RA.** Methylation at SLC6A4 is linked to family history of child abuse: An examination of the Iowa Adoptee sample. Am J Med Genet B Neuropsychiatr Genet. 2010;**153B**(2):710–3.
79. **Philibert R, Madan A, Andersen A, Cadoret R, Packer H, Sandhu H.** Serotonin transporter mRNA levels are associated with the methylation of an upstream CpG island. Am J Med Genet B Neuropsychiatr Genet. 2007;**144**(1):101–5.

80. **Duman EA, Canli T.** Influence of life stress, 5-HTTLPR genotype, and SLC6A4 methylation on gene expression and stress response in healthy Caucasian males. Biol Mood Anxiety Disord. 2015;**5**:2.

81. **Nikolova YS, Koenen KC, Galea S, Wang C-M, Seney ML, Sibille** E, et al. Beyond genotype: Serotonin transporter epigenetic modification predicts human brain function. Nat Neurosci. 2014;**17**(9):1153–5.

82. **Kaminsky ZA, Tang T, Wang S-C, Ptak C, Oh GHT, Wong AHC**, et al. DNA methylation profiles in monozygotic and dizygotic twins. Nat Gen. 2009;**41**(2):240–5.

83. **Weber-Stadlbauer U, Richetto J, Labouesse MA, Bohacek J, Mansuy IM, Meyer U.** Transgenerational transmission and modification of pathological traits induced by prenatal immune activation. Mol Psychiatry. 2017;**22**(1):102–12.

84. **Rodgers AB, Morgan CP, Leu NA, Bale TL.** Transgenerational epigenetic programming via sperm microRNA recapitulates effects of paternal stress. PNAS. 2015;**112**(44):13699–704.

85. **Gapp K, Jawaid A, Sarkies P, Bohacek J, Pelczar P, Prados** J, et al. Implication of sperm RNAs in transgenerational inheritance of the effects of early trauma in mice. Nat Neurosc. 2014;**17**(5):667–9.

86. **Reisinger SN, Kong E, Khan D, Schulz S, Ronovsky M, Berger** S, et al. Maternal immune activation epigenetically regulates hippocampal serotonin transporter levels. Neurobiol Stress. 2016;**4**:34–43.

Chapter 15

Electrocortical endophenotypes of anxiety

Erik M Mueller

15.1 Introduction

Some individuals experience fear and anxiety more often, in more different contexts, and more strongly than others. Fear and anxiety evolved to protect individuals from danger, and even today, being more fearful or anxious may be adaptive and related to reduced mortality in certain situations (1). On the other hand, extreme forms of dispositional fear or anxiety predispose individuals for anxiety disorders, which are among the most prevalent psychiatric disorders worldwide (2), dramatically impact the quality of life (3), and relate to other disorders like depression (4,5).

Although twin studies show that trait anxiety as measured with self-evaluative questionnaires and the propensity to develop anxiety disorders (4) is moderately heritable, the particular gene variants and mechanisms that make some individuals more anxious than others are not well understood. Due to the high complexity of this multidimensional, context-dependent, and long-evolved trait, probably hundreds or thousands of anxiety-related genes exist (6), each contributing to some kind of protein expression, which is somewhat relevant for some type of brain cell communication, which plays some role in brain networks that are also part of anxiety-related processes (e.g. 7) (Figure 15.1A and 15.1B). As a consequence, the direct association between an individual anxiety-related gene and overall dispositional anxiety is so small that it can hardly be detected with conventional sample sizes. To understand how single gene variants contribute to anxiety, it may thus be advisable to first isolate specific brain processes, which are involved in anxiety but which are less complex than the broad cognitive-affective-behavioral trait that is measured with questionnaires. These more specific brain processes (e.g. visual threat detection, fear conditioning, cardiovascular regulation, etc) are presumably affected by a smaller number of genes than the broad trait, and as a consequence, more strongly related to individual genes. This approach is termed the endophenotype or intermediate phenotype approach (9) because these specific brain processes reflect an intermediate node between genotype and observable phenotype.

Some of previously identified anxiety-related brain processes are subcortical (e.g. involving the amygdala; 10), others are cortical (e.g. involving the insula; 11), anterior cingulate cortex (12), prefrontal cortex (13), or sensory cortices (14). Some processes like worrying are slow and in the range of multiple seconds or minutes (15), while other processes like visual threat detection are extremely fast and in the range of tenths of a second (14) (Figure 15.2A). Whereas neuroimaging techniques are well suited to investigate slower anxiety-related brain processes, one of the best and most established methods to study fast cortical anxiety-related processes is electroencephalography (EEG), which has a much higher temporal resolution than functional Magnetic Resonance Imaging (fMRI) or other neuroimaging techniques. This chapter explores potential intermediate phenotypes of anxiety by introducing different electrocortical (EEG) markers of anxiety and by outlining how prior research has related them to molecular and behavioral genetics. As summarized in Table 15.1, the majority of anxiety-related EEG studies focuses on event-related potentials

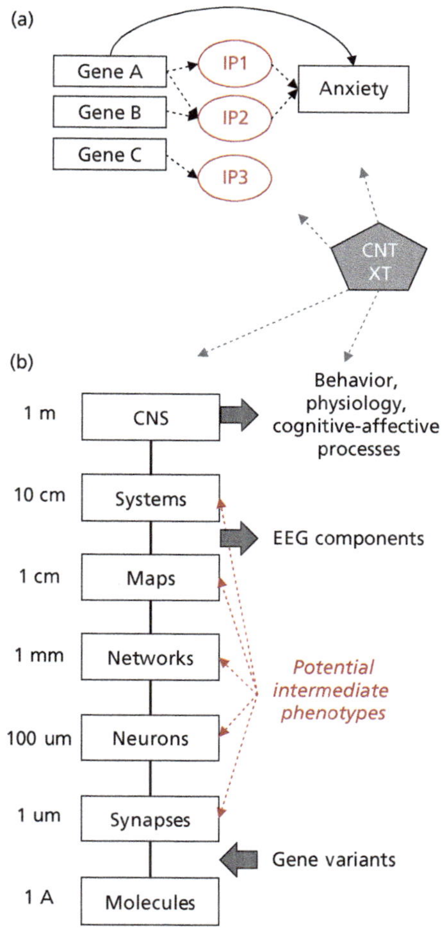

Figure 15.1 (A) Model of relationships between gene variants and dispositional anxiety. The direct association between a gene and anxiety (solid black line) is likely to be weak, as a complex trait like anxiety is composed of many different processes. The existence of intermediate phenotypes (IP) is proposed, which have a lower level of complexity than the complex trait (see also panel B). Importantly, IPs may be affected by one (IP1) or multiple (IP2) gene variants, IPs may be related to dispositional anxiety (IP1 and IP2) or not (IP3) and genes may be indirectly related to anxiety (Gene A and B) or not (Gene C). With an increasing number of involved genes and IPs the direct association between an individual gene variant and anxiety becomes increasingly smaller. In addition, the context in which an IP and dispositional anxiety are assessed (e.g. threat or non-threat context, type of specific threat, etc) may affect how strong genes, IPs and anxiety are related to each other. (B) Levels of organization in the nervous system and the spatial scale at which these levels can be identified. Genes in the size of several nanometers are to be placed at the lowest levels (between the Molecules and Synapses level). In contrast. anxiety-related behavior (e.g. avoidance), physiology (e.g. heart rate increase), and cognitive-affective processes (e.g. worrying, subjective feeling of anxiety) are dynamic, context-dependent adaptive processes that are emergent phenomena of the entire central nervous system, which reflects the highest level of complexity. Variation in complex CNS-level anxiety is caused by variations at all lower levels of organization. As all lower levels show a lower complexity they may potentially provide intermediate phenotypes for anxiety. EEG components reflect synchronized post-synaptic activity at patches of cortex in the range of centimeters (between Maps and Systems level) and may provide anxiety-related IPs that are closer to the gene variants and less complex than trait anxiety.

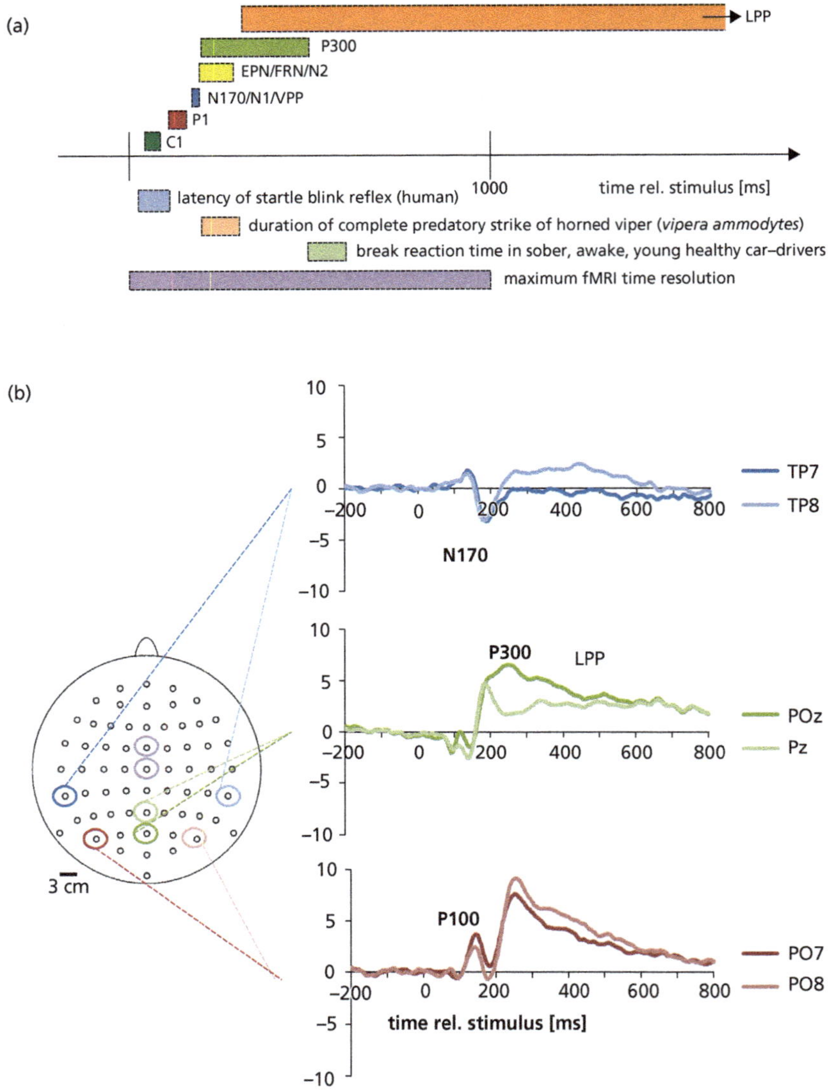

Figure 15.2 (A) Latencies of various threat-sensitive and/or anxiety related ERP-components (see text and table (1). Solid lines indicate that a genetic contribution has been shown in prior molecular genetic or behavioral genetic research. To illustrate the necessity of rapid brain processes in response to threat, latencies of some threat-related events are displayed below the time-axis including the latency of the human startle eye-blink reflex, the duration of a horned viper attack. To illustrate the advantage of EEG to measure rapid threat-related processes, the relatively poor temporal resolution of fMRI is also indicated. (B) Grand average visual evoked EEG potentials (VEPs) in response to a face picture displayed at 0 ms. Top row: VEP at temporo-parietal sensors TP7 and TP8 (referenced to average reference). A clear negative deflection at 170 ms can be seen (N170). Middle row: VEP at parietal midline sensors Pz and POz (referenced to mastoids). A positive deflection around 300 ms (P300) and an additional more sustained positivity (late positive potential, LPP) can be seen. Bottom row: VEP at lateral parieto-occipital sensors PO7 and PO8 (referenced to mastoids) showing a positive P100 component at about 130 ms. Right to the VEPs a 64 EEG channel layout is provided.

Source data from Janoo, A., & Gasc, J. P. (1992). High speed motion analysis of the predatory strike and fluorographic study of oesophageal deglutition in Vipera ammodytes: more than meets the eye. Amphibia-Reptilia, 13(4), 315–325 and Tashiro, M., Horikawa, E., Mochizuki, H., Sakurada, Y., Kato, M., Inokuchi, T., ... & Yanai, K. (2005). Effects of fexofenadine and hydroxyzine on brake reaction time during car-driving with cellular phone use. Human Psychopharmacology: Clinical and Experimental, 20(7), 501–509.

Table 15.1 Possible electrocortical endophenotypes of anxiety

General Approach	Specific Technique	Component	Time/Freq. range	Topography	Assumed Source	Anxiety Disorder	Trait Anxiety	Heritability	Gene variants
Event-related potentials	Visual evoked potentials	C1 [1]	50–100 ms	Parietal	V1 and other [2]	[3]	[4]	?	?
		P1 [5]	80–150 ms	Occipitotemporal	Ventral-lateral occipital cortex [5]	[6, 7] but [8]	[9, 10] but [11]	[12, 13]	?
		N1	160–180 ms	Temporal-Parietal	Ventral visual cortex	[14] (children)	?	[15]	?
		N170 [16]	160–180 ms	Temporal-Parietal	Ventral visual cortex	[17] but [7, 8, 18]	but [11]	[19]	?
		EPN[20]	180–280 ms?	Temporal-Occipital[20]	?	?	[21] but [9]	[22]	5-HTTLPR [23], TPH2[23]
		LPP [24]	300– > 2000 ms	Multiple visual cortical regions	Multiple	[6]	[9, 21, 25]	[22]	β1-receptor****[26]
	Task specific stimulus evoked potentials	NoGo-N2[27]	200–350 ms	Frontomedial	Anterior midcingulate	[28] but [29, 30]	[31–33] but [34]	[35]	BDNF*[36]
		FRN[37]	200–300 ms	Frontomedial	Anterior midcingulate	?	[38]	?	COMT**[39, 40]
		P300[41]	200–500 ms	Frontal ("P3a") and Parietal ("P3b")	Multiple [42]	[43, 44]	?	[22, 35, 45]	Non confirmed*****[46]
		NoGo-P300 anteriorization [47]	200-500 ms	Frontal vs. Parietal	Anterior midcingulate [48]	?	[49]	?	DTNBP1, 5-HTTLPR[50]
	Response evoked potentials	CRN/ERN [51, 52]	50–150 ms	Frontomedial	Anterior midcingulate	[53]	[38, 54, 55]	[56]	COMT** [57-59], Many others but nothing consistent [60]

(*continued*)

Table 15.1. Continued

General Approach	Specific Technique	Component	Time/Freq. range	Topography	Assumed Source	Anxiety Disorder	Trait Anxiety	Heritability	Gene variants
Other approaches	Time-Frequency analysis	Frontal theta	4–8 Hz	Frontomedial	Anterior midcingulate	[61]	[38, 62-64]	?	DRD4 [65], COMT** [58], HTR7*** [66]
	Cardio-EEG-Covariance Trace	N300H[67]	200500 ms	Centromedial	Midcingulate [68]	[43]	[69]	?	5-HTTLPR [69]

Not listed but with potential relevance for anxiety (among other components): CNV [70], N2PC [71], latency of threat-evoked components [8,11] and resting state frontal alpha asymmetry [72] [73] [74–76]; *BDNFVal66Met, **COMTVal158Met, *** rs7916403, ****G1165C, ***** although a large number of studies reported some kind of association between individual SNPs and P300 none of these associations could be confirmed in a very large sample of N = 4,211[46];

1. Jeffreys DA, Axford JG. Source locations of pattern-specific components of human visual evoked potentials. I. Component of striate cortical origin. Exp Brain Res. 1972:16:1-21.
2. Rauss K, Schwartz S, Pourtois G. Top-down effects on early visual processing in humans: A predictive coding framework. Neurosci Biobehav Rev. 2011;35(5):1237–53. doi:10.1016/j.neubiorev.2010.12.011.
3. Weymar M, Keil A, Hamm AO. Timing the fearful brain: unspecific hypervigilance and spatial attention in early visual perception. Soc Cogn Affect Neurosci. 2014;9(5):723–9. doi:10.1093/scan/nst044.
4. Eldar S, Yankelevitch R, Lamy D, Bar-Haim Y. Enhanced neural reactivity and selective attention to threat in anxiety. Biol Psychol. 2010;85(2):252–7. doi:10.1016/j.biopsycho.2010.07.010.
5. Clark, V. P., Hillyard, S. A. (1996). Spatial selective attention affects early extrastriate but not striate components of the visual evoked potential. J Cogn Neurosci, 8(5):387–402. doi:10.1162/jocn.1996.8.5.387
6. Weinberg A, Hajcak G. Electrocortical evidence for vigilance-avoidance in generalized anxiety disorder. Psychophysiology. 2011;48(6):842–51. doi:10.1111/j.1469-8986.2010.01149.x..
7. Mueller EM, Hofmann SG, Santesso DL, Meuret AE, Bitran S, Pizzagalli DA. Electrophysiological evidence of attentional biases in social anxiety disorder. Psychol Med. 2009b;39(7):1141–52. doi:10.1017/S0033291708004820.
8. Rossignol M, Philippot P, Douilliez C, Crommelinck M, and Campanella S. The perception of fearful and happy facial expression is modulated by anxiety: an event-related potential study. Neurosci Lett. (2005); 377, 115–20.
9. Holmes A, Nielsen MK, Green S. Effects of anxiety on the processing of fearful and happy faces: An event-related potential study. Biol Psychol. 2008;77(2):159–73. doi:10.1016/j.biopsycho.2007.10.003.
10. Li W, Zinbarg RE, Paller, KA. Trait anxiety modulates supraliminal and subliminal threat: Brain potential evidence for early and late processing influences. Cogn Affect Behav Neurosci. 2007:7:25–36.
11. Williams LM, et al. Neural biases to covert and overt signals of fear: dissociation by trait anxiety and depression. J Cogn Neurosci. 2007;19:1595–608.
12. Bulayeva KB, Pavlova TA, Guseynov GG. Visual evoked potentials: Phenotypic and genotypic variability. Behav Genet. 1993;23(5):443–7.
13. Buchsbaum MS. Average evoked response and stimulus intensity in identical and fraternal twins. Physiological Psychology. 1974;2(3):365–70.
14. Baving L, Rellum T, Laucht M, and Schmidt MH. Attentional enhancement to NoGo stimuli in anxious children. J Neural Transm (Vienna). (2004); 111, 985–99.
15. Smit DJ, Posthuma D, Boomsma DI, and De Geus EJ. Heritability of anterior and posterior visual N1. Int J Psychophysiol. (2007); 66, 196–204.
16. Bentin S, Allison T, Puce A, Perez E, McCarthy G. Electrophysiological Studies of Face Perception in Humans. J Cogn Neurosci. 1996;8(6):551–65. doi:10.1162/jocn.1996.8.6.551..
17. Kolassa IT, Miltner WH. Psychophysiological correlates of face processing in social phobia. Brain Res. 2006;1118(1):130–41. doi:10.1016/j.brainres.2006.08.019.
18. Rossignol M, et al. Enhanced perceptual responses during visual processing of facial stimuli in young socially anxious individuals. Neurosci Lett. 2012:526:68–73.
19. Shannon RW, Patrick CJ, Venables NC, He S. "Faceness" and affectivity: evidence for genetic contributions to distinct components of electrocortical response to human faces. Neuroimage. 2013;83:609–15. doi:10.1016/j.neuroimage.2013.06.014.
20. Eimer M, Holmes A, McGlone FP. The role of spatial attention in the processing of facial expression: an ERP study of rapid brain responses to six basic emotions. Cogn Affect Behav Neurosci. 2003;3:97–110.
21. Muhlberger A, Wieser MJ, Herrmann MJ, Weyers P, Troger C, Pauli P. Early cortical processing of natural and artificial emotional faces differs between lower and higher socially anxious persons. J Neural Transm (Vienna). 2009;116(6):735–46. doi:10.1007/s00702-008-0108-6.
22. Weinberg A, Venables NC, Proudfit GH, Patrick CJ. Heritability of the neural response to emotional pictures: evidence from ERPs in an adult twin sample. Soc Cogn Affect Neurosci. 2015;10(3):424–34. doi:10.1093/scan/nsu059.

23. Herrmann MJ, et al. Additive effects of serotonin transporter and tryptophan hydroxylase-2 gene variation on emotional processing. Cereb Cortex. 2007;17:1160–3.
24. Schupp HT, et al. Affective picture processing: the late positive potential is modulated by motivational relevance. Psychophysiology. 2000;37:257–61.
25. Moser JS, Huppert JD, Duval E, Simons RF. Face processing biases in social anxiety: an electrophysiological study. Biol Psychol. 2008;78(1):93–103. doi:10.1016/j.biopsycho.2008.01.005.
26. de Rover M, Brown SB, Boot N, Hajcak G, van Noorden MS, van der Wee NJ, Nieuwenhuis S. Beta receptor-mediated modulation of the late positive potential in humans. Psychopharmacology (Berl). 2012;219(4):971–9. doi:10.1007/s00213-011-2426-x.
27. Eimer M. Effects of attention and stimulus probability on ERPs in a Go/Nogo task. Biol Psychol. 1993;35(2):123–38.
28. Ruchsow M, Reuter K, Hermle L, Ebert D, Kiefer M, Falkenstein M. Executive control in obsessive–compulsive disorder: Event-related potentials in a Go/Nogo task. J Neural Transm (Vienna). 2007;114(12):1595–601. doi:10.1007/s00702-007-0779-4.
29. Herrmann MJ, Jacob C, Unterecker S, Fallgatter AJ. Reduced response-inhibition in obsessive-compulsive disorder measured with topographic evoked potential mapping. Psychiatry Res. 2003;120(3):265–71.
30. Kim MS, Kim YY, Yoo SY, Kwon JS. Electrophysiological correlates of behavioral response inhibition in patients with obsessive-compulsive disorder. Depress Anxiety. 2007;24(1):22–31. doi:10.1002/da.20195.
31. Sehlmeyer C, Konrad C, Zwitserlood P, Arolt V, Falkenstein M, Beste C. ERP indices for response inhibition are related to anxiety-related personality traits. Neuropsychologia. 2010;48(9):2488–95. doi:10.1016/j.neuropsychologia.2010.04.022.
32. Righi S, Mecacci L, Viggiano MP. Anxiety, cognitive self-evaluation and performance: ERP correlates. J Anxiety Disord. 2009;23(8):1132–38. doi:10.1016/j.janxdis.2009.07.018.
33. Amodio DM, Master SL, Yee CM, and Taylor SE. Neurocognitive components of the behavioral inhibition and activation systems: implications for theories of self-regulation. Psychophysiology. (2008); 45, 11–9.
34. Boksem MA, Tops M, Wester AE, Meijman TF, and Lorist MM. Error-related ERP components and individual differences in punishment and reward sensitivity. Brain Res. (2006); 1101, 92–101.
35. Anokhin AP, Heath AC, Myers E. Genetics, prefrontal cortex, and cognitive control: A twin study of event-related brain potentials in a response inhibition task. Neurosci Lett. 2004;368(3):314–8. doi:10.1016/j.neulet.2004.07.036.
36. Beste C, Baune BT, Domschke K, Falkenstein M, Konrad C. Paradoxical association of the brain-derived-neurotrophic-factor val66met genotype with response inhibition. Neuroscience. 2010b;166(1):178–84. doi:10.1016/j.neuroscience.2009.12.022.
37. Miltner WHR, Braun CH, Coles, MGH. Event-related brain potentials following incorrect feedback in a time-estimation task: Evidence for a "generic" neural system for error detection. J Cogn Neurosci. 1997;9:788–98.
38. Cavanagh JF, Shackman AJ. Frontal midline theta reflects anxiety and cognitive control: Meta-analytic evidence. J Physiol Paris. 29014; doi:10.1016/j.jphysparis.2014.04.003.
39. Marco-Pallares J, Cucurell D, Cunillera T, Kramer UM, Camara E, Nager W, et al. Genetic variability in the dopamine system (dopamine receptor D4, catechol-O-methyltransferase) modulates neurophysiological responses to gains and losses. Biol Psychiatry. 2009;66(2):154–61. doi:S0006-3223(09)00031-6 [pii] 10.1016/j.biopsych.2009.01.006.
40. Mueller EM, et al. The COMT Val158Met polymorphism regulates the effect of a dopamine antagonist on the feedback-related negativity. Psychophysiology. 2014;51:805–9.
41. Sutton, S., et al., Evoked-potential correlates of stimulus uncertainty. Science, 1965. 150: 1187–8.
42. Makeig S, Westerfield M, Jung TP, Enghoff S, Townsend J, Courchesne E, Sejnowski TJ. Dynamic brain sources of visual evoked responses. Science. 2002;295(5555):690–4. doi:10.1126/science.1066168295/5555/690 [pii].
43. Mueller EM, Panitz C, Nestoriuc Y, Stemmler G, Wacker J. Panic disorder and serotonin reuptake inhibitors predict coupling of cortical and cardiac activity. Neuropsychopharmacology. 2014b;39(2):507–14. doi:10.1038/npp.2013.224.
44. Clark CR, et al. Enlarged frontal P300 to stimulus change in panic disorder. Biol Psychiatry. 1996;39:845–56.
45. van Beijsterveldt CE, van Baal GC. Twin and family studies of the human electroencephalogram: a review and a meta-analysis. Biol Psychol. 2002;61(1-2):111–38.
46. Malone SM, Vaidyanathan U, Basu S, Miller MB, McGue M, Iacono WG. Heritability and molecular-genetic basis of the P3 event-related brain potential: a genome-wide association study. Psychophysiology. 2014;51(12):1246–58. doi:10.1111/psyp.12345..
47. Fallgatter AJ, Brandeis D, Strik WK. A robust assessment of the NoGo-anteriorisation of P300 microstates in a cued Continuous Performance Test. Brain Topogr. 1997;9:295–302.
48. Fallgatter AJ, Bartsch AJ, Herrmann MJ. Electrophysiological measurements of anterior cingulate function. J Neural Transm (Vienna). 2002;109:977–88.
49. Wacker J, Chavanon ML, Leue A, Stemmler G. Trait BIS predicts alpha asymmetry and P300 in a Go/No-Go task. European Journal of Personality. 2010;24(2):85–105..
50. Fallgatter AJ, Jatzke S, Bartsch AJ, Hamelbeck B, Lesch KP. Serotonin transporter promoter polymorphism influences topography of inhibitory motor control. Int J Neuropsychopharmacol. 1999;2(2):115–20. doi:10.1017/S1461145799001455..

(*continued*)

Table 15.1. Continued

51. Falkenstein M, et al. Effects of crossmodal divided attention on late ERP components. II. Error processing in choice reaction tasks. Electroencephalogr Clin Neurophysiol. 1991;78:447–55.
52. Gehring WJ, et al. A brain potential manifestation of error-related processing. Electroencephalogr Clin Neurophysiol Suppl. 1995;44:261–72.
53. Olvet DM, Hajcak G. The error-related negativity (ERN) and psychopathology: Toward an endophenotype. Clin Psychol Rev. 2008;28(8):1343–54. doi:S0272-7358(08)00109-8 [pii] 10.1016/j.cpr.2008.07.003.
54. Hajcak G, McDonald N, Simons RF. Error-related psychophysiology and negative affect. Brain Cogn. 2004;56(2):189–97.
55. Moser JS, Moran TP, Schroder HS, Donnellan MB, Yeung N. On the relationship between anxiety and error monitoring: A meta-analysis and conceptual framework. Front Hum Neurosci. 2013;7, 466. doi:10.3389/fnhum.2013.00466.
56. Anokhin AP, Golosheykin S, Heath AC. Heritability of frontal brain function related to action monitoring. Psychophysiology. 2008;45(4):524–34. doi:PSYP664 [pii] 10.1111/j.1469-8986.2008.00664.x.
57. Osinsky R, Hewig J, Alexander N, Hennig J. COMT Val158Met genotype and the common basis of error and conflict monitoring. Brain Res, 2012;1452, 108–18. doi:10.1016/j.brainres.2012.02.054.
58. Mueller EM, Makeig S, Stemmler G, Hennig J, and Wacker J. Dopamine effects on human error processing depend on catechol-*O*-methyltransferase VAL158MET genotype. J Neurosci. (2011); 31, 15818–25.
59. Kramer UM, Cunillera T, Camara E, Marco-Pallares J, Cucurell D, Nager W, Bauer P, Schule R, Schols L, Rodriguez-Fornells A, and Munte TF. The impact of catechol-*O*-methyltransferase and dopamine D4 receptor genotypes on neurophysiological markers of performance monitoring. J Neurosci. (2007); 27, 14190–8.
60. Ullsperger M. Genetic association studies of performance monitoring and learning from feedback: The role of dopamine and serotonin. Neurosci Biobehav Rev. 2010;34(5):649–59. doi:S0149-7634(09)00097-9 [pii] 10.1016/j.neubiorev.2009.06.009.
61. Riesel A, Endrass T, Auerbach LA, Kathmann N. Overactive performance monitoring as an endophenotype for obsessive-compulsive disorder: Evidence from a treatment study. Am J Psychiatry. 2015;172(7):665–73. doi:10.1176/appi.ajp.2014.14070886.
62. Mueller EM, Burgdorf C, Chavanon ML, Schweiger D, Wacker J, Stemmler G. Dopamine modulates frontomedial failure processing of agentic introverts versus extraverts in incentive contexts. Cogn Affect Behav Neurosci. 2014;14(2):756–68. doi:10.3758/s13415-013-0228-9.
63. Neo PS, McNaughton N. Frontal theta power linked to neuroticism and avoidance. Cogn Affect Behav Neurosci. 2011;11:396–403.
64. Neo PS, Thurlow JK, McNaughton N. Stopping, goal-conflict, trait anxiety and frontal rhythmic power in the stop-signal task. Cogn Affect Behav Neurosci. 2011;11:485–93.
65. Kramer UM, Rojo N, Schule R, Cunillera T, Schols L, Marco-Pallares J et al. ADHD candidate gene (DRD4 exon III) affects inhibitory control in a healthy sample. BMC Neurosci. 2009;10, 150. doi:10.1186/1471-2202-10-150..
66. Zlojutro, M., et al., Genome-wide association study of theta band event-related oscillations identifies serotonin receptor gene HTR7 influencing risk of alcohol dependence. Am J Med Genet B Neuropsychiatr Genet. 2011;156B:44–58.
67. Mueller EM, Stemmler G, Wacker J. Single-trial electroencephalogram predicts cardiac acceleration: A time-lagged P-correlation approach for studying neurovisceral connectivity. Neuroscience; 2009a:166(2):491–500. doi:10.1016/j.neuroscience.2009.12.051.
68. Panitz C, Wacker J, Stemmler G, Mueller EM. Brain–heart coupling at the P300 latency is linked to anterior cingulate cortex and insula—a cardio-electroencephalographic covariance tracing study. Biological Psychology. 2013;(94):185–91.
69. Mueller EM, Stemmler G, Hennig J, Wacker J. 5-HTTLPR and anxiety modulate brain–heart covariation. Psychophysiology. 2013;50:441–53.
70. Flor H, Birbaumer N, Roberts LE, Feige B, Lutzenberger W, Hermann C, and Kopp B. Slow potentials, event-related potentials, "gamma-band" activity, and motor responses during aversive conditioning in humans. Exp Brain Res. (1996); 112, 298–312.
71. Fox E, Derakshan N, Shoker L. Trait anxiety modulates the electrophysiological indices of rapid spatial orienting towards angry faces. Neuroreport. 2008;19:259–63.
72. Davidson RJ, Marshall JR, Tomarken AJ, and Henriques JB. While a phobic waits: Regional brain electrical and autonomic activity in social phobics during anticipation of public speaking. Biol Psychiatry. (2000); 47, 85–95.
73. Wacker J, Chavanon ML, Leue A, and Stemmler G. Is running away right? The behavioral activation–behavioral inhibition model of anterior asymmetry. Emotion. (2008); 8, 232–49.
74. Smit DJ, Posthuma D, Boomsma DI, and De Geus EJ. The relation between frontal EEG asymmetry and the risk for anxiety and depression. Biol Psychol. (2007); 74, 26–33.
75. Anokhin AP, Heath AC, Myers E. Genetic and environmental influences on frontal EEG asymmetry: A twin study. Biol Psychol. 2006;71:289–95.
76. Wacker J, Mueller EM, Pizzagalli DA, Hennig J, and Stemmler G. Dopamine D2 receptor blockade reverses the association between trait BAS and frontal asymmetry in an approach motivational context. Psychological Sci. (2013); 24, 489–97.

whereas other approaches including single-trial, frequency and time-frequency analyses are less common in this field. More complex approaches to isolate brain source activity such as independent component analysis (16) are not part of this chapter because to date they have been rarely used in the study of individual differences and molecular genetics (17).

15.2 Event-related potentials

In this approach, typically, the same (or similar) stimuli are presented multiple times and it is assumed that the critical brain reaction to these stimuli is more or less identical for each presentation. It is assumed that when averaging the EEG time-series across trials, any non-systematic fluctuations in the EEG (artifacts and "irrelevant" brain activity) are washed out and only the EEG-signature of the trial-invariant critical brain reaction remains (18). The resulting trial-averaged brain wave is termed event-related potential (ERP). Depending on the modality of the stimulus one can distinguish between visual-, auditory-, olfactory-, and somatosensory-evoked potentials. In addition, there are ERP characteristics that depend not so much on the sensory modality of a stimulus but more on their cognitive-affective implications in a given context or laboratory task (task-specific potentials). Moreover, there are also ERPs that are related to a behavioral response from the participant rather than an external stimulus (response-related potentials). The majority of ERP research on anxiety has been conducted with visual-evoked potentials, task-specific components, and response-evoked potentials. This research is outlined in the following sections.

15.2.1 Visual evoked potentials

C1. Like all other evoked potentials, visual-evoked potentials have a waveform shape with multiple troughs and peaks (Figure 15.2B). These troughs and peaks are often called *components*. The C1 component is one of the earliest components, which peaks between 50 and 100 ms at central parietal electrodes and is assumed to originate mainly but not exclusively from V1 (19). Importantly, the C1 has been shown to be enhanced for fear-evoking stimuli such as threat-conditioned gratings (20) or fearful faces (21). It has been shown that visual brain region EEG activity in this time window may be even enhanced for stimuli that have been fear conditioned about a year ago (22). Moreover, high levels of trait anxiety have been related to increased C1 amplitudes in response to threat-related (angry) faces (23) and spider phobics showed unspecific C1 enhancements in a task containing spider pictures (24). However, although the C1 may be related to an anxiety phenotype, behavioral, or molecular genetic studies on fear-related C1 have not yet been conducted. Accordingly, future work is needed to clarify whether C1 modulations may qualify as an intermediate phenotype for anxiety.

P100. The P100 is a positive deflection, which occurs over parieto-occipital electrodes about 100–130 ms after the presentation of visual stimuli (Figure 15.2B). The P100 amplitude is sensitive to attention and P100 enhancements may reflect increased sensory processing in extrastriate visual regions (25). Consistent with privileged processing of potential threats, the P100 is enhanced in response to danger-related faces or objects (26,27), and is enhanced in individuals with anxiety disorders (38). Moreover, when the amygdala, a well-known region implicated in rapid threat processing, is damaged, the P100 is no longer enhanced in response to threatening stimuli (29). Together, these findings suggest that P100 enhancements are a good marker for threat processing and anxiety. While the raw amplitude of the P100 is more similar in monozygotic versus dizygotic twins (30) and more similar in family members versus unrelated individuals (31), it is not yet known whether the actual threat-related enhancements of the P100 are heritable and/or whether they are related to specific gene variants.

N1 and N170. The N1 is a parieto-occipital negative deflection after the P1. It has been shown that this early negative component is particularly enhanced when faces rather than objects or other visual stimuli are shown and it peaks roughly around 170 ms (32). Therefore, this component is also termed N170 when elicited by facial stimuli (see Figure 15.1B). Whether N1 to non-face stimuli and the N170 to faces should be considered similar or equivalent is a matter of debate (33,34). Because some (35) but not others (21,36) found that the N170 is enhanced in response to threat-related faces, and some (37) but not others (38) found it to be enhanced in anxiety disorders, future work on its boundary conditions is needed before the N170 could be considered a potential intermediate phenotype for anxiety. Moreover, in spite of significant genetic contributions to individual differences in N1 and N170 as revealed by twin studies (35), single gene variants that relate to N1 or N170 amplitudes have not yet been identified.

15.2.2 Task specific and response evoked potentials

NoGo-N2. As states of anxiety are often accompanied with an inhibition of ongoing behavior (39), several studies investigated the electrophysiology of anxiety with so-called Go/NoGo tasks, which consist of frequent stimuli that signal the need for a behavioral response (Go-trials) and less frequent stimuli signaling that the prepotent behavioral response should be inhibited (NoGo-trials). NoGo versus Go-trials elicit a more negative frontomedial N2 amplitude at about 200–300 ms (NoGo-N2; 40,41), which presumably originates in medial frontal brain areas including the anterior cingulate cortex (42), a highly relevant region for anxiety (12). Because, in these trials, a prepotent behavioral response has to be suppressed it is assumed that the NoGo-N2 reflects the inhibition (43) or conflict (44) of motor plans. Consistent with the idea that behavioral inhibition or conflict detection are core features of anxiety (45), increased NoGo-N2 amplitudes have been observed in healthy individuals with high trait (78) and state (46) anxiety and anxiety-related disorders (47; but: 48,49). While twin studies indicate that interindividual variations in NoGo-N2 amplitude are genetically influenced (50) only few molecular genetic studies have been conducted. Because these studies focused on genes with different functions (dopamine D4 receptor: 51); brain-derived neurotrophic factor: 52); glutamate receptor: 53), tested relatively small samples (Nos: 20, 57, and 65, respectively), and have not yet revealed a consistent pattern across studies, more research is needed on the gene variants that contribute to anxiety-related variation in NoGo-N2.

Feedback-related and error-related negativity. The feedback-related negativity (FRN) is a negative frontomedial deflection in the stimulus-evoked ERP at about 250 ms, that is more negative after negative versus positive feedback. The FRN is often seen as a component similar to the error-related negativity (ERN). While the FRN emerges in the stimulus-evoked ERP, the ERN emerges in the response-evoked ERP. When individuals perform a response that they did not intend to undertake (e.g. accidentally press the left button even though the task required that the right button was pressed), the frontomedial response-evoked ERP amplitude within the first 100 ms after the button press is more negative than after a correct response. Like the NoGo-N2, both FRN (54,55) and ERN (56,57) have been source-localized to the anterior cingulate cortex. Because ERN and FRN seem to have the same brain source and because they are both characterized by more negative amplitudes whenever "events are worse than expected," they may share core aspects of their underlying neurobiology (58,59). Because animal work indicates that some dopaminergic neurons code for events that are worse than expected (or negative prediction errors; 60), the FRN and ERN have been theoretically linked to phasic dopamine changes (59) although this is not undisputed (61). Tonic dopamine (17,62) and other neurotransmitters including serotonin (63,64) are also assumed to be relevant.

The processing of unexpected negative feedback and of errors is of particular relevance for anxiety because such events often signal an unpredicted threat (65). Accordingly, relationships between ERN, FRN, and trait anxiety or anxiety disorders have been thoroughly investigated (for reviews, see 66,67). Many studies reported increased FRN and ERN amplitudes in individuals with high trait anxiety or related personality dimensions (68–70) although these correlations may depend on specific task characteristics like the presence or absence of trial-by-trial feedback (124), or type of task (71), and further aspects such as the particular facet of anxiety (72,73). Similarly, individuals who suffer from anxiety-related disorders (69,74–77,79) and moderate levels of depression (62,80,81,125) seem to show increased ERN/FRN amplitudes (but see, e.g. 82).

While the heritability of the FRN has not yet been investigated, one behavioral genetic study found considerably larger correlations of ERN amplitudes within monozygotic ($0.52 \leq r \leq 0.57$) versus dizygotic ($-0.08 \leq r \leq = 0.23$) twins, indicating that variability of ERN amplitudes is genetically influenced (e.g. estimated heritability: $a^2 = .47$; 83). Support for a genetic base of the ERN with relevance for anxiety also comes from studies by Riesel and colleagues who found (i) that not only patients with obsessive compulsive disorder (OCD) but also their first-degree relatives show increased ERN amplitudes (84), and (ii) that enhanced ERN amplitudes in OCD are unaffected by symptom reduction after cognitive behavioral therapy (76).

Because of their assumed association with dopamine (59), several researchers tested whether dopaminergic gene variants relate to FRN and ERN amplitudes. While most associations are based on single studies and/or seem difficult to replicate (64), several studies have now reported that the MET-allele of the catechol-*O*-methyl transferase (COMT) Val158Met polymorphism, associated with higher levels of prefrontal cortex dopamine, relates to reduced FRN and ERN amplitudes (62,85,86; but see 87). Intriguingly, this association was modulated for both FRN (62) and ERN (17) amplitudes by a dopamine D2 receptor blocker, which tentatively suggests that the link between COMT and individual differences in ERN/FRN is driven by dopaminergic mechanisms. Taken together, ERN and FRN seem to be promising anxiety- and depression-related endophenotypes that have been linked to dopamine-related gene variants. Future work is now needed to clarify whether ERN or FRN actually mediate a link between dopaminergic gene variants and anxiety. As the direct link between quantitative traits and common gene variants is so weak, very large samples will be needed to test for mediation effects.

P300. The P300 is one of the most prominent ERP components (Figure 15.1B). It peaks at 300–400 ms at frontal sites and somewhat later at posterior electrodes. In the time window of the P300, multiple neural generators are simultaneously active (97), which presumably relate to a number of different processes that occur in parallel. It has been assumed that the P300 reflects a noradrenaline-based brain response to enhance the processing of motivationally significant stimuli (88) like threat signals, and noradrenergic signaling is of potential relevance for panic (89,90). Consistently, some studies reported increased P300 amplitudes in panic disorder (91,98), suggesting that P300 may reflect a potential intermediate phenotype that links noradrenergic gene variants to panic-related traits or pathologies. However, although twin studies have estimated the heritability of the raw P300 to be in the range of 0.44–0.6 (99–101), numerous molecular genetic studies have failed to identify single SNPs that reliably relate to P300 amplitudes (102). A potential issue in this line of research is that individual raw P300 amplitudes in a given task at a single electrode are blurred by several sources of variance as they reflect a mixture of (i) several brain processes that occur in parallel, (ii) task-specific and general contributions to P300 amplitude, (iii) state and trait influences on P300, and (iv) brain and skull anatomy.

No-Go anteriorization. To the degree that these sources of variance are not affected by specific task conditions or do not vary across the scalp, difference scores that contrast conditions or topographical positions (i.e. P300 at electrode A minus P300 at electrode B), respectively, are useful to eliminate unwanted interindividual variation. Particularly promising appears the difference of frontal versus parietal P300 amplitudes during Go-NoGo tasks. In these tasks, NoGo trials produce a so-called *No-Go anteriorization* (larger P300 at frontal sites), which may reflect inhibitory processes that occur when a behavioral response must be suppressed. Of relevance, the No-Go anteriorization has not only been associated with dispositional behavioral inhibition/trait anxiety (103) but also with the short allele of the serotonin transporter polymorphism (104), which had been a widely discussed genetic predisposition for anxiety (105). Although behavioral genetic studies on the No-Go anteriorization have not yet been conducted, this EEG marker may be promising for future research on serotoninergic mechanisms of individual differences in anxiety.

Late positive potential. The late positive potential (LPP) (Figure 15.2B), a positive-going, centro-parietal slow wave that shows considerable temporal and morphological overlap with the P300 (101,106). In contrast to the P300, which only lasts for a few hundred milliseconds and is typically associated with oddball or Go-NoGo tasks, the LPP has a longer duration and is typically observed in studies where pictures with or without affective content are presented. Nevertheless, the brain processes that occur during the initial portion of the LPP may overlap to a large degree with the brain processes underlying P300. Like the P300, the LPP has been linked to noradrenaline (107) and may be another component with particular relevance for threat processing and anxiety. It is generally modulated not only by unpleasant but also by pleasant stimuli and presumably indicates sustained engagement and continued elaboration of affective content (101,108,109). The affective modulation of the LPP has been found to be related to state (110), trait (111–113), and pathological (114,115) anxiety, although some studies found this modulation to be increased in anxiety (110,112), while others found it to be decreased (111,115) or not significantly related to different types of anxiety disorders with and without comorbidities (116). Future work should therefore identify possible moderators of this association, which may include (i) the particular type of threat-related stimuli; (ii) the task in which these stimuli are presented, and/or (iii) other contextual factors that potentially influence whether anxious individuals show sustained engagement to presented stimuli or not.

Of relevance, a recent twin study found that the LPP amplitude (but not its modulation by valence) was significantly more similar for monozygotic versus dizygotic twins, indicating that individual differences in brain processes during the LPP time window are genetically determined. Moreover, a molecular genetic association of a beta-adrenergic receptor gene variant has been observed with the LPP amplitude but not with the affective modulation of the LPP (107). Although this pattern of findings may indicate that the LPP amplitude is a more promising endophenotype for anxiety than LPP modulation by valence, it should be noted that valence modulation of early proportions of the LPP (i.e. the P300) showed considerable heritability. Accordingly, future work in this field is clearly needed to improve understanding of the mechanisms of variation in LPP amplitude and in LPP modulation, and to determine unique and overlapping brain mechanisms that contribute to P300 and later LPP.

15.3 **Other approaches**

15.3.1 **Time-frequency approaches**

Phasic changes in oscillatory activity and frequency power may occur in response to an external stimulus, and if oscillations vary in phase from trial to trial, these effects are washed out and remain undetected with the event-related potential approach. Using complex mathematical analyses, however, time-frequency approaches assess how the power of oscillations across different frequencies changes over time irrespective of cross-trial phase inconsistencies. With this approach, it has been found that after individuals commit an error or receive negative feedback, a frontomedial burst in theta oscillations (4–8 Hz) can be observed (17,117). This burst is presumably related to the FRN, ERN, and NoGo-N2 (58), although it is not redundant with these time-domain measures (62,118). Of relevance, several studies suggest that frontomedial theta is linked to fear and anxiety (62,68,92–94), and associations with the COMT Val158Met polymorphism have been reported (17). As twin studies on the general heritability of variations in action-monitoring-related theta power have not yet been performed, we do not know whether phasic theta power may serve as an intermediate phenotype for anxiety.

15.3.2 **Brain–heart coupling**

Among the most prominent physiological components of fear and anxiety are changes in heart rate, which may serve to prepare the organism optimally to cope with potential threats. Of course, the heart does not change its speed autonomously but rather receives efferent commands from cortical and subcortical structures, which reach the sinoatrial node of the heart (the heart's pacemaker) via brainstem projections to the sympathetic and parasympathetic branches of the autonomic nervous system (95,96). The so called *cardio-EEG-covariance-tracing* (CECT) method has been developed to measure how strongly EEG and heart activity are coupled after the presentation of an external stimulus (67). This method uses the EEG magnitude of single trials at different time windows relative to stimulus onset and correlates this magnitude with subsequent changes in heart activity. Importantly, these correlations are performed across trials and within each subject, yielding an individual brain–heart coupling value for each participant. In general, it has been shown that the EEG activity 300 ms after the presentation of motivationally significant stimuli predicts how much the heart activity changes a few seconds later This phenomenon is termed N300H and is driven by serotonin (98,119) and brain activity in the anterior cingulate cortex (120) and varies between individuals. In line with the importance of brain–heart coupling for threat processing, the N300H is particularly strong in healthy anxious individuals (122), in subjects with panic disorder (98) and in response to fear-conditioned stimuli (121). Although its heritability has not yet been investigated with behavioral genetic approaches, the N300H has been found to differ between carriers of the short and long allele of the serotonin transporter polymorphism (122). We are currently investigating whether other neurotransmitter systems also contribute to variation in brain–heart coupling as measured with N300H.

15.4 **Conclusion**

Taken together, individual differences in anxiety-related brain activity are to some degree genetically determined. Because EEG has the perfect temporal resolution to study the rapid neural processing of threats it provides an ideal method for investigating intermediate phenotypes of fear and anxiety. Evidence for the genetic contribution to electrophysiological markers of anxiety comes from behavioral genetic twin studies and from molecular genetic studies investigating the contribution of individual SNPs. The general heritability has been established for a variety of critical indicators. While molecular genetic studies on the affective modulation of early visual components (e.g. C1, P100, N170) have not been strongly investigated, some individual gene variants have been tentatively linked to task-specific components (ERN, FRN, NoGo-N2, P300-anteriorization) and other anxiety-related EEG markers (stimulus-evoked theta, brain–heart coupling). Importantly, all of the observed links between a specific gene variant and a specific EEG component lack convincing replication and may be moderated by further variables such as the specific task or laboratory context in which EEG was collected. Moreover, environmental factors such as stressful life events may influence the effect of genes on neurobiological markers of anxiety (123) and gene–environment interactions should not be neglected on the search for EEG endophenotypes of anxiety. Studies that collected (i) genetic markers, (ii) critical EEG components, and (iii) individual differences in trait or clinical anxiety in the same dataset, and that were able to demonstrate a pathway from the gene over the brain process to the behavioral trait do not yet exist to my knowledge. Ultimately, however, it may be that it is these kind of multilevel studies, that shed light on the complex neurogenetic mechanisms, that lead to interindividual variation in anxiety.

Acknowledgements

I would like to thank Anna Weinberg and Anja Riesel for their helpful comments on this chapter.

References

1. **Lee WE, Wadsworth ME, Hotopf M.** The protective role of trait anxiety: A longitudinal cohort study. Psychol Med. 2006;**36**(3):345–51. doi:10.1017/S0033291705006847.
2. **Kessler RC, Berglund P, Demler O, Jin R, Merikangas KR, Walters EE.** Lifetime prevalence and age-of-onset distributions of DSM-IV disorders in the National Comorbidity Survey Replication. Arch Gen Psychiatry. 2005:**62**(6):593–602. doi:62/6/593 [pii] 10.1001/archpsyc.62.6.593.
3. **Lepine JP.** The epidemiology of anxiety disorders: prevalence and societal costs. J Clin Psychiatry. 2002;**63**(Suppl 14):4–8.
4. **Hettema JM, Neale MC, Myers JM, Prescott CA, Kendler KS.** A population-based twin study of the relationship between neuroticism and internalizing disorders. Am J Psychiatry. 2006;**163**(5):857–64. doi:163/5/857 [pii] 10.1176/appi.ajp.163.5.857.
5. **Watson D, Clark LA, Weber K, Assenheimer JS, Strauss ME, McCormick RA.** Testing a tripartite model: II. Exploring the symptom structure of anxiety and depression in student, adult, and patient samples. J Abnorm Psychol. 1995;**104**(1):15–25.
6. **Frazer KA, Murray SS, Schork NJ, Topol EJ.** Human genetic variation and its contribution to complex traits. Nat Rev Genet. 2009;**10**(4):241–51. doi:10.1038/nrg2554.
7. **Churchland PS, Sejnowski TJ.** Perspectives on cognitive neuroscience. Science. 1988 ;**242**(4879):741–5.
9. **Gottesman II, Gould TD.** The endophenotype concept in psychiatry: etymology and strategic intentions. Am J Psychiatry. 2003;**160**(4):636–45. doi:10.1176/appi.ajp.160.4.636.
10. **LeDoux JE.** Coming to terms with fear. PNAS. 2014;**111**(8):2871–78. doi:10.1073/pnas.1400335111.
11. **Paulus MP, Stein MB.** An insular view of anxiety. Biol Psychiatry. 2006:**60**(4):383–7. doi:10.1016/j.biopsych.2006.03.042.
12. **Shackman AJ, Salomons TV, Slagter HA, Fox AS, Winter JJ, Davidson RJ.** The integration of negative affect, pain and cognitive control in the cingulate cortex. Nat Rev Neurosci. 2001;**12**(3):154–67. doi:10.1038/nrn2994.
13. **Etkin A, Egner T, Kalisch R.** Emotional processing in anterior cingulate and medial prefrontal cortex. Trends Cogn Sci. 2011;**15**(2):85–93. doi:10.1016/j.tics.2010.11.004.
14. **Miskovic V, Keil A.** Acquired fears reflected in cortical sensory processing: A review of electrophysiological studies of human classical conditioning. Psychophysiology. 2012;**49**(9):1230–41. doi:10.1111/j.1469-8986.2012.01398.x.
15. **Hoehn-Saric R, Lee JS, McLeod DR, Wong DF.** Effect of worry on regional cerebral blood flow in nonanxious subjects. Psychiatry Res. 2005;**140**(3):259–69. doi:10.1016/j.pscychresns.2005.05.013.
16. **Makeig S, Bell AJ, Jung TP, Sejnowski TJ.** Independent component analysis of electroencephalographic data. Adv Neural Inf Process Syst. 1996;1996;**8**:145–51.
17. **Mueller EM, Makeig S, Stemmler G, Hennig J, Wacker J.** Dopamine effects on human error processing depend on catechol-*O*-methyltransferase Val158Met genotype. J Neurosci. 2011;**31**(44):15818–25. doi:10.1523/JNEUROSCI.2103-11.2011.
18. **Luck SJ.** *An Introduction to the Event-Related Potential Technique.* Cambridge, MA: MIT Press; 2014.

19. **Rauss K, Schwartz S, Pourtois G.** Top-down effects on early visual processing in humans: A predictive coding framework. Neurosci Biobehav Rev. 2011;**35**(5):1237–53. doi:10.1016/j.neubiorev.2010.12.011.
20. **Stolarova M, Keil A, Moratti S.** Modulation of the C1 visual event-related component by conditioned stimuli: evidence for sensory plasticity in early affective perception. Cereb Cortex. 2006;**16**(6):876–87. doi:10.1093/cercor/bhj031.
21. **Pourtois G, Grandjean D, Sander D, Vuilleumier P.** Electrophysiological correlates of rapid spatial orienting towards fearful faces. Cereb Cortex. 2004;**14**(6):619–33. doi:10.1093/cercor/bhh023.
22. **Mueller EM, Pizzagalli DA.** One-year-old fear memories rapidly activate human fusiform gyrus. Soc Cogn Affect Neurosci. 2016;**11**(2):308–16. doi:10.1093/scan/nsv122.
23. **Eldar S, Yankelevitch R, Lamy D, Bar-Haim Y.** Enhanced neural reactivity and selective attention to threat in anxiety. Biol Psychol. 2010;**85**(2):252–7. doi:10.1016/j.biopsycho.2010.07.010.
24. **Weymar M, Keil A, Hamm AO.** Timing the fearful brain: unspecific hypervigilance and spatial attention in early visual perception. Soc Cogn Affect Neurosci. 2014;**9**(5):723–9. doi:10.1093/scan/nst044.
25. **Clark VP, Hillyard SA.** Spatial selective attention affects early extrastriate but not striate components of the visual evoked potential. J Cogn Neurosci. 1996;**8**(5):387–402. doi:10.1162/jocn.1996.8.5.387.
26. **Santesso DL, Meuret AE, Hofmann SG, Mueller EM, Ratner KG, Roesch EB,** et al. Electrophysiological correlates of spatial orienting towards angry faces: a source localization study. Neuropsychologia. 2008:**46**(5):1338–48. doi:S0028-3932(07)00443-5 [pii] 10.1016/j.neuropsychologia.2007.12.013.
27. **Vuilleumier P, Pourtois G.** Distributed and interactive brain mechanisms during emotion face perception: Evidence from functional neuroimaging. Neuropsychologia. 2007;**45**(1):174–94. doi:10.1016/j.neuropsychologia.2006.06.003.
28. **Mueller EM, Stemmler G, Wacker J.** Single-trial electroencephalogram predicts cardiac acceleration: A time-lagged P-correlation approach for studying neurovisceral connectivity. Neuroscience; 2009a:**166**(2):491–500. doi:10.1016/j.neuroscience.2009.12.051.
29. **Rotshtein P, Richardson MP, Winston JS, Kiebel SJ, Vuilleumier P, Eimer M,** et al. Amygdala damage affects event-related potentials for fearful faces at specific time windows. Hum Brain Mapp. 2010;**31**(7):1089–105. doi:10.1002/hbm.20921.
30. **Buchsbaum MS.** Average evoked response and stimulus intensity in identical and fraternal twins. Physiological Psychology. 1974;**2**(3):365–70.
31. **Bulayeva KB, Pavlova TA, Guseynov GG.** Visual evoked potentials: Phenotypic and genotypic variability. Behav Genet. 1993;**23**(5):443–7.
32. **Bentin S, Allison T, Puce A, Perez E, McCarthy G.** Electrophysiological studies of face perception in humans. J Cogn Neurosci. 1996;**8**(6):551–65. doi:10.1162/jocn.1996.8.6.551.
33. **Carmel D, Bentin S.** Domain specificity versus expertise: Factors influencing distinct processing of faces. Cognition. 2002;**83**(1):1–29.
34. **Rossion B, Gauthier I, Tarr MJ, Despland P, Bruyer R, Linotte S,** et al. The N170 occipito-temporal component is delayed and enhanced to inverted faces but not to inverted objects: An electrophysiological account of face-specific processes in the human brain. Neuroreport. 2000;**11**(1):69–74.
35. **Shannon RW, Patrick CJ, Venables NC, He S.** "Faceness" and affectivity: evidence for genetic contributions to distinct components of electrocortical response to human faces. Neuroimage. 2013;**83**:609–15. doi:10.1016/j.neuroimage.2013.06.014.
36. **Pourtois G, Dan ES, Grandjean D, Sander D, Vuilleumier P.** Enhanced extrastriate visual response to bandpass spatial frequency filtered fearful faces: Time course and topographic evoked-potentials mapping. Hum Brain Mapp. 2005;**26**(1):65–79. doi:10.1002/hbm.20130.
37. **Kolassa IT, Miltner WH.** Psychophysiological correlates of face processing in social phobia. Brain Res. 2006;**1118**(1):130–41. doi:10.1016/j.brainres.2006.08.019.

38. **Mueller EM, Hofmann SG, Santesso DL, Meuret AE, Bitran S, Pizzagalli DA.** Electrophysiological evidence of attentional biases in social anxiety disorder. Psychol Med. 2009b;**39**(7):1141–52. doi:10.1017/S0033291708004820.
39. **Gray JA.** The Neuropsychology of Anxiety: An Enquiry into the Functions of the Septo-Hippocampal System: Oxford: Oxford University Press; 1982.
40. **Eimer M.** Effects of attention and stimulus probability on ERPs in a Go/Nogo task. Biol Psychol. 1993;**35**(2):123–38.
41. **Pfefferbaum A, Ford JM, Weller BJ, Kopell BS.** ERPs to response production and inhibition. Electroencephalogr Clin Neurophysiol. 1985;**60**(5):423–34.
42. **Bekker EM, Kenemans JL, Verbaten MN.** Source analysis of the N2 in a cued Go/NoGo task. Brain Res Cogn Brain Res. 2005;**22**(2):221–31. doi:10.1016/j.cogbrainres.2004.08.011.
43. **Falkenstein M, Hoormann J, Hohnsbein J.** ERP components in Go/Nogo tasks and their relation to inhibition. Acta Psychol (Amst). 1999;**101**(2-3):267–91.
44. **Nieuwenhuis S, Yeung N, van den Wildenberg W, Ridderinkhof KR.** Electrophysiological correlates of anterior cingulate function in a go/no-go task: Effects of response conflict and trial type frequency. Cogn Affect Behav Neurosci. 2003;**3**(1):17–26.
45. **Gray JA, McNaughton N.** *The Neuropsychology of Anxiety: An Enquiry into the Functions of the Septohippocampal System*, 2nd edn. Oxford: Oxford University Press; 2000.
46. **Righi S, Mecacci L, Viggiano MP.** Anxiety, cognitive self-evaluation and performance: ERP correlates. J Anxiety Disord. 2009;**23**(8):1132–38. doi:10.1016/j.janxdis.2009.07.018.
47. **Herrmann MJ, Jacob C, Unterecker S, Fallgatter AJ.** Reduced response-inhibition in obsessive-compulsive disorder measured with topographic evoked potential mapping. Psychiatry Res. 2003;**120**(3):265–71.
48. **Kim MS, Kim YY, Yoo SY, Kwon JS.** Electrophysiological correlates of behavioral response inhibition in patients with obsessive-compulsive disorder. Depress Anxiety. 2007;**24**(1):22–31. doi:10.1002/da.20195.
49. **Ruchsow M, Reuter K, Hermle L, Ebert D, Kiefer M, Falkenstein M.** Executive control in obsessive–compulsive disorder: Event-related potentials in a Go/Nogo task. J Neural Transm (Vienna). 2007;**114**(12):1595–601. doi:10.1007/s00702-007-0779-4.
50. **Anokhin AP, Heath AC, Myers E.** Genetics, prefrontal cortex, and cognitive control: A twin study of event-related brain potentials in a response inhibition task. Neurosci Lett. 2004;**368**(3):314–8. doi:10.1016/j.neulet.2004.07.036.
51. **Kramer UM, Rojo N, Schule R, Cunillera T, Schols L, Marco-Pallares J** et al. ADHD candidate gene (DRD4 exon III) affects inhibitory control in a healthy sample. BMC Neurosci. 2009;**10**, 150. doi:10.1186/1471-2202-10-150.
52. **Beste C, Baune BT, Domschke K, Falkenstein M, Konrad C.** Paradoxical association of the brain-derived-neurotrophic-factor val66met genotype with response inhibition. Neuroscience. 2010b;**166**(1):178–84. doi:10.1016/j.neuroscience.2009.12.022.
53. **Beste C, Baune BT, Domschke K, Falkenstein M, Konrad C.** Dissociable influences of NR2B-receptor related neural transmission on functions of distinct associative basal ganglia circuits. Neuroimage. 2010a;**52**(1):309–15. doi:10.1016/j.neuroimage.2010.04.022.
54. **Gehring WJ, Willoughby AR.** The medial frontal cortex and the rapid processing of monetary gains and losses. Science. 2002;**295**(5563):2279–82. doi:10.1126/science.1066893295/5563/2279 [pii].
55. **Walsh MM, Anderson JR.** Learning from experience: event-related potential correlates of reward processing, neural adaptation, and behavioral choice. Neurosci Biobehav Rev. 2012;**36**(8):1870–84. doi:10.1016/j.neubiorev.2012.05.008.
56. **Debener S, Ullsperger M, Siegel M, Fiehler K, von Cramon DY, Engel AK.** Trial-by-trial coupling of concurrent electroencephalogram and functional magnetic resonance imaging identifies the dynamics of performance monitoring. J Neurosci. 2005;**25**(50):11730–37. doi:25/50/11730 [pii] 10.1523/JNEUROSCI.3286-05.2005.

57. **Herrmann MJ, Rommler J, Ehlis AC, Heidrich A, Fallgatter AJ.** Source localization (LORETA) of the error-related-negativity (ERN/Ne) and positivity (Pe). Brain Res Cogn Brain Res. 2004;**20**(2):294–9. doi:10.1016/j.cogbrainres.2004.02.013.
58. **Cavanagh JF, Zambrano-Vazquez L, Allen JJ.** Theta lingua franca: a common mid-frontal substrate for action monitoring processes. Psychophysiology. 2012;**49**(2):220–38. doi:10.1111/j.1469-8986.2011.01293.x.
59. **Holroyd CB, Coles MG.** The neural basis of human error processing: Reinforcement learning, dopamine, and the error-related negativity. Psychol Rev. 2002;**109**(4):679–709.
60. **Schultz W.** Getting formal with dopamine and reward. Neuron. 2002;**36**(2):241–63. doi:S0896627302009674 [pii].
61. **Ullsperger M, Fischer AG, Nigbur R, Endrass T.** Neural mechanisms and temporal dynamics of performance monitoring. Trends Cogn Sci. 2014. doi:10.1016/j.tics.2014.02.009.
62. **Mueller EM, Burgdorf C, Chavanon ML, Schweiger D, Wacker J, Stemmler G.** Dopamine modulates frontomedial failure processing of agentic introverts versus extraverts in incentive contexts. Cogn Affect Behav Neurosci. 2014;**14**(2):756–68. doi:10.3758/s13415-013-0228-9.
63. **Fischer AG, Endrass T, Reuter M, Kubisch C, Ullsperger M.** Serotonin reuptake inhibitors and serotonin transporter genotype modulate performance monitoring functions but not their electrophysiological correlates. J Neurosci. 2015;**35**(21):8181–90. doi:10.1523/JNEUROSCI.5124-14.2015.
64. **Ullsperger M.** Genetic association studies of performance monitoring and learning from feedback: The role of dopamine and serotonin. Neurosci Biobehav Rev. 2010;**34**(5):649–59. doi:S0149-7634(09)00097-9 [pii] 10.1016/j.neubiorev.2009.06.009.
65. **Inzlicht M, Bartholow BD, Hirsh JB.** Emotional foundations of cognitive control. Trends Cogn Sci. 2015;**19**(3):126–32. doi:10.1016/j.tics.2015.01.004.
66. **Endrass T, Ullsperger M.** Specificity of performance monitoring changes in obsessive-compulsive disorder. Neurosci Biobehav Rev. 2014; doi:10.1016/j.neubiorev.2014.03.024.
67. **Olvet DM, Hajcak G.** The error-related negativity (ERN) and psychopathology: Toward an endophenotype. Clin Psychol Rev. 2008;**28**(8):1343–54. doi:S0272-7358(08)00109-8 [pii] 10.1016/j.cpr.2008.07.003.
68. **Cavanagh JF, Shackman AJ.** Frontal midline theta reflects anxiety and cognitive control: Meta-analytic evidence. J Physiol Paris. 2014; doi:10.1016/j.jphysparis.2014.04.003.
69. **Hajcak G, McDonald N, Simons RF.** Anxiety and error-related brain activity. Biol Psychol. 2003;**64**(1-2):77–90. doi:S0301051103001030 [pii].
70. **Hajcak G, McDonald N, Simons RF.** Error-related psychophysiology and negative affect. Brain Cogn. 2004;**56**(2):189–97.
71. **Grundler TO, Cavanagh JF, Figueroa CM, Frank MJ, Allen JJ.** Task-related dissociation in ERN amplitude as a function of obsessive-compulsive symptoms. Neuropsychologia. 2009;**47**(8-9):1978–87. doi:S0028-3932(09)00129-8 [pii]10.1016/j.neuropsychologia.2009.03.010.
72. **Moser JS, Moran TP, Schroder HS, Donnellan MB, Yeung N.** On the relationship between anxiety and error monitoring: A meta-analysis and conceptual framework. Front Hum Neurosci. 2013;**7**, 466. doi:10.3389/fnhum.2013.00466.
73. **Zambrano-Vazquez L, Allen JJ.** Differential contributions of worry, anxiety, and obsessive compulsive symptoms to ERN amplitudes in response monitoring and reinforcement learning tasks. Neuropsychologia. 2014;**61**:197–209. doi:10.1016/j.neuropsychologia.2014.06.023.
74. **Gehring WJ, Himle J, Nisenson LG.** Action-monitoring dysfunction in obsessive–compulsive disorder. Psychol Sci. 2000;**11**(1):1–6.
75. **Hajcak G, Simons RF.** Error-related brain activity in obsessive-compulsive undergraduates. Psychiatry Res. 2002;**110**(1):63–72. doi:S0165178102000343 [pii].
76. **Riesel A, Endrass T, Auerbach LA, Kathmann N.** Overactive performance monitoring as an endophenotype for obsessive-compulsive disorder: Evidence from a treatment study. Am J Psychiatry. 2015;**172**(7):665–73. doi:10.1176/appi.ajp.2014.14070886.

77. **Meyer A, Hajcak G, Torpey DC, Kujawa A, Kim J, Bufferd S,** et al. Increased error-related brain activity in six-year-old children with clinical anxiety. J Abnorm Child Psychol. 2013;**41**(8):1257–66. doi:10.1007/s10802-013-9762-8.

78. **Sehlmeyer C, Konrad C, Zwitserlood P, Arolt V, Falkenstein M, Beste C.** ERP indices for response inhibition are related to anxiety-related personality traits. Neuropsychologia. 2010;**48**(9):2488–95. doi:10.1016/j.neuropsychologia.2010.04.022.

79. **Weinberg A, Olvet DM, Hajcak G.** Increased error-related brain activity in generalized anxiety disorder. Biol Psychol. 2010;**85**(3):472–80. doi:10.1016/j.biopsycho.2010.09.011.

80. **Cavanagh JF, Bismark AJ, Frank MJ, Allen JJ.** Larger error signals in major depression are associated with better avoidance learning. Front Psychol. 2011;(2)331. doi:10.3389/fpsyg.2011.00331.

81. **Tucker DM, Luu P, Frishkoff G, Quiring J, Poulsen C.** Frontolimbic response to negative feedback in clinical depression. J Abnorm Psychol. 2003;**112**(4):667–78. doi:10.1037/0021-843X.112.4.667.

82. **Nieuwenhuis S, Nielen MM, Mol N, Hajcak G, Veltman DJ.** Performance monitoring in obsessive-compulsive disorder. Psychiatry Res. 2005;**134**(2):111–22. doi:S0165-1781(05)00056-9 [pii] 10.1016/j.psychres.2005.02.005.

83. **Anokhin AP, Golosheykin S, Heath AC.** Heritability of frontal brain function related to action monitoring. Psychophysiology. 2008;**45**(4):524–34. doi:PSYP664 [pii] 10.1111/j.1469-8986.2008.00664.x.

84. **Riesel A, Endrass T, Kaufmann C, Kathmann N.** Overactive error-related brain activity as a candidate endophenotype for obsessive-compulsive disorder: evidence from unaffected first-degree relatives. Am J Psychiatry, 2011;**168**(3):317–24. doi:10.1176/appi.ajp.2010.10030416.

85. **Marco-Pallares J, Cucurell D, Cunillera T, Kramer UM, Camara E, Nager W,** et al. Genetic variability in the dopamine system (dopamine receptor D4, catechol-O-methyltransferase) modulates neurophysiological responses to gains and losses. Biol Psychiatry. 2009;**66**(2):154–61. doi:S0006-3223(09)00031-6 [pii] 10.1016/j.biopsych.2009.01.006.

86. **Osinsky R, Hewig J, Alexander N, Hennig J.** COMT Val158Met genotype and the common basis of error and conflict monitoring. Brain Res, 2012;1452, 108–18. doi:10.1016/j.brainres.2012.02.054.

87. **Frank MJ, D'Lauro C, Curran T.** Cross-task individual differences in error processing: Neural, electrophysiological, and genetic components. Cogn Affect Behav Neurosci. 2007;**7**(4):297–308.

88. **Nieuwenhuis S, Aston-Jones G, Cohen JD.** Decision making, the P3, and the locus coeruleus-norepinephrine system. Psychol Bull. 2005;**131**(4):510–32. doi:2005-08334-002 [pii] 10.1037/0033-2909.131.4.510.

89. **Charney DS, Woods SW, Krystal JH, Nagy LM, Heninger GR.** Noradrenergic neuronal dysregulation in panic disorder: The effects of intravenous yohimbine and clonidine in panic disorder patients. Acta Psychiatr Scand. 1992;**86**(4):273–82.

90. **Morgan CA 3rd, Grillon C, Southwick SM, Nagy LM, Davis M, Krystal JH, Charney DS.** Yohimbine facilitated acoustic startle in combat veterans with post-traumatic stress disorder. Psychopharmacology (Berl). 1995;**117**(4):466–71.

91. **Clark CR, McFarlane AC, Weber DL, Battersby M.** Enlarged frontal P300 to stimulus change in panic disorder. Biol Psychiatry. 1996;**39**(10):845–56.

92. **Mueller EM, Panitz C, Hermann C, Pizzagalli DA.** Prefrontal oscillations during recall of conditioned and extinguished fear in humans. J Neurosci. 2014;**34**(21):7059–66. doi:10.1523/JNEUROSCI.3427-13.2014.

93. **Mueller EM, Pechtel P, Cohen AL, Douglas SR, Pizzagalli DA.** Potentiated processing of negative feedback in depression is attenuated by anhedonia. Depress Anxiety. 2015. doi:10.1002/da.22338.

94. **Riesel A, Kathmann N, Endrass T.** Overactive performance monitoring in obsessive-compulsive disorder is independent of symptom expression. Eur Arch Psychiatry Clin Neurosci. 2014;**264**(8):707–17. doi:10.1007/s00406-014-0499-3.

95. **Benarroch EE.** *Central Autonomic Network: Functional Organization and Clinical Correlations.* Armonk, NY: Futura Pub. Co.; 1994

96. **Thayer JF, Lane RD.** Claude Bernard and the heart-brain connection: Further elaboration of a model of neurovisceral integration. Neurosci Biobehav Rev. 209;**33**(2):81–8. doi:S0149-7634(08)00125-5 [pii] 10.1016/j.neubiorev.2008.08.004.

97. **Makeig S, Westerfield M, Jung TP, Enghoff S, Townsend J, Courchesne E, Sejnowski TJ.** Dynamic brain sources of visual evoked responses. Science. 2002;**295**(5555):690–4. doi:10.1126/science.1066168295/5555/690 [pii].

98. **Mueller EM, Panitz C, Nestoriuc Y, Stemmler G, Wacker J.** Panic disorder and serotonin reuptake inhibitors predict coupling of cortical and cardiac activity. Neuropsychopharmacology. 2014b;**39**(2):507–14. doi:10.1038/npp.2013.224.

99. **Anokhin AP, Golosheykin S, Heath AC.** Heritability of individual differences in cortical processing of facial affect. Behav Genet. 2010;**40**(2):178–85. doi:10.1007/s10519-010-9337-1.

100. **van Beijsterveldt CE, van Baal GC.** Twin and family studies of the human electroencephalogram: a review and a meta-analysis. Biol Psychol. 2002;**61**(1–2):111–38.

101. **Weinberg A, Venables NC, Proudfit GH, Patrick CJ.** Heritability of the neural response to emotional pictures: evidence from ERPs in an adult twin sample. Soc Cogn Affect Neurosci. 2015;**10**(3):424–34. doi:10.1093/scan/nsu059.

102. **Malone SM, Vaidyanathan U, Basu S, Miller MB, McGue M, Iacono WG.** Heritability and molecular-genetic basis of the P3 event-related brain potential: a genome-wide association study. Psychophysiology. 2014;**51**(12):1246–58. doi:10.1111/psyp.12345.

103. **Wacker J, Chavanon ML, Leue A, Stemmler G.** Trait BIS predicts alpha asymmetry and P300 in a Go/No-Go task. European Journal of Personality. 2010;**24**(2):85–105.

104. **Fallgatter AJ, Jatzke S, Bartsch AJ, Hamelbeck B, Lesch KP.** Serotonin transporter promoter polymorphism influences topography of inhibitory motor control. Int J Neuropsychopharmacol. 1999;**2**(2):115–20. doi:10.1017/S1461145799001455.

105. **Lesch KP, Bengel D, Heils A, Sabol SZ, Greenberg BD, Petri S,** et al. Association of anxiety-related traits with a polymorphism in the serotonin transporter gene regulatory region. Science. 1996;**274**(5292):1527–31.

106. **Proudfit GH, Bress JN, Foti D, Kujawa A, Klein DN.** Depression and Event-related Potentials: Emotional disengagement and reward insensitivity. Curr Opin Psychol. 2015;**4**, 110–3. doi:10.1016/j.copsyc.2014.12.018.

107. **de Rover M, Brown SB, Boot N, Hajcak G, van Noorden MS, van der Wee NJ, Nieuwenhuis S.** Beta receptor-mediated modulation of the late positive potential in humans. Psychopharmacology (Berl). 2012;**219**(4):971–9. doi:10.1007/s00213-011-2426-x.

108. **Azizian A, Polich J.** Evidence for attentional gradient in the serial position memory curve from event-related potentials. J Cogn Neurosci. 2007;**19**(12):2071–81. doi:10.1162/jocn.2007.19.12.2071.

109. **Dolcos F, Cabeza R.** Event-related potentials of emotional memory: Encoding pleasant, unpleasant, and neutral pictures. Cogn Affect Behav Neurosci. 2002;**2**(3):252–63.

110. **MacNamara A, Hajcak G.** Anxiety and spatial attention moderate the electrocortical response to aversive pictures. Neuropsychologia. 2009;**47**(13):2975–80. doi:10.1016/j.neuropsycholo gia.2009.06.026.

111. **Holmes A, Nielsen MK, Green S.** Effects of anxiety on the processing of fearful and happy faces: An event-related potential study. Biol Psychol. 2008;**77**(2):159–73. doi:10.1016/j.biopsycho.2007.10.003.

112. **Moser JS, Huppert JD, Duval E, Simons RF.** Face processing biases in social anxiety: an electrophysiological study. Biol Psychol. 2008;**78**(1):93–103. doi:10.1016/j.biopsycho.2008.01.005.

113. **Muhlberger A, Wieser MJ, Herrmann MJ, Weyers P, Troger C, Pauli P.** Early cortical processing of natural and artificial emotional faces differs between lower and higher socially anxious persons. J Neural Transm (Vienna). 2009;**116**(6):735–46. doi:10.1007/s00702-008-0108-6.

114. **MacNamara A, Hajcak G.** Distinct electrocortical and behavioral evidence for increased attention to threat in generalized anxiety disorder. Depress Anxiety. 2010;**27**(3):234–43. doi:10.1002/da.20679.

115. **Weinberg A, Hajcak G.** Electrocortical evidence for vigilance-avoidance in generalized anxiety disorder. Psychophysiology. 2011;**48**(6):842–51. doi:10.1111/j.1469-8986.2010.01149.x.
116. **Weinberg A, Perlman G, Kotov R, Hajcak G.** Depression and reduced neural response to emotional images: Distinction from anxiety, and importance of symptom dimensions and age of onset. J Abnorm Psychol. 2016;**125**(1):26–39. doi:10.1037/abn0000118.
117. **Luu P, Tucker DM, Derryberry D, Reed M, Poulsen C.** Electrophysiological responses to errors and feedback in the process of action regulation. Psychol Sci. 2003;**14**(1):47–53.
118. **Cohen MX, Elger CE, Ranganath C.** Reward expectation modulates feedback-related negativity and EEG spectra. Neuroimage. 2007;**35**(2):968–78. doi:10.1016/j.neuroimage.2006.11.056.
119. **Mueller EM, Evers EA, Wacker J, van der Veen FM.** Acute tryptophan depletion attenuates brain–heart coupling following external feedback. Front Hum Neurosci, 2012;**6**:77.
120. **Panitz C, Wacker J, Stemmler G, Mueller EM.** Brain–heart coupling at the P300 latency is linked to anterior cingulate cortex and insula—a cardio-electroencephalographic covariance tracing study. Biological Psychology. 2013;(94):185–91.
121. **Panitz C, Hermann C, Mueller EM.** Conditioned and extinguished fear modulate functional corticocardiac coupling in humans. Psychophysiology. 2015; doi:10.1111/psyp.12498.
122. **Mueller EM, Stemmler G, Hennig J, Wacker J.** 5-HTTLPR and anxiety modulate brain–heart covariation. Psychophysiology. 2013;**50**:441–53.
123. **Caspi A, Sugden K, Moffitt TE, Taylor A, Craig IW, Harrington H**, et al. Influence of life stress on depression: moderation by a polymorphism in the 5-HTT gene. Science. 2003;**301**(5631):386–9. doi:10.1126/science.1083968301/5631/386 [pii].
124. **Olvet DM, Hajcak G.** The effect of trial-to-trial feedback on the error-related negativity and its relationship with anxiety. Cogn Affect Behav Neurosci. 2009;**9**(4):427–33. doi:9/4/427 [pii] 10.3758/CABN.9.4.427
125. **Santesso DL, Steele KT, Bogdan R, Holmes AJ, Deveney CM, Meites TM, Pizzagalli DA.** Enhanced negative feedback responses in remitted depression. Neuroreport. 2008;**19**(10):1045–8. doi:10.1097/WNR.0b013e3283036e73

Chapter 16

Imaging genetics in depression

Ulrich Rabl and Lukas Pezawas

16.1 Introduction

Major depressive disorder (MDD) is highly prevalent and leads to increased mortality and disability worldwide (1). Unfortunately, first-line treatments result in a variable and modest treatment outcome leaving about two-thirds of patients with residual symptoms. Consequently, treatment augmentation or a drug change is frequently required, resulting in prolonged disability and suicidal behavior (2). Hence, the clinical demand to identify subgroups of MDD patients with respect to treatment outcome has fueled the search for clinical, genetic, and neural predictors in recent years (3–5). This unmet need asks for a better understanding of the causal mechanisms involved in MDD and antidepressant response, which will be crucial for the development of precision medicine approaches as well as new therapeutics. In response to this challenge, the last two decades witnessed an unprecedented effort towards understanding the biological foundation of depression. Around the turn of the millennium, imaging genetics evolved as a blend of methods from psychiatric genetics and clinical neuroimaging that have now matured into a distinct research field. Imaging genetics contributed significantly to our understanding of the links between genetic variation, neurotransmitter systems, brain circuits, behavior, and environmental effects involved in psychiatric illness (6). However, viable attempts to adopt these findings to the bedside are still mostly lacking, despite considerable progress being made (7). In this chapter, we provide an overview of the development and impact of imaging genetics research in the context of MDD and the serotonergic system and discuss potential strategies towards clinical application.

16.2 Historical overview

Following the observation of familial aggregation, twin and adoption studies provided the first estimates of environmental and genetic influence, providing heritability estimates ranging from around 80% in schizophrenia to around 30–40% in MDD. However, it also became obvious that psychiatric disorders do not "breed true," meaning that familial risk crosses traditional nosological boundaries such as the Kraepelinian divide. Even more, shared heritability between specific psychometric measures in healthy individuals and patients suggested a continuum between normal variation and illness (8). These findings stimulated the idea that finding the involved genes could give access to more meaningful diagnostics and treatments. In the 1970s, the highly successful gene hunt in Mendelian disorders stirred hope that the same genetic linkage techniques could be utilized on families with a pronounced load of psychiatric illness but eventually these methods failed to yield consistent results. At least, it became apparent that the genetic architecture of psychiatric risk had to be entirely distinct from monogenetic diseases. As the exact nature of psychiatry's genetic blueprint remained guesswork, a new research paradigm—association studies—was introduced to account in a better way for the combined effect of numerous genes with small individual impact (9). Since statistical power increases with limited search space, these studies focused on pre-selected candidate genes instead of the whole genome to detect genetic variants with modest individual effects. This method also allowed studying unrelated patients and controls instead of the hard-to-find pedigrees with substantial familial aggregation, although that advantage came at the high risk of bias due to population stratification. However, the most pressing problem was that association studies ultimately did not provide a solution for their primary raison d'être, namely the identification of unknown genes related to a specific phenotype. While linkage studies could identify entirely unknown risk loci across the genome by comparing their heritability patterns in affected families, association studies did not solve the "needle in the haystack" problem of gene discovery. Instead, these studies could only provide confirmatory evidence for candidates with sufficient prior evidence. Even more, selecting the right genes was not enough, but specific genetic markers—mostly single nucleotide polymorphisms (SNPs) and various number of tandem repeats (VNTRs)—needed to be determined *a priori*. More often than not, there were no good guesses where to look in the more than 20,000 genes of the human genome, and for most genetic markers, sufficient characterization concerning their impact on gene function was lacking. It also became evident that the human genome was significantly more diverse than anticipated from the first draft sequence of the human genome project. With genotyping costs becoming steadily cheaper, several large-scale initiatives such as the International HapMap Project or the 1000 Genomes Project were launched to reveal the true extent of genetic diversity, showing that the potential of single markers to sufficiently map the functional diversity of the genome was in fact fairly limited (10). Given these massive limitations, association studies focused on a small set of well-characterized candidates. Nevertheless, these studies are remarkable since they changed, for the first time, the direction of genetic research from a forward genetic approach that went from genetic discovery to detailed animal and *in vitro* research, to a reverse genetics mode, where the study of mechanisms was a prerequisite for population-level studies. Ultimately, association studies did not deliver on their promises and doubt about their initially promising results arose. One problem was that the many arbitrary decisions in these studies also increased their "researcher degrees of freedom," making them susceptible to sloppy data analyses and approximate replications. It also became undeniable that most studies did not have sufficient power for the expected effects, a situation that can massively distort effect size estimations in the presence of selective reporting (11). While some of these problems were widespread in the majority of the research environment, psychiatric genetics belonged to the primary suspects in sometimes heated methodological debates. Since association studies were imagined from the beginning as a workaround until genome-wide association became reality (9), geneticists emphasized the need for larger sample sizes and higher genetic resolution to achieve whole genome coverage. However, not all researchers were convinced that the lack of consistent results was simply due to low power and insufficient genomic coverage. Rather, a hypothesis gained traction that the phenotypes provided by the prevailing psychiatric nosology could be too blurry and heterogeneous to permit valid associations. This theory was already carved out by Gottesman and Shields, elaborating on a polygenetic theory of schizophrenia, hypothesizing that more confined measures of neuropsychiatric functioning could provide superior variables for genetic association than behavioral phenotypes. In 1973, this theory was cast in the concept of "endophenotypes," a term that was borrowed from a paper on the evolution of insects (12). The endophenotype theory suggested that complex behavioral disorders could be cut up into subcomponents that exhibit Mendelian-like heritability patterns and limit the influence of environmental variation. In this context, an endophenotype was defined as an internal phenotype that would be discoverable by a biological test or microscopic examination, thereby yielding more validity and objectivity than the subjective behavioral phenotypes. Endophenotypes are defined as heritable biological characteristics that are associated with the illness of interest in the population but manifest independently from illness activity in the individual. Within families, the endophenotype was expected to co-segregate with the illness and to be more prevalent in the non-affected family members when compared to the general population. After lying dormant for several years, the concept steadily gained momentum when the limited success of linkage and association studies became undeniable. The first attempts to translate the endophenotype concept into practice used measures derived from psychometric testing or electroencephalography, but these measures were found to be comparably polygenic as their behavioral counterparts. While some researchers, therefore, called their usefulness for risk gene discovery into question, others recognized that these intermediary measures could shed light on mechanisms of gene function that would remain hidden in clinical association studies (13). This shift to the elucidation of mechanisms was acknowledged with the alternative term "intermediate phenotype" as a substitute for "endophenotype" to de-emphasize the role of reduced genetic complexity. When modern neuroimaging techniques such as positron emission tomography (PET) and functional magnetic resonance imaging (fMRI) had technically matured enough to be applicable for clinical psychiatric research, they were natural candidates as intermediate phenotypes (14). In 1995, Small and colleagues used PET to study a genetic variant in the apolipoprotein E (*APOE*) gene that is known to potentiate risk for Alzheimer's disease, finding that non-demented carriers of the risk variant exhibit a similar but less pronounced pattern of parietal cortex metabolism as demented patients (15). Around the same time, encouraging results obtained by fMRI in schizophrenia patients prompted Daniel Weinberger to propose the use of fMRI, which employs the blood oxygen level-dependent (BOLD) signal as a proxy for neural activation, as a tool for genetic discovery. In 2000, Weinberger and colleagues published the first proof-of-concept study designed to discover unknown risk gene mechanisms. In this study, a polymorphism in the dopamine transporter gene *SLC6A3* was found to impact striatal dopamine transporter availability measured with single photon emission computed tomography (SPECT) (16). The first genetic studies using fMRI, which later became the prevailing method for imaging genetics research, followed soon thereafter, investigating again *APOE* and Val158Met, a polymorphism in the catechol-*O*-methyltransferase gene (*COMT*). The study on the *COMT* gene also included siblings of schizophrenia patients, showing intermediate effects in between patients and controls without familial risk. These findings suggested that brain-level intermediate phenotypes indeed fulfilled key criteria as proposed by Gottesman and Shields. It later became the standard approach to focus on healthy participants, which effectively eliminated the influence of confounders that are typical for clinical populations such as medication effects, cognitive deficits, smoking, or substance abuse (17). In the following years, *5-HTTLPR*, a VNTR in the serotonin transporter gene, *SLC6A4*, which had already been investigated by several clinical association studies since its discovery in 1996, became one of the best-studied polymorphisms in imaging genetics. Until the end of the 2000s, imaging genetics was characterized by the use of more complex models such as gene–gene interaction (epistasis), haplotypes, gene–environment (G×E) interaction, and the integration of new intermediate phenotypes such as functional connectivity or resting state fMRI (17). However, while these years were full of enthusiasm and the volume of literature exploded, the problems of association studies ultimately equally applied to imaging genetics studies. Around the same time, the first successful GWAS in somatic disorders were published and some geneticists began to lobby against imaging genetics research in the conviction that all candidate gene studies are likely to turn out as false positives, and that only GWAS will eventually deliver the genetic basis of psychiatric disorders (18). In 2005, Ioannidis even claimed that most published research findings are likely to be false positives (19). Given their sheer complexity, it became undeniable that the pioneering imaging genetics studies suffered both from the inherent problems of association studies and of the intricacies of neuroimaging research. Today, several initiatives are underway to standardize research and it is hoped that many of the proposed

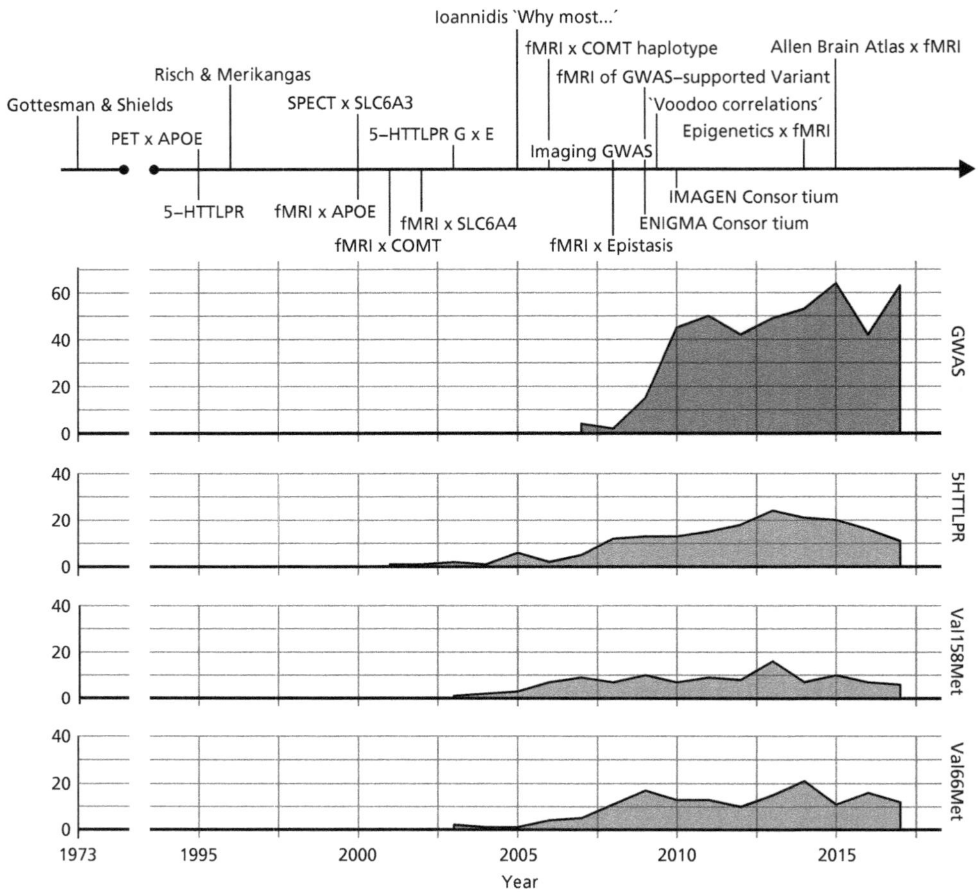

Figure 16.1 Timeline of the development of imaging genetics. Below are shown the number of neuroimaging publications per year for the most researched candidate polymorphisms (5-HTTLPR in the serotonin transporter gene *SLC6A4*, Val66Met in the *BDNF* gene, and Val158Met in the *COMT* gene) according to PubMed. While interest in these candidates has started to decline recently, interest in the combination of genome-wide association with neuroimaging is growing (search term: "x AND (fMRI OR MRI OR "magnetic resonance imaging" OR BOLD OR PET OR SPECT)" with x being either "Val158Met," "5-HTTLPR," "Val66Met," or "GWAS").

tools to reduce biases such as pre-registration are becoming standard practice in the future (20). The ice age between "endophenotypists" and the GWAS maximalists seems to have come to an end, and there are now several projects and groups underway such as the IMAGEN consortium that combine both approaches (21). Figure 16.1 provides a timeline of the development of imaging genetics.

16.3 **Brain circuits of major depressive disorder**

MDD is now increasingly viewed as a brain network disorder, a perspective that parallels a general trend in neuroimaging from the study of localized alterations of anatomy or function (sometimes derogatorily referred to as "blobology") towards alterations of connectivity, which studies the interactions between distinct brain areas (22). This increasing appreciation of network effects could, to some extent, be driven by methodological advances. While the application of brain localization techniques became more reliable due to an increased focus on methodological rigor, the basic principles and the range of problems that can be tested with these methods did not change significantly in the last two decades (20). In contrast, the variety of brain connectivity methods has seen massive growth and adoption over recent years. These techniques can be differentiated into measures of functional and effective connectivity. While functional connectivity assesses correlations between time-series derived from distinct regions of interest, effective connectivity measures such as Granger causality or dynamic causal modeling (DCM) are applied to deduce the directionality of network interactions (23). Recently, the transfer of graph theory, a broad mathematical field that studies the network properties of complex models of nodes and their relations, to brain imaging has added a whole new set of methods. These new measures quantify the macroscopic functional and structural properties of complex brain networks and provide entirely new potential intermediate phenotypes (24). While MDD research increasingly makes use of graph theory, the interpretation of these measures concerning their biological underpinnings and their application in imaging genetics is still in the early stages.

Beyond these technical advances, there has been a general shift in MDD research from a narrow focus on emotional processes to models that integrate mood, reward, and cognitive functions (25). Impairments of cognition and decision-making, in fact, rank among the most prevalent and debilitating symptoms of depression and are particularly challenging to treat (26).

In the following paragraphs, we introduce the primary nodes in the depression brain network and discuss the impact of genetic variation. This selection is admittedly subjective. However, meta-analytic attempts to reveal the core depression circuit also arrive only partly at overlapping conclusions depending on the technical approaches and the included data (27,28). On the other hand, human imaging genetics research almost neglected several brain regions likely to be involved in MDD such as the habenula, primarily due to the limited spatial and temporal resolution of human-applicable neuroimaging techniques.

16.3.1 Subgenual cingulate cortex

In 1997, a seminal study used PET to investigate potential metabolic abnormalities in bipolar and unipolar depressive patients, identifying an area in the prefrontal cortex ventral to the genu of the corpus callosum, mapping to Brodmann area 25, that showed decreased cerebral blood flow and glucose metabolism (29). This area is now known as subgenual or subcallosal anterior cingulate cortex (sACC) and was linked to MDD by a large body of studies. Beyond function, the sACC exhibits decreased gray matter and, in post-mortem studies, glial reductions in depressive patients. Also, it became a primary target for experimental deep brain stimulation in treatment-refractory patients, one of the first neuroscience-guided therapeutic approaches in psychiatry (30).

16.3.2 Amygdala

Hyperactivity of the amygdala has been one of the earliest neuroimaging findings in MDD patients, a result that is shared with the cluster of anxiety disorders (31). However, this finding has not been consistently replicated and therefore it is likely to depend on the specific paradigm and current state of the patient. Also, it has been suggested recently that a large vein in the immediate vicinity may potentially impact the amygdala fMRI signal, something which is still under debate (32). In a meta-analysis, a broader functional circuit, the salience network (SN), which comprises the anterior insula and the dorsal cingulate cortex and, depending on the used paradigm, extends to the amygdala, was found to be hyperactive in MDD (27). This network has been linked to error monitoring and physiological stress responses and may reflect a mode of constantly increased alertness in MDD (33). Partly overlapping with the salience network, the amygdala forms a network with the cingulate cortex that was first noted in animal work. Within this network, the anterior cingulate cortex performs evaluative processes in its more dorsal portion (dorsal anterior cingulate cortex, dACC) and emotion regulatory actions in the sACC, which are mediated by down-regulation of amygdala function (34). The integrity of this sACC-amygdala network also correlates with anxious traits and depressive symptoms (35).

16.3.3 Hippocampus

The hippocampus is a structure in the medial temporal lobe adjacent to the amygdala that is known best for its role in explicit memory formation and spatial navigation. Although the hippocampus most likely serves a general function in the interaction between internal states and the external environment, no established theory is available to reconcile its various functions into a general framework (36). Concerning MDD, the hippocampus' role in the regulation of allostasis—the process of achieving stability in the face of internal or external stressors through behavioral or physiological adaptions—is of specific importance. The hippocampus can impact the hormonal stress response by acting as an inhibitor of the hypothalamus–pituitary–adrenal axis, which is dysregulated in a subgroup of MDD patients (37). Further, hippocampal atrophy has been linked to MDD, an association that is subtle but consistent and moderated by clinical characteristics such as early age of onset and recurrent episodes (38). Studies on treatment mechanisms suggest that patients may undergo a waxing and waning of the hippocampus during the different states of the illness. Specifically, while hippocampal volume reduction is tied to chronically depressed states, volume growth was found during successful treatment by both animal and human studies (39).

16.3.4 Default mode network

Another important network that recently came into focus of MDD research is the default-mode network (DMN). The DMN is a functional network that is delineated based on its activation or deactivation during certain neuroimaging paradigms. Specifically, it becomes active during rest but is deactivated during goal-directed, attention-demanding, and non-self-referential behaviors (40). The DMN also exhibits state-dependent interactions with the hippocampus, which is likely to play an important role in sleep phase regulation and memory formation (41), and could be relevant in MDD, given the high prevalence of sleep and memory problems. In a meta-analysis, consistently increased connectivity between the DMN and the dACC, medial dorsal thalamus, and sACC was found, with the latter also predicting higher levels of rumination (42). Interestingly, alterations of the DMN seem even to extend to unmedicated remitted patients with subjects exhibiting increased functional connectivity with the dorsolateral prefrontal cortex and less deactivation of the anterior DMN during a cognitive task compared to healthy controls, which also correlated with the tendency for rumination (43). Also, remitted patients showed less hippocampal connectivity between the posterior DMN and the hippocampus, which were associated with cognitive reactivity and rumination (44). These findings suggest an increased integration of the anterior, but decoupling of the posterior DMN in MDD (44).

16.3.5 Genetic impact on depression brain circuits

While MDD only shows modest heritability (30–40%), compared to conditions such as schizophrenia or bipolar disorder (around 80%), the impact of genetic variants could, in fact, be much higher due to interactions with environmental factors and non-additive gene–gene interactions (epistasis). A seminal study on the *5-HTTLPR* genetic variant in *SLC6A4* provided an important proof of concept in this regard, showing that S-allele carriers of *5-HTTLPR* are at specifically high risk for depression in face of environmental adversity (45). One year earlier, one of the first imaging genetic studies using fMRI showed that the same variant, which was already linked to neuroticism, resulted in increased amygdala activation in S-carriers, providing the first evidence for a serotonin-related neural intermediate phenotype (46). A recent longitudinal study even extended this finding to epigenetic effects, finding changes in gene methylation in the *SLC6A4* promoter region that are associated with lower socioeconomic status impacting amygdala reactivity (47). Further, it could be shown that the S-allele impacts structure and function of the amygdala–cingulate network, which also is related to harm avoidance and risk-taking in decision-making (48,49). However, several negative results also exist regarding the initial association between *5-HTTLPR* and amygdala activation, suggesting that either the effect is much smaller than anticipated or conditional on not yet sufficiently understood factors (50). Interestingly, a recent study investigated the impact of *5-HTTLPR* during two fMRI paradigms, classic fear conditioning and instructed fear, and also found no effect in the amygdala, but 5-HTTLPR-dependent activation in the SN (51). This study is remarkable for its use of two independent samples to strengthen the conclusions, which are also supported by a meta-analysis of fear-conditioning studies that, in stark contrast to animal work, found no consistent amygdala activation but activation in the SN (52). However, one also has to keep in mind that the BOLD signal of the amygdala is especially difficult to obtain due to susceptibility artifacts and nearby vessels and that, similar to MDD, the effect is likely valence-specific (27,32). Other genes involved in serotonergic neurotransmission such as *HTR1A* have been less well researched, but were found to impact brain regions similar to *SLC6A4* (53). This limitation reflects the problem of candidate studies, which were restricted to a few well-supported candidates. In fact, only two candidate genes have been investigated with equal extent as *SLC6A4*, namely *BDNF* and *COMT*. In a first study investigating gene–gene interactions, termed epistasis, the *BDNF* Val66Met polymorphism was found to interact with *5-HTTLPR*. While the S-allele resulted in significant volume reduction in the presence of two Val alleles of Val66Met, no such effect was found in those S-carriers with one or two Met alleles (54). Val66Met was also found to impact hippocampal volume, although this effect was not replicated in larger cohorts and turned out to be small in a recent meta-analysis, putatively due to differences in measurement techniques (55). Beyond genetic main effects and gene–gene effects, several studies also addressed G×E interactions. These studies mostly focused on stressful life events or childhood maltreatment and primarily reported interaction effects between *BDNF*, *SLC6A4*, and *COMT* genetic variation and adversity on hippocampal volume, although effects on amygdala and cingulate cortex volume were also found. However, these effects as yet have only been reported in single studies and partly conflicting results have been reported (56). Recently, in the wake of low support of candidate variants in GWAS, the era of candidate imaging genetics studies seems, slowly, to be coming to an end (Figure 16.1). This decline of interest seems to especially concern *5-HTTLPR*, probably driven by increasing doubt on the highly publicized G×E interaction effect between childhood adversity and *5-HTTLPR*, which has been repeatedly called into question in meta-analyses (57). Similarly, several of the core findings of imaging genetics exhibited evidence for publication bias in meta-analyses (55,58). However, meta-analyses are also not without biases and specifically neuroimaging experiments tend to be very heterogeneous and might therefore be difficult to summarize (59). Novel work in primates, in fact, supports the conclusions of the initial human genetic studies on 5-HTTLPR (60).

More recently, genome-wide association is integrated into imaging genetics research with several strategies followed to date. Interesting first results stem from the investigation of gene candidates or polygenic risk scores derived from clinical GWAS (61,62). These results suggest that genetic risk highlighted in clinical studies is indeed linked to core networks of MDD, supporting the intermediate phenotype approach. Combined analyses of GWAS and imaging data are underway (63). Others started to integrate the Allen Brain Atlas post-mortem dataset to refine the search space in such analyses. Using this paradigm, a recent study found a set of 136 genes enriched for ion channels that showed a distribution of gene expression that followed the pattern of resting-state brain networks across space (64). Beyond suggesting a new paradigm for gene discovery, this study also provides another proof for the validity of brain imaging intermediate phenotypes.

16.4 Discussion and conclusion

The last two decades have witnessed considerable progress in our understanding of the biological foundations of MDD and other psychiatric disorders. Using normal genetic variation as a proxy for molecular manipulation, imaging genetics has proven useful as an explanatory bridge between population genetic approaches and the mechanism elucidation approaches of *in vitro* and animal studies (65). For that reason, imaging genetics has become one of the most favored approaches in the endeavor to map the involved neurocircuits and their modulation by genetic variants and environmental factors. Imaging genetics studies on the *BDNF*, *SLC6A4*, and *COMT* genes exemplify the "reverse genetics" mode that has been adopted in a majority of studies so far. In this approach, a study is conducted to delineate the neural phenotypes of known genetic variation of a gene that is already well characterized by animal or *in vitro* studies (66). While these studies provided important lessons, it has become clear now that the clinical impact of these variants on their own is negligible. Beyond the circularity of the candidate gene approach, several concerns are especially problematic in imaging research such as the problem of multiple comparisons, a statistical hitch that arises in situations where one searches for a small signal in an ocean of data. It has become clear that many earlier studies did not properly correct for multiple comparisons and that effect size estimates in imaging research can become distorted by "double dipping," a situation that occurs when the same dataset is used for both localization and the estimation of the size of the effect. Thanks to provocative papers dealing with "dead salmon" and "voodoo," the community is now well aware of these problems (67), although some of these problems are still heavily debated (68). For some of these statistical problems there seems to be no good workaround at all. We now know that the many post-hoc decisions made during data analysis can massively bias the results, a situation that has become known as "the garden of forking paths" or "researcher degrees of freedom." This bias does not necessarily occur with malicious intent ("p hacking" or "fishing expedition") but can occur subconsciously when there are multiple potential ways to undertake an analysis, and the data selection and analysis are not precisely pre-specified (69). Due to the sheer complexity of the data and the analysis pipelines, it is almost unavoidable for imaging researchers to avoid adjustments during data analysis entirely. Despite significant efforts to standardize analysis pipelines most projects still rely on individual data analysis steps, which also makes their summarization in meta-analyses problematic (70). We should, therefore, be aware that the majority of results in imaging genetics—and in many other fields—have been, in fact, only of exploratory, non-inferential nature, and they need validation by independent techniques. However, many of these problems can be overcome to some degree with pre-registration and larger sample sizes that allow higher power or even split-sample strategies (51.71). GWAS are a prime example of how dramatically the odds can change when studies are appropriately scaled. After years of fruitless attempts to unravel the genetics of schizophrenia, a large consortium pooling a mega-sample of 36,989 cases finally tracked down 108 statistically significant loci involved in the disorder (72). With regard to MDD, which has a much lower heritability, even studies exerting extreme measures to reduce genetic heterogeneity came up with few signals that could not be replicated in independent populations (73). Even here, however, size seems to trump everything, as demonstrated by an analysis of a dataset from 23andMe, which recently came up with 15 previously unknown loci (74). In contrast, next to none of the genes of the candidate gene era have been verified in these analyses, underscoring the value of a "forwards genetics" mode. However, how to proceed from here remains an unresolved question still. While imaging genetics has proven very useful in identifying the neural mechanisms of GWAS-derived candidates (75), doubts are now growing that the genes marked by GWAS provide little biological insight and are mostly unsuited as drug targets (76). Given the enormous efforts and cost of the last two decades of research, and the ever-growing consortia, the situation seems almost surreal when it is compared to the golden age of psychopharmacology, where an abundance of highly effective treatments was discovered more or less by serendipity. Probably, we should arrive at the realization that a search for "the" specific candidates for a highly polygenic trait has an absurdist element to it, and that to understand disease mechanisms, we need to get back to study paradigms that can elucidate mechanisms (62). However, given this rather pessimistic view about current research strategies, it must be said that it cannot be expected that the identification of a genotype must lead to a pharmacological treatment at all, even if the effect of the genotype is true. Pharmacological interventions and innovations are likely to happen somewhere in bottom-up molecular cascades and the bottle-necks involving such gene products. This has been shown to be a fruitful approach for Mendelian disorders like cystic fibrosis, where effective treatments (ivacaftor) have been established for certain genetic characteristics in a small subgroup of patients (77). Alternatively, it might also be reasonable to focus on environmental factors involved in psychiatric disorders since environmental effects are likely to be more accessible to intervention. Here, intermediate phenotypes can have the same role as a window into the brain as in genetics research (71). Given the increasing attention towards correct analytical strategies and increasing collaboration between clinicians, neuroimagists, geneticists, and biologists in recent years, one can be cautiously optimistic that the paths now taken will lead us to a translation into clinical practice.

References

1. **Becker AE, Kleinman A.** Mental health and the global agenda. N Engl J Med. 2013;**369**(1):66–73.
2. **Kupfer DJ, Frank E, Phillips ML.** Major depressive disorder: New clinical, neurobiological, and treatment perspectives. Lancet. 2012;**379**(9820):1045–55.
3. **Phillips ML, Chase HW, Sheline YI, Etkin A, Almeida JR, Deckersbach T**, et al. Identifying predictors, moderators, and mediators of antidepressant response in major depressive disorder: Neuroimaging approaches. Am J Psychiatry. 2015;**172**(2):124–138.
4. **Iniesta R, Malki K, Maier W, Rietschel M, Mors O, Hauser J**, et al. Combining clinical variables to optimize prediction of antidepressant treatment outcomes. J Psychiatr Res. 2016;78:94–102.
5. **Rabl U, Meyer BM, Pezawas L.** Neuroimaging Genetics: Principles and Practices. In: Bigos KL, Hariri AR, Weinberger DR (eds).Variability of Antidepressant Drug Response Contribution of Imaging Genetics Studies. Oxford University Press; 2016.
6. **Meyer-Lindenberg A, Weinberger DR.** Intermediate phenotypes and genetic mechanisms of psychiatric disorders. Nat Rev Neurosci. 2006;**7**(10):818–27.
7. **Weinberger DR, Radulescu E.** Finding the elusive psychiatric "lesion" with 21st-century neuroanatomy: A note of caution. Am J Psychiatry. 2016;**173**(1):27–33.
8. **Smoller JW.** Disorders and borders: psychiatric genetics and nosology. Am J Med Genet B Neuropsychiatr Genet. 2013;**162B**(7):559–78.
9. **Risch N, Merikangas K.** The future of genetic studies of complex human diseases. Science. 1996;**273**(5281):1516–7.
10. **Altshuler D, Durbin RM, Abecasis GR, Bentley DR, Chakravarti A, Clark AG**, et al. A map of human genome variation from population-scale sequencing. Nature. 2010;**467**(7319):1061–73.

11. **Nord CL, Valton V, Wood J, Roiser JP.** Power-up: A reanalysis of "power failure" in neuroscience using mixture modeling. J Neurosci. 2017;**37**(34):8051–61.
12. **Gottesman, II, Shields J.** Genetic theorizing and schizophrenia. Br J Psychiatry. 1973;**122**(566):15–30.
13. **Flint J, Timpson N, Munafo M.** Assessing the utility of intermediate phenotypes for genetic mapping of psychiatric disease. Trends Neurosci. 2014;**37**(12):733–41.
14. **Raichle ME.** A brief history of human brain mapping. Trends Neurosci. 2009;**32**(2):118–26.
15. **Small GW, Mazziotta JC, Collins MT, Baxter LR, Phelps ME, Mandelkern MA,** et al. Apolipoprotein E type 4 allele and cerebral glucose metabolism in relatives at risk for familial Alzheimer disease. JAMA. 1995;**273**(12):942–7.
16. **Heinz A, Goldman D, Jones DW, Palmour R, Hommer D, Gorey JG,** et al. Genotype influences in vivo dopamine transporter availability in human striatum. Neuropsychopharmacology. 2000;**22**(2):133–9.
17. **Bigos KL, Weinberger DR.** Imaging genetics--days of future past. Neuroimage. 2010;**53**(3):804–9.
18. **Abbott A.** Psychiatric genetics: The brains of the family. Nature. 2008;**454**(7201):154–7.
19. **Ioannidis JP.** Why most published research findings are false. PLoS Med. 2005;**2**(8):e124.
20. **Poldrack RA, Baker CI, Durnez J, Gorgolewski KJ, Matthews PM, Munafò MR,** et al. Scanning the horizon: Towards transparent and reproducible neuroimaging research. Nat Rev Neurosci. 2017;**18**(2):115–26.
21. **Geschwind DH, Flint J.** Genetics and genomics of psychiatric disease. Science. 2015;**349**(6255):1489–94.
22. **Kaiser RH, Andrews-Hanna JR, Wager TD, Pizzagalli DA.** Large-scale network dysfunction in major depressive disorder: A meta-analysis of resting-state functional connectivity. JAMA Psychiatry. 2015;**72**(6):603–11.
23. **Smith SM.** The future of FMRI connectivity. Neuroimage. 2012;**62**(2):1257–66.
24. **Bullmore E, Sporns O.** Complex brain networks: Graph theoretical analysis of structural and functional systems. Nat Rev Neurosci. 2009;**10**(3):186–98.
25. **Rock PL, Roiser JP, Riedel WJ, Blackwell AD.** Cognitive impairment in depression: A systematic review and meta-analysis. Psychol Med. 2014;**44**(10):2029–40.
26. **Shilyansky C, Williams LM, Gyurak A, Harris A, Usherwood T, Etkin A.** Effect of antidepressant treatment on cognitive impairments associated with depression: A randomised longitudinal study. Lancet Psychiatry. 2016;**3**(5):425–35.
27. **Hamilton JP, Etkin A, Furman DJ, Lemus MG, Johnson RF, Gotlib IH.** Functional neuroimaging of major depressive disorder: A meta-analysis and new integration of base line activation and neural response data. Am J Psychiatry. 2012;**169**(7):693–703.
28. **Diener C, Kuehner C, Brusniak W, Ubl B, Wessa M, Flor H.** A meta-analysis of neurofunctional imaging studies of emotion and cognition in major depression. Neuroimage. 2012;**61**(3):677–85.
29. **Drevets WC, Price JL, Simpson JR, Jr., Todd RD, Reich T, Vannier M,** et al. Subgenual prefrontal cortex abnormalities in mood disorders. Nature. 1997;**386**(6627):824–7.
30. **Hamani C, Mayberg H, Stone S, Laxton A, Haber S, Lozano AM.** The subcallosal cingulate gyrus in the context of major depression. Biol Psychiatry. 2011;**69**(4):301–8.
31. **Domschke K, Dannlowski U.** Imaging genetics of anxiety disorders. Neuroimage. 2010;**53**(3):822–31.
32. **Boubela RN, Kalcher K, Huf W, Seidel EM, Derntl B, Pezawas L,** et al. fMRI measurements of amygdala activation are confounded by stimulus correlated signal fluctuation in nearby veins draining distant brain regions. Sci Rep. 2015;**5**:10499.
33. **Uddin LQ.** Salience processing and insular cortical function and dysfunction. Nat Rev Neurosci. 2015;**16**(1):55–61.
34. **Etkin A, Egner T, Kalisch R.** Emotional processing in anterior cingulate and medial prefrontal cortex. Trends Cogn Sci. 2011;**15**(2):85–93.
35. **Kim MJ, Whalen PJ.** The structural integrity of an amygdala-prefrontal pathway predicts trait anxiety. J Neurosci. 2009;**29**(37):11614–8.

36. **Lisman J, Buzsaki G, Eichenbaum H, Nadel L, Ranganath C, Redish AD.** Viewpoints: How the hippocampus contributes to memory, navigation and cognition. Nat Neurosci. 2017;**20**(11):1434–47.
37. **McEwen BS.** Allostasis and the epigenetics of brain and body health over the life course: The brain on stress. JAMA Psychiatry. 2017;**74**(6):551–2.
38. **Schmaal L, Veltman DJ, van Erp TG, Sämann PG, Frodl T, Jahanshad N,** et al. Subcortical brain alterations in major depressive disorder: Findings from the ENIGMA major depressive disorder working group. Mol Psychiatry. 2016;**21**(6):806–12.
39. **Arnone D, McKie S, Elliott R, Juhasz G, Thomas EJ, Downey D,** et al. State-dependent changes in hippocampal grey matter in depression. Mol Psychiatry. 2013;**18**(12):1265–72.
40. **Raichle ME.** The brain's default mode network. Annu Rev Neurosci. 2015;**38**:433–47.
41. **Mitra A, Snyder AZ, Hacker CD, Pahwa M, Tagliazucchi E, Laufs H,** et al. Human cortical-hippocampal dialogue in wake and slow-wave sleep. PNAS. 2016;**113**(44):E6868–E6876.
42. **Hamilton JP, Farmer M, Fogelman P, Gotlib IH.** Depressive rumination, the default-mode network, and the dark matter of clinical neuroscience. Biol Psychiatry. 2015;**78**(4):224–30.
43. **Bartova L, Meyer BM, Diers K, Rabl U, Scharinger C, Popovic A,** et al. Reduced default mode network suppression during a working memory task in remitted major depression J Psychiatr Res. 2015;**64**:9–18.
44. **Figueroa CA, Mocking RJT, van Wingen G, Martens S, Ruhe HG, Schene AH.** Aberrant default-mode network-hippocampus connectivity after sad memory-recall in remitted-depression. Soc Cogn Affect Neurosci. 2017;**12**(11):1803–13.
45. **Caspi A, Sugden K, Moffitt TE, Taylor A, Craig IW, Harrington H,** et al. Influence of life stress on depression: moderation by a polymorphism in the 5-HTT gene. Science. 2003;**301**(5631):386–9.
46. **Hariri AR, Mattay VS, Tessitore A, Kolachana B, Fera F, Goldman D,** et al. Serotonin transporter genetic variation and the response of the human amygdala. Science. 2002;**297**(5580):400–3.
47. **Swartz JR, Hariri AR, Williamson DE.** An epigenetic mechanism links socioeconomic status to changes in depression-related brain function in high-risk adolescents. Mol Psychiatry. 2017;**22**(2):209–14.
48. **Roiser JP, de Martino B, Tan GC, Kumaran D, Seymour B, Wood NW,** et al. A genetically mediated bias in decision making driven by failure of amygdala control. J Neurosci. 2009;**29**(18):5985–91.
49. **Pezawas L, Meyer-Lindenberg A, Drabant EM, Verchinski BA, Munoz KE, Kolachana BS,** et al. 5-HTTLPR polymorphism impacts human cingulate–amygdala interactions: A genetic susceptibility mechanism for depression. Nat Neurosci. 2005;**8**(6):828–34.
50. **Bastiaansen JA, Servaas MN, Marsman JB, Ormel J, Nolte IM, Riese H,** et al. Filling the gap: Relationship between the serotonin-transporter-linked polymorphic region and amygdala activation. Psychol Sci. 2014;**25**(11):2058–66.
51. **Klumpers F, Kroes MC, Heitland I, Everaerd D, Akkermans SE, Oosting RS,** et al Dorsomedial prefrontal cortex mediates the impact of serotonin transporter linked polymorphic region genotype on anticipatory threat reactions. Biol Psychiatry. 2015;**78**(8):582–9.
52. **Fullana MA, Harrison BJ, Soriano-Mas C, Vervliet B, Cardoner N, Àvila-Parcet A,** et al. Neural signatures of human fear conditioning: An updated and extended meta-analysis of fMRI studies. Mol Psychiatry. 2016;**21**(4):500–8.
53. **Fakra E, Hyde LW, Gorka A, Fisher PM, Muñoz KE, Kimak M,** et al. Effects of HTR1A C(-1019)G on amygdala reactivity and trait anxiety. Arch Gen Psychiatry. 2009;**66**(1):33–40.
54. **Pezawas L, Meyer-Lindenberg A, Goldman AL, Verchinski BA, Chen G, Kolachana BS,** et al. Evidence of biologic epistasis between BDNF and SLC6A4 and implications for depression. Mol Psychiatry. 2008;**13**(7):709–6.
55. **Harrisberger F, Spalek K, Smieskova R, Schmidt A, Coynel D, Milnik A,** et al. The association of the BDNF Val66Met polymorphism and the hippocampal volumes in healthy humans: A joint meta-analysis of published and new data. Neurosci Biobehav Rev. 2014;**42**:267–78.
56. **Bogdan R, Pagliaccio D, Baranger DA, Hariri AR.** Genetic moderation of stress effects on corticolimbic circuitry. Neuropsychopharmacology. 2016;**41**(1):275–96.

57. **Culverhouse RC, Saccone NL, Horton AC, Ma Y, Anstey KJ, Banaschewski T**, et al. Collaborative meta-analysis finds no evidence of a strong interaction between stress and 5-HTTLPR genotype contributing to the development of depression. Mol Psychiatry. 2018;**23**(1):133–42.
58. **Murphy SE, Norbury R, Godlewska BR, Cowen PJ, Mannie ZM, Harmer CJ**, et al. The effect of the serotonin transporter polymorphism (5-HTTLPR) on amygdala function: A meta-analysis. Mol Psychiatry. 2013;**18**(4):512–20.
59. **Ioannidis JP.** The mass production of redundant, misleading, and conflicted systematic reviews and meta-analyses. **Milbank Q.** 2016;**94**(3):485–514.
60. **Santangelo AM, Ito M, Shiba Y, Clarke HF, Schut EH, Cockcroft G**, et al. Novel primate model of serotonin transporter genetic polymorphisms associated with gene expression, anxiety and sensitivity to antidepressants. Neuropsychopharmacology. 2016;**41**(9):2366–76.
61. **Schott BH, Assmann A, Schmierer P, Soch J, Erk S, Garbusow M**, et al. Epistatic interaction of genetic depression risk variants in the human subgenual cingulate cortex during memory encoding. Transl Psychiatry. 2014;**4**:e372.
62. **Yüksel D, Dietsche B, Forstner AJ, Witt SH, Maier R, Rietschel M**, et al. Polygenic risk for depression and the neural correlates of working memory in healthy subjects. Prog Neuropsychopharmacol Biol Psychiatry. 2017;**79**(Pt B):67–76.
63. **Thompson PM, Stein JL, Medland SE, Hibar DP, Vasquez AA, Renteria ME**, et al. The ENIGMA Consortium: Large-scale collaborative analyses of neuroimaging and genetic data. Brain Imaging Behav. 2014;**8**(2):153–82.
64. **Richiardi J, Altmann A, Milazzo AC, Chang C, Chakravarty MM, Banaschewski T**, et al. Brain Networks. Correlated gene expression supports synchronous activity in brain networks. Science. 2015;**348**(6240):1241–4.
65. **Tabery J.** Beyond versus: The struggle to understand the interaction of nature and nurture. 2014.
66. **Meyer-Lindenberg A.** Intermediate or brainless phenotypes for psychiatric research? Psychol Med. 2010;**40**(7):1057–62.
67. **Vul E, Harris C, Winkielman P, Pashler H.** Puzzlingly High Correlations in fMRI Studies of Emotion, Personality, and Social Cognition. Perspect Psychol Sci. 2009;**4**(3):274–90.
68. **Eklund A, Nichols TE, Knutsson H.** Cluster failure: Why fMRI inferences for spatial extent have inflated false-positive rates. PNAS. 2016;**113**(28):7900–5.
69. **Gelman A, Loken E.** The garden of forking paths: Why multiple comparisons can be a problem, even when there is no “fishing expedition” or “p-hacking” and the research hypothesis was posited ahead of time. 2013.
70. **Carp J.** The secret lives of experiments: Methods reporting in the fMRI literature. Neuroimage. 2012;**63**(1):289–300.
71. **Lederbogen F, Kirsch P, Haddad L, Streit F, Tost H, Schuch P**, et al. City living and urban upbringing affect neural social stress processing in humans. Nature. 2011;**474**(7352):498–501.
72. **Ripke S, Neale BM, Corvin A, Walters JT, Farh KH, Holmans PA**, et al. Schizophrenia Working Group of the Psychiatric Genomics C. Biological insights from 108 schizophrenia-associated genetic loci. Nature. 2014;**511**(7510):421–7.
73. **Cai N, Bigdeli TB, Kretzschmar W, Li Y, Liang J, Song L**, et al. CONVERGE consortium. Sparse whole-genome sequencing identifies two loci for major depressive disorder. Nature. 2015;**523**(7562):588–91.
74. **Hyde CL, Nagle MW, Tian C, Chen X, Paciga SA, Wendland JR**, et al. Identification of 15 genetic loci associated with risk of major depression in individuals of European descent. Nat Genet. 2016;**48**(9):1031–6.
75. **Esslinger C, Walter H, Kirsch P, Erk S, Schnell K, Arnold C**, et al. Neural mechanisms of a genome-wide supported psychosis variant. Science. 2009;**324**(5927):605.
76. **Boyle EA, Li YI, Pritchard JK.** An expanded view of complex traits: From polygenic to omnigenic. Cell. 2017;**169**(7):1177–86.
77. **Lin L, Yee SW, Kim RB, Giacomini KM.** SLC transporters as therapeutic targets: Emerging opportunities. Nat Rev Drug Discov. 2015;**14**(8):543–60.

Chapter 17

Psychosocial stress and telomere regulation

Idan Shalev and Waylon J Hastings

17.1 Introduction

"He who is of calm and happy nature will hardly feel the pressure of age, but to him who is of an opposite disposition youth and age are equally a burden," Plato

Stress is universal; everyone experiences it. But as Plato hinted, perception and disposition are important factors in determining the magnitude of our response, noting that not everyone is equally burdened. More than 2,000 years after Plato, scientists are now seeking to answer, not only how stressful disposition affects our mind, but also how adverse challenges impact health and aging processes by looking at microscopic changes inside the cell. On a cellular level, stress is considered to be a disruption to homeostasis, the relatively stable equilibrium maintained across the multitude of processes within cells. As we get older, continued stress exposure can cause increased damage to individual cells, and eventually induce wear and tear on physiological systems. One factor that is both stress-responsive and has also received growing attention as a marker for cellular aging is telomere length (TL). In fact, the very discovery of telomeres was a result of experiments simulating cellular stress using radiation.

During the 1930s, Herman Muller and Barbara McClintock found they could force breakage and subsequent reformation of new chromosome products by irradiating fruit flies' DNA with X-rays (1, 2), a stressor which can result in DNA damage response and cell death. Interestingly, the products never contained end to end fusions of the original chromosomes, leading Muller to the following hypothesis: "The terminal gene must have a special function, that of sealing the end of the chromosome, so to speak, and that for some reason a chromosome cannot persist indefinitely without having its ends thus sealed. This gene may accordingly be distinguished by a special term, the telomere" (i.e. "end part" from Greek) (3). Subsequent work in the 1950s by Rebecca Gerschman and colleagues showed the same DNA damage induced by X-rays could also be produced through oxidative stress (i.e. the accumulation of highly reactive, oxygen-containing-molecules which are naturally produced by mitochondria during cellular respiration), resulting in decreased survival in mice (4).

This prompted biologists to consider how stress may contribute to aging at the cellular level. Denham Harmon followed Gerschman's work, proposing that cell death is the result of pathways triggered by accumulated oxidative damage to DNA (5). Later in the 1960s Hayflick and Moorhead discovered that cells have an inherent limit to their replicative capacity, the number of times a cell can divide. This threshold was later named the "Hayflick limit." They theorized that the limited capacity for normal cells to divide is an expression of aging, and that it determines the longevity of the organism (6), building on an idea first theorized in the late nineteenth century by August Weissman (112). In the early 1970s, Olovnikov and Watson implicated telomeres in this process in what they independently formalized as the "end replication problem," wherein 12–30 bp of telomeric DNA are lost with each division due to replication inefficiency (7, 8). Further division would be halted when telomeres become critically short, preventing additional loss of critical DNA when cells enter a state of replicative arrest (i.e. senescence). While most cells conform to the Hayflick limit of finite cell divisions, there were multiple reports that certain cells, such as stem cells, seemingly divided indefinitely, somehow circumventing the Hayflick limit. Thus, a protective mechanism must be at work at telomeric ends, at least in some cells.

The first clue came a decade later when Blackburn and Gall, working with the ciliate organism *Tetrahymena*, showed the end regions of chromosomes were not a gene as Muller hypothesized, but rather a series of 5'-TTGGGG-3' tandem repeats (9) (5'-TTAGGG-3' in humans). Later in the 1980s, Blackburn and Greider discovered a unique protein, telomerase, which adds telomeric repeats to the 3' ends of chromosomes during each cell division, providing a viable solution to the end replication problem (10). As opposed to stem cells, most somatic cells lack sufficient telomerase and as a consequence, telomeres progressively shorten with each cell division.

While no empirical evidence existed to implicate telomere shortening or oxidative damage as the cause of cellular senescence, two seminal studies in the 1990s provided some clarity. In the first, Harley and colleagues confirmed shortening of telomeres with age in human connective tissue cells (11). Once von Zglinicki's work showed exposure to higher levels of oxidative stress increases the rate of telomeric shortening *in vitro*, there was compelling evidence implicating both telomeres and cellular stress in the aging process (12).

Growing body of research over the last 20 years has continued to explore the methods and means of telomere regulation in response to stress, including stress induced by psychosocial factors in humans. The results of these studies are presented in this chapter. We begin with a brief overview of the link between telomeres and health, followed by a state of the science review of the underlying biological mechanisms of telomere regulation. We conclude with some of the open questions in the field and discuss future research directions.

17.2 General impacts of stress on telomeres and health

It is generally accepted that telomere length decreases with chronological age. However, the rate of shortening varies across the lifespan. Although we currently lack within-individual evidence, it is suggested that the greatest loss occurs in the first years of life, corresponding to the rapid growth rates and high production and turnover of cells, followed by a stable plateau into childhood and young adulthood and a gradual decline into old age (13). Beyond this natural erosion of telomeres by age, recent research has shown that the rate of shortening is influenced by physical, psychological, and social conditions, which invites the question; how strong are telomeres as an indicator and predictor of healthspan and lifespan?

Debate still rages on this question, particularly in the realm of gerontology. Based on measurements in peripheral blood leukocytes, some (14), but not all (15), studies have found inverse relationships between TL and mortality, such that shorter telomeres are associated with higher mortality rates. Still others find the association holds, but significantly weakens when controlling for inflammatory markers (16). Possible explanations for these mixed results are varied TL assays and DNA extraction procedures (17, 18), sample sizes, population composition, and statistical adjustments used in the analyses. In a recent population study, including 64,637 individuals, the authors reported significant association between short TL and all-cause mortality (19). Furthermore, the exceptionally long telomeres found in centenarians are associated with above average longevity and increased cognitive function (20). These individuals also display decreased incidence of physically taxing conditions such as obesity, diabetes, and heart disease.

Incidence rates of aging-related health conditions (e.g. cardiovascular disease, diabetes, and stroke) have consistently been shown to be associated with shorter TL, independent of conventional risk factors (21, 22). Further, poor lifestyle behaviors can lead to and intensify negative health outcomes. Smoking is one such example that has been associated with shorter telomeres in cross-sectional studies (23, 24). Another detrimental outcome associated with smoking, and particularly related to telomeres, is cancer. Two large-scale meta-analysis found shorter telomeres to be significantly associated with specific types of cancer (25, 26). It should be noted, however, that findings in this area are mixed. Retrospective studies tend to report shorter telomeres in leukocytes of those with a history or current diagnosis of cancer. By contrast, Mendelian randomization studies report increased cancer risk for those carrying variants associated with longer leukocyte telomere length (113, 114). Taken together, this may imply that longer telomeres predispose one to develop cancer, and that the bout with cancer prompts telomere shortening in turn.

While unhealthy behaviors and exposures can damage telomeres, positive health behaviors, on the other hand, can mitigate telomere loss. For example, endurance training and mindfulness meditation is associated with increased telomerase activity (27, 28, 115), suggesting a protective effect on the rate of telomere erosion. The association between mindfulness and telomerase activity was mediated by changes in feelings of perceived control and neuroticism, highlighting the relationship between psychological and physical well-being. A foundational work in this area was by Epel and colleagues in 2004, showing a relationship between the burden of caregiving and shorter telomeres, which highlighted for the first time the link between psychological stress and TL (29). Follow-up studies provided further support that caregivers of Alzheimer's patients and sisters of women with breast cancer displayed similarly shortened TL (30, 31).

A population which has been a focus of research regarding the effects of stress on telomeres are individuals who were exposed to early-life adversity. This population is of particular importance as early adversity is known to propagate across the lifespan and to powerfully affect health in later life (32, 33). Thus, there is an interest in the role of telomere erosion as a potential mediator of such adverse effects. Beginning with work in 2010 (34), numerous studies have reported the negative impacts of early-life stress on TL in adults (reviewed in 35, 36). In the first study of children, greater exposure to (poor quality) institutional care was associated with shorter TL in middle childhood (37). However, conclusions from these studies have been limited by their cross-sectional and retrospective design. Subsequent longitudinal research involving repeated telomere measurements by Shalev and colleagues provided critical prospective evidence Children exposed to multiple kinds of violence between age 5 and 10 years showed significantly more telomere erosion over time (i.e. by age 10) than did other children (38).

The fact that telomeres prove sensitive to adversity and predictive of health has resulted in their regard as a "biological clock" for studying accumulated cellular aging throughout the life course. Intriguingly, recent data suggest that this accelerated aging process begins very early in life. One small-sample study linked greater prenatal stress and shorter cord blood TL at birth (39), a finding which was recently replicated in another study using a prospective design over the whole course of gestation (40). This developmental origin of health and disease is known to have long-lasting effects on health and aging processes (41). In a cohort study by Shalev and colleagues, perinatal complications at birth were linked with two aging indicators in midlife, TL and perceived facial aging, independent of family history and social risks present before birth, and of life-course health (42). These associations raise questions of underlying mechanisms of how stress "gets under the skin" at the cellular level. In the sections to follow, we first review basic telomere regulation by intrinsic systems, before discussing stress-related mechanisms and their impact on telomere stability and regulation.

17.3 **Telomere length regulation by intrinsic systems**

The preceding section highlighted the versatile nature of telomeres as a biomarker for understanding the deleterious impact of various stressors on health and well-being. Given telomeres' association with these processes, extensive research has been conducted to elucidate biological mechanisms responsible for telomere regulation. In order to understand the biological embedding of stress, it is useful to first review the basic biology of the telomere system at the cellular level before turning our attention to regulation of telomeres by stress-related processes. Thus, in the ensuing section we highlight key *intrinsic regulators* of TL, beginning with baseline determinants (Figure 17.1).

17.3.1 **Baseline determinants**

Despite being composed of DNA bases, telomeres are not like genes. Their length varies between chromosomes and between cells (43), and continually changes within individuals across the lifespan. Even at birth, TL varies significantly across individuals as a result of heritable factors. A recent meta-analysis of 19,713 individuals found TL heritability level of 0.70 based on twins and family members' comparisons (44). A similar heritability level, 64%, was calculated in another recent twin study, with shared environmental experiences accounting for an additional 22% of TL variation (45). Studies are only beginning to examine which specific genetic variations can

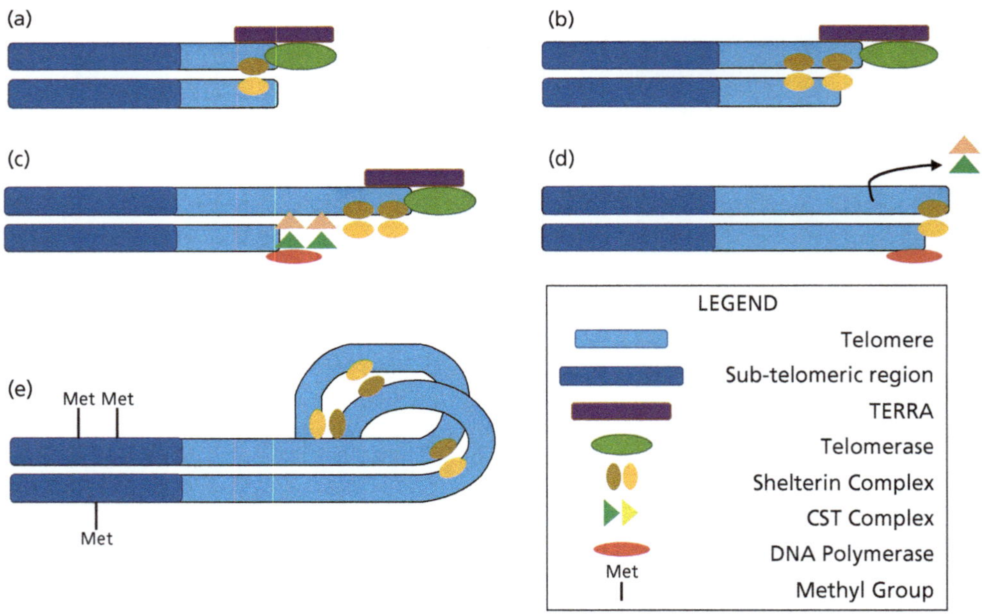

Figure 17.1 Schematic representation of telomere length regulation by intrinsic mechanisms. (A) TERRA assists in recruitment of telomerase to telomere while shelterin stabilizes DNA. (B) Telomerase elongates telomeres through addition of TTAGGG repeats to the leading strand. (C) CST complex competitively excludes shelterin complex by binding to single-stranded DNA and recruits DNA polymerase. (D) DNA polymerase elongates lagging strand of telomeric DNA, leaving a small 3' overhang to which shelterin proteins remain bound. (E) Shelterin proteins induce conformation of a T-loop structure to stabilize telomere ends. DNA is methylated in sub-telomeric regions, preventing further transcription of TERRA.

account for this relatively high heritability level of TL. For example, one study showed that variations in seven autosomal genes are associated with TL and disease (46). Interestingly, three of these genes are directly involved in the formation and activity of telomerase, while another two have been implicated in the function of telomere system proteins. Intriguingly, all seven genes are on different chromosomes, highlighting the complex genomic nature of TL regulation.

Parents influence TL as well, albeit in different ways. Studies reported significant correlation between mother and offspring TL (44, 47), suggesting a strong maternal influence on TL at birth. Paternal influence is more subtle. The strongest evidence has been for a paternal age at conception (PAC) effect, such that increased paternal age at conception is associated with longer telomeres in offspring (44, 48–50). This effect may even be cumulative across generations, as paternal grandfather's age at the time of father's conception is also significantly related to TL, independent of the paternal age effect (51). One potential explanation for this phenomenon is the observation of increased TL in sperm cells of older males (52). Despite genetic influences on TL *before* birth, twin-based studies have provided empirical support for environmental factors as the dominant influence *across the lifespan* (53). These environmental factors are translated to changes in TL through modulation of intrinsic regulation and stress-related processes.

17.3.2 **Telomerase and the telomeric repeat containing RNA (TERRA)**

As mentioned in section 17.1, telomerase is the master regulator of TL. The telomerase enzyme is a complex protein composed of four subunits (i.e. TERT, TERC, DKC1, and TEP1), which are encoded by four genes located on different chromosomes. All play important roles in telomere maintenance. Although telomerase can protect telomeres from erosion, most somatic cells lack sufficient expression of telomerase. The predominant cells which constitutively express telomerase are stem cells, which need be maintained throughout life to serve as a source for newly differentiated cells, and sperm cells, whose telomeres can actually elongate with age (52). In humans, telomerase adds TTAGGG repeats to the leading strand upon each round of cell division, allowing DNA polymerase a longer consensus sequence to replicate and elongate chromosome ends (54).

For many years the repeat rich telomeric sequence was thought to be a noncoding region similar to other "junk" regions in the genome. Yet, telomeres are transcribed into what is known as the telomeric repeat containing RNA (TERRA), which helps recruit telomerase to the shortest telomeres (55, 56). This ensures that the shortest telomeres in a given cell are prioritized for elongation. Recent work in mouse brain tissue and human lymphocyte cells has elaborated on the importance of TERRA, including a role in the immune response (57). When exported extracellularly, TERRA stimulates the expression of inflammatory cytokines and may cue cell degradation when unable to bind to critically short telomeres in the nucleus (57).

17.3.3 **Shelterin and CST complexes**

In addition to telomerase and TERRA, telomere intrinsic regulation involves several other proteins, namely the shelterin and CST complexes. Both complexes stabilize and cap the chromosome at various points during the cell division cycle. Specifically, the shelterin complex is a six-protein system (i.e. RAP1, TRF1, TRF2, TIN2, POT1, and TPP1) whose concerted activity promotes accessibility for telomerase and prevents end-to-end fusions of chromosomes (58). To accomplish this, shelterin proteins bind to and stabilize the local area of already replicated double-stranded DNA until telomerase can bind, while additional proteins are bound to the single-stranded DNA overhang to prevent fusion with other chromosomes during cell division (59). As cell division nears conclusion, shelterin complex proteins are competitively excluded by binding of the three-protein CST complex (i.e. CTC1, STN1, and TEN1), which prevents further elongation by telomerase (60). Additional function of the CST complex is the recruitment of DNA polymerase and ligase for elongation of the lagging strand, complementary to the telomerase-added repeats (61). Once elongation is complete, TERRA interacts with shelterin proteins to induce formation of a stabilizing structure called the T-loop (62). This folded-back structure prevents end-to-end fusions of chromosomes by enzymes responsible for normal repair of double-stranded DNA breaks (63).

17.3.4 Epigenetics

Recent research has shown increased impact of epigenetic regulation on TL. Epigenetic control refers to such processes as methylation and acetylation of DNA and histones, which in turn alters chromatin structure, transcription factor access, and subsequent gene expression. Epigenetic maintenance of TL is a dynamic process involving the regulation of TERRA levels and interactions with other stabilizing proteins in the shelterin and CST complexes (64). Importantly, epigenetic control is one pathway to produce drastically elongated telomeres, most likely through interactions between histones and proteins in the shelterin complex (64). Specifically, increased binding of shelterin proteins TRF1 and TRF2 to histone tails could increase recruitment of telomerase to telomeres (65). While telomere elongation may be perceived as a positive outcome, elongated telomeres cannot effectively form stabilizing T-loop structures, and thus are more prone to breakage or recombination with other chromosome ends. In fact, recombination-mediated telomere elongation (i.e. alternative lengthening of telomeres [ALT]) is a hallmark of certain cancer cells (66). By contrast, DNA methylation in the subtelomeric regions decreases TL, perhaps by inhibiting access to promoter regions to decrease transcription of TERRA (67). Furthermore, loss of methylation occurs in response to critically short telomeres, which may explain the age-associated prevalence of cancer as stem cells are near their Hayflick limit (68).

Future work will continue to elucidate the connection between epigenetics and telomeres. While these intrinsic processes, including variations in specific genes or the mode of parental inheritance can influence TL at birth, substantial influence results from environmental stressors across the lifespan. In the next section, we detail stress-related mechanisms that can further shorten telomeres and impair health.

17.4 Telomere length regulation by stress-related systems

Under normal conditions, the aforementioned intrinsic processes work in concert to maintain and protect telomeres. However, stress, by its very definition, is a disruption to the internal regulation of an organism, and the mechanisms governing TL are no exception. Multiple mechanisms are hypothesized to embed stress biologically in a manner that shortens telomeres. These embedding mechanisms involve complex interactions between stress-related systems and factors which, in turn, can increase the rate of telomere erosion and cellular senescence and promote aging-related diseases (Figure 17.2).

17.4.1 Neuroendocrine systems

Evidence linking shorter TL and increased telomere erosion with psychosocial stressors (35, 69) and internalizing disorders (24, 70, 71) calls attention, when trying to understand the process of telomere regulation, to the physiological stress systems, in particular the sympathetic–adrenal–medulla (SAM) axis of the autonomic nervous system and the hypothalamic–pituitary–adrenal (HPA) axis of the neuroendocrine system. Although the mechanistic link between systemic

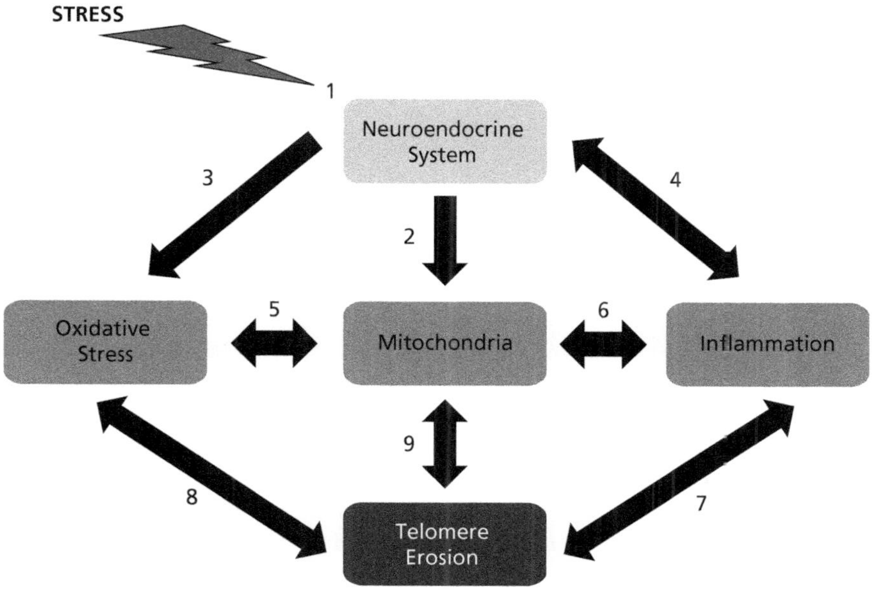

Figure 17.2 Schematic representation of telomere length regulation by stress-related mechanisms. (1) The human stress response initiated via the sympathetic–adrenal–medulla and hypothalamic–pituitary–adrenal axes stimulate the release of catecholamines and glucocorticoids into the periphery. (2) Cortisol release stimulates cellular metabolism through interaction with receptors in mitochondria. (3) Cortisol diminishes antioxidant production, disrupting their balance with free radicals, leading to oxidative stress within cells. (4) Chronic release of cortisol results in over-activation of the inflammatory response. Higher levels of inflammatory cytokines as a result of chronic inflammation also dampen diurnal rhythm of cortisol. (5) Increased cell metabolism increases production of reactive oxygen species (ROS) catalyzed by mitochondria. These can induce oxidative damage to cellular DNA, including mitochondrial DNA. (6) Under severe stress, damaged mitochondria release pro-inflammatory markers. Pro-inflammatory markers, in turn, can suppress mitochondrial function. (7) Inflammatory factors mark cells/tissues for degradation by macrophages. Damaged tissues are replenished by juvenile stem cells, decreasing TL as they divide and differentiate. Critically short telomeres activate senescent pathways, including release of inflammatory markers. (8) Free radicals permeate the nucleus to cause DNA damage, to which telomeres are especially susceptible. If left unrepaired, DNA damage triggers cell death pathways, further increasing stem-cell proliferation and TL erosion. Telomere-induced senescence increase levels of ROS. (9) During times of oxidative stress, TERT, the catalytic subunit of telomerase, localizes within mitochondria which can impair TL elongation. By contrast the age-associated decrease in mitochondrial activity can be rescued by enhancing TERT activity.

dysregulation of the HPA axis and TL is not entirely clear (36), empirical evidence suggests that chronic stress-induced secretion of the glucocorticoid cortisol down-regulates the activity of telomerase in lymphocyte cells, while increasing oxidative stress through mitochondrial dysregulation, which in turn leads to more rapid erosion of telomeres and, eventually, cellular senescence (72–75). Several studies of humans document significant associations between HPA axis indices, including physiological stress reactivity in children, and shorter TL (76–79). Furthermore, dysregulation along three stress pathways, namely the immune–inflammatory pathway, HPA axis, and autonomic nervous system were each associated with shorter TL in an additive manner (80). To understand further the mechanisms integrating these systems, we need to focus on processes in the periphery, specifically those associated with stress-related cellular metabolism and inflammation.

17.4.2 Inflammation

An important feature of the immune system implicated in the telomere erosion process is inflammation. The inflammatory response increases flow of leukocytes to injured tissues to degrade damaged cells or digest pathogens. Similarly, individually damaged cells release inflammatory cytokines to mark themselves for degradation. As the damaged tissue is removed, stem cells divide and differentiate to give rise to new cells, decreasing TL in the body's whole-cell reserve while simultaneously introducing naïve cells to the system with relatively long telomeres.

Glucocorticoids typically suppress the immune response during stress, redistributing energy to processes more vital to immediate survival. Part of this suppression involves increased release of anti-inflammatory markers, which also down-regulate the overall stress response as part of a negative feedback loop (81). This negative feedback is important to re-establish immune support for an organism's defense. However, when an individual is exposed to chronic stress, the immune system overcompensates, becoming overactive and resistant to normal regulatory measures, which can in turn prompt higher levels of inflammation, increased cell death, and stem cell depletion (82). For example, early-life stress, a "toxic" exposure linked to shorter TL, has been associated with elevated inflammation in adulthood (83). This heightened inflammation occurs in spite of higher than average levels of baseline cortisol, which should suppress inflammatory activity.

Models using human cell lines have also shown decreased telomerase activity in T-cells as an ancillary function of cortisol (73), an effect which likely holds *in vivo* (77). Therefore, chronic inflammation produces a "double-up" effect on TL erosion wherein cells are stimulated to divide as part of the inflammatory response, but have decreased telomerase activity to maintain TL. Moreover, an important feature of senescent cells, apart from growth arrest, is their increased secretion of inflammatory factors, such as interleukins 6 and 8. This feature is known as the senescence-associated secretory phenotype (SASP) (84). Thus, increased senescence rate resulting from accelerated telomere erosion goes on to increase the level of inflammatory markers, making clear that the inflammation–telomere–erosion relation is reciprocal rather than unidirectional. Over time, chronic low-grade inflammation, termed "inflammaging" (116), can damage tissues and accelerate aging.

The inflammation–telomere–erosion process is also influenced by the telomere position effect (TPE) (85). First observed in yeast, and later in human cells in 2001, TPE describes the reversible silencing of genes as a function of TL (86). While DNA damage responses can be triggered by critically short telomeres, TPE does not rely on very short TL. TPE can occur on multiple chromosomes affecting genes near and farther away from telomeres, and importantly, changes as a function of TL (87). Notably, these effects can influence inflammation as the *ISG15* gene, a contributor to the inflammatory response, is within the subset of genes found to be regulated by TPE (87).

17.4.3 Oxidative stress

While mobilization of energy increases an organism's ability to respond to adversity, reactive oxygen species (ROS) generated as a result of cellular metabolism expose the cell to oxidative stress. The G-rich DNA sequences common in telomeres are especially susceptible to oxidative cleavage as demonstrated by *in vitro* experiments showing increased telomere erosion when cells were exposed to high concentrations of ROS (88, 89). ROS production is also increased in senescent cells, providing more fuel for cellular damage and telomere erosion (90). Adding to this effect, oxidative damage also deters the renewal ability of stem cells (91).

To prevent this damage and potential cell death signaling cascades, cells generate antioxidants, such as glutathione, which reduce ROS in the cytosol, mitochondria, and nucleus to prevent DNA damage (92, 93). However, these defensive processes may be impaired when they are needed most. Specifically, animal models have demonstrated the detrimental effect of stress-induced glucocorticoids on antioxidants' ability to reduce reactive oxygens, which can further accumulate through interactions between glucocorticoids and mitochondria as discussed more broadly later in the chapter (94). As such, the balance between antioxidants and oxidative stress is an important measure of stress response efficacy. For example, Epel and colleagues (2004) used the ratio of ROS metabolites to the antioxidant vitamin E in their foundational work investigating TL in caregivers (29).

17.4.4 **Mitochondria**

Mitochondria are other important mediators to consider as these organelles participate in the stress response, in part by sensing levels of glucocorticoids (95). Glucocorticoids permeate the cell membrane, bind to intracellular receptors, and translocate within mitochondria to stimulate ATP production by oxidative phosphorylation. Although ROS are only produced in 1% to 3% of reactions catalyzed in mitochondria, any increase in activity, such as that associated with aging or the stress response, will increase ROS accumulation (96). While this alone can damage telomeres, recent studies provide empirical evidence for a more complex interaction between mitochondria, telomeres, and aging. For example, mitochondrial DNA exhibits a characteristic 5 kb deletion with increased age, and overall mitochondrial respiratory activity is decreased in aging tissues and in individuals with age-related disorders (97–100). Further research has shown higher mitochondrial DNA copy numbers, as well as shorter telomeres, co-occur in individuals with a history of childhood maltreatment (101).

In vitro studies on the oxidative stress response of human fibroblasts have further shown that TERT, the catalytic subunit of telomerase, is excluded from the nucleus and subsequently co-localizes with mitochondria (102). Similarly, the age-associated decrease in mitochondrial activity can be rescued by enhancing TERT activity *in vitro* (103). Taken together, these results imply that TERT may protect mitochondria during stress to allow production of needed ATP, but as a result can decrease the normative telomere elongation. This is in line with a life-history evolutionary perspective of favoring immediate over long-term survival in response to stress. Notwithstanding, protection of mitochondria is nevertheless important for long-term survival. Damaged mitochondria release markers triggering inflammatory and cell death responses similar to those resulting from nuclear DNA damage (74). As a result of their connection to each other and to stress-associated networks, a new theory has emerged purporting a mitochondrial–telomere axis of aging to describe the interconnectedness of the two structures within the aging process (104).

In sum, stress is a complex process with impacts that radiate throughout the human body. While stress functions as an adaptive response to promote short-term survival over gross longevity, under chronic conditions stress can increase stem cell division and differentiation, shorten TL, and eventually promote aging-related disorders such as cancer, diabetes, and cardiovascular disease. Here, we highlighted mechanistic pathways of several key components, including the neuroendocrine response, inflammation, oxidative stress, and mitochondrial control, and their respective impacts on telomere length and erosion.

17.5 Conclusion and future directions

The emergent field of telomere science has opened exciting new opportunities to study stress-related lifelong aging processes. Notably, stress in its various forms is suggested to influence the rate of telomere erosion throughout life and thus, elucidating the underlying mechanisms is of great interest. A better understanding can inform clinical treatments and intervention efforts to reverse the damaging effects of stress on aging processes. Considerable progress in the past two decades has increased our knowledge on the link between TL and health, as well as factors known to influence telomeres. These include intrinsic mechanisms (e.g. heritability, telomerase, TERRA, and epigenetics, to name a few), and biological embedding mechanisms, most likely through complex interaction between the neuroendocrine system, inflammation, oxidative stress, and mitochondrial regulation, although much remains to be illuminated about the mechanisms highlighted here. To put it in historical context, these intrinsic and stress related regulators of telomere length have been discovered in a time frame shorter than that between Muller's and Blackburn's seminal publications. Informed by the literature described herein, research over the next decade will undoubtedly uncover even more knowledge concerning stress, telomeres, and aging.

Future progress will require advancing methodological measurement techniques currently employed by scientists investigating telomeres. The current predominant technique for TL measurement is the quantitative PCR (qPCR) that provides an average length of the telomeric region across all cells in a given sample (105), but suffers from greater measurement error compared with other techniques (106). Other methods such as the "gold standard" Southern blot analysis can measure the distribution of TL within samples (107). Nevertheless, as TL can vary between the *same* chromosome and between *different* chromosomes of the same cell (43), and as telomere-induced senescence may rely on a few (or even single) critically short telomeres (108), future research will benefit from high-resolution and ideally high-throughput techniques. Further, while empirical evidence suggests TL to be correlated across different cell types of the same individual, new knowledge can emerge by measuring TL within specific types of cells and on specific chromosomes (109, 110). For example, how epigenetic processes influence TL regulation may vary across tissues, which inherently express different genes as a function of cell type. Moreover, the observation of "pseudo-lengthening" of telomeres, likely resulting from higher proportions of newly differentiated stem-cells, highlights the need for increased specificity when measuring TL *in vivo* (111). Furthermore, given the diversity of health outcomes associated with early-life adversity, and telomeres' ability to modulate gene expression in a chromosome-specific manner through TPE, future longitudinal studies may reveal associations between specific types of biopsychosocial stressors and chromosome-specific TL erosion. Finally, advancements in nanotechnology for nucleic acid analysis and DNA sequencing techniques will put us closer to finding factors underpinning Plato's old observation of inequality in stress and aging.

References

1. **Muller H.** The remaking of chromosomes. Collecting net. 1938;**13**(198):181–95.
2. **McClintock B.** The stability of broken ends of chromosomes in zea mays. Genetics. 1941;**26**(2):234–82.
3. **Muller HJ.** Studies in Genetics. Bloomington, IN: Indiana University Press; 1962.
4. **Gerschman R, Gilbert DL, Nye SW, Dwyer P, Fenn WO.** Oxygen poisoning and x-irradiation: a mechanism in common. Science. 1954;**119**(3097):623–6.
5. **Harman D.** Aging—a theory based on free-radical and radiation-chemistry. J Gerontol. 1956;**11**(3):298–300.
6. **Hayflick L, Moorhead PS.** The serial cultivation of human diploid cell strains. Exp Cell Res. 1961;**25**:585–621.

7. **Olovnikov AM.** Principle of marginotomy in template synthesis of polynucleotides. Dokl Akad Nauk SSSR. 1971;**201**(6):1496–9.
8. **Watson JD.** Origin of concatemeric T7 DNA. Nature-New Biol. 1972;**239**(94):197–201.
9. **Blackburn EH, Gall JG.** A tandemly repeated sequence at the termini of the extrachromosomal ribosomal RNA genes in tetrahymena. J Mol Biol. 1978;**120**(1):33–53.
10. **Greider CW, Blackburn EH.** Identification of a specific telomere terminal transferase activity in tetrahymena extracts. Cell. 1985;**43**(2 Pt 1):405–13.
11. **Harley CB, Futcher AB, Greider CW.** Telomeres shorten during aging of human fibroblasts. Nature. 1990;**345**(6274):458–60.
12. **von Zglinicki T, Saretzki G, Docke W, Lotze C.** Mild hyperoxia shortens telomeres and inhibits proliferation of fibroblasts—a model for senescence. Experimental Cell Research. 1995;**220**(1):186–93.
13. **Frenck RW, Jr., Blackburn EH, Shannon KM.** The rate of telomere sequence loss in human leukocytes varies with age. PNAS. 1998;**95**(10):5607–10.
14. **Cawthon RM, Smith KR, O'Brien E, Sivatchenko A, Kerber RA.** Association between telomere length in blood and mortality in people aged 60 years or older. Lancet. 2003;**361**(9355):393–5.
15. **Martin-Ruiz CM, Gussekloo J, van Heemst D, von Zglinicki T, Westendorp RGJ.** Telomere length in white blood cells is not associated with morbidity or mortality in the oldest old: A population-based study. Aging Cell. 2005;**4**(6):287–90.
16. **Glei DA, Goldman N, Weinstein M, Risques RA.** Shorter ends, faster end? Leukocyte telomere length and mortality among older Taiwanese. J Gerontol A Biol Sci Med Sci. 2015;**70**(12):1490–8.
17. **Cunningham JM, Johnson RA, Litzelman K, Skinner HG, Seo S, Engelman CD,** et al. Telomere length varies by DNA extraction method: Implications for epidemiologic research. Cancer Epidemiology Biomarkers & Prevention. 2013;**22**(11):2047–54.
18. **Hofmann JN, Hutchinson AA, Cawthon R, Liu C-S, Lynch SM, Lan Q,** et al. Telomere length varies by DNA extraction method: Implications for epidemiologic research—letter. Cancer Epidemiology Biomarkers & Prevention. 2014;**23**(6):1129–30.
19. **Rode L, Nordestgaard BG, Bojesen SE.** Peripheral blood leukocyte telomere length and mortality among 64,637 individuals from the general population. J Natl Cancer Inst. 2015;**107**(6):djv074.
20. **Atzmon G, Cho M, Cawthon RM, Budagov T, Katz M, Yang X,** et al. Evolution in health and medicine Sackler colloquium: Genetic variation in human telomerase is associated with telomere length in Ashkenazi centenarians. PNAS. 2010;**107** Suppl 1:1710–7.
21. **D'Mello MJJ, Ross SA, Briel M, Anand SS, Gerstein H, Pare G.** Association between shortened leukocyte telomere length and cardiometabolic outcomes systematic review and meta-analysis. Circ-Cardiovasc Gene. 2015;**8**(1):82.
22. **Haycock PC, Heydon EE, Kaptoge S, Butterworth AS, Thompson A, Willeit P.** Leucocyte telomere length and risk of cardiovascular disease: A systematic review and meta-analysis. BMJ. 2014;**349**:g4227.
23. **Valdes AM, Andrew T, Gardner JP, Kimura M, Oelsner E, Cherkas LF,** et al. Obesity, cigarette smoking, and telomere length in women. Lancet. 2005;**366**(9486):662–4.
24. **Shalev I, Moffitt TE, Braithwaite AW, Danese A, Fleming NI, Goldman-Mellor S,** et al. Internalizing disorders and leukocyte telomere erosion: A prospective study of depression, generalized anxiety disorder and post-traumatic stress disorder. Mol Psychiatry. 2014;**19**(11):1163–70.
25. **Ma H, Zhou Z, Wei S, Liu Z, Pooley KA, Dunning AM,** et al. Shortened telomere length is associated with increased risk of cancer: A meta-analysis. PLoS One. 2011;**6**(6):e20466.
26. **Wentzensen IM, Mirabello L, Pfeiffer RM, Savage SA.** The association of telomere length and cancer: A meta-analysis. Cancer Epidemiol Biomarkers Prev. 2011;**20**(6):1238–50.
27. **Jacobs TL, Epel ES, Lin J, Blackburn EH, Wolkowitz OM, Bridwell DA,** et al. Intensive meditation training, immune cell telomerase activity, and psychological mediators. Psychoneuroendocrinology. 2011;**36**(5):664–81.
28. **Schutte NS, Malouff JM.** A meta-analytic review of the effects of mindfulness meditation on telomerase activity. Psychoneuroendocrinology. 2014;**42**:45–8.

29. **Epel ES, Blackburn EH, Lin J, Dhabhar FS, Adler NE, Morrow JD**, et al. Accelerated telomere shortening in response to life stress. PNAS. 2004;**101**(49):17312–5.
30. **Damjanovic AK, Yang Y, Glaser R, Kiecolt-Glaser JK, Nguyen H, Laskowski B**, et al. Accelerated telomere erosion is associated with a declining immune function of caregivers of Alzheimer's disease patients. J Immunol. 2007;**179**(6):4249–54.
31. **Parks CG, Miller DB, McCanlies EC, Cawthon RM, Andrew ME, DeRoo LA**, et al. Telomere length, current perceived stress, and urinary stress hormones in women. Cancer Epidemiol Biomarkers Prev. 2009;**18**(2):551–60.
32. **Heim C, Nemeroff CB.** The role of childhood trauma in the neurobiology of mood and anxiety disorders: Preclinical and clinical studies. Biol Psychiatry. 2001;**49**(12):1023–39.
33. **Miller GE, Chen E, Parker KJ.** Psychological stress in childhood and susceptibility to the chronic diseases of aging: Moving toward a model of behavioral and biological mechanisms. Psychol Bull. 2011;**137**(6):959–97.
34. **Tyrka AR, Price LH, Kao HT, Porton B, Marsella SA, Carpenter LL.** Childhood maltreatment and telomere shortening: Preliminary support for an effect of early stress on cellular aging. Biol Psychiatry. 2010;**67**(6):531–4.
35. **Price LH, Kao HT, Burgers DE, Carpenter LL, Tyrka AR.** Telomeres and early-life stress: An overview. Biol Psychiatry. 2013;**73**(1):15–23.
36. **Shalev I.** Early life stress and telomere length: Investigating the connection and possible mechanisms: a critical survey of the evidence base, research methodology and basic biology. Bioessays. 2012;**34**(11):943–52.
37. **Drury SS, Theall K, Gleason MM, Smyke AT, De Vivo I, Wong JY**, et al. Telomere length and early severe social deprivation: Linking early adversity and cellular aging. Mol Psychiatry. 2012;**17**(7):719–27.
38. **Shalev I, Moffitt TE, Sugden K, Williams B, Houts RM, Danese A**, et al. Exposure to violence during childhood is associated with telomere erosion from 5 to 10 years of age: A longitudinal study. Mol Psychiatry. 2013;**18**(5):576–81.
39. **Entringer S, Epel ES, Lin J, Buss C, Shahbaba B, Blackburn EH**, et al. Maternal psychosocial stress during pregnancy is associated with newborn leukocyte telomere length. Am J Obstet Gynecol. 2013;**208**(2):134 e1–7.
40. **Marchetto NM, Glynn RA, Ferry ML, Ostojic M, Wolff SM, Yao R**, et al. Prenatal stress and newborn telomere length. Am J Obstet Gynecol. 2016;**215**(1):94–e1.
41. **Gluckman PD, Hanson MA, Cooper C, Thornburg KL.** Effect of in utero and early-life conditions on adult health and disease. N Engl J Med. 2008;**359**(1):61–73.
42. **Shalev I, Caspi A, Ambler A, Belsky DW, Chapple S, Cohen HJ**, et al. Perinatal complications and aging indicators by midlife. Pediatrics. 2014;**134**(5):e1315–23.
43. **Lansdorp PM, Verwoerd NP, van de Rijke FM, Dragowska V, Little MT, Dirks RW**, et al. Heterogeneity in telomere length of human chromosomes. Hum Mol Genet. 1996;**5**(5):685–91.
44. **Broer L, Codd V, Nyholt DR, Deelen J, Mangino M, Willemsen G**, et al. Meta-analysis of telomere length in 19,713 subjects reveals high heritability, stronger maternal inheritance and a paternal age effect. Eur J Hum Genet. 2013;**21**(10):1163–8.
45. **Hjelmborg JB, Dalgard C, Moller S, Steenstrup T, Kimura M, Christensen K**, et al. The heritability of leucocyte telomere length dynamics. J Med Genet. 2015;**52**(5):297–302.
46. **Codd V, Nelson CP, Albrecht E, Mangino M, Deelen J, Buxton JL**, et al. Identification of seven loci affecting mean telomere length and their association with disease. Nat Genet. 2013;**45**(4):422–7.
47. **Nawrot TS, Staessen JA, Gardner JP, Aviv A.** Telomere length and possible link to X chromosome. Lancet. 2004;**363**(9408):507–10.
48. **De Meyer T, Rietzschel ER, De Buyzere ML, De Bacquer D, Van Criekinge W, De Backer GG**, et al. Paternal age at birth is an important determinant of offspring telomere length. Hum Mol Genet. 2007;**16**(24):3097–102.

49. **Nordfjall K, Svenson U, Norrback KF, Adolfsson R, Roos G.** Large-scale parent-child comparison confirms a strong paternal influence on telomere length. Eur J Hum Genet. 2010;**18**(3):385–9.
50. **Unryn BM, Cook LS, Riabowol KT.** Paternal age is positively linked to telomere length of children. Aging Cell. 2005;**4**(2):97–101.
51. **Eisenberg DT, Hayes MG, Kuzawa CW.** Delayed paternal age of reproduction in humans is associated with longer telomeres across two generations of descendants. PNAS. 2012;**109**(26):10251–6.
52. **Kimura M, Cherkas LF, Kato BS, Demissie S, Hjelmborg JB, Brimacombe M,** et al. Offspring's leukocyte telomere length, paternal age, and telomere elongation in sperm. PLoS Genet. 2008;**4**(2):e37.
53. **Bakaysa SL, Mucci LA, Slagboom PE, Boomsma DI, McClearn GE, Johansson B,** et al. Telomere length predicts survival independent of genetic influences. Aging Cell. 2007;**6**(6):769–74.
54. **Harrington L.** Biochemical aspects of telomerase function. Cancer Lett. 2003;**194**(2):139–54.
55. **Arora R, Brun CM, Azzalin CM.** TERRA: Long noncoding RNA at eukaryotic telomeres. Prog Mol Subcell Biol. 2011;**51**:65–94.
56. **Cusanelli E, Romero CA, Chartrand P.** Telomeric noncoding RNA TERRA is induced by telomere shortening to nucleate telomerase molecules at short telomeres. Mol Cell. 2013;**51**(6):780–91.
57. **Wang Z, Deng Z, Dahmane N, Tsai K, Wang P, Williams DR,** et al. Telomeric repeat-containing RNA (TERRA) constitutes a nucleoprotein component of extracellular inflammatory exosomes. PNAS. 2015;**112**(46):e6293–300.
58. **Liu D, O'Connor MS, Qin J, Songyang Z.** Telosome, a mammalian telomere-associated complex formed by multiple telomeric proteins. J Biol Chem. 2004;**279**(49):51338–42.
59. **Smogorzewska A, van Steensel B, Bianchi A, Oelmann S, Schaefer MR, Schnapp G,** et al. Control of human telomere length by TRF1 and TRF2. Mol Cell Biol. 2000;**20**(5):1659–68.
60. **Chen LY, Redon S, Lingner J.** The human CST complex is a terminator of telomerase activity. Nature. 2012;**488**(7412):540–4.
61. **Wu P, Takai H, de Lange T.** Telomeric 3' overhangs derive from resection by Exo1 and Apollo and fill-in by POT1b-associated CST. Cell. 2012;**150**(1):39–52.
62. **Flynn RL, Centore RC, O'Sullivan RJ, Rai R, Tse A, Songyang Z,** et al. TERRA and hnRNPA1 orchestrate an RPA-to-POT1 switch on telomeric single-stranded DNA. Nature. 2011;**471**(7339):532–6.
63. **Griffith JD, Comeau L, Rosenfield S, Stansel RM, Bianchi A, Moss H,** et al. Mammalian telomeres end in a large duplex loop. Cell. 1999;**97**(4):503–14.
64. **Galati A, Micheli E, Cacchione S.** Chromatin structure in telomere dynamics. Front Oncol. 2013;**3**:46.
65. **Galati A, Micheli E, Alicata C, Ingegnere T, Cicconi A, Pusch MC,** et al. TRF1 and TRF2 binding to telomeres is modulated by nucleosomal organization. Nucleic Acids Res. 2015;**43**(12):5824–37.
66. **Cesare AJ, Reddel RR.** Alternative lengthening of telomeres: models, mechanisms and implications. Nat Rev Genet. 2010;**11**(5):319–30.
67. **Ng LJ, Cropley JE, Pickett HA, Reddel RR, Suter CM.** Telomerase activity is associated with an increase in DNA methylation at the proximal subtelomere and a reduction in telomeric transcription. Nucleic Acids Res. 2009;**37**(4):1152–9.
68. **Benetti R, Garcia-Cao M, Blasco MA.** Telomere length regulates the epigenetic status of mammalian telomeres and subtelomeres. Nat Genet. 2007;**39**(2):243–50.
69. **Shalev I, Entringer S, Wadhwa PD, Wolkowitz OM, Puterman E, Lin J,** et al. Stress and telomere biology: A lifespan perspective. Psychoneuroendocrinology. 2013;**38**(9):1835–42.
70. **Simon NM, Smoller JW, McNamara KL, Maser RS, Zalta AK, Pollack MH,** et al. Telomere shortening and mood disorders: Preliminary support for a chronic stress model of accelerated aging. Biol Psychiatry. 2006;**60**(5):432–5.
71. **Wolkowitz OM, Mellon SH, Epel ES, Lin J, Dhabhar FS, Su Y,** et al. Leukocyte telomere length in major depression: Correlations with chronicity, inflammation and oxidative stress--preliminary findings. PLoS One. 2011;**6**(3):e17837.
72. **Behl C, Lezoualc'h F, Trapp T, Widmann M, Skutella T, Holsboer F.** Glucocorticoids enhance oxidative stress-induced cell death in hippocampal neurons in vitro. Endocrinology. 1997;**138**(1):101–6.

73. **Choi J, Fauce SR, Effros RB.** Reduced telomerase activity in human T lymphocytes exposed to cortisol. Brain Behav Immun. 2008;**22**(4):600–5.
74. **Picard M, Juster RP, McEwen BS.** Mitochondrial allostatic load puts the "gluc" back in glucocorticoids. Nat Rev Endocrinol. 2014;**10**(5):303–10.
75. **Vartak S, Deshpande A, Barve S.** Reduction in the telomere length in human t-lymphocytes on exposure to cortisol. Curr Res Med Med Sci. 2014;**4**(2):20–5.
76. **Epel ES, Lin J, Wilhelm FH, Wolkowitz OM, Cawthon R, Adler NE**, et al. Cell aging in relation to stress arousal and cardiovascular disease risk factors. Psychoneuroendocrinology. 2006;**31**(3):277–87.
77. **Tomiyama AJ, O'Donovan A, Lin J, Puterman E, Lazaro A, Chan J**, et al. Does cellular aging relate to patterns of allostasis? An examination of basal and stress reactive HPA axis activity and telomere length. Physiology & Behavior. 2012;**106**(1):40–5.
78. **Kroenke CH, Epel E, Adler N, Bush NR, Obradovic J, Lin J**, et al. Autonomic and adrenocortical reactivity and buccal cell telomere length in kindergarten children. Psychosomatic Medicine. 2011;**73**(7):533–40.
79. **Gotlib IH, LeMoult J, Colich NL, Foland-Ross LC, Hallmayer J, Joormann J**, et al. Telomere length and cortisol reactivity in children of depressed mothers. Mol Psychiatry. 2015;**20**(5):615–20.
80. **Revesz D, Verhoeven JE, Milaneschi Y, de Geus EJ, Wolkowitz OM, Penninx BW.** Dysregulated physiological stress systems and accelerated cellular aging. Neurobiol Aging. 2014;**35**(6):1422–30.
81. **Tian R, Hou G, Li D, Yuan TF.** A possible change process of inflammatory cytokines in the prolonged chronic stress and its ultimate implications for health. Scientific World Journal. 2014;**2014**:780616.
82. **Schuettpelz LG, Link DC.** Regulation of hematopoietic stem cell activity by inflammation. Front Immunol. 2013;**4**:204.
83. **Danese A, Pariante CM, Caspi A, Taylor A, Poulton R.** Childhood maltreatment predicts adult inflammation in a life-course study. PNAS. 2007;**104**(4):1319–24.
84. **Merino A, Buendia P, Martin-Malo A, Aljama P, Ramirez R, Carracedo J.** Senescent CD14+CD16+ monocytes exhibit proinflammatory and proatherosclerotic activity. J Immunol. 2011;**186**(3):1809–15.
85. **Gottschling DE, Aparicio OM, Billington BL, Zakian VA.** Position effect at S. cerevisiae telomeres: reversible repression of Pol II transcription. Cell. 1990;**63**(4):751–62.
86. **Baur JA, Zou Y, Shay JW, Wright WE.** Telomere position effect in human cells. Science. 2001;**292**(5524):2075–7.
87. **Robin JD, Ludlow AT, Batten K, Magdinier F, Stadler G, Wagner KR**, et al. Telomere position effect: Regulation of gene expression with progressive telomere shortening over long distances. Genes Dev. 2014;**28**(22):2464–76.
88. **Henle ES, Han Z, Tang N, Rai P, Luo Y, Linn S.** Sequence-specific DNA cleavage by Fe2+-mediated fenton reactions has possible biological implications. J Biol Chem. 1999;**274**(2):962–71.
89. **von Zglinicki T.** Oxidative stress shortens telomeres. Trends Biochem Sci. 2002;**27**(7):339–44.
90. **Passos JF, Simillion C, Hallinan J, Wipat A, von Zglinicki T.** Cellular senescence: Unravelling complexity. Age (Dordr). 2009;**31**(4):353–63.
91. **Naka K, Muraguchi T, Hoshii T, Hirao A.** Regulation of reactive oxygen species and genomic stability in hematopoietic stem cells. Antioxid Redox Signal. 2008;**10**(11):1883–94.
92. **Masella R, Di Benedetto R, Vari R, Filesi C, Giovannini C.** Novel mechanisms of natural antioxidant compounds in biological systems: Involvement of glutathione and glutathione-related enzymes. J Nutr Biochem. 2005;**16**(10):577–86.
93. **Kurz DJ, Decary S, Hong Y, Trivier E, Akhmedov A, Erusalimsky JD.** Chronic oxidative stress compromises telomere integrity and accelerates the onset of senescence in human endothelial cells. J Cell Sci. 2004;**117**(11):2417–26.
94. **McIntosh LJ, Hong KE, Sapolsky RM.** Glucocorticoids may alter antioxidant enzyme capacity in the brain: Baseline studies. Brain Res. 1998;**791**(1-2):209–14.

95. **Manoli I, Alesci S, Blackman MR, Su YA, Rennert OM, Chrousos GP.** Mitochondria as key components of the stress response. Trends Endocrin Met. 2007;**18**(5):190–8.

96. **Valko M, Leibfritz D, Moncol J, Cronin MT, Mazur M, Telser J.** Free radicals and antioxidants in normal physiological functions and human disease. Int J Biochem Cell Biol. 2007;**39**(1):44–84.

97. **Arnold AS, Egger A, Handschin C.** PGC-1alpha and myokines in the aging muscle—a mini-review. Gerontology. 2011;**57**(1):37–43.

98. **Yen TC, Su JH, King KL, Wei YH.** Ageing-associated 5 kb deletion in human liver mitochondrial DNA. Biochem Biophys Res Commun. 1991;**178**(1):124–31.

99. **Cortopassi GA, Arnheim N.** Detection of a specific mitochondrial DNA deletion in tissues of older humans. Nucleic Acids Res. 1990;**18**(23):6927–33.

100. **Lin MT, Beal MF.** Mitochondrial dysfunction and oxidative stress in neurodegenerative diseases. Nature. 2006;**443**(7113):787–95.

101. **Tyrka AR, Parade SH, Price LH, Kao HT, Porton B, Philip NS,** et al. Alterations of mitochondrial DNA copy number and telomere length with early adversity and psychopathology. Biol Psychiatry. 2016;**79**(2):78–86.

102. **Ahmed S, Passos JF, Birket MJ, Beckmann T, Brings S, Peters H,** et al. Telomerase does not counteract telomere shortening but protects mitochondrial function under oxidative stress. J Cell Sci. 2008;**121**(7):1046–53.

103. **Sahin E, Colla S, Liesa M, Moslehi J, Muller FL, Guo M,** et al. Telomere dysfunction induces metabolic and mitochondrial compromise. Nature. 2011;**470**(7334):359–65.

104. **Sahin E, DePinho RA.** Axis of ageing: telomeres, p53 and mitochondria. Nat Rev Mol Cell Biol. 2012;**13**(6):397–404.

105. **Cawthon RM.** Telomere length measurement by a novel monochrome multiplex quantitative PCR method. Nucleic Acids Res. 2009;**37**(3):e21.

106. **Aviv A, Hunt SC, Lin J, Cao X, Kimura M, Blackburn E.** Impartial comparative analysis of measurement of leukocyte telomere length/DNA content by Southern blots and qPCR. Nucleic Acids Res. 2011:gkr634.

107. **Kimura M, Stone RC, Hunt SC, Skurnick J, Lu X, Cao X,** et al. Measurement of telomere length by the Southern blot analysis of terminal restriction fragment lengths. Nature Protocols. 2010;**5**(9):1596–607.

108. **Abdallah P, Luciano P, Runge KW, Lisby M, Geli V, Gilson E,** et al. A two-step model for senescence triggered by a single critically short telomere. Nature Cell Biology. 2009;**11**(8):988–93.

109. **Daniali L, Benetos A, Susser E, Kark JD, Labat C, Kimura M,** et al. Telomeres shorten at equivalent rates in somatic tissues of adults. Nat Commun. 2013;**4**:1597.

110. **Friedrich U, Griese E, Schwab M, Fritz P, Thon K, Klotz U.** Telomere length in different tissues of elderly patients. Mech Ageing Dev. 2000;**119**(3):89–99.

111. **Epel E.** How "reversible" is telomeric aging? Cancer Prev Res (Phila). 2012;**5**(10):1163–8.

112. **Weismann A.** Über Leben und Tod; eine biologische Untersuchung (2d. Aufl. ed.). Jena: G. Fischer; 1892.

113. **Haycock, PC, Burgess S, Nounu A, Zheng J, Okoli GN, Bowden J,** et al. Association between telomere length and risk of cancer and non-neoplastic diseases: A Mendelian randomization study. JAMA Oncology. 2017;**3.5**: 636–51. doi: 10.1001/jamaoncol.2016.5945.

114. **Zhang C, Doherty JA, Burgess S, Hung RJ, Lindstrom S, Kraft P,** et al. Genetic determinants of telomere length and risk of common cancers: A mendelian randomization study. Hum Mol Genet. 2015;**24**: 5356–66. doi: 10.1093/hmg/ddv252.

115. **Werner CM, Hecksteden A, Morsch A, Zundler J, Wegmann M, Kratzsch J,** et al. Differential effects of endurance, interval, and resistance training on telomerase activity and telomere length in a randomized, controlled study. European Heart Journal. 2018. doi:10.1093/eurheartj/ehy585.

116. **Franceschi C, Bonafè M, Valensin S, Olivieri F, De Luca M, Ottaviani E,** et al. Inflamm-aging: An evolutionary perspective on immunosenescence. Ann N Y Acad Sci. 2000;**908**(1), 244–54.

Chapter 18

Genetic effects on peripheral psychophysiological measures of emotion processing

Annette Conzelmann, Paul Pauli,
Alexander Prehn-Kristensen, and Tobias Renner

18.1 Chapter overview

The chapter begins with a description of two selected peripheral psychophysiological indicators of emotion processing which have been recently approached in genetic studies. First, the startle reflex and its affect modulation will be described, followed by the skin conductance level and skin conductance responses. Their suitability to detect genetic effects on emotion processing will be discussed, and studies of genetic effects on emotion processing assessed with startle reactivity and electrodermal activity will then be summarized. The chapter will end with an overview of implications for clinical research. The whole chapter focuses on studies in humans, and a detailed overview on animal studies can be obtained from other literature reviews (1).

18.2 Startle reflex and electrodermal activity to assess valence and arousal effects of emotion processing

18.2.1 Startle reflex

The startle reflex is a defensive reaction following a sudden short threatening stimulus. This reaction consists of a forward thrusting of the head and a descending flexor wave reaction. The most stable component is the blink reflex of the eyelids. In research, this blink reflex is usually elicited by a loud abrupt short noise and can be measured 30–50 ms thereafter via electromyography of the orbicularis oculi muscle. The startle reflex is based on a short brain circuit from the nucleus cochlearis to the nucleus reticularis pontis caudalis and then to spinal motor neurons. Psychophysiological evidence has indicated that the amplitude of the startle reflex is related to the individual's emotional state. The more the subject is in a state of negative emotion engaging the defensive system, the more the startle reflex is increased (2, 3). Accordingly, this reflex is an index of the organism's valence disposition. It is increased in anxiety patients probably because of their increased amygdala activity, which has been involved in startle modulation (4).

Furthermore, this reflex can be modulated by emotional foreground stimuli, a phenomenon which is called the affect modulation of the startle reflex. Reflexes with the same valence as an ongoing emotional state are augmented and inhibited during a mismatching emotional state, such that the defensive startle reflex is potentiated during unpleasant stimuli engaging the defensive motivational system as compared to neutral stimuli (startle potentiation). To the contrary, the startle reflex is inhibited during pleasant stimuli compared to neutral stimuli (startle inhibition) because the mismatching appetitive system is engaged. Mostly, pictures are used as foreground stimuli (3, 5). Furthermore, conditioned stimuli (CSs) are used, for example, in fear-conditioning paradigms where the subsequent startle potentiation is called fear-potentiated startle reaction (6). Startle inhibition by pleasant stimuli is influenced by the nucleus accumbens and startle potentiation by unpleasant stimuli by the amygdala (7, 8). This affect modulation of the startle reflex has been widely assessed and replicated (4). It is stronger when emotional stimuli are more arousing (9). Affect modulation of the startle reflex has been used to detect emotional deficits in patient samples with disturbances in the startle and emotion brain circuit, for example, within the amygdala, nucleus accumbens or the prefrontal cortex (4, 10–12). Accordingly, during unpleasant stimuli, anxious patients show an increased, and psychopaths a reduced startle potentiation (13, 14). Attention deficit hyperactive disorder (ADHD) patients are characterized by a reduced startle inhibition during the presentation of pleasant stimuli (10).

18.2.2 Electrodermal activity

Electrodermal activity or skin conductance reflects the arousal dimension of emotion processing. Skin conductance is controlled by the sympathetic nervous system, the hypothalamus, the brainstem, and especially the reticular formation and the nucleus coeruleus. Afferent neurons from the sympathetic axis innervate eccrine sweat glands, and their activity modulates the conductance of an externally applied current (15, 16). Skin conductance is increased when the sympathetic nervous system leads to increased secretion in the sweat glands. In most studies, skin conductance is measured with electrodes placed on the thenar and hypothenar areas of the palm. Skin conductance can be obtained during baseline measurements without any emotional stimulation and indicates the emotional arousal of an individual. This tonic level of electrodermal activity can be assessed as skin conductance level (SCL) or non-specific skin conductance responses (NS.SCRs) in response-free recording intervals (15). Elevated skin conductance baseline activity was associated with a worse course of schizophrenia and decreased activity was found in aggressive children and criminals (17–19).

Furthermore, we can assess phasic skin conductance responses to stimuli (SCRs) (17, 20). It has been widely shown that SCRs during emotionally pleasant and unpleasant stimuli are higher than during neutral stimuli because emotional stimuli are more arousing than neutral stimuli (21). SCRs are also increased during exposure to aversively conditioned stimuli in fear conditioning paradigms, as these elicit arousal reactions previously associated with the unconditioned stimulus (22). When the effects of stimuli on skin conductance are assessed in parallel with brain activity, higher-order brain structures are involved such as the amygdala, anterior cingulate cortex, hippocampus, insula, and dorsolateral and ventromedial prefrontal cortex (16). SCRs are higher in stressful or fearful situations (23) and SCRs are reduced in ADHD patients (24).

18.2.3 Usability of startle reflex and electrodermal activity to investigate gene effects on emotion processing

Although we have to be cautious and keep in mind that the affect-modulated startle may only be a valid measure for specific emotional stimuli (9, 25), both measures, the startle reflex and electrodermal activity, are prominent indicators of valence and arousal dimensions of emotion processing. Indeed, they are considered as important markers of psychiatric disorders (26). They might be less biased by social requirements than subjective reports of participants. They can be easily and cheaply assessed in large samples of healthy volunteers as well as patients with mental disorders. The latter is important given that we need large sample sizes to assess genetic effects on emotion processing. Both measures can be obtained in children and adults (4, 27–29).

Additionally, we have animal models for the startle reflex modulation (1). These measurements are also sensitive to pharmacological influences (30, 31). Furthermore, for both measures the brain circuits are known, which could help in understanding the underlying neural basis of potential genetic effects.

Therefore, the startle reflex and electrodermal activity might be well suited to investigate genetic effects on emotional processing. Electrodermal activity has been found to be stable and heritable (29, 32), especially the number of SCRs (33). However, for the startle reflex, there has been a discussion of whether this reflex is appropriate to detect genetic effects, and whether it is stable and heritable. On the one hand, there are studies revealing a moderate stability and heritability of baseline and the affect-modulated startle (34, 35). On the other hand, there are findings pointing to heritability of baseline startle response, but not its affect modulation (36). Heritability of the affect-modulated startle may be expected on the basis of animal studies (37), and because of its demonstrated association with trait factors and psychiatric disorders (4). Nevertheless, baseline startle reactivity without any further stimulation is considered to be a more stable indicator of genetic influences on emotion, as compared to affective stimulus-driven startle modulation (38).

Looking at the neurobiological basis, genotypes that modulate transmitter systems such as dopamine and serotonin for startle reactivity, or noradrenalin for electrodermal activity, might be promising. However, as indicated in section 18.3, other neurotransmitter systems may also be relevant. In the following, studies of the effects of genetic polymorphisms on startle reactivity, affect-modulated startle, and skin conductance responses are summarized.

18.3 **Genetic effects on emotion processing assessed with the startle reflex and electrodermal activity**

18.3.1 **Baseline startle reflex**

There are several studies on the serotonergic system. Concerning *5-HTT*, it was reported that carriers of the low-expressing S-allele showed increased amygdala activity during the processing of fearful stimuli (39) and also exhibited stronger overall startle responses across conditions than L-allele homozygotes (40, 41). In another study, smokers with the L/L genotype showed greater inhibition of the startle response when provided with nicotine versus placebo than those with the S/S or S/L genotypes (42). However, there are also conflicting results, which take into account an interaction effect with environmental stress (43). There was also an effect of *TPH2*, a gene regulating serotonin synthesis. It was found that in male T-allele carriers, startle amplitude was higher compared to male G/G homozygotes, while in females the effect was reversed and depended on the menstrual cycle. Interestingly, this was only found for younger adults, but not for children and older adults (44).

Other studies concentrated on the dopaminergic system. A study investigating the effects of the catechol-*O*-methyl-transferase polymorphism (*COMT*) in adults with the 22q11 deletion syndrome revealed that Met homozygotes (higher dopamine levels in the prefrontal cortex) showed reduced startle responses compared with Val homozygotes (45). Additionally, pregnant women with increased estradiol levels were found to show a higher startle reactivity if they were Val/Val homozygotes as compared to heterozygotes (46).

Furthermore, a polymorphism in the brain-derived neurotrophic factor gene (*BDNF*), which was associated with neural organization, depression, and anxiety (10, 47–49), was linked to startle reactivity, with reduced responses in Met-carriers (50).

Another study investigated effects of the gene coding for stathmin (*STMN1*). There was a STMN1 SNP1 × gender interaction effect with the female C/C-carriers showing stronger overall startle reactivity than the male C/C-carriers. Additionally, there was a significant STMN1 SNP2 × gender interaction effect on startle reactivity with the female T/T-carriers showing larger overall startle reactivity than the male T/T-carriers (51).

Schizophrenia patients carrying the neuropeptide S receptor gene (*NPSR1*) Ile107 variant, associated with higher neuropeptide S transmission, showed significantly reduced startle amplitudes compared to patients homozygous for the Asn107 variant and compared to healthy controls (52).

However, there are also negative results for genetic effects on startle reactivity for the transcription factor 4 (*TCF4*) gene, the serotonin-2A receptor gene polymorphism (*5-HT2AR*), *COMT*, adenosine A2A receptor gene (*ADORA2A*), Neuregulin 1 (*NRG1*), proline dehydrogenase (*PRODH*), and *5-HTT* (43, 53-58).

18.3.2 **Affect-modulated startle**

In relation to *5-HTT*, there is some indication that the 5-HTT S-allele might be associated with increased startle responses to threat (59). Another study investigated the impact of *5-HTT* and stressful life events on the affect-modulated startle response but revealed no genetic effect, and no gene–environment (G×E) interaction (40). One study showed that carriers of the *5-HTT* S-allele exhibited stronger fear potentiation of the startle reflex as compared to L/L-carriers (6).

Some studies on the dopaminergic system concentrated on the dopamine D4 receptor polymorphism (*DRD4*). Carriers of the 7R-allele associated with sensation seeking (60) exhibited blunted affect-modulated startle in comparison with the 7R-absent group (57, 61) associated with anxiety disorders (62). Concerning *COMT*, one study demonstrated that Met homozygotes exhibited a stronger startle potentiation towards unpleasant pictures (63). However, there are also other results reporting startle potentiation towards unpleasant stimuli only in Val-allele homozygotes, but not in Met-allele homozygotes. Met-allele homozygotes but not Val homozygotes displayed startle inhibition towards pleasant stimuli. Additionally, there was a G×E effect. Only COMT Val/Val homozygotes scoring high in childhood maltreatment showed startle potentiation towards unpleasant pictures and startle inhibition towards pleasant pictures (64). In a fear-conditioning paradigm, Met homozygotes showed more pronounced fear potentiation than Val-allele carriers, during the extinction phase (6). Another study investigated *COMT* effects on affect-modulated startle reactivity under L-dopa intervention. The Val-allele was associated with an increased startle potentiation by unpleasant stimuli independently of L-dopa or placebo intervention. Met homozygotes showed startle potentiation by unpleasant pictures only under L-dopa administration, but not in the placebo condition (65). Additionally, *BDNF* Met-carriers showed decreased fear-potentiated startle responses (22).

One study investigated the interaction between *NPSR* and the adenosine system (related to arousal and anxiety) manipulated through the administration of caffeine as adenosine A2A receptor antagonist, or placebo. AA/AT-carriers did not show differences in the affect-modulated startle response, but the more active TT variant exhibited increased startle amplitude in response to neutral stimuli and a decreased startle amplitude in response to unpleasant stimuli in the caffeine condition compared to the placebo condition (66). Furthermore, investigating the *ADORA2A* polymorphism, in female TT risk genotype carriers, the highest startle magnitudes were observed after caffeine administration in response to unpleasant stimuli (58).

The endocannabinoid (eCB) system is involved in various psychiatric disorders, such as mood disorders, addiction, pain, and anxiety disorders (67, 68). A study on the fatty acid amide hydrolase (*FAAH*) polymorphism revealed that A-carriers compared to C/C homozygotes were characterized by an increased startle potentiation towards unpleasant picture stimuli and reduced startle inhibition in response to pleasant pictures, possibly reflecting a drift into negative emotionality (69).

Another candidate gene implicated in risk for affective and anxiety disorders is the monoamine oxidase A (*MAOA*) gene. Women with the G/G genotype showed sustained reactivity to unpleasant stimuli during picture-offset (70).

There are also studies investigating gene × gene (G×G) effects in fear-conditioning paradigms. The *5-HTT* and the neuropeptide S receptor (*NPSR1*) polymorphisms were studied. Only carriers of both risk alleles (*5-HTT* S-allele and *NPSR1* T-allele) showed higher startle responses in the threat as compared to the safe condition (71). Another study investigated effects of *5-HTT* and the corticotropin-releasing hormone receptor 1 (*CRHR1*) polymorphisms. G-allele carriers showed reduced acquisition of cue-specific fear-conditioned responses compared with C/C-carriers. In addition, reduced cue potentiation of the G-allele carriers was strongest in the *5-HTT*/rs25531 group (72).

However, it must be noted that there are also results showing no genetic effects on the affect-modulated startle response or fear conditioning for the following genes: *DAT, DRD4, COMT, 5-HTT, TPH2, STMN1, DRD2, ANKK1 Taq I* (10, 40, 41, 44, 46, 50, 51, 61, 73, 74).

18.3.3 **Skin conductance responses**

There are fewer candidate gene studies investigating SCRs. With regards to the serotonergic system, *5-HTT* effects did not reach significance in two separate samples, but a post-hoc analysis across two samples indicated higher SRCs towards threat in the S/S group (59). Another study found that the extinction of conditioned fear assessed with SCRs was reduced in S-carriers of the *5-HTT* genotype (75).

With respect to the dopaminergic system, the D2 dopamine receptor gene (*DRD2*) was associated with differential conditioning of SCRs and an aversive priming effect (76). Another study on *BDNF* employing SCRs revealed that the Met-allele was associated with impaired fear extinction (77). In an anorexia subgroup analysis, *BDNF* Met-carriers had a significant increased frequency of SCRs during the processing of stimuli presenting women with underweight, in comparison to Val/Val homozygotes (78).

Concerning the oxytocin receptor polymorphism (*OXTR*), G/G individuals showed increased SCRs compared to A-allele carriers (79).

Additionally, there are results that indicate G×E effects. Although, SCRs towards stress were not affected by *5-HTT*, there was a significant interaction effect of *5-HTT* and attachment security on electrodermal skin conductance level. Children with a secure attachment and two L-alleles revealed a lower skin conductance level during a stress test (80). Furthermore, there are reports that S-allele carriers with a higher number of traumatic life events compared with L homozygotes showed stronger SCRs later during fear acquisition. Furthermore, *TPH2* T-allele carriers with a higher number of traumatic events showed stronger first interval electrodermal responses during the extinction phase, compared to non-carriers (81).

On the other hand, there are reports of no effects of gene polymorphisms on SCRs. There was no influence of *COMT, BDNF, 5-HTT,* NPSR1, the ankyrin repeat and kinase domain 1 (*ANKK1*) polymorphism and type 1 receptor adenylate cyclase activating polypeptide polymorphism (*ADCYAP1R1*) on SRCs during fear-conditioning paradigms (6, 22, 23, 46, 74, 76, 82). There were no *5-HTT* effects or interaction with life events on SCRs during a fear conditioning paradigm (83). *NPSR1* showed no impact on the skin conductance level during an emotional *n*-back task (84).

18.3.4 **Conclusion**

This review of the literature shows that there are promising results indicating genetic effects on baseline startle activity and affect-modulated startle, as well as electrodermal reactivity. Notably, studies on the former measure found relatively more genetic effects. However, there are also controversial results, which could be attributed to the use of small sample sizes, which may have contributed to the failure to detect small genetic effects (26). It must be also noted that there are only a few studies carried out in humans, which suggests that this remains a relatively unexplored research field, thus far. Furthermore, studies are often conducted by the same work groups and should be replicated in independent laboratories. Therefore, it is not clear whether we have a publication bias in candidate gene studies available to date. It would be important to replicate previous findings in large new samples and also to report non-significant findings. One way to achieve this goal is through multicenter studies. One other restriction is that the affect-modulated startle depends on the stimuli used. It is very important to use paradigms which show the expected valence modulation (9), which was not always the case in genetic studies of the affect-modulated startle response (23, 41, 44, 63, 70, 73). Studies should also consider the potential effects of gender, menstrual cycle, and age (44, 51). In addition, we are also far from an informative picture of G×G, G×E effects, and epigenetic effects. Of course, this requires very large samples and long-term data collection. Genome-wide association studies with large sample sizes would also be extremely helpful in detecting candidate genes. Indeed, using such an approach, one study detected one gene that might be promising for affect-modulated startle studies, the poly ADP-ribose polymerase 14 gene (*PARP14*), and another one for baseline startle activity (glutamate receptor, ionotropic, kainate 3, *GRIK3*) (38). Additionally, there is a large number of animal studies on startle, which could offer a starting point for future research designed to test pharmacogenetic effects or to describe the brain mechanisms underlying genetic influences (1). Animal startle studies have also allowed the investigation of gene knockout models (85). For future studies assessing startle or electrodermal activity, hypotheses about candidate polymorphisms could be derived based on studies of genetic effects on other dependent variables related to emotions (86). To sum up, the affect-modulated startle and skin conductance could be very useful in future genetic studies. However, we are currently still at the beginning of exploring this important scientific topic and therefore, future research is warranted.

18.4 Implications for the understanding of psychiatric disorders

As described earlier, startle reactivity and electrodermal activity are promising markers for emotion processing and their underlying genetics. These measures offer us the opportunity of understanding more about the effect of genes on emotion and their neurobiological mechanisms. In addition, the startle reflex and electrodermal activity have been found to be markers of psychiatric disorders such as psychopathy, anxiety disorders, or ADHD (4, 10, 24, 33). Therefore, these paradigms are ideally suited as phenotypic assessments that could show sensitivity to genetic effects in psychiatric disorders, and could also be investigated in interaction with psychopharmacological interventions. Furthermore, measures such as the startle reflex and electrodermal activity can be easily assessed, which is a great advantage for research in large samples of patients. To date, as indicated earlier, there are too few studies including patient samples. Hypotheses about potentially interesting genetic effects on emotion processing in psychiatric disorders, which could be tested using startle and electrodermal activity, could be derived by considering disturbances in emotion-related neurotransmitter systems and known candidate genes in these disorders. Complementary to this, it may also be possible to derive new hypotheses about candidate genes for psychiatric disorders when a gene is found to influence a specific disorder-associated pattern of startle or electrodermal response, such as exaggerated startle activity in anxiety disorders. In the future, longitudinal research in large sample of patients, based on multicentric efforts, may shed light in G×E effects on emotional problems and their psychophysiological correlates.

References

1. **Zhang L, Hu XZ, Li H, Li X, Smerin S, Benedek DM**, et al. Startle response related genes. Med Hypotheses. 2011;77(4):685–91.
2. **Cook EW, 3rd, Davis TL, Hawk LW, Spence EL, Gautier CH.** Fearfulness and startle potentiation during aversive visual stimuli. Psychophysiol. 1992;**29**:633–45.
3. **Lang PJ, Bradley MM, Cuthbert BN.** Emotion, attention, and the startle reflex. Psychol Rev. 1990;**97**:377–98.
4. **Grillon C, Baas J.** A review of the modulation of the startle reflex by affective states and its application in psychiatry. Clin Neurophysiol. 2003;**114**(9):1557–79.
5. **Lang PJ, Bradley MM, Cuthbert BN.** A motivational analysis of emotion: Reflex–cortex connections. Psychol Sci. 1992;**3**:44–9.
6. **Lonsdorf TB, Weike AI, Nikamo P, Schalling M, Hamm AO, Ohman A.** Genetic gating of human fear learning and extinction: Possible implications for gene–environment interaction in anxiety disorder. Psychol Sci. 2009;**20**(2):198–206.
7. **Koch M.** The neurobiology of startle. Prog Neurobiol. 1999;**59**:107–28.
8. **Koch M, Schmid A, Schnitzler HU.** Pleasure-attenuation of startle is disrupted by lesions of the nucleus accumbens. Neuroreport. 1996;**7**(8):1442–6.
9. **Cuthbert BN, Bradley MM, Lang PJ.** Probing picture perception: Activation and emotion. Psychophysiology. 1996;**33**(2):103–11.
10. **Conzelmann A, Mucha RF, Jacob CP, Weyers P, Romanos J, Gerdes AB**, et al. Abnormal affective responsiveness in attention-deficit/hyperactivity disorder: subtype differences. Biol Psychiatr. 2009;**65**(7):578–85.
11. **Plessen KJ, Bansal R, Zhu H, Whiteman R, Amat J, Quackenbush GA**, et al. Hippocampus and amygdala morphology in attention-deficit/hyperactivity disorder. Arch Gen Psychiatr. 2006;**63**(7):795–807.
12. **Scheres A, Milham MP, Knutson B, Castellanos FX.** Ventral striatal hyporesponsiveness during reward anticipation in attention-deficit/hyperactivity disorder. Biol Psychiatr. 2007;**61**(5):720–4.
13. **Patrick CJ.** Emotion and psychopathy: Startling new insights. Psychophysiology. 1994;**31**(4):319–30.
14. **Globisch J, Hamm AO, Esteves F, Öhman A.** Fear appears fast: Temporal course of startle reflex potentiation in animal fearful subjects. Psychophysiology. 1999;**36**:66–75.
15. **Boucsein W.** Electrodermal Activity. New York, NY: Plenum Press; 1992.
16. **Critchley HD.** Electrodermal responses: What happens in the brain. Neuroscientist. 2002;**8**(2):132–42.
17. **Schell AM, Dawson ME, Rissling A, Ventura J, Subotnik KL, Gitlin MJ**, et al. Electrodermal predictors of functional outcome and negative symptoms in schizophrenia. Psychophysiology. 2005;**42**(4):483–92.
18. **Posthumus JA, Böcker KBE, Raaijmakers MAJ, Van Engeland H, Matthys W.** Heart rate and skin conductance in four-year-old children with aggressive behavior. Biol Psychol. 2009;**82**:164–8.
19. **Raine A, Venables PH, Williams M.** Relationships between central and autonomic measures of arousal at age 15 years and criminality at age 24 years. Arch Gen Psychiatr. 1990;**47**(11):1003–7.
20. **De Pascalis V, Valerio E, Santoro M, Cacace I.** Neuroticism-anxiety, impulsive-sensation seeking and autonomic responses to somatosensory stimuli. Int J Psychophysiol. 2007;**63**(1):16–24.
21. **Lang PJ, Greenwald MK, Bradley MM, Hamm HO.** Looking at pictures: Affective, facial, visceral, and behavioral reactions. Psychophysiology. 1993;**30**:781–5.
22. **Lonsdorf TB, Weike AI, Golkar A, Schalling M, Hamm AO, Ohman A.** Amygdala-dependent fear conditioning in humans is modulated by the BDNFval66met polymorphism. Behav Neurosci. 2010;**124**(1):9–15.
23. **Pohlack ST, Nees F, Ruttorf M, Cacciaglia R, Winkelmann T, Schad LR**, et al. Neural mechanism of a sex-specific risk variant for posttraumatic stress disorder in the type i receptor of the pituitary adenylate cyclase activating polypeptide. Biol Psychiatr. 2015;**78**(12):840–7.

24. **Conzelmann A, Gerdes AB, Mucha RF, Weyers P, Lesch KP, Bahne CG**, et al. Autonomic hypoactivity in boys with attention-deficit/hyperactivity disorder and the influence of methylphenidate. World J Biol Psychiatr. 2014;**15**(1):56–65.

25. **Bernat E, Patrick CJ, Benning SD, Tellegen A.** Effects of picture content and intensity on affective physiological response. Psychophysiology. 2006;**43**(1):93–103.

26. **Vrieze SI, Malone SM, Pankratz N, Vaidyanathan U, Miller MB, Kang HM**, et al. Genetic associations of nonsynonymous exonic variants with psychophysiological endophenotypes. Psychophysiology. 2014;**51**(12):1300–8.

27. **van Goozen SH, Snoek H, Matthys W, van Rossum I, van Engeland H.** Evidence of fearlessness in behaviourally disordered children: A study on startle reflex modulation. J Child Psychol Psychiatry & Allied Disciplines. 2004;**45**(4):884–92.

28. **Lazzaro I, Gordon E, Li W, Lim CL, Plahn M, Whitmont S**, et al. Simultaneous EEG and EDA measures in adolescent attention deficit hyperactivity disorder. Int J Psychophysiol. 1999;**34**(2):123–34.

29. **Schell AM, Dawson ME, Nuechterlein KH, Subotnik KL, Ventura J.** The temporal stability of electrodermal variables over a one-year period in patients with recent-onset schizophrenia and in normal subjects. Psychophysiology. 2002;**39**(2):124–32.

30. **Kokkinidis L, MacNeill EP.** Potentiation of d-amphetamine and L-dopa-induced acoustic startle activity after long-term exposure to amphetamine. Psychopharmacology. 1982;**78**(4):331–5.

31. **Davis M.** Cocaine: Excitatory effects on sensorimotor reactivity measured with acoustic startle. Psychopharmacology. 1985;**86**(1-2):31–6.

32. **Crider A, Kremen WS, Xian H, Jacobson KC, Waterman B, Eisen SA**, et al. Stability, consistency, and heritability of electrodermal response lability in middle-aged male twins. Psychophysiology. 2004;**41**(4):501–9.

33. **Isen JD, Iacono WG, Malone SM, McGue M.** Examining electrodermal hyporeactivity as a marker of externalizing psychopathology: A twin study. Psychophysiology. 2012;**49**(8):1039–48.

34. **Carlson SR, Katsanis J, Iacono WG, McGue M.** Emotional modulation of the startle reflex in twins: Preliminary findings. Biol Psychol. 1997;**46**(3):235–46.

35. **Larson CL, Ruffalo D, Nietert JY, Davidson RJ.** Temporal stability of the emotion-modulated startle response. Psychophysiology. 2000;**37**(1):92–101.

36. **Anokhin AP, Golosheykin S, Heath AC.** Genetic and environmental influences on emotion-modulated startle reflex: A twin study. Psychophysiology 2007;**44**(1):106–12.

37. **Anisman H, Kelly O, Hayley S, Borowski T, Merali Z, McIntyre DC.** Acoustic startle and fear-potentiated startle in rats selectively bred for fast and slow kindling rates: Relation to monoamine activity. Eur J Neurosci 2000;**12**(12):4405–16.

38. **Vaidyanathan U, Malone SM, Miller MB, McGue M, Iacono WG.** Heritability and molecular genetic basis of acoustic startle eye blink and affectively modulated startle response: A genome-wide association study. Psychophysiology. 2014;**51**(12):1285–99.

39. **Hariri AR, Mattay VS, Tessitore A, Kolachana B, Fera F, Goldman D**, et al. Serotonin transporter genetic variation and the response of the human amygdala. Science. 2002;**297**(5580):400–3.

40. **Armbruster D, Moser DA, Strobel A, Hensch T, Kirschbaum C, Lesch KP**, et al. Serotonin transporter gene variation and stressful life events impact processing of fear and anxiety. Int J Neuropsychopharmacol. 2009;**12**(3):393–401.

41. **Brocke B, Armbruster D, Müller J, Hensch T, Jacob CP, Lesch KP**, et al. Serotonin transporter gene variation impacts innate fear processing: Acoustic startle response and emotional startle. Mol Psychiatr. 2006;**11**(12):1106–12.

42. **Minnix JA, Robinson JD, Lam CY, Carter BL, Foreman JE, Vandenbergh DJ**, et al. The serotonin transporter gene and startle response during nicotine deprivation. Biol Psychol. 2011;**86**(1):1–8.

43. **Verona E, Joiner TE, Johnson F, Bender TW.** Gender specific gene–environment interactions on laboratory-assessed aggression. Biological Psychol. 2006;**71**(1):33–41.

44. **Armbruster D, Mueller A, Strobel A, Kirschbaum C, Lesch KP, Brocke B.** Influence of functional tryptophan hydroxylase 2 gene variation and sex on the startle response in children, young adults, and older adults. Biol Psychol. 2010;**83**(3):214–21.

45. **de Koning MB, Boot E, Bloemen OJ, van Duin ED, Abel KM, de Haan L**, et al. Startle reactivity and prepulse inhibition of the acoustic startle response are modulated by catechol-*O*-methyl-transferase Val(158) Met polymorphism in adults with 22q11 deletion syndrome. J Psychopharmacol. 2012;**26**(12):1548–60.

46. **Comasco E, Hellgren C, Sundstrom-Poromaa I.** Influence of catechol-*O*-methyltransferase Val158Met polymorphism on startle response in the presence of high estradiol levels. Eur Neuropsychopharmacol. 2013;**23**(7):629–35.

47. **Huang EJ, Reichardt LF.** Neurotrophins: Roles in neuronal development and function. Annu Rev Neurosci. 2001;**24**:677–736.

48. **Rattiner LM, Davis M, Ressler KJ.** Brain-derived neurotrophic factor in amygdala-dependent learning. Neuroscientist. 2005;**11**(4):323–33.

49. **Verhagen M, van der Meij A, van Deurzen PA, Janzing JG, Arias-Vasquez A, Buitelaar JK**, et al. Meta-analysis of the BDNF Val66Met polymorphism in major depressive disorder: Effects of gender and ethnicity. Mol Psychiatr. 2010;**15**(3):260–71.

50. **Armbruster D, Muller-Alcazar A, Strobel A, Lesch KP, Kirschbaum C, Brocke B.** BDNF val(66)met genotype shows distinct associations with the acoustic startle reflex and the cortisol stress response in young adults and children. Psychoneuroendocrinology. 2016;**66**:39–46.

51. **Brocke B, Lesch KP, Armbruster D, Moser DA, Muller A, Strobel A**, et al. Stathmin, a gene regulating neural plasticity, affects fear and anxiety processing in humans. Am J Med Genet B Neuropsychiatr Genet. 2010;**153B**(1):243–51.

52. **Lennertz L, Quednow BB, Schuhmacher A, Petrovsky N, Frommann I, Schulze-Rauschenbach S**, et al. The functional coding variant Asn107Ile of the neuropeptide S receptor gene (NPSR1) is associated with schizophrenia and modulates verbal memory and the acoustic startle response. Int J Neuropsychopharmacol. 2012;**15**(9):1205–15.

53. **Quednow BB, Schmechtig A, Ettinger U, Petrovsky N, Collier DA, Vollenweider FX**, et al. Sensorimotor gating depends on polymorphisms of the serotonin-2A receptor and catechol-*O*-methyltransferase, but not on neuregulin-1 Arg38Gln genotype: A replication study. Biol Psychiatr. 2009;**66**(6):614–20.

54. **Quednow BB, Wagner M, Mossner R, Maier W, Kuhn KU.** Sensorimotor gating of schizophrenia patients depends on Catechol-*O*-methyltransferase Val158Met polymorphism. Schizophr Bull. 2010;**36**(2):341–6.

55. **Quednow BB, Kuhn KU, Mossner R, Schwab SG, Schuhmacher A, Maier W**, et al. Sensorimotor gating of schizophrenia patients is influenced by 5-HT2A receptor polymorphisms. Biol Psychiatr. 2008;**64**(5):434–7.

56. **Quednow BB, Ettinger U, Mossner R, Rujescu D, Giegling I, Collier DA**, et al. The schizophrenia risk allele C of the TCF4 rs9960767 polymorphism disrupts sensorimotor gating in schizophrenia spectrum and healthy volunteers. J Neurosci. 2011;**31**(18):6684–91.

57. **Roussos P, Giakoumaki SG, Bitsios P.** Cognitive and emotional processing in high novelty seeking associated with the L-DRD4 genotype. Neuropsychologia. 2009;**47**(7):1654–9.

58. **Domschke K, Gajewska A, Winter B, Herrmann MJ, Warrings B, Muhlberger A**, et al. ADORA2A Gene variation, caffeine, and emotional processing: A multi-level interaction on startle reflex. Neuropsychopharmacol. 2012;**37**(3):759–69.

59. **Klumpers F, Kroes MC, Heitland I, Everaerd D, Akkermans SE, Oosting RS**, et al. Dorsomedial prefrontal cortex mediates the impact of serotonin transporter linked polymorphic region genotype on anticipatory threat reactions. Biol Psychiatr. 2015;**78**(8):582–9.

60. **Benjamin J, Li L, Patterson C, Greenberg BD, Murphy DL, Hamer DH.** Population and familial association between the D4 dopamine receptor gene and measures of novelty seeking. Nat Genet. 1996;**12**:81–4.

61. **Pauli P, Conzelmann A, Mucha RF, Weyers P, Baehne CG, Fallgatter AJ**, et al. Affect-modulated startle reflex and dopamine D4 receptor gene variation. Psychophysiology 2010:25–33.
62. **Tochigi M, Hibino H, Otowa T, Kato C, Marui T, Ohtani T**, et al. Association between dopamine D4 receptor (DRD4) exon III polymorphism and Neuroticism in the Japanese population. Neurosci Lett. 2006;**398**(3):333–6.
63. **Montag C, Buckholtz JW, Hartmann P, Merz M, Burk C, Hennig J**, et al. COMT genetic variation affects fear processing: Psychophysiological evidence. Behav Neurosci. 2008;**122**(4):901–9.
64. **Klauke B, Winter B, Gajewska A, Zwanzger P, Reif A, Herrmann MJ**, et al. Affect-modulated startle: Interactive influence of catechol-*O*-methyltransferase Val158Met genotype and childhood trauma. PloS One. 2012;7(6):e39709.
65. **Domschke K, Winter B, Gajewska A, Unterecker S, Warrings B, Dlugos A**, et al. Multilevel impact of the dopamine system on the emotion-potentiated startle reflex. Psychopharmacology (Berl). 2015;**232**(11):1983–93.
66. **Domschke K, Klauke B, Winter B, Gajewska A, Herrmann MJ, Warrings B**, et al. Modification of caffeine effects on the affect-modulated startle by neuropeptide S receptor gene variation. Psychopharmacology (Berl). 2012;**222**(3):533–41.
67. **Parolaro D, Realini N, Vigano D, Guidali C, Rubino T.** The endocannabinoid system and psychiatric disorders. Exp Neurol. 2010;**224**(1):3–14.
68. **Sipe JC, Chiang K, Gerber AL, Beutler E, Cravatt BF.** A missense mutation in human fatty acid amide hydrolase associated with problem drug use. PNAS. 2002;**99**(12):8394–9.
69. **Conzelmann A, Reif A, Jacob C, Weyers P, Lesch KP, Lutz B**, et al. A polymorphism in the gene of the endocannabinoid-degrading enzyme FAAH (FAAH C385A) is associated with emotional-motivational reactivity. Psychopharmacology (Berl). 2012;**224**(4):573–9.
70. **Larson CL, Taubitz LE, Robinson JS.** MAOA T941G polymorphism and the time course of emotional recovery following unpleasant pictures. Psychophysiology. 2010;**47**(5):857–62.
71. **Glotzbach-Schoon E, Andreatta M, Reif A, Ewald H, Troger C, Baumann C**, et al. Contextual fear conditioning in virtual reality is affected by 5HTTLPR and NPSR1 polymorphisms: Effects on fear-potentiated startle. Front Behav Neurosci. 2013;**7**:31.
72. **Heitland I, Groenink L, van Gool JM, Domschke K, Reif A, Baas JM.** Human fear acquisition deficits in relation to genetic variants of the corticotropin-releasing hormone receptor 1 and the serotonin transporter—revisited. Genes Brain Behav. 2016;**15**(2):209–20.
73. **Nederhof E, Creemers HE, Huizink AC, Ormel J, Oldehinkel AJ.** L-DRD4 genotype not associated with sensation seeking, gambling performance and startle reactivity in adolescents: The TRAILS study. Neuropsychologia. 2011;**49**(5):1359–62.
74. **Torrents-Rodas D, Fullana MA, Arias B, Bonillo A, Caseras X, Andion O**, et al. Acquisition and generalization of fear conditioning are not modulated by the BDNF-val66met polymorphism in humans. Psychophysiology. 2012;**49**(5):713–9.
75. **Agren T, Furmark T, Eriksson E, Fredrikson M.** Human fear reconsolidation and allelic differences in serotonergic and dopaminergic genes. Transl Psychiatr. 2012;**2**:e76.
76. **Huertas E, Ponce G, Koeneke MA, Poch C, Espana-Serrano L, Palomo** T, et al. The D2 dopamine receptor gene variant C957T affects human fear conditioning and aversive priming. Genes Brain Behav. 2010;**9**(1):103–9.
77. **Soliman F, Glatt CE, Bath KG, Levita L, Jones RM, Pattwell SS**, et al. A genetic variant BDNF polymorphism alters extinction learning in both mouse and human. Science. 2010;**327**(5967):863–6.
78. **Clarke J, Ramoz N, Fladung AK, Gorwood P.** Higher reward value of starvation imagery in anorexia nervosa and association with the Val66Met BDNF polymorphism. Transl Psychiatr. 2016;**6**(6):e829.
79. **Smith KE, Porges EC, Norman GJ, Connelly JJ, Decety J.** Oxytocin receptor gene variation predicts empathic concern and autonomic arousal while perceiving harm to others. Soc Neurosci. 2014;**9**(1):1–9.

80. **Gilissen R, Bakermans-Kranenburg MJ, van Ijzendoorn MH, Linting M.** Electrodermal reactivity during the Trier Social Stress Test for children: Interaction between the serotonin transporter polymorphism and children's attachment representation. Dev Psychobiol. 2008;**50**(6):615–25.
81. **Hermann A, Kupper Y, Schmitz A, Walter B, Vaitl D, Hennig J**, et al. Functional gene polymorphisms in the serotonin system and traumatic life events modulate the neural basis of fear acquisition and extinction. Plos One. 2012;**7**(9):e44352.
82. **Raczka KA, Gartmann N, Mechias ML, Reif A, Buchel C, Deckert J**, et al. A neuropeptide S receptor variant associated with overinterpretation of fear reactions: A potential neurogenetic basis for catastrophizing. Mol Psychiatr. 2010;**15**(11):1045, 67–74.
83. **Klucken T, Alexander N, Schweckendiek J, Merz CJ, Kagerer S, Osinsky R**, et al. Individual differences in neural correlates of fear conditioning as a function of 5-HTTLPR and stressful life events. Soc Cogn Affect Neurosci. 2013;**8**(3):318–25.
84. **Guhn A, Domschke K, Muller LD, Dresler T, Eff F, Kopf J**, et al. Neuropeptide S receptor gene variation and neural correlates of cognitive emotion regulation. Soc Cogn Affect Neurosci. 2015;**10**(12):1730–7.
85. **Fujioka R, Nii T, Iwaki A, Shibata A, Ito I, Kitaichi K**, et al. Comprehensive behavioral study of mGluR3 knockout mice: Implication in schizophrenia related endophenotypes. Mol Brain. 2014;**7**:31.
86. **Gatt JM, Nemeroff CB, Dobson-Stone C, Paul RH, Bryant RA, Schofield PR**, et al. Interactions between BDNF Val66Met polymorphism and early life stress predict brain and arousal pathways to syndromal depression and anxiety. Mol Psychiatr. 2009;**14**(7):681–95.

Part 4

Disorders and therapy

Chapter 19

The genetics of personality/psychopathology: A brief review of constructs, results, approaches, and implications

Thomas J Bouchard, Jr, Wendy Johnson, and Irving I Gottesman[1]

"We shall never, probably, disentangle the inextricable web of affinities between the members of any one class; but when we have a distinct object in view, and we do not look to some unknown plan of creation, we may hope to make sure but slow progress."

Darwin, 1859

19.1 Introduction

This chapter selectively reviews recent work on the genetics, both quantitative and molecular, of personality/psychopathology. We use the conjunction "personality/psychopathology" because an often useful way to think of "psychiatric disorders" is as extremes of continuously measured (not necessarily normally distributed) personality traits (2). Put simply, we treat psychopathology as part of the personality "trait sphere" (3, 4). We do not deny that there may be some specific, rare behavioral traits that are discrete classes (5), but most currently offered examples do not hold up to closer scrutiny (6), and proof of existence of a taxon is a very difficult enterprise (7).

19.2 A brief exposition on psychometric traits and quantitative genetic methods

Quantitative measures of personality and psychopathology abound. Some instruments make use of brief statements describing a person's rather general psychological status, the Minnesota Multiphasic Personality Inventory (MMPI: 8, 9) being a prime example. Other measures make use of specific symptom counts gathered by interview, such as the Diagnostic Interview Schedule (10). Scores and scales can be derived in many ways (11) and the number of different scales derived from different combinations of items from the single MMPI item pool runs into the hundreds. Scales can also be constructed at several levels (facet, trait, higher-order factor) depending on the intended use (12). All such scales show evidence of genetic influence. This is usually quantified using heritability estimates. Until modern genotyping technology became available in the early 2000s such estimates were based on kinship studies for humans and breeding studies for other organisms.

1. Prof. Gottesman passed away on June 29, 2016, before the completion of this manuscript.

The heritability estimates for personality and psychiatric traits constitute standards against which estimates of genetic influence based on aggregating estimates of effects of individual genetic units (alleles) using modern genotyping techniques are compared. Currently the latter estimates fall far short of the former, leading to the so-called puzzle of "missing heritability" (13). In reality, what we have is not missing heritability but a new level of analysis and the new questions this always raises. There is now consensus that complex traits/diseases such as personality and psychopathology are highly polygenic, influenced by many genes of very small effect (14), with some genes contributing positively and others contributing negatively, sometimes even for the same trait in different individuals. In addition, the same genes may well contribute differently to different traits, and different genes often fill the same biological roles. As Wray and Maier (15) pointed out:

> Consideration of these factors can quickly lead to philosophical musings of the definition of disease, since even for a single genetic disease under a polygenic model of disease, each individual could carry a unique portfolio of risk loci. In the genomics era, a disease definition may be at the pathway level, whereby a single genetic disease considers different portfolios of risk loci impacting the same pathway, or, more practically, the class of individuals who respond to the same treatment. (p. 225)

The concept of heritability is often misunderstood. Heritability is *not* an intrinsic feature of a trait or condition. Without going into details, heritability provides an estimate of the relative importance of genetic and environmental influences on trait variation in the very broadly construed overall environmental conditions experienced by the particular *population* under study. It is a population statistic involving relative proportions of variance, with no relevance to mean trait levels in that population, to any particular individuals within that population, or to absolute magnitudes of population variance (which also can and do vary from sample to sample). Most studies of genetic influence have involved populations living in what might be called "average expectable environments" or normal ranges of environments in so-called developed nations. Consequently, findings reported apply to those environments and may not apply to more extreme or qualitatively different environments. High heritability does not imply genetic determination. Johnson, Penke, and Spinath (16) provide a thoughtful discussion of numerous misconceptions regarding heritability.

19.3 **Quantitative genetic analysis of measures of personality/psychopathology**

Recently, Polderman and colleagues (17) carried out a meta-analysis of the heritability of human traits (physical and psychological) based on 50 years of twin studies (1958–2012; among 3,404 traits, 764 studies of traits labeled "Psychiatric" and 1,774 traits, 280 studies of traits labeled "Temperament and Personality Functions"). Meta-analysis is intended to hone in on "the" size of some quantitative factor, so this study was in some ways misuse of the technique. Even individual traits do not have intrinsic heritabilities, and, even if they did, there would be no reason to suspect that groupings of different traits should have similar or the same heritabilities. Still, the results offer a comprehensive catalog of the ranges of heritabilities commonly observed across many different traits.

This catalog suggests considerable consistency for human traits. Across the Personality/ Psychiatric category, Polderman and colleagues (17) reported an average heritability (h^2) of 0.41 and a common (shared) family environmental influence (c^2) of 0.16 for females and values of 0.41 and 0.17 for males. Due to the large sample sizes, the standard errors of these estimates were tiny. Separate compilations by sex were very similar for all human trait categories. However, such overall averages, or even within-trait category averages, cannot tell us much that really matters. What would be of primary importance in understanding their relevance to the underlying transactions between genetic and environmental influences on trait development would be the sample-to-sample and study-to-study variances in these statistics and any specific genetic factors and/or environmental circumstances that contributed systematically to these variances. If these were small, we would infer something quite different about consistency of manifestation of genetic similarities than if they were large. The study did not offer any information about this.

The compilations in the Polderman and colleagues (17) paper were also limited in focusing on samples of twins reared together. Estimates of genetic and environmental variance components from such studies rely on specific assumptions about degrees of genetic and environmental relationships between mono- and dizygotic co-twins, independence of genetic and environmental influences, and degrees to which twins and their environments can be considered typical of those of much more commonly occurring singleton births. A good way to check their validities is to estimate the same quantities in samples from the same populations with different degrees of genetic and environmental relationships, although doing this is rare. One place where it has been done, however, is Minnesota. Table 19.1 provides a summary of results from two large Minnesota projects making use of the Multidimensional Personality Questionnaire (18).

Table 19.1 Heritability estimates from a multiple group design by gender and intraclass correlations for monozygotic and dizygotic twins reared apart (MZA, DZA) and reared together (MZT, DZT).

	Heritabilities from multiple group behavior genetic design			**Reared Apart twin intralcass correlations**		**Reared Together twin intraclass correlations**	
Scales							
	Women	**Men**	**Mean**	**MZA**	**DZA**	**MZT**	**DZT**
	@5,000	**@3,400**	**@8,500**	***N* = 74**	***N* = 54**	***N* = 626**	***N* = 631**
Primary scales							
Well-being	0.40	0.40	0.40	0.50	0.20	0.45	0.11
Social potency	0.54	0.53	0.54	0.54	0.02	0.59	0.27
Achievement	0.38	0.32	0.35	0.33	0.16	0.37	0.13
Social closeness	0.47	0.44	0.46	0.44	0.29	0.49	0.21
Stress reaction	0.45	0.43	0.44	0.47	0.42	0.45	0.17
Alienation	0.39	0.61	0.50	0.35	0.26	0.45	0.15
Aggression	0.39	0.35	0.37	0.40	0.22	0.38	0.13
Control	0.33	0.47	0.40	0.46	–0.03	0.41	0.19
Harm avoidance	0.45	0.46	0.46	0.45	0.21	0.47	0.21
Traditionalism	0.55	0.52	0.54	0.52	0.32	0.61	0.33
Absorption	0.44	0.26	0.35	0.56	0.24	0.42	0.15
Mean	0.44	0.44	0.44	0.46	0.21	0.46	0.19
Higher-order factors							
Positive emotionality	0.48	0.53	0.51	0.43	-0.02	0.55	0.17
Negative emotionality	0.42	0.48	0.45	0.47	0.40	0.44	0.13
Constraint	0.53	0.50	0.52	0.58	-0.01	0.56	0.28
Mean	0.48	0.50	0.49	0.50	0.12	0.52	0.19

Note: The DZA and DZT samples include both same sex and opposite sex twins. The data were derived from multiple tables in multiple studies conducted in Minnesota.

Besides 1,252 monozygotic twins and 1,263 dizygotic twins reared together (MZT, DZT individuals), the Finkel and McGue (19) study included 495 parents, 333 siblings, 1,690 spouses, and 535 adult offspring of the twins. Estimates of genetic and environmental variance components using the MZT and DZT correlations in these data were very consistent in aggregate with those from the monozygotic twin correlations in the Minnesota Study of Twins Reared Apart (MISTRA) (20). The monozygotic twin reared apart (MZA) correlation directly estimates heritability though it relies on somewhat different (though no less specific) assumptions (21). This consistency suggests that the assumptions underlying both study designs do not distort the estimates.

In contrast with Polderman and colleagues' (17) summaries for a mixture of psychopathological and more ordinary personality characteristics, which reported some common family environmental influences, the Minnesota MPQ results with "normal range" personality traits suggested very modest, if any, such influences. This has been typical of normal range personality results. Perhaps measures of psychopathology contain more such influence than measures of normal range personality. The heritability of major depression, no matter how it is assessed, however, is generally reported to be about 0.40 with very little common environmental influence (22). A recent large twin study replicated the absence of common environmental influences for its full sample (heritability of 0.52), but application of more sophisticated statistical methods suggested that heritabilities differed by sex, with males showing common environmental influences of 0.22 and heritability of 0.35, and females showing values of 0.01 and 0.54 (23). This study demonstrated that quantitative genetic methods continue to evolve, that individual estimates can deviate considerably from the Polderman and colleagues (17) reported averages, and that many findings in the field should be considered tentative.

19.4 The psychopathology "*p*" factor

One of the most interesting and perplexing problems in psychiatric diagnosis has been the common occurrence of comorbidity (24, 25), the apparent appropriateness of multiple psychiatric diagnoses for many patients. Recently the categorical approach to psychopathology that generates this comorbidity has yielded to a more empirical, quantitative, scale-based factor approach (26) that articulates "dimensions" or "axes" along which dispositional vulnerabilities vary subject to underlying genetic and environmental influences. Under this newer model comorbidity arises because these psychopathological dimensions or axes are themselves interrelated (correlated). Caspi and colleagues (67) and Krueger and Markon (68) have argued that, parallel to the *g* factor in cognitive abilities, there is a general *p* factor in psychopathology. As with the *g* factor, the existence of a general *p* factor does not negate the importance of more specific, lower-order factors in addressing important scientific and applied problems such as variations in efficacies of psychopharmacological agents and treatments based on behavior among patients exhibiting similar clinical features. The suggested *p* factor does not fully explain either overall rates of comorbidity or their tendencies to involve some groupings of disorders more than others; residuals specific to individual scales or diagnoses also offer explanation. Caspi and colleagues studied symptoms gathered from longitudinal interviews while Krueger and Markon studied multiple independent replications of cross-sectional questionnaire data, but their resulting models were very similar. Their *p* factors contributed directly to two major factors both groups labeled Internalizing (*Int*) and Externalizing (*Ext*). Both projects also found evidence for a thought-disorder factor, but it tended to collapse into the other domains. We ignore this important but unresolved issue here. There is now a wide consensus regarding the validity and utility of the *p* factor (27, 28). A symptom/network theory consistent with the *p* factor has been put forward by Borsboom (29, p. 7) and by McNally's group (30).

19.5 Molecular genetics: Genome-wide association studies (GWAS) and "missing heritability"

Given pervasive evidence of genetic influence on personality/psychopathology-related traits, their impact on people's lives, and burgeoning availability of cost-effective technology to examine the genomes of large samples of people directly, there is high interest in identifying the specific genetic variants that contribute to these traits. The method most commonly used in recent years to do this has been the Genome-Wide Association Study (GWAS). A GWAS correlates hundreds of thousands of single-nucleotide polymorphisms (SNPs, sequences of DNA varying relatively frequently in human populations) with the traits or diagnostic measures of interest across the genomes of population samples often running into the tens and even hundreds of thousands. Results from these studies have overwhelmingly fed the emerging consensus that these traits are influenced by very large numbers of genes with very small effect sizes and that the same genes influence or are involved in many different disorders. This is consistent with the notion that the diagnostic psychopathological categories are in reality dimensional (31). Typically, the associations (links between the SNPs or nearby genes and the traits) suggested in these studies involve genes expressed in the brain, suggesting plausible biological mechanisms (32, 33), although effect sizes for individual genes are always so small that the possibility that any indicated biological mechanism is necessary to cause the trait or condition is very remote.

As noted earlier, there are marked differences between quantitative estimates of heritability based on twin studies and GWAS estimates, the so-called missing heritability. The GWAS methods assume additive effects. Genes underlying GWAS markers form "genetic interaction networks" (34, Figure 4, provides a nice example), and Mackay (35) has shown "that additivity can be an emergent property of underlying genetic epistatic and interaction networks" (35, p. 22). Under these models, traits manifested by the simulated networks appear to be additive, but the underlying processes that were used to generate them involve epistatic and interactive aspects. Adding interaction terms to equations based on weighting associations identified in GWAS studies could make their estimates of genetic variance more consistent with those from quantitative genetic studies. There has been modest success in this area using pathway polygenic risk scores (PPRS) that weight the individual genes according not to the strengths of their statistical associations with the trait but according to the relative importance of the biological pathways they implicate. Genome-wide polygenic scores (GPS) are now approaching effect sizes found in some of the social and biological sciences (36, 37), although they involve such large numbers of individual genetic markers that most of them did not reach statistical significance in any sample and they tell us little more than do family-based "traditional" quantitative genetic studies

19.6 Genes and biological pathways (mechanisms)

The long-term goal of understanding how genes influence complex phenotypic personality traits will require specifying precisely how particular genes transact with environments to influence development of the mechanisms/pathways underlying traits. All the traits discussed here are also clearly influenced by environments. Studies of model organisms offer many examples of dependence of genetic expression on environmental circumstances, so understanding the natures of these gene–environment transactions in trait development and expression is critical.[2] Two theories of psychopathology offer examples of attempts to accomplish this: the "Feinberg Hypothesis"

2. The focus of this chapter has precluded discussion of the complexities introduced by both environmental influences and development. These topics have been dealt with elsewhere (64, 65).

involving neural mechanisms, and the Vascular Hypothesis involving pathophysiology. We chose these two examples not to pit them against each other but to emphasize that there are very likely multiple causal mechanisms influencing any existing *p* factor and they are likely biologically heterogeneous. Some causal mechanisms probably involve neural networks but coexisting ones may influence systems that support very general bodily functions (e.g. blood vessels, hormones, etc.). For example, genes influencing well-being (sometimes considered reversed *Int*) are expressed in the central nervous system as well as in adrenal and pancreatic tissues (38). Genes and environments may also become linked through repeated environmental exposures (self-selected or otherwise; passive or active gene–environment correlation). Moreover, just like *g*, *p* could be emergent rather than latent, which requires a very different theoretical framework within which to understand genetic influences (39).

The Feinberg Hypothesis suggests that schizophrenia is in part caused by a fault in the synaptic pruning normally occurring during adolescence. This is the critical period during which symptoms often first appear (40). Progressive cortical thinning, as psychosis develops, has been observed (41), and genes involved in this thinning have now been associated with schizophrenia (42) as well as bipolar disorder (43). There is also evidence that *de novo*, newly appearing, mutations influence genetic risk for schizophrenia by influencing synaptic networks (44).

The Vascular Hypothesis posits that abnormalities in genetic polymorphisms that regulate inflammatory responses interfere with "exquisitely precise regulation of the delivery of energy and oxygen required for normal brain function" (45). A corollary of this hypothesis is that environmental agents trigger inflammatory responses. Inflammatory influences on the brain, measured via retinal imaging, have now been related to both the *p* factor and IQ (46, 47). Genes regulating inflammatory response (among others) have also been related to risk of schizophrenia (48).

Potential involvement in specific personality/psychopathology traits of the pathophysiological mechanisms underlying these hypotheses can be investigated in considerable detail (49). Even if these mechanisms influenced behavior in only individual patients and only to small degrees, they would be causal mechanisms in the fundamental sense of the word. There may well be many coexisting mechanisms of this sort, each influenced by numerous genes, and the particular ones involved may differ among traits. Methods facilitating discovery and roles played by genes underlying complex traits are rapidly proliferating (50–52).

The Feinberg and Vascular Hypotheses were proposed before GWAS came into regular use. GWAS have merely offered evidence potentially consistent with them. Moreover, the manner in which they have done so suggests rather strongly that it is not any particular genes that matter in generating psychopathologies, but genes' aggregate involvement in emergent processes that somehow send biological pathways "off-course," with the "somehow" remaining a wide-open space in which environmental input is important. In addition, though we know many so-called risk genes are expressed in the human brain, we do not know the neuronal subtypes influenced by those genes (53).

19.7 **Closing thoughts from an evolutionary perspective: Complexity as a consequence of "descent with modification"**

Four large facts stand out from the body of research we have briefly reviewed. First, all complex traits are heritable. Second is pervasive comorbidity, both at the phenotypic and molecular levels. Third, these traits are influenced by a very large number of genes of very small effect. Finally, there is pleiotropy/epistasis, individual genes having multiple influences and interacting with other genes, in simultaneous transaction with the environment. How do these facts intermesh with evolutionary processes? Nature is a tinkerer and the brain is not an elegantly designed and optimized adaptive machine. The brain is made up of multiple layers of more recently evolved systems overlaid on older systems (54, chapter 3, 55). This creates problems of coordination among systems, but also a level of redundancy that can maintain phenotypic function despite disruption of individual operative pathways. Jacob (56) likened the human brain to a jet engine (cortex) mounted on an old horse cart (subcortical regions) and noted, "It is not surprising ... that accidents, difficulties, and conflicts, can occur" (56, p. 1166). Consequently, complex biological organisms and their brains lie somewhere between "design" and "bricolage" (57). That "somewhere" is a "kluge," a less-than-elegant mechanism, but "good enough" (58), at least for some period of time in some environments. Systems biology and analysis of modular networks (59) are being used to explicate brain networks. The connectome project's goal, for example, is to identify and map the networks that underlie brain disorders (60). It has recently indicated that the human cerebral cortex is structurally more complex than previously believed, articulating 97 new areas as well as the previously known 83 (61). These networks are also likely to be kludgy. Weiss and Buchanan (62) summarized the situation as follows;

> Much of life seems to be characterized by ad hoc, ephemeral, contextual probabilism without proper underlying distributions. To the extent that this is true, causal effects are not asymptotically predictable, and new ways of understanding life may be required. (62, p. 761).

Turkheimer (63) came to a similar "gloomy" conclusion for behavioral traits.

> What we will see instead is a proliferation of small, diverse, contingent findings that do not accumulate into coherent scientific theories. These will not be robust findings with large effect sizes; they will be the signature of a complex problem being addressed at the wrong level of analysis. They will be the keyless sidewalk under the genomic streetlight. (63, p. 28).

We do not disagree that the problem is complex and we may need "new ways of understanding life." We, however, side with Darwin and "hope to make sure but slow progress." There is no reason to doubt that how genes influence the multi-various pathways (networks/structures, etc.) that underlie "traits" can be worked out.[3] The various solutions will be messy (kludgy). One consequence of this messiness is that a given value on any particular trait (e.g. IQ) is likely to be achievable via different genetic and environmental pathways (equifinality) due to unique portfolios of loci living in unique environments. In an important sense any individual gene per se may or may not be involved, its role depending on a variety of other things (contextual probability). From this point of view, the status of every individual on every trait is to some extent emergent.

References

1. **Darwin C.** On the origin of the species by means of natural selection, or the preservation of favored races in the struggle for life: A facsimile of the first edition with an introduction by Ernst Mayr (1964). Cambridge, MA: Harvard University Press; 1859.
2. **Pettersson E, Mendle J, Turkheimer E, Horn EE, Ford DC, Simms LJ, Clark LA.** Do maladaptive behaviors exist at one or both ends of personality traits? Psychol Assess. 2014;**26**(2):433–46. doi:10.1037/a0035587.

3. Most of these pathways will be initially worked out using lower animals where powerful experimental manipulations can be implemented and causal mechanisms explicated (34, 66).

3. **Cattell RB.** The description of personality: Basic traits resolved into clusters. J Abnorm Soc Psychol. 1943;**38**:476–506. doi:10.1037/h0054116.
4. **Markon KE, Chmielewski M, Miller CJ.** The reliability and validity of discrete and continuous measures of psychopathology: A quantitative review. Psychol Bull. 2011;**137**(5):856–79. doi:10.1037/a0023678.
5. **Norris AL, Marcus DK, Green BA.** Homosexuality as a discrete class. Psychol Sci. 2015;**26**(12):1843–53. doi:10.1177/0956797615598617.
6. **Haslam N, Holland E, Kuppens P.** Categories versus dimensions in personality and psychopathology: A quantitative review of taxometric research. Psychol Med. 2012;**42**:903–20. doi:10.1017/S0033291711001966.
7. **Wilmot MP.** A contemporary taxometric analysis of the latent structure of self-monitoring. Psychol Assess. 2015;**27**(2):353–64. doi:10.1037/pas0000030.
8. **Tellegen A, Ben-Porath YS, McNulty JL, Arbisi PA, Graham JR, Kaemmer B.** MMPI-2 Restructured Clinical (RC) Scales: Development, Validation, and Interpretation. Minneapolis, MN: University of Minnesota Press; 2003.
9. **Trumbetta SL, Bolinskey PK, Gottesman I.** The MMPI-ARC and MMPI-AF scales: Psychometric properties of the MMPI-2-RF restructured clinical (RC) and higher order factor scales when adapted for use with archival adolescent MMPI data. Archives of Assessment Psychology; 2013;**3**:23–36.
10. **Robins LN, Cottler L, Bucholz KK, Compton W.** Diagnostic Interview Schedule for DSM-IV. St. Louis, MO: Washington University School of Medicine; 1995.
11. **Simms LJ.** Classical and modern methods of psychological scale construction. Soc Personal Psychol Compass. 2008;**2**(1):414–33. doi:10.1111/j.1751-9004.2007.00044.x.
12. **Condon DM.** An organizational framework for the psychological individual differences: Integrating the affective, cognitive, and conative domains. (PhD), Northwestern University, Evanston, IL; 2014.
13. **Nolte IM, van der Most PJ, Alizadeh BZ, de Bakker PI, Boezen HM, Bruinenberg M**, et al. Missing heritability: is the gap closing? An analysis of 32 complex traits in the Lifelines Cohort Study. Eur J Hum Genet. 2017. doi:10.1038/ejhg.2017.50.
14. **Chabris CF, Lee JJ, Cesarini D, Benjamin DJ, Laibson DI.** The fourth law of behavior genetics. Curr Dir Psychol Sci. 2015;**24**(4):304–12. doi:10.1177/0963721415580430.
15. **Wray NR, Maier R.** Genetic basis of complex genetic disease: The contribution of disease heterogeneity to missing heritability. Curr Epidemiol Rep. 2014;**1**(4):220–7. doi:10.1007/s40471-014-0023-3.
16. **Johnson W, Penke L, Spinath FM.** Heritability in the era of molecular genetics: Some thoughts for understanding genetic influences on behavioral traits. Eur J Personality. 2011;**25**:254–66. doi:10.1002/per.836.
17. **Polderman TJC, Benyamin B, De Leeuw CA, Sullivan PF, van Bochoven A, Visscher PM**, et al. Meta-analysis of the heritability of human traits based on fifty years of twin studies. Nat Genet. 2015;**47**, 702–9. doi:10.1038/ng.3285.
18. **Tellegen A, Waller NG.** Exploring personality through test construction: Development of the Multidimensional Personality Questionnaire. In: Boyle GJ, Matthews G, Saklofske D (eds). The SAGE Handbook of Personality Theory and Assessment: Vol. 2—Personality Measurement and Testing. Los Angeles, CA: Sage; 2008.
19. **Finkel D, McGue M.** Sex differences and nonadditivity in heritability of the Multidimensional Personality Questionnaire scales. J Pers Soc Psychol. 1997;**72**(4), 929–38.
20. **Segal NL.** Born Together-Reared Apart: The Landmark Minnesota Twin Study. Cambridge, MA: Harvard University Press; 2012.
21. **Bouchard TJ Jr, Lykken DT, McGue M, Segal NL, Tellegen A.** When kin correlations are not squared. Science. 1990;**250**:1498.
22. **Flint J, Kendler KS.** The genetics of major depression. Neuron. 2014;**81**:484–503. doi:10.1016/j.neuron.2014.01.027.

23. **Molenar D, Middeldorp CM, Willemsen G, Ligthart L, Nivard MG, Boomsma DI.** Evidence for gender-dependent genotype by environment interaction in adult depression. Behav Genet. 2016;**46**:59–71. doi:10.1007/s10519-015-9752-4.

24. **Bebbington P.** Categories, continua and the growth of psychiatric knowledge. Soc Psychiatry Psychiatr Epidemiol. 2015;**50**:507–10. doi:10.1007/s00127-015-1033-2.

25. **Moffitt TE, Caspi A, Taylor A, Kokaua J, Milne BJ, Polanczyk G, Poulton R.** How common are common mental disorders? Evidence that lifetime prevalence rates are doubled by prospective versus retrospective ascertainment. Psychol Med. 2010. **40**:899–909. doi:10.1017/s0033291709991036.

26. **Lahey BB, Krueger RF, Rathouz PJ, Waldman ID, Zald DH.** A hierarchical causal taxonomy of psychopathology across the life span. Psychol Bull. 2017;**143**(2): 142–86. doi:10.1037/bul0000069.

27. **Kim H, Eaton NR.** A hierarchical integration of person-centered comorbidity models: Structure, stability, and transition over time. Clin Psychol Sci. 2017;**5**(4):595–612. doi:10.1177/2167702617704018.

28. **Kotov R, Krueger RF, Watson D, Achenbach TM, Althoff RR, Bagby RM,** et al. The Hierarchical Taxonomy of Psychopathology (HiTOP): A dimensional alternative to traditional nosologies. J Abnorm Psychol. 2017;**126**(4):454–77. doi:10.1037/abn0000258.

29. **Borsboom D.** A network theory of mental disorders. World Psychiatry. 2017;**16**(1):5–13. doi:10.1002/wps.20375.

30. **Jones PJ, Heeren A, McNally RJ.** Commentary: A network theory of mental disorders. Frontiers in Psychology. 2017;8. doi:10.3389/fpsyg.2017.01305.

31. **Lee JJ, Vattikuti S, Chow CC.** Uncovering the genetic architectures of quantitative traits. Comput Struct Biotechnol J. 2016;**14**:28–34. doi:10.1016/j.csbj.2015.10.002.

32. **Hibar DP, Stein JL, Renteria ME, Arias-Vasquez A, Desrivieres S, Jahanshad N** et al. Common genetic variants influence human subcortical brain structures. Nature. 2015;**520**(7546):224–9. doi:10.1038/nature14101.

33. **Lee JJ, McGue M.** Why behavioral genetics matters: Comment on Plomin et al. (2016). Perspect Psychol Sci. 2016;**11**(1):29–30. doi:10.1177/1745691615611932.

34. **Shorter J, Couch C, Huang W, Carbone MA, Peiffer J, Anholt RR,** et al. Genetic architecture of natural variation in Drosophila melanogaster aggressive behavior. PNAS. 2015;**112**(27):E3555–63. doi:10.1073/pnas.1510104112.

35. **Mackay TF.** Epistasis and quantitative traits: using model organisms to study gene-gene interactions. Nat Rev Genet. 2014;**15**(1):22–33. doi:10.1038/nrg3627.

36. **Bouchard TJ,** Jr. Genes and human psychological traits. In: Carruthers P, Laurence S, Stich S (eds). The Innate Mind: Foundations for the Future, Vol. **3**. Oxford: Oxford University Press; 2007.

37. **Selzam S, Krapohl E, von Stumm S, O'Reilly PF, Rimfeld K, Kovas Y,** e al. Predicting educational achievement from DNA. Mol Psychiatry, 2016;**22**(2):267–72. doi:10.1038/mp.2016.107.

38. **Okbay A, Baselmans BM, De Neve JE, Turley P, Nivard MG, Fontana MA,** et al. Genetic variants associated with subjective well-being, depressive symptoms, and neuroticism identified through genome-wide analyses. Nat Genet. 2016;**48**(6):624–33. doi:10.1038/ng.3552.

39. **van der Maas HL, Dolan CV, Grasman RP, Wicherts JM, Huizenga HM, Raijmakers ME.** A dynamical model of general intelligence: the positive manifold of intelligence by mutualism. Psychol Rev. 2006;**114**(4):842–61. doi:10.1037/0033-295X.113.4.842.

40. **Feinberg I.** Schizophrenia: Caused by a fault in programmed synaptic elimination during adolescence? J Psychiatric Res. 1982–1983;**17**:3119–334.

41. **Cannon TD, Chung Y, He G, Sun D, Jacobson A, van Erp TG,** et al. Progressive reduction in cortical thickness as psychosis develops: A multisite longitudinal neuroimaging study of youth at elevated clinical risk. Biol Psychiatry. 2015;**77**(2):147–57. doi:10.1016/j.biopsych.2014.05.023.

42. **Sekar A, Bialas AR, de Rivera H, Davis A, Hammond TR, Kamitaki N,** et al. Schizophrenia risk from complex variation of complement component 4. Nature. 2016;**530**(7589):177–83. doi:10.1038/nature16549.

43. **Hanford LC, Nazarov A, Hall GB, Sassi RB.** Cortical thickness in bipolar disorder: a systematic review. Bipolar Disorders. 2016;**18**(1):4–18. doi:10.1111/bdi.12362.
44. **Fromer MP, Pocklington A, Kavanagh AJ, Williams HJ, Dwyer S, Gormley P**, et al. De novo mutations in schizophrenia implicate synaptic networks. Nature. 2014;**506**:179–84. doi:10.1038/nature12929.
45. **Hanson DR, Gottesman II.** Theories of schizophrenia: A genetic-inflammatory-vascular synthesis. BMC Medical Genetics. 2015;**6**:7. doi:10.1186/1471-2350-6-7.
46. **Meier MH, Shalev I, Moffitt TE, Kapur S, Keefe RS, Wong TY**, et al. Microvascular abnormality in schizophrenia as shown by retinal imaging. Am J Psychiatry. 2013;**170**(12):1451–9. doi:10.1176/appi.ajp.2013.13020234.
47. **Shalev I, Moffitt TE, Wong TY, Meier MH, Houts RM, Ding J**, et al. Retinal vessel caliber and lifelong neuropsychological functioning: Retinal imaging as an investigative tool for cognitive epidemiology. Psychol Sci. 2013;**24**(7):1198–207. doi:10.1177/0956797612470959.
48. **Andreassen OA, Harbo HF, Wang Y.** Genetic pleiotropy between multiple sclerosis and schizophrenia but not bipolar disorder: Differential involvement of immune-related gene loci. Mol Psychiatry. 2015;**20**:207–14. doi:10.1038/mp.2013.195.
49. **Ruzzo EK, Geschwind DH.** Schizophrenia genetics complements its mechanistic understanding. Nat Neurosci. 2016;**19**(4):523–5. doi:10.1038/nn.4277.
50. **Gamazon ER, Wheeler HE, Shah KP, Mozaffari SV, Aquino-Michaels K, Im HK**, et al. A gene-based association method for mapping traits using reference transcriptome data. Nat Genet. 2015;**47**:1091–8. doi:10.1038/ng.3367.
51. **Potkin SG, van Erp TGM, Ling S, Macciardi E, Xie X.** Identifying unanticipated genes and mechanisms in serious mental illness: GWAS-based imaging genetic strategies. In: KL Bigos, AR Hariri, DR Weinberger (eds), Neuroimaging Genetics: Principles and Practices. New York, NY: Oxford University Press; 2016 (141–156).
52. **Wang Y, Thompson WK, Schork AJ, Holland D, Chen CH, Bettella F**, et al. Leveraging genomic annotations and pleiotropic enrichment for improved replication rates in schizophrenia GWAS. PLoS Genetics. 2016;**12**(1):e1005803. doi:10.1371/journal.pgen.1005803.
53. **Lake BB, Ai R, Kaeser GE, Salathia NS, Yung YC, Liu R**, et al. Neuronal subtypes and diversity revealed by single-nucleus RNA sequencing of the human brain. Science. 2016;**352**(6293):1586–90. doi:10.1126/science.aaf1204.
54. **Allman JM.** Evolving Brains. New York, NY: Scientific American Library; 2000.
55. **Panksepp J.** Cross-species affective neuroscience decoding of the primal affective experiences of humans and related animals. PLoS One. 2011;**6**(9):e21236. doi:10.1371/journal.pone.0021236.
56. **Jacob F.** Evolution and tinkering. Science. 1977;**196**(4295):1161–6.
57. **Wilkins AS.** Between "design" and "bricolage": Genetic networks, levels of selection, and adaptive evolution. PNAS. 2007;**104**:8590–6. doi:10.1073pnas.0701044104.
58. **Markus G.** Kluge: The Haphazard Construction of the Human Mind. Boston, MA: Houghton Mifflin; 2008.
59. **Sporns O, Betzel RF.** Modular brain networks. Ann Rev Psycholog. 2016; **67**:613–40. doi:10.1146/annurev-psych-122414-033634.
60. **Fornito A, Zalesky A, Breakspear M.** The connectomics of brain disorders. Nat Rev Neurosci. 2015;**16**:159–72. doi:10.1038/nrn3901.
61. **Glasser MF, Coalson TS, Robinson EC, Hacker CD, Harwell J, Yacoub** E et al. A multi-modal parcellation of human cerebral cortex. Nature. 2016;**536**(7615):171–8. doi:10.1038/nature18933.
62. **Weiss KM, Buchanan AV.** Is life law-like? Genetics. 2011;**188**(4):761–71. doi:10.1534/genetics.111.131318.
63. **Turkheimer E.** Weak genetic explanation 20 years later: Reply to Plomin et al. (2016). Perspect Psychol Sci. 2016;**11**(1):24–8. doi:10.1177/1745691615617442.

64. **Bouchard TJ, Jr.** Experience producing drive theory: Personality "Writ Large." Personality and Individual Differences. 2016;**90**:302–14. doi:10.1016/j.paid.2015.11.007.
65. **Johnson W.** Developing Difference. Basingstoke: Palgrave Macmillan; 2014.
66. **Kukekova AV, Temnykh SV, Johnson JL, Trut LN, Acland GM.** Genetics of behavior in the silver fox. Mammalian Genome. 2012;**23**(1–2):164–77. doi:10.1007/s00335-011-9373-z.
67. **Caspi A, Houts RM, Belsky DW, Goldman-Mellor SJ, Harrington** H, et al. The p Factor: One general psychopathology factor in the structure of psychiatric disorders? Clin Psychol Sci. 2014; *2*(2): 119–37. doi:10.1177/2167702613497473.
68. **Krueger RF, Markon KE.** The role of the DSM-5 personality trait model in moving toward a quantitative and empirically based approach to classifying personality and psychopathology. Ann Rev Clin Psychol. 2014;**10**: 7.1–7.20. doi:10.1146/annurev-clinpsy-032813-153732.

Chapter 20

Resilience

Rebecca Alexander and Justine Megan Gatt

20.1 Introduction

The neurobiology of resilience is a field of surging importance in the move towards optimizing mental health and well-being. Resilience refers to the process of adaptive recovery from adversity and the ability to maintain optimal levels of well-being (e.g. life satisfaction, optimism, positivity, mastery) in the face of dynamic challenges to psychological resources (1). A trajectory of low resilience would place an individual at higher risk of developing psychiatric problems following adversity, with depression and anxiety being the most common disorders (2). From a neurobiological perspective, resilience has been defined as the process of adaptive and constructive use of the stress response system in response to a stressor (3, 4). However, the process of resilience is complex and dynamic (5, 6), and it is likely to involve the interaction of several neurobiological systems including those that drive more pleasurable states and executive control. In addition, the genetic endowment of an individual, as well as developmental and environmental influences, all interplay in an intricate manner to impact the process (7). This chapter aims to illustrate the broad scope of multidisciplinary research in this area and progress to date. It will focus attention on the neurobiological and developmental literature, as well as on recent genetic and epigenetic work. The current challenges in this field will be highlighted including issues pertaining to conceptual clarity and measurement methods, and how these may be best addressed in future work.

20.2 Concepts and definitions

Resilience is currently recognized as the process of adaptive recovery following adversity or trauma (1). However, it has also been separately defined as either the presence of minimal psychopathology following trauma (8), or a trait or attribute (e.g. positive affect, self-esteem, internal control, and hardiness) (9). These ambiguities in definition have hampered progress in understanding resilience because they often focus on a single aspect rather than the dynamic interplay between the person, their environment, and the challenges they face (10, 11). From this perspective, resilience needs to be understood as a contextual process by which an individual is able to buffer and overcome adversity via successful adaptation to stressors as a product of socio-ecological and psychological resources available to them at any given time (11). It is likely that genetic factors influencing brain function provide a biological context upon which environmental effects unfold over time to determine resilience in the face of adversity. Future research needs to focus on these dynamic interactions between genetic and environmental factors, and their modulatory effects on brain circuitry and cognitive processing in the resilience process.

An additional challenge is the lack of conceptual delineation between the constructs of *resilience* and *mental well-being* (12). Mental health encapsulates both mental illness and mental well-being, with well-being defining a state that could vary from languishing up to flourishing levels (Figure 20.1). Resilience, on the other hand, describes the process of returning to the optimal

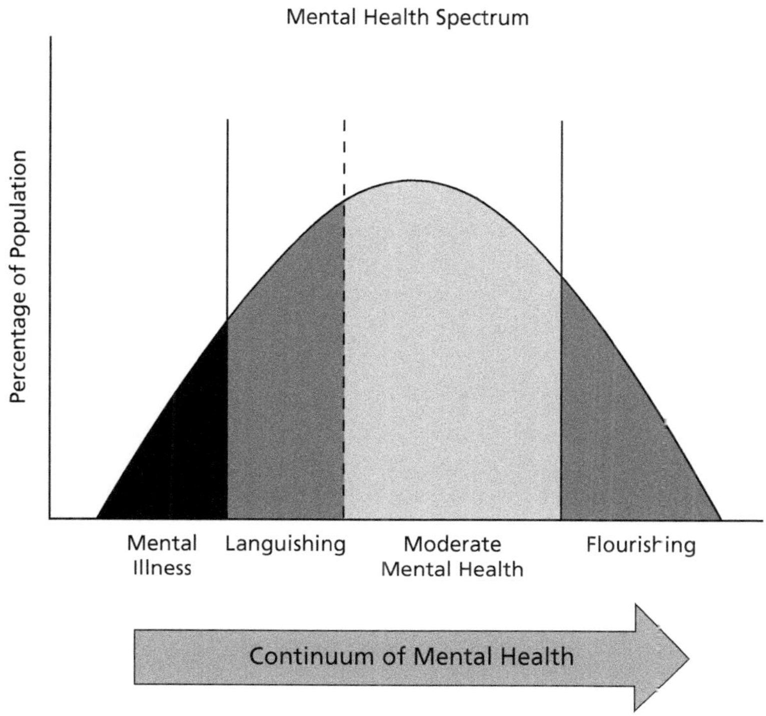

Figure 20.1 Mental health encapsulates both mental illness and mental well-being, with mental well-being spanning a continuum of states ranging from languishing to flourishing levels of functioning.

Reproduced with permission from Huppert FA, Baylis N, Keverne B (eds). The Science of Well-being. Oxford: Oxford University Press; 2005.

mental state following adversity or trauma exposure. Ambiguities between the concepts emerge when one measure is considered a prerequisite or component of the other, or an equivalent proxy measure, without taking into consideration adversity exposure or change over time (12). In addition, research in well-being has also been hampered by a lack of universal definition and agreed measurement. Two similar yet separate categories of definition have commonly been used in the well-being literature: subjective well-being (SWB) and psychological well-being (PWB) (12). SWB is a measure of hedonia or positive and negative affect and life satisfaction (13, 14), whereas PWB measures eudaimonia or an individual's human potential including autonomy, mastery, life purpose, and the capacity for personal growth (15). Scientists are increasingly recognizing the value in adopting a more "composite" approach that incorporates both forms of subjective and psychological well-being, given that the constructs are so strongly associated with one another (16–19). One such example is the recently developed COMPAS-W Wellbeing Scale (16), which provides scores for total mental well-being as well as sub-scores for indices of SWB (e.g. positivity and satisfaction) and PWB (e.g. composure, own-worth, mastery, and achievement). Adopting a more comprehensive measure of well-being within the context of the resilience process will aid understanding and promotion of these mechanisms in mental health. A focus on the positive, adaptive qualities of mental well-being will enable researchers to harness the health-enhancing qualities of this unique construct in the resilience process. In section 20.3, we discuss a neural framework of resilience, drawing on the clinical and healthy population literature.

20.3 **Neural circuits of resilience**

Scientists frequently refer to the same networks as a point of exploration in the neuroscience field of resilience. In mood and affective disorders, there is much focus on the structural and functional integrity of the brain and relevant networks, particularly in relation to stress (20–22). Human and animal studies suggest similar networks in stress resilience and vulnerability (23). In particular, it is proposed that the threat and reward networks implicated in mood and anxiety disorders are also involved in adaptive stress responses and resilience (23, 24). In parallel, the executive control network is another key network modulating attentional processes, response inhibition, and self-regulatory behaviors, and interacting with the threat and reward networks towards adaptive (or maladaptive) stress responses (21, 22).

20.3.1 **Neural circuitry of threat**

The neural regions involved in the threat response include the amygdala, hippocampus, medial prefrontal cortex (mPFC), nucleus accumbens, ventromedial hypothalamus, and a number of brain stem nuclei (25–27). These regions interact in response to potentially threatening cues in the environment for processes of threat perception, emotional response, fear conditioning, modulation of fear memories, and extinction learning (28). Successful adaptation to stress requires the capacity to avoid over-generalization of threat stimuli and to engage adaptive reconsolidation and extinction of fear memories (24).

The administration of threat stimuli during neuroimaging studies in humans and animals help identify correlates of the threat network. Human imaging studies in patients with post-traumatic stress disorder (PTSD) and other anxiety disorders have identified a number of potential biomarkers for a heightened threat response. These include greater activation of the amygdala and insula to negative emotional stimuli (29); increased psychophysiological threat responses measured by skin conductance (30); and hypo-activation of dorsal and rostral anterior cingulate cortices and ventromedial PFC, indicative of reduced top-down inhibition of the amygdala (28). In contrast, other studies suggest that an optimally functioning threat network may be a factor that underpins resilience to psychopathology in trauma-exposed individuals. For instance, in one study, healthy individuals with a history of childhood maltreatment showed an absence of fear bias during an emotional face processing task that is normally found in clinical patients (31). The researchers suggested that the failure to develop a negative fear bias may be due to optimal functioning of the threat circuitry. It is important to note that human imaging studies of the threat network such as these are somewhat limited in advancing the knowledge base of resilience. The comparison of trauma-exposed groups with and without psychopathology is not equivalent to investigating the threat network in a resilient population that is trauma-exposed yet flourishing. This needs to be the focus in future imaging studies of resilience.

Animal studies have shown that impaired synaptic plasticity may underlie maladaptive fear conditioning and extinction (32). Inhibiting brain-derived neurotrophic factor (BDNF) signaling in the amygdala of rodents results in impaired fear conditioning and extinction (33, 34). Additionally, contextual fear learning which is inhibited in BDNF knockout mice has been shown to be rescued following an infusion of BDNF protein into the hippocampus (35). Prelimbic and infralimbic BDNF in mPFC is also required for consolidation and extinction of learned fear (36, 37). There are also several other signaling pathways that have been implicated in fear consolidation and extinction learning including glutamate and γ-aminobutyric acid (GABA) (32, 38–40), dopamine, acetylcholine, norepinephrine, nitric oxide, and endocannabinoids (32). These pathways may provide neurochemical targets for pharmacological intervention, and may help parse some of the specific neurobiological features that uniquely contribute to the resilience process in response to threat.

20.3.2 Neural circuitry of reward

A functional reward network is also instrumental to successful adaptation and possibly resilience to stress (6). The mesolimbic dopamine pathway is pivotal in the reward network and involves dopamine signaling from the ventral tegmental area to the nucleus accumbens; the primary reward structure in the brain that regulates responses to positive stimuli and drugs of addiction. This signaling pathway also extends to regions such as the amygdala, hippocampus, and mPFC (23, 41, 42). In rodent studies, firing patterns of mesolimbic dopamine neurons have been shown to mediate resilience to chronic stress and antidepressant response (43). In contrast, impaired reward functioning is associated with anhedonia and depression-like behaviors in both rodents (44, 45) and humans (46–49).

Neuroimaging studies in patients with major depressive disorder and PTSD show evidence of reduced striatal activity and reward responsiveness during reward-based tasks (50, 51). Alterations to the white matter microstructure of reward circuits are associated with familial predisposition to depression (52), whereas continuity of neural firing in reward centers has been associated with resilience in Special Forces Soldiers operating in a high-reward condition (53). Converging evidence is reported in studies measuring reward sensitivity using behavioral measures or self-report. For instance, a recent study of healthy college students showed that self-report measures of high reward sensitivity were associated with higher positive affect following experimental stress (54). High positive affect before and after stress mediated the relationship between reward sensitivity and trait resilience.

Like all aspects of neuropsychological functioning, the reward network is developmentally sensitive, with higher reward sensitivity noted during adolescence (6). Therefore, factors that may impact reward functioning or sensitivity to rewarding stimuli may have a more prominent effect during development, particularly when the onset of depression is more likely. Consistent with this, increased reactivity in mPFC and decreased activation in striatum have been linked to adolescent depression (55). Future studies need to delineate the specific role of the reward network in resilience, and whether it is the total magnitude and response time and/or the sensitivity to specific positive stimuli that is more important, and how this varies with age. In addition, an understanding of how these processes may counter the impact of threat or stress will be useful in the development of biologically based psycho-educational therapies, designed to increase resilience capacities in vulnerable populations.

20.3.3 Executive control networks

The executive control network underpins core processes such as inhibition, working memory, and attention, with evidence accumulating for its role in both susceptibility to psychopathology and resilience (56–60). Pre-trauma deficits in response inhibition and attention regulation, linked to dysfunction in the dorsal prefrontal networks, have been reported to predict subsequent PTSD in vulnerable individuals (56). Similarly, in remitted patients with major depression, improvements in executive and attentional functioning were found post-remittance (61). In terms of resilience, neurocognitive studies in humans are just beginning to focus on the role of executive control in this process. For instance, in a recent study of male soldiers administered the Connor–Davidson Resilience Scale (CD-RISC), the Attentional Control Scale (ACS), and a dot-probe task with happy, neutral, and threatening faces; a positive relationship between attentional control and trait resilience was reported (62). This effect was in the absence of a direct association between trait resilience and attentional biases to specific emotions, yet when attentional control was high, attentional bias to threatening faces was associated with higher trait resilience (though not to positive or neutral faces). These findings suggest a possible modulating effect between the threat and executive networks in male soldier resilience. These findings require replication in typical male/female populations, and for other aspects of executive functioning such as working memory, set shifting, and response inhibition. Each of these processes involves a precise set of functions that may differentially interact with associated threat and reward networks in both risk and resilient groups.

Overall, the neuro-scientific evidence indicates possible key roles for the threat, reward, and executive networks in risk and resilience. Beyond these functional networks, scientists are also starting to look at other networks during rest such as the default mode and salience networks, which interact with task-activated networks to guide attentional control and initiate action in response to environmental stimuli (63). This type of work is still very new and requires further focus. Across the neuroimaging paradigms however, most work has focused on psychopathology and risk, or resilience as defined by an absence of psychopathology given trauma exposure or by a trait measure. Studies need to start incorporating all key aspects of the resilience process: presence of trauma exposure, absence of psychopathology, and presence of elevated positive functioning (or mental well-being). Based on the task-activated functional magnetic resonance imaging (fMRI) paradigms alone, we would predict that optimal functioning of the threat network (i.e. conditioning, threat perception, extinction, and modulation of fear memories) along with consistent and timely activation of the reward network would be imperative to resilience. The executive control network most likely modulates the threat and reward networks, which consequently leads to an adaptive stress and emotional response. In section 20.4, we consider the likely impact of early life experiences on brain function, which may help clarify the diverging trajectories that lead to altered network function and risk or resilience pathways.

20.4 **Early life experiences and developmental factors**

Our environment plays a central role in influencing adaptive stress responses across the lifespan (64). The impact of environment can include familial experiences during development (e.g. familial structure, rearing, parenting, and family break-up), and more unique sporadic experiences that may be specific to the individual (e.g. illness or injury, environmental disaster, bullying, or abuse). Early life experiences are particularly influential in shaping later developmental outcomes, both positive and negative, as the brain is still undergoing development and is highly plastic during infancy, childhood, and adolescence. In particular, development is sensitive to both the timing and nature of experienced stressors.

Prenatal stress is one of the first core factors that may influence brain development and later adaptive functions. In animal studies, the offspring of rats exposed to prenatal chronic mild stress (PCMS) or chronic unpredictable stress (CUS) show more depression-like symptoms under stress (65, 66). In particular, rodent offspring showed elevated amygdala corticotropin-releasing hormone (CRH) expression and decreased hippocampal glucocorticoid receptor (GR) expression following prenatal stress, which has previously been linked to the pathogenesis of stress-related disorders (67). Evidence from animal studies also supports the detrimental impact of exposure to trauma during early development. Numerous studies have indicated that offspring abused by their mothers in early infancy show reduced independence and lower stress regulation skills in adulthood (23). These animals exhibit high anxiety, increased HPA axis activation, and high basal CRH levels in cerebrospinal fluid (68–70). Studies of non-human primates have shown that abused animals were more likely to abuse their own offspring, perpetuating an intergenerational cycle of abuse (71).

Similar impact of abuse has been reported in humans. Prenatal stress and childhood trauma are associated with the dysregulation of the HPA axis and depressive symptomology. Prenatal stress has been linked to hyper-activation of the HPA axis and risk of hypercortisclemia in adulthood (72). Additionally, antenatal stress at developmentally sensitive times has been associated with hypo-activation of the HPA axis and blunted daytime cortisol levels in depressed adolescents (73). Supporting evidence for the dysregulation of the HPA axis also comes from a study that found blunted cortisol levels following the Trier Social Stress Tests (TSST) in adolescent females with a history of childhood maltreatment (74). Early-life stress and childhood maltreatment also lead to reductions in hippocampal volume and increased activation of the amygdala (75–77). Further, research suggests that exposure to early-life trauma is associated with higher self-reported depression and anxiety scores in the healthy population, and that exposure to adult trauma alone may not be as influential (78). In particular, early-life trauma characterized by interpersonal violation (e.g. sexual abuse, emotional abuse, and bullying/peer rejection) was the strongest predictor of depression and anxiety scores, when compared with non-interpersonal trauma (e.g. illness, family break-up, and natural disaster) (78). These findings suggest the nature and timing of trauma are critical to later mental well-being outcomes, and that infancy and early childhood are of particular importance in shaping these outcomes.

The impact of early-life stress may however be modulated by the controllability of the stressor as well as the availability of protective buffers. In animal studies, uncontrollable early-life stress can lead to "learned helplessness" in which case the animal displays heightened fear responses without attempting to escape from the stimulus (79). Learned helplessness has been linked to dysregulation of the serotonergic neurons in dorsal raphe nuclei (80) and reduced hippocampal cell production (81). On the other hand, mild to moderate controllable stress can lead to "stress inoculation," whereby the animal develops increased tolerance to stress (7). For example, infant monkeys exposed to short periods of separation from their mothers during early life experienced less anxiety and lower hormonal stress response on separation at nine months of age than non-inoculated monkeys (82). A number of other protective factors that promote stress adaptability have also been identified in studies of maltreated and trauma-exposed children (6). Among these were enriched and stimulating environments (83), positive and secure relationships with caregivers, social support from teachers and peers, strong community values, religious beliefs that emphasized meaning through humor and prosocial behavior (84, 85), and positive opportunities at major life transition points, such as a supportive marital partner or friend (8).

Together, these findings highlight the crucial role of the environment in shaping risk and resilience processes. It is important to remember that these modulatory processes are likely to interact with an individual's unique genetic make-up or genome in shaping the outcome. In section 20.5, we consider recent genetic research and the implications these findings may have for future knowledge advancement.

20.5 **Genetic mechanisms**

The genetics of resilience is another blossoming field with the potential to offer great insight into the biological mechanisms that underpin vulnerability versus adaptability to stress. Twin studies suggest that both genetics and environment play a role in mental illness risk and mental health. Specifically, the comparison of identical to non-identical twins reveals small to moderate genetic heritability estimates of 29–42% for anxiety and depression symptoms (86–88) and 35–64% for well-being and trait resilience (16, 89). Yet, most psychiatric genetic research to date has focused on identifying genetic alleles of risk within the diathesis-stress framework, in which individuals are viewed as more or less vulnerable to environmental stressors and adversity as an effect of their genome. More recently there has been a move to reconsider "vulnerability" genes as "plasticity" genes, where genetically "vulnerable" individuals may instead be perceived as more sensitive to environmental influences, regardless of whether the environmental cue is positive or negative (90). Refocusing attention on the plasticity of the genome in relation to the environment requires a broader research agenda in which the same genes are investigated for their capacity to confer risk or protection depending on the environmental conditions facing the individual.

20.5.1 Genome-wide association studies (GWAS)

Most GWAS conducted to date in the psychiatric field have been in psychiatric disorders, particularly in those with a strong genetic origin, such as schizophrenia (91). While resilience per se is yet to be targeted as a specific phenotype, a recent study has identified a number of significant genes associated with subjective well-being and the associated phenotypes of depression and neuroticism (92). Three independent single nucleotide polymorphisms (SNPs) reached genome-wide significance for subjective well-being (rs3756290, rs2075677, rs4958581); these were independent of the SNPs identified for depressive symptoms and neuroticism. Similarly, in a previous epigenome-wide association study (EWAS), differentially methylated sites were associated with well-being (93). Two sites (cg10845147, cg01940273) reached genome-wide significance after stringent correction. The closest gene to site cg10845147 is *NEURL1B*, which is involved in regulation of the Notch pathway and therefore the regulation of cell survival and cell proliferation (93). For cg01940273, the closest gene is alkaline phosphatase, placental-like 2 (*ALLPL2*), which is responsible for the dephosphorylation of proteins, nucleotides, or alkaloids, and is sensitive to environmental and disease impacts such as smoking (94). Together, these studies highlight the importance of genetic processes underpinning cell survival and proliferation and normal dephosphorylation in well-being, and how they may be moderated by environmental factors prenatally and during early or later development. It is important to examine whether these same genes may modulate resilience outcomes in youth and adulthood when taking into consideration early-life adversity and stress.

20.5.2 Candidate gene association studies

The majority of candidate gene studies in psychiatry tend to focus on identifying risk alleles for major psychopathological disorders, targeting genes known to modulate stress reactivity and emotional functioning such as the serotonergic system, HPA axis, neuropeptide Y (NPY), noradrenergic, and dopaminergic systems, and brain-derived neurotrophic factor (BDNF) (3, 6, 23). Table 20.1 presents some common genetic risk variants frequently reported in genetic association studies of specific disorders. A review of meta-analyses of such association studies found evidence for a potential pleiotropic effect of specific genetic risk variants across a number of mental disorders (i.e. major depressive disorder [MDD], anxiety disorder [AD], schizophrenia [SZ], bipolar disorder [BPD], and attention deficit hyperactivity disorder [ADHD]) (91). Specifically, the review identified 134 genes (206 variants) that were significant risk variants for MDD, AD, SZ, BPD, and ADHD, of which 13 genetic variants were shared in common across two or more disorders (91). It is quite possible that the opposing "protective" variants may indeed be predictive of resilience, but this still requires confirmation.

In contrast, genetic association studies in resilience or well-being are very sparse. So far, the most common approach is to test the protective association of opposing risk variants in individuals with an absence of pathology following trauma, or with trait resilience measures. For instance, gene–environment interactions have been identified for a number of SNPs of the CHR type 1 receptor gene (*CRHR1*) which has been found to moderate the risk of depressive

Table 20.1 Summary of genetic risk variants and associated neurotransmitter systems implicated in psychopathology

Neurotransmitter/ system	Associated brain regions	Gene	Risk variants	Associated pathology	References
Serotonergic	Amygdala, nucleus accumbens and PFC	*SLC6A3*	Short- (S) allele of 5-HTTLPR	MDD (depression, negative mood, emotion dysregulation)	Feder et al., 2009 (23) Karg, Burmeister, Shedden, and Sen, 2011 (119)
		HTR1A *HTR3A* *HTR3C*	G variant of *HTR1A* C-1019G C variant of *HTR2C* rs6318 C variant of *HTR3A* 42C > T rs1062613		Kim et al., 2011 (120) Brummett et al., 2012 (121) Bhatnagar et al., 2004 (122) Gatt et al., 2010 (123) Harrell & Allan, 2003 (124) Niesler et al., 2001 (125)
Noradrenergic and dopaminergic	1. LC-NE system (PFC, amygdala, hypothalamus andhippocampus) 2. Mesolimbic-dopamine pathway (ventral tegmental area, nucleus accumbens, hippocampus, amygdala, and medial PFC)	*DRD2* *DRD4* *DAT1*	GG genotype of *DRD2* rs1076560 Long- (L) allele of DRD4 (7R or 8R) SLC6A3 3 variable number tandem repeat (VNTR)	PTSD (abnormal stress responsivity, anxiety, negative mood, emotion processing deficits, limbic responsivity to unpleasant stimuli)	Blasi et al., 2009 (126) Dragan and Oniszczenko, 2009 (127) Ptáček, Kuželová, & Stefano, 2011 (128) Segman et al., 2002 (129)
		COMT Val^{158}*Met*	Met-allele carriers		Heinz and Smolka, 2006 (130) Kolassa et al., 2010 (131) Williams et al., 2010 (132)

(*continued*)

Table 20.1 Continued

Neurotransmitter/ system	Associated brain regions	Gene	Risk variants	Associated pathology	References
Hypothalamic–pituitary–adrenal axis (HPA axis)	CRH secretion from hypothalamus signals release of ACTH from pituitary and activates secretion of glucocorticoids	*CRHR1* *FKBP5*	CC genotype of *CRHR1* rs7209436 GG genotype of *CRHR1* rs110402 AG & GG genotypes of *FKBP5* rs9296158 CC, CT, and TT genotypes of *FKBP5* rs1360780 AA, AC & CC genotypes of *FKBP5* rs3800373 CC, CT, TT genotypes of *FKBP5* rs9470080 AA genotype of *FKBP5* rs4713916	MDD PTSD/MDD MDD only	Bradley, Binder, and Epstein, 2008 (95) Binder et al., 2008 (133) Zimmermann et al., 2011 (134)
Brain-derived neurotrophic factor (BDNF)	Hippocampus, amygdala, PFC, and basal forebrain	BDNF Val66Met	Met/Met homozygotes	Anxiety Unipolar depression (reduced hippocampal volume & function)	Chen et al., 2006 (135) Egan et al., 2003 (136) Gatt et al., 2008; Gatt et al., 2009 (137, 138) Videbech & Ravnkilde, 2004 (139)
Neuropeptide Y (NPY)	Amygdala, hippocampus, hypothalamus, and locus coeruleus	*NPY* haplotypes H3 and H4	H3/H3, H4/H4, H3/H4 carriers for rs16142, rs2023890, rs17374047	Anxiety disorder (e.g. PD, GAD, Phobias)	Donner et al., 2012 (140) Zhang et al., 2012 (141) Zhou et al., 2008 (142)

psychopathology in individuals with a background of child abuse (95). Individuals homozygous for the rare TT- or AA-alleles of two SNPs of the *CRHR1* gene (rs110402, rs7209436) showed lower depression scores despite evidence of moderate to severe early childhood abuse, demonstrating a protective role for these alleles. In another study, Rana and colleagues examined the association between 65 candidate gene SNPs and 31 ancestry informative marker SNPs with dispositional optimism (LOT) (96) and trait resilience (CD-RISC) (97) in 935 older adults (98). No significant genetic associations were found when using a corrected p threshold of 0.00089, but several associations emerged at the 0.05 level; namely associations between optimism with monoamine oxidase A (*MAOA*) (rs6323), interleukin-10 (*IL10*) (rs1800896), and fibrinogen gamma chain (*FGG*) (rs1800792), and between resilience and *MAOA* (rs6323) (98). Similarly, in a recent study of physical resilience in aging (99), small yet significant associations were found between physical resilience and two SNPs of the serotonin gene *SLC64A* (rs25533, rs1042173). Across both studies, effects were small and are most likely polygenic, but they do support the role of protective genetic variants in positive mental health outcomes.

There are also some other studies that have focused on identifying candidate genetic variants that are protective and that may underpin quantitative measures of mental well-being. For instance, in another aging study, measures of subjective well-being (SWB) were associated with the *COMT* (rs4680) gene. Male carriers of the *COMT* Met-allele reported significantly better SWB than male/female carriers of the Val-allele or female carriers of the Met-allele (100). Male Met-allele carriers also showed an association with physical activity; thereby the role of *COMT* in well-being may be associated with or mediated by levels of physical activity. The results are, however, surprising given that *COMT* Met-allele carriers have also been shown to experience greater negative affect than Val carriers (101–104). It is possible that age mitigates this contrary result. Turan and colleagues have suggested that the *COMT* Met-allele carriers are less likely to experience negative affect with age. In a longitudinal study, participants provided ecological momentary assessments (EMA) of emotional experience over one week on three occasions at five-yearly intervals. Results indicated that *COMT* Met-allele carriers experienced greater negative affect at an earlier age, but there was no effect of *COMT* on affective experience in older age (105). Earlier research has indicated that in fact *COMT*-Met allele carriers have higher levels of cognitive functioning (106, 107). The authors suggested that *COMT* Met-allele carriers may harness this cognitive functioning to hone their emotion regulation skills across time and promote improved affective experience in later life. Clearly, further genetic and neuroscientific research would be required to establish the veracity of this hypothesis.

20.5.3 **Epigenetic mechanisms**

Epigenetics refers to heritable changes in gene expression that may be independent of a corresponding change to the underlying DNA sequence (108). Epigenetic modifications impact phenotypic variation through mechanisms such as DNA methylation and demethylation, and histone modifications such as methylation, acetylation, and phosphorylation (6). These processes are forms of chromatin regulation which impact gene transcription by altering the state of gene regulatory elements in a highly stable manner (109). One of the most researched epigenetic processes is DNA methylation, in which a methyl group is added to DNA. This occurs most commonly at CpG sites, where cytosine and guanine base pairs are linked consecutively. If methylation occurs in the promoter region of the gene, it has the capacity to block critical binding factors that regulate gene expression, thereby reducing the expression of the gene (110). Acetylation is another epigenetic process involved in gene transcription and expression whereby an acetyl group is added to DNA. For example, histone acetylation is a process by which an acetyl group is added to a histone N-terminal tail allowing for chromatin reorganization and efficient gene transcription. Histones are alkaline proteins located in eukaryotic cell nuclei that function to organize DNA into structured elements called nucleosomes. Histones act as coils around which DNA winds and are active in gene regulation (111).

Epigenetic processes such as methylation and acetylation can be activated via environmental exposure to critical events at sensitive developmental periods in the lifespan. Early enriching life experiences may therefore have the power to modify genetic mechanisms towards resilience to stress, whereas exposure to stress and poor rearing may elicit genomic changes that contribute to increased risk of psychiatric symptoms. For instance, rat pups of mothers who provided high levels of maternal care showed increased hippocampal GR expression and enhanced glucocorticoid negative feedback sensitivity, and a lower HPA arousal pattern in response to stress compared with pups from mothers showing low levels of maternal care (112). This suggests that effective caregiving may confer protection against stress via decreased sensitivity of the HPA axis. Another animal study found that histone methyltransferase was upregulated in the nucleus accumbens (NAc) of resilient mice following chronic social defeat stress (113); suggesting histone methylation processes may confer protection against the effects of chronic stress and subsequent depression. In humans, the majority of research to date has examined the influence of DNA methylation on genes that impact the stress response (108). Post-mortem examination of suicide victims with a history of child abuse indicate increased methylation of a GR (*NR3C1*) promoter in hippocampus, and therefore decreased hippocampal GR expression, when compared to suicide victims or control subjects without abuse (114). More broadly, DNA methytransferase (DNMT) expression was altered in a number of post-mortem brain regions of suicide victims when compared with controls who died of other causes (115); specifically, increases in DNMT-3 expression in the frontal cortex and in DNA methylation in the γ-aminobutyric acid (GABA) A receptor α1 subunit promoter region.

More recently, there has been a number of studies geared towards identifying epigenetic mechanisms that impact stress vulnerability during critical developmental periods (108). For instance, prenatal chronic and war-related stress has significant impacts on HPA axis gene methylation in newborns at transcription factor binding (TFB) sites for genes regulating HPA function, including *CRH, CRHBP, NR3C1*, and *FKBP5* (116). Adversity during early childhood affects methylation of the glucocorticoid receptor (GR) gene, *NR3C1*, and internalizing behavior in pre-schoolers (117). Even in previously abused adults, epigenetic mechanisms are apparent for the oxytocin receptor (*OXTR*) towards risk of psychopathology (110). Together, these studies suggest an important role for epigenetic mechanisms in modulating the effects of environmental stressors on subsequent behavioral outcomes across the lifespan. Future research will need to consider the manner in which similar epigenetic processes influence well-being outcomes and resilience.

20.6 **Conclusion**

Resilience is a complex and dynamic construct, with proposed neuroscience and genetic frameworks providing some guidance towards future research initiatives. From a neuroscience perspective, resilience involves the adaptive ability to harness the stress response system to promote healthy emotional and cognitive functioning. Emerging theory and evidence supports the role for the threat and reward networks in the brain in connection with executive control networks. Developmental research provides context for how these neural systems develop and are modified by the environment across different life stages. Genetic work has revealed numerous genetic variants that influence key neurochemical systems in the brain to promote risk and resilience to stress throughout development. With the advances in knowledge, scientists do, however, need to be mindful in how they are defining resilience and its measurement. Much neuroscientific and genetic research in "resilience" is conducted in trauma-exposed individuals or analogous animal models with no overt psychopathology. While this method certainly has its merits, the findings from such studies can only really inform mechanisms of risk and vulnerability rather than resilience, which is a process that also encapsulates optimal mental well-being and flourishing. Recognition and adoption of these key differences in future studies will aid knowledge development and progress, and thereby enable new approaches to psychotherapeutic and pharmacological interventions designed to build resilience and promote well-being in vulnerable populations.

References

1. **APA.** The road to resilience. Washington DC; 2016 [cited 2016 28 January]. Available from: <http://www.apa.org/helpcenter/road-resilience.aspx>.
2. **Kessler RC, McLaughlin KA, Green JG, Gruber MJ, Sampson NA, Zaslavsky AM** et al. Childhood adversities and adult psychopathology in the WHO World Mental Health Surveys. BJ Psychiatry. 2010;**197**(5):378–85.
3. **Russo SJ, Murrough JW, Han M-H, Charney DS, Nestler EJ.** Neurobiology of resilience. Nat Neurosci. 2012;**15**(11):1475–84.
4. **Southwick SM, Charney DS.** Ready for anything. Scientific American Mind. 2013;**24**(3):32–41.
5. **Ungar M, Ghazinour M, Richter J.** Annual Research Review: What is resilience within the social ecology of human development? J Child Psychol Psychiatry. 2013;**54**(4):348–66.
6. **Wu G, Feder A, Cohen H, Kim JJ, Calderon S, Charney DS**, et al. Understanding resilience. Front Behav Neurosci. 2013;30.
7. **Southwick SM, Charney DS.** The science of resilience: Implications for the prevention and treatment of depression. Science. 2012;**338**(6103):79–82.
8. **Werner EE.** Resilience in development. Curr Dir Psychol Sci. 1995;**4**(3):81–5.
9. **Friborg O, Hjemdal O, Rosenvinge JH, Martinussen M.** A new rating scale for adult resilience: What are the central protective resources behind healthy adjustment? Int J Methods Psychiatr Res. 2003;**12**(2):65–76.
10. **Kumpfer KL.** Outcome measures of interventions in the study of children of substance-abusing parents. Pediatrics. 1999;**103**(Supplement 2):1128.
11. **Stein DJ.** The psychobiology of resilience. CNS spectrums. 2009;**14**(2 Suppl 3):41–7.
12. **Schultze-lutter F, Schimmelmann BG, Schmidt SJ.** Resilience, risk, mental health and well-being: Associations and conceptual differences. Eur Child Adolesc Psychiatry. 2016;**25**(5):459–66.
13. **Diener E, Emmons RA, Larsen RJ, Griffin S.** The satisfaction with life scale. J Personal Assess. 1985;**49**(1):71–5.
14. **Diener E, Suh EM, Lucas RE, Smith HL.** Subjective well-being: Three decades of progress. Psychol Bull. 1999;**125**(2):276–302.
15. **Ryff CD.** Happiness is everything, or is it? Explorations on the meaning of psychological well-being. J Personal Soc Psychol. 1989;**57**(6):1069–81.
16. **Gatt JM, Burton KLO, Schofield PR, Bryant RA, Williams LM.** The heritability of mental health and wellbeing defined using COMPAS-W, a new composite measure of wellbeing. Psychiatry Research. 2014;**219**(1):204–13.
17. **Keyes CL.** Promoting and protecting mental health as flourishing: A complementary strategy for improving national mental health. American Psychologist. 2007;**62**(2):95.
18. **Keyes CLM, Shmotkin D, Ryff CD.** Optimizing well-being: The empirical encounter of two traditions. J Personal Soc Psychol. 2002;**82**(6):1007–22.
19. **Seligman M.** Flourishing. A New Understanding of Happiness and Well-being and How to Achieve Them. London: Free Press; 2011.
20. **Franklin TB, Saab BJ, Mansuy IM.** Neural mechanisms of stress resilience and vulnerability. Neuron. 2012;**75**(5):747–61.
21. **Williams LM.** Precision psychiatry: A neural circuit taxonomy for depression and anxiety. Lancet Psychiatry. 2016;**3**(5):472–80.
22. **Williams LM.** Defining biotypes for depression and anxiety based on large-scale circuit dysfunction: A theoretical review of the evidence and future directions for clinical translation. Depress Anxiety. 2017;**34**(1):9–24.
23. **Feder A, Nestler EJ, Charney DS.** Psychobiology and molecular genetics of resilience. Nat Rev Neurosci. 2009;**10**(6):446–57.

24. **Charney DS.** Psychobiological mechanisms of resilience and vulnerability: Implications for successful adaptation to extreme stress. Am J Psychiatry. 2004;**161**(2):195–216.
25. **Davis M.** Neural systems involved in fear and anxiety measured with fear-potentiated startle. American Psychologist. 2006;**61**(8):741–56.
26. **Maren S.** Pavlovian fear conditioning as a behavioral assay for hippocampus and amygdala function: Cautions and caveats. Eur J Neurosci. 2008;**28**(8):1661–6.
27. **Quirk GJ, Mueller D.** Neural mechanisms of extinction learning and retrieval. Neuropsychopharmacology. 2008;**33**(1):56–72.
28. **Shin LM, Liberzon I.** The neurocircuitry of fear, stress, and anxiety disorders. Neuropsychopharmacology. 2010;**35**(1):169–91.
29. **Etkin A, Wager TD.** Functional neuroimaging of anxiety: A meta-analysis of emotional processing in PTSD, social anxiety disorder, and specific phobia. Am J Psychiatry. 2007;**164**(10):1476–88.
30. **MacNamara A, Rabinak CA, Fitzgerald DA, Zhou XJ, Shankman SA, Milad MR,** et al. Neural correlates of individual differences in fear learning. Behav Brain Res. 2015;**287**:34–41.
31. **Suzuki A, Poon L, Kumari V, Cleare AJ.** Fear biases in emotional face processing following childhood trauma as a marker of resilience and vulnerability to depression. Child Maltreatment. 2015;**20**(4):240–50.
32. **Mahan AL, Ressler KJ.** Fear conditioning, synaptic plasticity and the amygdala: Implications for posttraumatic stress disorder. Trends Neurosci. 2012;**35**(1):24–35.
33. **Chhatwal JP, Stanek-Rattiner L, Davis M, Ressler KJ.** Amygdala BDNF signaling is required for consolidation but not encoding of extinction. Nat Neurosci. 2006;**9**(7):870–2.
34. **Rattiner LM, Davis M, French CT, Ressler KJ.** Brain-derived neurotrophic factor and tyrosine kinase receptor B involvement in amygdala-dependent fear conditioning. J Neurosci. 2004;**24**(20):4796–806.
35. **Liu IY, Lyons WE, Mamounas LA, Thompson RF.** Brain-derived neurotrophic factor plays a critical role in contextual fear conditioning. J Neurosci. 2004;**24**(36):7958–63.
36. **Choi DC, Maguschak KA, Ye K, Jang S-W, Myers KM, Ressler KJ,** et al. Prelimbic cortical BDNF is required for memory of learned fear but not extinction or innate fear. PNAS. 2010;**107**(6):2675–80.
37. **Peters J, Dieppa-Perea LM, Melendez LM, Quirk GJ.** Induction of fear extinction with hippocampal-infralimbic BDNF. Science. 2010;**328**(5983):1288–90.
38. **Dalton GL, Wang YT, Floresco SB, Phillips AG.** Disruption of AMPA receptor endocytosis impairs the extinction, but not acquisition of learned fear. Neuropsychopharmacology. 2008;**33**(10):2416–26.
39. **Liu JL, Li M, Dang XR, Wang ZH, Rao ZR, Wu SX,** et al. A NMDA receptor antagonist, MK-801 impairs consolidating extinction of auditory conditioned fear responses in a Pavlovian model. PLoS One. 2009;**4**(10):e7548.
40. **Zimmerman JM, Maren S.** NMDA receptor antagonism in the basolateral but not central amygdala blocks the extinction of Pavlovian fear conditioning in rats. Eur J Neurosci. 2010;**31**(9):1664–70.
41. **Berridge KC, Kringelbach ML.** Building a neuroscience of pleasure and well-being. Psychology of Well-Being. 2011;**1**(1):1–26.
42. **Nestler EJ, Carlezon Jr WA.** The mesolimbic dopamine reward circuit in depression. Biol Psychiatry. 2006;**59**(12):1151–9.
43. **Cao JL, Covington HE, 3rd, Friedman AK, Wilkinson MB, Walsh JJ, Cooper DC,** et al. Mesolimbic dopamine neurons in the brain reward circuit mediate susceptibility to social defeat and antidepressant action. J Neurosci. 2010;**30**(49):16453–8.
44. **Francis TC, Chandra R, Friend DM, Finkel E, Dayrit G, Miranda J,** et al. Nucleus accumbens medium spiny neuron subtypes mediate depression-related outcomes to social defeat stress. Biol Psychiatry. 2015;**77**(3):212–22.
45. **Heshmati M, Russo SJ.** Anhedonia and the brain reward circuitry in depression. Curr Behav Neurosci Rep. 2015;**2**(3):146–53.

46. **Keedwell PA, Andrew C, Williams SCR, Brammer MJ, Phillips ML.** The neural correlates of anhedonia in major depressive disorder. Biol Psychiatry. 2005;**58**(11):843–53.

47. **McMakin DL, Olino TM, Porta G, Dietz LJ, Emslie G, Clarke G,** et al. Anhedonia predicts poorer recovery among youth with selective serotonin reuptake inhibitor treatment-resistant depression. J Am Acad Child Adolesc Psychiatry. 2012;**51**(4):404–11.

48. **Vrieze E, Demyttenaere K, Bruffaerts R, Hermans D, Pizzagalli DA, Sienaert P,** et al. Dimensions in major depressive disorder and their relevance for treatment outcome. J Affect Disord. 2014;**155**:35–41.

49. **Vrieze E, Pizzagalli DA, Demyttenaere K, Hompes T, Sienaert P, de Boer P,** et al. Reduced reward learning predicts outcome in major depressive disorder. Biol Psychiatry. 2013;**73**(7):639–45.

50. **Pizzagalli DAP, Holmes AJAM, Dillon DGP, Goetz ELBA, Birk JLBA, Bogdan RAM,** et al. Reduced caudate and nucleus accumbens response to rewards in unmedicated individuals with major depressive disorder. Am J Psychiatry. 2009;**166**(6):702–10.

51. **Sailer U, Robinson S, Fischmeister FPS, König D, Oppenauer C, Lueger-Schuster B,** et al. Altered reward processing in the nucleus accumbens and mesial prefrontal cortex of patients with posttraumatic stress disorder. Neuropsychologia. 2008;**46**(11):2836–44.

52. **Bracht T, Linden D, Keedwell P.** A review of white matter microstructure alterations of pathways of the reward circuit in depression. J Affect Disord. 2015;**187**:45–53.

53. **Vythilingam M, Nelson EE, Scaramozza M, Waldeck T, Hazlett G, Southwick SM,** et al. Reward circuitry in resilience to severe trauma: An fMRI investigation of resilient special forces soldiers. Psychiatry Research. 2009;**172**(1):75–7.

54. **Corral-Frías NS, Nadel L, Fellous JM, Jacobs WJ.** Behavioral and self-reported sensitivity to reward are linked to stress-related differences in positive affect. Psychoneuroendocrinology. 2016;**66**:205–13.

55. **Forbes EE, Dahl RE.** Research Review: Altered reward function in adolescent depression: What, when and how? J Child Psychol Psychiatry. 2012;**53**(1):3–15.

56. **Aupperle RL, Melrose AJ, Stein MB, Paulus MP.** Executive function and PTSD: Disengaging from trauma. Neuropharmacology. 2012;**62**(2):686–94.

57. **Fertuck EA, Lenzenweger MF, Clarkin JF.** The association between attentional and executive controls in the expression of borderline personality disorder features: A preliminary study. Psychopathology. 2005;**38**(2):75–81.

58. **Gazzaley A, Nobre AC.** Top-down modulation: Bridging selective attention and working memory. Trends Cogn Sci. 2012;**16**(2):129–35.

59. **Paelecke-Habermann Y, Pohl J, Leplow B.** Attention and executive functions in remitted major depression patients. J Affect Disord. 2005;**89**(1–3):125–35.

60. **Routledge K, Burton KLO, Williams LM, Harris A, Schofield PR, Clark CR, Gatt JM.** The shared and unique genetic relationship between mental wellbeing, depression and anxiety symptoms and cognitive function in healthy twins. Cognition and Emotion. 2017;**31**(7):1465–79.

61. **Gruber S, Rathgeber K, Bräunig P, Gauggel S.** Stability and course of neuropsychological deficits in manic and depressed bipolar patients compared to patients with major depression. J Affect Disord. 2007;**104**(1–3):61–71.

62. **Schäfer J, Wittchen HU, Höfler M, Heinrich A, Zimmermann P, Siegel S,** et al. Is trait resilience characterized by specific patterns of attentional bias to emotional stimuli and attentional control? J Behav Ther Exp Psychiatry. 2015;**48**:133–9.

63. **Kong F, Wang X, Hu S, Liu J.** Neural correlates of psychological resilience and their relation to life satisfaction in a sample of healthy young adults. NeuroImage. 2015;**123**:165–72.

64. **Rende R.** Behavioral resilience in the post-genomic era: Emerging models linking genes with environment. Front Hum Neurosci [Internet]. 2012; **6**(50).

65. **Mueller BR, Bale TL.** Sex-specific programming of offspring emotionality after stress early in pregnancy. J Neurosci. 2008;**28**(36):9055–65.

66. **Wang Y, Ma Y, Hu J, Cheng W, Jiang H, Zhang X**, et al. Prenatal chronic mild stress induces depression-like behavior and sex-specific changes in regional glutamate receptor expression patterns in adult rats. Neuroscience. 2015;**301**:363–74.
67. **Musazzi L, Treccani G, Mallei A, Popoli M.** The action of antidepressants on the glutamate system: Regulation of glutamate release and glutamate receptors. Biol Psychiatry. 2013;**73**(12):1180–8.
68. **Claes S.** Corticotropin-releasing hormone (CRH) in psychiatry: From stress to psychopathology. Ann Med. 2004;**36**(1):50–61.
69. **McCormack K, Sanchez MM, Bardi M, Maestripieri D.** Maternal care patterns and behavioral development of rhesus macaque abused infants in the first 6 months of life. Dev Psychobiol. 2006;**48**(7):537–50.
70. **Strome EM, Wheler GT, Higley JD, Loriaux DL, Suomi SJ, Doudet DJ.** Intracerebroventricular corticotropin-releasing factor increases limbic glucose metabolism and has social context-dependent behavioral effects in nonhuman primates. PNAS. 2002;**99**(24):15749–54.
71. **Maestripieri D, Lindell SG, Higley JD.** Intergenerational transmission of maternal behavior in rhesus macaques and its underlying mechanisms. Dev Psychobiol. 2007;**49**(2):165–71.
72. **Heim C, Shugart M, Craighead WE, Nemeroff CB.** Neurobiological and psychiatric consequences of child abuse and neglect. Dev Psychobiol. 2010;**52**(7):671–90.
73. **Van den Bergh BRH, Van Calster B, Smits T, Van Huffel S, Lagae L.** Antenatal maternal anxiety is related to HPA-axis dysregulation and self-reported depressive symptoms in adolescence: A prospective study on the fetal origins of depressed mood. Neuropsychopharmacology. 2008;**33**(3):536–45.
74. **MacMillan HL, Georgiades K, Duku EK, Shea A, Steiner M, Niec A**, et al. Cortisol response to stress in female youths exposed to childhood maltreatment: Results of the Youth Mood Project. Biol Psychiatry. 2009;**66**(1):62–8.
75. **Dannlowski U, Stuhrmann A, Beutelmann V, Zwanzger P, Lenzen T, Grotegerd D**, et al. Limbic scars: Long-term consequences of childhood maltreatment revealed by functional and structural magnetic resonance imaging. Biol Psychiatry. 2012;**71**(4):286–93.
76. **Davidson RJ, McEwen BS.** Social influences on neuroplasticity: Stress and interventions to promote well-being. Nat Neurosci. 2012;**15**(5):689–95.
77. **Janssen J, Pol HEH, De Leeuw F-E, Schnack HG, Lampe IK, Kok RM**, et al. Hippocampal volume and subcortical white matter lesions in late life depression: Comparison of early and late onset depression. J Neurol Neurosurg Psychiatry. 2007;**78**(6):638–40.
78. **Chu DA, Williams LM, Harris AWF, Bryant RA, Gatt JM.** Early life trauma predicts self-reported levels of depressive and anxiety symptoms in nonclinical community adults: Relative contributions of early life stressor types and adult trauma exposure. J Psychiatr Res. 2013;**47**(1):23–32.
79. **Overmier JB, Seligman ME.** Effects of inescapable shock upon subsequent escape and avoidance responding. J Comp Physiol Psychol. 1967;**63**(1):28–33.
80. **Greenwood BN, Foley TE, Day HE, Campisi J, Hammack SH, Campeau S**, et al. Freewheel running prevents learned helplessness/behavioral depression: Role of dorsal raphe serotonergic neurons. J Neurosci. 2003;**23**(7):2889–98.
81. **Ho Y, Wang S.** Adult neurogenesis is reduced in the dorsal hippocampus of rats displaying learned helplessness behavior. Neuroscience. 2010;**171**(1):153–61.
82. **Parker KJ, Buckmaster CL, Schatzberg AF, Lyons DM.** Prospective investigation of stress inoculation in young monkeys. Arch Gen Psychiatr. 2004;**61**(9):933–41.
83. **Green T, Gehrke B, Bardo M.** Environmental enrichment decreases intravenous amphetamine self-administration in rats: Dose-response functions for fixed-and progressive-ratio schedules. Psychopharmacology. 2002;**162**(4):373–8.
84. **Masten AS, Best KM, Garmezy N.** Resilience and development: Contributions from the study of children who overcome adversity. Devel Psychopathol. 1990;**2**(04):425–44.
85. **Werner EE.** Children and war: Risk, resilience, and recovery. Dev Psychopathol. 2012;**24**(2):553–8.

86. **Burton KLO, Williams LM, Richard Clark C, Harris A, Schofield PR, Gatt JM.** Sex differences in the shared genetics of dimensions of self-reported depression and anxiety. J Affect Disord. 2015;**188**:35–42.

87. **Johnson W, McGue M, Gaist D, Vaupel JW, Christensen K.** Frequency and heritability of depression symptomatology in the second half of life: Evidence from Danish twins over 45. Psychol Med. 2002;**32**(07):1175–85.

88. **Stein MB, Jang KL, Livesley WJ.** Heritability of social anxiety-related concerns and personality characteristics: A twin study. J Nerv Ment Dis. 2002;**190**(4):219–24.

89. **Hansson K, Cederblad M, Lichtenstein P, Reiss D, Pedersen N, Belderhiser J,** et al. Individual resiliency factors from a genetic perspective: Results from a twin study. Family Process. 2008;**47**(4):537–51.

90. **Belsky J, Jonassaint C, Pluess M, Stanton M, Brummett B, Williams R.** Vulnerability genes or plasticity genes? Mol Psychiatry. 2009;**14**(8):746–54.

91. **Gatt JM, Burton KLO, Williams LM, Schofield PR.** Specific and common genes implicated across major mental disorders: A review of meta-analysis studies. J Psychiatr Res. 2015;**60**:1–13.

92. **Okbay A, Baselmans BM, De Neve J-E, Turley P, Nivard MG, Fontana MA,** et al. Genetic variants associated with subjective well-being, depressive symptoms, and neuroticism identified through genome-wide analyses. Nat Genet. 2016;**48**:624–33.

93. **Baselmans BM, van Dongen J, Nivard MG, Lin BD, Zilhão NR, Boomsma DI,** et al. Epigenome-wide association study of wellbeing. Twin Res Hum Genet. 2015;**18**(6):710–9.

94. **Koshida K, Stigbrand T, Munck-Wikland E, Hisazumi H, Wahren B.** Analysis of serum placental alkaline phosphatase activity in testicular cancer and cigarette smokers. Urol Res. 1990;**18**(3):169–73.

95. **Bradley RG, Binder EB, Epstein MP.** Influence of child abuse on adult depression: Moderation by the corticotropin-releasing hormone receptor gene. Arch Gen Psychiatr. 2008;**65**(2) 190–200.

96. **Scheier MF, Carver CS.** Optimism, coping, and health: Assessment and implications of generalized outcome expectancies. Health Psychol. 1985;**4**(3):219–47.

97. **Connor KM, Davidson JR.** Development of a new resilience scale: The Connor–Davidson Resilience Scale (CD-RISC). Depress Anxiety. 2003;**18**(2):76–82.

98. **Rana BK, Darst BF, Bloss C, Shih P-aB, Depp C, Nievergelt CM,** et al. Candidate SNP associations of optimism and resilience in older adults: Exploratory study of 935 community-dwelling adults. Am J Geriatr Psych. 2014;**22**(10):997–1006.e5.

99. **Resnick B, Klinedinst NJ, Yerges-Armstrong L, Choi EY, Dorsey SG.** The impact of genetics on physical resilience and successful aging. J Aging Health. 2015;**27**(6):1084–104.

100. **Olsson LA, Nilsson TK.** The triangulation of COMT polymorphism rs4680, physical activity, and subjective well-being in retired community-dwelling elderly subjects Örebro University, Sweden. Örebro University; 2015. Available from: <http://www.diva- portal.org/smash/record.jsf?pid=diva2%3A792545&dswid=-5306>.

101. **Enoch M-A, Xu K, Ferro E, Harris CR, Goldman D.** Genetic origins of anxiety in women: A role for a functional catechol-*O*-methyltransferase polymorphism. Psychiat Genet. 2003 **13**(1):33–41.

102. **Jabbi M, Kema IP, van der Pompe G, te Meerman GJ, Ormel J, den Boer JA.** Catechol-*o*-methyltransferase polymorphism and susceptibility to major depressive disorder modulates psychological stress response. Psychiat Genet. 2007;**17**(3):183–93.

103. **Perroud N, Jaussent I, Guillaume S, Bellivier F, Baud P, Jollant F,** et al. COMT but not serotonin-related genes modulates the influence of childhood abuse on anger traits. Genes Brain Behav. 2010;**9**(2):193–202.

104. **Smolka MN, Schumann G, Wrase J, Grüsser SM, Flor H, Mann K,** et al. Catechol-*O*-methyltransferase Val158Met genotype affects processing of emotional stimuli in the amygdala and prefrontal cortex. J Neurosci. 2005;**25**(4):836–42.

105. **Turan B, Sims T, Best SE, Carstensen LL.** Older age may offset genetic influence on affect: The COMT polymorphism and affective well-being across the life span. Psychol Aging. 2016;**31**(3):287.

106. **Egan MF, Goldberg TE, Kolachana BS, Callicott JH, Mazzanti CM, Straub RE,** et al. Effect of COMT Val108/158 Met genotype on frontal lobe function and risk for schizophrenia. PNAS. 2001;**98**(12):6917–22.

107. **Goldberg TE, Egan MF, Gscheidle T, Coppola R, Weickert T, Kolachana BS,** et al. Executive subprocesses in working memory: Relationship to catechol-*O*-methyltransferase Val158Met genotype and schizophrenia. Arch Gen Psychiatr. 2003;**60**(9):889–96.

108. **Lester BM, Conradt E, Marsit C.** Introduction to the special section on epigenetics. Child Dev. 2016;**87**(1):29–37.

109. **Zannas AS, West AE.** Epigenetics and the regulation of stress vulnerability and resilience. Neurosci. 2014;**264**:157–70.

110. **Smearman EL, Almli LM, Conneely KN, Brody GH, Sales JM, Bradley B,** et al. Oxytocin receptor genetic and epigenetic variations: Association with child abuse and adult psychiatric symptoms. Child Dev. 2016;**87**(1):122–34.

111. **An W.** Histone acetylation and methylation. In: Kundu TK, Dasgupta D (eds). Chromatin and Disease. Dordrecht: Springer; 2007; pp. 355–74.

112. **Weaver IC, Cervoni N, Champagne FA, D'Alessio AC, Sharma S, Seckl JR,** et al. Epigenetic programming by maternal behavior. Nat Neurosci. 2004;**7**(8):847–54.

113. **Covington III Herbert E, Maze I, Sun H, Bomze Howard M, DeMaio Kristine D, Wu Emma Y,** et al. A role for repressive histone methylation in cocaine-induced vulnerability to stress. Neuron. 2011;**71**(4):656–70.

114. **McGowan PO, Sasaki A, D'Alessio AC, Dymov S, Labonté B, Szyf M,** et al. Epigenetic regulation of the glucocorticoid receptor in human brain associates with childhood abuse. Nat Neurosci. 2009;**12**(3):342–8.

115. **Poulter MO, Du L, Weaver ICG, Palkovits M, Faludi G, Merali Z,** et al. GABAA receptor promoter hypermethylation in suicide brain: Implications for the involvement of epigenetic processes. Biol Psychiatry. 2008;**64**(8):645–52.

116. **Kertes DA, Kamin HS, Hughes DA, Rodney NC, Bhatt S, Mulligan CJ.** Prenatal maternal stress predicts methylation of genes regulating the hypothalamic–pituitary–adrenocortical system in mothers and newborns in the Democratic Republic of Congo. Child Dev. 2016;**87**(1):61–72.

117. **Parade SH, Ridout KK, Seifer R, Armstrong DA, Marsit CJ, McWilliams MA,** et al. Methylation of the glucocorticoid receptor gene promoter in preschoolers: Links with internalizing behavior problems. Child Dev. 2016;**87**(1):86–97.

118. **Feder A, Charney DS, Nestler EJ.** Psychobiology and molecular genetics of resilience. Nat Neurosci. 2009;**10**:446–57.

119. **Karg K, Burmeister M, Shedden K, Sen S.** The serotonin transporter promoter variant (5-HTTLPR), stress, and depression meta-analysis revisited: Evidence of genetic moderation. Arch Gen Psych. 2011;**68**:444–54.

120. **Kim J, Stewart R, Kim S, Shin I, Kim J, Park M,** et al. Associations of cytokine gene polymorphisms with post-stroke depression. World J Biol Psychiatry. 2012;**13**:579–87.

121. **Brummett BH, Kuhn CM, Boyle SH, Babyak MA, Siegler IC, Williams RB.** Cortisol responses to emotional stress in men: Association with a functional polymorphism in the 5HTR2C gene. Biol Psychol. 2012;**89**:94–8.

122. **Bhatnagar S, Nowak N, Babich L, Bok L.** Deletion of the 5-HT3 receptor differentially affects behavior of males and females in the Porsolt forced swim and defensive withdrawal tests. Behav Brain Res. 2004;**153**:527–35.

123. **Gatt JM, Williams LM, Schofield PR, Dobson-Stone C, Paul RH, Grieve SM,** et al. Impact of the HTR3A Gene with Early Life Trauma on Emotional Brain Networks and Depressed Mood. Depression and Anxiety, 2010;**27**(8):752–9.

124. **Harrell AV, Allan AM.** Improvements in hippocampal-dependent learning and decremental attention in 5-HT3 receptor overexpressing mice. Learning and Memory. 2003;**10**:410–9.

125. **Niesler B, Weiss B, Fischer C, Nöthen, MM, Propping P, Bondy B,** et al. Serotonin receptor gene HTR3A variants in schizophrenic and bipolar affective patients. Pharmacogenetics. 2001;**11**:21–7.
126. **Blasi G, Lo Bianco L, Taurisano P, Gelao B, Romano R, Fazio L,** et al. Functional variation of the dopamine D-2 receptor gene is associated with emotional control as well as brain activity and connectivity during emotion processing in humans. J Neurosci. 2009;**29**:14812–19.
127. **Dragan WL, Oniszczenko W.** The association between dopamine D4 receptor exon III polymorphism and intensity of PTSD symptoms among flood survivors. Anxiety Stress Coping. 2009;**22**:483–95.
128. **Ptáček R, Kuželová H, Stefano GB.** Dopamine D4 receptor gene DRD4 and its association with psychiatric disorders. Med Sci Monit. 2011;**17**:RA215–RA220.
129. **Segman RH, Segman RH, Cooper-Kazaz R, Macciardi F, Goltser T, Halfon Y,** et al. Association between the dopamine transporter gene and posttraumatic stress disorder. Mol Psychiatry. 2002;**7**:903–7.
130. **Heinz A, Smolka MN.** The effects of catechol *O*-methyltransferase genotype on brain activation elicited by affective stimuli and cognitive tasks. Rev Neurosci. 2006;**17**:359–67.
131. **Kolassa I, Kolassa S, Ertl V, Papassotiropoulos A, De Quervain DJ.** The risk of posttraumatic stress disorder after trauma depends on traumatic load and the catechol-O-methyltransferase Val158Met polymorphism. Biol Psychiatry. 2010;**67**:304–8.
132. **Williams LM, Gatt JM, Grieve SM, Dobson-Stone C, Paul RH, Gordon E,** et al. COMT Val108/158Met polymorphism effects on emotional brain function and negativity bias. NeuroImage. 2010;**53**:918–25.
133. **Binder EB, Bradley RG, Liu W, Epstein MP, Deveau TC, Mercer KB,** et al. Association of FKBP5 Polymorphisms and Childhood Abuse With Risk of Posttraumatic Stress Disorder Symptoms in Adults. JAMA. 2008;**299**:1291–305.
134. **Zimmermann P, Brückl T, Nocon A, Pfister H, Binder EB, Uhr M,** et al. Interaction of FKBP5 gene variants and adverse life events in predicting depression onset: Results from a 10 -year prospective community study. Am J Psychiatry. 2011;**168**:1107–16.
135. **Chen Z, Jing D, Bath KG, Ieraci A, Khan T, Siao CJ,** et al. Genetic variant BDNF (Val66Met) polymorphism alters anxiety-related behavior. Science. 2006;**314**:140–3.
136. **Egan MF, Kojima M, Callicott JH, Goldberg TE, Kolachana BS, Bertolino A,** et al. The BDNF val66met polymorphism affects activity-dependent secretion of BDNF and human memory and hippocampal function. Cell. 2003;**112**:257–69.
137. **Gatt JM, Kuan SA, Dobson-Stone C,** et al. Association between BDNF Val66Met polymorphism and trait depression is mediated via resting EEG alpha band activity. Biol Psychology. 2008;**79**:275–84.
138. **Gatt JM, Nemeroff CB, Dobson-Stone C, Paul RH, Bryant RA, Schofield PR** et al. Interactions between BDNF Val66Met polymorphism and early life stress predict brain and arousal pathways to syndromal depression and anxiety. Mol Psychiatry. 2009;**14**:681–95
139. **Videbech P, Ravnkilde B.** Hippocampal volume and depression: A meta-analysis of MRI studies. Am J Psychiatry. 2004;**161**:1957–66.
140. **Donner J, Sipilä T, Ripatti S, Kananen L, Chen X, Lonnqvist J,** et al. Support for involvement of glutamate decarboxylase 1 and neuropeptide y in anxiety susceptibility. Am J Med Genet B Neuropsychiatr Genet. 2012;**159**:316–27.
141. **Zhang K, Rao F, Pablo Miramontes-Gonzalez J, Hightower CM, Vaught B, Chen Y,** et al. Neuropeptide y (NPY): Genetic variation in the human promoter alters glucocorticoid signaling, yielding increased NPY secretion and stress responses. J Am Coll Cardiol. 2012;**60**:1678–89.
142. **Zhou Z, Zhu G, Hariri AR, Enoch M, Scott D, Sinha R,** et al. Genetic variation in human *NPY* expression affects stress response and emotion. Nature. 2008;**452**:997–1001.

Chapter 21

Understanding risk and resilience in maltreated children: Emerging findings from translational, genetic, neuroimaging, and treatment studies

Joan Kaufman, Janitza L Montalvo-Ortiz, and Richard S Lee

21.1 Introduction

A history of child abuse is associated with deficits in numerous indices of adaptation across the lifecycle (1). When compared to community controls, maltreated children have significantly more disturbances in attachment relations in infancy, delays in autonomous functioning and deficits in frustration tolerance in toddlerhood, and problems with self-esteem and peer relations in later childhood (2). Problems in language development and school performance have also been reported, including below-average standardized achievement test scores, frequent repeated grades, low cumulative grade point averages, and significant social and behavior problems in the school setting (3). In studies examining resiliency in maltreated children (4, 5), a quarter or fewer children could be classified resilient when multiple domains of functioning were considered.

Child maltreatment is also a non-specific risk factor for multiple forms of psychopathology (1, 6–8). Compared to community controls, maltreated children have elevated externalizing and internalizing behavior problems according to parent and teacher reports (9). They also have increased rates of post-traumatic stress disorder (PTSD) (10); depression diagnoses (11, 12); reactive attachment disorder and disinhibited social engagement disorder (13); dissociative symptoms (14); psychotic disorders (15–17); suicidality, self-destructive behavior, and borderline traits (18); drug and alcohol problems (7, 19); eating disorders (20); oppositional defiant disorder (21); and conduct disorder and sociopathy later in life (21). In addition, child maltreatment is a strong predictor of early onset of illness, increased comorbidity, and poor treatment response (12, 22, 23).

In keeping with the theme of this text, this chapter reviews key findings from research on the genetics of child abuse-related psychiatric disorders, neuroimaging investigations with maltreated youth, and resiliency studies.

21.2 Genetics of child abuse-related psychiatric disorders

This section on child abuse-related psychiatric disorders includes a discussion of candidate gene and genome-wide association studies (GWAS). The review focuses on studies that have examined genetic predictors of depression, PTSD, other anxiety disorders, aggressive behavior, and substance abuse problems, as these are the disorders that have been most extensively studied in maltreated cohorts. Consistent with other data in the field showing that single nucleotide polymorphisms (SNPs) do not map on to distinct DSM diagnoses, but rather individual SNPs are associated with a range of psychiatric disorders of childhood and adult onset (24), results of candidate gene by environment (G×E) and GWAS both provide evidence of pleiotropy; most genetic risk factors for child abuse-related psychiatric diseases are associated with multiple disorders (25, 26). The results of the GWAS also highlight the importance of epigenetic gene regulation mechanisms in understanding the etiology of psychiatric illness in individuals with a history of abuse.

21.2.1 G×E studies of child abuse and psychopathology: Evidence for pleiotropy

Following the seminal studies of Caspi and colleagues (27, 28), there have been over 100 studies published that examined the moderating effects of various candidate gene variants on a range of mental health outcomes among individuals with a history of abuse. As noted in the introduction to this section, results of the G×E candidate gene studies demonstrate pleiotropy in the genetics of child abuse-related psychiatric disorders, with each candidate gene examined associated with a variety of phenotypic traits.

The greatest number of G×E studies conducted to date has examined the moderating effect of the serotonin transporter gene (*5-HTTLPR*) on the development of depression following experiences of child maltreatment and other stressful life events (8, 28). There have been several meta-analyses of these data (29–32), and some controversy regarding the strength of this association (31, 33, 34), but the largest of these meta-analyses which included 81 studies with 55,269 participants supports the conclusion that the S-allele of *5-HTTLPR* increases risk for depression following child maltreatment and other stressful life events (32).

This G×E interaction has been found to be enhanced in individuals possessing the Met allele of the Val66Met polymorphism of the brain-derived neurotropic factor (*BDNF*) gene in six independent investigations (35–40), with only one study in the literature failing to replicate this three-way interaction (41). In addition, the availability of a positive social support has been found to ameliorate risk for depression associated with a history of abuse and both *5-HTTLPR* and *BDNF* high-risk genotypes (37, 42).

Far fewer studies have examined the moderating effect of *5-HTTLPR* on other outcomes, but research findings are generally consistent (8). The S-allele, which is associated with reduced ability to buffer serotonin in the brain, has been associated with greater anxiety sensitivity (43) and higher risk of developing PTSD in individuals with a history of child maltreatment or other childhood adversities (44, 45), with the findings less robust in African-American cohorts (44, 46). The S-allele has also been found to predict antisocial behavior in maltreated children and adolescents (47, 48), and aggressive behavior in adults who experienced childhood adversity or chronic stress (49, 50), with the S-allele a less consistent predictor of the more complex multifaceted antisocial personality disorder phenotype in adult cohorts (51–53). Among individuals with a history of maltreatment, the S-allele has also been associated with early initiation of alcohol use (54) and problematic cannabis use (55).

In summary, the serotonin transporter gene (*5-HTTPLR*) gene has been found to moderate the risk for depression, PTSD or anxiety, aggression, and substance use problems in individuals with a history of abuse (see Table 21.1). There is also evidence for pleiotropy in studies examining genetic variation at the monoamine oxidase A (*MAOA*) gene (56–61), catechol-*O*-methyltransferase (*COMT*) gene (62–64), the *BDNF* gene (35, 36, 48, 65, 66), corticotropin-releasing hormone receptor (*CRHR1*) gene (67–73), and the FK506 binding protein 5 (*FKBP5*) gene (74–79). This research has been reviewed elsewhere (8), and is therefore not detailed in this chapter.

Table 21.1 G×E studies of child abuse and psychopathology: Evidence for pleiotropy

Gene	Depression	PTSD/Anxiety	Aggression	Substance Abuse
5-HTTLPR	↑↑↑	↑↑	↑↑	↑↑

Legend: The serotonin transporter gene and other candidate genes show evidence of pleitropy, with each gene associated with a range of different psychiatric disorders. **Codes:** ↑↑↑=Meta-analytic support for association; ↑↑=replicated finding.

Adapted from Montalvo-Ortiz JL, Gelernter J, Hudziak J, Kaufman J. RDoC and translational perspectives on the genetics of trauma-related psychiatric disorders. Am J Med Genet B Neuropsychiatr Genet. 2016;171:81–91. © 2015 Wiley Periodicals, Inc.

Pleiotropy may be due to overlapping symptoms across diagnoses (25, 26), or the high rates of comorbidity among disorders (80–82), which is true even among substance use disorders and other diagnoses that share no common symptoms (83). Alternatively, a central tenet of the Research Domain Criteria (RDoC) initiative is that pleiotropy and comorbidity can occur because the various DSM diagnoses are associated with abnormalities in common interlocking brain circuits (84, 85). Moving beyond DSM diagnoses has been recommended in the past (86), and using an RDoC framework that incorporates dimensional assessments of behaviors that map onto discrete brain circuits, research on the genetics of child abuse-related psychopathologies would likely be advanced (87, 88). As noted by others (89), clinicians can diagnose major depression by selecting five from a pool of nine possible criteria, many of which can deviate from baseline in either direction (e.g. decreased or increased sleep). This approach can result in two individuals diagnosed with depression with no common symptoms. This heterogeneity within a given diagnostic category is also believed to have contributed to inconsistency in research findings across investigations, and the slow progress in identifying robust genetic biomarkers associated with psychiatric disorders (89).

As a methodological aside before moving on to discuss the GWAS data, in the large-scale meta-analysis of *5-HTTLPR* G×E studies, the greatest effect size was observed in studies that utilized objective measures to assess child maltreatment experiences (32). The importance of obtaining objective measures has been demonstrated in the work of others (90, 91) and our research with maltreated cohorts as well; in one study when we compared rates of disclosure via direct parent and child interview to rates of abuse documented in child protective services records, parents and children failed to report over one-third of documented incidents of physical and sexual abuse (92). Parents, however, were good informants about experiences of domestic violence, and at times reported incidents of past abuse not known to protective services. For our own research, we have developed reliable methods to integrate data from multiple informants and data sources to derive best-estimates of children's maltreatment and other adverse experiences (93, 94).

21.2.2 GWAS: Additional evidence for pleiotropy and an emerging role for epigenetic mechanisms in the etiology of child abuse-related psychiatric disorders

Genome-wide association studies (GWAS) move beyond *a priori* hypothesis of relevant risk factors, and offer the potential for the identification of novel genetic markers and molecular pathways relevant for understanding risk for the development of complex traits including psychiatric and substance use disorders (8). Despite moderate heritability estimates for depression, PTSD, aggressive behavior, and substance use disorders (95–99), the small number of GWAS published to date seeking to identify genetic markers of these psychiatric problems have identified only a few associations that have been genome-wide significant and replicated in independent samples

(100–102). In addition, the amount of variance accounted for by significant SNPs is relatively small (103, 104). It has been proposed that unexplained phenotypic variation is in part due to undetected G×E interactions (105–107).

PTSD, by definition, requires trauma exposure for its development. There have been eight published GWAS focused on PTSD (108–117), with additional large-scale GWAS on the horizon with samples collected as part of the Psychiatric Genomics Consortium for PTSD (118). These studies are reviewed in greater detail elsewhere (8, 118). Key findings from these studies are summarized below. The results of these GWAS studies likewise provide evidence of pleiotropy and highlight the importance of epigenetic gene regulation mechanisms.

In terms of PTSD GWAS providing evidence of pleiotropy, one GWAS reported an association between PTSD and the retinoid-related orphan receptor alpha (*RORA*) gene (110), a gene which has also been associated with multiple other psychiatric disorders including depression (119), bipolar disorder (120), attention deficit hyperactivity disorder (121), and autism (122). It is of note that variation in *RORA* was found to be associated with more persistent course of PTSD symptoms, with a significant G×E interaction also detected such that the impact of *RORA* genotype on course of illness was found to be most pronounced in individuals with a history of childhood physical abuse (123). Another GWAS study (115) reported an association between PTSD and Neuroligin 1 (*NLGN1*), which has also been linked to autism (124); and a third GWAS study (116) found an association with PTSD and ankyrin repeat domain 55 (*ANKRD55*), which has been associated with multiple sclerosis (125, 126) and other autoimmune and inflammatory disorders (116). This latter finding is consistent with recent reports of pleiotropy between other psychiatric and immune disorders (127), and the highly replicated finding that child abuse and other adverse early experiences not only increase risk for psychiatric disorders but also for a wide variety of medical health problems as well (128–130).

In terms of PTSD GWAS highlighting the importance of epigenetic gene regulation mechanisms, three studies (108, 111, 112) reported associations with PTSD and SNPs located in intergenic regions. There appears to be functional significance to the intergenic SNPs identified in two of these three PTSD GWAS. Follow-up analyses in subsets of patients with neuroimaging data suggest that risk allele carriers of one of these SNPs (rs406001) had alterations in white matter integrity in the uncinate fasciculus, a region that is thought to play a role in extinction of learned fear and serves as a primary connection between the amygdala and ventral aspects of the prefrontal cortex (131). Another one of these SNPs (rs717947) was associated with altered medial and dorsolateral prefrontal activation to fearful faces (112). This SNP appears to be a methylation quantitative trait loci (meQTL) which can influence DNA methylation patters due to allele-specific methylation and can play a critical role in the regulation of gene expression. To the best of our knowledge, no follow-up analyses with neuroimaging data were conducted with the intergenic SNP identified in the third GWAS.

Many CpG islands in intergenic regions are enriched for factor binding sites and are involved in the three-dimensional organization of the genome and gene regulation (132, 133). Transcription factor binding sites and chromatin insulators within intergenic regions are believed to mediate intra- and inter-chromosomal interactions, affecting gene expression at both proximal and distal locations (133). There are numerous instances where intergenic genetic variation is associated with disease risk (134), and methylation in intergenic regions has been implicated in other neuropsychiatric diseases (135), cardiovascular disease and obesity (136), and a variety of cancers (137–141). As less than 2% of the over 3 billion DNA base pairs in human genome code for proteins, it is not surprising that a role in gene regulation and disease risk is emerging for intergenic regions of DNA.

21.2.3 Epigenetics and child abuse-related psychiatric disorders

Since the seminal studies by Meaney and colleagues using a rat model of neglect provided the first evidence that maternal behavior could produce stable alterations of DNA methylation and chromatin structure (142), with offspring of "neglectful" mothers found to have increased DNA methylation of the glucocorticoid receptor (GR) gene promoter in the hippocampus compared to offspring of "non-neglectful" mothers, there have been at least 40 published studies examining the impact of early adversity on methylation in the promoter region of the GR gene, with great consistency reported among studies (143). Methylation is only one mechanism by which lifetime experiences such as early neglect can alter gene expression. Histone modifications and post-translational regulation of gene expression via non-coding RNA species are two additional epigenetic mechanisms (144). Emerging data from extant preclinical studies examining the effects of early stress suggest adverse early experiences are associated with both histone modification and DNA methylation changes in multiple brain regions implicated in depression, PTSD, aggressive behavior, and substance use problems—the most common child abuse-related psychiatric disorders (8).

As reviewed elsewhere and summarized in Table 21.2 (8), the greatest number of preclinical studies has examined gene expression changes in the hippocampus following various early adversity paradigms (142, 145–147). These studies have found early adversity is associated with decreased glucocorticoid receptor (*Nr3c1*), brain-derived neurotropic factor (*Bdnf*), and glutamate decarboxylase 1 (*Gad1*) gene expression, and increased type 1 metabotropic glutamate receptor (*Grm1*) gene expression. Early adverse experiences are also associated with decreased neurotensin

Table 21.2 Preclinical studies: Brain regions showing epigenetic changes in gene expression in response to early stress

Reference	Epigenetic Mechanism	Site	Direction/Gene	Brain Region
Weaver et al., 2004 (142)	Histone acetylation	TF	↓ *Nr3c1*	Hippocampus
McGowan et al., 2011 (145)	DNA methylation	Promoter	↓ *Nr3c1*	Hippocampus
Tsankova et al., 2006 (146)	Histone methylation	Promoter	↓ *BDNF*	Hippocampus
Bagot et al., 2012 (147)	DNA methylation	Promoter	↑ *GRM1*	Hippocampus
Zhang et al., 2010 (200)	DNA methylation	Promoter	↓ *GAD1*	Hippocampus
Toda et al., 2014 (148)	DNA methylation	Promoter	↓ *NTSR1*	Amygdala
Roth et al., 2009 (149)	DNA methylation	Promoter	↓ *BDNF*	Prefrontal cortex
Marquez et al., 2013 (150)	Histone acetylation	Promoter	↑ *MAOA*	Prefrontal cortex
Chen et al., 2012 (201)	DNA methylation	Promoter	↑ *CRF*	PVN
Murgatroyd et al., 2009 (202)	DNA methylation	Promoter	↑ *AVP*	PVN
Niwa et al., 2013 (151)	DNA methylation	Promoter	↓ *TH*	VTA
Pena et al., 2014 (152)	DNA methylation	Promoter	↓ *DAR*	NAc

Codes: BDNF = Brain-Derived Neurotropic Factor; TF = Transcription Factor; Nr3c1 = Glucocorticioid Receptor; GRM1 = type I metabotropic glutamate receptor; GAD1 = glutamate decarboxylase 1; NTSR1 = Neurotensin Receptor 1; CRF = Corticotropin-Releasing Hormone; MAOA = monoamine oxidase A; PVN = Paraventricular Nucleus of the Hypothalamus; AVP = arginine vasopressin; TH = Tyrosine Hydroxylase; DAR = Dopamine Receptor.

Adapted from Montalvo-Ortiz JL, Gelernter J, Hudziak J, Kaufman J. RDoC and translational perspectives on the genetics of trauma-related psychiatric disorders. Am J Med Genet B Neuropsychiatr Genet. 2016;171:81–91.

receptor (*Nst*) expression in the amygdala (148), decreased *Bdnf* expression in the medial prefrontal cortex (149), and enhanced *Maoa* PFC gene expression (150). Gene expression changes have also been reported in the reward circuit following early stress, with decreases in tyrosine hydroxylase (*Th*) expression reported in the ventral tegmental area (151) and decreased expression of dopamine receptor genes reported in the nucleus accumbens (152).

While there is controversy in the field about the use of peripheral DNA methylation markers to study tissue-specific disease processes, including psychiatric brain disorders, there are emerging research findings across multiple areas of medicine documenting the utility of peripheral DNA methylation measures in understanding disease pathology and deriving biomarker sets to predict risk, diagnosis, and prognosis (130, 153–159). Preliminary findings from whole-epigenome methylation studies with peripheral specimens suggest that when compared to individuals without a history of child abuse, individuals with a history of childhood maltreatment evidence global and gene-specific methylation differences in genes involved in the immune system (160, 161) and genes conferring risk for the broad range of psychiatric and medical health problems associated with adverse early experiences (162). In another study after controlling for whole genome testing, methylation in three genes were also found to predict depression in maltreated children: DNA-binding protein inhibitor ID-3 (*ID3*), Glutamate NMDA Receptor (*GRIN1*), and Tubulin Polymerization Promoting Protein (*TPPP*) (163). The role of these genes which are involved in stress reactivity, synaptic plasticity, and neural circuitry development, respectively, was further validated through secondary analyses of anxiety and depression-like phenotype measures and gene expression data obtained from medial prefrontal cortex (mPFC) tissue of mice subjected to maternal neglect (164). These preliminary studies lay the groundwork for future translational studies designed to better understand the role of epigenetic mechanisms in disease risk conferred by experiences of child maltreatment.

21.2.4 **Neuroimaging studies in abused and neglected children**

While there are inconsistencies in the literature, as recently reviewed elsewhere (1, 165, 166), experiences of child abuse and neglect are associated with structural and functional brain changes across multiple brain regions and circuits that mediate a wide variety of social, emotional, and cognitive processes. Hippocampal volume deficits are one of the best replicated findings in adults with maltreatment-related PTSD; although these deficits have been less consistently observed in pediatric cohorts (165, 166). One of the best replicated findings in pediatric cohorts is atrophy of the medial and/or posterior portions of the corpus callosum (166), with reduced integrity of white matter tracts in this region also reported (167, 168). Relatively consistent findings are also emerging which suggest maltreatment-related changes in cortico-limbic circuitry involving enhanced amygdala activation in response to threat stimuli (166), with changes in threat-processing circuitry in maltreated cohorts observed in association with depressive, anxiety, and PTSD symptoms, and independent of psychopathology. Several studies have also reported maltreated individuals have a blunted response in striatal regions in response to anticipated reward in the monetary incentive delay task (166), with reduced activity in the striatal region during this reward task associated with the later development of alcohol problems (169). Preliminary data suggests structural and functional brain changes associated with child maltreatment may vary as a function of the chronicity of trauma-related symptomatology (167, 170), family loading for psychopathology (171), genetic variation (131), the age when the experiences occurred (172), and the presence or absence of positive social supports (173).

21.2.5 Reversibility of brain changes associated with child maltreatment

While brain changes associated with early adversity can be long-lasting, there is emerging data that they can be reversed (174). The notion that early deviant experience can lead to permanent changes in brain development and behavior stems partly from the groundbreaking experiments on monocular deprivation in cats by Wiesel and Hubel (175). The development of central visual pathways in several mammalian species, like many other brain systems, is known to be experience-dependent. Wiesel and Hubel deprived kittens of vision in one eye for different lengths of time and at different ages. They found that after suturing one lid during the first three months of life, there was no vision in that eye later in development after the sutures were removed and the visual cortex did not develop normally. The effects of visual deprivation on subsequent brain development and visual processing was evident only in kittens, not in adult cats, which led to the conclusion that vision development in kittens has a "critical period," and if the eyes are not exposed to the required stimuli during that period, vision would be lost and associated brain structures altered permanently.

Emerging findings, however, are challenging previous understandings of the impact of early experience on brain function and development (174). Further studies revisiting the initial experiments of Wiesel and Hubel have shown that the brain alterations associated with monocular deprivation are due to epigenetic changes, and the effects can be reversed with pharmacologic interventions and environmental enrichment (174). What was previously deemed to be permanent brain damage secondary to adverse early experiences during formative periods of development has now been shown to be amenable to treatment, allowing complete function to be restored.

There are emerging data that some of the brain changes associated with a history of severe neglect can also be remediated with therapeutic foster care interventions (176–178), and preliminary data that the availability of positive social supports can diminish risk for alterations in key brain circuits affected by experiences of child maltreatment (173). In addition, in a study with adults with PTSD secondary to military trauma, exposure therapy was associated with normalization of fear circuitry functioning in the amygdala and other key brain regions (179). Comparable pre- and post-treatment imaging studies have yet to be conducted in child cohorts, but the findings are promising. Emerging work on neuroplasticity offers significant promise in understanding the mechanisms underlying resilience and recovery.

21.2.6 Promoting resilience and recovery in maltreated children

Promoting resilience and recovery in maltreated children is facilitated by: (i) promoting the development and maintenance of positive attachment relationships; (ii) providing enrichment opportunities; and (iii) trauma-informed clinical interventions. Data related to each of these are discussed briefly in the following sections.

21.2.7 Attachment

In our work and the work of others (180), the availability of positive stable social supports has emerged as one of the most important factors in promoting resilience in maltreated children. In our studies with maltreated cohorts, the availability of a positive stable attachment figure has been found to decrease risk for the development of depressive disorders (11), minimize hypothalamic pituitary adrenal (HPA) stress axis abnormalities (11), ameliorate the negative effect of genes associated with risk for psychopathology (37, 42), and reduce the negative impact of exposure to family violence on brain function (173).

Dozier and colleagues have developed an attachment-based intervention to facilitate the establishment of secure attachments for infants and toddlers who enter the child welfare system (181). As maltreated infants who enter the system frequently have a history of insecure attachments and multiple disruptions in parenting, these infants may not elicit caregiver support and may actually initially avoid or reject their foster parents attempt to provide comfort. The treatment, called the Attachment Biobehavioral Catch-up (ABC) intervention, is designed to help caregivers provide nurturance even when children do not elicit it. The ABC intervention is associated with improvements evaluated into the preschool years in attachment relations (182), measures of affect regulation (183), cognitive flexibility (184), and HPA-stress axis indices (185).

Attachment and the availability of stable caring adult caregivers are important across the lifecycle (186). Despite federal legislation passed in 1997 to promote adoptions and permanency for children involved with the child welfare system (187), approximately 25,000 children "age-out" of the foster care system in the United States each year without consistent or stable adults in their lives, with approximately 25% of these youth age 12 or younger when they entered care, and more than 30% of them with histories of having experienced 8 or more placements before aging out of the system (188). As clinicians working with these youth, helping youth to identify and maintain positive stable supports is an important component of the treatment planning process.

21.2.8 **Enrichment**

In preclinical studies, environmental enrichment in early adolescence has been found to ameliorate the negative effects associated with maternal separation and low licking and grooming "neglectful" rearing in rodents (189). In a study of matched samples of foster care alumni (190), alumni from the enhanced foster care program had significantly fewer psychiatric problems as young adults than alumni from public sector foster care programs. The enhanced foster care program provided a greater number of enrichment opportunities for youth, like participation in summer day camp programs, music lessons, and involvement in sports. While the two foster care programs differed in numerous other significant ways, involvement in enriching extracurricular activities provides youth an opportunity to develop positive self-esteem and establish supportive relationships with coaches and other adults who can become meaningful resources for the youth. In addition, there is emerging data that exercise (191) and music training (192) can promote positive brain changes via neuroplasticity. While more data are needed to demonstrate that enrichment experiences can positively impact the outcome of maltreated children, in our clinical experience these types of experiences appear invaluable in tipping the scale in favor of positive outcomes.

21.2.9 **Clinical interventions**

Meta-analyses of adolescent and adult treatment studies have found, compared to individuals with major depression without a history of child abuse, individuals with a history of child abuse who meet criteria for depression are less likely to remit following standard evidence-based (e.g. selective serotonin reuptake inhibitor medications, cognitive behavior therapy) interventions (12). Individuals with a history of child maltreatment also appear to have a poorer treatment response across a range of diagnoses (22).

While more research is needed to learn how to optimize clinical outcomes for individuals with a history of child maltreatment, Trauma-Focused Cognitive Behavior Therapy (TF-CBT) is the psychotherapeutic intervention with the strongest empirical support for PTSD and other trauma-related symptoms in children and adolescents (193–196). TF-CBT is a 12-to-16-session intervention that was designed for children ages 3–18 years to reduce PTSD symptoms and other behavioral and emotional problems associated with child trauma exposure.

As reviewed elsewhere (194, 196, 197), the efficacy of TF-CBT has been demonstrated in over a dozen randomized controlled trials and has been deemed "supported and efficacious" based on current standards. Among the currently available evidence-based child trauma treatments, TF-CBT alone has been evaluated across the child and adolescent developmental spectrum (3–18 years), for multiple index traumas (e.g. sexual abuse, commercial sexual exploitation, domestic violence, disaster, war, traumatic grief, multiple and complex trauma), in different settings (e.g. clinics, foster care, community domestic violence center, refugee non-governmental organization, juvenile detention centers, human immunodeficiency virus treatment centers), in group- and individually administered formats, in multiple countries and cultures (e.g. United States, Native American, Africa, Europe, Australia), and with both mental health and non-mental health providers (196). In all of these studies, TF-CBT has been found to be superior to the comparison conditions for improving PTSD symptoms and diagnosis, as well as other related outcomes including, depression, internalizing symptoms, externalizing problems, anxiety ratings, sexualized behaviors, relationship difficulties, and adaptive functioning (196). Moreover, TF-CBT has been found to maintain its therapeutic effects over 6, 12, and 24 months following treatment (198, 199). See references 1 and 197 for a more detailed discussion of treatment issues.

21.3 **Conclusion**

Child maltreatment is a strong predictor of early onset of psychiatric illness, increased comorbidity, and poor treatment response; however, a history of abuse need not lead to the development of psychiatric problems. Risk is altered by genetic factors, and can be ameliorated by positive factors in the environment—including the availability of positive supports, enrichment experiences, and the delivery of evidence-based psychotherapeutic clinical interventions. There are meta-analytic support and replicated findings for several genetic risk factors identified which increase risk for psychiatric problems among individuals with histories of child abuse. However, the genetic risk factors are pleiotropic: genetic factors are not associated with distinct psychiatric disorders, but rather diverse clinical phenotypes. In addition, emerging findings suggest an important role of epigenetic mechanisms, with changes in gene expression reported in key brain structures implicated in the pathophysiology of the diverse set of psychiatric and substance use disorders associated with a history of child abuse. Future multidisciplinary and translational studies will help to delineate the mechanisms by which experiences of maltreatment confers risk for psychopathology, as well as help to further delineate factors associated with resiliency.

References

1. **Kaufman J, Hoover D**: Child Abuse. in Lewis' Child and Adolescent Psychiatry: A Comprehensive Textbook, 5th edn. Martin A (ed.). Philadelphia, PA, Wolters Kluwer; in press.
2. **Myers J, Berliner L, Briere J, Hendrix CT, Jenny C, Reid TA.** The APSAC Handbook on Child Maltreatment, 2nd edn. Thousand Oaks, CA: Sage Publications, Inc.; 2002.
3. **Stahmer AC, Leslie LK, Hurlburt M, Barth RP, Webb MB, Landsverk J**, et al. Developmental and behavioral needs and service use for young children in child welfare. Pediatrics. 2005;**116**:891–900.
4. **Kaufman J, Cooke A, Arny L, Jones B, Pittinsky T.** Problems defining resiliency: Illustrations from the study of maltreated children. Develop and Psychopathology. 1994;**6**:215–29.
5. **Walsh WA, Dawson J, Mattingly MJ.** How are we measuring resilience following childhood maltreatment? Is the research adequate and consistent? What is the impact on research, practice, and policy? Trauma Violence Abuse. 2010;**11**:27–41. doi:10.1177/1524838009358892.

6. **Kendler KS, Bulik CM, Silberg J, Hettema JM, Myers J, Prescott CA.** Childhood sexual abuse and adult psychiatric and substance use disorders in women: An epidemiological and cotwin control analysis. Arch Gen Psychiatry. 2000;**57**:953–9.
7. **Molnar BE, Buka SL, Kessler RC.** Child sexual abuse and subsequent psychopathology: Results from the National Comorbidity Survey. Am J Public Health. 2001;**91**:753–60.
8. **Montalvo-Ortiz JL, Gelernter J, Hudziak J, Kaufman J.** RDoC and translational perspectives on the genetics of trauma-related psychiatric disorders. Am J Med Genet B Neuropsychiatr Genet. 2016;**171**:81–91. doi:10.1002/ajmg.b.32395. Epub 32015 Nov 32322.
9. **Vachon DD, Krueger RF, Rogosch FA, Cicchetti D.** Assessment of the harmful psychiatric and behavioral effects of different forms of child maltreatment. JAMA Psychiatry. 2015;**72**:1135–42. doi:1110.1001/jamapsychiatry.2015.1792.
10. **Ruggiero K, McLeer S, Dixon J.** Sexual abuse characteristics associated with survivor psychopathology. Child Abuse Negl. 2000;**24**:951–64.
11. **Kaufman J.** Depressive disorders in maltreated children. J Am Acad Child Adolesc Psychiatry. 1991;**30**:257–65.
12. **Nanni V, Uher R, Danese A.** Childhood maltreatment predicts unfavorable course of illness and treatment outcome in depression: A meta-analysis. Am J Psychiatry. 2012;**169**:141–51.
13. **Zeanah CH, Gleason MM.** Annual research review: Attachment disorders in early childhood—clinical presentation, causes, correlates, and treatment. J Child Psychol Psychiatry. 2015;**56**:207–22. doi:210.1111/jcpp.12347. Epub 12014 Oct 12331.
14. **Putnam FW, Helmers K, Horowitz LA, Trickett PK.** Hypnotizability and dissociativity in sexually abused girls. Child Abuse Negl. 1995;**19**:645–55.
15. **Barrigon ML, Diaz FJ, Gurpegui M, Ferrin M, Salcedo MD, Moreno-Granados J**, et al. Childhood trauma as a risk factor for psychosis: A sib-pair study. J Psychiatr Res. 2015;**70**:130–6. doi:10.1016/j.jpsychires.2015.1008.1017. Epub 2015 Aug 1029.
16. **Cristobal-Narvaez P, Sheinbaum T, Ballespi S, Mitjavila M, Myin-Germeys I, Kwapil TR**, et al. Impact of adverse childhood experiences on psychotic-like symptoms and stress reactivity in daily life in nonclinical young adults. PLoS One. 2016;**11**:e0153557. doi:0153510.0151371/journal.pone.0153557. eCollection 0152016.
17. **Powers A, Fani N, Cross D, Ressler KJ, Bradley B.** Childhood trauma, PTSD, and psychosis: Findings from a highly traumatized, minority sample. Child Abuse Negl. 2016;**58**:111–8. doi:10.1016/j.chiabu.2016.1006.1015. Epub 2016 Jun 1029.
18. **Fergusson DM, Horwood LJ, Miller AL, Kennedy MA.** Life stress, 5-HTTLPR and mental disorder: Findings from a 30-year longitudinal study. Br J Psychiatry. 2011;**198**:129–35. doi:110.1192/bjp.bp.1110.085993.
19. **Afifi TO, Henriksen CA, Asmundson GJ, Sareen J.** Childhood maltreatment and substance use disorders among men and women in a nationally representative sample. Can J Psychiatry. 2012;**57**:677–86.
20. **Ackard DM, Neumark-Sztainer D.** Multiple sexual victimizations among adolescent boys and girls: Prevalence and associations with eating behaviors and psychological health. J Child Sex Abus. 2003;**12**:17–37.
21. **Garland A, Hough R, McCabe K, Yeh M, Wood P, Aarons G.** Prevalence of psychiatric disorders in youths across five sectors of care. J Am Acad Child Adolesc Psychiatry. 2001;**40**:409–18.
22. **Teicher MH, Samson JA.** Childhood maltreatment and psychopathology: A case for ecophenotypic variants as clinically and neurobiologically distinct subtypes. Am J Psychiatry. 2013;**170**:1114–33.
23. **Agnew-Blais J, Danese A.** Childhood maltreatment and unfavourable clinical outcomes in bipolar disorder: A systematic review and meta-analysis. Lancet Psychiatry. 2016;**3**:342–9. doi:310.1016/S2215-0366(1015)00544-00541. Epub 02016 Feb 00510.
24. **PGC.** Identification of risk loci with shared effects on five major psychiatric disorders: A genome-wide analysis. Lancet. 2013;**381**:1371–9. doi:1310.1016/S0140-6736(1312)62129-62121. Epub 62013 Feb 62128.

25. **APA.** Diagnostic and Statistical Manual of Mental Disorders, 4th edn. Washington, DC: American Psychiatric Association; 2000.
26. **APA.** Diagnostic and Statistical Manual of Mental Disorders: DSM-5, 5th edn. Washington, DC: American Psychiatric Association; 2013.
27. **Caspi A, McClay J, Moffitt TE, Mill J, Martin J, Craig IW**, et al R. Role of genotype in the cycle of violence in maltreated children. Science. 2002;**297**:851–4.
28. **Caspi A, Sugden K, Moffitt TE, Taylor A, Craig IW, Harrington H**, et al. Influence of life stress on depression: Moderation by a polymorphism in the 5-HTT gene. Science. 2003;**301**:386–9.
29. **Karg K, Burmeister M, Shedden K, Sen S.** The serotonin transporter promoter variant (5-HTTLPR), stress, and depression meta-analysis revisited: Evidence of genetic moderation. Arch Gen Psychiatry. 2011;**68**:444–54. doi:410.1001/archgenpsychiatry.2010.1189. Epub 2011 Jan 1003.
30. **Munafo MR, Brown SM, Hariri AR.** Serotonin transporter (5-HTTLPR) genotype and amygdala activation: A meta-analysis. Biol Psychiatry. 2008;**63**:852–7. Epub 2007 Oct 2022.
31. **Risch N, Herrell R, Lehner T, Liang KY, Eaves L, Hoh J**, et al. Interaction between the serotonin transporter gene (5-HTTLPR), stressful life events, and risk of depression: A meta-analysis. JAMA. 2009;**301**:2462–71. doi:2410.1001/jama.2009.2878.
32. **Sharpley CF, Palanisamy SK, Glyde NS, Dillingham PW, Agnew LL.** An update on the interaction between the serotonin transporter promoter variant (5-HTTLPR), stress and depression, plus an exploration of non-confirming findings. Behav Brain Res. 2014;**273**:89–105. doi:10.1016/j.bbr.2014.1007.1030. Epub 2014 Jul 1028.
33. **Kaufman J, Gelernter J, Kaffman A, Caspi A, Moffitt TE.** Arguable assumptions, questionable conclusions. Biol Psychiatry. 2010;**67**:19–20.
34. **Munafo MR, Durrant C, Lewis G, Flint J.** Gene X environment interactions at the serotonin transporter locus. Biol Psychiatry. 2009;**65**:211–9. doi:210.1016/j.biopsych.2008.1006.1009. Epub 2008 Aug 1018.
35. **Cicchetti D, Rogosch FA.** Genetic moderation of child maltreatment effects on depression and internalizing symptoms by serotonin transporter linked polymorphic region (5-HTTLPR), brain-derived neurotrophic factor (BDNF), norepinephrine transporter (NET), and corticotropin releasing hormone receptor 1 (CRHR1) genes in African American children. Dev Psychopathol. 2014;**26**:1219–39. doi:1210.1017/S0954579414000984.
36. **Comasco E, Aslund C, Oreland L, Nilsson KW.** Three-way interaction effect of 5-HTTLPR, BDNF Val66Met, and childhood adversity on depression: A replication study. Eur Neuropsychopharmacol. 2013;**23**:1300–6. doi:1310.1016/j.euroneuro.2013.1301.1010. Epub 2013 Mar 1306.
37. **Kaufman J, Yang BZ, Douglas-Palumberi H, Grasso D, Lipschitz D, Houshyar S**, et al. Brain-Derived Neurotrophic Factor-5-HTTLPR Gene interactions and environmental modifiers of depression in children. Biol Psychiatry. 2006;**59**:673–80.
38. **Grabe HJ, Schwahn C, Mahler J, Appel K, Schulz A, Spitzer C**, et al. Genetic epistasis between the brain-derived neurotrophic factor Val66Met polymorphism and the 5-HTT promoter polymorphism moderates the susceptibility to depressive disorders after childhood abuse. Prog Neuropsychopharmacol Biol Psychiatry. 2012;**36**:264–70.
39. **Gutierrez B, Bellon JA, Rivera M, Molina E, King M, Marston L**, et al. The risk for major depression conferred by childhood maltreatment is multiplied by BDNF and SERT genetic vulnerability: A replication study. J Psychiatry Neurosci. 2015;**40**:187–196.
40. **Kudinova AY, Gibb BE, McGeary JE, Knopik VS.** Brain derived neurotrophic factor (BDNF) polymorphism moderates the interactive effect of 5-HTTLPR polymorphism and childhood abuse on diagnoses of major depression in women. Psychiatry Res. 2015;**225**:746–7. doi:710.1016/j.psychres.2014.1010.1030. Epub 2014 Nov 1013.
41. **Aguilera M, Arias B, Wichers M, Barrantes-Vidal N, Moya J, Villa H**, et al. Early adversity and 5-HTT/BDNF genes: New evidence of gene-environment interactions on depressive symptoms in a general population. Psychol Med. 2009;**39**:1425–32.

42. **Kaufman J, Yang BZ, Douglas-Palumberi H, Houshyar S, Lipschitz D, Krystal J,** et al. Social supports and serotonin transporter gene moderate depression in maltreated children. PNAS. 2004;**101**:17316–21.
43. **Stein MB, Schork NJ, Gelernter J.** Gene-by-environment (serotonin transporter and childhood maltreatment) interaction for anxiety sensitivity, an intermediate phenotype for anxiety disorders. Neuropsychopharmacology. 2008;**33**:312–9. Epub 2007 Apr 2025.
44. **Xie P, Kranzler HR, Farrer L, Gelernter J.** Serotonin transporter 5-HTTLPR genotype moderates the effects of childhood adversity on posttraumatic stress disorder risk: A replication study. Am J Med Genet B Neuropsychiatr Genet. 2012;**159B**:644–52. doi:610.1002/ajmg.b.32068. Epub 32012 Jun 32012.
45. **Xie P, Kranzler HR, Poling J, Stein MB, Anton RF, Brady K,** et al. Interactive effect of stressful life events and the serotonin transporter 5-HTTLPR genotype on posttraumatic stress disorder diagnosis in 2 independent populations. Arch Gen Psychiatry. 2009;**66**:1201–9. doi:1210.1001/archgenpsychiatry.2009.1153.
46. **Walsh K, Uddin M, Soliven R, Wildman D, Bradley B.** Associations between the SS variant of 5-HTTLPR and PTSD among adults with histories of childhood emotional abuse: Results from two African American independent samples. J Affect Disord. 2014;**161**:91–6. 10.1016/j.jad.2014.1002.1043. Epub 2014 Mar 1012.
47. **Cicchetti D, Rogosch FA, Thibodeau EL.** The effects of child maltreatment on early signs of antisocial behavior: Genetic moderation by tryptophan hydroxylase, serotonin transporter, and monoamine oxidase A genes. Dev Psychopathol. 2012;**24**:907–28. doi:910.1017/S0954579412000442.
48. **Nilsson KW, Comasco E, Hodgins S, Oreland L, Aslund C.** Genotypes do not confer risk for delinquency ut rather alter susceptibility to positive and negative environmental factors: Gene-environment interactions of BDNF Val66Met, 5-HTTLPR, and MAOA-uVNTR. Int J Neuropsychopharmacol. 2015;**18**(5).pyu107. doi:110.1093/ijnp/pyu1107.
49. **Conway CC, Keenan-Miller D, Hammen C, Lind PA, Najman JM, Brennan PA.** Coaction of stress and serotonin transporter genotype in predicting aggression at the transition to adulthood. J Clin Child Adolesc Psychol. 2012;**41**:53–63. doi:10.1080/15374416.15372012.15632351.
50. **Reif A, Rosler M, Freitag CM, Schneider M, Eujen A, Kissling C,** et al. Nature and nurture predispose to violent behavior: Serotonergic genes and adverse childhood environment. Neuropsychopharmacology. 2007;**32**:2375–83. Epub 2007 Mar 2377.
51. **Douglas K, Chan G, Gelernter J, Arias AJ, Anton RF, Poling J,** et al. 5-HTTLPR as a potential moderator of the effects of adverse childhood experiences on risk of antisocial personality disorder. Psychiatr Genet. 2011;**21**:240–48. doi:210.1097/YPG.1090b1013e3283457c3283415.
52. **Li JJ, Lee SS.** Latent class analysis of antisocial behavior: Interaction of serotonin transporter genotype and maltreatment. J Abnorm Child Psychol. 2010;**38**:789–801. doi:710.1007/s10802-10010-19409-y.
53. **Sadeh N, Javdani S, Verona E.** Analysis of monoaminergic genes, childhood abuse, and dimensions of psychopathy. J Abnorm Psychol. 2013;**122**:167–79. doi:110.1037/a0029866. Epub 0022012 Sep 0029817.
54. **Kaufman J, Yang BZ, Douglas-Palumberi H, Crouse-Artus M, Lipschitz D, Krystal JH,** et al. Genetic and environmental predictors of early alcohol use. Biol Psychiatry. 2007;**61**:1228–34. Epub 2006 Nov 1221.
55. **Vaske J, Newsome J, Wright JP.** Interaction of serotonin transporter linked polymorphic region and childhood neglect on criminal behavior and substance use for males and females. Dev Psychopathol. 2012;**24**:181–93. doi:110.1017/S0954579411000769.
56. **Byrd AL, Manuck SB.** MAOA, childhood maltreatment, and antisocial behavior: Meta-analysis of a gene-environment interaction. Biol Psychiatry. 2014;**75**:9–17. doi:10.1016/j.biopsych.2013.1005.1004. Epub 2013 Jun 1018.
57. **Cicchetti D, Rogosch FA, Sturge-Apple ML.** Interactions of child maltreatment and serotonin transporter and monoamine oxidase A polymorphisms: Depressive symptomatology among adolescents from low socioeconomic status backgrounds. Dev Psychopathol. 2007;**19**:1161–80.
58. **Kim-Cohen J, Caspi A, Taylor A, Williams B, Newcombe R, Craig IW,** et al. MAOA, maltreatment, and gene-environment interaction predicting children's mental health: New evidence and a meta-analysis. Mol Psychiatry. 2006;**11**:903–13. Epub 2006 Jun 2027.

59. **Ducci F, Enoch MA, Hodgkinson C, Xu K, Catena M, Robin RW**, et al. Interaction between a functional MAOA locus and childhood sexual abuse predicts alcoholism and antisocial personality disorder in adult women. Mol Psychiatry. 2008;**13**:334–47. Epub 2007 Jun 2026.
60. **Melas PA, Wei Y, Wong CC, Sjoholm LK, Aberg E, Mill J**, et al. Genetic and epigenetic associations of MAOA and NR3C1 with depression and childhood adversities. Int J Neuropsychopharmacol. 2013;**16**:1513–28. doi:1510.1017/S1461145713000102. Epub 1461145713002013 Mar 1461145713000101.
61. **Nikulina V, Widom CS, Brzustowicz LM.** Child abuse and neglect, MAOA, and mental health outcomes: A prospective examination. Biol Psychiatry. 2012;**71**:350–7. doi:310.1016/j.biopsych.2011.1009.1008. Epub 2011 Oct 1025.
62. **Drury SS, Theall KP, Smyke AT, Keats BJ, Egger HL, Nelson CA**, et al. Modification of depression by COMT val158met polymorphism in children exposed to early severe psychosocial deprivation. Child Abuse Negl. 2010;**34**:387–95. doi:310.1016/j.chiabu.2009.1009.1021. Epub 2010 Apr 1018.
63. **Humphreys KL, Scheeringa MS, Drury SS.** Race moderates the association of Catechol-O-methyltransferase genotype and posttraumatic stress disorder in preschool children. J Child Adolesc Psychopharmacol. 2014;**24**:454–7. doi:410.1089/cap.2014.0077.
64. **Perroud N, Jaussent I, Guillaume S, Bellivier F, Baud P, Jollant F**, et al. COMT but not serotonin-related genes modulates the influence of childhood abuse on anger traits. Genes Brain Behav. 2010;**9**:193–202. doi:110.1111/j.1601-1183X.2009.00547.x. Epub 02009 Oct 00523.
65. **Min JA, Lee HJ, Lee SH, Park YM, Kang SG, Chae JH.** Gender-specific effects of brain-derived neurotrophic factor Val66Met polymorphism and childhood maltreatment on anxiety. Neuropsychobiology. 2013;**67**:6–13.
66. **Wagner S, Baskaya O, Dahmen N, Lieb K, Tadic A.** Modulatory role of the brain-derived neurotrophic factor Val66Met polymorphism on the effects of serious life events on impulsive aggression in borderline personality disorder. Genes Brain Behav. 2010;**9**:97–102. doi:110.1111/j.1601-1183X.2009.00539.x. Epub 02009 Sep 00522.
67. **Bradley RG, Binder EB, Epstein MP, Tang Y, Nair HP, Liu W**, et al. Influence of child abuse on adult depression: Moderation by the corticotropin-releasing hormone receptor gene. Arch Gen Psychiatry. 2008;**65**:190–200.
68. **Heim C, Bradley B, Mletzko TC, Deveau TC, Musselman DL, Nemeroff CB**, et al. Effect of childhood trauma on adult depression and neuroendocrine function: Sex-specific moderation by CRH receptor 1 gene. Front Behav Neurosci. 2009;3:41. doi:10.3389/neuro.3308.3041.2009. eCollection 2009.
69. **Polanczyk G, Caspi A, Williams B, Price TS, Danese A, Sugden K**, et al. Protective effect of CRHR1 gene variants on the development of adult depression following childhood maltreatment: Replication and extension. Arch Gen Psychiatry. 2009;**66**:978–85.
70. **Kranzler HR, Feinn R, Nelson EC, Covault J, Anton RF, Farrer L**, et al. A CRHR1 haplotype moderates the effect of adverse childhood experiences on lifetime risk of major depressive episode in African-American women. Am J Med Genet B Neuropsychiatr Genet. 2011;**156B**:960–8. doi:910.1002/ajmg.b.31243. Epub 32011 Oct 31213.
71. **Laucht M, Treutlein J, Blomeyer D, Buchmann AF, Schmidt MH, Esser G**, et al. Interactive effects of corticotropin-releasing hormone receptor 1 gene and childhood adversity on depressive symptoms in young adults: Findings from a longitudinal study. Eur Neuropsychopharmacology. 2013;**23**:358–67. doi:310.1016/j.euroneuro.2012.1006.1002. Epub 2012 Jun 1028.
72. **Nelson EC, Agrawal A, Pergadia ML, Wang JC, Whitfield JB, Saccone FS**, et al. H2 haplotype at chromosome 17q21.31 protects against childhood sexual abuse-associated risk for alcohol consumption and dependence. Addict Biol. 2010;**15**:1–11.
73. **Grabe HJ, Schwahn C, Appel K, Mahler J, Schulz A, Spitzer C**, et al. Childhood maltreatment, the corticotropin-releasing hormone receptor gene and adult depression in the general population. Am J Med Genet B Neuropsychiatr Genet. 2010;**153B**:1483–93. doi:1410.1002/ajmg.b.31131. Epub 32010 Oct 31118.

74. **Appel K, Schwahn C, Mahler J, Schulz A, Spitzer C, Fenske K**, et al. Moderation of adult depression by a polymorphism in the FKBP5 gene and childhood physical abuse in the general population. Neuropsychopharmacology. 2011;**36**:1982–91.
75. **Bevilacqua L, Carli V, Sarchiapone M, George DK, Goldman D, Roy A**, et al. Interaction between FKBP5 and childhood trauma and risk of aggressive behavior. Arch Gen Psychiatry. 2012;**69**:62–70.
76. **Xie P, Kranzler HR, Poling J, Stein MB, Anton RF, Farrer LA**, et al. Interaction of FKBP5 with childhood adversity on risk for post-traumatic stress disorder. Neuropsychopharmacology. 2010;**35**:1684–92.
77. **Binder EB, Bradley RG, Liu W, Epstein MP, Deveau TC, Mercer KB**, et al. Association of FKBP5 polymorphisms and childhood abuse with risk of posttraumatic stress disorder symptoms in adults. JAMA. 2008;**299**:1291–305. doi:1210.1001/jama.1299.1211.1291.
78. **Levran O, Peles E, Randesi M, Li Y, Rotrosen J, Ott J**, et al. Stress-related genes and heroin addiction: A role for a functional FKBP5 haplotype. Psychoneuroendocrinology. 2014;**45**:67–76. doi:10.1016/j.psyneuen.2014.1003.1017. Epub 2014 Apr 1016.
79. **van Zuiden M, Geuze E, Willemen HL, Vermetten E, Maas M, Amarouchi K**, et al. Glucocorticoid receptor pathway components predict posttraumatic stress disorder symptom development: A prospective study. Biol Psychiatry. 2012;**71**:309–16. doi:310.1016/j.biopsych.2011.1010.1026. Epub 2011 Dec 1012.
80. **Kaufman J, Charney D.** Comorbidity of mood and anxiety disorders. Depression and Anxiety. 2000;**12**:69–76.
81. **Kessler RC, Avenevoli S, McLaughlin KA, Green JG, Lakoma MD, Petukhova M**, et al. Lifetime co-morbidity of DSM-IV disorders in the US National Comorbidity Survey Replication Adolescent Supplement (NCS-A). Psychol Med. 2012;**42**:1997–2010.
82. **Kessler RC, McGonagle KA, Zhao S, Nelson CB, Hughes M, Eshleman S**, et al. Lifetime and 12-month prevalence of DSM-III-R psychiatric disorders in the United States. Results from the National Comorbidity Survey. Arch Gen Psychiatry. 1994;**51**:8–19.
83. **Kessler RC, Crum RM, Warner LA, Nelson CB, Schulenberg J, Anthony JC.** Lifetime co-occurrence of DSM-III-R alcohol abuse and dependence with other psychiatric disorders in the National Comorbidity Survey. Arch Gen Psychiatry. 1997;**54**:313–21.
84. **Etkin A, Cuthbert B.** Beyond the DSM: Development of a transdiagnostic psychiatric neuroscience course. Acad Psychiatry. 2014;**38**:145–50.
85. **Insel TR.** The NIMH Research Domain Criteria (RDoC) Project: Precision medicine for psychiatry. Am J Psychiatry. 2014;**171**:395–7.
86. **Gottesman, II, Gould TD.** The endophenotype concept in psychiatry: Etymology and strategic intentions. Am J Psychiatry. 2003;**160**:636–45.
87. **Cuthbert BN.** The RDoC framework: Facilitating transition from ICD/DSM to dimensional approaches that integrate neuroscience and psychopathology. World Psychiatry. 2014;**13**:28–35.
88. **Ford JM, Morris SE, Hoffman RE, Sommer I, Waters F, McCarthy-Jones S**, et al. Studying hallucinations within the NIMH RDoC Framework. Schizophr Bull. 2014;21.
89. **Kaffman A, Krystal JH.** New frontiers in animal research of psychiatric illness. Methods Mol Biol. 2012;**829**:3–30.
90. **Widom CS, Shepard RL.** Accuracy of adult recollections of childhood victimization, Part 1: Childhood physical abuse. Psychological Assessment. 1996;**84**:412–21.
91. **Widom CS, Morris S.** Accuracy of adult recollections of childhood victimization, Part II: Childhood sexual abuse. Psychological Assessment. 1997;**9**:34–46.
92. **Grasso D, Boonsiri J, Lipschitz D, Guyer A, Houshyar S, Douglas-Palumberi H**, et al. Posttraumatic stress disorder: The missed diagnosis. Child Welfare. 2009;**88**:157–76.
93. **Holbrook H, O'Loughlin K, Althoff R, Douglas-Palumberi H, Kaufman J, Hudziak J.** The Yale-Vermont Adversity in Childhood Scale: A quantitative approach to adversity assessment. In: American Academy of Child and Adolescent Psychiatry's 61st Annual Meeting. San Diego, CA; 2015.

94. **Kaufman J, Jones B, Steiglitz E, Vitulano L, Mannarino A.** The use of multiple informants to assess children's maltreatment experiences. Journal of Family Violence. 1994;**9**:227–48.
95. **Coccaro EF, Bergeman CS, McClearn GE.** Heritability of irritable impulsiveness: A study of twins reared together and apart. Psychiatry Res. 1993;**48**:229–42.
96. **True WR, Rice J, Eisen SA, Heath AC, Goldberg J, Lyons MJ**, et al. A twin study of genetic and environmental contributions to liability for posttraumatic stress symptoms. Arch Gen Psychiatry. 1993;**50**:257–64.
97. **Bierut LJ, Heath AC, Bucholz KK, Dinwiddie SH, Madden PA, Statham DJ**, et al. Major depressive disorder in a community-based twin sample: Are there different genetic and environmental contributions for men and women? Arch Gen Psychiatry. 1999;**56**:557–63.
98. **Maes HH, Woodard CE, Murrelle L, Meyer JM, Silberg JL, Hewitt JK**, et al. Tobacco, alcohol and drug use in eight- to sixteen-year-old twins: The Virginia Twin Study of Adolescent Behavioral Development. J Stud Alcohol. 1999;**60**:293–305.
99. **Pickens RW, Svikis DS, McGue M, Lykken DT, Heston LL, Clayton PJ.** Heterogeneity in the inheritance of alcoholism. A study of male and female twins. Arch Gen Psychiatry. 1991;**48**:19–28.
100. **Lee SH, Wray NR, Goddard ME, Visscher PM.** Estimating missing heritability for disease from genome-wide association studies. Am J Hum Genet. 2011;**88**:294–305.
101. **Maher B.** Personal genomes: The case of the missing heritability. Nature. 2008;**456**:18–21.
102. **Thapar A, Harold G.** Editorial perspective: Why is there such a mismatch between traditional heritability estimates and molecular genetic findings for behavioural traits? J Child Psychol Psychiatry. 2014;**55**:1088–91.
103. **Bierut LJ.** Nicotine dependence and genetic variation in the nicotinic receptors. Drug Alcohol Depend. 2009;**104** Suppl 1:S64–69.
104. **Rietschel M, Treutlein J.** The genetics of alcohol dependence. Ann N Y Acad Sci. 2013;**1282**:39–70.
105. **Bookman EB, McAllister K, Gillanders E, Wanke K, Balshaw D, Rutter J**, et al. Gene–environment interplay in common complex diseases: Forging an integrative model-recommendations from an NIH workshop. Genet Epidemiol. 2011;**9**:20571.
106. **Hutter CM, Mechanic LE, Chatterjee N, Kraft P, Gillanders EM.** Gene–environment interactions in cancer epidemiology: A National Cancer Institute Think Tank report. Genet Epidemiol. 2013;**37**:643–57.
107. **Mechanic LE, Chen HS, Amos CI, Chatterjee N, Cox NJ, Divi RL**, et al. Next generation analytic tools for large scale genetic epidemiology studies of complex diseases. Genet Epidemiol. 2012;**36**:22–35.
108. **Guffanti G, Galea S, Yan L, Roberts AL, Solovieff N, Aiello AE**, et al. Genome-wide association study implicates a novel RNA gene, the lincRNA AC068718.1, as a risk factor for post-traumatic stress disorder in women. Psychoneuroendocrinology. 2013;**38**:3029–38. doi:3010.1016/j.psyneuen.2013.3008.3014. Epub 2013 Sep 3028.
109. **Liberzon I, King AP, Ressler KJ, Almli LM, Zhang P, Ma ST**, et al. Interaction of the ADRB2 gene polymorphism with childhood trauma in predicting adult symptoms of posttraumatic stress disorder. JAMA Psychiatry. 2014;**71**:1174–82. doi:1110.1001/jamapsychiatry.2014.1999.
110. **Logue MW, Baldwin C, Guffanti G, Melista E, Wolf EJ, Reardon AF**, et al. A genome-wide association study of post-traumatic stress disorder identifies the retinoid-related orphan receptor alpha (RORA) gene as a significant risk locus. Mol Psychiatry. 2013;**18**:937–42. doi:910.1038/mp.2012.1113. Epub 2012 Aug 1037.
111. **Xie P, Kranzler HR, Yang C, Zhao H, Farrer LA, Gelernter J.** Genome-wide association study identifies new susceptibility loci for posttraumatic stress disorder. Biol Psychiatry. 2013;**74**:656–63. doi:610.1016/j.biopsych.2013.1004.1013. Epub 2013 May 1028.
112. **Almli LM, Stevens JS, Smith AK, Kilaru V, Meng Q, Flory J**, et al. A genome-wide identified risk variant for PTSD is a methylation quantitative trait locus and confers decreased cortical activation to fearful faces. Am J Med Genet B Neuropsychiatr Genet. 2015;**168**:327–36. doi:310.1002/ajmg.b.32315. Epub 32015 May 32318.

113. **Nievergelt CM, Maihofer AX, Mustapic M, Yurgil KA, Schork NJ, Miller MW**, et al. Genomic predictors of combat stress vulnerability and resilience in U.S. Marines: A genome-wide association study across multiple ancestries implicates PRTFDC1 as a potential PTSD gene. Psychoneuroendocrinology. 2015;**51**:459–71. doi:10.1016/j.psyneuen.2014.1010.1017. Epub 2014 Oct 1030.

114. **Solovieff N, Roberts AL, Ratanatharathorn A, Haloosim M, De Vivo I, King AP**, et al. Genetic association analysis of 300 genes identifies a risk haplotype in SLC18A2 for post-traumatic stress disorder in two independent samples. Neuropsychopharmacology. 2014;**39**:1872–9. doi:1810.1038/npp.2014.1834. Epub 2014 Feb 1814.

115. **Kilaru V, Iyer SV, Almli LM, Stevens JS, Lori A, Jovanovic T**, et al. Genome-wide gene-based analysis suggests an association between Neuroligin 1 (NLGN1) and post-traumatic stress disorder. Transl Psychiatry. 2016;**6**:e820. doi:10.1038/tp.2016.1069.

116. **Stein MB, Chen CY, Ursano RJ, Cai T, Gelernter J, Heeringa SG**, et al. Genome-wide association studies of posttraumatic stress disorder in 2 cohorts of US Army soldiers. JAMA Psychiatry. 2016;**73**:695–704. doi:610.1001/jamapsychiatry.2016.0350.

117. **Ashley-Koch AE, Garrett ME, Gibson J, Liu Y, Dennis MF, Kimbrel NA**, et al. Genome-wide association study of posttraumatic stress disorder in a cohort of Iraq-Afghanistan era veterans. J Affect Disord. 2015;**184**:225–34. doi:10.1016/j.jad.2015.1003.1049.

118. **Logue MW, Amstadter AB, Baker DG, Duncan L, Koenen KC, Liberzon I**, et al. The Psychiatric Genomics Consortium Posttraumatic Stress Disorder Workgroup: Posttraumatic stress disorder enters the age of large-scale genomic collaboration. Neuropsychopharmacology. 2015 118.

119. **Terracciano A, Tanaka T, Sutin AR, Sanna S, Deiana B, Lai S**, et al. Genome-wide association scan of trait depression. Biol Psychiatry. 2010;**68**:811–7. doi:810.1016/j.biopsych.2010.1006.1030. Epub 2010 Aug 1025.

120. **Le-Niculescu H, Patel SD, Bhat M, Kuczenski R, Faraone SV, Tsuang MT**, et al. Convergent functional genomics of genome-wide association data for bipolar disorder: Comprehensive identification of candidate genes, pathways and mechanisms. Am J Med Genet B Neuropsychiatr Genet. 2009;**150B**:155–81. doi:110.1002/ajmg.b.30887.

121. **Neale BM, Lasky-Su J, Anney R, Franke B, Zhou K, Maller JB**, et al. Genome-wide association scan of attention deficit hyperactivity disorder. Am J Med Genet B Neuropsychiatr Genet. 2008;**147B**:1337–44. doi:1310.1002/ajmg.b.30866.

122. **Sarachana T, Xu M, Wu RC, Hu VW.** Sex hormones in autism: Androgens and estrogens differentially and reciprocally regulate RORA, a novel candidate gene for autism. PLoS One. 2011;**6**:e17116. doi:17110.11371/journal.pone.0017116.

123. **Lowe SR, Meyers JL, Galea S, Aiello AE, Uddin M, Wildman DE**, et al. RORA and posttraumatic stress trajectories: Main effects and interactions with childhood physical abuse history. Brain Behav. 2015;**5**:e00323. doi:00310.01002/brb00323.00323. Epub 02015 Mar 00328.

124. **Glessner JT, Wang K, Cai G, Korvatska O, Kim CE, Wood S**, et al. Autism genome-wide copy number variation reveals ubiquitin and neuronal genes. Nature. 2009;**459**:569–73. doi:510.1038/nature07953. Epub 02009 Apr 07928.

125. **Alloza I, Otaegui D, de Lapuente AL, Antiguedad A, Varade J, Nunez C**, et al. ANKRD55 and DHCR7 are novel multiple sclerosis risk loci. Genes Immun. 2012;**13**:253–7. doi:210.1038/gene.2011.1081. Epub 2011 Dec 1031.

126. **Lill CM, Schjeide BM, Graetz C, Liu T, Damotte V, Akkad DA**, et al. Genome-wide significant association of ANKRD55 rs6859219 and multiple sclerosis risk. J Med Genet. 2013;**50**:140–3. doi:110.1136/jmedgenet-2012-101411. Epub 102013 Jan 101412.

127. **Wang Q, Yang C, Gelernter J, Zhao H.** Pervasive pleiotropy between psychiatric disorders and immune disorders revealed by integrative analysis of multiple GWAS. Hum Genet. 2015;**134**:1195–209. doi:1110.1007/s00439-00015-01596-00438. Epub 02015 Sep 00434.

128. **Anda RF, Croft JB, Felitti VJ, Nordenberg D, Giles WH, Williamson DF**, et al. Adverse childhood experiences and smoking during adolescence and adulthood. JAMA. 1999;**282**:1652–8.

129. **Felitti VJ, Anda RF, Nordenberg D, Williamson DF, Spitz AM, Edwards V**, et al. Relationship of childhood abuse and household dysfunction to many of the leading causes of death in adults. The Adverse Childhood Experiences (ACE) Study. Am J Prev Medicine. 1998;**14**:245–58.
130. **Yang B-Z, Zhang H, Ge W, Weder N, Douglas-Palumberi H, Perepletchikova F**, et al. Child abuse and epigenetic mechanisms of disease risk. Am J Prevent Med. 2013;**44**:101–7.
131. **Almli LM, Srivastava A, Fani N, Kerley K, Mercer KB, Feng H**, et al. Follow-up and extension of a prior genome-wide association study of posttraumatic stress disorder: Gene × environment associations and structural magnetic resonance imaging in a highly traumatized African-American civilian population. Biol Psychiatry. 2014;**76**:e3–4. doi:10.1016/j.biopsych.2014.1001.1017. Epub 2014 Jan 1028.
132. **Hodges E, Molaro A, Dos Santos CO, Thekkat P, Song Q, Uren PJ**, et al. Directional DNA methylation changes and complex intermediate states accompany lineage specificity in the adult hematopoietic compartment. Mol Cell. 2011;**44**:17–28.
133. **Yang J, Corces VG.** Chromatin insulators: A role in nuclear organization and gene expression. Adv Cancer Res. 2011;**110**:43–76.
134. **Cotnoir-White D, Laperriere D, Mader S.** Evolution of the repertoire of nuclear receptor binding sites in genomes. Mol Cell Endocrinol. 2011;**334**:76–82.
135. **Qureshi IA, Mattick JS, Mehler MF.** Long non-coding RNAs in nervous system function and disease. Brain Res. 2010;**1338**:20–35.
136. **Hanson M, Godfrey KM, Lillycrop KA, Burdge GC, Gluckman PD.** Developmental plasticity and developmental origins of non-communicable disease: Theoretical considerations and epigenetic mechanisms. Prog Biophys Mol Biol. 2011;**106**:272–80.
137. **Dedeurwaerder S, Fumagalli D, Fuks F.** Unravelling the epigenomic dimension of breast cancers. Curr Opin Oncol. 2011;**23**:559–65.
138. **Ogoshi K, Hashimoto S, Nakatani Y, Qu W, Oshima K, Tokunaga K**, et al. Genome-wide profiling of DNA methylation in human cancer cells. Genomics. 2011;**98**:280–7.
139. **Watanabe Y, Maekawa M.** Methylation of DNA in cancer. Adv Clin Chem. 2010;**52**:145–67.
140. **Wilson AS, Power BE, Molloy PL.** DNA hypomethylation and human diseases. Biochim Biophys Acta. 2007;**1775**:138–62.
141. **Yegnasubramanian S, Wu Z, Haffner MC, Esopi D, Aryee MJ, Badrinath R**, et al. Chromosome-wide mapping of DNA methylation patterns in normal and malignant prostate cells reveals pervasive methylation of gene-associated and conserved intergenic sequences. BMC Genomics. 2011;**12**:313.
142. **Weaver IC, Cervoni N, Champagne FA, D'Alessio AC, Sharma S, Seckl JR**, et al. Epigenetic programming by maternal behavior. Nat Neurosci. 2004;**7**:847–54.
143. **Turecki G, Meaney MJ.** Effects of the social environment and stress on glucocorticoid receptor gene methylation: A systematic review. Biol Psychiatry. 2014:022.
144. **Turecki G, Ota V, Belangero S, Jackowski A, Kaufman J.** Early life adversity, genomic plasticity, and psychopathology. Lancet Psychiatry. 2014;**1**:461–6.
145. **McGowan PO, Suderman M, Sasaki A, Huang TC, Hallett M, Meaney MJ**, et al. Broad epigenetic signature of maternal care in the brain of adult rats. PLoS One. 2011;6:e14739. doi:14710.11371/journal.pone.0014739.
146. **Tsankova NM, Berton O, Renthal W, Kumar A, Neve RL, Nestler EJ.** Sustained hippocampal chromatin regulation in a mouse model of depression and antidepressant action. Nat Neurosci. 2006;**9**:519–25. Epub 2006 Feb 2026.
147. **Bagot RC, Zhang TY, Wen X, Nguyen TT, Nguyen HB, Diorio J**, et al. Variations in postnatal maternal care and the epigenetic regulation of metabotropic glutamate receptor 1 expression and hippocampal function in the rat. PNAS. 2012;**109**:17200–7. doi:17210.11073/pnas.1204599109. Epub 1204592012 Oct 1204599108.
148. **Toda H, Boku S, Nakagawa S, Inoue T, Kato A, Takamura N**, et al. Maternal separation enhances conditioned fear and decreases the mRNA levels of the neurotensin receptor 1 gene with hypermethylation of this gene in the rat amygdala. PLoS One. 2014;**9**:e97421.
149. **Roth TL, Zoladz PR, Sweatt JD, Diamond DM.** Epigenetic modification of hippocampal Bdnf DNA in adult rats in an animal model of post-traumatic stress disorder. J Psychiatr Res. 2011;**45**:919–26.

150. **Marquez C, Poirier GL, Cordero MI, Larsen MH, Groner A, Marquis J,** et al. Peripuberty stress leads to abnormal aggression, altered amygdala and orbitofrontal reactivity and increased prefrontal MAOA gene expression. Transl Psychiatry. 2013;**3**:e216.

151. **Niwa M, Jaaro-Peled H, Tankou S, Seshadri S, Hikida T, Matsumoto Y,** et al. Adolescent stress-induced epigenetic control of dopaminergic neurons via glucocorticoids. Science. 2013;**339**:335–9. doi:310.1126/science.1226931.

152. **Pena CJ, Neugut YD, Calarco CA, Champagne FA.** Effects of maternal care on the development of midbrain dopamine pathways and reward-directed behavior in female offspring. Eur J Neurosci. 2014;**39**:946–56. doi:910.1111/ejn.12479. Epub 12014 Jan 12421.

153. **Brennan K, Garcias-Closas M, Orr N, Fletcher O, Jones M, Ashworth A,** et al. Intragenic ATM methylation in peripheral blood DNA as a biomarker of breast cancer risk. Cancer Res. 2012;**28**:28.

154. **Sapienza C, Lee J, Powell J, Erinle O, Yafai F, Reichert J,** et al. DNA methylation profiling identifies epigenetic differences between diabetes patients with ESRD and diabetes patients without nephropathy. Epigenetics. 2011;**6**:20–8.

155. **Suijkerbuijk KP, van Diest PJ, van der Wall E.** Improving early breast cancer detection: Focus on methylation. Ann Oncol. 2011;**22**:24–9.

156. **Dempster EL, Pidsley R, Schalkwyk LC, Owens S, Georgiades A, Kane F,** et al. Disease-associated epigenetic changes in monozygotic twins discordant for schizophrenia and bipolar disorder. Hum Mol Genet. 2011;**20**:4786–96.

157. **Kaminsky Z, Tochigi M, Jia P, Pal M, Mill J, Kwan A,** et al. A multi-tissue analysis identifies HLA complex group 9 gene methylation differences in bipolar disorder. Mol Psychiatry. 2011;**7**:7.

158. **Klengel T, Pape J, Binder EB, Mehta D.** The role of DNA methylation in stress-related psychiatric disorders. Neuropharmacology. 2014;**80**:115–32. doi:10.1016/j.neuropharm.2014.1001.1013. Epub 2014 Jan 1019.

159. **Smith AK, Kilaru V, Klengel T, Mercer KB, Bradley B, Conneely KN,** et al. DNA extracted from saliva for methylation studies of psychiatric traits: Evidence tissue specificity and relatedness to brain. Am J Med Genet B Neuropsychiatr Genet. 2015;**168B**:36–44. doi:10.1002/ajmg.b.32278. Epub 32014 Oct 32229.

160. **Naumova OY, Lee M, Koposov R, Szyf M, Dozier M, Grigorenko EL.** Differential patterns of whole-genome DNA methylation in institutionalized children and children raised by their biological parents. Dev Psychopathol. 2011;**29**:1–13.

161. **Smith AK, Conneely KN, Kilaru V, Mercer KB, Weiss TE, Bradley B,** et al. Differential immune system DNA methylation and cytokine regulation in post-traumatic stress disorder. Am J Med Genet B Neuropsychiatr Genet. 2011;**156B**:700–8. doi:710.1002/ajmg.b.31212. Epub 32011 Jun 31228.

162. **Yang X, Ewald ER, Huo Y, Tamashiro KL, Salvatori R, Sawa A,** et al. Glucocorticoid-induced loss of DNA methylation in non-neuronal cells and potential involvement of DNMT1 in epigenetic regulation of Fkbp5. Biochem Biophys Res Commun. 2012;**420**:570–5. doi:510.1016/j.bbrc.2012.1003.1035. Epub 2012 Mar 1016.

163. **Weder N, Zhang H, Jensen K, Yang BZ, Simen S, Jackowski J,** Et al. Child abuse, depression, and methylation in genes involved with stress, neural plasticity, and brain circuitry. J Am Acad Child Adolsc Psychiatry. 2014;**53**:417–24.

164. **Montalvo-Ortiz JL, Bordner KA, Carlyle BC, Gelernter J, Simen AA, Kaufman J.** The role of genes involved in stress, neural plasticity, and brain circuitry in depressive phenotypes: Convergent findings in a mouse model of neglect. Behav Brain Res. 2016;**315**:71–4. doi:10.1016/j.bbr.2016.1008.1010. Epub 2016 Aug 1016.

165. **Bick J, Nelson CA.** Early adverse experiences and the developing brain. Neuropsychopharmacology. 2016;**41**:177–96.

166. **Teicher MH, Samson JA.** Annual Research Review: Enduring neurobiological effects of childhood abuse and neglect. J Child Psychol Psychiatry. 2016;**57**:241–66. doi:210.1111/jcpp.12507. Epub 12016 Feb 12501.

167. **De Bellis MD, Hooper SR, Chen SD, Provenzale JM, Boyd BD, Glessner CE,** et al. Posterior structural brain volumes differ in maltreated youth with and without chronic posttraumatic stress disorder. Dev Psychopathol. 2015;**27**:1555–76. doi:1510.1017/S0954579415000942.

168. **Jackowski A, Douglas-Palumberi H, Jackowski M, Win L, Schultz RT, Staib LH,** et al. Corpus Callosum in maltreated children with PTSD: A diffusion tensor imaging study. Psychiatry research. Neuroimaging. 2008;**162**:256–61.

169. **Nees F, Tzschoppe J, Patrick CJ, Vollstadt-Klein S, Steiner S, Poustka L,** et al. Determinants of early alcohol use in healthy adolescents: The differential contribution of neuroimaging and psychological factors. Neuropsychopharmacology. 2012;**37**:986–95. doi:910.1038/npp.2011.1282. Epub 2011 Nov 1023.

170. **Morey RA, Haswell CC, Hooper SR, De Bellis MD.** Amygdala, hippocampus, and ventral medial prefrontal cortex volumes differ in maltreated youth with and without chronic posttraumatic stress disorder. Neuropsychopharmacology. 2015:205.

171. **Carballedo A, Lisiecka D, Fagan A, Saleh K, Ferguson Y, Connolly G,** et al. Early life adversity is associated with brain changes in subjects at family risk for depression. World J Biol Psychiatry. 2012;**13**:569–78. doi:510.3109/15622975.15622012.15661079. Epub 15622012 Apr 15622920.

172. **Pechtel P, Lyons-Ruth K, Anderson CM, Teicher MH.** Sensitive periods of amygdala development: The role of maltreatment in preadolescence. Neuroimage. 2014;**97**:236–44. doi:10.1016/j.neuroimage.2014.1004.1025. Epub 2014 Apr 1013.

173. **Orr CA, Hudziak J, Albaugh MD, O'Loughlin K, D'Alberto N, Holbrook H,** et al. Relations between childhood adversity, post-traumatic symptoms, and neural activity. In: Annual Meeting of the Society for Research in Child Development. Philadelphia, PA:2015.

174. **Weder N, Kaufman J.** Critical periods revisited: Implications for intervention with traumatized children. J Am Acad Child Adolesc Psychiatry. 2011;**50**:1087–9.

175. **Wiesel TN, Hubel DH.** Effects of visual deprivation on morphology and physiology of cells in the cats lateral geniculate body. J Neurophysiol. 1963;**26**:978–93.

176. **Bick J, Fox N, Zeanah C, Nelson CA.** Early deprivation, atypical brain development, and internalizing symptoms in late childhood. Neuroscience. 2015:026.

177. **Bruce J, McDermott JM, Fisher PA, Fox NA.** Using behavioral and electrophysiological measures to assess the effects of a preventive intervention: A preliminary study with preschool-aged foster children. Prev Sci. 2009;**10**:129–40. doi:110.1007/s11121-11008-10115-11128.

178. **McDermott JM, Troller-Renfree S, Vanderwert R, Nelson CA, Zeanah CH, Fox NA.** Psychosocial deprivation, executive functions, and the emergence of socio-emotional behavior problems. Front Hum Neurosci. 2013;7:167. doi:10.3389/fnhum.2013.00167. eCollection 02013.

179. **Roy MJ, Costanzo ME, Blair JR, Rizzo AA.** Compelling evidence that exposure therapy for PTSD normalizes brain function. Stud Health Technol Inform. 2014;**199**:61–65.

180. **Domhardt M, Munzer A, Fegert JM, Goldbeck L.** Resilience in survivors of child sexual abuse: A systematic review of the literature. Trauma Violence Abuse. 2015;**16**:476–93. doi:410.1177/1524838014557288. Epub 1524838014552014 Nov 1524838014557210.

181. **Dozier M, Peloso E, Lewis E, Laurenceau JP, Levine S.** Effects of an attachment-based intervention on the cortisol production of infants and toddlers in foster care. Dev Psychopathol. 2008;**20**:845–59.

182. **Bernard K, Dozier M, Bick J, Lewis-Morrarty E, Lindhiem O, Carlson E.** Enhancing attachment organization among maltreated children: Results of a randomized clinical trial. Child Dev. 2012;**83**:623–36. doi:610.1111/j.1467-8624.2011.01712.x. Epub 02012 Jan 01712.

183. **Lind T, Bernard K, Ross E, Dozier M.** Intervention effects on negative affect of CPS-referred children: Results of a randomized clinical trial. Child Abuse Negl. 2014;**38**:1459–67. doi:1410.1016/j.chiabu.2014.1404.1004. Epub 2014 May 1410.

184. **Lewis-Morrarty E, Dozier M, Bernard K, Terracciano SM, Moore SV.** Cognitive flexibility and theory of mind outcomes among foster children: Preschool follow-up results of a randomized clinical trial. J Adolesc Health. 2012;**51**:S17–22. doi:10.1016/j.jadohealth.2012.1005.1005.

185. **Bernard K, Hostinar CE, Dozier M.** Intervention effects on diurnal cortisol rhythms of child protective services-referred infants in early childhood: Preschool follow-up results of a randomized clinical trial. JAMA Pediatr. 2015;**169**:112–9.

186. **Dozier M, Kaufman J, Kobak R, O'Connor T, Sagi-Schwartz A, Scott S**, et al. Consensus statement on group care. In: Applying Research in Child and Adolescent Development to Child Welfare Placement Practices Meeting Participants. Annie E. Casey Foundation (ed.). YLC. New York, NY; 2012.

187. **Dozier M, Kaufman J, Kobak R, O'Connor TG, Sagi-Schwartz A, Scott S, Shauffer C, Smetana J, van IJzendoorn MH, Zeanah CH.** Consensus statement on group care for children and adolescents: A statement of policy of Orthopsychiatric Association. American Journal of Orthopsychiatry; 2014:**84**(3):219–25. PMID: 24827016 PMID: 18996506. https://www.acf.hhs.gov/cb/resource/afcars-report-25.

188. DHHS. The AFCARS Report. in Adoption and Foster Care Analysis and Reporting System (AFCARS) FY 2013 data, U.S. Department of Health and Human Services, Administration for Children and Families, Administration on Children, Youth and Families, Children's Bureau, programs/cb; 2014.

189. **Bredy TW, Humpartzoomian RA, Cain DP, Meaney MJ, Morley-Fletcher S, Rea M**, et al. Partial reversal of the effect of maternal care on cognitive function through environmental enrichment. Eur J Neurosci. 2003;**18**:571–576.

190. **Kessler RC, Pecora PJ, Williams J, Hiripi E, O'Brien K, English D**, et al. Effects of enhanced foster care on the long-term physical and mental health of foster care alumni. Arch Gen Psychiatry. 2008;**65**:625–33.

191. **Davidson RJ, McEwen BS.** Social influences on neuroplasticity: Stress and interventions to promote well-being. Nat Neurosci. 2012;**15**:689–95.

192. **Hudziak JJ, Albaugh MD, Ducharme S, Karama S, Spottswood M, Crehan E**, et al. Cortical thickness maturation and duration of music training: Health-promoting activities shape brain development. J Am Acad Child Adolesc Psychiatry. 2014;**53**:1153–61 e1152.

193. **Cohen JA, Mannarino AP, Deblinger E.** Treating Trauma and Traumatic Grief in Children and Adolescents. New York, NY: Guilford Press; 2006.

194. **Cohen JA, Bukstein O, Walter H, Benson SR, Chrisman A, Farchione TR**, et al. Practice parameter for the assessment and treatment of children and adolescents with posttraumatic stress disorder. J Am Acad Child Adolesc Psychiatry. 2010;**49**:414–30.

195. **Cohen JA, Mannarino A.** Trauma-Focused CBT for Children and Adolescents.Treatment Applications. New York, NY: Guilford Press; 2012.

196. **Cohen JA, Mannarino AP.** Trauma-focused cognitive behavior therapy for traumatized children and families. Child Adolesc Psychiatr Clin N America. 2015;**24**:557–70.

197. **Hoover D, Kaufman J.** Posttraumatic stress disorder. In: Martin A (ed.). Lewis' Child and Adolescent Psychiatry: A Comprehensive Textbook, 5th edn. Philadelphi, PA, Wolters Kluwer; 2017.

198. **Deblinger E, Mannarino AP, Cohen JA, Steer RA.** A follow-up study of a multisite, randomized, controlled trial for children with sexual abuse-related PTSD symptoms. J Am Acad Child Adolesc Psychiatry. 2006;**45**:1474–84.

199. **Deblinger E, Steer RA, Lippmann J.** Two-year follow-up study of cognitive behavioral therapy for sexually abused children suffering post-traumatic stress symptoms. Child Abuse Negl. 1999;**23**:1371–8.

200. **Zhang TY, Hellstrom IC, Bagot RC, Wen X, Diorio J, Meaney MJ.** Maternal care and DNA methylation of a glutamic acid decarboxylase 1 promoter in rat hippocampus. J Neurosci. 2010;**30**(39): 13130–13137. doi: 13110.11523/JNEUROSCI.11039-13110.12010.

201. **Chen J, Evans AN, Liu Y, Honda M, Saavedra JM, Aguilera G.** Maternal deprivation in rats is associated with corticotrophin-releasing hormone (CRH) promoter hypomethylation and enhances CRH transcriptional responses to stress in adulthood. J Neuroendocrinol. 2012;**24**(7):1055–1064. doi: 1010.1111/j.1365-2826.2012.02306.x.

202. **Murgatroyd C, Patchev AV, Wu Y**, et al., Dynamic DNA methylation programs persistent adverse effects of early-life stress. Nat Neurosci. 2009;**12**(12):1559–1566.

Chapter 22

Animal models of post-traumatic stress disorder: Towards understanding of individual differences

Lisa Heltzel and Judith R Homberg

22.1 Introduction

Fear responses are normal and essential processes in an organism. Fear learning is needed to respond appropriately to cues and contexts that are associated with danger (1, 2). However, sometimes fear learning goes awry, leading to intrusive fear memory failing to extinguish and eliciting an intense, distressing and persisting response to an external trauma, which is a real or perceived threat to the individuals life or lives of others close to the individual (3, 4). These are core features of anxiety disorders, specifically post-traumatic stress disorder (PTSD) (Box 22.1). In the Diagnostic and Statistical Manual of Mental Disorders V (DSM-V), PTSD is listed as a trauma or stressor-related trauma (5, 6). The eight criteria of PTSD according to DSM-V are as follows: (A) Stressor must initiate the syndrome and symptoms; (B) Re-experience of the event must be present; (C) Subjects must display increased avoidance of trauma-associated stimuli; (D) Negative changes in cognition and mood must be present after the event; (E) Alterations in arousal and reactivity must occur after the traumatic event; (F) Displayed symptoms must be persistent for more than one month; (G) Symptoms must significantly impair the individual's functioning; and (H) Other factors—such as medication, substance use or other illness—that may cause the symptoms must be excluded. Criterion (B) is particularly related to impaired fear learning and extinction, causing PTSD-patients recurrently to revive the traumatic event in response to cues related to the trauma (7). To date, the pathophysiological basis of PTSD is poorly understood, making current treatments symptomatic and only effective for fewer than half the patients (8). Furthermore, side effects and residual symptoms following treatment are the rule rather than exception. Hence, there is an urgent need to advance the understanding of its pathophysiology.

Interestingly, there are substantial individual differences in long-term responses to trauma among individuals; in humans the probability of developing PTSD is estimated to be in range of 15–20%, depending on the type of trauma (3, 9, 10). However, in addition, the symptoms among PTSD patients can be quite diverse: they can be mild or extreme, and under the latter condition patients show hyper-reactivity to traumatic events, have delayed extinction, and/or fail to extinguish memories related to negative events (11, 12). On the other hand, there is the ability to adapt to traumatic events (i.e. resilience), which is defined as low reactivity and fast recovery from the threat during extinction (11, 13, 14). Understanding individual differences in resilience versus vulnerability may be key to the development of new treatment options.

Dissociating resilience and vulnerability is not trivial, since data from patients are generally obtained *following* the diagnosis of PTSD. This makes it practically impossible to determine whether the observed differences between psychiatric patients and healthy controls are related to respective causes (i.e. vulnerability) or consequences (i.e. symptoms) of PTSD. Furthermore, trauma exposure cannot be controlled in humans and prospective studies are often complicated by confounding factors. Another limitation of human studies is that only non-invasive techniques can be used, hindering the investigation of detailed mechanistic underpinnings. Animal studies controlling vulnerability factors as well as trauma exposure can overcome these limitations, and allow controlled post-mortem and *in vivo* non-invasive experiments to link behavior to neural mechanisms. Animal models are also challenging as it is essential that they have translational value. In this chapter, we present currently available behavioral methods to measure resilience versus vulnerability in rodents and animal models that have the potential to elucidate key factors of vulnerability, which are biological factors, genotype, age, and gender.

22.2 Behavioral methods to measure PTSD in rodents

22.2.1 Pavlovian fear conditioning and extinction

Since intrusive extinction-resistant fear-related memories form the core aspect of PTSD, the Pavlovian fear conditioning and extinction paradigm is used to understand neural mechanisms related to PTSD. The process involves pairing an emotionally neutral stimulus such as a tone or light (conditioned stimulus; CS) with a fear-inducing stimulus, typically a mild footshock or loud noise (unconditioned stimulus; US) (15). The combination of these stimuli will condition the animal to a point where the tone alone is enough to elicit a conditioned fear response (CR). Repeated presentation of the CS in the absence of the US leads to fear extinction, whereby the CR gradually weakens. Besides the fear elicited by the CS (cued fear conditioning), rodents can have fearful responses to the context or place where the fear was learned (contextual fear conditioning) (9, 16).

Fear conditioning is studied in boxes where most conditions can be controlled, such as light, sound, reward, or punishment. These boxes can be closed to prevent escape and are soundproof to decrease environmental noise. The floor of shock chambers consists of a metal grid floor, through which an electrical footshock (US) can be applied. Additionally, a speaker is present producing a tone (CS). While for testing contextual fear memory and its extinction (recall) the animals are placed in the same boxes as used during conditioning, for testing cued fear memory and its extinction (recall) a novel context is used; a chamber that is different regarding visual, olfactory, and tactile cues to avoid generalization (17).

Prior to the conditioning procedure, animals receive habituation sessions by placing them into conditioning chambers such that they get familiar with the environment. Those sessions can take for example 10 minutes in 1 day (18, 19) or can consist of two 15-minutes sessions on 2 uninterrupted days (20). Minimally 24 hours after the habituation sessions, the fear-conditioning procedure begins. It is important that the chambers are thoroughly cleaned after each animal, using 70% ethanol.

Cued fear conditioning (Figure 22.1A) begins with a short habituation period, during which animals explore the environment for two minutes. Then a consistent 30-second 85 dB 2.8 kHz tone (CS) is presented, which is followed by an electrical 1 second 0.6 mA footshock (US). This tone–shock pairing will be repeated five times with an inter-trial interval of one minute (19). The animals learn to associate the CS with the US, the CS acquires a predictive property and elicits freezing behavior (CR). Freezing behavior is defined as the absence of any movement except for those that are needed for breathing (21). The retention of the fear memory and fear extinction are usually measured 24 hours after conditioning. The retention or recall of the extinction memory is measured 24 hours or more after the extinction session. Cue extinction (recall) sessions consist of a 2-minute habituation period, followed by 5 to 25 CS presentations in the absence of footshocks. In the absence of the US, the CR (i.e. freezing) will decrease with repeated CS presentation as the animals learn that the CS now predicts safety. When measuring contextual fear memory retention and extinction (Figure 22.1B), animals are returned to the context in which they were exposed to the US during conditioning. The session lasts for ten minutes without footshock delivery and the total duration of freezing is measured (17).

An important phenomenon related to conditioned fear involves fear generalization. Fear generalization is defined as fear responses to non-dangerous events that resemble the threatening ones. For cued fear generalization, animals are stressed with two different stimuli in the same context (Figure 22.1C), a loud tone (stimulus A, the CS) and footshock on day 1 versus a mild tone (stimulus B, the generalization stimulus; GS) and nofoot shock on day 2. The GS is thus perceptually comparable but not identical to the CS. In a novel context, 24 hours later, fear extinction is measured and stimulus A and B are randomly presented. In the case of fear generalization, the animal will freeze in response to stimulus B as well as stimulus A. Contextual fear generalization (Figure 22.1D) is assessed by stressing the animals in two distinct conditioning chambers (context A on day 1 versus context B on day 2) in combination with two stressors that are slightly different (e.g. intense and mild foot shock, respectively). Twenty-four hours later, fear extinction sessions are performed; animals return successively to condition chamber A and B in the absence of the footshocks and their freezing response is measured. It is expected that animals freeze more in context A where they received the intense footshocks compared to context B where they received the mild footshocks. Upon fear generalization, freezing will be increased in context B relative to the amount of freezing originally measured during conditioning. In the fear generalization paradigm, the fear response is always more intense in response to the CS (the initial trauma) compared to the GS, and with increasing dissimilarity of the GS to the CS the response will be less intense. Failure to extinguish the CR and fear generalization are correlated to PTSD (21).

While the Pavlovian fear conditioning and extinction paradigm has offered us extensive understanding of the neural mechanisms—specifically regarding the communication between the prefrontal cortex and amygdala—underlying fear conditioning, the retention of the fear memory, fear extinction, and the recall of the fear extinction memory, it is now commonly agreed upon that these are normal "healthy" processes. Indeed, individual differences are seen without any stressor or trauma, whereas exposure to a stressor or trauma is a necessary component of the DSM-V criteria for PTSD (see criterion A). Accordingly, throughout the years, various stressors have been used to mimic PTSD-inducing trauma. It is important to keep in mind that each model elucidates some, but not necessarily all, PTSD-related criteria. In Table 22.1 we give a short overview of widely accepted stressors that can be applied in animal (i.e. mouse and rat) research to induce PTSD-like symptoms, specifically maladaptive fear learning (DSM-V criterion B). We subdivided the stressors into physical, psychological, and social ones and also indicate the associated DSM-V criteria (labelled A to H). For a more detailed description and relevant information of the models, see Borghans and Homberg (5). In the next sections these stressors will be addressed in the behavioral approaches to measure individual differences in resilience and vulnerability to PTSD.

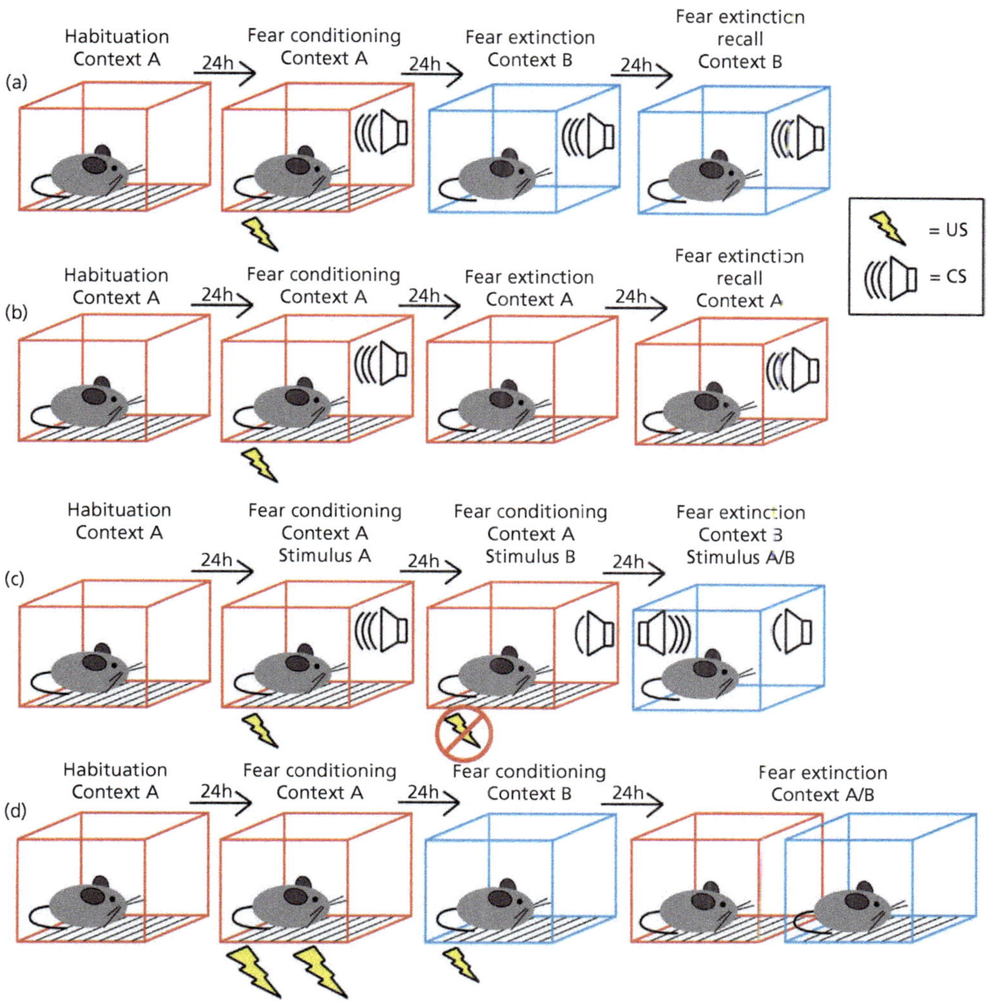

Figure 22.1 Schematic outline of Pavlovian fear conditioning and extinction paradigms. (A) Cued fear conditioning. (B) Contextual fear conditioning. (C) Cued fear generalization. (D) Contextual fear generation.

22.3 **Measuring individual differences in PTSD symptomatology**

Individual differences in resilience versus vulnerability to PTSD can be mediated by biological factors, genetic factors, and age as well as gender. When considering investigation of individual differences it is essential to dissociate cause (i.e. pre-existing or inborn mechanisms shaping resilience versus vulnerability) and consequence (i.e. mechanisms that become overt or are triggered by trauma exposure), which requires careful experimental design.

The most widely used design to study individual differences in resilience and vulnerability to PTSD involves the comparison of trauma exposed animals to controls. Half of the animals from a genetically homogeneous group undergo a certain stress procedure, while the

Table 22.1 Rodent models for post-traumatic stress disorder subdivided according to the nature of the stressors (physical, psychological, and social) and the associated DSM-V criteria

Physical stressors (unpleasant stimuli)	
Rodent model for PTSD	*Associated DSM-V criteria*
Single-prolonged stress	A,B,C,D,E,F,G,H
Restraint stress	A,B,C,D,E,F,G,H
Foot shock	A,B,C,E,F,G,H
Stress-enhanced fear learning	A,B,C,E,F,G,H
Underwater trauma	A,B,E,F,G,H
Social stressors (natural social behavior of animals)	
Rodent model for PTSD	*Associated DSM-V criteria*
Housing instability	A,B,E,G,H
Social instability	A,B,E,F,G,H
Early life stress	A,B,C,D,E,F,G,H
Social defeat	A,B,C,E,F,G,H
Physiological stressors (instinctive behavior to natural predators)	
Rodent model for PTSD	*Associated DSM-V criteria*
Predator-based physiological stress/predator scent stress	A,B,C,D,E,F,G,H

other animals receive a sham procedure. Afterwards, PTSD-related symptoms and associated mechanisms are assessed. Whereas the observed differences can be directly linked to the effects of stress, they do not necessarily inform us about pre-existing mechanisms that increase stress-/trauma-induced PTSD symptomatology. Another design involves a procedure in which all animals from a group undergo the same stress procedure and are tested on stress-related symptomatology afterwards. Comparing the behaviorally (most) resilient animals to the behaviorally (most) vulnerable animals allows one to distinguish potential adaptive from maladaptive neuroadaptations due to stress exposure. More so than in the stressed versus control contrast, this comparison links mechanisms directly to the behavioral phenotype (i.e. psychopathology). However, the observed neuroadaptive changes are not necessarily linked to any alteration induced by the stress procedure in itself as the animals' mechanistic make-up might have already differed before the procedure (and reflect pre-existing inborn factors). Hence, in order to identify pre-existing/inborn mechanisms determining resilience or vulnerability to PTSD requires the pre-screening of animals or the use of animals with pre-defined behavioral characteristics before a stress procedure is applied, without any interference with the stress effects themselves. This illustrates the fact that investigating individual differences in resilience and vulnerability to PTSD is a tedious task when one aims to collect mechanistic information that is at the core of individual differences. In the following, we set out paradigms reported to study individual differences related to biological factors, genotype, age, and gender that use one of these approaches or combines them in order to assess changes in fear learning, memory, and extinction as well as other PTSD symptoms and related mechanisms that shape resilience and vulnerability.

22.3.1 Biology-dependent individual differences

PTSD is characterized by dysregulation of the central stress response. Therefore, vulnerability to PTSD can result from alterations and dysfunctions of the hypothalamus–pituitary–adrenal axis (HPA axis). This axis consists of the paraventricular nucleus of the hypothalamus (PVN), the pituitary gland, and the adrenal gland, which produces cortisol (humans) or corticosterone (animals). When the PVN is activated by a stressor it releases corticotropin-releasing factor (CRF). The released CRF stimulates the pituitary gland, which produces adrenocorticotropic hormone (ACTH). ACTH is released into the bloodstream and reaches the adrenal glands where it stimulates the production and release of cortisol/corticosterone into the bloodstream (22). It has been proposed that PTSD is related to lower basal levels of cortisol because of a stronger negative feedback over the HPA axis and decreased cortisol levels following stress (23, 24). Remarkably, increased CRF levels in the CSF and a reduced ACTH response were found in PTSD patients (25). Accordingly, HPA axis components present relevant targets for individual differences in vulnerability to PTSD.

Recently, a mouse model for PTSD induction was established (26, 27), revealing individual differences in PTSD vulnerability and underlying mechanisms in relation to the HPA axis. In this model, a group of 60 mice was first exposed to a severe stressor (i.e. intense electric footshock) in a certain context (context A), followed by a mild stressor (i.e. weak electric footshock) the next day in a distinct context (context B), which was different in terms of spatial, auditory, and olfactory cues. By stressing the animals in two distinct contexts, this protocol actively stimulates the process of fear generalization (of context A to B) (28, 29). One week after stressing the animals, they were tested in five multiple behavioral paradigms (Table 22.2) to examine if PTSD-like symptomatology—hypervigilance, compulsivity, impaired attention, compromised risk assessment, and insomnia (DSM-V)—was induced. Thereafter, the animals were subdivided into extreme groups showing resilience to PTSD or PTSD-like behavior.

Measuring the *level of anxiety* was done using the light/dark transfer test. A box was divided into a dark and an illuminated area, connected by a small passage in the center of the dividing wall. Mice were placed in the dark area, and the animal's movements were recorded in five-minute sessions. Behavioral measures were: time spent in the light arena, number of visits to the light arena, and total distance travelled in the light arena. A mouse was considered less anxious when it explored the light area more, and/or more time was spent in the light area, and/or more frequently the light area was visited (30). To determine *risk assessment behavior*, time spent and the number of visits in a defined risky area (an additional area surrounding the opening of the illuminated box) were measured. The percentage risk assessment was the time spent in the risk area divided by the time spent in the illuminated area (which excludes the risk zone) (26). *Startle and prepulse inhibition* measurements were performed on a vibration-sensitive platform with a Plexiglas and an integrated sensor for movement detection, and two high-frequency speakers for the audio stimuli. Throughout all sessions there was white background noise (65 dB). The acoustic startle response (ASR) session began with a five-minute habituation period. Then, 32 × 120dB startle stimuli (40 ms duration) were presented with a randomly varying inter-trial interval of 12–30 ms. The stimuli were alternated with 40 additional startle stimuli, randomly preceded by prepulses of either 74dB (40 ms), 78dB (40 ms), or 82 dB (40ms) (26, 30). The latency (ms) to peak startle amplitude was measured in response to startle stimuli and startle stimuli preceded by prepulses. Prepulse inhibition (PPI) disruption occurs when the response to a startling stimulus with a prepulse stimulus is less inhibited. To measure *total locomotor activity during light cycles*, mice were housed individually for 72 hours in the home cage. First the mice were habituated to the individual housing condition for 24 hours. Locomotor activity was measured in two light and two dark cycles at 10 min intervals during the previous 48 hours (26). For the *marble burying* test, mice were placed in a cage (30 × 27 × 26 cm) that was filled with 5 cm of autoclaved bedding with 20 marbles centrally arranged 4 × 5 and filmed for 30 minutes. After 25 minutes the unburied marbles were counted (31).

After the behavioral tests, mice were categorized into resilient and PTSD-like groups of mice' behavior using a scoring method. Behavioral data of the individual rats were ranked from highest to lowest; for the behavioral measures of risk assessment and the latency to peak startle amplitude a score of 3 points was given to the lowest 20% results, and a score of 2 points was given for less percentage PPI. A score of 1 point was given to mice belonging to the upper 20% with the highest activity in the light cycle (i.e. non-active phase) and percentage most marbles buried data (see Table 22.2). If the total score was 5 or higher, mice were categorized as PTSD-like. Mice that had a score lower than 1 (showing no extreme behavior) were considered to be resilient. Henckens and co-workers as well as Lebow and colleagues (26, 27) have shown that 20% of mice indeed developed PTSD-like symptomatology, whereas 20% of the other extreme did not show any of these behaviors and were resilient. This observation perfectly mimics the human situation in which only a relatively small fraction (15–20%) of the individuals exposed to a traumatic event develop PTSD (29).

What are the potential mechanisms that dissociate the resilient and PTSD-like mice? CRF can bind to two different CRF receptors: CRFR1 and CRFR2 (32). Both CRFR1 and CRFR2 activation regulates anxiety-like behavior in rodents, dependent on the brain region where the receptors are expressed. In most brain regions, CRFR1 activation leads to anxiety-like behavior. However, less is known about CRFR2 activation. It is possible that CRFR2 compensates stress responses. CRFR2 has been examined in the bed nucleus of the stria terminalis (BNST), a stress-related area (26). In the BSNT of mice with a PTSD-like phenotype compared with resilient mice (characterized as described earlier after stress exposure), CRFR2 expression levels were increased (26, 27) without changes in CRFR1 expression levels and CRF plasma levels. Henckens and colleagues (27) used optogenetic activation and inhibition of BNST GABAergic CRFR2-expressing neurons in order to study its association with PTSD susceptibility. Activation of these neurons reduced anxiety-like behavior. Furthermore, activation immediately after stress exposure resulted in a reduction of PTSD-symptom development compared to control mice, while inhibition resulted in opposite behavior. These findings suggest that CRFR2-expressing neurons in the BSNT enhance stress recovery following trauma exposure, and thereby modulate PTSD-behavior. However other studies

Table 22.2 Scores to behavioral tests and their PTSD-like attributes.

Behavioral test	PTSD-like attributes	Behavioral measure	Score	
Risk assessment behavior	Judgement of danger	% Risk assessment behavior	Bottom 20%	3
Latency to peak startle response	Alertness	Latency to peak startle amplitude	Bottom 20%	3
Pre-pulse inhibition (PPI)	Sensorimotor gating	% Total PPI disruption	Bottom 20%	2
Locomotor activity in the non-active phase of the light/dark cycle	Sleep disturbances	Total light locomotor activity	Top 20%	1
Marble burying	Hypervigilance	% Marbles buried	Top 20%	1
		PTSD-like	≥ 5	
		Resilient	< 1	

reported contrary findings; by using a BNST-CRFR2-specific lentiviral knockdown (KD), Lebow and Neufeld-Cohen (26) sought to determine the effect of reduced CRFR2 levels on the development of PTSD-like behavior. All mice (KD and controls) were tested in the behavioral tests as described earlier, and no differences were found between the groups. However, stress exposed CRFR2-KD mice displayed less PTSD-like symptoms compared to controls, suggesting that low BSNT-CRF2 receptor levels reduce the development of PTSD-like behavior.

22.3.2 Genotype-dependent individual differences

One genetic factor widely studied in relation to PTSD vulnerability involves the serotonin transporter (5-HTT) linked polymorphic region (5-HTTLPR) (33, 34). Neuroimaging studies have revealed that the low activity short (S-) allelic variant of this polymorphism is associated with amygdala hyperactivity (35) as well as reduced prefrontal cortical control over the amygdala (36, 37). The low-expressing polymorphic variant is more frequently found in patients with PTSD than in controls (38). Since there is a striking correspondence between this genetic brain phenotype and the brain pathology that has been linked to PTSD (21, 39), it is likely that 5-HTT genotype affects fear learning, memory, and extinction.

The 5-HTTLPR in humans is mimicked by serotonin transporter knockout in rodents (40). It has been reported that 5-HTT knockout (5-HTT$^{-/-}$) mice and rats show a deficiency in fear extinction (41, 42) or the recall of the fear extinction memory (43–45). These rodent findings seem to match those found in humans, as human 5-HTTLPR S-allele carriers show increased acquisition of fear memory (46), as well as reduced fear extinction (47). The extinction (recall) deficit in 5-HTT$^{-/-}$ rodents is accompanied by morphological abnormalities in the basolateral amygdala (BLA) and in the infralimbic cortex (IL) of the ventro-medial prefrontal cortex (vmPFC) (44). Wellman and Izquierdo (44) tested 5-HTT$^{-/-}$ and wildtype (WT) mice in a Pavlovian fear paradigm (cued fear conditioning), and after the completion of the tests, the brains (dendrites and spines) were analysed. This analysis revealed increased spine density in the BLA and increased dendritic length in the IL of 5-HTT$^{-/-}$ mice. A plausible explanation could be that the increased spine density in BLA dendrites might result in increased excitatory output of these neurons resulting in amygdala hyperactivity, whereas the prolonged IL dendrites function as an adaptive response to control this activity.

Another genetic factor that has been reported in relation to PTSD susceptibility is brain-derived neurotrophic factor (BDNF), which is a growth factor that regulates fear learning and memory by synaptic plasticity (48, 49). The BDNF gene harbours the Val66Met polymorphism, a substitution of valine to methionine at codon 66 (50). In humans the BDNF-Met-allele leads to decreased BDNF secretion and presumably to impairments in learning and altered stress responses (48, 51), which are affected in PTSD. This polymorphism can be mimicked in an inbred knockin mouse strain that expresses the variant BDNF-allele (BDNF$^{Met/Met}$) (52). Soliman and Glatt (49) investigated mice and humans with all the variants of BDNF-alleles (Val/Val, Val/Met, and Met/Met) and the relation to fear learning and extinction by using cued fear conditioning and extinction suitable for mouse or human. They showed that Met-allele carriers had impaired fear extinction while fear conditioning was intact. Neuroimaging studies were performed to correlate brain activity to these behavioral effects. These studies revealed that Met-allele carriers, during fear extinction, exhibited reduced activity in the vmPFC and amygdala hyperactivity (49). Interestingly, these brain regions were similarly affected by the 5-HTTLPR S-allele. These findings have a strong biological foundation, as mice carrying the Met-allele also displayed impaired contextual fear conditioning and increased anxiety. Worth knowing, the serotonergic and BDNF systems functionally overlap and interact. BDNF is a downstream target of serotonin (5-HT) (53). A double-mutant mouse model was created to investigate the behavioral consequences of altered BDNF secretion along with decreased 5-HT transport (54). Anxiety- and stress-related effects were studied in relation to four genotypes in mice ($SERT^{+/+}/BDNF^{+/+}$ (WT); $SERT^{+/+}/BDNF^{+/-}$; $SERT^{-/-}/BDNF^{+/+}$; and $SERT^{-/-}/BDNF^{+/-}$ (double-mutant)). Anxiety-like behavior was increased in $SERT^{-/-}/BDNF^{+/-}$ mice compared to the other three genotypes, tested on the elevated plus maze (test described in a following paragraph; EPM). To examine stress-related behavior, mice were injected with saline (mild-stressor), and the plasma ACTH response (determined by radioimmunoassay) was compared to non-injected mice. All genotypes, except for WT, had increased plasma ACTH levels; the double-mutant even had a fourfold higher response. These results reveal a BDNF–SERT genotype interaction affecting anxiety levels and stress responsivity (54, 55), as well as vulnerability to psychopathology.

Another polymorphism related to vulnerability to PTSD involves the Taq1A polymorphism in the dopamine receptor 2 (DRD2) gene (50). DRD2 belongs to the class of the G-protein-coupled receptor family that inhibits adenylyl cyclase; this receptor is associated with pleasure, reward circuitry, and learning mechanisms. A reduced number of DRD2 in the human brain has been found in TaqA1-allele carriers (50). The study by Huertas and colleagues (56) showed no effect of the Taq1A polymorphism on human fear conditioning, but another DRD2 polymorphism (C957T) could be related to learning deficits. Animal studies focusing merely on DRD2 polymorphisms and its effects on fear learning and PTSD are not very common nowadays. There is evidence from several human studies that there is an association between the A1-allele and PTSD in a subset of patients (e.g. "harmful alcohol drinkers") (57), while others (58) did not find such an association. Interestingly, in a human population at high risk for developing stress-related-disorders such as PTSD, Hemmings and colleagues (50) found a gene–gene interaction between DRD2 Taq1A and BDNF Val66Met variants. The Davidson Trauma Scale (DTS) was used to measure the severity of PTSD symptoms and therefore the susceptibility to PTSD (59). In subjects with a Val/Val BDNF genotype, the DTS increased with the addition of the DRD2 Taq1A A1-allele. Whereas in Met-allele carriers of the BDNF gene, the DTS decreased with the addition of the DRD2 Taq1A A1-allele. The optimal concentration level of dopamine function is between baseline levels, depletion, and over-dosing have both negative effects dependent on task and brain area (60). Dopamine concentrations are influenced by DRD2 activation, in addition, DRD2 activation is exerted in an optimal range (61). It is known that BDNF interacts with the dopaminergic system (62), but the mechanism of action of the interaction between BDNF and DRD2 remains to be determined. This interaction probably affects the available dopamine levels, too low and too high, and impacts PTSD susceptibility (50). Animal studies focused on fear conditioning and extinction in combination with double-mutants (gene–gene interaction) are lacking, but could contribute to the understanding of PTSD susceptibility.

22.3.3 Age-dependent individual differences

Adolescence is a period of physical and brain maturation that is characterized by an increase in pervasive fears and risky behaviors, which coincides with the emergence of anxiety and other affective disorders (63–69). Recent data suggest that adolescence is associated with the organizational changes of the cognitive control circuitry regulating emotional behaviors. More specifically, there is evidence for immaturity of the prefrontal cortex (PFC) and its top-down control over subcortical areas mediating emotion and motivation such as the amygdala. According to the developmental imbalance hypothesis, the maturation of the PFC is delayed in comparison to the amygdala, resulting in a temporary imbalance between emotion and its regulatory processes (70). Given that fear learning, memory, and extinction are dependent on the prefrontal cortex and amygdala, age, and specifically adolescence, reflects an important factor influencing resilience and vulnerability to PTSD.

Regarding fear extinction, it has been reported that pre-adolescent (i.e. postnatal day [*P*] 24), adolescent (*P*35), and adult (*P*70) rats express identical extinction acquisition following conditioned stimulus (CS) and shock pairings. However, when tested the next day for extinction recall, adolescent rats showed an almost complete failure to maintain extinction of CS-elicited conditioned freezing compared with *P*24 and *P*70 rats (71, 72). It has also been reported that *P*29 adolescent mice show a temporary suppression of contextual fear due a retrieval failure (73). Luna and Sweeney (74) suggested that developmental changes in cognition during adolescence and infancy are different, as adolescent changes involve the improvement of existing capacities rather than the acquisition of new ones. As such, fear extinction in *P*17 rats appears to rely on memory erasure instead of new learning (75). This is reflected by the findings that extinction in adult and *P*24 rats, but not in *P*17 rats, is GABA- and *N*-methyl-D-aspartate-dependent (NMDA) (75–78). Accordingly, the mPFC is thought to be critical for long-term extinction in *P*24 and adult rats, but not in *P*17 rats (79, 80). Interestingly, it has been suggested that the neural mechanisms underlying extinction in P24 and adult rats are similar (80). Therefore, changes in the neural circuitry underlying extinction up to pre-adolescence may have a "development" nature, whereas changes from *P*24 to adulthood may be considered as "maturation." Maturation may go along with physiological and molecular changes, as the suppression of contextual fear was associated with a blunted synaptic potentiation response in the amygdala (73). Furthermore, levels of phosphorylated mitogen-activated protein kinase (pMAPK) in the infralimbic (IL) cortex (corresponding to the vmPFC in humans) were decreased in adolescent rats compared to *P*24 and *P*70 rats in association with impaired fear extinction recall. Extensive extinction training in adolescents increased pMAPK levels (72), as well as the recall of the extinction memory (71, 72), indicating that adolescents are less efficient in utilizing prefrontal areas. Levels of pAkt and pERK2, the downstream targets of the MAPK/ERK pathway, were decreased in the hippocampi of *P*29 mice during suppression of contextual fear, indicating that adolescence is associated with a lack of memory retrieval-associated signalling in the hippocampus (73). The MAPK/ERK pathway is part of the intracellular cascade that is involved in producing long-lasting changes in synaptic efficacy (81), and indeed there is evidence for changes in the molecular composition of synapses and the synaptic release machinery of the medial prefrontal cortex in adolescent rats (82).

22.3.4 Gender-dependent individual differences

Women have been considered to be more susceptible to stress-related psychopathology, such as PTSD, than men. Several reports show a higher lifetime prevalence of PTSD in women than in men (8.5% versus 3.4%, respectively) (83). Ironically, most studies use exclusively male subjects; however, recent studies include female subjects and reveal sex-specific differences in responding to fear and stress (84). As described earlier in the Pavlovian paradigm, the most common measured fear response is freezing behavior. However, Gruene and colleagues (85) identified an alternative fear response in rats; a rapid, forward movement across the chamber that resembled escape-like behavior (e.g. "darting"). Adult (8–10 weeks) male and female rats underwent habituation, cued fear conditioning, and extinction (recall) (Figure 22.1A). Besides freezing, locomotor activity was also recorded and analysed during all procedures. In traces of each animal's velocity, the darting responses are identified and measured as discrete events. Darts are defined as all movements at the velocity rate threshold or higher. The threshold can be determined by cross-referencing velocity data with experimenter scoring of darting behavior. It was reported (85) that females are more likely to display darting behavior and less freezing during fear conditioning and extinction than men. The possibility of an effect of the oestrous cycle on darting was excluded. These findings suggest that there are different behaviors of responding to stress and fear, that they are gender related, and that measuring freezing alone may not be sufficient enough to determine fear responses. Female rats can therefore be incorrectly said to be resilient, because of their low freezing levels, while in fact they display darting behavior.

To explore gender-related differences in baseline and post-stress anxiety-like behavior, an animal model (86) that focused on individual responses in males and females was established. The stress paradigm that was used was the predator scent stress (PSS), in which the animals were confronted with the odor of one of their natural predators (5). In this study the researchers exposed rats to used cat litter for 10 min. Before (baseline) and after PSS exposure, male and female rats were tested in the elevated plus maze (EPM) and acoustic startle response (ASR) to measure anxiety-like behaviors, related to PTSD symptomatology (i.e. anxiety, fearfulness, and hypervigilance). The EPM test was performed on a plus-shaped platform maze, with two opposing open and two opposing closed arms (surrounded by 40-cm high walls on three sides). Rats were placed in the center of the platform facing an open arm and explored the maze for five minutes (while recorded). Behavioral measures were: time spent in open and closed arms; number of entries in open and closed arms (an arm is entered when all four paws are in); total entries into all arms (total exploration) (87). These measures were included in the "Anxiety Index," which can be calculated as follows:

$$Anxiety\,index = 1 - \left[\frac{\dfrac{time\,spent\,in\,the\,open\,arms}{total\,time\,on\,the\,maze} + \dfrac{number\,of\,entries\,to\,the\,open\,arms}{total\,exploration\,on\,the\,maze}}{2} \right]$$

A value of 0 was related to no anxiety-like behavior, and an increase reflected an increase in anxiety (from 0 to 1) (86). ASR measurements were performed on a vibration-sensitive platform with a Plexiglas and an integrated sensor for movement detection, and two high-frequency speakers for the audio stimuli. Throughout all sessions there was white background noise (68 dB). Sessions began with a five-minute habituation period. Then, 30 × 110dB startle stimuli (40 ms duration) were presented with a randomly varying inter-trial interval of 30 or 45 s, in 6 blocks. The mean startle amplitude (over all 30 trials) and the percentage of startle response habituation was measured in response to acoustic stimuli as follows:

$$Percent\,habituation = 100 \times \frac{[average\,startle\,amplitude\,in\,block\,1 - average\,startle\,amplitude\,in\,block\,5]}{average\,startle\,amplitude\,in\,block\,1}$$

Anxiety-like behavior was characterized by restrained exploratory behavior on the maze, and a higher startle response in the ASR paradigm that doesn't undergo adaptation to acoustic stimuli (88). Using cut-off behavioral criteria (CBCs) (86–88) animals were categorized on their behavioral phenotype. Animals that (i) spent 5 min in the closed arms and didn't enter the open arms in the EPM, and (ii) had a mean amplitude of the startle response (at 110dB) > 1200 units and no adaptation over time, were categorized as PTSD-like animals displaying extreme behavior). Animals that (i) spent 0–1 min in the closed arms of the EPM and entered the open-arm ≥ 8 times, and (ii) had a mean amplitude of the startle response (at 110dB) < 700 units and showed adaptation over time, were considered as resilient animals.

Females were found to display a higher level of anxiety-like behavior at baseline compared to males. After stress exposure, both male and female showed changes in anxiety-like behavior (compared to baseline), while males showed a larger change in response to PSS exposure. Regarding prevalence of PTSD-like behavior, no gender difference was found, although females were more susceptible to develop extreme behavior in terms of intensity of the response (4, 86). As these studies show, males and females differ in baseline anxiety and that responses upon stress exposure are gender-dependent.

22.4 Discussion and conclusion

Individual differences in vulnerability to PTSD and other anxiety-related disorders are a key clinical observation, but it is as yet poorly understood. As we have set out here, animal models are very helpful to improve this understanding on the premise that they have high translational value. More specifically, the genetic rodent models revealed, amongst others, that the 5-HTTLPR and BDNF Val66Met polymorphisms lead to impairments in fear extinction (recall), as is also observed in humans carrying these polymorphisms (47, 49). Since these polymorphisms have been associated with PTSD, these findings imply that genotype-driven changes in learning and memory contribute to PTSD vulnerability. Given the translational value of the genetic animal models, the specific nature of the learning and memory impairments can be elucidated and other phenotypes displayed by the animals can be explored to inform new or adjust existing therapies to treat PTSD, in a personalized manner. For instance, while 5-HTT knockout rats show an impairment in fear extinction, they also are behaviorally flexible in response to cues predicting reward (89). Accordingly, a behavioral distraction therapy in which a cue predicting reward was presented simultaneously with the cue predicting shock, was able to improve extinction specifically in 5-HTT knockout rats (42). Studies focusing on age-dependent changes in fear learning and extinction demonstrated that adolescence is an important factor in the vulnerability to develop PTSD or resiliency. Adolescence rodents tend to have more fear extinction recall failures, due to developmental changes in the PFC and amygdala resulting in temporary memory suppression and/or removal. This was also found in humans (39). Understanding of the molecular and cellular changes in this developmental phase (discussed in the previous paragraph) are of great importance because the occurrence of anxiety disorders enlarges among youth. As mentioned, extinction causes an increase of the NMDA receptors in the IL in preadolescent and adult mice, but not in adolescent animals, resulting in impaired fear extinction (2). Accordingly, the NMDA2B receptor partial agonist d-cycloserine (DCS) was able to alleviate the impairment in cued fear-extinction retention (71). Clinical studies showed that DCS intervention to patients with anxiety disorders, facilitates fear extinction (90, 91). This is promising for the treatment of the impaired fear learning and extinction in PTSD patients (92). Finally, rodent male–female comparisons in fear learning and extinction paradigms have revealed that there are gender-specific responses to fear and stress. Besides the well-known freezing behavior, darting (a rapid, forward movement across the chamber that resembles escape-like behavior) is described as an alternative fear response. While males display more freezing behavior, females show darting behavior as stress response (85). For early diagnosis of PTSD it is thus worthwhile to measure gender-specific fear responses. In addition, the animal model of Mazor and colleagues (86) showed that baseline anxiety is dependent on gender; female rodents are more anxious than males. After trauma, rodent females respond in a more extreme manner than males, but PTSD-like incidence was not affected in the rat population. In humans there is a higher prevalence of PTSD in women compared with men, but this is possibly due to the type of trauma and gender differences in stress coping (83, 93). Animal studies thus help to refine the understanding of gender-specific fear and coping responses. Importantly, genotype, age, and gender are not independent acting biological factors; they strongly interact. For instance, the fear extinction (recall) deficit in both 5-HTT$^{-/-}$ and Val66Met rodents is influenced by age; during adolescence the impairments are transiently normalized for both genetic models (48, 94). Whether gender interferes with these findings remains to be established. Nonetheless, this example illustrates how critical outcomes are dependent on combinations of factors. While it is possible to investigate genotype × age × gender interactions in humans, this is highly challenging in terms of the numbers of subjects required. In rodents, a systematic analysis of factor interactions is more feasible, and given the high translational value of the genetic animal models and the fear conditioning/extinction paradigm, resulting data are likely to be relevant for humans.

Although the fear conditioning/extinction paradigm is well-validated and considered to have high translational value, it does not fully recapitulate the symptoms of PTSD. As discussed in detail, the studies of Henckens (27) and Lebow (26) used five behavioral paradigms (Table 22.2) to measure PTSD-like symptomatology: hypervigilance, compulsivity, impaired attention, compromised risk assessment, and insomnia (DSM-V). The study of Mazor (86), which also discussed regarding gender differences, used the elevated plus maze and acoustic startle response to measure anxiety-like behavior related to PTSD symptoms (i.e. anxiety, fearfulness, and hypervigilance). However, PTSD is clinically also characterized by avoidance and memory of trauma-related stimuli. Toledano and colleagues (10) used the single prolonged stress (SPS) paradigm to investigate the avoidance of trauma-associated cues. When SPS is given with additional cues (olfactory and/or auditory), these can later be used as trauma reminders. Avoidance can then be measured with an odor discrimination test on a Y-maze, and the reactivity (freezing) to the trauma-associated tone, in order to more completely capture PTSD symptomatology.

Modern technologies, such as functional magnetic resonance imaging (fMRI), can be used to further dissect how fear and related pathologies (e.g. PTSD) are processed in the brain. fMRI is a non-invasive technique which indirectly measures neural activity across the whole brain through the blood oxygen level-dependent (BOLD) signal (95, 96). In rodents, one can study pharmacological, genetic, and environmental manipulations and its effects on brain function (97). fMRI in rodents is mainly performed while the animals were anaesthetized or resting, which limits it as a translational tool for fear pathologies (98). Nevertheless, there is a growing need for improved translation of preclinical imaging studies. Several studies (95–100) successfully developed protocols for fMRI scanning in awake, genetically modified rodents to investigate brain responses and activation to fear. Studies combining fear conditioning with fMRI use a flashing light as CS and a footshock as US. Then, during extinction training, the CS can be presented in the fMRI scanner allowing the measurement of brain activation while the animals experience fear (96–98). Presenting a fear-inducing scent of a predator (PSS) while the rodents are in the fMRI elicits also a stress response. An odor that is frequently used is 2,4,5-trimethylthiazoline (TMT; a chemical extract of fox feces) (100), but also an airflow containing cat fur has been used (99). For both paradigms it is important that the fear responses are confirmed in additional behavioral tests. An alternative methodology to map fear-related brain activity is manganese-enhanced MRI (MEMRI) (101). Manganese is easily traceable with MRI and visualizes neuronal functional connectivity, but toxicity remains a challenge (102). Besides the normal daily handling to minimize stress, rodents have to be habituated to the restraining device of the fMRI scanner (a so-called mock scanner) to acclimatize to the scanning environment. The animals also have to be lightly anesthetized before placing them in and out of the restrainer. Rodents have to be restrained during fMRI experiments because they must remain motionless, but this also prevents them from fully eliciting fear responses, reactions, and/or escape (99, 103). The whole-brain scanning is an advantage of fMRI, but the restraining and the non-invasive use make it difficult to use it in freely moving rodents. Barkus and colleagues (104) used a method analogous to fMRI, *in vivo* voltammetric monitoring,

Box 22.1 Post-traumatic stress disorder (PTSD) related numbers and facts

Estimates are that up to 90% of all people will be exposed to a severe traumatic event during their life time (105), of which a substantial part (15–20%) ultimately develops PTSD (106). The quality of life of an estimated 7.7 million European citizens is presently severely affected by PTSD's symptomatology (involved annual costs: €9 billion) (107), and numbers are rising; the number of veterans diagnosed with PTSD almost tripled, suicide attempts doubled, and the money spent on their care increased by tenfold over the past decade (103). Current medication falls short as treatments are symptomatic, only effective for fewer than half of the patients (8), and side effects and residual symptoms following treatment are rule rather than exception. To improve treatment, a better understanding of PTSD pathophysiology is urgently needed.

to study fear behavior in freely moving rodents. This method allows the measurement of region-specific (but not whole-brain) changes in brain tissue oxygen during behavior (103), creating possibilities to use it during behavioral experiments.

In conclusion, we strongly encourage the use of translational rodent models to increase our understanding of individual differences in vulnerability to PTSD, which ultimately will help to refine existing or develop novel personalized strategies to prevent or treat PTSD.

References

1. **King EC, Pattwell SS, Glatt CE, Lee FS.** Sensitive periods in fear learning and memory. Stress. 2014;**17**(1):13–21.
2. **Pattwell SS, Duhoux S, Hartley CA, Johnson DC, Jing D, Elliott MD,** et al. Altered fear learning across development in both mouse and human. PNAS. 2012;**109**(40):16318–23.
3. **Ritov G, Boltyansky B, Richter-Levin G.** A novel approach to PTSD modeling in rats reveals alternating patterns of limbic activity in different types of stress reaction. Mol Psychiatry. 2016;**21**(5):630–41.
4. **Daskalakis NP, Yehuda R, Diamond DM.** Animal models in translational studies of PTSD. Psychoneuroendocrinology. 2013;**38**(9):1895–911.
5. **Borghans B, Homberg JR.** Animal models for posttraumatic stress disorder: An overview of what is used in research. World journal of psychiatry. 2015;**5**(4):387.
6. **American Psychiatric Association.** Diagnostic and Statistical Manual of Mental Disorders, 5th edn. Arlington, VA: American Psychiatric Association; 2013.
7. **Marin MF, Camprodon JA, Dougherty DD, Milad MR.** Device-based brain stimulation to augment fear extinction: Implications for PTSD treatment and beyond. Depress Anxiety. 2014;**31**(4):269–78.
8. **Maxmen JS, Ward NG.** Psychotropic Drugs: Fast Facts, 3rd edn. New York, NY: W.W. Norton; 2002.
9. **Johnson LR, McGuire J, Lazarus R, Palmer AA.** Pavlovian fear memory circuits and phenotype models of PTSD. Neuropharmacology. 2012;**62**(2):638–46.
10. **Toledano D, Gisquet-Verrier P.** Only susceptible rats exposed to a model of PTSD exhibit reactivity to trauma-related cues and other symptoms: An effect abolished by a single amphetamine injection. Behav Brain Res. 2014;**272**:165–74.
11. **Coyner J, McGuire JL, Parker CC, Ursano RJ, Palmer AA, Johnson LR.** Mice selectively bred for high and low fear behavior show differences in the number of pMAPK (p44/42 ERK) expressing neurons in lateral amygdala following Pavlovian fear conditioning. Neurobiol Learn Mem. 2014;**112**:195–203.

12. **Norrholm SD, Jovanovic T, Olin IW, Sands LA, Bradley B, Ressler KJ.** Fear extinction in traumatized civilians with posttraumatic stress disorder: Relation to symptom severity. Biol Psychiatry. 2011;**69**(6):556–63.
13. **Bush DE, Sotres-Bayon F, LeDoux JE.** Individual differences in fear: Isolating fear reactivity and fear recovery phenotypes. J Traum Stress. 2007;**20**(4):413–22.
14. **Yehuda R.** Risk and resilience in posttraumatic stress disorder. J Clin Psychiatry. 2004.
15. **Pavlov IP.** Conditioned Reflexes. An Investigation of the Physiological Activity of the Cerebral Cortex. Oxford: Oxford University Press; 1927.
16. **Galatzer-Levy IR, Bonanno GA, Bush DE, LeDoux J.** Heterogeneity in threat extinction learning: Substantive and methodological considerations for identifying individual difference in response to stress. Front Behav Neurosci. 2013;7:55.
17. **Castro-Gomes V, Bergstrom HC, McGuire JL, Parker CC, Coyner J, Landeira-Fernandez J,** et al. A dendritic organization of lateral amygdala neurons in fear susceptible and resistant mice. Neurobiol Learn Mem. 2016;**127**:64–71.
18. **Schipper P, Henckens MJ, Borghans B, Hiemstra M, Kozicz T, Homberg JR.** Prior fear conditioning does not impede enhanced active avoidance in serotonin transporter knockout rats. Behav Brain Res. 2017;**326**:77–86.
19. **Schipper P, Henckens MJ, Lopresto D, Kozicz T, Homberg JR.** Acute inescapable stress alleviates fear extinction recall deficits caused by serotonin transporter abolishment. Behav Brain Res. 2018;**346**:16–20.
20. **Ponder C, Kliethermes C, Drew M, Muller J, Das K, Risbrough V,** et al. Selection for contextual fear conditioning affects anxiety-like behaviors and gene expression. Genes, Brain, Behav. 2007;**6**(8):736–49.
21. **Lopresto D, Schipper P, Homberg JR.** Neural circuits and mechanisms involved in fear generalization: Implications for the pathophysiology and treatment of posttraumatic stress disorder. Neurosci Biobehav Rev. 2016;**60**:31–42.
22. **Smith SM, Vale WW.** The role of the hypothalamus–pituitary–adrenal axis in neuroendocrine responses to stress. Dialogues Clin Neurosci. 2006;**8**(4):383.
23. **Henckens MJ, Klumpers F, Everaerd D, Kooijman SC, van Wingen GA, Fernández G.** Interindividual differences in stress sensitivity: Basal and stress-induced cortisol levels differentially predict neural vigilance processing under stress. Soc Cogn Affect Neurosci. 2016c;**11**(4):663–73.
24. **Whitaker AM, Gilpin NW.** Blunted hypothalamus–pituitary–adrenal axis response to predator odor predicts high stress reactivity. Physiol Behav. 2015;**147**:16–22.
25. **Bremner JD, Licinio J, Darnell A, Krystal JH, Owens MJ, Southwick SM,** et al. Elevated CSF corticotropin-releasing factor concentrations in posttraumatic stress disorder. Am J Psychiatry. 1997;**154**:624–9.
26. **Lebow M, Neufeld-Cohen A, Kuperman Y, Tsoory M, Gil S, Chen A.** Susceptibility to PTSD-like behavior is mediated by corticotropin-releasing factor receptor type 2 levels in the bed nucleus of the stria terminalis. J Neurosci. 2012;**32**(20):6906–16.
27. **Henckens M, Printz Y, Shamgar U, Dine J, Lebow M, Drori Y,** et al. CRF receptor type 2 neurons in the posterior bed nucleus of the stria terminalis critically contribute to stress recovery. Mol Psychiatry. 2017;**22**(12):1691–700.
28. **Jovanovic T, Kazama A, Bachevalier J, Davis M.** Impaired safety signal learning may be a biomarker of PTSD. Neuropharmacology. 2012;**62**(2):695–704.
29. **Kessler RC, Chiu WT, Demler O, Walters EE.** Prevalence, severity, and comorbidity of 12-month DSM-IV disorders in the National Comorbidity Survey Replication. Arch Gen Psychiatry. 2005;**62**(6):617–27.
30. **Neufeld-Cohen A, Tsoory MM, Evans AK, Getselter D, Gil S, Lowry CA,** et al. A triple urocortin knockout mouse model reveals an essential role for urocortins in stress recovery. PNAS. 2010;**107**(44):19020–5.

31. **Sztainberg Y, Kuperman Y, Justice N, Chen A.** An anxiolytic role for CRF receptor type 1 in the globus pallidus. J Neurosci. 2011;**31**(48):17416–24.

32. **Henckens MJ, Deussing JM, Chen A.** Region-specific roles of the corticotropin-releasing factor-urocortin system in stress. Nat Rev Neurosci. 2016b;**17**(10):636–51.

33. **Gressier F, Calati R, Balestri M, Marsano A, Alberti S, Antypa N,** et al. The 5-HTTLPR polymorphism and posttraumatic stress disorder: A meta-analysis. J Traum Stress. 2013;**26**(6):645–53.

34. **Miller R, Wankerl M, Stalder T, Kirschbaum C, Alexander N.** The serotonin transporter gene-linked polymorphic region (5-HTTLPR) and cortisol stress reactivity: A meta-analysis. Mol Psychiatry. 2013;**18**(9):1018–24.

35. **Hariri AR, Mattay VS, Tessitore A, Kolachana B, Fera F, Goldman D,** et al. Serotonin transporter genetic variation and the response of the human amygdala. Science. 2002;**297**(5580):400–3.

36. **Volman I, Verhagen L, den Ouden HE, Fernández G, Rijpkema M, Franke B,** et al. Reduced serotonin transporter availability decreases prefrontal control of the amygdala. J Neurosci. 2013;**33**(21):8974–9.

37. **Pezawas L, Meyer-Lindenberg A, Drabant EM, Verchinski BA, Munoz KE, Kolachana BS,** et al. 5-HTTLPR polymorphism impacts human cingulate-amygdala interactions: A genetic susceptibility mechanism for depression. Nat Neurosci. 2005;**8**(6):828–34.

38. **Lee HJ, Lee MS, Kang RH, Kim H, Kim SD, Kee BS,** et al. Influence of the serotonin transporter promoter gene polymorphism on susceptibility to posttraumatic stress disorder. Depress Anxiety. 2005;**21**(3):135–9.

39. **Morey RA, Haswell CC, Hooper SR, De Bellis MD.** Amygdala, hippocampus, and ventral medial prefrontal cortex volumes differ in maltreated youth with and without chronic posttraumatic stress disorder. Neuropsychopharmacology. 2016;**41**(3):791–801.

40. **Caspi A, Hariri AR, Holmes A, Uher R, Moffitt TE.** Genetic sensitivity to the environment: The case of the serotonin transporter gene and its implications for studying complex diseases and traits. Focus. 2010;**8**(3):398–416.

41. **Pang RD, Wang Z, Klosinski LP, Guo Y, Herman DH, Celikel T,** et al. Mapping functional brain activation using [14 C]-iodoantipyrine in male serotonin transporter knockout mice. PLoS One. 2011;**6**(8):e23869.

42. **Nonkes LJ, de Pooter M, Homberg JR.** Behavioural therapy based on distraction alleviates impaired fear extinction in male serotonin transporter knockout rats. J Psychiatry Neurosci. 2012;**37**(4):224.

43. **Narayanan V, Heiming RS, Jansen F, Lesting J, Sachser N, Pape H-C,** et al. Social defeat: Impact on fear extinction and amygdala-prefrontal cortical theta synchrony in 5-HTT deficient mice. PLoS One. 2011;**6**(7):e22600.

44. **Wellman C, Izquierdo A, Garrett J, Martin K, Carroll J, Millstein R,** et al. Impaired stress-coping and fear extinction and abnormal corticolimbic morphology in serotonin transporter knock-out mice. J Neurosci. 2007;**27**(3):684–91.

45. **Schipper P, Kiliaan AJ, Homberg JR.** A mixed polyunsaturated fatty acid diet normalizes hippocampal neurogenesis and reduces anxiety in serotonin transporter knockout rats. Behav Pharmacol. 2011;**22**(4):324–34.

46. **Garpenstrand H, Annas P, Ekblom J, Oreland L, Fredrikson M.** Human fear conditioning is related to dopaminergic and serotonergic biological markers. Behav Neurosci. 2001;**115**(2):358.

47. **Klucken T, Alexander N, Schweckendiek J, Merz CJ, Kagerer S, Osinsky R,** et al. Individual differences in neural correlates of fear conditioning as a function of 5-HTTLPR and stressful life events. Soc Cogn Affect Neurosci. 2012:nss005.

48. **Dincheva I, Pattwell SS, Tessarollo L, Bath KG, Lee FS.** BDNF modulates contextual fear learning during adolescence. Dev Neurosci. 2014;**36**(3-4):269–76.

49. **Soliman F, Glatt CE, Bath KG, Levita L, Jones RM, Pattwell SS,** et al. A genetic variant BDNF polymorphism alters extinction learning in both mouse and human. Science. 2010;**327**(5967):863–6.

50. **Hemmings SM, Martin LI, Klopper M, van der Merwe L, Aitken L, de Wit E,** et al. BDNF Val66Met and DRD2 Taq1A polymorphisms interact to influence PTSD symptom severity: A preliminary investigation in a South African population. Prog Neuro-Psychopharmacol Biol Psychiatry. 2013;**40**:273–80.
51. **Frielingsdorf H, Bath KG, Soliman F, DiFede J, Casey B, Lee FS.** Variant brain-derived neurotrophic factor Val66Met endophenotypes: Implications for posttraumatic stress disorder. Ann N Y Acad Sci. 2010;**1208**(1):150–7.
52. **Chen Z-Y, Jing D, Bath KG, Ieraci A, Khan T, Siao C-J,** et al. Genetic variant BDNF (Val66Met) polymorphism alters anxiety-related behavior. Science. 2006;**314**(5796):140–3.
53. **Homberg JR, Molteni R, Calabrese F, Riva MA.** The serotonin–BDNF duo: Developmental implications for the vulnerability to psychopathology. Neurosci Biobehav Rev. 2014;**43**:35–47.
54. **Ren Patterson RF, Cochran LW, Holmes A, Sherrill S, Huang SJ, Tolliver T,** et al. Loss of brain-derived neurotrophic factor gene allele exacerbates brain monoamine deficiencies and increases stress abnormalities of serotonin transporter knockout mice. J Neurosci Res. 2005;**79**(6):756–71.
55. **Gardier AM.** Mutant mouse models and antidepressant drug research: Focus on serotonin and brain-derived neurotrophic factor. Behav Pharmacol. 2009;**20**(1):18–32.
56. **Huertas E, Ponce G, Koeneke M, Poch C, Espana-Serrano L, Palomo T,** et al. The D2 dopamine receptor gene variant C957T affects human fear conditioning and aversive priming. Genes, Brain Behav. 2010;**9**(1):103–9.
57. **Young RM, Lawford BR, Noble EP, Kann B, Wilkie A, Ritchie T,** et al. Harmful drinking in military veterans with post-traumatic stress disorder: Association with the D2 dopamine receptor A1 allele. Alcohol Alcohol. 2002;**37**(5):451–6.
58. **Bailey JN, Goenjian AK, Noble EP, Walling DP, Ritchie T, Goenjian HA.** PTSD and dopaminergic genes, DRD2 and DAT, in multigenerational families exposed to the Spitak earthquake. Psychiatry Res. 2010;**178**(3):507–10.
59. **Davidson JR, Book S, Colket J, Tupler L, Roth S, David D,** et al. Assessment of a new self-rating scale for post-traumatic stress disorder. Psychol Med. 1997;**27**(01):153–60.
60. **Cools R, D'Esposito M.** Inverted-U–shaped dopamine actions on human working memory and cognitive control. Biol Psychiatry. 2011;**69**(12):e113-e25.
61. **Monte-Silva K, Kuo M-F, Thirugnanasambandam N, Liebetanz D, Paulus W, Nitsche MA.** Dose-dependent inverted U-shaped effect of dopamine (D2-like) receptor activation on focal and nonfocal plasticity in humans. J Neurosci. 2009;**29**(19):6124–31.
62. **Hünnerkopf R, Strobel A, Gutknecht L, Brocke B, Lesch KP.** Interaction between BDNF Val66Met and dopamine transporter gene variation influences anxiety-related traits. Neuropsychopharmacology. 2007;**32**(12):2552–60.
63. **Pine DS, Coplan JD, Papp LA, Klein RG, Martinez JM, Kovalenko P,** et al. Ventilatory physiology of children and adolescents with anxiety disorders. Arch Gen Psychiatry. 1998;**55**(2):123–9.
64. **Somerville LH, Kelley WM, Heatherton TF.** Self-esteem modulates medial prefrontal cortical responses to evaluative social feedback. Cerebral Cortex. 2010c;**20**(12):3005–13.
65. **Somerville LH, Casey B.** Developmental neurobiology of cognitive control and motivational systems. Curr Opin Neurobiol. 2010a;**20**(2):236–41.
66. **Somerville LH, Jones RM, Casey B.** A time of change: Behavioral and neural correlates of adolescent sensitivity to appetitive and aversive environmental cues. Brain Cogn. 2010b;**72**(1):124–33.
67. **Steinberg L.** Cognitive and affective development in adolescence. Trends Cogn Sci. 2005;**9**(2):69–74.
68. **Dahl RE.** Adolescent brain development: A period of vulnerabilities and opportunities. Keynote address. Ann N Y Acad Sci. 2004;**1021**(1):1–22.
69. **Britton JC, Lissek S, Grillon C, Norcross MA, Pine DS.** Development of anxiety: The role of threat appraisal and fear learning. Depress Anxiety. 2011;**28**(1):5–17.
70. **Casey B, Jones RM, Hare TA.** The adolescent brain. Ann N Y Acad Sci. 2008;**1124**(1):111–26.

71. **McCallum J, Kim JH, Richardson R.** Impaired extinction retention in adolescent rats: Effects of D-cycloserine. Neuropsychopharmacology. 2010;**35**(10):2134–42.
72. **Kim JH, Li S, Richardson R.** Immunohistochemical analyses of long-term extinction of conditioned fear in adolescent rats. Cerebral Cortex. 2010:bhq116.
73. **Pattwell SS, Bath KG, Casey B, Ninan I, Lee FS.** Selective early-acquired fear memories undergo temporary suppression during adolescence. PNAS. 2011;**108**(3):1182–7.
74. **Luna B, Garver KE, Urban TA, Lazar NA, Sweeney JA.** Maturation of cognitive processes from late childhood to adulthood. Child Dev. 2004;**75**(5):1357–72.
75. **Kim JH, Richardson R.** The effect of temporary amygdala inactivation on extinction and reextinction of fear in the developing rat: Unlearning as a potential mechanism for extinction early in development. J Neurosci. 2008;**28**(6):1282–90.
76. **Harris JA, Westbrook RF.** Evidence that GABA transmission mediates context-specific extinction of learned fear. Psychopharmacology. 1998;**140**(1):105–15.
77. **Santini E, Muller RU, Quirk GJ.** Consolidation of extinction learning involves transfer from NMDA-independent to NMDA-dependent memory. J Neurosci. 2001;**21**(22):9009–17.
78. **Langton JM, Kim JH, Nicholas J, Richardson R.** The effect of the NMDA receptor antagonist MK-801 on the acquisition and extinction of learned fear in the developing rat. Learning & Memory. 2007;**14**(10):665–8.
79. **Sierra-Mercado D, Corcoran KA, Lebrón-Milad K, Quirk GJ.** Inactivation of the ventromedial prefrontal cortex reduces expression of conditioned fear and impairs subsequent recall of extinction. Eur J Neurosci. 2006;**24**(6):1751–8.
80. **Kim JH, Hamlin AS, Richardson R.** Fear extinction across development: The involvement of the medial prefrontal cortex as assessed by temporary inactivation and immunohistochemistry. J Neurosci. 2009;**29**(35):10802–8.
81. **English JD, Sweatt JD.** A requirement for the mitogen-activated protein kinase cascade in hippocampal long term potentiation. J Biol Chem. 1997;**272**(31):19103–6.
82. **Counotte DS, Li KW, Wortel J, Gouwenberg Y, Van Der Schors RC, Smit AB**, et al. Changes in molecular composition of rat medial prefrontal cortex synapses during adolescent development. Eur J Neurosci. 2010;**32**(9):1452–60.
83. **McLean CP, Asnaani A, Litz BT, Hofmann SG.** Gender differences in anxiety disorders: Prevalence, course of illness, comorbidity and burden of illness. J Psychiatric Res. 2011;**45**(8):1027–35.
84. **Bangasser DA, Wicks B.** Sex-specific mechanisms for responding to stress. J Neurosci Res. 2017;**95**(1-2):75–82.
85. **Gruene TM, Flick K, Stefano A, Shea SD, Shansky RM.** Sexually divergent expression of active and passive conditioned fear responses in rats. Elife. 2015;**4**:e11352.
86. **Mazor A, Matar MA, Kaplan Z, Kozlovsky N, Zohar J, Cohen H.** Gender-related qualitative differences in baseline and post-stress anxiety responses are not reflected in the incidence of criterion-based PTSD-like behaviour patterns. World J Biol Psychiatry. 2009;**10**(4-3):856–69.
87. **Cohen H, Zohar J, Matar MA, Zeev K, Loewenthal U, Richter-Levin G.** Setting apart the affected: The use of behavioral criteria in animal models of post traumatic stress disorder. Neuropsychopharmacology. 2004;**29**(11).
88. **Cohen H, Zohar J.** An animal model of posttraumatic stress disorder: The use of cut-off behavioral criteria. Ann N Y Acad Sci. 2004;**1032**(1):167–78.
89. **Nonkes LJ, Maes JH, Homberg JR.** Improved cognitive flexibility in serotonin transporter knockout rats is unchanged following chronic cocaine self-administration. Addiction Biol. 2011, 2013;**18**(3):434–40.
90. **Ressler KJ, Rothbaum BO, Tannenbaum L, Anderson P, Graap K, Zimand E**, et al. Cognitive enhancers as adjuncts to psychotherapy: Use of D-cycloserine in phobic individuals to facilitate extinctionof fear. Arch Gen Psychiatry. 2004;**61**(11):1136–44.

91. **Hofmann SG, Meuret AE, Smits JA, Simon NM, Pollack MH, Eisenmenger K**, et al. Augmentation of exposure therapy with D-cycloserine for social anxiety disorder. Arch Gen Psychiatry. 2006;**63**(3):298–304.
92. **de Kleine RA, Smits JA, Hendriks G-J, Becker ES, van Minnen A.** Extinction learning as a moderator of d-cycloserine efficacy for enhancing exposure therapy in posttraumatic stress disorder. J Anxiety Disord. 2015;**34**:63–7.
93. **Olff M, Langeland W, Draijer N, Gersons BP.** Gender differences in posttraumatic stress disorder. Psychol Bull. 2007;**133**(2):183.
94. **Schipper P.** Smart coping: Serotonin transporter gene variation and environment interact in determining the behavioral adaptation to stress: [Sl: sn]; 2017.
95. **Febo M, Ferris CF.** Oxytocin and vasopressin modulation of the neural correlates of motivation and emotion: Results from functional MRI studies in awake rats. Brain Res. 2014;**1580**:8–21.
96. **Harris A, Lennen R, Brydges N, Jansen M, Pernet C, Whalley H**, et al. The role of brain-derived neurotrophic factor in learned fear processing: An awake rat fMRI study. Genes, Brain Behav. 2016;**15**(2):221–30.
97. **Harris AP, Lennen RJ, Marshall I, Jansen MA, Pernet CR, Brydges NM**, et al. Imaging learned fear circuitry in awake mice using fMRI. Eur J Neurosci. 2015;**42**(5):2125–34.
98. **Brydges NM, Whalley HC, Jansen MA, Merrifield GD, Wood ER, Lawrie SM**, et al. Imaging conditioned fear circuitry using awake rodent fMRI. PLoS One. 2013;**8**(1):e54197.
99. **Reed MD, Price KE, Archbold J, Moffa A, Febo M.** Predator odor-evoked BOLD activation in the awake rat: Modulation by oxytocin and V 1a vasopressin receptor antagonists. Brain Res. 2013;**1494**:70–83.
100. **Febo M, Pira AS.** Increased BOLD activation to predator stressor in subiculum and midbrain of amphetamine-sensitized maternal rats. Brain Res. 2011;**1382**:118–27.
101. **Chen W, Tenney J, Kulkarni P, King JA.** Imaging unconditioned fear response with manganese-enhanced MRI (MEMRI). Neuroimage. 2007;**37**(1):221–9.
102. **Wang L, Lu H, Brown PL, Rea W, Vaupel B, Yang Y**, et al. Manganese-enhanced MRI reflects both activity-independent and activity-dependent uptake within the rat habenulomesencephalic pathway. PloS One. 2015;**10**(5):e0127773.
103. **McHugh SB, Fillenz M, Lowry JP, Rawlins JNP, Bannerman DM.** Brain tissue oxygen amperometry in behaving rats demonstrates functional dissociation of dorsal and ventral hippocampus during spatial processing and anxiety. Eur J Neurosci. 2011;**33**(2):322–37.
104. **Barkus C, Line SJ, Huber A, Capitao L, Lima J, Jennings K**, et al. Variation in serotonin transporter expression modulates fear-evoked hemodynamic responses and theta-frequency neuronal oscillations in the amygdala. Biol Psychiatry. 2014;**75**(11):901–8.
105. **Kilpatrick DG, Resnick HS, Milanak ME, Miller MW, Keyes KM, Friedman MJ.** National estimates of exposure to traumatic events and PTSD prevalence using DSM-IV and DSM-5 criteria. J Traum Stress. 2013;**26**(5):537–47.
106. **Santiago PN, Ursano RJ, Gray CL, Pynoos RS, Spiegel D, Lewis-Fernandez R**, et al. A systematic review of PTSD prevalence and trajectories in DSM-5 defined trauma exposed populations: Intentional and non-intentional traumatic events. PloS One. 2013;**8**(4):e59236.
107. **Olesen J, Gustavsson A, Svensson M, Wittchen HU, Jönsson B.** The economic cost of brain disorders in Europe. Eur J Neurol. 2012;**19**(1):155–62.
108. **Medicine Io.** Treatment for Posttraumatic Stress Disorder in Military and Veteran Populations: Final Assessment. Washington, DC: The National Academies Press; 2014.

Chapter 23

Genetics of impulsivity, anger, and aggression as risk factors for suicidal behavior

Dan Rujescu and Ina Giegling

23.1 Personality as risk factors of suicidal behavior

Suicidal behavior is a complex interplay of many different risk factors like social (e.g. lack of social support, social isolation), socio-economic (e.g. unemployment), medical (e.g. mental disorders, chronic pain), personality (e.g. impulsivity, anger, and aggression), and biological (e.g. genetics) characteristics (1, 2). Being unmarried, having a psychiatric illness, feeling hopeless, and having recently experienced a stressful life event were factors for an increased risk of suicidal behavior. Personality in particular seems to have a high impact on suicidal behavior (3). This chapter will focus on the genetics of impulsivity, anger, and aggression as risk factors for suicidal behavior.

Such intermediate phenotypes are a valuable tool in the search for the genetics of complex traits. As suicidal behavior is an interplay of many risk factors, a decomposition into intermediate phenotypes (e.g. impulsivity, anger, or aggression) can be very useful. The expectation is that such genetic factors contributing to intermediate phenotypes like personality will be easier to identify. Suicide-related behavior is a highly heterogeneous trait implying that the current clinically defined phenotype might not be optimal for genetic studies. Therefore, simpler, quantifiable measures of functioning may be more useful as they could be less complex compared to clinical phenotypes. The assumption is that the number of genes required to produce variations in these traits may be fewer than those involved in producing a diagnostic entity. Such an intermediate phenotype should be state-independent, associated with the illness, and co-segregate in families with the disease (4). Personality traits, which are partly under genetic control, may be intermediate phenotypes for the genetic component of suicidal behavior (5, 6).

23.1.1 Impulsivity

Impulsivity is defined as the inability to resist impulses, which refers to explosive and instantaneous, automatic, or semi-automatic psychomotor actions (7). The International Society for Research on Impulsivity defines impulsivity as "a behavior without adequate thought, the tendency to act with less forethought than do most individuals of equal ability and knowledge, or a predisposition toward rapid, unplanned reactions to internal or external stimuli without regard to the negative consequences of these reactions" (<http://www.impulsivity.org>/; 8).

There are many facets of impulsivity such as motor impulsivity (a tendency to pre-potent motor disinhibition); decision-making impulsivity (a tendency towards decision-making); choice impulsivity (difficulty in delaying gratification and choosing immediate small rewards despite negative long-term effects), or reflection impulsivity (insufficient information sampling before making a choice (9)). Interestingly, impulsivity may predispose to suicidality (10). A variety of studies demonstrated an association between impulsive personality traits and suicidal behavior (11–15). On the other hand, there are reports indicating a weak correlation between treat impulsivity and suicidal behavior (16). Beside the correlation between traits and suicidal behavior, extensive research focused on special subtypes of suicide or suicide attempts, namely, impulsive attempts. Such premeditation is a critical factor in the clinical evaluation of suicide attempt. Different interventions may be necessary in patients who have carefully planned the attempt or are liable to act out suddenly in response to circumstances. A considerable proportion of attempts is made without premeditation (17).

Rimkeviciene and colleagues (18) provided a meta-analysis on definitions, characteristics and risk factors for impulsive suicide attempts. Three major different research criteria emerged: time-related criteria; absence of proximal planning/preparations; and presence of suicide plan in lifetime/previous year. No risk factors were identified uniformly related to suicide attempt impulsivity across all criteria groups but several relationships emerged between separate criteria and specific characteristics of suicide attempters. Nevertheless, more standardized definition criteria for impulsive suicide attempts or suicides are necessary to make different studies comparable.

23.1.2 Anger

Anger is a further risk factor for suicidality. Spielberger suggested that anger can be defined "as a psychobiological state or condition consisting of subjective feelings that vary in intensity, from mild irritation or annoyance to intense fury and rage, with concomitant activation or arousal of the autonomic nervous system" (19). In several studies on suicidality, anger is measured through, for example, the "State–Trait Anger eXpression Inventory" (STAXI; 20, 21) consisting of subscales like "state anger" (measuring the respondent's current feelings of anger); "trait anger" (expression of anger without provocation; with two subscales: "angry temperament" and "angry reaction"); "anger-in" (determining the frequency with which the respondent holds in or suppresses his/her anger); "anger-out" (measuring the frequency with which the respondent expresses his/her anger to other people or objects); and "anger control" (assesses the degree to which the respondent attempts to control his/her expression of anger). Trait anger has been associated with a previous history of attempted suicide in different samples (3, 22, 23). In univariate analyses controlled for age and sex, late-life suicide attempters were lower in warmth and positive emotions than non-attempters. However, only lower anxiety was associated with attempt status when all seven emotion traits were included as predictors. Of the attempters, those who had made a greater number of attempts reported lower positive emotions and higher anger/ hostility and guilt, though only lower positive emotions had a significant effect independent of the other emotions (24). Furthermore, the association between anger and suicidality has been demonstrated in different psychiatric disorders like depression (25, 26), eating disorders (27, 233), alcohol use disorders (28, 29), and multiple psychiatric disorders that include anger as a symptom (e.g. borderline personality disorder, posttraumatic stress disorder) (30, 31).

Studies in adolescent samples reveal a strong correlation of anger with self-harm (32) and attempted suicide (22, 33–35). A study by Hawkins and Cougle (36) found that anger experience and expression uniquely predicted suicidal ideation, plans, and attempts, even after controlling for demographics and psychiatric comorbidity (36). Further, among individuals reporting lifetime suicidal ideation, anger experience and expression were unique predictors of suicide attempts. Anger experience and expression also demonstrated independent and additive effects on risk of suicide.

Interestingly, a newer study by Hawkins and colleagues (37) showed that anger is associated with suicidal ideation and behavior via perceived burdensomeness and with greater acquired capability for suicide via experiences with painful and provocative events.

23.1.3 Aggression

Aggression has been constantly associated with suicidal behavior for several decades (38). Since then, a large number of studies replicated this link (39–42). Aggressive behavior frequently occurs together with anger. Several typologies exist, but for psychiatry the most relevant distinction is that between reactive and proactive aggression. Reactive aggression or "emotional/hostile aggression" is accompanied by visible signs of anger and occurs in response to a frustrating event or a perceived threat. Typical manifestations include anger expressions, temper tantrums, and vengeful hostility (43, 44). Proactive aggression, or instrumental aggression, is designed to attain a goal such as social dominance, including bullying or domination. In psychiatry, aggression has mostly been measured by self-ratings including, for example, aspects like "spontaneous aggression," "reactive aggression," "self-aggression," or "aggression inhibition." One popular instrument is the Brown–Goodwin Inventory of Aggression (45), showing significant associations with suicide attempts in different psychiatric patients.

For example, aggression has been related to suicidality in mood disorders (46). Mann and colleagues (47) found higher rates of lifetime aggression in suicide attempters compared to non-suicidal patients. Oquendo and colleagues (41) and Grunebaum and co-workers (48) reported a correlation of lifetime aggression with suicide attempts in patients with mood disorders. A regression tree analysis in bipolar patients identified current depression and aggressive traits as indicators of a remote suicide attempt (49). Interestingly, bipolar I and II depressed patients had more life-time impulsivity and aggression/hostility than unipolar patients. In addition, bipolar I and II patients also differed on these trait characteristics (50). A link between aggression and suicidal behavior has also been shown in schizophrenia (12, 48, 51–53), borderline personality disorder (54–56), substance use disorders (57–60), post-traumatic stress disorder (61), and in non-clinical samples (62, 63).

23.2 Genetics of personality as risk factors of suicidal behavior

23.2.1 Heritability

A genetic component of suicidal behavior is supported by adoption and family studies. Mittendorfer-Rutz and colleagues (64) reported that the risk of a suicide attempt increases when the biological parents had committed a suicide attempt (factor 4.2 for the mother and 3.3 for the father). This risk increases to 7.3 when two or more family members are affected (64). Twin studies allow separating effects due to shared environment from genetic factors. The largest meta-analysis by Voracek and Loibl (65), including publications between 1812 and 2006, showed that 23 of the 168 monozygotic pairs (13.7%) were concordant for suicidal behavior whereas this was only the case in 5 of the 347 dizygotic twin pairs (1.4%). Similar results indicated a concordance rate of 11.3% in monozygotic and 1.8% in dizygotic twins (66, 67, 234). The heritability of serious suicide attempts is estimated to be 55% (68) or 58% in a newer study by Thornton and colleagues (69). Adoption studies also provided evidence for a genetic component of suicidal behavior (70–72). Schulsinger and colleagues (71) showed that the biological relatives of index cases showed a sixfold higher suicide rate (4.46%) than the biological relatives of a matched, non-suicidal adoptee control group (0.74%). Interestingly, there was no suicide among the adoptive relatives of the index cases. Von Borczyskowski and colleagues (73) support the role of a genetic risk by presenting a study based on the Swedish registry with a total of 2,471,496 people, including 27,600 adoptees.

Beside suicidality, impulsivity, anger, and aggression also have a clear genetic component. Studies on heritability rely on the kind of measurement, for example, through self-ratings or neuropsychological experiments, as well as on other ratings (e.g. patient him/herself at different time points in his or her course of the disorder; parents or teachers in the case of children, etc). There are some well-studied behavioral models of impulsivity like delay discounting, a decline in the subjective value of reward with increasing delay until its receipt (74). A commonly used computerized procedure involving choice between varying amounts of money available immediately and a standard amount presented at variable delays was administered to a population-based sample of twins aged 16 and 18 ($n = 560$; 134 monozygotic and 142 dizygotic pairs). A generalized estimate of heritability for the entire sample was 57% (95% CI: 47% to 66%), suggesting that most of the observed individual differences in the degree of delay discounting can be attributed to genetic factors (74). Regarding anger, there are several different estimates of heritability depending on the rating scales performed. Coccaro and co-workers (75) presented results from a twin study based on the Affective Lability Scale (ALS) and Affect Intensity Measure (AIM) questionnaires. Both were administered to 796 male twins in the Vietnam Era Twin Registry and subjected to twin and model-fitting analyses. Complete data were available from 182 monozygotic and 119 dizygotic twin pairs. Additive genetic influence accounted for 40% of the variance in affect intensity and 25% of the variance in the ALS subscale assessing anxiety–depression mood shifts. Non-additive genetic influence was indicated for ALS subscales measuring shifts between normal mood and depression (29%) and anger (27%), respectively (75). In contrast to impulsivity and anger, aggression is studied mostly using animal models. For psychiatric phenotypes, mostly self-ratings are applied. There is a high heritability of aggression, and evidence indicates that especially physical and proactive aggression showed the highest heritability. There are several twin studies providing high heritability estimates, such as those of Burt and colleagues (76), which reported a heritability of 65%, or Tuvblad and Baker (77), which found a heritability of 50%. There are also sex differences, with boys showing higher heritability, especially around and after adolescence (78, 79; for review see 80).

23.2.2 **Molecular genetics**

23.2.2.1 Serotonergic system

There is a long history of genetic association studies in the context of impulsivity, anger, and aggression focusing on genes which are involved in serotonergic neurotransmission. Serotonergic function was regarded as crucial for the regulation of impulsive and aggressive behavior, which in turn has been demonstrated to correlate with suicidal behavior in various studies (81). These studies are mostly based on research on low cerebrospinal fluid (CSF) concentrations of the serotonin metabolite 5-hydroxyindoleacetic acid (5-HIAA) regarded as a putative indicator for low serotonin turnover (38, 82, 83). Animal studies showed that low CSF 5-HIAA is associated with aggressiveness, low social affiliation, high-risk behavior, and premature mortality (84, 85, 220–222). In humans, 5-HIAA in the CSF correlates negatively with various aggressive behaviors, as demonstrated in healthy and psychiatric samples throughout the lifespan. Although there are some negative reports (86, 87), the vast majority of early findings suggested that lowered 5-HIAA is related to the vulnerability for aggressive behavior (82, 83, 88) which, beside other factors, led to major scientific work on serotonin and aggression.

For suicidal behavior, many studies investigated the post-mortem brains of suicide victims which have shown alterations of the serotonergic system, especially in the prefrontal cortex (for review, see 235). Furthermore, clinical studies found a relationship between lower levels of 5-HIAA and suicidal behavior (83), and there also seems to be a relationship between lower serotonergic function and aggressive, impulsive, and suicidal behavior. Therefore, early genetic studies investigated the serotonergic system in suicidal behavior and personality traits from the very beginning.

23.2.2.2 Tryptophan hydroxylases 1 and 2

The enzyme tryptophan hydroxylase (TPH) converts L-tryptophan, in a rate-limiting step, into 5-hydroxytryptophan. Therefore, single nucleotide polymorphisms (SNPs) within this gene seemed promising in genetic studies on aggression and suicidal behavior. The first meta-analysis (89) did not find an association of one of these SNPs within intron 7 A218C (TPH1), but a second meta-analysis focusing only on Caucasians reported a higher frequency of the A218-allele in patients with suicidal behavior (90). Further meta-analyses (91, 92) and studies (93–95, 223) confirmed this association.

This "A-" allele was also associated with higher aggression, a tendency to experience unprovoked anger, and a more outwardly expression of anger (130). Rujescu and colleagues (131) replicated these findings in showing association of A-carrier and higher scores on the Trait Anger Scale of the STAXI, especially on the subscale "Angry Temperament". Similar results were provided by Baud and colleagues (132), who reported that suicide attempters carrying the AA genotype scored significantly lower on the Anger Control subscale involving a reduced capacity to control anger. Interestingly, a study by Ruocco and colleagues (133) examined the relationship of two SNPs in the TPH1 gene (rs1799913 and rs4537731) to prefrontal cortex (PFC) activation on a response inhibition task. Participants carrying the risk alleles showed less activity primarily in bilateral inferior frontal gyri and medial PFC under conditions of response inhibition (133).

In addition to TPH1, there is a second tryptophan hydroxylase isoform (134, 135). The first investigation of SNPs in the TPH2 gene in a sample of suicide victims and ethnically matched healthy controls showed an association of one SNP with completed suicide (136). Since then, several case-control studies were published, reporting both positive and negative results (137–139; for review, see 140). The most recent study by Pawlak and colleagues (95) supports the role of TPH2 in suicidal behavior.

Interestingly, polymorphisms in the TPH2 gene were also associated with personality traits like aggression or impulsivity. For instance, Kulikov and colleagues (96, 97) reported that the C1473G polymorphism in the tryptophan hydroxylase-2 gene is associated with intermale aggression in mice (see also 98, 99). Perez-Rodriguez and colleagues (100) found an association between TPH2 and borderline personality disorder, affective lability, suicidal/parasuicidal behavior, and aggression scores. For overt aggressive impulsivity, TPH2 showed an effect in attention-deficit hyperactivity disorder children (101). TPH2 was also associated with eating disorders correlated with perfectionism and impulsivity, and seemed to affect impulsivity which, in turn, might predispose to the self-induced vomiting phenotype (102). Furthermore, TPH2 was also associated with risk-taking behavior in a gambling task (103), executive function (104), and response inhibition (105).

Interestingly, mice lacking Tph2 show intense compulsive and impulsive behaviors that include extreme aggression. The impulsivity was motor rather than cognitive, considering that Tph2$^{-/-}$ mice showed normal acquisition and reversal learning on a spatial learning task. Restoration of 5HT levels by treatment of Tph2$^{-/-}$ mice with its immediate precursor, 5-hydroxytryptophan, attenuated compulsive and impulsive-aggressive behaviors (106). Mosienko and colleagues (107) reported that Tph2($^{-/-}$) mice exhibited increased depression-like behavior and decreased anxiety-like behavior accompanied by strong aggressiveness (107). Furthermore, Tph2($^{-/-}$) males displayed increased impulsivity and high aggressiveness. Tph2($^{-/-}$) females displayed greater emotional reactivity to aversive conditions as reflected by changes in behavior at baseline, including increased freezing and decreased locomotion in novel environments (108, 109). A recent study showed that loss of Nesp55, which is co-localized with serotonin, alters novelty-related behavior.

The $Nespm^{/+}$ mice showed that loss of Nesp55 leads to impulsive choices as measured by a delayed reinforcement task, whereby $Nespm^{/+}$ mice were less willing to wait for a delayed, larger reward, preferring instead to choose an immediate, smaller reward. Furthermore, in midbrain regions there were reductions in mRNA expression of the serotonin-specific genes Tph2 and Slc6a4, but not the dopamine-specific gene Th in $Nespm^{/+}$ mice, suggesting an altered serotonergic system could contribute, in part, to changes in impulsive behavior (224).

23.2.2.3 Monoamine oxidase A

Monoamine oxidase A (MAOA) is one of the key enzymes in the metabolism of serotonin. Studies investigating the possible association of MAOA polymorphisms and suicidal behavior yielded inconsistent results (236, 110–113). Ho and colleagues (114) examined polymorphisms in a sample of bipolar patients. They found a VNTR variant to be associated with a history of suicide attempts, especially in females. The Fnu4H1 RFLP only showed significant differences in allele frequencies for female subjects, but not in the total sample. Another study showed a strong association of the high-activity-related EcoRV allele and depressed suicide in male subjects, but not in females or the total sample (115). A study examining serotonergic genes and suicidality, psychopathology and aggression in adolescents gives further support for an association between MAOA and suicidal behavior (116). A study including 432 community controls, 385 patients with MDD who had not attempted suicide, 96 community subjects without mental disorders who had attempted suicide, and 109 patients with MDD who had attempted suicide, showed that the MAOA-uVNTR 4R-allele is associated with enhanced vulnerability to suicide in depressed males, but not in community subjects (117). A meta-analysis on MAOA-uVNTR polymorphism in suicidal behavior reported no significant association. Studies on gene–environment interactions or genetic risk factors for endophenotypes of suicidal behaviors are warranted.

Some of these endophenotypes are personality parameters like impulsivity, anger, and aggression. Brunner and colleagues (118, 119) reported a large Dutch family with a new form of X-linked mild mental retardation, with all affected males showing aggressive, impulsive, and sometimes violent behavior. A second investigation showed that each of five affected males had a point mutation in the MAOA gene, which changes a glutamine to a termination codon (118–120).

Several studies in humans followed based on these results. Analyses between the MAOA VNTR and ADHD analysis showed an association with a subgroup of children with comorbid conduct problems (121). The lower expression allele was also associated with a history of abuse before 15 years of age in male subjects and with higher impulsivity in males but not females (111, 122).

Another study found an association with cluster B personality, which is highly correlated with impulsivity and aggressiveness (123), or alcoholism and antisocial behavior, disorders which are also highly correlated with these personality traits (124–128). Stetler and colleagues (129) revealed a robust association between low-activity MAOA-uVNTR alleles and violent crime. This association was replicated in the group of Caucasian violent offenders.

Manuck and colleagues studied a four repeat variant of the MAOA-uVNTR polymorphism in context of aggressiveness and impulsivity and reported that men carrying the one-quarter-allele scored significantly lower on aggressiveness and impulsivity than men in the two-thirds-allele group (237). Also, Zalsman and colleagues (116) reported an association with low-activity MAOA genotypes and clinical traits of impulsive–aggression and suicidality in adolescents.

Beside the VNTR, the functional SNP rs6323 within MAOA was associated with higher levels of anger expressed outwards (STAXI "anger-out" subscale) in suicidal males as well as females (141).

There are also a few imaging or neuropsychological studies providing support for the association between MAOA and impulsivity/aggression. Meyer-Lindenberg and colleagues (142) show that the low expression variant of the MAOA VNTR was associated with increased risk of violent behavior, and predicted pronounced limbic volume reductions and hyperresponsive amygdala during emotional arousal, with diminished reactivity of regulatory prefrontal regions, compared with the high expression allele. In men, the low expression allele was also associated with changes in orbitofrontal volume, amygdala and hippocampus hyperreactivity during aversive recall, and impaired cingulate activation during cognitive inhibition.

Decision-making impairment in suicide attempters is another endophenotype for suicidality. Genetic factors have been associated with the vulnerability to suicidal behavior. The Iowa Gambling Task was used to assess decision-making in 168 patients with a personal history of attempted suicide. Patients carrying the MAOA-HH (in women) genotypes significantly improved their performance during the task, suggesting a genetic modulation of the learning process required for advantageous decision-making (143). Next, an experiment was reported, drawing from work in psychology and behavioral economics, which demonstrated that aggression occurs with greater intensity and frequency as provocation is experimentally manipulated upwards, especially among low-activity MAOA (MAOA-L) subjects (144). Kuepper and colleagues (145) performed a modified version of a competitive reaction time task—a commonly used tool for assessing aggressiveness. A main effect of MAOA-uVNTR was found on a measure of reactive aggressiveness for both men and women, whereby the low-activity alleles of the MAOA-uVNTR were associated with substantially increased aggressive reactions. This effect was unique for reactive aggressiveness. This indicates that the low-functional alleles of the MAOA-uVNTR are not associated with increased aggressive behavior per se, but rather with an increased aggressive reactivity to provocation.

More recently, 20 years after the "Brunner" publication, Piton and colleagues (146) performed a targeted high-throughput sequencing of 220 genes, including MAOA, in patients with undiagnosed intellectual disability. They identified a c.797_798delinsTT (p.C266F) missense mutation in MAOA in a boy with autism spectrum disorder, attention deficit and autoaggressive behavior. Two maternal uncles carried the mutation and had severe intellectual disability, with a history of maltreatment in early childhood. This novel missense mutation decreased MAOA enzymatic activity, leading to abnormal levels of urinary monoamines.

23.2.2.4 Serotonin transporter

The serotonin transporter (5-HTT) is responsible for the reuptake of released serotonin from the synaptic cleft. The 5′-promotor region of the 5-HTT gene contains a functional insertion/deletion variant (5-HTTLPR) with initially two and later three common alleles ("short" (S) and "long" (L_A and L_G). The 5-HTT transcription and serotonin reuptake are higher in cells containing the homozygous LL genotype, compared with cells having the LS or SS forms (147). There are several studies on the association of the 5-HTTLPR and suicidal behavior. A meta-analysis conducted by Anguelova and colleagues (148), including 12 studies with 10 Caucasian populations, 1 US population, and 1 Chinese sample pooled a total number of 1,168 cases (suicide completers and suicide attempters) and 1371 controls. They found a significant association of the S-allele with suicidal behavior as a whole in the subsample of suicide attempters but not suicide completer. A second meta-analysis, including 18 studies with 1,521 suicide attempters or completers and 2,429 controls found no overall association, but the S-allele was associated with violent suicidal behavior, which is mainly characterized by the use of highly lethal and violent methods like hanging or shooting, but not with non-violent suicide. This could suggest a link with higher impulsive and aggressive behavior traits (149). Li and He (150) reviewed 39 studies that examined the association between

5-HTTLPR and suicide showing an association with the S allele. Another meta-analysis on 31 primary case–control studies (cases $n = 6{,}324$; controls $n = 10{,}285$) supports the need to distinguish between groups of suicide completions and suicide attempts. When only the attempted suicide case–control studies were analyzed ($n = 25$), the pooled OR reached actual significance (1.13; CI1.05–1.21, $p = 0.001$). The pooled completed suicide case-control studies (n = 6) were nonsignificant (93).

In addition to suicidal behavior, there are many studies concentrating on the association of the 5-HTTLPR and impulsivity or aggression. Lesch and colleagues (147) found that individuals carrying one or two short 5-HTTLPR-alleles had significantly higher scores for NEO Neuroticism and lower scores for Agreeableness in a large, predominantly male cohort (147). These findings were robustly replicated by Greenberg and colleagues (151), who also reported associations with subscales like N2-Angry hostility and N5-Impulsiveness. Several other studies were performed since then. For instance, Gerra and colleagues (152) reported that the short-short (SS) genotype frequency was significantly higher among aggressive/novelty seeker (NS) experimenters with poor school achievements, compared with drugs experimenters without aggressiveness and school failure. When evaluated in the entire sample, BDHI mean total scores, TPQ NS scores, and school failure frequency were significantly higher in SS individuals, in comparison with LL subjects (for review, see 153–155). Other studies were done in children with ADHD and/or conduct disorders. The most recent meta-analysis on 19 studies found no association (156). Vassos and colleagues (157) reported a meta-analysis on the 5-HTTLPR and aggression, in which 31 studies were included. Although there was only a trend for an overall association with aggression, follow-up analyses in samples of age > 16 and < 65, and substance users, found significant associations (157). Another recent meta-analysis tested the interaction of the 5-HTTLPR genotype and environmental adversities on antisocial behavior. Eight studies, comprising 12 independent samples, totaling 7,680 subjects, with an effective sample size of 6,724, were included. A significant interaction effect between the 5-HTTLPR genotype and environmental adversities on antisocial behavior was detected (158)

23.2.2.5 HTR1A receptor

Serotonin-1A (HTR1A) receptors play a role in impulsivity-related behavior. The C-1019G functional polymorphism (rs6295) regulates the HTR1A receptor gene expression in presynaptic raphe neurons (increased receptor concentration and reduced neuronal firing) (159). Lemonde and colleagues (160) examined the C-1019G SNP in the promoter of the HTR1A gene in a sample of suicide victims and healthy controls. They found the G-allele to be significantly over-represented in the suicide group, and the homozygous GG genotype was four times more frequent among suicide completers. This association was not replicated by Huang and co-workers (161), whereas Sawiniec and colleagues (162) showed a positive association. Serretti and co-workers (163) did not reveal any significant association with suicidal behavior for the rs6295. Only marginally significant associations were found with the STAXI or the Questionnaire for Measuring Factors of Aggression (FAF) whereas there was no association with the Temperament and Character Inventory (TCI) (164). Strobel and colleagues (165) previously reported a significant effect of the HTR1A-1019 polymorphism on NEO Neuroticism (especially the Anxiety and Depression facets), with carriers of the G-allele showing higher scores. Carriers of the G-allele also exhibited higher TPQ Harm Avoidance scores. Angles and colleagues (166) performed a meta-analysis on the rs6295 polymorphism within 5-HT1A and suicidal behavior. Four studies including 957 patients and 957 controls were eligible. No association was detected, which was also the case in the meta-analysis by Schild and co-workers (167). Benko and colleagues (159) studied the relationship between impulsivity and the C-1019G polymorphism of the HTR1A in a population sample of 725 volunteers using the Impulsiveness subscale (IVE-I) of the Eysenck Impulsiveness, Venturesomeness, and Empathy scale, and also the Barratt Impulsiveness Scale (BIS-11). Subjects carrying GG genotype showed significantly higher impulsiveness scores for the IVE-I scale, for the Motor, Cognitive Impulsiveness, and for the BIS total, but not for the Nonplanning Impulsiveness subscale of the BIS-11.

23.2.2.6 HTR1B receptor

Although, mostly, there is no conclusive evidence of the association between the HTR1B gene and suicidal behavior (52, 139, 168–177), some weak positive associations have been reported (178, 179). However, a meta-analysis showed no overall association (180).

On the other hand, there is substantial evidence for the involvement of HTR1B in anger, aggression, and impulsivity. First, studies on serotonin 1B knockout mice showed an increase in aggressive behavior relative to wildtype mice (181). Lappalainen and colleagues (182) reported the involvement of the HTR1B gene in alcoholism, with aggressive and impulsive behavior represented by antisocial personality disorder and intermittent explosive disorder comorbid with alcoholism. Interestingly, Jensen and colleagues (183) reported that a SNP in the serotonin receptor 1B mRNA moderates regulation by miR-96 and is associated with aggressive human behaviors. Furthermore, Conner and co-workers (184) showed that functional polymorphisms within HTR1B predicted self-reported anger and hostility among young men. Stoltenberg and colleagues (2012) performed a cross-sectional study and examined whether polymorphisms in six serotonin system candidate genes and the experience of early-life trauma (age 0–12) were associated with individual differences in impulsivity in a non-clinical sample of Caucasian university students ($N = 424$). An association with HTR1B was shown. Another positive association study was reported by Varga and team (185). One HTR1B SNP was associated with trait impulsivity assessed with the Barratt Impulsiveness Scale (BIS-11) in a sample of 687 Caucasian young adults. The involvement of HTR1B in 695 adults with ADHD (266 and 396 subjects with and without comorbid substance use disorder, respectively) was showed by Sánchez-Mora and colleagues (186). Another study examined whether HTR1B is related to hostility, anger, and aggressive behavior phenotypes over the lifespan and whether it modifies the connection between childhood aggressive behavior and adulthood hostility and anger. The participants were 967 women and men from a large population based sample (the Young Finns Study) with a 27-year follow-up. The HTR1B SNP rs6296 was associated with childhood aggressive behavior, but not with adulthood anger or hostility, and moderated the association between childhood aggressive behavior and adulthood hostility (187). Interestingly, an association of variants of the HTR1B gene and whole-brain analyses demonstrated large-scale neural activation differences in the inferior and medial frontal and temporal/parietal regions of the response inhibition. Activation in these regions was significantly associated with stop-task performance, but not with ADHD diagnosis or severity (188).

23.2.2.7 HTR2A receptor

The increased density of brain and platelet serotonin 2A (HTR2A) receptors in subjects with suicidal behavior is supported by several studies (189). Thus, the HTR2A receptor gene has been regarded as a major candidate for the genetic susceptibility to this behavior (190). However, a meta-analysis pooling nine studies with altogether 596 suicide completers or attempters, and 1,003 healthy controls could not find any association with the T102C SNP (148). Furthermore, a later meta-analysis of 25 studies did not find any association with this SNP, but interestingly, the meta-analysis of a promoter SNP in this gene (A-1438G; rs2007004O) showed association with suicidal behavior (92). A meta-analysis by Wang and colleagues (191) on the 102T/C polymorphism demonstrated no association. However, in patients with schizophrenia, the C/C genotype may increase the risk of attempted suicide. One approach studied the contributions of six CpG sites in HTR2A exon I in schizophrenia patients assessed for lifetime suicide attempt. Potential methylation in 25 CpG SNPs across the entire HTR2A gene was analyzed considering their direct involvement in the process of methylation. When methylation was compared between attempters and nonattempters, the polymorphism T102C (rs6313) was significantly different between the two groups. Furthermore, in the methylation analysis, nominal association was found with suicide attempt for six of the 25 SNPs analyzed (192).

Regarding impulsivity, anger, and aggression, platelet serotonin 2A receptor sites have been found to be associated with impulsivity and aggression in a Positron Emission Tomography (PET) study (193). Moreover, the SNP rs6311 was shown to be associated with impulsive traits in alcohol dependents, measured with the Baratt Impulsiveness Scale (BIS) (194, 195). This result was partially confirmed (196, 197). Giegling and colleagues (198) demonstrated the association of SNPs in the HTR2A gene with anger- and aggression-related personality traits. CC homozygotes for the functional SNP rs6311 reported more anger-related traits in general and had higher scores on the subscale "Trait Anger", with its "Angry Reaction" component, and lower scores on the subscale "Anger Out". HTR2A SNPs have also been found to be associated with low anxiety-related traits, which are negatively related to impulsive traits (199, 200). In addition, TCI Self-Transcendence was related to the rs6311 SNP (201). Serretti and colleagues (163) investigated a few SNPs in relation to personality traits assessed with the TCI, in three independent samples including healthy subjects and patients. Although the SNP rs594242 showed an association with Self-Directedness in the German sample, and rs6313 was marginally associated with Novelty Seeking in the Italian sample, this study did not support a major effect of these SNPs on temperament (202). The interaction between COMT Val158Met and HTR2A T102C polymorphisms was found to be associated with subscale impulsiveness. TT carriers of HTR2A T102C polymorphism, who also had the Met/Met genotype of COMT Val158Met SNP, scored significantly higher on impulsiveness compared to Val-allele carriers (203). Also, Zalsman and colleagues (116) reported an association in adolescents. Homozygosity for the T-allele of the HTR2A 102T/C polymorphism was associated with lower impulsivity and aggression.

In addition to self-report measures like the Barratt Impulsiveness Scale (BIS-11), the stop-signal task was also used as a method of estimating the level of behavioral impulsivity. This measure is an interesting endophenotype which was investigated, for instance, in a sample of 304 alcohol-dependent patients. A significant association between high levels of behavioral impulsivity and the CC genotype of rs6313 in alcohol-dependent patients was detected (204). Pinto and colleagues (205) also studied endophenotypes of ADHD, such as performance in a go/no-go task, in a twin sample. The strongest association was observed for the rs7984966 SNP and reaction time variability.

23.2.2.8 Genome-wide association studies

In addition to studies on candidate genes related to the serotonergic system, several such studies have been performed, for example, for genes from the dopaminergic and noradrenergic neurotransmitter systems. Most of the effects in these studies, like those of SNPs within the serotonergic system, did not survive correction for multiple testing in genome-wide association studies, mostly due to small samples sizes. For suicidal behavior per se, there is no conclusive genome-wide association result surviving correction for multiple testing (206–211). There are also no positive findings for Treatment Emergent Suicidal Ideation (TESI) 238, 212, 213).

Regarding anger, aggression, and impulsivity, most genome-wide associations studies undertaken thus far were done in correlated phenotypes like conduct disorder (214), ADHD (215) adult antisocial behavior (216), alcohol and substance abuse disorders (for review, see 217), aggressive behavior in chicken (218), and related personality constructs such as extraversion (219) or excitement seeking. Results have drawn attention to the CTNNA2 catenin cadherin-associated protein alpha 2 gene (225). Interestingly, a genome-wide association study by Ehlers and colleagues (226) on impulsivity replicated this association for a variant 50 kb upstream from CTNNA2, a neuronal-specific catenin. A second locus upstream of nei endonuclease VIII-like 3 (NEIL3) on chromosome 4 also achieved genome-wide significance.

Other genome-wide association studies, for example, on behavioral disinhibition (227), inattention, and hyperactivity–impulsivity (228) or aggressiveness in attention-deficit hyperactivity disorder (229) detected no genome wide significant effects.

An association very near to genome-wide significance (p = 2.9E-08) was reported by Mick and colleagues (230) who studied dysregulated emotional control, including clinically significant anger, hostility, and aggression. A genome-wide association study of proneness to anger using the Spielberger State–Trait Anger Scale was conducted (n = 8,747). An association was found on chromosome 6q21 in the gene coding for the non-receptor protein-tyrosine kinase, Fyn. Fyn interacts with NDMA receptors and inositol-1,4,5-trisphosphate (IP3)-gated channels to regulate calcium influx and intracellular release in the post-synaptic density. Another larger (n = 18,988) genome-wide association study on aggressive behavior using well-validated parent-reported questionnaires was presented by Pappa and colleagues (231). This is the largest sample exploring children's aggressive behavior to date, with measures in two developmental stages (early childhood and middle childhood/early adolescence). In the total sample, one region in chromosome 2 (2p12) was identified at near genome-wide significance (top SNP rs11126630, $p = 5.30 \times 10(-8)$). The separate meta-analyses in the two developmental stages revealed suggestive evidence of association at the same locus. Recently, Rautiainen and colleagues (232) reported a genome-wide association study and a replication analysis of Finnish criminal offenders fulfilling DSM-IV criteria for antisocial personality disorder. Results indicated suggestive associations of two clusters at 6p21.2 and at 6p21.32 at the human leukocyte antigen (HLA) region. Imputation of HLA-alleles revealed an independent nominal association with DRB1*01:01. Two polymorphisms at 6p21.2 LINC00951-LRFN2 gene region were replicated in a separate data set, and rs4714329 reached genome-wide significance ($p = 1.6 \times 10(-9)$. The risk allele also associated with antisocial features in the general population conditioned for severe problems in childhood family (p = 0.012).

23.3 Conclusion

The last few years have brought an enormous amount of new knowledge on genetics of impulsivity, anger, and aggression. Genome-wide association studies only just started, and exome and whole-genome sequencing projects are already underway. All these new techniques will help us further to characterize genetic risk factors of suicidal behavior and personality traits like impulsivity, anger, or aggression. These genetic variants have raised a number of fascinating new clinical and scientific questions, especially concerning the phenotypic boundaries between mental disorders as they are currently classified, their modes of inheritance, as well as the implications of this new information for diagnosis and genetic counselling.

References

1. **Wasserman D, Rihmer Z, Rujescu D, Sarchiapone M, Sokolowski M, Titelman D**, et al. The European Psychiatric Association (EPA) guidance on suicide treatment and prevention. Eur Psychiatry. 2012;**27**:129–41.
2. **Martin-Carrasco M, Evans-Lacko S, Dom G, Christodoulou NG, Samochowiec J, González-Fraile** E, et al. EPA guidance on mental health and economic crises in Europe. Eur Arch Psychiatry Clin Neurosci. 2016 Mar;**266**(2):89–24.

3. **Brezo J, Paris J, Turecki G.** Personality traits as correlates of suicidal ideation, suicide attempts, and suicide completions: A systematic review. Acta Psychiatrica Scandinavica. 2006;**113**:180–206.
4. **Gottesman II, Gould TD.** The endophenotype concept in psychiatry: Etymology and strategic intentions. Am J Psychiatry. 2003 Apr;**160**(4):636–45.
5. **Baud P.** Personality traits as intermediary phenotypes in suicidal behavior: Genetic issues. Am J Med Genet. Part C, Seminars in Medical Genetics. 2005;**133**:34–42.
6. **Chistiakov DA, Kekelidze ZI, Chekhonin VP.** Endophenotypes as a measure of suicidality. J Appl Genet. 2012 Nov;**53**(4):389–413.
7. **Kempf E.** Psychopathology. New York, NY: Arno Press; 1976.
8. **Moeller FG, Barratt ES, Dougherty DM, Schmitz JM, Swann AC.** Psychiatric aspects of impulsivity. Am J Psychiatry. 2001;**158**:1783–93.
9. **Fineberg NA, Chamberlain SR, Goudriaan AE, Stein DJ, Vanderschuren LJ, Gilla CM,** et al. New developments in human neurocognition: Clinical, genetic, and brain imaging correlates of impulsivity and compulsivity. CNS Spectr. 2014 Feb;**19**(1):69–89.
10. **Glick AR.** The role of serotonin in impulsive aggression, suicide, and homicide in adolescents and adults: A literature review. Int J Adolesc Med Health. 2015 May;**27**(2):143–50.
11. **McGirr A, Turecki G.** The relationship of impulsive aggressiveness to suicidality and other depression-linked behaviors. Curr Psychiatry Reports 2007;**9**:460–6.
12. **McGirr A, Turecki G.** What is specific to suicide in schizophrenia disorder? Demographic, clinical and behavioural dimensions. Schizophrenia Research 2008;**98**:217–24.
13. **Turecki G.** Dissecting the suicide phenotype: The role of impulsive–aggressive behaviours. J Psychiatry Neurosci. 2005;**30**:398–408.
14. **Zouk H, Tousignant M, Seguin M, Lesage A, Turecki G.** Characterization of impulsivity in suicide completers: Clinical, behavioral and psychosocial dimensions. J Affect Disord. 2006;**92**:195–204.
15. **Menon V, Sarkar S, Kattimani S, Mathan K.** Do personality traits such as impulsivity and hostility-aggressiveness predict severity of intent in attempted suicide? Findings from a record based study in South India. Indian J Psychol Med. 2015 Oct-Dec;**37**(4):393–8.
16. **Anestis MD, Soberay KA, Gutierrez PM, Hernández TD, Joiner TE.** Reconsidering the link between impulsivity and suicidal behavior. Pers Soc Psychol Rev. 2014 Nov;**18**(4):366–86.
17. **Wyder M, De Leo D.** Behind impulsive suicide attempts: Indications from a community study. J Affect Disord. 2007;**104**:167–73
18. **Rimkeviciene J, O'Gorman J, De Leo D.** Impulsive suicide attempts: A systematic literature review of definitions, characteristics and risk factors. J Affect Disord. 2015 Jan 15;**171**:93–104.
19. **Spielberger CD, Reheiser EC, Sydeman SJ.** Measuring the experience, expression, and control of anger. Issues in Comprehensive Pediatric Nursing. 1995;**18**:207–232.
20. **Spielberger CD.** State-Trait Anxiety Inventory: Bibliography, 2nd edn. Palo Alto, CA: Consulting Psychologists Press; 1989.
21. **Schwenkmezger P, Hodapp V.** A questionnaire for assessing anger and expression of anger. Z Klin Psychol Psychopathol Psychother. 1991;**39**:63–68
22. **Stein D, Apter A, Ratzoni G, Har-Even D, Avidan G.** Association between multiple suicide attempts and negative affects in adolescents. J Am Acad Child Adolesc Psychiatry. 1998 May;**37**(5):488–94.
23. **Gvion Y, Horesh N, Levi-Belz Y, Apter A.** A proposed model of the development of suicidal ideations. Compr Psychiatry. 2015 Jan;**56**:93–102.
24. **Seidlitz L, Conwell Y, Duberstein P, Cox C, Denning D.** Emotion traits in older suicide attempters and non-attempters. J Affec Disord. 2001;**66**:123–31.
25. **Painuly N, Sharan P, Mattoo SK.** Antecedents, concomitants and consequences of anger attacks in depression. Psychiatry Research. 2007;**153**:39–45.
26. **Bowen R, Balbuena L, Peters EM, Leuschen-Mewis C, Baetz M.** The relationship between mood instability and suicidal thoughts. Arch Suicide Res. 2015;**19**(2):161–71.

27. **Nickel MK, Simek M, Lojewski N, Muehlbacher M, Fartacek R, Kettler C**, et al. Familial and sociopsychopathological risk factors for suicide attempt in bulimic and in depressed women: Prospective study. Int J Eat Disord. 2006;**39**:410–7.
28. **Haw C, Houston K, Townsend E, Hawton K.** Deliberate self-harm patients with alcohol disorders: Characteristics, treatment, and outcome. Crisis. 2001;**22**:93–101.
29. **Hellmuth JC, Stappenbeck CA, Hoerster KD, Jakupcak M.** Modeling PTSD symptom clusters, alcohol misuse, anger, and depression as they relate to aggression and suicidality in returning U.S. veterans. J Trauma Stress. 2012 Oct;**25**(5):527–34.
30. **Nepon J, Belik SL, Bolton J, Sareen J.** The relationship between anxiety disorders and suicide attempts: Findings from the National Epidemiologic Survey on Alcohol and Related Conditions. Depress Anxiety. 2010 Sep;**27**(9):791–8.
31. **Nock MK, Hwang I, Sampson NA, Kessler RC.** Mental disorders, comorbidity and suicidal behavior: Results from the National Comorbidity Survey Replication. Mol Psychiatry. 2010 Aug;**15**(8):868–76.
32. **Hawton K, Kingsbury S, Steinhardt K, James A, Fagg J.** Repetition of deliberate selfharm by adolescents: The role of psychological factors. J Adolesc. 1999;**22**:369–78.
33. **Cautin RL, Overholser JC, Goetz P.** Assessment of mode of anger expression in adolescent psychiatric inpatients. Adolescence. 2001;**36**:163–70.
34. **Esposito C, Spirito A, Boergers J, Donaldson D.** Affective, behavioral, and cognitive functioning in adolescents with multiple suicide attempts. Suicide Life Threat Behav. 2003;**33**:389–99.
35. **Kirkcaldy BD, Brown J, Siefen RG.** Disruptive behavioural disorders, self harm and suicidal ideation among German adolescents in psychiatric care. Int J Adolesc Med Health. 2006 Oct–Dec;**18**(4):597–614.
36. **Hawkins KA, Cougle JR.** A test of the unique and interactive roles of anger experience and expression in suicidality: Findings from a population-based study. J Nerv Ment Dis. 2013 Nov;**201**(11):959–63.
37. **Hawkins KA, Hames JL, Ribeiro JD, Silva C, Joiner TE, Cougle JR.** An examination of the relationship between anger and suicide risk through the lens of the interpersonal theory of suicide. J Psychiatr Res. 2014 Mar;**50**:59–65.
38. **Asberg M, Traskman L, Thoren P.** 5-HIAA in the cerebrospinal fluid. A biochemical suicide predictor? Arch Gen Psychiatry. 1976; **33**(10):1193–7.
39. **Corbitt EM, Malone KM, Haas GL, Mann JJ.** Suicidal behavior in patients with major depression and comorbid personality disorders. J Affect Disord. 1996 Jun 20;**39**(1):61–72.
40. **Malone KM, Haas GL, Sweeney JA, Mann JJ.** Major depression and the risk of attempted suicide. J Aff Disord. 1995; **34**:173–85.
41. **Oquendo MA, Mann JJ.** The biology of impulsivity and suicidality. Psychiatr Clin North Am. 2000 Mar;**23**(1):11–25.
42. **Oquendo MA, Russo SA, Underwood MD, Kassir SA, Ellis SP, Mann JJ**, et al. Higher ost-mortem prefrontal 5-HT2A receptor binding correlates with lifetime aggression in suicide. Biol Psychiatry. 2006 Feb 1;**59**(3):235–43.
43. **Berkowitz L.** Aversively stimulated aggression. Some parallels and differences in research with animals and humans. Am Psychol. 1983 Nov;**38**(11):1135–44.
44. **Price JM, Dodge KA.** Reactive and proactive aggression in childhood: Relations to peer status and social context dimensions. J Abnorm Child Psychol. 1989 Aug;**17**(4):455–71.
45. **Brown GL, Goodwin FK, Ballenger JC, Goyer PF, Major LF.** Aggression in humans correlates with cerebrospinal fluid amine metabolites. Psychiatry Res. 1979;**1**:131–9.
46. **Straub R, Wolfersdorf M, Keller F, Hole G.** Personality, motivation and affect as modulating factors in suicidal behavior of depressed women. Fortschritte der Neurologie-Psychiatrie 1992;**60**:45–53.
47. **Mann JJ, Waternaux C, Haas GL, Malone KM.** Toward a clinical model of suicidal behavior in psychiatric patients. Am J Psychiatry. 1999;**156**:181–9.

48. **Grunebaum MF, Ramsay SR, Galfalvy HC, Ellis SP, Burke AK, Sher L**, et al. Correlates of suicide attempt history in bipolar disorder: A stress-diathesis perspective. Bipolar Disorders. 2006;**8**:551–7.
49. **Mann JJ, Ellis SP, Waternaux CM, Liu X, Oquendo MA, Malone KM**, et al. Classification trees distinguish suicide attempters in major psychiatric disorders: A model of clinical decision making. J Clin Psychiatry. 2008;**69**:23–31.
50. **Dervic K, Garcia-Amador M, Sudol K, Freed P, Brent DA, Mann JJ**, et al. Bipolar I and II versus unipolar depression: Clinical differences and impulsivity/aggression traits. Eur Psychiatry. 2015 Jan;**30**(1):106–13.
51. **Mork E, Walby FA, Harkavy-Friedman JM, Barrett EA, Steen NE, Lorentzen S**, et al. Clinical characteristics in schizophrenia patients with or without suicide attempts and non-suicidal self-harm—a cross-sectional study. BMC Psychiatry. 2013 Oct 9;**13**:255.
52. **Hong CJ, Pan GM, Tsai SJ.** Association study of onset age, attempted suicide, aggressive behavior, and schizophrenia with a serotonin 1B receptor (A-161T) genetic polymorphism. Neuropsychobiology. 2004;**49**(1):1–4.
53. **Malone KM, Waternaux C, Haas GL, Cooper TB, Li S, Mann JJ.** Cigarette smoking, suicidal behavior, and serotonin function in major psychiatric disorders. Am J Psychiatry. 2003;**160**:773–9.
54. **Baus N, Fischer-Kern M, Naderer A, Klein J, Doering S, Pastner B**, et al. Personality organization in borderline patients with a history of suicide attempts. Psychiatry Res. 2014;Aug 15;**218**(1-2):129–33.
55. **Brodsky BS, Groves SA, Oquendo MA, Mann JJ, Stanley B.** Interpersonal precipitants and suicide attempts in borderline personality disorder. Suicide and Life- Threatening Behavior. 2006;**36**:313–22.
56. **Horesh N, Orbach I, Gothelf D, Efrati M, Apter A.** Comparison of the suicidal behavior of adolescent inpatients with borderline personality disorder and major depression. J Nerv Ment Dis. 2003;**191**:582–8.
57. **Khemiri L, Jokinen J, Runeson B, Jayaram-Lindström N.** Suicide risk associated with experience of violence and impulsivity in alcohol dependent patients. Sci Rep. 2016 Jan 19;**6**:19373.
58. **Sher L, Oquendo M, Galfalvy H, Grunebaum M, Burke A, Zalsman G**, et al. The relationship of aggression to suicidal behavior in depressed patients with a history of alcoholism. Addictive Behaviors. 2005;**30**:1144–53.
59. **Sher L, Stanley BH, Harkavy-Friedman JM, Carballo JJ, Arendt M, Brent DA**, et al. Depressed patients with co-occurring alcohol use disorders: A unique patient population. J Clin Psychiatry. 2008:e1–9.
60. **Tremeau F, Darreye A, Staner L, Correa H, Weibel H, Khidichian F**, et al. Suicidality in opioid-dependent subjects. Am J Addictions. 2008;**17**:187–94.
61. **Fanning JR, Lee R, Coccaro EF.** Comorbid intermittent explosive disorder and posttraumatic stress disorder: Clinical correlates and relationship to suicidal behavior. Compr Psychiatry. 2016 Oct;**70**:125–33.
62. **Ille R, Huber HP, Zapotoczky HG.** Aggressive and suicidal behavior. A cluster analysis study of suicidal and nonclinical subjects. Psychiatrische Praxis. 2001;**28**:24–8.
63. **Hong L, Guo L, Wu H, Li P, Xu Y, Gao X**, et al. Bullying, depression, and suicidal ideation among adolescents in the Fujian province of China: A cross-sectional study. Medicine (Baltimore). 2016 Feb;**95**(5):e2530.
64. **Mittendorfer-Rutz E, Rasmussen F, Wasserman D.** Familial clustering of suicidal behaviour and psychopathology in young suicide attempters. A register-based nested case control study. Soc Psychiatry Psychiatr Epidemiol. 2008 Jan;**43**(1):28–36.
65. **Voracek M, Loibl LM.** Genetics of suicide: a systematic review of twin studies. Wien Klin Wochenschr. 2007;**119**(15-16):463–75.
66. **Roy A, Linnoila M.** Alcoholism and suicide. Suicide Life Threat Behav. 1986; **16**(2):244–73.
67. **Roy A, Segal NL, Centerwall BS, Robinette CD.** Suicide in twins. Arch Gen Psychiatry. 1991; **48**(1):29–32.

68. **Statham DJ, Heath AC, Madden PA, Bucholz KK, Bierut L, Dinwiddie SH,** et al. Suicidal behaviour: An epidemiological and genetic study. Psychol Med. 1998; **28**(4):839–55.

69. **Thornton LM, Welch E, Munn-Chernoff MA, Lichtenstein P, Bulik CM.** Anorexia nervosa, major depression, and suicide attempts: Shared genetic factors. Suicide Life Threat Behav. 2016 Oct;**46**(5):525–34.

70. **Kety SS, Rosenthal D, Wender PH, Schulsinger F.** The types and prevalence of mental illness in the biological and adoptive families of adopted schizophrenics. J Psychiatric Res. 1968;**6**:345–62.

71. **Schulsinger F.** A family study of suicide. In: Schov M, Stromgren G (eds). Origin, Prevention and Treatment of Affective Disorders. London: Academic Press; 1979, pp. 277–87.

72. **Wender PH, Kety SS, Rosenthal D, Schulsinger F, Ortmann J, Lunde I.** Psychiatric disorders in the biological and adoptive families of adopted individuals with affective disorders. Arch Gen Psychiatry. 1986 Oct;**43**(10):923–9.

73. **von Borczyskowski A, Lindblad F, Vinnerljung B, Reintjes R, Hjern A.** Familial factors and suicide: An adoption study in a Swedish National Cohort. Psychol Med. 2011 Apr;**41**(4):749–58.

74. **Anokhin AP, Golosheykin S, Mulligan RC.** Long-term test-retest reliability of delayed reward discounting in adolescents. Behav Processes. 2015 Feb;**111**:55–9.

75. **Coccaro EF, Ong AD, Seroczynski AD, Bergeman CS.** Affective intensity and lability: Heritability in adult male twins. J Affect Disord. 2012 Feb;**136**(3):1011–6.

76. **Burt SA.** Are there meaningful etiological differences within antisocial behavior? Results of a meta-analysis. Clin Psychol Rev. 2009 Mar;**29**(2):163–78.

77. **Tuvblad C, Baker LA.** Human aggression across the lifespan: Genetic propensities and environmental moderators. Adv Genet. 2011;**75**:171–214.

78. **Baker LA, Raine A, Liu J, Jacobson KC.** Differential genetic and environmental influences on reactive and proactive aggression in children. J Abnorm Child Psychol. 2008 Nov;**36**(8):1265–78.

79. **Wang P, Niv S, Tuvblad C, Raine A, Baker LA.** The genetic and environmental overlap between aggressive and non-aggressive antisocial behavior in children and adolescents using the Self-Report Delinquency Interview (SR-DI). Journal Criminal Justice. 2013 Sep;**41**(5):277–84.

80. **Waltes R, Chiocchetti AG, Freitag CM.** The neurobiological basis of humanaggression: A review on genetic and epigenetic mechanisms. Am J Med Genet B Neuropsychiatr Genet. 2016 Jul;**171**(5):650–75.

81. **Mann JJ.** Neurobiology of suicidal behaviour. Nat Rev Neurosci. 2003; **4**(10):819–28.

82. **Asberg M, Schalling D, Traeskman-Bendz L, Waegner A.** Psychobiology of suicide, impulsivity, and related phenomena. In: Meltzer HY (ed.). Psychopharmacology. The Third Generation of Progress. New York, NY: Raven Press; 1987, pp 655–68.

83. **Asberg M.** Neurotransmitters and suicidal behavior. The evidence from cerebrospinal fluid studies. Ann N Y Acad Sci. 1997;**836**:158–81.

84. **Mehlman PT, Higley JD, Faucher I, Lilly AA, Taub DM, Vickers J,** et al. Low CSF 5-HIAA concentrations and severe aggression and impaired impulse control in nonhuman primates. Am J Psychiatry 1994; **151**(10):1485–91.

85. **Westergaard GC, Suomi SJ, Higley JD, Mehlman PT.** CSF 5-HIAA and aggression in female macaque monkeys: Species and interindividual differences. Psychopharmacology (Berl) 1999;**146**(4):440–6.

86. **Roggenbach J, Müller-Oerlinghausen B, Franke L.** Suicidality, impulsivity and aggression: Is there a link to 5HIAA concentration in the cerebrospinal fluid? Psychiatry Res. 2002 Dec 15;**113**(1-2):193–206.

87. **Duke AA, Bègue L, Bell R, Eisenlohr-Moul T.** Revisiting the serotonin-aggression relation in humans: A meta-analysis. Psychol Bull. 2013 Sep;**139**(5):1148–72.

88. **Lee R, Coccaro E.** Post-mortem pharmacologycology of criminality and aggression. Can J Psychiatry. 2001; **46**(1):35–44.

89. **Lalovic A, Turecki G.** Meta-analysis of the association between tryptophan hydroxylase and suicidal behavior. Am J Med Genet. 2002; **114**(5):533–540.

90. **Rujescu D, Giegling I, Sato T, Hartmann AM, Moller HJ.** Genetic variations in tryptophan hydroxylase in suicidal behavior: Analysis and meta-analysis. Biol Psychiatry 2003; **54**(4):465–73.
91. **Bellivier F, Chaste P, Malafosse A.** Association between the TPH gene A218C polymorphism and suicidal behavior: A meta-analysis. Am J Med Genet B Neuropsychiatr Genet. 2004;**124B**(1):87–91.
92. **Li D, He L.** Meta-analysis shows association between the tryptophan hydroxylase (TPH) gene and schizophrenia. Hum Genet. 2006 Aug;**120**(1):22–30.
93. **Clayden RC, Zaruk A, Meyre D, Thabane L, Samaan Z.** The association of attempted suicide with genetic variants in the SLC6A4 and TPH genes depends on the definition of suicidal behavior: A systematic review and meta-analysis. Transl Psychiatry. 2012 Oct 2;**2**:e166.
94. **Antypa N, Serretti A, Rujescu D.** Serotonergic genes and suicide: A systematic review. Eur Neuropsychopharmacol. 2013 Oct;**23**(10):1125–42.
95. **Pawlak J, Dmitrzak-Weglarz M, Wilkosc M, Szczepankiewicz A, Leszczynska-Rodziewicz A, Zaremba D**, et al. Suicide behavior as a quantitative trait and its genetic background. J Affect Disord. 2016 Jul 19;**206**:241–50.
96. **Kulikov AV, Osipova DV, Naumenko VS, Popova NK.** The C1473G polymorphism in the tryptophan hydroxylase-2 gene and intermale aggression in mice. Dokl Biol Sci. 2005 May–Jun;**402**:208–10.
97. **Kulikov AV, Osipova DV, Naumenko VS, Popova NK.** Association between Tph2 gene polymorphism, brain tryptophan hydroxylase activity and aggressiveness in mouse strains. Genes Brain Behav. 2005 Nov;**4**(8):482–5.
98. **Osipova DV, Kulikov AV, Popova NK.** C1473G polymorphism in mouse tph2 gene is linked to tryptophan hydroxylase-2 activity in the brain, intermale aggression, and depressive-like behavior in the forced swim test. J Neurosci Res. 2009 Apr;**87**(5):1168–74.
99. **Osipova DV, Kulikov AV, Mekada K, Yoshiki A, Moshkin MP, Kotenkova EV**, et al. Distribution of the C1473G polymorphism in tryptophan hydroxylase 2 gene in laboratory and wild mice. Genes Brain Behav. 2010 Jul;**9**(5):537–43.
100. **Perez-Rodriguez MM, Weinstein S, New AS, Bevilacqua L, Yuan Q, Zhou Z**, et al Tryptophan-hydroxylase 2 haplotype association with borderline personality disorder and aggression in a sample of patients with personality disorders and healthy controls. J Psychiatr Res. 2010 Nov;**44**(15):1075–81.
101. **Oades RD, Lasky-Su J, Christiansen H, Faraone SV, Sonuga-Barke EJ, Banaschewski T**, et al. The influence of serotonin- and other genes on impulsive behavioral aggression and cognitive impulsivity in children with attention-deficit/hyperactivity disorder (ADHD): Findings from a family-based association test (FBAT) analysis. Behav Brain Funct. 2008 Oct 20;**4**:48.
102. **Slof-Op't Landt MC, Bartels M, Middeldorp CM, van Beijsterveldt CE, Slagboom PE, Boomsma DI**, et al. Genetic variation at the TPH2 gene influences impulsivity in addition to eating disorders. Behav Genet. 2013 Jan;**43**(1):24–33.
103. **Juhasz G, Downey D, Hinvest N, Thomas E, Chase D, Toth ZG**, et al. Risk-taking behavior in a gambling task associated with variations in the tryptophan hydroxylase 2 gene: Relevance to psychiatric disorders. Neuropsychopharmacology. 2010 Apr;**35**(5):1109–19.
104. **Reuter M, Ott U, Vaitl D, Hennig J.** Impaired executive control is associated with a variation in the promoter region of the tryptophan hydroxylase 2 gene. J Cogn Neurosci. 2007 Mar;**19**(3):401–8.
105. **Stoltenberg SF, Glass JM, Chermack ST, Flynn HA, Li S, Weston ME**, et al. Possible association between response inhibition and a variant in the brain-expressed tryptophan hydroxylase-2 gene. Psychiatr Genet. 2006 Feb;**16**(1):35–8.
106. **Angoa-Pérez M, Kane MJ, Briggs DI, Sykes CE, Shah MM, Francescutti DM**, et al. Genetic depletion of brain 5HT reveals a common molecular pathway mediating compulsivity and impulsivity. J Neurochem. 2012 Jun;**121**(6):974–84
107. **Mosienko V, Bert B, Beis D, Matthes S, Fink H, Bader M**, et al. Exaggerated aggression and decreased anxiety in mice deficient in brain serotonin. Transl Psychiatry. 2012 May 29;**2**:e122.

108. **Gutknecht L, Popp S, Waider J, Sommerlandt FM, Göppner C, Post A,** et al. Interaction of brain 5-HT synthesis deficiency, chronic stress and sex differentially impact emotional behavior in Tph2 knockout mice. Psychopharmacology (Berl). 2015 Jul;**232**(14):2429–41.

109. **Waider J, Araragi N, Gutknecht L, Lesch KP.** Tryptophan hydroxylase-2 (TPH2) in disorders of cognitive control and emotion regulation: A perspective. Psychoneuroendocrinology. 2011 Apr;**36**(3):393–405.

110. **Ono H, Shirakawa O, Nishiguchi N, Nishimura A, Nushida H, Ueno Y,** et al. No evidence of an association between a functional monoamine oxidase a gene polymorphism and completed suicides. Am J Med Genet. 2002; **114**(3):340–42.

111. **Huang YY, Cate SP, Battistuzzi C, Oquendo MA, Brent D, Mann JJ.** An association between a functional polymorphism in the monoamine oxidase a gene promoter, impulsive traits and early abuse experiences. Neuropsychopharmacology. 2004; **29**(8):1498–505.

112. **Linkowska K, Daca P, Sykutera M, Pufal E, Bloch-Bogusławska E, Grzybowski T.** Search for association between suicide and 5-HTT, MAOA and DAT polymorphism in Polish males. Arch Med Sadowej Kryminol. 2010 Apr-Sep;**60**(2-3):112–7.

113. **Buttenschøn HN, Flint TJ, Foldager L, Qin P, Christoffersen S, Hansen NF,** et al. An association study of suicide and candidate genes in the serotonergic system. J Affect Disord. 2013 Jun;**148**(2–3):291–8.

114. **Ho LW, Furlong RA, Rubinsztein JS, Walsh C, Paykel ES, Rubinsztein DC.** Genetic associations with clinical characteristics in bipolar affective disorder and recurrent unipolar depressive disorder. Am J Med Genet. 2000;**96**(1):36–42.

115. **Du L, Faludi G, Palkovits M, Sotonyi P, Bakish D, Hrdina PD.** High activity-related allele of MAO-A gene associated with depressed suicide in males. Neuroreport 2002; **13**(9):1195–8.

116. **Zalsman G, Patya M, Frisch A, Ofek H, Schapir L, Blum I,** et al. Association of polymorphisms of the serotonergic pathways with clinical traits of impulsive-aggression and suicidality in adolescents: A multi-center study. World J Biol Psychiatry. 2011 Feb;**12**(1):33–41.

117. **Lung FW, Tzeng DS, Huang MF, Lee MB.** Association of the MAOA promoter uVNTR polymorphism with suicide attempts in patients with major depressive disorder. BMC Med Genet. 2011 May 24;**12**:74.

118. **Brunner HG, Nelen M, Breakefield XO, Ropers HH, van Oost BA.** Abnormal behavior associated with a point mutation in the structural gene for monoamine oxidase A. Science. 1993 Oct 22;**262**(5133):578–80.

119. **Brunner HG, Nelen MR, van Zandvoort P, Abeling NG, van Gennip AH, Wolters EC,** et al. X-linked borderline mental retardation with prominent behavioral disturbance: phenotype, genetic localization, and evidence for disturbed monoamine metabolism. Am J Hum Genet. 1993 Jun;**52**(6):1032–9.

120. **Brunner HG.** MAOA deficiency and abnormal behaviour: Perspectives on an association. Ciba Found Symp. 1996;**194**:155–64; discussion 164–7.

121. **Lawson DC, Turic D, Langley K, Pay HM, Govan CF, Norton N,** et al. Association analysis of monoamine oxidase A and attention deficit hyperactivity disorder. Am J Med Genet B Neuropsychiatr Genet. 2003 Jan 1;**116B**(1):84–9.

122. **Kinnally EL, Huang YY, Haverly R, Burke AK, Galfalvy H, Brent DP,** et al. Parental care moderates the influence of MAOA-uVNTR genotype and childhood stressors on trait impulsivity and aggression in adult women. Psychiatr Genet. 2009 Jun;**19**(3):126–33.

123. **Jacob CP, Müller J, Schmidt M, Hohenberger K, Gutknecht L, Reif A,** et al. Cluster B personality disorders are associated with allelic variation of monoamine oxidase A activity. Neuropsychopharmacology. 2005 Sep;**30**(9):1711–8.

124. **Saito T, Lachman HM, Diaz L, Hallikainen T, Kauhanen J, Salonen JT,** et al. Analysis of monoamine oxidase A (MAOA) promoter polymorphism in Finnish male alcoholics. Psychiatry Res. 2002 Mar 15 ;**109**(2):113–9.

125. **Contini V, Marques FZ, Garcia CE, Hutz MH, Bau CH.** MAOA-uVNTR polymorphism in a Brazilian sample: Further support for the association with impulsive behaviors and alcohol dependence. Am J Med Genet B Neuropsychiatr Genet. 2006 Apr 5;**141B**(3):305–8.

126. **Tikkanen R, Sjöberg RL, Ducci F, Goldman D, Holi M, Tiihonen J**, et al. Effects of MAOA-genotype, alcohol consumption, and aging on violent behavior. Alcohol Clin Exp Res. 2009 Mar;**33**(3):428–34.

127. **Tikkanen R, Ducci F, Goldman D, Holi M, Lindberg N, Tiihonen J**, et al. MAOA alters the effects of heavy drinking and childhood physical abuse on risk for severe impulsive acts of violence among alcoholic violent offenders. Alcohol Clin Exp Res. 2010 May;**34**(5):853–60.

128. **Laqua C, Zill P, Koller G, Preuss U, Soyka M.** [Association between the MAOA-uVNTR polymorphism and antisocial personality traits in alcoholic men]. Fortschr Neurol Psychiatr. 2015 Mar;**83**(3):162–9.

129. **Stetler DA, Davis C, Leavitt K, Schriger I, Benson K, Bhakta S**, et al. Association of low-activity MAOA allelic variants with violent crime in incarcerated offenders. J Psychiatr Res. 2014 Nov;**58**:69–75.

130. **Manuck SB, Flory JD, Ferrell RE, Dent KM, Mann JJ, Muldoon MF.** Aggression and anger-related traits associated with a polymorphism of the tryptophan hydroxylase gene. Biol Psychiatry. 1999; **45**(5):603–14.

131. **Rujescu D, Giegling I, Bondy B, Gietl A, Zill P, Moller HJ.** Association of anger-related traits with SNPs in the TPH gene. Mol Psychiatry. 2002;**7**:1023–9.

132. **Baud P, Perroud N, Courtet P, Jaussent I, Relecom C, Jollant F**, et al. Modulation of anger control in suicide attempters by TPH-1. Genes Brain Behav. 2009;Feb;**8**(1):97–100.

133. **Ruocco AC, Rodrigo AH, Carcone D, McMain S, Jacobs G, Kennedy JL.** Tryptophan hydroxylase 1 gene polymorphisms alter prefrontal cortex activation during response inhibition. Neuropsychology. 2016 Jan;**30**(1):18–27.

134. **Walther DJ, Peter JU, Bashammakh S, Hortnagl H, Voits M, Fink H**, et al. Synthesis of serotonin by a second tryptophan hydroxylase isoform. Science. 2003; **299**(5603):76.

135. **Zhang X, Beaulieu JM, Sotnikova TD, Gainetdinov RR, Caron MG.** Tryptophan hydroxylase-2 controls brain serotonin synthesis. Science. 2004 Jul 9;**305**(5681):217.

136. **Zill P, Buttner A, Eisenmenger W, Moller HJ, Bondy B, Ackenheil M.** Single nucleotide polymorphism and haplotype analysis of a novel tryptophan hydroxylase isoform (TPH2) gene in suicide victims. Biol Psychiatry 2004; **56**(8):581–6.

137. **Lopez de Lara C, Brezo J, Rouleau G, Lesage A, Dumont M, Alda M**, et al. Effect of tryptophan hydroxylase-2 gene variants on suicide risk in major depression. Biol Psychiatry. 2007 Jul 1;**62**(1):72–80.

138. **Mouri K, Hishimoto A, Fukutake M, Shiroiwa K, Asano M, Nagasaki Y**, et al. TPH2 is not a susceptibility gene for suicide in Japanese population. Prog Neuropsychopharmacol Biol Psychiatry. 2009 Nov 13;**33**(8):1546–50.

139. **Zupanc T, Pregelj P, Tomori M, Komel R, Paska AV.** No association between polymorphisms in four serotonin receptor genes, serotonin transporter gene and alcohol-related suicide. Psychiatr Danub. 2010 Dec;**22**(4):522–7.

140. **Tsai SJ, Hong CJ, Liou YJ.** Recent molecular genetic studies and methodological issues in suicide research. Prog Neuropsychopharmacol Biol Psychiatry. 2011 Jun 1;**35**(4):809–17.

141. **Antypa N, Giegling I, Calati R, Schneider B, Hartmann AM, Friedl M**, et al. MAOA and MAOB polymorphisms and anger-related traits in suicidal participants and controls. Eur Arch Psychiatry Clin Neurosci. 2013 Aug;**263**(5): 393–403.

142. **Meyer-Lindenberg A, Buckholtz JW, Kolachana BR, Hariri A, Pezawas L, Blasi G**, et al. Neural mechanisms of genetic risk for impulsivity and violence in humans. PNAS. 2006 Apr 18;**103**(16):6269–74.

143. **Jollant F, Buresi C, Guillaume S, Jaussent I, Bellivier F, Leboyer M**, et al. The influence of four serotonin-related genes on decision-making in suicide attempters. Am J Med Genet B Neuropsychiatr Genet. 2007 Jul 5;**144B**(5):615–24.

144. **McDermott R, Tingley D, Cowden J, Frazzetto G, Johnson DD.** Monoamine oxidase A gene (MAOA) predicts behavioral aggression following provocation. PNAS. 2009 Feb 17;**106**(7):2118–23.

145. **Kuepper Y, Grant P, Wielpuetz C, Hennig J.** MAOA-uVNTR genotype predicts interindividual differences in experimental aggressiveness as a function of the degree of provocation. Behav Brain Res. 2013 Jun 15;**247**:73–8.

146. **Piton A, Poquet H, Redin C, Masurel A, Lauer J, Muller J**, et al. 20 ans après: A second mutation in MAOA identified by targeted high-throughput sequencing in a family with altered behavior and cognition. Eur J Hum Genet. 2014 Jun;**22**(6):776–83.

147. **Lesch KP, Bengel D, Heils A, Sabol SZ, Greenberg BD, Petri S**, et al. Association of anxiety-related traits with a polymorphism in the serotonin transporter gene regulatory region. Science. 1996; **274**(5292):1527–31.

148. **Anguelova M, Benkelfat C, Turecki G.** A systematic review of association studies investigating genes coding for serotonin receptors and the serotonin transporter: II. Suicidal behavior. Mol Psychiatry. 2003; **8**(7):646–53.

149. **Lin PY, Tsai G.** Association between serotonin transporter gene promoter polymorphism and suicide: Results of a meta-analysis. Biol Psychiatry 2004; **55**(10):1023–30.

150. **Li D, He L.** Meta-analysis supports association between serotonin transporter (5-HTT) and suicidal behavior. Mol Psychiatry. 2007 Jan;**12**(1):47–54. Epub 2006 Sep 12.

151. **Greenberg BD, Li Q, Lucas FR, Hu S, Sirota LA, Benjamin J**, et al. Association between the serotonin transporter promoter polymorphism and personality traits in a primarily female population sample. Am J Med Genet. 2000 Apr 3;**96**(2):202–16.

152. **Gerra G, Garofano L, Castaldini L, Rovetto F, Zaimovic A, Moi G**, et al. Serotonin transporter promoter polymorphism genotype is associated with temperament, personality traits and illegal drugs use among adolescents. J Neural Transm (Vienna). 2005 Oct;**112**(10):1397–410.

153. **Murphy DL, Lesch KP.** Targeting the murine serotonin transporter: Insights into human neurobiology. Nat Rev Neurosci. 2008 Feb;**9**(2):85–96.

154. **Lesch KP.** Linking emotion to the social brain. The role of the serotonin transporter in human social behaviour. EMBO Rep. 2007 Jul;8 Spec No:S24–9. Review.

155. **Canli T, Lesch KP.** Long story short: I serotonin transporter in emotion regulation and social cognition. Nat Neurosci. 2007 Sep;**10**(9):1103–9. Review.

156. **Lee YH, Song GG.** Meta-analysis of case-control and family-based associations between the 5-HTTLPR L/S polymorphism and susceptibility to ADHD. J Atten Disord. 2015 Jun 5. Pii:1087054715587940. [Epub ahead of print]

157. **Vassos E, Collier DA, Fazel S.** Systematic meta-analyses and field synopsis of genetic association studies of violence and aggression. Mol Psychiatry. 2014 Apr;**19**(4):471–7.

158. **Tielbeek JJ, Karlsson Linnér R, Beers K, Posthuma D, Popma A, Polderman TJ.** Meta-analysis of the serotonin transporter promoter variant (5-HTTLPR) in relation to adverse environment and antisocial behavior. Am J Med Genet B Neuropsychiatr Genet. 2016 Jul;**171**(5):748–60.

159. **Benko A, Lazary J, Molnar E, Gonda X, Tothfalusi L, Pap D**, et al. Significant association between the C(-1019)G functional polymorphism of the HTR1A gene and impulsivity. Am J Med Genet B Neuropsychiatr Genet. 2010 Mar 5;**153B**(2):592–9.

160. **Lemonde S, Turecki G, Bakish D, Du L, Hrdina PD, Bown CD**, et al. Impaired repression at a 5-hydroxytryptamine 1A receptor gene polymorphism associated with major depression and suicide. J Neurosci 2003; **23**(25):8788–99.

161. **Huang YY, Battistuzzi C, Oquendo MA, Harkavy-Friedman J, Greenhill L, Zalsman G**, et al. Human 5-HT1A receptor C(-1019)G polymorphism and psychopathology. Int J Neuropsychopharmacol. 2004; **7**(4):441–51.

162. **Sawiniec J, Borkowski K, Ginalska G, Lewandowska-Stanek H.** Association between 5-hydroxytryptamine 1A receptor gene polymorphism and suicidal behavior. Przegl Lek. 2007;**64**(4-5):208–11.

163. **Serretti A, Mandelli L, Giegling I, Schneider B, Hartmann AM, Schnabel A**, et al. HTR2C and HTR1A gene variants in German and Italian suicide attempters and completers. Am J Med Genet B Neuropsychiatr Genet. 2007 Apr 5;**144B**(3):291–9.
164. **Serretti A, Calati R, Giegling I, Hartmann AM, Möller HJ, Rujescu D.** Serotonin receptor HTR1A and HTR2C variants and personality traits in suicide attempters and controls. J Psychiatr Res. 2009 Feb;**43**(5):519–25.
165. **Strobel A, Gutknecht L, Rothe C, Reif A, Mössner R, Zeng Y**, et al. Allelic variation in 5-HT1A receptor expression is associated with anxiety-and depression-related personality traits. J Neural Transm (Vienna). 2003 Dec;**110**(12):1445–53.
166. **Angles MR, Ocaña DB, Medellín BC, Tovilla-Zárate C.** No association between the HTR1A gene and suicidal behavior: A meta-analysis. Rev Bras Psiquiatr. 2012 Mar;**34**(1):38–42.
167. **Schild AH, Pietschnig J, Tran US, Voracek M.** Genetic association studies between SNPs and suicidal behavior: A meta-analytical field synopsis. Prog Neuropsychopharmacol Biol Psychiatry. 2013 Oct 1;**46**:36–42.
168. **Huang YY, Grailhe R, Arango V, Hen R, Mann JJ.** Relationship of psychopathology to the human serotonin1B genotype and receptor binding kinetics in ost-mortem brain tissue. Neuropsychopharmacology. 1999; **21**(2):238–46.
169. **New AS, Gelernter J, Goodman M, Mitropoulou V, Koenigsberg H, Silverman J**, et al. Suicide, impulsive aggression, and HTR1B genotype. Biol Psychiatry 2001; **50**(1):62–65.
170. **Huang YY, Oquendo MA, Friedman JM, Greenhill LL, Brodsky B, Malone KM**, et al. Substance abuse disorder and major depression are associated with the human 5-HT1B receptor gene (HTR1B) G861C polymorphism. Neuropsychopharmacology. 2003; **28**(1):163–9.
171. **Nishiguchi N, Shirakawa O, Ono H, Nishimura A, Nushida H, Ueno Y**, et al. No evidence of an association between 5HT1B receptor gene polymorphism and suicide victims in a Japanese population. Am J Med Genet 2001; **105**(4):343–5.
172. **Arango V, Huang YY, Underwood MD, Mann JJ.** Genetics of the serotonergic system in suicidal behavior. J Psychiatr Res. 2003 ; **37**(5):375–86.
173. **Turecki G, Sequeira A, Gingras Y, Seguin M, Lesage A, Tousignant M**, et al. Suicide and serotonin: Study of variation at seven serotonin receptor genes in suicide completers. Am J Med Genet B Neuropsychiatr Genet. 2003; **118B**(1):36–40.
174. **Pooley EC, Houston K, Hawton K, Harrison PJ.** Deliberate self-harm is associated with allelic variation in the tryptophan hydroxylase gene (TPH A779C), but not with polymorphisms in five other serotonergic genes. Psychol Med 2003; **33**(5):775–83.
175. **Rujescu D, Giegling I, Sato T, Moller HJ.** Lack of association between serotonin 5-HT1B receptor gene polymorphism and suicidal behavior. Am J Med Genet B Neuropsychiatr Genet. 2003; **116B**(1):69–71.
176. **Stefulj J, Buttner A, Skavic J, Zill P, Balija M, Eisenmenger W**, et al. Serotonin 1B (5HT-1B) receptor polymorphism (G861C) in suicide victims: Association studies in German and Slavic population. Am J Med Genet B Neuropsychiatr Genet. 2004;**127B**(1):48–50.
177. **Tsai SJ, Hong CJ, Yu YW, Chen TJ, Wang YC, Lin WK.** Association study of serotonin 1B receptor (A-161T) genetic polymorphism and suicidal behaviors and response to fluoxetine in major depressive disorder. Neuropsychobiology. 2004; **50**(3):235–8.
178. **Wang S, Zhang K, Xu Y, Sun N, Shen Y, Xu Q.** An association study of the serotonin transporter and receptor genes with the suicidal ideation of major depression in a Chinese Han population. Psychiatry Res. 2009 Dec 30;**170**(2-3):204–7.
179. **Murphy TM, Ryan M, Foster T, Kelly C, McClelland R, O'Grady J**, et al. Risk and protective genetic variants in suicidal behaviour: Association with SLC1A2, SLC1A3, 5-HTR1B &NTRK2 polymorphisms. Behav Brain Funct. 2011 Jun 28;7:22.
180. **Kia-Keating BM, Glatt SJ, Tsuang MT.** Meta-analyses suggest association between COMT, but not HTR1B, alleles, and suicidal behavior. Am J Med Genet B Neuropsychiatr Genet. 2007 Dec 5;**144B**(8):1048–53.
181. **Saudou F, Amara DA, Dierich A, LeMeur M, Ramboz S, Segu L**, er al. Enhanced aggressive behavior in mice lacking 5-HT1B receptor. Science. 1994 Sep 23;**265**(5180):1875–8.

182. **Lappalainen J, Long JC, Eggert M, Ozaki N, Robin RW, Brown GL**, et al. Linkage of antisocial alcoholism to the serotonin 5-HT1B receptor gene in 2 populations. Arch Gen Psychiatry. 1998 Nov;**55**(11):989–94.

183. **Jensen KP, Covault J, Conner TS, Tennen H, Kranzler HR, Furneaux HM.** A common polymorphism in serotonin receptor 1B mRNA moderates regulation by miR-96 and associates with aggressive human behaviors. Mol Psychiatry. 2009 Apr;**14**(4):381–9.

184. **Conner TS, Jensen KP, Tennen H, Furneaux HM, Kranzler HR, Covault J.** Functional polymorphisms in the serotonin 1B receptor gene (HTR1B) predict self-reported anger and hostility among young men. Am J Med Genet B Neuropsychiatr Genet. 2010 Jan 5;**153B**(1):67–78.

185. **Varga G, Szekely A, Antal P, Sarkozy P, Nemoda Z, Demetrovics Z**, et al. Additive effects of serotonergic and dopaminergic polymorphisms on trait impulsivity. Am J Med Genet B Neuropsychiatr Genet. 2012 Apr;**159B**(3):281–8.

186. **Sánchez-Mora C, Ramos-Quiroga JA, Garcia-Martínez I, Fernàndez-Castillo N, Bosch R, Richarte V**, et al. Evaluation of single nucleotide polymorphisms in the miR-183-96-182 cluster in adulthood attention-deficit and hyperactivity disorder (ADHD) and substance use disorders (SUDs). Eur Neuropsychopharmacol. 2013 Nov;**23**(11):1463–73.

187. **Hakulinen C, Jokela M, Hintsanen M, Merjonen P, Pulkki-Råback L, Seppälä I**, et al. Serotonin receptor 1B genotype and hostility, anger and aggressive behavior through the lifespalhe Young Finns study. J Behav Med. 2013 Dec;**36**(6):583–90.

188. **van Rooij D, Hartman CA, van Donkelaar MM, Bralten J, von Rhein D, Hakobjan M**, et al. Variation in serotonin neurotransmission genes affects neural activation during response inhibition in adolescents and young adults with ADHD and healthy controls. World J Biol Psychiatry. 2015;**16**(8):625–34.

189. **Bachus SE, Hyde TM, Akil M, Weickert CS, Vawter MP, Kleinman JE.** Neuropathology of suicide. A review and an approach. Ann N Y Acad Sci.1997;**836**:201–219.

190. **Mann JJ.** The neurobiology of suicide. Nat Med. 1998; **4**(1):25–30.

191. **Wang JY, Jia CX, Lian Y, Sun SH, Lyu M, Wu A.** Association of the HTR2A 102T/C polymorphism with attempted suicide: A meta-analysis. Psychiatr Genet. 2015 Aug;**25**(4):168–77.

192. **Bani-Fatemi A, Howe AS, Matmari M, Koga A, Zai C, Strauss J, De Luca V.** Interaction between methylation and CpG single-nucleotide polymorphisms in the HTR2A gene: Association analysis with suicide attempt in schizophrenia. Neuropsychobiology. 2016;**73**(1):10–5.

193. **Coccaro EF, Kavoussi RJ, Sheline YI, Berman ME, Csernansky JG.** Impulsive aggression in personality disorder correlates with platelet 5-HT2A receptor binding. Neuropsychopharmacology. 1997 Mar;**16**(3):211–6

194. **Preuss UW, Koller G, Bahlmann M, Soyka M, Bondy B.** No association between suicidal behavior and 5-HT2A-T102C polymorphism in alcohol dependents. Am J Med Genet 2000; **96**(6): 877–8.

195. **Hill EM, Stoltenberg SF, Bullard KH, Li S, Zucker RA, Burmeister M.** Antisocial alcoholism and serotonin-related polymorphisms: Association tests. Psychiatr Genet. 2002 Sep;**12**(3): 143–53.

196. **Ni X, Bismil R, Chan K, Sicard T, Bulgin N, McMain S, Kennedy JL.** Serotonin 2A receptor gene is associated with personality traits, but not to disorder, in patients with borderline personality disorder. Neurosci Lett. 2006;**408**: 214–9.

197. **Nomura M.** Involvement of a polymorphism in the 5-HT2A receptor gene in impulsive behavior. Nippon Yakurigaku Zasshi. 2006;**127**: 9–13.

198. **Giegling I, Hartmann AM, Moller HJ, Rujescu D.** Anger- and aggression-related traits are associated with polymorphisms in the 5-HT-2A gene. J Affect Disord. 2006; **96**(1-2):75–81.

199. **Golimbet VE, Alfimova MV, Mitiushina NG.** Polymorphism of the serotonin 2A receptor gene (5HTR2A) and personality traits. Mol Biol (Mosk). 2004;**38**:404–12.

200. **Rybakowski F, Slopien A, Dmitrzak-Weglarz M, Czerski P, Rajewski A, Hauser J.** The 5-HT2A-1438 A/G and 5-HTTLPR polymorphisms and personality dimensions in adolescent anorexia nervosa: Association study. Neuropsychobiology. 2006;**53**:33–9.

201. **Ham BJ, Kim YH, Choi MJ, Cha JH, Choi YK, Lee MS.** Serotonergic genes and personality traits in the Korean population. Neurosci Lett. 2004;**354**:2–5.

202. **Serretti A, Calati R, Giegling I, Hartmann AM, Möller HJ, Colombo C,** et al. 5-HT2A SNPs and the Temperament and Character Inventory. Prog Neuropsychopharmacol Biol Psychiatry. 2007 Aug 15;**31**(6):1275–81.

203. **Salo J, Pulkki-Råback L, Hintsanen M, Lehtimäki T, Keltikangas-Järvinen L.** The interaction between serotonin receptor 2A and catechol-*O*-methyltransferase gene polymorphisms is associated with the novelty-seeking subscale impulsiveness. Psychiatr Genet. 2010 Dec;**20**(6):273–81.

204. **Jakubczyk A, Wrzosek M, Lukaszkiewicz J, Sadowska-Mazuryk J, Matsumoto H, Sliwerska E,** et al. The CC genotype in HTR2A T102C polymorphism is associated with behavioral impulsivity in alcohol-dependent patients. J Psychiatr Res. 2012 Jan;**46**(1):44–9.

205. **Pinto R, Asherson P, Ilott N, Cheung CH, Kuntsi J.** Testing for the mediating role of endophenotypes using molecular genetic data in a twin study of ADHD traits. Am J Med Genet B Neuropsychiatr Genet. 2016 Oct;**171**(7):982–92.

206. **Perlis RH, Huang J, Purcell S, Fava M, Rush AJ, Sullivan PF,** et al. Genome-wide association study of suicide attempts in mood disorder patients. Am J Psychiatry. 2010 Dec;**167**(12):1499–507.

207. **Schosser A, Butler AW, Ising M, Perroud N, Uher R, Ng MY,** et al. Genomewide association scan of suicidal thoughts and behaviour in major depression. PloS One. 2011;**6**(7):e20690.

208. **Willour VL, Seifuddin F, Mahon PB, Jancic D, Pirooznia M, Steele J,** et al. A genome-wide association study of attempted suicide. Mol Psychiatry. 2012 Apr;**17**(4):433–44.

209. **Mullins N, Perroud N, Uher R, Butler AW, Cohen-Woods S, Rivera M,** et al. Genetic relationships between suicide attempts, suicidal ideation and major psychiatric disorders: A genome-wide association and polygenic scoring study. Am J Med Genet B Neuropsychiatr Genet. 2014 Jul;**165B**(5):428–37.

210. **Zai CC, Gonçalves VF, Tiwari AK, Gagliano SA, Hosang G, de Luca V,** et al. A genome-wide association study of suicide severity scores in bipolar disorder. J Psychiatr Res. 2015 Jun;**65**:23–9.

211. **Galfalvy H, Haghighi F, Hodgkinson C, Goldman D, Oquendo MA, Burke A,** et al. A genome-wide association study of suicidal behavior. Am J Med Genet B Neuropsychiatr Genet. 2015 Oct;**168**(7):557–63.

212. **Perroud N, Uher R, Ng MY, Guipponi M, Hauser J, Henigsberg N,** et al. Genome-wide association study of increasing suicidal ideation during antidepressant treatment in the GENDEP project. Pharmacogenomics J. 2012 Feb;**12**(1):68–77.

213. **Menke A, Domschke K, Czamara D, Klengel T, Hennings J, Lucae S,** et al. Genome-wide association study of antidepressant treatment-emergent suicidal ideation. Neuropsychopharmacology. 2012 Feb;**37**(3):797–807.

214. **Dick DM, Aliev F, Krueger RF, Edwards A, Agrawal A, Lynskey M,** et al. Genome-wide association study of conduct disorder symptomatology. Mol Psychiatry. 2011;**16**(8):800–808.

215. **Lesch KP, Selch S, Renner TJ, Jacob C, Nguyen TT, Hahn T,** et al. Genome-wide copy number variation analysis in attention-deficit/hyperactivity disorder: Association with neuropeptide Y gene dosage in an extended pedigree. Mol Psychiatry. 2011 May;**16**(5):491–503.

216. **Tielbeek JJ, Medland SE, Benyamin B, Byrne EM, Heath AC, Madden PA,** et al. Unraveling the genetic etiology of adult antisocial behavior: A genome-wide association study. PloS One. 2012;**7**(10):e45086.

217. **Schuckit MA.** A brief history of research on the genetics of alcohol and other drug use disorders. J Stud Alcohol Drugs Suppl. 2014;**75** Suppl 17:59–67.

218. **Li Z, Zheng M, Abdalla BA, Zhang Z, Xu Z, Ye Q,** et al. Genome-wide association study of aggressive behaviour in chicken. Sci Rep. 2016 Aug 3;**6**:30981.

219. **van den Berg SM, de Moor MH, Verweij KJ, Krueger RF, Luciano M, Arias Vasquez A,** et al. Meta-analysis of genome-wide association studies for extraversion: Findings from the Genetics of Personality Consortium. Behav Genet. 2016 Mar;**46**(2):170–82.

220. **Higley JD, Mehlman PT, Taub DM, Higley SB, Suomi SJ, Vickers JH,** et al. Cerebrospinal fluid monoamine and adrenal correlates of aggression in free-ranging rhesus monkeys. Arch Gen Psychiatry. 1992;**49**(6):436–41.

221. **Higley JD, Mehlman PT, Poland RE, Taub DM, Vickers J, Suomi SJ**, et al. CSF testosterone and 5-HIAA correlate with different types of aggressive behaviors. Biol Psychiatry.1996;**40**(11):1067–82.

222. **Higley JD, Mehlman PT, Higley SB, Fernald B, Vickers J, Lindell SG**, et al. Excessive mortality in young free-ranging male nonhuman primates with low cerebrospinal fluid 5-hydroxyindoleacetic acid concentrations. Arch Gen Psychiatry. 1996;**53**(6):537–43.

223. **González-Castro TB, Juárez-Rojop I, López-Narváez ML, Tovilla-Zárate CA.** Association of TPH-1 and TPH-2 gene polymorphisms with suicidal behavior: A systematic review and meta-analysis. BMC Psychiatry. 2014 Jul 8;**14**:196.

224. **Dent CL, Humby T, Lewis K, Plagge A, Fischer-Colbrie R, Wilkins JF**, et al. Impulsive choices in mice lacking imprinted Nesp55. Genes Brain Behav. 2016 Aug 10. Doi:10.1111/gbb.12316. [Epub ahead of print] PubMed PMID: 27509352.

225. **Terracciano A, Esko T, Sutin AR, de Moor MH, Meirelles O, Zhu G**, et al. Meta-analysis of genome-wide association studies identifies common variants in CTNNA2 associated with excitement-seeking. Transl Psychiatry. 2011 Oct 18;**1**:e49.

226. **Ehlers CL, Gizer IR, Bizon C, Slutske W, Peng Q, Schork NJ**, et al. Single nucleotide polymorphisms in the REG-CTNNA2 region of chromosome 2 and NEIL3 associated with impulsivity in a Native American sample. Genes Brain Behav. 2016 Jul;**15**(6):568–77.

227. **Derringer J, Corley RP, Haberstick BC, Young SE, Demmitt BA, Howrigan DP**, et al. Genome-wide association study of behavioral disinhibition in a selected adolescent sample. Behav Genet. 2015 Jul;**45**(4):375–81.

228. **Ebejer JL, Duffy DL, van der Werf J, Wright MJ, Montgomery G, Gillespie NA**, et al. Genome-wide association study of inattention and hyperactivity-impulsivity measured as quantitative traits. Twin Res Hum Genet. 2013 Apr;**16**(2):560–74.

229. **Brevik EJ, van Donkelaar MM, Weber H, Sánchez-Mora C, Jacob C, Rivero O**, et al. Genome-wide analyses of aggressiveness in attention-deficit hyperactivity disorder. Am J Med Genet B Neuropsychiatr Genet. 2016 Jul;**171**(5):733–47.

230. **Mick E, McGough J, Deutsch CK, Frazier JA, Kennedy D, Goldberg RJ.** Genome-wide association study of proneness to anger. PLoS One. 2014 Jan 28;**9**(1):e87257.

231. **Pappa I, St Pourcain B, Benke K, Cavadino A, Hakulinen C, Nivard MG**, et al. A genome-wide approach to children's aggressive behavior: The EAGLE consortium. Am J Med Genet B Neuropsychiatr Genet. 2016 Jul;**171**(5):562–72.

232. **Rautiainen MR, Paunio T, Repo-Tiihonen E, Virkkunen M, Ollila HM, Sulkava S**, et al. Genome-wide association study of antisocial personality disorder. Transl Psychiatry. 2016 Sep 6;**6**(9):e883.

233. **Iliceto P, Pompili M, Candilera G, Natali MA, Stefani H, Lester D, Serafini G, Girardi P.** Gender-related differences concerning anger expression and interpersonal relationships in a sample of overweight/obese subjects. Clin Ter. 2012;**163**(5):e279–85.

234. **Brent DA, Mann JJ.** Family genetic studies, suicide, and suicidal behavior. Am J Med Genet C Semin Med Genet. 2005 Feb;**133C**(1):13–24.

235. **Arango V, Underwood MD, Mann JJ.** Postmortem findings in suicide victims. Implications for in vivo imaging studies. Ann N Y Acad Sci. 1997 Dec;**836**:269–87.

236. **Kunugi H, Ishida S, Kato T, Tatsumi M, Sakai T, Hattori M, Hirose T, Nanko S.** A functional polymorphism in the promoter region of monoamine oxidase-A gene and mood disorders. Mol Psychiatry. 1999 Jul;**4**(4):393–5.

237. **Manuck SB, Flory JD, Ferrell RE, Mann JJ, Muldoon MF.** A regulatory polymorphism of the monoamine oxidase-A gene may be associated with variability in aggression, impulsivity, and central nervous system serotonergic responsivity. Psychiatry Res. 2000 Jul 24;**95**(1):9–23.

238. **Laje G, Allen AS, Akula N, Manji H, John Rush A, McMahon FJ.** Genome-wide association study of suicidal ideation emerging during citalopram treatment of depressed outpatients. Pharmacogenet Genomics. 2009 Sep;**19**(9):666–74.

Chapter 24

Causes of distress-induced emotional eating

Tatjana van Strien

24.1 Introduction

Negative mood or distress is associated with both increased and decreased food intake (1), with eating less being the typical and predominant response (2). Negative mood or distress is normally associated with a hyperactive hypothalamic–pituitary–adrenal (HPA) axis, with physiological adaptations that are biologically designed to prepare the individual for a fight-or-flight response (2). These adaptations include inhibition of gastric motility (3) and the release of sugar into the bloodstream (4), thereby suppressing feelings of hunger. However, so-called emotional eaters show an atypical response to distress and eat the same amount or even more after distress (5, 6).

The present chapter explores possible causes of distress-induced emotional eating in terms of etiology and mechanisms. After a paragraph on the assessment of emotional eating, studies on possible mechanisms will be presented and we will examine whether this opposite response to distress of emotional eating is the result of a failure to experience the typical post-stress reduction of hunger, poor interoceptive awareness, or alexithymia, and of changes in the stress reactivity of the HPA axis as indicated by cortisol stress responses. Also, the emergence of emotional eating in adolescence will be addressed by presenting two prospective studies on increases of emotional eating in association with inadequate parenting and depressive feelings in interaction with genetic vulnerability (e.g. the dopamine D2 Receptor (*DRD2*) gene and the serotonin transporter gene (*SCL6A4/5-HTT*)). Finally, emotional eating as a mediator between depression and both body mass index and weight gain will be examined, and suggestions will be given for obesity interventions and future research.

24.2 Assessment of emotional eating

Emotional eating can be assessed using the Emotional Eating Scale (87), the Emotional Appetite Questionnaire (7), three items from the revised 18-item Three-Factor Eating Questionnaire (8), six items from the further refined 21-item version (TFEQ-R21; 9), or the 13-item emotional eating scale of the Dutch Eating Behavior Questionnaire (DEBQ) (DEBQ-E) (10).

Most studies in the present chapter measured emotional eating with the DEBQ-E, the very first measure of emotional eating. The DEBQ-E was developed by Van Strien and colleagues (10) in a representative Dutch community sample. Apart from a scale on emotional eating, the DEBQ also has a scale on restrained eating (eating less than desired to maintain or lose body weight) and external eating (i.e. eating in response to external food-related cues such as sight and smell of attractive food).

The DEBQ-E only has items on eating in response to negative emotions because of evidence that eating in response to positive and to negative emotions refer to different constructs (11–13). Furthermore, only eating in response to negative emotions was shown to be an "obese eating style," an eating style that is associated with weight gain and obesity (13).

The DEBQ has been rated as "up to the mark" or "good" by the Dutch Committee on Tests and Testing (COTAN) on all EFPA (European Federation of Psychologists' Associations) criteria (e.g. norms, reliability (internal consistency, test–retest), and validity (dimensional validity, construct validity, and criterion validity)) (14). The three DEBQ scales showed, for example, a robust dimensional validity, not only in the original Dutch version (15), but also in other translations (e.g. Chinese (16), English (17), French (18), German (19), Italian (20), Japanese (21), Spanish (22), and Turkish (23)). The DEBQ-E has repeatedly been shown to predict distress-induced food intake, with low emotional eaters showing the typical stress response of eating less and high emotional eaters showing the atypical stress response of eating similar amounts or more after stress (5, 24, 25). The DEBQ-E was also shown to measure a different construct than external eating (26, 27).

24.3 **Mechanisms**

24.3.1 **Hunger**

Emotional eating has been classified as eating in the absence of hunger because the typical response to negative emotions or stress is loss of appetite, and high emotional eaters are thought to eat despite their presumed loss of appetite. Support for the contention that loss of appetite is indeed the typical stress response was obtained in a study with the Trier Social Stress Test (TSST: 28, 29). Participants with extremely high or low scores on an emotional eating questionnaire were subjected to both a control condition and a stress condition on two consecutive days. On the control day, participants had to rate six different fabrics (wool, fur, felt, silk, linen, and cotton) on various attributes (e.g. softness, pleasantness, warmth). On the distress day, the participants were subjected to the TSST, a distress manipulation that involves public speaking before a jury, coupled with an arithmetic challenge. The TSST thereby induces social-evaluative (ego) threat and is generally associated with an activation of the HPA axis, including an increased secretion of cortisol (28). Van Strien and colleagues (29) examined the mean hunger values in the control and the stress condition before and at three time points after the stress versus the control manipulation. There was a significant moderator effect of the stress manipulation on the hunger values over time, with the mean hunger values being indeed significantly lower after the stress than after the control manipulation (see 29, Figure 2). A further interesting finding, which contradicts the common view that emotional eating is eating in the absence of hunger, was that high emotional eaters did not show the typical reduction of reports of hunger in the stress condition. There was a significant moderator effect of emotional eating on self-reported hunger over time in the stress condition, with high emotional eaters showing a trend of having higher hunger ratings in the stress condition than the low emotional eaters (see Figure 24.1). A similar finding was found by Gluck and colleagues (30) for patients with binge eating disorder (a condition that is closely related to emotional eating).

Interestingly, Raspopow and colleagues (31) found differences between high and low emotional eaters in their ghrelin responses following food intake, with ghrelin levels declining in the low emotional eaters but not in the high emotional eaters. Ghrelin is the so-called hunger hormone which regulates appetite and the distribution and rate of use of energy.

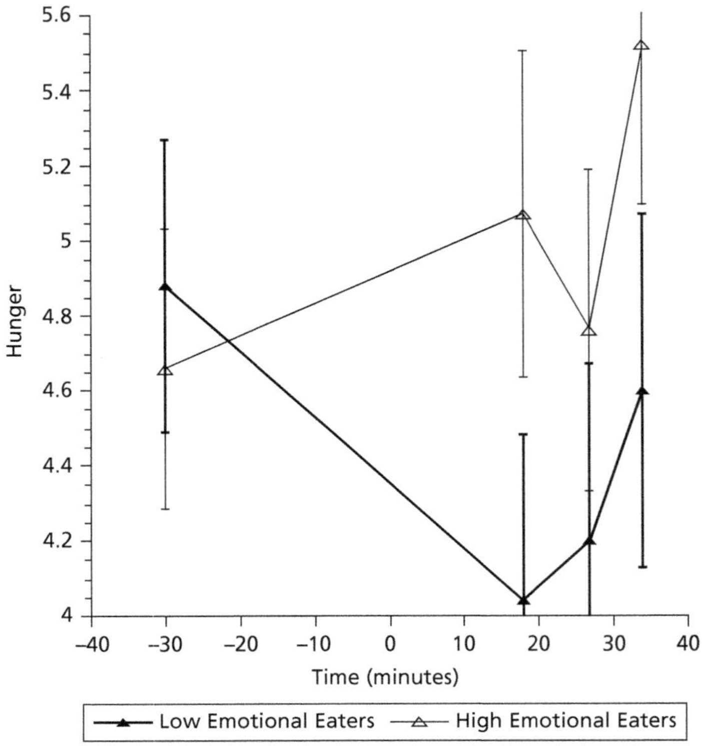

Figure 24.1 The moderator effect of emotional eating (low versus high) on the hunger values in the stress condition, showing that high emotional eaters had higher hunger ratings in the stress condition than low motional eaters (but only a borderline significant difference ($p = 0.06$) at T2 (18 min)).

Reprinted from *Appetite*, 79 (1), Tatjana van Strien, Machteld A Ouwens, Carmen Engel, and Carolina de Weerth, Hunger, inhibitory control and distress-induced emotional eating, Figure 3, p. 129, doi.org/10.1016/j.appet.2014.04.020

24.3.2 **Poor interoceptive awareness and alexithymia**

The atypical stress response of high emotional eaters of eating similar amounts or more after stress is, according to Bruch (32), the result of poor interoceptive awareness as an outcome of adverse early learning experiences in which there was insufficient regard for the real needs of a child. If the response of the caregiver is continuously inappropriate, be it neglectful or overly permissive, the outcome will be that the child is not sufficiently able to recognize whether it is hungry, or satiated, or suffering from some other discomfort. It may start to overeat in response to any arousal state: emotional overeating.

Poor interoceptive awareness which comprises both deficits in awareness and identification of both emotions, and hunger, and satiety (33), is closely associated with alexithymia (34, 35), a concept that specifically refers to difficulty in identifying feelings and describing feelings to other people.

Poor interoceptive awareness and alexithymia were both found to be positively associated with the DEBQ-E (36, 37). Further, in an experimental study of 86 women, alexithymia, as measured with the Toronto Alexithymia Scale (38), moderated the relationship between distress and food consumption in an ad lib taste task (39). Whereas the women with low alexithymia (good ability to recognize and describe emotions) showed the typical response of eating less after distress, the high alexithymic females ate the same or even a little more, thus showing the "biological unnatural" and atypical stress response. Distress was manipulated in this study by letting the participants believe that they had to give a speech in front of an evaluative audience versus that they had to touch a variety of fabrics. The former mood manipulation was found to be effective in inducing higher tension.

24.3.3 Blunted HPA axis

An entirely different explanation for the atypical stress responses of emotional eaters may be that high emotional eaters may not *have* the typical post-stress reduction of hunger, because chronic early-life stress caused their HPA-axis stress response to be reversed. Instead of responding to stress with hyperactivation and the typical neurovegetative symptoms of loss of appetite, hypophagia, and weight loss, these high emotional eaters may respond to stress with a hypoactivation of the HPA axis and the atypical neurovegetative symptoms of increased appetite, hyperphagia, and weight gain (2).

Adverse rearing experiences early in life were indeed found to have lasting effects on stress-responsive neurobiological systems, particularly when parent–infant relationships are involved (40, 41). This line of research was inspired by the famous observations by Spitz in small infants in understaffed orphanages (42), followed up by (controlled) studies on animals and humans (e.g. 43). Although high emotional eaters in some studies also responded to stress with *increased* cortisol responses (31, 44), Tomiyama and colleagues (45) found that women who experienced chronic stress (caregivers of chronically ill children) showed both higher scores on emotional eating and a dampened HPA-axis activity in comparison of women who did not experience chronic stress (caregivers of healthy children). This resulted in the hypothesis by Van Strien and colleagues (46) that high emotional eaters with a blunted HPA-axis stress reactivity (blunted cortisol stress responses) would show the highest food intake after stress.

This hypothesis was indeed supported. High emotional eaters with a blunted cortisol stress response showed a higher food intake after the stressor than those with the typical elevated cortisol stress response. In this study, participants with extremely high or low scores on the DEBQ-E were subjected to both a control condition and the stress condition (TSST) on two consecutive days. The distress manipulation was effective in inducing more negative mood in the distress condition. Saliva samples for cortisol determination were on both days collected at five fixed time points. On both days, the participants were invited to help themselves to water and to food (four bowls filled with white grapes, pieces of carrot, M&Ms, and pieces of buttercake) after the mood manipulation (further details of the procedure can be found in 46).

The finding that high emotional eaters with a *blunted* cortisol response ate more food after distress than those with an elevated cortisol stress response (see 47, for similar findings), suggests that emotional overeating is indeed associated with a lowered HPA-axis functioning.

It is, however, as yet unclear whether this blunted cortisol response is secondary (i.e. the result of an adaptive down regulation) or primary to emotional eating (see 46). According to the chronic stress-response network model developed by Dallman (48), highly stressed people may have coped with high levels of stress by engaging in stress eating, thereby developing blunted HPA-axis responses as a reflection of an adaptive downregulation secondary to emotional overeating. Alternatively, a blunted cortisol response may also have been primary, instead of secondary, to emotional eating. A blunted cortisol response has been related to adverse early life experiences (49) and could point at a failure to recover from stress. In this line of thought, emotional eating would be the consequence of a lowered HPA-axis functioning, which would explain why emotional eaters are more receptive to the reinforcing value of food and use food as "self-medication" to blunt effects of negative emotions.

24.3.4 Blunted HPA axis and alexithymia

Studies on the association between HPA-axis reactivity and alexithymia show conflicting results (50, 51). In the study by Hua and colleagues (50), with the TSST as stressor, the alexithymia aspect of difficulty in identifying feelings was associated with an elevated cortisol stress response. Härtwig and colleagues (51), by contrast, found evidence for alexithymia being associated with blunted cortisol responses. Hua and co-workers (50) explain this discrepancy in results by the age difference of the participants in the two studies. The older adults in the study by Härtwig and colleagues (51) may have experienced more cumulative life stress than the college students in the study by Hua and co-workers (50). Alexithymia is associated with chronic stress (52), and chronic stress, in turn, may ultimately result in hypoactivity of the HPA axis (53).

24.4 Emergence of emotional eating in adolescence

Emotional eating is highly prevalent in adults, particularly in (female) adults who are overweight or obese (54, 55). In children, in contrast, the prevalence of emotional eating is rather low (56, 57). This suggests that most young children show the natural biological distress response of losing appetite and that emotional eating emerges in the transition between childhood and adulthood. the period of adolescence. It has been suggested that the biologically atypical response of emotional eating is acquired as a result of inadequate parenting or adverse experiences early in life (32, 58), although emotional eating may not become manifest until late adolescence or adulthood.

In two prospective studies in Dutch families with two adolescent children, the interaction between gene and environment (G×E) on the emergence of emotional eating in adolescence was addressed (59, 60).

24.4.1 Parenting × dopamine D2 receptor gene (*DRD2*)

In the prospective study by Van Strien and colleagues (59), the emergence of emotional eating was assessed in relation to the inadequate parenting practice of psychological control, and the dopaminergic (reward sensitivity) brain system, namely polymorphisms in the dopamine D2 receptor gene (*DRD2*). Parental psychological control (e.g. "My father (mother) makes me feel guilty when I fail at school") can be seen as an aversive rearing experience, because it is a manipulative and suppressive type of parental control which includes guilt induction and the withdrawal of love, thereby intruding into the psychological and emotional development of the child (61). In an earlier study on young adolescents, both maternal and paternal psychological control were found to be associated with emotional eating, although the effects were small (58). This indicates that not all people who experienced parental psychological control develop emotional eating, suggesting a possible role for genetics. In a subsequent prospective study with a four-year follow-up, it was tested whether emotional eating develops as a result of an interaction between parental psychological control and polymorphisms in the dopamine D2 receptor gene (*DRD2*). Carrying the A1-allele of the *DRD2* gene Taq1A polymorphism (rs1800497) is associated with reduced *DRD2* receptor availability in the brain (62). It was hypothesized that the relationship between psychological control and emotional eating after four years would be stronger for adolescents who carried at least one A1-allele. *DRD2* genotype was indeed found to moderate the relationship between both maternal and paternal psychological control and increases in emotional eating in both sexes. Adolescents showed an increase in emotional eating in relation to high psychological control only if they carried at least one *DRD2* A1-allele (see Figure 24.2).

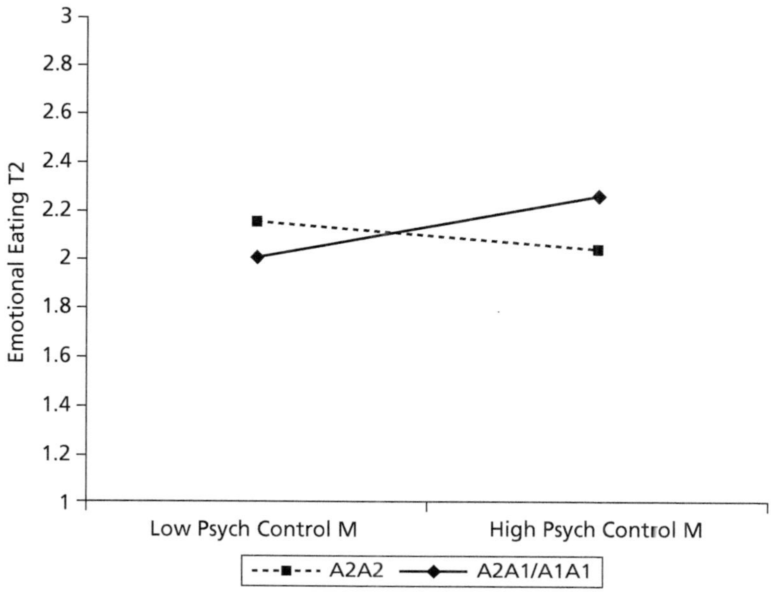

Figure 24.2 Two-way interaction between DRD2 genotype (A1A2/A1A1 versus A2A2) and maternal psychological control on emotional eating at T2. The group A1A//A1A1 is the group at risk because the possession of at least one A1-allele is associated with reduced D2 receptor availability in the brain.

Reprinted from *Appetite*, 54 (2), Tatjana van Strien, Harriëtte M Snoek, Carmen S van der Zwaluw, and Rutger C.M.E. Engels, Parental control and the dopamine D2 receptor gene (DRD2) interaction on emotional eating in adolescence, Figure 1, p. 258, doi.org/10.1016/j.appet.2009.11.006,

24.4.2 **Depressive feelings × serotonin transporter (5-HTT) gene (*SLC6A4*)**

The prospective study by van Strien and colleagues (60) tested whether emotional eating develops as a result of an interaction between depressive feelings and a serotonergic brain system, namely polymorphisms in the serotonin transporter (5-HTT) gene (*SLC6A4*). The short allele of the *5-HTTLPR* polymorphism in the serotonin transporter gene is associated with lower serotonin activity (63). Low levels of serotonin activity have been associated with a more reactive arousal system, including increased appetite and body weight (64). It was hypothesized that adolescents with depressive feelings would show higher increase in emotional eating if they carried the short allele of the *5-HTTLPR* polymorphism in the serotonin transporter gene. We further expected this to hold true particularly for females because of the higher prevalence of depression and emotional eating in women (65). The serotonin transporter gene was indeed found to moderate the relation between depressive feelings and the increase in emotional eating in both sexes, in the youngest siblings (13 years at baseline (n = 286) (Figure 24.3). In the older siblings (15 years at baseline; (n = 298)), this specific moderation effect was only found in girls. Younger adolescents and older adolescent girls showed a higher increase in emotional eating when they carried the *5-HTTLPR* short allele.

24.4.3 **Plasticity genes**

It should be noted that the two studies by Van Strien and colleagues (59, 60) did not show main effects of the polymorphisms. Only in interaction with an environmental factor such as parental

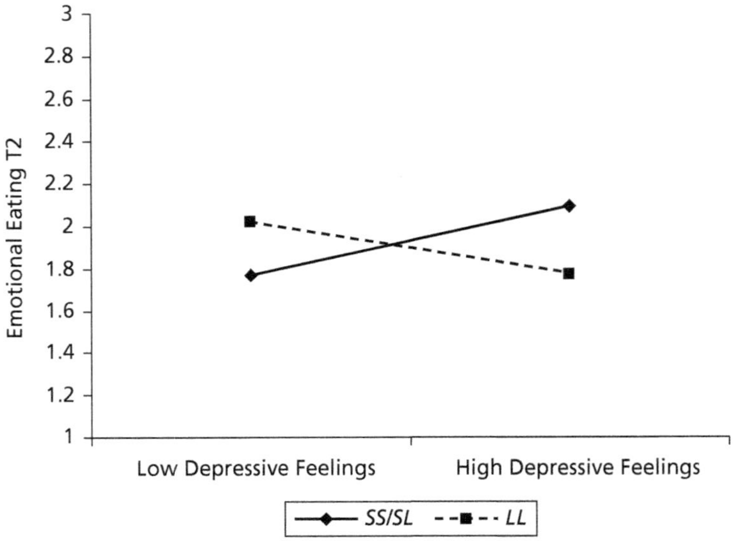

Figure 24.3 Youngest siblings. 2-way interaction between 5-HTTLPR genotype (SS/SL vs LL) and depressive feelings on increase in emotional eating at T2. Individuals who carried at least one copy of the s allele (SS/SL) are the individuals at risk.

Reprinted from *Journal of Psychiatric Research*, 44 (15), Tatjana van Strien, Carmen S van der Zwaluw, and Rutger CME Engels, Emotional eating in adolescents: A gene (SLC6A4/5-HTT)—Depressive feelings interaction analysis, Figure 1, p. 1039, doi.org/10.1016/j.jpsychires.2010.03.012,

control or with a negative mood such as depressive feelings did the polymorphisms predict future increases in emotional eating. It has been suggested that both gene polymorphisms function as "plasticity" genes, rather than "vulnerability" genes (e.g. 66), meaning that these genes are associated with increased sensitivity to the environment.

Reinspection of the results of the two prospective studies by van Strien and colleagues (59, 60) revealed support for Belsky and co-workers' (66) framework of "plasticity genes" (67, p. 52). Adolescents with at least one risk gene (at least one *DRD2* A1-allele) who grew up in the more favorable rearing condition of having parents who did not use psychological control showed the lowest increase of emotional eating after four years and seemed to fare even better than adolescents carrying the A2A2 genotype. The same held true for young adolescents (13 years) who combined carrying at least one *5-HTTLPR* short allele with low depressive feelings in Van Strien and colleagues' study (60). Similar results were obtained in the older adolescent girls (15 years). In contrast, different results were obtained in the older adolescent boys. The association of depressive feelings with emotional eating after four years was even negative in boys with at least one *5-HTTLPR* short allele (see 60, Figure 2).

24.4.4 Sex differences and pubertal hormones

The sex difference in the interaction between *5-HTTLPR*, depressive feelings, and increase in emotional eating in the study by Van Strien and colleagues (60) is consistent with the conclusions of a systematic review (68). A general outcome of this review was that sex differences in the association of *5-HTTLPR* with affective disorders started with adolescence but disappeared among the elderly. This would suggest a possible role of hormonal fluctuation during puberty. In support, a review of human and animal studies on puberty as a critical risk period for eating disorders concluded that part of the larger role of puberty in girls is to be attributed to genetic factors associated with estrogen activation at puberty (69). During puberty, estrogen may activate genetic risk for emotional eating by regulating gene transcription (turning genes on and off); for example, the transcription of genes related to serotonin and dopamine (see further 69, 70, p. 905).

24.5 Depression, obesity, and therapy

Depression and obesity are common conditions with severe medical consequences and high costs for society. A meta-analysis by Luppino and colleagues (71) showed that depression predicted later weight gain and obesity. Feeling depressed is normally associated with loss of appetite and subsequent weight loss. There exists, however, a subtype of depression that is characterized by the atypical features of increased appetite and subsequent weight gain (72). Emotional eating has been considered a marker of this atypical depression subtype (60) because it shares with this depression subtype the atypical feature of increased appetite in response to distress such as feelings of depression.

24.5.1 Emotional eating as mediator

In various cross-sectional studies, emotional eating acted as a mediator between depression and obesity (73–75). In a prospective study on the parents of the adolescents in the Van Strien and co-workers' studies (60), emotional eating was found to act as a mediator between maternal depression and weight gain after five years. Depressive symptoms were related to higher emotional eating and emotional eating predicted greater increases in BMI independently of depression (76). No causal chain between depression, emotional eating, and weight gain was found in the fathers. This sex difference in the mediation effect of emotional eating is consistent with the female preponderance in atypical depression and emotional eating (60).

A further study by van Strien and colleagues (77) assessed whether emotional eating also acted as mediator between depression and body mass index in two European countries with a different prevalence of depression, namely Denmark (a high prevalence of depression) and Spain (a low prevalence of depression). In both countries, there was a mediation effect of emotional eating. In Denmark, this mediation effect was stronger for participants with increased appetite and for females than for participants with decreases/no change in appetite and for males. This supports depression with atypical features as possible explanation for the mediation effect of emotional eating between depression and BMI. In Spain, there was no support for depression with atypical features as explanation of the mediation effect of emotional eating, because the strength of the mediation effect of emotional eating was neither influenced by change in appetite, nor by gender. Instead, post-hoc analyses suggested that the strength of the chain was influenced by "stress of unemployment," with stronger mediation effects of emotional eating between depression and BMI for unemployed than for employed people.

24.5.2 Northern latitude hypotheses

It should be noted that most studies that found mediation effects for emotional eating have been carried out in countries that are located in northern latitudes, where atypical depression is more common. An exception is a study in females by Clum and colleagues (73), which was conducted one year after a stressful life event, namely Hurricane Katrina in the Greater New Orleans area (30°00′ northern latitude). Also, Denmark has a high northern latitude (56°00′), whereas mass unemployment was an ego-threatening stressor in Spain (26°00′–44°00′ northern latitude). It would be of interest to assess in future studies whether the finding by Van Strien and colleagues (77) that depression with atypical features is a possible underlying mechanism in the mediation effect of emotional eating is specific for countries with a higher northern latitude.

24.5.3 Therapy

The finding that emotional eating acted as mediator between depression and weight gain and between depression and body mass index is also of interest for obesity interventions and interventions on atypical depression. A reduction of emotional eating may be a target to prevent future weight gain, for example by teaching high emotional eaters emotion regulation skills. A pilot study of a group dialectical behavior therapy (DBT) adapted for obese emotional eaters demonstrated DBT to be a highly acceptable and effective intervention for reducing eating related pathology (e.g. emotional eating and also depression) at both the end of treatment and at the six months follow-up (78). Mindfulness training may be a further option for treatment. A brief mindfulness training was found to enhance the cortisol response to a stressor (79), and to change the cortisol awakening response and to reduce abdominal fat (80).

24.6 Future perspectives and conclusions

In future research, possible differences between the sexes and the two depression subtypes should be taken into account. An important research question for the future would be, for example, to disentangle whether the blunted HPA-axis response in the study by Van Strien and colleagues (46) is primary, or rather secondary to emotional overeating, as a result of adaptive downregulation: the chronic stress response network by Dallman (48). The same holds true for the question whether emotional eating reflects a deficit of reward (a primary low baseline dopamine D2 receptor availability in the brain (81) or a heightened sensitivity of reward, resulting in an adaptive downregulation of the dopaminergic system (88).

It would also be of interest to find out whether the mediation effect of emotional eating between depression and weight gain (76) is moderated by 5-HTTLPR genotype or by cortisol stress reactivity. In the study by Van Strien and colleagues (76), 5-HTTLPR did not moderate the mediation effect of emotional eating between depressive feelings and future weight gain, perhaps because larger samples are needed to find such moderated mediation effects (82). Further, although there has been evidence that cortisol acts as a mediator between depression and obesity, the results were contradictory. A distinction between the two depression subtypes (with, i.e. emotional eating) could in the future clarify this inconsistency (83).

We also do not yet have conclusive evidence that eating comfort foods (sweet, fatty, and energy dense) helps emotional eaters to feel better. Results of a functional magnetic resonance imaging study by Bohon and colleagues (84) on consumatory food reward during negative mood in high versus low emotional eaters suggest a negative answer. This result is in line with studies showing that the improvement in mood is at best only short lived (85), and may even become worse after some time (86). However, more research in this respect is needed. See Gibson (89) for further suggestions for future research on the psychobiology of comfort eating.

In conclusion, possible mechanisms of distress-induced emotional eating are an atypical hunger stress response, poor interoceptive awareness, high alexithymia, and a blunted HPA-axis stress response (as indicated by a blunted cortisol stress response). Inadequate parenting and a high degree of depressive feelings may, in interaction with a genetic vulnerability, be associated with increases in emotional eating during adolescence. Emotional eating may also act as a mediator between depression and body mass index or weight gain. This finding is of interest for both obesity interventions and interventions for atypical depression.

References

1. **Greeno CG, Wing RR.** Stress-induced eating. Psychol Bull. 1994;**115**:444–64.
2. **Gold PW, Chrousos GP.** Organization of the stress system and its deregulation in melancholic and atypical depression: High vs low CRH/NE states. Mol Pyschiatry. 2002;**7**:254–75.
3. **Carlson AJ.** The Control of Hunger in Health and Disease. Chicago, IL: University of Chicago Press; 1916.
4. **Cannon WB.** Bodily Changes in Pain, Hunger, Fear and Rage, 2nd edn. New York, NY: Appleton; 1915.
5. **Oliver G, Wardle J, Gibson EL.** Stress and food choice. A laboratory study. Psychosom Med. 2000;**62**:853–65.
6. **Van Strien T, Ouwens MA.** Counteregulation in female obese emotional eaters. Schachter, Goldman and Gordon's (1968) test of psychosomatic theory revisited. Eating Behav. 2003;**3**:329–40.
7. **Geliebter A, Aversa A.** Emotional eating in overweight, normal weight, and underweight individuals. Eating Behav. 2004;**3**:341–7.
8. **Karlsson J, Persson LO, Sjöström L, Sullivan M.** Psychometric properties and factor structure of the Three-Factor Eating Questionnaire (TFEQ) in obese men and women. Results from the Swedish Obese Subjects (SOS) study. Int J Obes. 2000;**24**:1715–25.
9. **Tholin S, Rasmussen F, Tynelius P, Karlsson J.** Genetic and environmental influences on eating behavior: The Swedish Young Male Twins Study. Am J Clin Nutr. 2005;**81**:564–69.
10. **Van Strien T, Frijters JER, Bergers GPA, Defares PB.** The Dutch Eating Behavior Questionnaire (DEBQ) for assessment of restrained, emotional and external eating behavior. Int J Eat Disord. 1986;**5**:295–315.
11. **Nolan LJ, Halparin LB, Geliebter A.** Emotional appetite questionnaire. Construct validity and relationship with BMI. Appetite. 2010;**54**:314–19.
12. **Van Strien T, Cebolla A, Etchemendy E, Gutiërez-maldonado J, Ferrer-Garcia M, Botella C,** et al. Emotional eating and food intake after sadness and joy. Appetite. 2013;**66**:20–25.
13. **Van Strien T, Donker MH, Ouwens MA.** Is desire to eat in response to positive emotions an "obese" eating style: Is Kummerspeck for some people a misnomer? Appetite. 2016:**100**:225–35.
14. **COTAN.** Beoordeling Nederlandse Vragenlijst voor Eetgedrag, NVE, 2013 [Review Dutch Eating Behaviour Questionnaire (DEBQ, 2013)]. Available at: <http://www.cotandocumentatie.nl/test_details.php?id=848>.
15. **Barrada JR, Van Strien T, Cebolla A.** Internal structure and measurement invariance of the Dutch Eating Behavior Questionnaire (DEBQ) in a (nearly) representative Dutch community sample. Eur Eat Disord Rev. 2016:**24**(6):503–9.
16. **Wu S, Cai T, Luo X.** Validation of the Dutch Eating Behavior Questionnaire (DEBQ) in a sample of Chinese adolescents. Psychol Health Med. 2016. doi:10.1080/13548506.2016.1173712.
17. **Wardle J.** Eating style: A validation study of the Dutch Eating Behaviour Questionnaire in normal subjects and women with eating disorders. J Psychosom Res. 1987;**31**:161–9.
18. **Lluch A, Kahn JP, Stricker-Krongrad A, Ziegler O, Drouin P, Méjean L.** Internal validation of a French version of the Dutch Eating Behaviour Questionnaire. Eur Psychiatry. 1996;**11**:198–203.
19. **Nagi M, Hilbert A, de Zwaan M, Braehler E, Kersting A.** The German version of the Dutch Eating Behavior Questionnaire. Psychometric properties, measurement invariance and population-based norms. PloS One. 2016:**11**(9):e0162510.
20. **Dakanalis A, Zanetti MA, Clerici M, Madeddu F, Riva G, Caccialanza R.** Italian version of the Dutch Eating Behavior Questionnaire. Psychometric proprieties and measurement invariance across sex, BMI-status and age. Appetite. 2013;**71**:187–95.
21. **Imada S.** Psychological studies about eating behavior (3): A Japanese version of the Dutch Eating Behavior Questionnaire (DEBQ). Studies in the Humanities & Sciences. 1994;**34**:281–91.
22. **Cebolla A, Barrada JR, van Strien T, Oliver E, Baños R.** Validation of the Dutch Eating Behavior Questionnaire (DEBQ) in a sample of Spanish women. Appetite. 2014;**73**:58–64.

23. **Bozan N, Bas M, Asci FH.** Psychometric properties of the Turkish version of Dutch Eating Behaviour Questionnaire (DEBQ). Preliminary results. Appetite. 2011;**56**:564–6.
24. **Van Strien T, Herman CP, Anschutz D, Engels RCME, de Weerth C.** Moderation of distress-induced eating by emotional eating scores. Appetite. 2012;**58**:277–84.
25. **Wallis DJ, Hethrington MM.** Stress and eating. The effects of ego-threat and cognitive demand on food intake in restrained and emotional eaters. Appetite. 2004;**43**:39–46.
26. **Ouwens MA, Van Strien T, Van Leeuwe JFJ.** Possible pathways between depression, emotional and external eating: A structural equation model. Appetite. 2009;**53**:245–8.
27. **Van Strien T, Herman CP, Anschutz DJ.** The predictive validity of the DEBQ- external eating scale for eating in response to food commercials while watching television. Int J Eat Disord. 2012;**45**:257–62.
28. **Kirschbaum C, Pirke KM, Hellhammer DH.** The Trier Social Stress Test: A tool for investigating psychosocial stress responses in a laboratory setting. Neuropsychobiology. 1993;**28**:76–81.
29. **Van Strien T, Ouwens MA, Engel C, de Weerth C.** Hunger, inhibitory control and distress-induced emotional eating. Appetite. 2014;**79**:124–33.
30. **Gluck ME, Geliebter A, Hung J, Yahav E.** Cortisol, hunger, and desire to binge eat following a cold stress test in women with binge eating disorder. Psychosom Med. 2004;**66**:876–81.
31. **Raspopow K, Abizaid A, Mathesen K, Anisman H.** Psychological stressor effects on cortisol and ghlerin in emotional and non-emotional eaters: Influence on anger and shame. Horm Behav. 2010;**58**:677–84.
32. **Bruch H.** Hunger Awareness and Individuation. Eating Disorders, Obesity, Anorexia Nervosa and the Person Within. New York, NY. Basic Books, Inc.
33. **Koch A, Pollatos O.** Interoceptive sensitivity, body weight and eating behavior in children: A prospective study. Front Psychol. 2014;**5**:1–11. doi:10.3389/fpsyg.2014.01003s.
34. **Garner DM.** Eating Disorder Inventory: 2 Manual. Odessa, FL: Psychological Assessment Resources; 1991.
35. **Taylor GJ, Parker JDA, Bagby RM, Bourke MP.** Relationships between alexithymia and psychological characteristics associated with eating disorders. Journal of Psychosom Res. 1996;**412**:561–8.
36. **Larsen JK, Van Strien T, Eisenga R, Engels RCME.** Gender differences in the association between alexithymia and emotional eating in obese individuals. J Psychosom Res. 2006;**60**:237–43.
37. **Van Strien T, Engels RCME, Van Leeuwe J, Snoek HM.** The Stice model of overeating: Tests in clinical and non-clinical samples. Appetite. 2005;**45**:205–13.
38. **Bagby R, Parker JDA, Taylor GJ.** The twenty-item Toronto Alexithymia Scale: I Item selection and cross-validation of the factor structure. J Psychosom Res. 1994;**38**:23–32.
39. **Van Strien T, Ouwens MA.** Effects of distress, alexithymia and impulsivity on eating. Eating Behav. 2007;**8**:251–7.
40. **Cicchetti D, Rogisch FA.** Diverse patterns of neuroendocrine activity in maltreated children. Dev Psychopathol. 2001;**13**:677–93.
41. **Yehuda R, Bierer LM, Schmeidler J, Aferiat DH, Breslau I, Dolan S.** Low cortisol and risk for PTSD in adult offspring of holocaust survivors. Am J Psychiatry. 2000;**157**:1252–9.
42. **Emde RN, Polak PR, Spitz RA.** Anaclitic depression in an infant raised in an institution. J Am Acad Child Adolesc Psychiatry. 1965;**4**:545–53.
43. **Barr CS, Newman TK, Schwandt M, Coutney C, Dvoskin RI, Lindell SG,** et al. (2004). Sexual dichotomy of an interaction between early adversity and the serotonin transporter gene promoter variant in rhesus macaques. PNAS. 2004;12358–63.
44. **Epel E, Jimenez S, Brownell K, Stroud L, Stoney C, Niaura R.** Are stress eaters at risk for the metabolic syndrome? Ann N Y Acad Sci. 2004;**1032**:208–10.
45. **Tomiyama AJ, Dallman MF, Epel ES.** Comfort food is comforting to those most stressed: Evidence of the chronic stress response network in high stress women. Psychoneuroendocrinology. 2011;**36**:1513–19.

46. **Van Strien T, Roelofs K, de Weerth C.** Cortisol reactivity and distress-induced emotional eating. Psychoneuroendocrinology. 2013;**38**(5): 677–84.
47. **Tryon MS, DeCant R, Laugero KD.** Having your cake and eating it too: A habit of comfort food may link chronic social stress exposure and acute stress-induced cortisol hyporesponsiveness. Physiol Behav. 2013 Apr 10;114–15:32–7. doi: 10.1016/j.physbeh.2013.02.018.
48. **Dallman MF.** Stress induced obesity and the emotional nervous system. Trends Endocrin Metab. 2010;**21**:159–65.
49. **Elzinga BM, Roelofs K, Tollenaar MS, Bakvis P, Van Pelt J, Spinhoven P.** Diminished cortisol responses to psychosocial stress associated with life time events. A study among healthy young subjects. Psychoneuroendocrinology. 2008;**33**:227–37.
50. **Hua J, Le Scanff C, Larue J, José F, Martin J, Devillers L**, et al. Global stress response during a social stress test: Impact of alexithymia and its subfactors. Psychoneuroendocrinology. 2014;**50**:53–61.
51. **Härtwig EA, Aust S, Heuser I.** HPA system activity in alexithymia: A cortisol awakening response study. Psychoneuroendocrinology. 2013;**38**:2121–6.
52. **Aust S, Härtwig EA, Heuser I, Bajbouj M.** The role of early emotional neglect in alexithymia. Psychol Trauma US. 2013;**5**(3):225–32.
53. **Fries E, Hesse J, Hellhammer J, Hellhammer DH.** A new view on hypocortisoism. Psychoneuroendocrinology. 2005;**30**:1010–16.
54. **Van Strien T, Herman CP, Verheijden MW.** Eating style, overeating and overweight in a representative Dutch sample. Does external eating play a role? Appetite. 2009;**52**:380–7.
55. **Van Strien T, Herman CP, Verheijden MW.** Eating style, overeating and weight gain. A prospective 2-year follow-up study in a representative Dutch sample. Appetite. 2012;**59**(3):782–9. doi:10.1016/j.appet.2012.08.009.
56. **Sleddens EF, Kremers SP, Thijs C.** The Children's Eating Behaviour questionnaire. Factorial validity and association with Body Mass Index in Dutch children aged 6–7. Int J Behav Nutr Phys Act. 2008;**5**:49. <http://dx.doi.org/10.1186/1479-5868-5-49>.
57. **Van Strien T, Oosterveld P.** The Children's DEBQ (DEBQ-C) for assessment of restrained, emotional and external eating in 7–12 year old children. Int J Eat Disord. 2008;**41**:72–81.
58. **Snoek HM, Engels RC, Janssens JM, van Strien T.** Parental behaviour and adolescents emotional eating. Appetite. 2007;**49**:223–30.
59. **Van Strien T, Snoek HM, van der Zwaluw CS, Engels RCME.** Parental control and the dopamine D2 receptor gene (*DRD2*) interaction on emotional eating in adolescence. Appetite. 2010a;**54**:255–61.
60. **Van Strien T, van der Zwaluw CS, Engels, RCME.** Emotional eating in adolescents: A gene (SLC6A4/5-HTT)-depressive feelings interaction analysis. J Psychiatr Res. 2010b;**44**:1035–42.
61. **Barber BK.** Parental psychological control: Revisiting a neglected construct. Child Develop. 1996;**67**:3296–319.
62. **Gluskin BS, Mickey BJ.** Genetic variation and dopamine D2 receptor availability: A systematic review and meta-analysis of human in vivo molecular imagining studies. Transl Psychiatry. 2016;**6**:e747; doi:10.1038/tp.2016.22.
63. **Hariri A, Holmes A.** Genetics of emotional regulation: The role of the serotonin transporter in neural function. Trends Cogn Sci. 2005;**10**:182–91.
64. **Leibovitz SF, Alexander JT.** Hypothalamic serotonin in control of eating behavior, meal size and body weight. Biol Psychiatry. 1998;**44**:851–64.
65. **Cyranowski JM, Frank E, Young E, Shear MK.** Adolescent onset of the gender difference in lifetime rates of major depression. Arch Gen Psychiatry. 2000;**57**:21–7.
66. **Belsky J, Jonassaint C, Pluess M, Stanton M, Brummett B, Williams R.** Vulnerability genes or plasticity genes. Mol Psychiatry. 2009;**14**:746–54.
67. **Van Strien T, Levitan RD, Engels RCME, Homberg JR.** Season of birth, the dopamine D4 receptor gene (DRD4) and emotional eating in males and females: Evidence of a genetic plasticity factor? Appetite. 2015;**90**:51–7. doi:10.1016/j.appet.2015.02.024.

68. **Gressier F, Calati R, Serretti A.** 5-HTTLPR and gender differences in affective disorders: A systematic review. J Affect Disord. 2016;**190**:193–207.
69. **Klump KL.** Puberty as a critical risk period for eating disorders: A review of human and animal studies. Horm Behav. 2013;**64**:399–410.
70. **Klump KL, O'Connor SM, Hildebrandt BA, Keel PK, Neale M, Sisk CL**, et al. Differential effects of estrogen and progesterone on genetic and environmental risk for emotional eating in women. Clin Psychol Sci. 2016;**4**:895–908.
71. **Luppino FS, de Wit LM, Bouvy PF, Stijnen T, Cuijpers P, Peninx B**, et al. Overweight, obesity and depression. A systematic review and meta-analysis of longitudinal studies. Arch Gen Psychiatry. 2010;**67**:220–9s.
72. **American Psychiatric Association,** (2013). Diagnostic and Statistical Manual of Mental Disorders, 5th edn. Washington, DC: American Psychiatric Publication; 2013.
73. **Clum GA, Rice JC, Broussard M, Johnson CC, Webber LS.** Associations between depressive symptoms, self-efficacy, eating styles, exercise and body mass index in women. J Behav Med. 2014;**37**(4):577–86.
74. **Goldschmidt AB, Crosby RD, Engel SG, Crow SJ, Cao L, Peterson CB**, et al. Affect and eating behavior in obese adults with and without elevated depression symptoms. Int J Eat Disord. 2014;**47**(3):281–6.
75. **Konttinen H, Männistö S, Sarlio-Lähteenkorva S, Silventoinen K, Haukkala A.** Emotional eating, depressive symptoms and self-reported food consumption. A population-based study. Appetite. 2010;**54**(3):473–9. doi:<http://dx.doi.org/10.1016/j.appet.2010.01.014>.
76. **Van Strien T, Konttinen H, Homberg JR, Engels RCME, Winkens LHH.** Emotional eating as a mediator between depression and weight gain. Appetite. 2016;**100**:216–24.
77. **Van Strien T, Winkens LHH, Broman Toft M, Pedersen S, Brouwer IA, Visser M**, et al. The mediation effect of emotional eating between depression and body mass index in Denmark and Spain. Appetite. 2016;**105**:500–8.
78. **Roosen MA, Safer DL, Adler S, Cebolla A, Van Strien T.** Group dialectical therapy adapted for obese emotional eaters: A pilot study. Nutrición Hospitalaria. 2012;**27**:1125–31.
79. **Creswell JD, Pacilio LE, Lindsay EK, Brown KW.** Brief mindfulness meditation training alters psychological and neuroendocrine responses to social evaluative stress. Psychoneuroendocrinology. 2014;**44**:1–12.
80. **Daubenmier J, Kristeller J, Hecht FM, Maninger N, Kuwata M, Jhaveri K**, et al. Mindfulness intervention for stress eating to reduce cortisol and abdominal fat among overweight and obese women: An exploratory randomized controlled study. J Obes. 2011. (Article ID 651936).
81. **Volkow ND, Wang GJ, Maynard L, Jayne M, Fowler JS, Zhu W.** Brain dopamine is associated with eating behaviours in humans. Int J Eat Disord. 2003;**33**(2):136–42.
82. **Rozenblat V, Ong D, Fuller-Tyszkiewicz M, Akkermann K, Collier D, Engels R**, et al. A meta-analysis of gene (5-HTT) x environment interactions in eating pathology using secondary data analyses. J Psychiatric Res. 2017;**84**:62–72.
83. **Pervanidou P, Chrousos GP.** Neuroendocrinology of post-traumatic stress disorder. Prog Brain Res. 2010;**182**:149–60.
84. **Bohon C, Stice E, Spoor S.** Female emotional eaters show abnormalities in consummatory and anticipatory food reward: A functional magnetic resonance imaging study. Int J Eat Disord. 2009;**42**:210–21.
85. **Macht M, Mueller J.** Immediate effects of chocolate on experimentally induced mood states. Appetite. 2007;**49**:667–74.
86. **Haedt-Matt A, Keel PK, Racine SE, Burt SA, Hu JY, Boker S**, et al. Do emotional eating urges regulate affect? Concurrent and prospective associations and implications for risk models of binge eating. Int J Eat Disord. 2014;**47**:847–77.

87. **Arnow B, Kenardy J, Agras WS.** The Emotional Eating Scale: The development of a measure to assess coping with negative affect by eating. Int J Eat Disord. 1995;**18** (1):79–90.
88. **Davis C, Strachan S, Berkson M.** Sensitivity to reward: Implications for overeating and overweight. Appetite. 2004;**42**(2):131–8.
89. **Gibson EL.** The psychobiology of comfort eating: Implications for neuropharmacological interventions. Behav Pharmacol. 2012;**23**:442–60.

Chapter 25

Genetics of obsessive–compulsive disorder and Tourette's syndrome

Nuno R Zilhão, Dorret I Boomsma, Dirk JA Smit, and Danielle C Cath

25.1 Introduction

The first scientific descriptions of obsessive–compulsive disorder (OCD) and Tourette Syndrome (TS) as psychiatric disorders stem from medical research in the nineteenth century, starting with the works of Jean Itard (1825) and George Gilles de la Tourette (1885) (on tic disorders), and of Pierre Janet (1900) (on OCD). Conceptualization of the two disorders over time reflected the evolving views regarding their underlying causes and mechanisms. The Freudian psychogenic view prevailed for many decades, explaining both TS/tics and OCD symptoms as manifestations of inner psychological or psychosexual conflicts that result from environmental adversities (predominantly stemming from early-life traumatic events and dysfunctional upbringing). From the second half of the twentieth century onwards, advances in pharmacotherapeutics and neurobiology (1–3) questioned the psychoanalytical explanations of tics and OC symptoms, and provided new findings into the neural origin of these diseases.

Developments within the field of twin and family studies, following modern applications of structural equation modeling (4), have advanced the field of psychiatric genetic research. Both family-based and epidemiological twin studies have sought to estimate the contribution of genetic and environmental influences to variation in the phenotype, often specified as the heritability of the liability to develop the disease. Collaborative psychiatric genetic research further contributed to insight into the genetic basis of both OCD and TS, as well as their shared underlying etiology (5–8). This rapidly growing field now also aims to find the genetic DNA variants that help to identify pathways of disease development. The success of these efforts will aid in understanding the biological mechanisms underlying the disorders. Here, we provide an overview of the genetic studies of OCD and TS, and of the genetic overlap between these disorders.

25.2 The symptom characteristics of OCD and TS

25.2.1 Clinical definition and symptomatology

The current definition of OCD was introduced in the 1980s in the Diagnostic and Statistical Manual of Mental Disorders, 3rd edition (9). Until then, OCD was commonly referred to as "obsessive–compulsive neurosis," or psychasthenia. OCD had been considered a sub-dimension of depression (10). According to DSM-IV (11), OCD was classified as an anxiety disorder, characterized by persistent, repetitive, and intrusive thoughts/images (obsessions) that cause anxiety and/or tension and are in > 80% of the cases followed by anxiety/tension-reducing actions and ritualized behaviors (compulsions). Relief from carrying out compulsions is usually brief and the repetitive behavior by and in itself fuels subsequent repetitive behavior (12), with the person getting caught in an ongoing cycle of fear, doubt, worry, and distress, leading to high levels of social impairment. Global burden of disease (GBD) estimates have found that OCD accounted for 2.2% of all years lost to disability (YLD), (13), placing it in the eleventh position of causes of non-fatal disease burden worldwide (14). Whereas in DSM-IV OCD was classified among the anxiety disorders, the new edition (DSM-5) includes OCD in its own category ("OCD and related disorders") alongside hoarding, excoriating disorder, body dysmorphic disorder, and trichotillomania (15).

The course of the disease is heterogeneous: around 50% of cases remit with time, and 50% of persons with OCD run a chronic course (16). Recent studies from the Netherlands Twin Register showed that while OCD diagnosis as well as symptom severity on average decreases from age 35 onwards, it shows an increase after age 60 that is predominantly driven by checking behavior (17). In all, OCD is considered as a lifelong disorder, which varies with regard to symptomatology. The most frequently reported obsessions are fear of contamination, a need for order and symmetry, and persistent and unwanted aggressive thoughts of causing harm to oneself or others (15). Compulsions appear as repetitive behaviors or rituals in an attempt to decrease anxiety or tension as a result of the obsessions. The most frequently reported compulsions include washing and cleaning, checking, counting, symmetry, and hoarding behavior (18, 19).

According to the DSM-5 (2013), three categories of tic disorders are recognized: TS, chronic tic disorder (motor or vocal), and provisional tic disorder (15). They are classified as neurodevelopmental disorders according to DSM-IV and DSM-5, and characterized by sudden, repetitive, and unwanted motor movements or sounds (tics) (20). Tics can be either motor/vocal and simple/complex. Simple motor tics include brief, abrupt movements, involving only single muscle groups (e.g. eye blinking and rolling, nose wrinkling, head jerk/nodding, shoulder shrugs). Complex motor tics are performed as a sequence of simpler movements (e.g, touching objects, hopping/jumping, squatting). Simple vocal tics include simple meaningless sounds and noises such as grunting, sniffing, throat clearing, coughing, and snorting. Complex vocalizations entail words/syllables or making of animal sounds, echolalia (repeating another person's words), or coprolalia (obscene words) (20).

Tics have a fluctuating course with time. Strikingly, over 70% of TS children experience a decrease in tic symptoms in adolescence, starting at age 12, most likely as a result of the maturation of the frontal lobes (21). Tics are usually more acute and expressive in periods of stress and anticipation, and tend to be reduced when the individual is mentally absorbed, focused and concentrated on activities. During adolescence, in up to 80% of tic-affected children, symptom severity decreases to an extent that the individuals no longer experience any distress and run a course that is indistinguishable from individuals with normal development (22, 23). However, in a small number of individuals, tics persist into adulthood and sometimes even increase in severity. To what extent this differential course (remitting versus persistent) reflects differential genetic architecture, differences in environmental influences, or differences in both has not been studied to date.

2.2.2 Diagnosis and comorbidity

OCD is frequently accompanied by psychiatric comorbidities, mostly with concurrent major depression (31%), social or specific phobias (11%), anxiety disorders (25%), or bipolar disorder (7%), and 20% of those affected have comorbid TS or tics (24–29).

For people with TS, most frequently occurring comorbid psychiatric disorders include attention deficit hyperactivity disorder (ADHD) and OCD (30, 31). Between 20% and 89% of individuals with TS exhibit OC behavior (32, 33), and between 50% and 60% are diagnosed with ADHD (34). Furthermore, increased rates of anxiety and depression have been reported in individuals with TS (35). Hirschtritt and colleagues published a comprehensive analysis on the characteristics of ADHD and OCD as comorbid disorders in TS, reporting a mean of 2.1 comorbid disorders in TS patients and 1.6 when excluding OCD and ADHD (36).

25.3 Population prevalence and clinical course

Lifetime and one-year prevalence of OCD is estimated at 2.3% and 1.2%, respectively (37). Moreover, many individuals experience OC symptoms without the full diagnosis. Hence, the prevalence of OC symptoms and subthreshold OCD (defined as having OC symptoms, but either less than one hour per day, although with significant suffering or distress, or more than one hour per day, but without suffering or distress) is likely to be much higher (up to 6%) in the general population (38–40).

For TS, Robertson and colleagues (41) suggested an overall prevalence estimate of 1% in the general population, and Scahill and colleagues reported prevalence rates between 0.5– 0.7% in children between age 6 and 18 (42). For tic disorders, prevalence rates range between 0.3% and 0.8% for chronic motor tic disorder, whereas for vocal tic disorder no reliable estimate is currently available. In adults, tic prevalence rates are considerably lower than in children. Zilhao and colleagues (43) found prevalence rates varying from 0.3% (TS) to 4.5% (transient tic disorder), in a population-based twin family study from the Netherlands, using self-reported data in adults.

OCD shows a slight preponderance for women and a bimodal pattern for the age of onset. Boys typically experience an onset before 10 years of age and more often have tic-related symptoms, while girls experience a later onset (44, 45). With respect to TS, males are more commonly diagnosed with TS in male to female ratios of 3– 4:1 (46, 47). The mean age of onset of TS is estimated at 7 years ranging from 2 to 21 years of age (48). Typically, tics start at 5–7 years of age (range between 3 and 8 years), and worsen at the age of 12 years, followed by a decline in severity during adolescence. By adulthood, it is estimated that roughly one-third of children with TS may be tic-free. The onset of vocal tics usually occurs around 2 years later than motor tics.

25.4 Genetic studies of TS and OCD: Heritability and linkage

Family studies investigate clustering of a trait within families, that is, in biological relatives, and provide a first insight into a possible contribution of genetic factors to the etiology of complex disorders. However, increased familial aggregation is in itself not sufficient to prove a genetic origin of a disorder because phenotypic resemblance between family members may also be due to shared or common environment.

The first family studies reported on OCD were based solely on family history data, without directly interviewing the relatives of patients themselves. Although these studies tend to underestimate the real disease rate within families, they have generated consistent conclusions on the familial nature of OCD (49–51). A total of 15 family study reports have appeared in the literature, of which 8 were in adults and 7 in children or adolescents, all using direct interviews with at least one family member, in some instances combined with family history data (44, 52). Overall, the results showed an up to tenfold increased rate of OCD in first-degree relatives of children/adolescents with OCD and a twofold increase in relatives of adults with OCD. Interestingly, within families of affected male probands, male relatives tend to report more tic symptoms, while female relatives more often present obsessive–compulsive behavior (45).

In contrast to family studies, twin studies can make a distinction between genetic and shared environmental factors by modeling the differences in correlations between monozygotic(MZ) and dizygotic (DZ) twins (53). Twin studies of OCD go back to 1929, with the first report on OC symptom pathology of twins by Lange. In the following 40 years a number of twin studies were published on OCD (for a historical overview, see 51). Two of the largest clinical twin studies performed within the period of the DSM-III diagnostic system (54, 55) reported concordance rates for OCD of 45% in MZ pairs and 15% in DZ pairs (54), or, according to another study, 33% in MZ pairs and 7% in DZ pairs (55), indicating a genetic background for OCD.

Other studies on the heritability of OCD focused on OC symptoms (OCS) instead of OC disorder. By using continuous data from self-report questionnaires of a variety of OC symptoms, and population-based epidemiological samples as opposed to clinically based samples, larger twin samples could be recruited (56). Figure 25.1 summarizes the heritability estimates obtained from twin studies in OCD, using DSM diagnostic criteria. The first relatively large study performed on OC symptoms in an epidemiological sample of 419 twin pairs was conducted by Clifford and colleagues (57). Using genetic structural equation modeling (SEM), the study obtained heritability estimates for obsessional traits and OC symptoms of, respectively, 44% and 47%. Subsequent studies, using larger epidemiological sample sets, provided heritability estimates ranging between 26% and 65%, depending on age of the sample and sex (49, 50, 58, 59). Bolton and colleagues estimated the heritability of OC symptoms in a population-based sample of 854 6-year old twins to be 29% (60). In recent years three studies have added a longitudinal perspective to the twin design in cohorts of children, adolescents, and adults (61–63). Stability as expressed by longitudinal phenotypic correlations was between 0.50 and 0.63, with substantial genetic contributions to symptom persistence (34–56%), indicating not only that OC symptoms are influenced by genetic factors but also that these factors are highly stable over time in adults. In children, stability of OC symptoms was estimated at between 35% and 51% for boys and between 28% and 34% for girls (59).

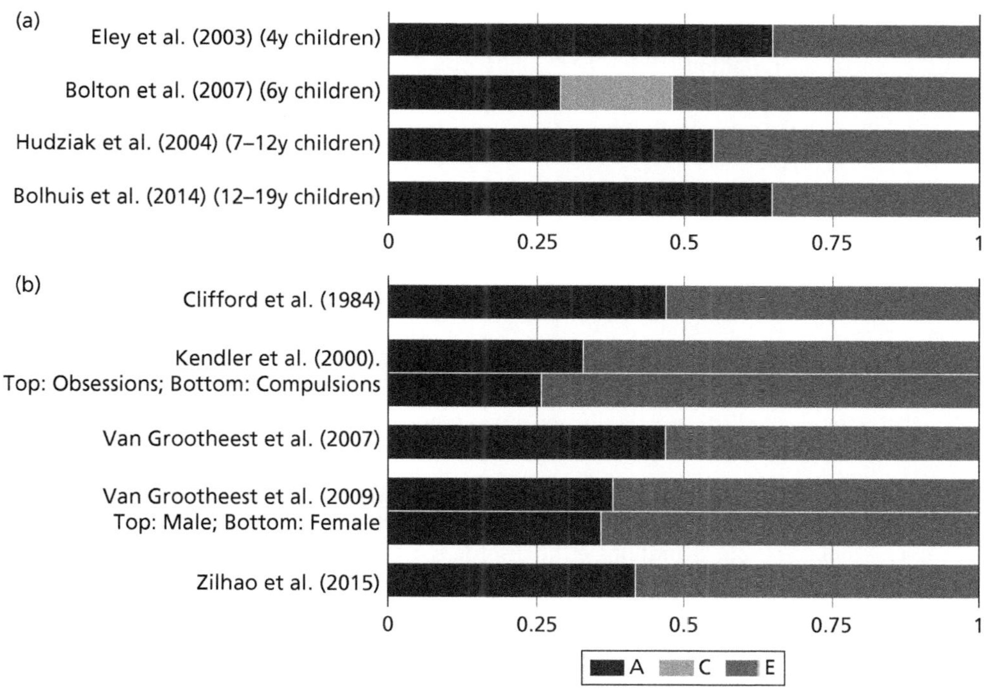

Figure 25.1 Heritability estimates from published twin studies in OCD: (a) studies in children samples; (b) studies in adult samples.

Note: Overlapping bars indicate when separate estimates resulted from one single study.

The first studies of familial effects in TS (3) described an increased risk of tics in families with TS probands. However, no distinction was made between recurrent childhood tics or newly onset adult tics in these reports. The first studies on newly developed tics in relatives of TS probands (64, 65) reported increased frequencies of tics/TS in these families when compared to the general population. The first large-scale family study utilizing direct interviews with patients and family members and employing a control group was published in 1991, with rates of reported TS being significant higher among relatives of probands than among relatives of controls (66). Overall, family studies in TS show an overall figure of a 10- to 100-fold increase among first-degree relatives of affected family members. Further, increased prevalences of tic disorders, ADHD, and OCD were found in parents and siblings of 4,479 school-age children who were assessed for TS and tic disorders (67). Stewart and colleagues assessed (direct interview) 692 relatives of 239 patients diagnosed for TS, ADHD, or TS and ADHD, with the aim to explore familial associations between TS and ADHD (68). This case-control study found an increased risk of comorbid TS and ADHD among relatives of cases diagnosed for one of these two disorders. A large multigenerational family sample from Sweden assessed 4,826 individuals for TS and chronic tic disorder between the periods of 1969 and 2009. There was an increased risk for tic disorders proportional to the degree of genetic relatedness, with first-degree relatives having a risk of 1:18.69 (odds ratio), and heritability of tic disorder estimated at 0.77 (69). A slightly different approach was taken by de Haan and colleagues (70) who estimated heritability of different tic symptom factors, carrying out a factor analysis in a sample of probands (N = 494) and their family members (N = 351). Three factors (complex vocal tics and obscene behavior; body tics; head/neck tics) formed a core etiological entity, with heritabilities between 0.19 and 0.25. Recent factor analytic studies of symptoms tried to uncover tic-based subtypes or TS-related phenotypes including both tics and co-morbidities. Hirschtritt and colleagues analyzed item-level data from 1191 TS probands and 2,303 first-degree relatives and identified a sub-phenotype characterized by high rates of social disinhibition, with a heritability of 0.53 (36). Darrow and colleagues reported similar results for 3,494 individuals assessed for TS, OCD, and ADHD symptoms—two cross-disorder (TS-related) phenotypes were identified: disinhibition (heritability 0.35) and symmetry (heritability 0.39) (71).

Fewer twin studies have been performed in TS than in OCD. The available evidence points to a genetic basis of the disorder. Figure 25.2 summarizes the heritability estimates obtained from twin studies of TS/tic disorders. The largest *clinical* study (MZ twins: N = 60, DZ twins: N = 26) found concordance rates in MZ and DZ twins of 77% and 23% (65). Another clinical study analyzed the concordance rates for TS and tic disorders among MZ pairs (N = 16 pairs) in relation to intra-pair differences in birth weight. Results showed 56% concordance rates of TS and 94% of tic disorder. The results suggested that prenatal events could be affecting different phenotypic expression for TS—that is, intra-pair birth-weight difference was found to be a predictor of the intra-pair difference for tic scoring (72).

Heritability estimates from epidemiological studies range between 0.28 and 0.56 (60, 73, 74). An extended twin-family study in adults from the Netherlands (N = 8,323 twins; 7,164 family members) looked at four definitions of tic disorders, ranging from a lenient definition (any tic reported at any lifetime point) to the strictest definition including the most severe manifestation (TS or chronic tic disorder). Heritability estimates depended on the definition of TS: heritability ranged from 0.37 for any tic reported to 0.25 for TS or chronic tic disorder (43).

Linkage studies seek to find genetic markers co-segregating with the disease/ phenotype of interest that are possibly "linked" to the causal genetic variants. The linkage approach involves scanning large family pedigrees with known disease transmission patterns, and narrowing down regions of the genome that segregate in association with the disorder. So far, five studies

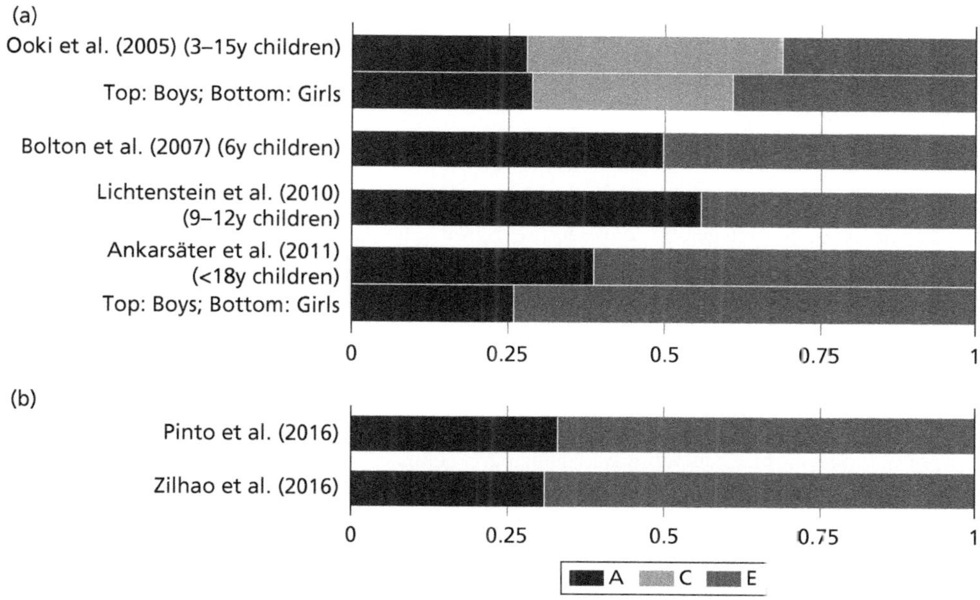

Figure 25.2 Heritability estimates from published twin studies in TS/tic disorders: (a) studies in children samples; (b) studies in adult samples.

Note: Overlapping bars indicate when separate estimates resulted from one single study.

on OCD have been performed (75–77). Results implicated genomic regions on chromosomes 1q, 3q, 6q, 6p, 6q, 7p, 9p, 10p, 11p, 14q, 15q, and 19q. However, none of these results reached genome-wide significance. Hanna and collaborators conducted a study in 66 individuals from seven families scanned for childhood OCD, and found region 9p24 to be a candidate region with linkage (75), a finding that was replicated by Willour and colleagues (78). A suggestive signal (LOD score = 3.13—usually the threshold for genome-wide linkage significance is 3.6) was reported in three Costa Rican families in region 15q14 (79), overlapping with findings from a previous study (77). The strongest signal so far (LOD score = 3.77) was recently found on the 1p36 region, by Mathews and colleagues in 33 Caucasian families (n = 245 individuals) with childhood onset OCD (80).

In TS, five genome-wide linkage studies have been performed with inconsistent results (81–85). In 1997, the Tourette Syndrome Association International Consortium for Genetics (TSAIC), published a linkage scan on 110 sib pairs, and found suggestive evidence for linkage in regions 4q and 8p (84). Simonic and colleagues studied an Afrikaner population using linkage transmission disequilibrium tests (TDT), and replicated previous findings for linkage in three previously identified regions, 2p11, 8q22, and 11q23-24 (83, 86). Interestingly, for this last region (11q23-24), another study from a French-Canadian population of 127 families found the marker D11S1377 to be statistically significant for linkage (LOD score = 3.18) (82). The largest linkage study thus far undertaken was from TSAIC within a collaborative effort comprising 2,040 individuals from 238 nuclear families and 18 multigenerational families (85). Statistically significant linkage was found for marker D2S144 on chromosome 2p32.2.

Linkage studies were the primary method for genetic analysis until the era of large-scale genome-wide association studies (GWAS). Because linkage studies require large family pedigrees and several affected generations with a disease of interest, or large numbers of affected sibling pairs, they became less popular. When dense genotyping became feasible at reasonable costs, GWAS in unrelated subjects quickly lead to successes for numerous complex traits and diseases, including genetic disorders (87).

25.5 **Genetic association studies**

Genetic association studies can be performed at a population level, in large samples of unrelated persons. Candidate gene studies require *a priori* (biological) hypotheses, on a gene(s) that has a potential functional relevance to the disease under study (gathered either from pharmacological studies, animal/theoretical models, or linkage studies). With the advent of GWAS and the decreasing costs in determining genetic variants, candidate gene studies have become less important. Here we summarize the genes found thus far through candidate gene studies, that is, those reported more than once to be implicated in either OCD or TS/tic disorders. Most studies searching for candidate genes associated with OCD and TS focused on neurotransmitters, or neurodevelopmental pathways. Particularly genes in the serotonergic, dopaminergic, and glutamatergic systems, that are also often the pharmacological targets of prescribed drugs, have been studied. Results have implied the serotonin transporter *SLC6A4, MAOA* gene encoding the enzyme monoamine oxidase A, the *COMT* gene, encoding catechol-*O*-methyl transferase affecting dopaminergic functioning, and *DRD4*.

A *de novo* mutation in the serotonin transporter *SLC6A4* gene has been implicated in OCD and comorbid disorders (anorexia nervosa, autism) in a study in two unrelated families (88). This finding was replicated in two family studies in OCD (89, 90). A recent meta-analysis of a total of 113 genetic association studies in OCD identified *COMT* and *MAOA* polymorphisms associated with OCD, in males (91). Still, the only consistent replication so far, has been associated with the glutamate transporter gene *SLC1A1* (89, 90, 92).

For TS, attention has largely focused on the *SLITRK1* gene, a member of the SLIT and TRK family proteins, involved in the control of neurite outgrowth. In 2005, Abelson and colleagues mapped an inversion and two subsequent mutations at 13q33.1, close to the gene *SLITRK1* (13q31.1), in two patients with TS (93). Subsequently, a Canadian-based family study identified one polymorphism tagging the major haplotype of *SLITRK1*, and these findings were subsequently replicated (94, 95). With respect to the dopamine receptor gene *DRD4*, Liu and collaborators studied a tandem repeat polymorphism in *DRD4* in a population-based family study of a large Han Chinese sample. Results suggested a protective role for the 2-repeat allele and a deleterious role for the 4-repeat allele (96). Finally, a mutation in the *HDC* gene (involved in the neuronal histaminergic pathways) was described in a multigenerational linkage family study where the father and all his eight children were diagnosed with TS (97). A subsequent study in 520 nuclear families with TS found a significant association of SNPs rs854150 and rs1894236 (intronic SNPs in gene *HDC*), providing supportive evidence for the role of this gene in the development of TS (98). SNPs (single nucleotide polymorphisms) represent common inter-individual genetic variation across the human genome. Due to methodological advances in the genomic era, scientists are able to determine, through direct and indirect means, millions of these common genetic SNP variants in the human genome.

The focus of genetic research has recently shifted from candidate gene studies to GWAS that seek to establish an association between common genetic variants of small effect, and a trait of interest. Through large international collaborations (e.g. the OCF-genetics consortium, OC GAS consortium, TSA genetic consortium, and TIC-genetics, all collated within the Psychiatric Genomics Consortium), and high-throughput analysis techniques, very large datasets of cases and controls became available. Participants in these studies are genotyped for large number of

SNPs and the genotype data are imputed to even larger numbers based on a common reference set, allowing for meta-analyses in large numbers of subjects.

GWAS studies are largely hypothesis-free and exploratory. They entail scanning the entire genome for SNPs that are associated with the phenotype of interest, tha is those that vary in allele frequency between cases and controls. The rationale behind these studies is that the collective variation embedded in all these SNPs can account for a statistically significant effect and play an important role in the development of disorders. The exploratory nature of GWAS, involving large numbers of statistical tests, has forced a stringent control of type I error rate by setting very low p-value thresholds (the generally accepted genome-wide threshold is $5x10^{-8}$ (99–101), prompting the need for large datasets. Unlike candidate gene studies, genome-wide data in GWAS additionally allow controlling for population stratification effects—the systematic difference in allele frequencies between subpopulations (or cases/controls). Here we briefly overview the knowledge gathered thus far in this field.

Four GWAS have been conducted for OCD and TS. The International OCD Foundation Genetics Consortium (IOCDF-GC) performed the first GWAS in OCD, involving 1,465 cases, 5,557 controls, and 400 trios (7). Two separate case-control analyses were performed with and without the trios with a total of 469,410 autosomal SNPs and 9,657 X-chromosome SNPs. The study found two trend-significant SNPs ($P = 2.49 \times 10^{-6}$, $P = 3.44 \times 10^{-6}$) located in the *DLGAP1* gene in the case-control analysis, and a significant SNP ($P = 3{,}84 \times 10^{-8}$) near the *BTBD3* gene. However, in the subsequent meta-analysis combining the trios and the case-control samples, no genome-wide significant SNPs were obtained.

A second GWAS was published recently by the OCD Collaborative Genetics Association Study (OCGAS) consisting of 1,406 individuals with OCD among a total sample of 5,061 individuals from 1,065 families (6). The authors failed to find any genome-wide significant result, but their follow-up gene-based analysis revealed a significant association of OCD with *IQCK, C16orf88*, and *OFC11*, all protein-coding genes. Recently, den Braber and colleagues reported a series of genetic analyses on OC symptoms from a population-based sample in the Netherlands (5). One SNP (rs8100480), located within the muscle-specific gene *MEF2BNB* was genome-wide significantly associated with OCS. Also, in this same work, a predictive Polygenic Risk Score (PRS) analysis was carried out using additional data from the previous GWAS IOCDF-GC. In this method, the samples are divided into discovery and target datasets (for each trait) and, in each of these, a risk score is calculated for every subject using the total number of risk alleles. Interestingly, in the study by den Braber and colleagues, the PRS from the clinical sample (IOCDF-GC) predicted OC scores in the population-based twin family sample (5).

In TS, one GWAS study has been performed so far, published in 2012 by the Tourette Syndrome Association International Consortium for Genetics (TSAICG), in a sample of 1,285 cases and 4,964 ancestry-matched controls (8). The top signal, an SNP located in the *COL27A1* gene, did not reach genome-wide significance ($P = 1.85 \text{ x } 10^{-6}$), which might be the result of power issues. Currently, several GWAS replication efforts are underway through worldwide consortia (TSAICG replication effort, EM-TICS, TS-EUROTRAIN).

Copy number variants (CNV) entail a form of variation where segments of DNA may be duplicated or deleted, with segments of differing length present in different people. The first GWAS on copy number variants in TS was performed in an ascertained sample from Latin America (210 cases and 285 controls) (102). Results showed an increase of large CNVs among cases compared to controls with four duplications and two deletions were located in the *CCL8A1* and *NRXN1* gene, respectively. McGrath and colleagues performed a cross-disorder genome-wide CNV analysis on TS and OCD (103). Large CNVs (> 500 kb) with frequency < 1% were assessed in a case-control design (2,699 cases; 1,789 controls), ascertained for OCD ($N = 1{,}613$) and TS ($N = 1{,}086$).

A 3.3-fold increase of large deletions was observed among OCD/TS cases compared to controls, mostly located in the 16p13.11 locus, a region previously linked to other neurodevelopmental disorders, with weaker evidence in the TS cases than in the OCD cases (103).

Bertelsen screened 188 TS cases from a Danish cohort, and identified seven patients with intronic deletions in the *IMMP2L* gene, with a significant higher frequency when compared to both the Danish control population and the genotyping reference panel (Affymetrix) cohort (104).

25.6 SNP heritability and genetic correlations

GWAS studies focus on the identification of common genetic variants. Other techniques focus on the estimation of the heritability of a trait as a result of all measured and imputed SNPS (referred to as SNP-based heritability), that is the amount of heritability explained by the joint effect of common variants across the genome. This method—referred to as Genome-Wide Complex Trait Analysis (GCTA)—calculates the genetic similarity between individuals based on the identified SNPs, and estimates how much of this similarity explains their phenotypic similarity. In the bivariate situation, genetic relatedness can be used to explain the genetic correlation between the phenotypes. In the study by den Braber and colleagues, GCTA was performed and the SNP-based heritability rate for OC scores was estimated at 14% (5). In an attempt to assess the degree of shared heritability between OCD and TS (105), Davis and colleagues performed GCTA with data collected in the scope of the consortia eTSAICG and IOCFGC. In this study SNP heritabilities for OCD and TS were reported at 0.37 and 0.58, respectively. Interestingly, it was found that for TS, 21% of the heritability was explained by rare SNPs (with a frequency in the population of less than 0.05), whereas for OCD, the entire heritability was attributable to common SNPs. The results of the bivariate analysis combining both datasets (OCD and TS) revealed a genetic correlation of 0.41 (se = 0.15, $p = 0.002$). This reflects the genetic overlap, that is, the degree of shared heritability between these two disorders. Following on their first collaborative effort on GWAS in TS and OCD (7, 8), a PRS analysis was carried out in the same groups. The combined GWAS analysis sought to unravel functional variants shared between these two disorders (106). The aggregated risk score in the discovery sample was then evaluated for predicting the disease status on the target sample. This analysis failed to detect a polygenic signal, and the overall conclusion from this work pointed towards specific, rather than shared, genetic effects underlying the etiology of OCD and TS.

Finally, in a study from a population-based twin family sample ($N = 5393$ individuals), Zilhao et al. found that 61.1% of the phenotypic correlation between OS symptoms (OCS) and TS/tic disorders was due to genetic factors. Furthermore, the genetic correlation between OCS and TS/tic disorders was modest—0.37 (107).

25.7 Conclusions and directions for future research

Heritability estimates of OCD and TS/tic disorders range between 0.27 and 0.65 for OCD and between 0.25 and 0.77 for TS. With respect to genetic factors underlying the etiology for these disorders, we expect the genetic liability to be highly polygenic, with small effect sizes for individual SNPs. With the increase in sample size and the continued collaborations in the field of psychiatric genetics, we also expect that new genetic variants will soon be uncovered.

An open issue is the discrepancy found between heritability estimates as found in epidemiological twin studies (between 0.25 and 0.37) versus clinical TS samples/ TS families using either SEM- or SNP-based methods (between 0.25 and 0.77). Explanations might be (i) a larger amount of measurement error in the epidemiological samples (as a result of, for instance, retrospective self-report measures used on lifetime tic occurrence). Generally, measurement error adds to inflated estimates of unique environmental influence, and lower heritability estimates. Alternatively,

(ii) the discrepancy between clinical and epidemiological samples might reflect phenotypic differences. TS family studies might have an over-representation of participants with persistent TS, which might be associated with increased genetic load. By contrast, the epidemiological studies also included subjects with childhood-only tic disorders and may be genetically more heterogeneous, or may represent a group in which environmental factors play a larger role.

To conclude, we have witnessed a remarkable progress in our conceptualization of TS and OCD in the past decades. In 1986 TS was conceptualized as an autosomal dominant genetic disorder (1), with only a few genes with a large effect contributing to its expression. Moreover, within TS families, comorbid OCD was considered as an alternate expression of the same phenotype. It is largely consensual now that TS is highly heterogeneous, mediated by large numbers of SNPs of small effect, with influences of CNVs and rare variants, and that TS and OCD only partly share genetic architecture. Current research technology on the genetics of OCD and TS offers interesting perspectives for new discoveries and understanding of these two disorders in the near future. First, a number of international consortia and collaborative efforts aim to standardize phenotyping and diagnostic criteria. This will allow researchers to increase sample sizes, and combine information from different populations worldwide while properly accounting for population stratification and cultural differences.

Second, we expect that both OCD and TS will undergo a similar scenario to that of other psychiatric disorders such as schizophrenia, for which, within less than a decade after the first GWAS published, more than 100 loci were identified as a result of huge sample sizes (n = 37,000 cases and 113,000 controls) in collaborative GWAS meta-analytic efforts (108). Shortly following this report, the involvement of the major histocompatibility complex (MHC) locus was confirmed, specifically the role of complement component 4 (C4) genes in the reduced number of synapses in the brains of schizophrenic individuals (109). These genes regulate the expression of the human C4 protein, which has a central role in the classical complement pathway, part of the innate immune system. This seems to confirm hypotheses that schizophrenia may in part be mediated by an affected immune system. In the near future, largely increased sample sizes due to worldwide consortia might equally yield results with respect to gene pathways involved in TS and OCD.

Third, the fast development of other genetic techniques in targeting other genetic variation than SNPs, including CNVs, *de novo* mutations or rare variants and micro RNAs supplemented by pathway and network approaches suggests that new insightful results will soon be generated. There is overall consensus that the emerged paradigm of "missing heritability" should be interpreted as *hidden* and not missing. Part of this genetic variation, which has not been captured so far, may indeed be obscured in other forms of genetic variation than the currently measured SNPs.

Finally, it is expected that similar developments in other fields such as epigenetics, animal models, and systems biology (proteomics, metabolomics, and imaging genetics) will lead to new insights. In the first epigenome-wide association study on TS, we (110) found potential enrichment for genes previously associated with psychiatric disorders. In sum, there are many scientific roads that can be followed in understanding how OCD and TS develop, how genes express themselves to cause liability for these diseases, and possibly provide new drug targets or treatment options.

References

1. **Pauls DL, Leckman JF.** The inheritance of Gilles de la Tourette's syndrome and associated behaviors. Evidence for autosomal dominant transmission. N Engl J Med. 1986 Oct 16;**315**(16):993–7.
2. **Peterson B, Riddle MA, Cohen DJ, Katz LD, Smith JC, Hardin MT**, et al. Reduced basal ganglia volumes in Tourette's syndrome using three-dimensional reconstruction techniques from magnetic resonance images. Neurology. 1993 May;**43**(5):941–9.

3. **Shapiro AK, Shapiro E**, Gerald Young J, Feinberg TE. Gilles De La Tourette Syndrome, 2nd edn. New York, NY: Raven Press Ltd.; 1988.
4. **Martin NG, Eaves LJ, Kearsey MJ, Davies P.** The power of the classical twin study. Heredity (Edinb). 1978 Feb;**40**(1):97–116.
5. **den Braber A, Zilhão NR, Fedko IO, Hottenga JJ, Pool R, Smit DJ**, et al. Obsessive–compulsive symptoms in a large population-based twin-family sample are predicted by clinically based polygenic scores and by genome-wide SNPs. Transl Psychiatry. 2016;6.
6. **Mattheisen M, Samuels JF, Wang Y, Greenberg BD, Fyer a J, McCracken JT**, et al. Genome-wide association study in obsessive–compulsive disorder: Results from the OCGAS. Mol Psychiatry. 2014 May 13;(October 2013):1–8.
7. **Stewart SE, Yu D, Scharf JM, Neale BM, Fagerness J, Mathews C**, et al. Genome-wide association study of obsessive–compulsive disorder. Mol Psychiatry. 2013 Jul;**18**(7):788–98.
8. **Scharf JM, Yu D, Mathews C a, Neale BM, Stewart SE, Fagerness J**, et al. Genome-wide association study of Tourette's syndrome. Mol Psychiatry. 2013;**18**(6):721–8.
9. **American Psychiatric Association**. Diagnostic and Statistical Manual of Mental Disorders, 3rd edn. Washington, DC: American Psychiatric Association; 1980.
10. **World Health Organization**. The ICD-10 Classification of Mental and Behavioural Disorders. Int Classif. 1992;**10**:1–267.
11. **American Psychiatric Association**. Diagnostic and Statistical Manual of Mental Disorders, Fourth edn, 4th edn. Washington, DC: American Psychiatric Association; 2000.
12. **Van Den Hout M, Kindt M.** Repeated checking causes memory distrust. Behav Res Ther. 2003;**41**(3):301–16.
13. **Ustun TB.** Global burden of depressive disorders in the year 2000. Br J Psychiatry. 2004 May 1; **184**(5):386–92.
14. **WHO**. World Health Statistics 2007. Geneva: WHO 2007.
15. **American Psychiatric Association**. Diagnostic and Statistical Manual of Mental Disorders, 5th edn; DSM-5. Washington, DC: American Psychiatric Association. 2013.
16. **Skoog G, Skoog I.** A 40-year follow-up of patients with obsessive–compulsive disorder. Arch Gen Psychiatry. 1999;**56**:121–7.
17. **Cath DC, Nizar K, Boomsma D, Mathews CA.** Age-specific prevalence of hoarding and obsessive compulsive disorder: A population-based study. Am J Geriatr Psychiatry. 2017;**25**(3):245–55.
18. **Katerberg H, Delucchi KL, Stewart SE, Lochner C, Denys DAJP, Stack DE**, et al. Symptom dimensions in OCD: Item-level factor analysis and heritability estimates. Behav Genet. 2010 Jul;**40**(4):505–17.
19. **Mataix-Cols D, Do Rosario-Campos MC, Leckman JF.** A multidimensional model of obsessive–compulsive disorder. Am J Psychiatry. 2005;162(2):228–38.
20. **Cath DC, Hedderly T, Ludolph AG, Stern JS, Murphy T, Hartmann A**, et al. European clinical guidelines for Tourette Syndrome and other tic disorders. Part I: Assessment. Eur Child Adolesc Psychiatry. 2011;**20**:155–71.
21. **Felling RJ, Singer HS.** Neurobiology of tourette syndrome: current status and need for further investigation. J Neurosci. 2011;**31**(35):12387–95.
22. **Burd LK.** Long-term follow-up of an epidemiologically defined cohort of patients with Tourette syndrome. J Child Neurol. 2001;**16**:431–7.
23. **Leckman JF, Zhang H, Vitale A, Lahnin F, Lynch K, Bondi C**, et al. Course of tic severity in Tourette syndrome: The first two decades. Pediatrics. 1998 Jul;**102**(1 Pt 1):14–9.
24. **Angst J, Gamma A, Endrass J, Hantouche E, Goodwin R, Ajdacic V**, et al. Obsessive–compulsive syndromes and disorders: Significance of comorbidity with bipolar and anxiety syndromes. Eur Arch Psychiatry Clin Neurosci. 2005;**255**(1):65–71.
25. **Canavera KE, Ollendick TH, Ehrenreich May JT, Pincus DB.** Clinical correlates of comorbid obsessive–compulsive disorder and depression in youth. Child Psychiatry Hum Dev. 2010;**41**(6):583–94.

26. **Geller DA.** Obsessive–compulsive and spectrum disorders in children and adolescents. Psychiatr Clin North Am. 2006;**29**(2):353–70.

27. **Langley AK, Lewin AB, Bergman RL, Lee JC, Piacentini J.** Correlates of comorbid anxiety and externalizing disorders in childhood obsessive compulsive disorder. Eur Child Adolesc Psychiatry. 2010;**19**(8):637–45.

28. **Pallanti S, Grassi G, Sarrecchia ED, Cantisani A, Pellegrini M.** Obsessive–compulsive disorder comorbidity: Clinical assessment and therapeutic implications. Front Psychiatr. 2011.

29. **Storch EA, Lewin AB, Geffken GR, Morgan JR, Murphy TK.** The role of comorbid disruptive behavior in the clinical expression of pediatric obsessive–compulsive disorder. Behav Res Ther. 2010;**48**(12):1204–10.

30. **Freeman RD, Fast DK, Burd L, Kerbeshian J, Robertson MM, Sandor P.** An international perspective on Tourette syndrome: Selected findings from 3,500 individuals in 22 countries. Dev Med Child Neurol. 2000;**42**(October 1997):436–47.

31. **Freeman RD, Aabech H, Allaghband-Rad J, Berlin C, Bruun R, Budman C,** et al. Tic disorders and ADHD: Answers from a world-wide clinical dataset on Tourette syndrome. Eur Child Adolesc Psychiatry. 2007;**16**(Suppl. 1):15–23.

32. **Karno M, Golding JM, Sorenson SB, Burnam M a.** The epidemiology of obsessive–compulsive disorder in five US communities. Arch Gen Psychiatry. 1988;**45**(12):1094–9.

33. **Weissman MM, Bland RC, Canino GJ, Greenwald S, Hwu HG, Lee CK,** et al. The cross national epidemiology of obsessive compulsive disorder. The Cross National Collaborative Group. J Clin Psychiatry. 1994;**55** Suppl:5–10.

34. **Polanczyk G, De Lima MS, Horta BL, Biederman J, Rohde LA.** The worldwide prevalence of ADHD: A systematic review and metaregression analysis. Am J Psychiatry. 2007;**164**(6):942–8.

35. **Cath DC, Ludolph AG.** Other psychiatric comorbidities in Tourette Syndrome. In: D Martino, JF Leckman (eds). Tourette Syndrome. Oxford University Press; 2013, pp. 74–106.

36. **Hirschtritt ME, Lee PC, Pauls DL, Dion Y, Grados MA, Illmann C,** et al. Lifetime prevalence, age of risk, and genetic relationships of comorbid psychiatric disorders in Tourette syndrome. JAMA Psychiatry. 2015;**72**(4):325–33.

37. **Ruscio A, Stein D, Chiu W, Kessler R.** The epidemiology of obsessive–compulsive disorder in the National Comorbidity Survey Replication. Mol Psychiatry. 2010;**15**:53–63.

38. **Adam Y, Meinlschmidt G, Gloster AT, Lieb R.** Obsessive–compulsive disorder in the community: 12-month prevalence, comorbidity and impairment. Soc Psychiatry Psychiatr Epidemiol. 2012;**47**(3):339–49.

39. **de Bruijn C, Beun S, de Graaf R, ten Have M, Denys D.** Subthreshold symptoms and obsessive–compulsive disorder: Evaluating the diagnostic threshold. Psychol Med. 2010;**40**(6):989–97.

40. **Fineberg N, Hengartner MP, Bergbaum C, Gale T, Rössler W, Angst J.** Lifetime comorbidity of obsessive–compulsive disorder and sub-threshold obsessive–compulsive symptomatology in the community: Impact, prevalence, socio-demographic and clinical characteristics. Int J Psychiatry Clin Pract. 2013 Aug;**17**(3):188–96.

41. **Robertson MM.** The prevalence and epidemiology of Gilles de la Tourette syndrome. Part 1: The epidemiological and prevalence studies. J Psychosom Res. 2008:65(5):461–72.

42. **Scahill L, Dalsgaard S, Bradbury K.** The prevalence of tourette syndrome and its relationship to clinical features. In: D Martino, JF Leckman (eds). Tourette Syndrome. Oxford: Oxford University Press; 2013. pp. 121–33.

43. **Zilhão NR, Olthof MC, Smit DJA, Cath DC, Ligthart L, Mathews CA,** et al. Heritability of tic disorders: A twin-family study. Psychol Med. 2016;1–12.

44. **Nestadt G.** A family study of obsessive–compulsive disorder. Arch Gen Psychiatry. 2000: 358–63.

45. **Pauls DL, Alsobrook JP, Goodman W, Rasmussen S, Leckman JF.** A family study of obsessive–compulsive disorder. Am J Psychiatry. 1995;**152**:76–84.

46. **Cavanna AE, Rickards H.** The psychopathological spectrum of Gilles de la Tourette syndrome. Neurosci Biobehav Rev. 2013. p. 1008–15.

47. **Robertson MM.** Tourette syndrome, associated conditions and the complexities of treatment. Brain. 2000;**123**(3):425–62.

48. **Leckman JF, Bloch MH, King RA.** Symptom dimensions and subtypes of obsessive–compulsive disorder: A developmental perspective. 2009;21–33.

49. **Eley TC, Bolton D, O'Connor TG, Perrin S, Smith P, Plomin R.** A twin study of anxiety-related behaviours in pre-school children. J Child Psychol Psychiatry Allied Discip. 2003;**44**:945–60.

50. **Hudziak JJ, Van Beijsterveldt CEM, Althoff RR, Stanger C, Rettew DC, Nelson EC**, et al. Genetic and environmental contributions to the Child Behavior Checklist Obsessive–Compulsive Scale: A cross-cultural twin study. Arch Gen Psychiatry. 2004;**61**:608–16.

51. **van Grootheest DS, Cath DC, Beekman AT, Boomsma DI.** Twin studies on obsessive–compulsive disorder: a review. Twin Res Hum Genet. 2005;**8**(5):450–8.

52. **Grados MA, Riddle MA, Samuels JF, Liang KY, Hoehn-Saric R, Bienvenu OJ**, et al. The familial phenotype of obsessive–compulsive disorder in relation to tic disorders: The Hopkins OCD family study. Biol Psychiatry. 2001;**50**:559–65.

53. **Boomsma D, Busjahn A, Peltonen L.** Classical twin studies and beyond. Nat Rev Genet. 2002;**3**(11):872–82.

54. **Carey, G., & Gottesman II.** Twin and familiy studies of anxiety, phobic, and obsessive disorders. In: DF Klein, J Rabkin (eds). Anxiety: New research and changing concepts. New York, NY: Raven Press; 1981, pp. 117–36.

55. **Skre I, Onstad S, Torgersen S, Lygren S, Kringlen E.** A twin study of DSM-III-R anxiety disorders. Acta Psychiatr Scand. 1993;(88):85–92.

56. **Macdonald AM, Murray RM, Clifford CA.** The contribution of heredity to obsessional disorder and personality: A review of family and twin study evidence. In: MT Tsuang, KS Kendler, **MJ Lyons** (eds). Genetic issues in Psychosocial Epidemiology. New Brunswick, NJ: Rutgers University Press; 1991; pp. 191–212.

57. **Clifford CA, Murray RM, Fulker DW** (1984). Genetic and environmental influences on obsessional traits and symptoms. Psychol Med. 1984;(14):791–800.

58. **Jonnal AH, Gardner CO, Prescott CA, Kendler KS.** Obsessive and compulsive symptoms in a general population sample of female twins. Am J Med Genet—Neuropsychiatr Genet. 2000;**96**:791–6.

59. **van Grootheest, Bartels M, Cath DC, Beekman AT, Hudziak JJ, Boomsma DI.** Genetic and environmental contributions underlying stability in childhood obsessive–compulsive behavior. Biol Psychiatry. 2007 Feb 1;**61**(3):308–15.

60. **Bolton D, Rijsdijk F, O'Connor TG, Perrin S, Eley TC.** Obsessive–compulsive disorder, tics and anxiety in 6-year-old twins. Psychol Med. 2007;**37**:39–48.

61. **Bolhuis K, McAdams T a, Monzani B, Gregory a M, Mataix-Cols D, Stringaris a**, et al. Aetiological overlap between obsessive–compulsive and depressive symptoms: A longitudinal twin study in adolescents and adults. Psychol Med. 2014 May;**44**(7):1439–49.

62. **van Grootheest, Cath D, Hottenga JJ, Beekman AT, Boomsma DI.** Genetic factors underlie stability of obsessive–compulsive symptoms. Twin Res Hum Genet. 2009 Oct;**12**(5):411–9.

63. **Zilhao NR, Smit DJ a, Den Braber A, Dolan C V, Willemsen G, Boomsma DI**, et al. Genetic and environmental contributions to stability in adult obsessive compulsive behavior. Twin Res Hum Genet. 2014;**18**(1):52–60.

64. **Pauls DL, Cohen DJ, Heimbuch R, Detlor J, Kidd KK.** Familial pattern and transmission of Gilles de la Tourette syndrome and multiple tics. Arch Gen Psychiatry. 1981;**38**(10):1091–3.

65. **Price RA, Kidd KK, Cohen DJ, Pauls DL, Leckman JF.** A twin study of Tourette syndrome. Arch Gen Psychiatry. 1985;**42**:815–20.

66. **Pauls DL, Raymond CL, Stevenson JM, Leckman JF.** A family study of Gilles de la Tourette syndrome. Am J Hum Genet. 1991 Jan;**48**(1):154–63.

67. **Khalifa N, von Knorring A-L.** Tourette syndrome and other tic disorders in a total population of children: Clinical assessment and background. Acta Paediatr. 2005;**94**:1608–14.
68. **Stewart SE, Illmann C, Geller D a, Leckman JF, King R, Pauls DL.** A controlled family study of attention-deficit/hyperactivity disorder and Tourette's disorder. J Am Acad Child Adolesc Psychiatry. 2006;**45**(11):1354–62.
69. **Mataix-Cols D, Isomura K, Pérez-Vigil A, Chang Z, Rück C, Larsson KJ,** et al. Familial risks of Tourette syndrome and chronic tic disorders: A population-based cohort study. JAMA psychiatry. 2015;
70. **de Haan MJ, Delucchi KL, Mathews CM, Cath DC.** Tic symptom dimensions and their heritabilities in Tourette's syndrome. Psychiatr Genet. 2015;**25**:112–8.
71. **Darrow SM, Hirschtritt ME, Davis LK, Illmann C, Osiecki L, Grados M,** et al. Identification of two heritable cross-disorder endophenotypes for Tourette syndrome. Am J Psychiatry. 2017;**174**(4):387–96.
72. **Hyde TM, Aaronson BA, Randolph C, Rickler KC, Weinberger DR.** Relationship of birth weight to the phenotypic expression of Gilles de la Tourette's syndrome in monozygotic twins. Neurology. 1992;**42**:652–8.
73. **Anckarsäter H, Lundström S, Kollberg L, Kerekes N, Palm C, Carlström E,** et al. The Child and Adolescent Twin Study in Sweden (CATSS). Twin Res Hum Genet. 2011;**14**:495–508.
74. **Pinto R, Monzani B, Leckman JF, Christian R, Serlachius E, Lichtenstein P,** et al. Understanding the covariation of tics, attention-deficit/hyperactivity, and obsessive–compulsive symptoms: A population-based adult twin study. Am J Med Genet B Neuropsychiatr Genet. 2016;**171**(7):938–47
75. **Hanna GL, Veenstra-VanderWeele J, Cox NJ, Boehnke M, Himle J, Curtis GC,** et al. Genome-wide linkage analysis of families with obsessive–compulsive disorder ascertained through pediatric probands. Am J Med Genet. 2002 Jul 8;**114**(5):541–52.
76. **Hanna GL, Veenstra-Vanderweele J, Cox NJ, Van Etten M, Fischer DJ, Himle J,** et al. Evidence for a susceptibility locus on chromosome 10p15 in early-onset obsessive–compulsive disorder. Biol Psychiatry. 2007 Oct 15;**62**(8):856–62.
77. **Shugart YY, Samuels J, Willour VL, Grados M a, Greenberg BD, Knowles J a,** et al. Genomewide linkage scan for obsessive–compulsive disorder: Evidence for susceptibility loci on chromosomes 3q, 7p, 1q, 15q, and 6q. Mol Psychiatry. 2006 Aug;**11**(8):763–70.
78. **Willour VL, Shugart YY, Samuels J, Grados M, Cullen B, Iii OJB,** et al. Replication study supports evidence for linkage to 9p24 in obsessive–compulsive disorder. 2004;508–13.
79. **Ross J, Badner J, Garrido H, Sheppard B, Chavira D a, Grados M,** et al. Genomewide linkage analysis in Costa Rican families implicates chromosome 15q14 as a candidate region for OCD. Hum Genet. 2011 Dec;**130**(6):795–805.
80. **Mathews C a, Badner J a, Andresen JM, Sheppard B, Himle J a, Grant JE,** et al. Genome-wide linkage analysis of obsessive–compulsive disorder implicates chromosome 1p36. Biol Psychiatry. Elsevier Inc.; 2012 Oct 15;**72**(8):629–36.
81. **Barr CL, Wigg KG, Zovko E, Sandor P, Tsui LC.** Linkage study of the dopamine D5 receptor gene and Gilles de la Tourette syndrome. Am J Med Genet. 1997;**74**:58–61.
82. **Mérette C, Brassard A, Potvin A, Bouvier H, Rousseau F, Emond C,** et al. Significant linkage for Tourette syndrome in a large French Canadian family. Am J Hum Genet. 2000;**67**:1008–13.
83. **Simonic I, Gericke GS, Ott J, Weber JL.** Identification of genetic markers associated with Gilles de la Tourette syndrome in an Afrikaner population. Am J Hum Genet. 1998;**63**:839–46.
84. **The Tourette Syndrome Association International Consortium for Genetics.** A complete genome screen in sib pairs affected by Gilles de la Tourette Syndrome. Am J Hum Genet. 1999 Nov;**65**(5):1428–36.
85. **The Tourette Syndrome Association International Consortium for Genetics.** Genome scan for Tourette disorder in affected-sibling-pair and multigenerational families. Am J Hum Genet. 2007 Feb;**80**(2):265–72.
86. **Simonic I, Nyholt DR, Gericke GS, Gordon D, Matsumoto N, Ledbetter DH,** et al. Rapid publication further evidence for linkage of Gilles de la Tourette Syndrome (GTS) Susceptibility Loci on Chromosomes 2p11, 8q22, and 11q23-24 in South African Afrikaners. 2001;**167**(February):163–7.

87. **Lee SH, Ripke S, Neale BM, Faraone S V, Purcell SM, Perlis RH,** et al. Genetic relationship between five psychiatric disorders estimated from genome-wide SNPs. Nat Genet. 2013;**45**(9):984–94.
88. **Ozaki N, Goldman D, Kaye WH, Plotnicov K, Greenberg BD, Lappalainen J,** et al. Serotonin transporter missense mutation associated with a complex neuropsychiatric phenotype. Mol Psychiatry. 2003;**8**(11):933–6.
89. **Hu X-Z, Lipsky RH, Zhu G, Akhtar LA, Taubman J, Greenberg BD,** et al. Serotonin transporter promoter gain-of-function genotypes are linked to obsessive–compulsive disorder. Am J Hum Genet. 2006;**78**(5):815–26.
90. **Voyiaziakis E, Evgrafov O, Li D, Yoon H-J, Tabares P, Samuels J,** et al. Association of SLC6A4 variants with obsessive–compulsive disorder in a large multicenter US family study. Mol Psychiatry. 2011;**16**(1):108–20.
91. **Taylor S.** Molecular genetics of obsessive–compulsive disorder: A comprehensive meta-analysis of genetic association studies. Mol Psychiatry. 2013 Jul;**18**(7):799–805.
92. **Stewart SE, Fagerness JA, Platko J, Smoller JW, Scharf JM, Illmann C,** et al. Association of the SLC1A1 glutamate transporter gene and obsessive–compulsive disorder. Am J Med Genet. 2007;**144B**(8):1027–33.
93. **Abelson JF, Kwan KY, O'Roak BJ, Baek DY, Stillman AA, Morgan TM,** et al. Sequence variants in SLITRK1 are associated with Tourette's syndrome. Science. 2005:317–20.
94. **Karagiannidis I, Rizzo R, Tarnok Z, Wolanczyk T, Hebebrand J, Nöthen MM,** et al. Replication of association between a SLITRK1 haplotype and Tourette syndrome in a large sample of families. Mol Psychiatry. 2012;**17**(7):665–8.
95. **Miranda DM, Wigg K, Kabia EM, Feng Y, Sandor P, Barr CL.** Association of SLITRK1 to Gilles de la Tourette syndrome. Am J Med Genet B Neuropsychiatr Genet. 2009;**150B**:483–6.
96. **Liu S, Cui J, Zhang X, Wu W, Niu H, Ma X,** et al. Variable number tandem repeats in dopamine receptor D4 in Tourette's syndrome. Mov Disord. 2014;**29**(13):1687–91.
97. **Ercan-Sencicek AG, Stillman AA, Ghosh AK, Bilguvar K, O'Roak BJ, Mason CE,** et al. L-histidine decarboxylase and Tourette's syndrome. N Engl J Med. 2010;**362**:1901–8.
98. **Karagiannidis I, Dehning S, Sandor P, Tarnok Z, Rizzo R, Wolanczyk T,** et al. Support of the histaminergic hypothesis in Tourette syndrome: Association of the histamine decarboxylase gene in a large sample of families. J Med Genet. 2013;**50**(11):760–4.
99. **Hoggart CJ, Clark TG, De Iorio M, Whittaker JC, Balding DJ.** Genome-wide significance for dense SNP and resequencing data. Genet Epidemiol. 2008;**32**(2):179–85.
100. **International T, Consortium H.** A haplotype map of the human genome. Nature. 2005;**437**(7063):1299–320.
101. **Risch N, Merikangas K.** The future of genetic studies of complex human diseases. Science. 1996;**273**(5281):1516–7.
102. **Nag A, Bochukova EG, Kremeyer B, Campbell DD, Muller H, Valencia-Duarte AV,** et al. CNV analysis in Tourette syndrome implicates large genomic rearrangements in COL8A1 and NRXN1. PLoS One. 2013;**8**(3).
103. **McGrath LM, Yu D, Marshall C, Davis LK, Thiruvahindrapuram B, Li B,** et al. Copy number variation in obsessive–compulsive disorder and tourette syndrome: A cross-disorder study. J Am Acad Child Adolesc Psychiatry. 2014 Aug;**53**(8):910–9.
104. **Bertelsen B, Melchior L, Jensen LR, Groth C, Glenthøj B, Rizzo R,** et al. Intragenic deletions affecting two alternative transcripts of the IMMP2L gene in patients with Tourette syndrome. Eur J Hum Genet. 2014;**22**(11):1283–9.
105. **Davis LK, Yu D, Keenan CL, Gamazon ER, Konkashbaev AI, Derks EM,** et al. Partitioning the heritability of Tourette syndrome and obsessive compulsive disorder reveals differences in genetic architecture. PLoS Genet. 2013 Oct;**9**(10):e1003864.
106. **Yu D, Mathews CA, Neale BM, Ph D, Davis LK, Gamazon ER,** et al. Cross-disorder genome-wide analyses suggest a complex genetic relationship between Tourette's syndrome and OCD. Am J Psychiatry. 2014;1–12.

107. **Zilhao NR, Smit DJ, Boomsma DI, Cath DC.** Cross-disorder genetic analysis of tic disorders, obsessive–compulsive, and hoarding symptoms. Front Psychiatry. 2016;7(Jun).

108. **Ripke S, O'Dushlaine C, Chambert K, Moran JL, Kähler AK, Akterin S**, et al. Genome-wide association analysis identifies 13 new risk loci for schizophrenia. Nat Genet. 2013 Oct 25;**45**(10):1150–9.

109. **Sekar A, Bialas AR, de Rivera H, Davis A, Hammond TR, Kamitaki N**, et al. Schizophrenia risk from complex variation of complement component 4. Nature. 2016;**530**(7589):177–83.

110. **Zilhão NR, Padmanabhuni SS, Pagliaroli L, Barta C, Smit DJ a., Cath D**, et al. Epigenome-wide association study of tic disorders. Twin Res Hum Genet. 2015;**18**(6):1–11.

Chapter 26

Therapygenetics: Predicting psychological treatment response from genetic markers

Jonathan RI Coleman, Kathryn J Lester, and Thalia C Eley

26.1 Introduction

Emotional disorders represent the majority of mental illness diagnoses worldwide and account for a large economic and emotional burden (1–2). In addition to pharmacological treatment, a number of psychological treatment modalities have been developed, of which a major contributor is evidence-based cognitive behavior therapy (CBT; 3–4). CBT works by providing a supportive environment in which individuals with emotional disorders identify and explore the negative thought processes and behavioral patterns that underlie their distress. The therapist guides the recipient through cognitive and behavioral interventions that allow these negative schemas to manifest without adverse consequences. From a cognitive perspective, the participant is thus able to adjust negative biases in their perception. The behavioral component is allied to classical conditioning learning processes. Exposure therapy allows the participant to confront afflictive stimuli without negative consequences, which diminishes their association. This extinction is reinforced by the development of coping strategies, with the eventual aim of reduced distress and impairment (4).

Psychological therapies are effective, and their use is widespread. For example, the UK's Increasing Access to Psychological Treatment program aims to offer therapy to at least 15% of all individuals suffering from emotional disorders (5–9). Compared to waitlisted groups, CBT shows good efficacy in adults (10) and in children (11). There is little evidence of significant differences between different psychological therapies, potentially suggesting a general mechanism by which response occurs (11–13). However, not all recipients respond equally, with around 40% to 50% of recipients retaining a clinically significant level of impairment or distress following treatment (11, 14). Psychological therapy involves considerable financial and emotional investment from both provider and recipient, and, in the absence of guidance as to the optimal treatment for any given individual, allocation to specific treatment types is usually dictated by availability and instinct. Therefore, there is clear value in identifying predictors of response, which could ultimately be used to inform treatment choice.

Identifying such predictors has been a main area of focus in the study of psychological therapies, and a wide variety of potential demographic and clinical factors have been proposed to have an effect, but the evidence basis for such factors is mixed. Higher disorder severity at the start of therapy has repeatedly been implicated in poorer treatment response (15). However, results vary between studies, not least because response is inconsistently defined such that initial disorder severity is part of the definition in some investigations but not in others (14). Similarly inconsistent findings have been reported for the negative effect of psychiatric comorbidities on therapy response (16, 17). Studies of therapist characteristics strongly suggest that response to psychological therapy does not just result from factors related to the recipient but also from those acting on the therapist, and from the relationship between the therapist and the patient. Although evidence for the effect of specific therapist behaviors such as adherence to therapeutic manuals and competence is inconsistent, there does appear to be a robust positive correlation between therapeutic alliance and treatment outcome (18, 19).

Given the difficulty of defining influences on treatment response in psychological therapies, DNA provides a useful potential substrate for identifying predictors. Genetic variation is effectively uniform across all tissues in the body, in contrast to alternative sources of biological variation, such as gene expression or DNA methylation. Furthermore, DNA variation is not altered by environmental influences (including treatment) or time, comparing favorably to other (including non-biological) sources of variation (20).

26.2 Review of therapygenetic studies to date

We performed a systematic review of therapygenetic studies to date. This extends previous reviews that focused on CBT (21) and on anxiety disorders (22). The literature search was conducted by browsing Web of Science with the inquiries ("therapygenetics"), ("genetics" AND ("cognitive behavior therapy" OR "cognitive behavior therapy" OR "cognitive behavioral therapy" OR "cognitive behavioral therapy") AND "response"), (genes AND psychotherapy), and (gene AND psychological AND therapy), including all studies listed up to July 28, 2016. Additional studies were identified from the reference sections of the included studies and by consulting previous reviews (21, 22). This yielded 1,567 records, of which 12 were duplicated due to the multiple search strategies. Following removal of records that were non-empirical (241 records), did not report therapy response as an outcome (1,149 records), did not assess genetic variants (21 records), or focused on phenotypes other than emotional disorders (109 records), 35 studies were identified for inclusion in the review (Table 26.1) (23–56). Two studies of the same cohort (investigating different aspects of response to therapy) reported results from the same analysis and are combined in Table 26.1 (35, 44). One conference publication (57) reported in a previous review (21) was superseded by a journal publication (23). Finally, one study assessed change in disorder severity between treated and non-treated groups, stratified by a genetic variant (28). Although this study differs from all others in that it uses a group-differences design rather than an individual-differences one, it is included as this approach could be informative about the role of genetic variation in response to psychological treatments, particularly in the development of stratified medicine.

26.3 Serotonergic system genes: Serotonin transporter (*SLC6A4*), serotonin receptor 1A (*HTR1A*), serotonin receptor 2A (*HTR2A*), and tryptophan hydroxylase 2 (*TPH2*)

The most studied gene in the field of therapygenetics is *SLC6A4*, which encodes the serotonin transporter, responsible for the reuptake of serotonin from the synaptic cleft (58, 59). The serotonergic system is of considerable interest in the study of emotional disorders, not least because the serotonin transporter is targeted by selective serotonin reuptake inhibitors, which are effective antidepressant and anxiolytic drugs (58, 59). A considerable amount of research has investigated the promoter repeat polymorphism of the gene (5HTTLPR), as the short allele (which reduces the expression of the gene; 60) has been implicated in the differential development of depressive outcomes in individuals exposed to early-life trauma (61, 62).

Table 26.1 Therapygenetic studies to date

Gene	Authors	Ref.	Diagnosis	Sample	Treatment	Results
SLC6A4 (5HTTPLR)	Di Bella et al. (2002)	31	AN + BN	*N* = 158, 100% female	4–8 wk CBT ± serotonin reuptake inhibitor or placebo	Genotypic model (LL × SL × SS), without rs25531. A significant association with the LL genotype and improved response was observed in AN participants, but was specific to the CBT + SRI group. No significant effects were observed in the BN group.
	Steiger et al. (2008)	52	BN	*N* = 98, Caucasian, M_{age} = 27, 100% female	16–32 wk multimodal therapy based on individual CBT for all participants, with additional group CBT and medication	Dominant models with and without rs25531 (without: LL × SL/SS; with: L'L' × S'L'/S'S', where L_A = L', L_G or S –> S') S allele carriers showed less reduction in binge episodes following 32 wks of therapy (biallelic/ triallelic model), and poorer response after 16 wks on measures of anxiety and depression.
	Wang et al. (2009)	54	PTSD	*N* = 35	12 wk prolonged exposure therapy or 12 wk Escitalopram (*N* = 20)	Recessive model (LL/SL × SS) without rs25531. No significant association with treatment response in exposure therapy group.
	Bryant et al. (2010)	26	PTSD	*N* = 45, Caucasian, M_{age} = ~ 43 yrs, 33% female	8 × 90min weekly individual CBT	Dominant model with rs25531 (L'L' × S'L'/S'S', where L_A = L', L_G or S –> S') No significant association at post-treatment. At 6mth follow-up, higher % of S-allele than L-allele carriers met PTSD criteria and had significantly higher symptom scores.
	Lonsdorf et al. (2010)	43	PD ± Ag	*N* = 69, Caucasian, M_{age} = ~ 35 yrs, 62% female	10 × 2hr weekly group (*N* = 38) or Internet-delivered (*N* = 31) CBT	Dominant (LL × SL/SS) and genotypic model (LL × SL × SS) without rs25531 and dominant (L'L' × S'L'/S'S') model with rs25531 No significant association with treatment response.

	Kohen et al. (2011)	38	DEP (post-stroke)	*N* = 61, mixed ethnicity	9 × positive problem solving plus antidepressant vs usual care plus antidepressant	Additive (number of S alleles) and genotypic (LL × SL × SS) models, without rs25531 SS and SL carriers had a significantly greater mean percentage reduction in depression ratings and more likely to be in remission at nine-week follow-up than those in the control group.
	Sakolsky et al. (2011)	49	SEP; GAD; SAD	*N* = 211, Caucasian, 7–17 yrs	Sertraline, 14 sessions of CBT, combination therapy or 12 wk placebo	Additive (# S alleles), dominant (LL × SL/SS) and recessive (LL/SL × SS) models, without rs25531, and Additive (# S′ alleles), dominant (L′L′ × S′L′/S′S′) and recessive (L′L′/S′L′ × S′S′) models with rs25531. No significant association with treatment response.
	Eley et al. (2012)	32	ANX	*N* = 359, Caucasian, M_{age} = 9.44 yrs, 49% female	10–12 session group or individual CBT or guided self-help	Genotypic (LL × SL × SS) and recessive (LL/SL × SS) models, without rs25531 No significant association at post treatment. At follow-up, higher % of SS genotype carriers free of anxiety diagnoses than SL/LL genotype carriers. SS genotype carriers had significantly greater reduction in symptom severity scores.
	Hedman et al. (2012)	36	SAD	*N* = 126, 98% Caucasian, M_{age} = ~ 35 yrs, 36% female	15 × 2.5hr weekly group (*N* = 62) or Internet-delivered (*N* = 64) CBT	Additive (# S alleles) and genotypic (LL × SL × SS) models without rs25521 and genotypic (L_AL_A × L_GL_A × SL_A × SL_G × SS) model with rs25531 No significant association with treatment response.
	Castellini et al. (2012)	27	AN + BN	*N* = 201, Caucasian, M_{age} = 27 yrs, 97% female	19–40 × 1hr individual CBT	Dominant model (LL × SL/SS) without rs25531 No significant association with recovery at six-year follow-up, although significant associations with the S-allele were identified for increased depressive comorbidity and greater diagnostic instability.

(*continued*)

Table 26.1 Continued

Gene	Authors	Ref.	Diagnosis	Sample	Treatment	Results
	Bockting et al. (2013)	25	Recurrent DEP	*N* = 180, Caucasian, M_{age} = 45 yrs, 74% female	Brief CBT vs. treatment as usual	Recessive (LL/SL × SS) model, without rs25531 No significant association with time to recurrence.
	Andersson et al. (2013)	23	SAD	*N* = 314, M_{age} = 37 yrs, 50% female (Overlaps with Hedman et al.)	Trial 1 (*N* = 126): As Hedman, Trial 2 (*N* = 204): 9 wks of Internet-delivered CBT	Dominant (LL × SL/SS) and genotypic (LL × SL × SS) models without rs25531. No significant effect on treatment response at long-term follow-up.
	Felmingham et al. (2013)	33	PTSD	*N* = 55, Caucasian, M_{age} = ~ 47 yrs, 36% female (Overlaps with Bryant et al.)	8 wk exposure therapy	Dominant (LL × SL/SS) model, unclear if with or without rs25531 No significant association with treatment response.
	Knuts et al. (2014)	37	Ag	*N* = 99, Caucasian, M_{age} = ~ 39 yrs, 74% female	5 session, 1 week *in vivo* exposure therapy	Genotypic model with rs25531 (L'L' × S'L' × S'S') SS/SL carriers showed better treatment response two wks after therapy.
	Gloster et al. (2015) and Lueken et al. (2015)	35, 44	PD ± Ag	*N* = 228, Caucasian, M_{age} = 36.2 yrs, 75% female (Overlaps with Lonsdorf et al.)	Manualized exposure-based CBT, 12 × twice-weekly sessions	Additive (# S' alleles [Lueken]) and dominant (L'L' × S'L'/S'S' [Gloster]) models with rs25331. No significant associations with treatment response.
	Lester et al. (2016b)	42	ANX	*N* = 829, mixed ethnicity, M_{age} = 10 yrs	Manualized CBT (individual, group, or guided self-help), varying (4–52 wk) duration	Recessive model (LL/SL × SS) without rs25531. No significant association at post treatment or follow-up. In mega-analysis with Eley et al. (2012), higher % of SS genotype carriers free of anxiety diagnoses than SL/LL genotype carriers for all anxiety disorders (no significant effect for primary anxiety disorder).

	Cicchetti et al. (2015)	28	DEP	*N* = 58, 54% African American, 22% Caucasian, 24% Hispanic/biracial, M_{age} = 25 yrs, 100% female* (* includes additional group, *N* = 68, randomized to treatment as usual).	Manualized interpersonal psychotherapy, 14 wks or treatment as usual	Dominant model (LL × SL/SS) without rs25531 Women with LL genotype who participated in IPT reported significantly fewer depressive symptoms at post-intervention (but not at eight-month follow-up) compared to those who participated in treatment as usual. No differences in depressive symptoms were found between intervention groups for SL/SS genotypes. The results were primarily found in African-American participants.
SLC6A4 (STin2 VNTR)	Kohen et al. (2011)	38	DEP (post-stroke)	*N* = 64, mixed ethnicity	9 × positive problem solving plus antidepressant vs. usual care plus antidepressant	9/12 and 12/12 genotype carriers in intervention group had a significantly greater mean percentage reduction in depression scores and greater likelihood of remission at 9 wk follow-up than controls.
	Sakolsky et al. (2011)	49	SEP; GAD; SAD	*N* = 211, Caucasian, 7–17 yrs	Sertraline, 14 sessions of CBT, combination therapy or 12 wk placebo	At 12-week assessment STin2 12-copy variant carriers showed significantly greater improvement.
SLC6A4 (rs421417)	Serretti et al. (2013)	51	DEP	*N* = 43, M_{age} = 51 yrs*, 66% female* (* From sample with antidepressants, *N* = 160)	6-week interpersonal counselling	No significant association with response to IPC.
SLC6A4 (rs8076005)	Matsumoto et al. (2014)	45	DEP	*N* = 40, M_{age} = 44 yrs, 60% female (Overlaps with Serretti et al.)	6-week interpersonal counselling	No significant association with response to IPC.
SLC6A4 (rs25531)	Hedman et al. (2012)	36	SAD	*N* = 126, 98% Caucasian, M_{age} = ~ 35 yrs, 36% female	15 × 2.5hr weekly group (*N* = 62) or Internet-delivered (*N* = 64) CBT	No significant association with treatment response.

(continued)

Table 26.1 Continued

Gene	Authors	Ref.	Diagnosis	Sample	Treatment	Results
HTR1A (rs6295)	Serretti et al. (2013)	51	DEP	$N = 43$, $M_{age} = 51$ yrs*, 66% female* (* From sample with antidepressants, $N = 160$)	6-week interpersonal counselling	No significant association with response to IPC.
	Straube et al. (2014)	53	PD + Ag	$N = 171$ $M_{age} = 36$ yrs, 70% female, Caucasian (Overlaps with Reif et al.)	12 × twice weekly exposure-based CBT	No significant association with response to exposure-based CBT.
HTR2A (rs6311/ G1438A)	Steiger et al. (2008)	52	BN	$N = 98$, Caucasian, $M_{age} = 27$, 100% female	16–32 wk multimodal therapy based on individual CBT for all participants, with additional group CBT and medication	G allele carriers showed less reduction in binge episodes following 32 wks of therapy.
HTR2A (rs7997012)	Kotte et al. (2007)	39	DEP	$N = 58$, 100% male	16 wk group CBT	G allele predicted significantly larger reduction in BDI scores across treatment compared to A-allele carriers.
HTR2A (rs7997012, rs2224721)	Serretti et al. (2013)	51	DEP	$N = 43$, $M_{age} = 51$ yrs*, 66% female* (* From sample with antidepressants, $N = 160$)	6 wk interpersonal counselling	No significant association with response to IPC.
HTR2A (rs6314)	Matsumoto et al. (2014)	45	DEP	$N = 40$, $M_{age} = 44$ yrs, 60% female (Overlaps with Serretti et al.)	6 wk interpersonal counselling	No significant association with response to IPC.

TPH2 (rs4570625/ G703T)	Andersson et al. (2013)	23	SAD	*N* = 314, M_{age} = 37 yrs, 50% female (Overlaps with Hedman et al.)	Trial 1 (*N* = 126): As Hedman, Trial 2 (*N* = 204): 9 wks of Internet delivered CBT	Effects of risk genotype on treatment response at post-treatment, but inconsistent direction of effect across trials. No significant effect on treatment response at long-term follow-up.
TPH2 (rs11179023)	Matsumoto et al. (2014)	45	DEP	*N* = 40, M_{age} = 44 yrs, 60% female (Overlaps with Serretti et al.)	6 wk interpersonal counselling	No significant association with response to IPC.
COMT (rs4680 val158met)	Lonsdorf et al. (2010)	43	PD ± Ag	*N* = 69, Caucasian, M_{age} = ~ 35 yrs, 62% female	10 × 2hr weekly group (*N* = 38) or Internet-delivered (*N* = 31) CBT	No significant effect of COMTVal158Met genotype on change in anxiety or depression scores across cognitive modules (wks 1–3). Met/Met genotype carriers had significantly smaller reduction in anxiety scores across exposure modules (wks 4–9) compared to Val carriers.
	Hedman et al. (2012)	36	SAD	*N* = 126, 98% Caucasian, M_{age} = ~ 35 yrs, 36% female	15 × 2.5hr weekly group (*N* = 62) or Internet-delivered (*N* = 64) CBT	No significant association with treatment response.
	Andersson et al. (2013)	23	SAD	*N* = 314, M_{age} = 37 yrs, 50% female (Overlaps with Hedman et al.)	Trial 1 (*N* = 126): As Hedman, Trial 2 (*N* = 204): 9 wks of Internet-delivered CBT	No significant effect on treatment response at long-term follow-up.
	Serretti et al. (2013)	51	DEP	*N* = 43, M_{age} = 51 yrs*, 66% female* (* From sample with antidepressants, *N* = 160)	6-week interpersonal counselling	No significant association with response to IPC.

(continued)

Table 26.1 Continued

Gene	Authors	Ref.	Diagnosis	Sample	Treatment	Results
	Bakker et al. (2014)	24	DEP	*N* = 126, Caucasian, M_{age} = 45 yrs, 78% female	8 × 2.5 hour weekly sessions of mindfulness-based CBT vs treatment as usual	No significant association with treatment response.
DRD4 (rs936461)	Bakker et al. (2014)	24	DEP	*N* = 126, Caucasian, M_{age} = 45 yrs, 78% female	8 × 2.5 hour weekly sessions of mindfulness-based CBT vs treatment as usual	Significant association in genotype × group × time interaction, but effect driven by changes in the control group.
DRD2 (rs6276 and rs6277)	Bakker et al. (2014)	24	DEP	*N* = 126, Caucasian, M_{age} = 45 yrs, 78% female	8 × 2.5 hour weekly sessions of mindfulness-based CBT vs treatment as usual	No significant association with treatment response.
SLC6A3/DAT1 (rs6347)	Bakker et al. (2014)	24	DEP	*N* = 126, Caucasian, M_{age} = 45 yrs, 78% female	8 × 2.5 hour weekly sessions of mindfulness-based CBT vs treatment as usual	No significant association with treatment response.
MAOA (u VNTR)	Reif et al. (2014)	46	PD ± Ag	*N* = 288, Caucasian (Overlaps with Lonsdorf et al.)	12 × twice weekly CBT	Carriers of the long, higher activity allele had significantly worse treatment outcome, elevated heart rate, greater fear and panic attacks during a behavioral avoidance task and failure to habituate during repetitive exposure.
GRIN2B (rs1019385)	Sakolsky et al. (2010)	48	SEP; GAD; SAD	*N* = 211, Caucasian, 7–17 yrs (Overlaps with Sakolsky et al.)	Sertraline, 14 sessions of CBT, combination therapy or 12 wk placebo	No significant association with treatment response.

GRIK4 (rs1954787)	Sakolsky et al. (2010)	48	SEP; GAD; SAD	*N* = 211, Caucasian, 7–17 yrs (Overlaps with Sakolsky et al.)	Sertraline, 14 sessions of CBT, combination therapy or 12 wk placebo	Significant association with treatment response at 12 week assessment.
CHRM2 (rs324650, rs1824024)	Bakker et al. (2014)	24	DEP	*N* = 126, Caucasian, M_{age} = 45 yrs, 78% female	8 × 2.5 hour weekly sessions of mindfulness-based CBT vs treatment as usual	Significant association with rs1824024 in genotype × group × time interaction, increasing positive affect from baseline to post-treatment in the treatment group.
OPRM1 (rs495491, rs609148)	Bakker et al. (2014)	24	DEP	*N* = 126, Caucasian, M_{age} = 45 yrs, 78% female	8 × 2.5 hour weekly sessions of mindfulness-based CBT vs treatment as usual	Significant associations with both genotypes in genotype × group × time interaction, increasing positive affect from baseline to post-treatment in the treatment group.
BDNF (rs6265 val66met)	Sakolsky et al. (2010)	48	SEP; GAD; SAD	*N* = 211, Caucasian, 7–17 yrs (Overlaps with Sakolsky et al.)	Sertraline, 14 sessions of CBT, combination therapy or 12 wk placebo	No significant association with treatment response.
	Lester et al. (2012)	40	ANX	*N* = 374, Caucasian, M_{age} = 9.46 yrs, 49% female (Overlaps with Eley et al.)	10–12 session group or individual CBT or guided self-help	No significant association with treatment response.
	Hedman et al. (2012)	36	SAD	*N* = 126, 98% Caucasian, M_{age} = ~ 35 yrs, 36% female	15 × 2.5hr weekly group (*N* = 62) or Internet-delivered (*N* = 64) CBT	No significant association with treatment response.
	Fullana et al. (2012)	34	OCD	*N* = 106, Caucasian, M_{age} = 33 yrs, 50% female	20 × 45min weekly exposure-based CBT plus SSRI	Met allele carriers significantly less likely to respond to treatment than non-Met-allele carriers. Genotype predicted response only and not change in severity scores.

(continued)

Table 26.1 Continued

Gene	Authors	Ref.	Diagnosis	Sample	Treatment	Results
	Felmingham et al. (2013)	33	PTSD	*N* = 55, Caucasian, M_{age} = ~ 47yrs, 36% female. (Overlaps with Bryant et al.)	8 wk exposure therapy	Met allele associated with poorer response to CBT post-treatment.
	Matsumoto et al. (2014)	45	DEP	*N* = 40, M_{age} = 44 yrs, 60% female (Overlaps with Serretti et al.)	6 wk interpersonal counselling	No significant association with response to IPC.
	Santacana et al. (2016)	50	PD ± Ag	*N* = 97, Spanish Caucasian, M_{age} = 36 years, 63% female	9 wk exposure-based group CBT	rs6265 genotype did not predict membership of a high response vs low response trajectory.
BDNF (rs6265, rs11030101, rs11030102)	Bakker et al. (2014)	24	DEP	*N* = 126, Caucasian, M_{age} = 45 yrs, 78% female	8 × 2.5 hour weekly sessions of mindfulness-based CBT vs treatment as usual	Significant association with rs11030101 in genotype × group × time interaction, but effect driven by changes in the control group.
NGF (rs6330)	Lester et al. (2012)	40	ANX	*N* = 374, Caucasian, M_{age} = 9.46 yrs, 49% female (Overlaps with Eley et al)	10–12 session group or individual CBT or guided self-help	No significant association with treatment response at post treatment. At follow-up, children with one or more copies of T allele were significantly more likely to be free of anxiety diagnosis.
NR3C1/GR (rs4142324)	Yehuda et al. (2014)	56	PTSD	*N* = 25, 57% Caucasian, 43% Hispanic, M_{age} = 49 yrs, 11% female* (* includes control group, *N* = 12)	Prolonged exposure treatment 12 × 1.5 hour weekly sessions, vs minimal attention	Responders (post-treatment) were significantly more likely to carry GG-/CG- than CC-alleles.

NR3C1/GR (rs6195, rs4142324)	Roberts et al. (2015)	47	ANX	*N* = 1000, Mixed ancestry 64.3% Caucasian, M_{age} = 9.9 yrs, 51% female (Overlaps with Eley et al. demographics from full sample *N* = 1152)	Manualized CBT (individual, group, or guided self-help), varying (4–52 wk) duration	No significant association with treatment response.
CRHR1 (rs7209436, rs110402, rs242924 TAT haplotype)	Cicchetti et al. (2015)	28	DEP	*N* = 58, 54% African-American, 22% Caucasian, 24% Hispanic/biracial, M_{age} = 25 yrs, 100% female* (* includes control group *N* = 68)	Manualized interpersonal psychotherapy, 14 wks or treatment as usual	Women with 0 copies of the TAT haplotype who participated in IPT reported significantly fewer depressive symptoms at post-intervention and eight-month follow-up compared to those who participated in treatment as usual. No difference was observed between treatment groups in individuals with 1 or 2 copies of the TAT haplotype.
FKBP5 (rs1360780)	Wilker et al. (2014)	55	PTSD	*N* = 43, Ugandan, M_{age} = 31.91 yrs, 67% female	12 biweekly sessions of narrative exposure therapy	No genotype-dependent differences in therapy outcome at four-month follow-up. Carriers of the risk (T) allele were at increased risk of symptom relapse at ten-month follow-up.
FKBP5 (Single region containing rs1360780, rs3800373, rs9296158, rs9470080, rs4713916)	Roberts et al. (2015)	47	ANX	*N* = 924, Mixed ancestry 64.3% Caucasian, M_{age} = 9.9 yrs, 51% female (Overlaps with Eley et al., demographics from full sample *N* = 1152)	Manualized CBT (individual, group, or guided self-help), varying (4–52 wk) duration	No significant association with treatment response, although CpG-specific methylation was associated with poorer treatment response in risk allele carriers.
CREB1 (rs2253206)	Matsumoto et al. (2014)	45	DEP	*N* = 40, M_{age} = 44 yrs, 60% female (Overlaps with Serretti et al.)	6 wk interpersonal counselling	No significant association with response to IPC.

(continued)

Table 26.1 Continued

Gene	Authors	Ref.	Diagnosis	Sample	Treatment	Results
CNR1 (15 regions containing 130 variants)	Lester et al. (2016a)	41	ANX	*N* = 920–1,175, Mixed ancestry (70% Caucasian), M_{age} = 9.8 yrs, 52% female (Overlaps with Lester et al., 2016b)	Manualized CBT (individual, group, or guided self-help), varying (4–52 wk) duration	rs806365-T was associated with less reduction in anxiety severity from post-treatment to a six-month follow-up in individuals with fear-based anxiety disorders specifically (*N* = 399).
CNR2 (2 regions containing 160 variants)	Lester et al. (2016a)	41	ANX	*N* = 929, Mixed ancestry (70% Caucasian), M_{age} = 9.8 yrs, 52% female (Overlaps with Lester et al., 2016b)	Manualized CBT (individual, group, or guided self-help), varying (4–52 wk) duration	No significant association with CBT response.
FAAH (13 regions containing 319 variants)	Lester et al. (2016a)	41	ANX	*N* = 939, Mixed ancestry (70% Caucasian), M_{age} = 9.8 yrs, 52% female (Overlaps with Lester et al., 2016b)	Manualized CBT (individual, group, or guided self-help), varying (4–52wk) duration	No significant association with CBT response.
GWAS (3017604 variants)	Coleman et al. (2016b)	29	ANX	*N* = 980, Mixed ancestry (92% Caucasian); M_{age} = 9.9 yrs, 55% female (Overlaps with Lester et al., 2016b)	Manualized CBT (individual, group, or guided self-help), varying (4–52 wk) duration	No variants were significantly associated with treatment response at a genome-wide level of significance ($p = 5 \times 10^{-8}$).
	Coleman et al. (2016a)	30	SP; PD + Ag	*N* = 181, Caucasian, M_{age} = 39 yrs, 64% female	Exposure therapy, varying duration (5–30 wk)	No variants were significantly associated with treatment response at a genome-wide level of significance ($p = 5 \times 10^{-8}$).

Diagnoses: AN = anorexia nervosa, BN = bulimia nervosa, PTSD = post-traumatic stress disorder, PD = panic disorder, Ag = agoraphobia, DEP = depression, SEP = separation anxiety, GAD = generalized anxiety disorder, SAD = social anxiety disorder, ANX = mixed anxiety disorders, OCD = obsessive compulsive disorder, SP = specific phobia.

Results from therapygenetic studies of the 5HTTLPR have been equivocal, with associations in both directions as well as null results reported (Table 26.1). Furthermore, the interpretation of these results is confounded by the biology of the variant and differences in the analytical model between studies. Specifically, the 5HTTLPR is linked to a single nucleotide polymorphism (rs25531), the G-allele of which reduces the expression of the serotonin transporter irrespective of the 5HTTLPR-alleles carried (63). This complexity is not accounted for in all studies of the 5HTTLPR. Furthermore, the model tested between studies is not consistent, with additive (number of S-alleles), dominant (LL versus SL/SS), recessive (LL/SL versus SS), and genotypic (LL versus SL versus SS) models used. As such, although the 5HTTLPR has been the most studied variant in therapygenetic research, methodological differences reduce the comparability of different studies and prevent strong conclusion regarding the effect of the 5HTTLPR on the outcome.

The serotonin transporter gene is affected by variants other than the 5HTTLPR. Two studies have explored the effect of a variable number tandem repeat in the second intron of the gene (STin2; 38, 49). Although the studies identified an improvement in response to psychotherapy associated with carrying the 12-copy tandem repeat, both were performed in medicated populations. As such, it is unclear whether the association between the variant and improved outcome was specific to psychotherapy. Furthermore, the sparsity of studies and the differences between them limit the strength of conclusion.

Interest in the serotonergic system in emotional disorders has extended beyond the serotonin transporter. There are 14 known serotonin receptors, of which *HTR2A* and *HTR1A* have been examined in a therapygenetic context (64). Both receptor types are expressed broadly across the brain, with particular density in cortical regions and the raphe nuclei (65–67). Activation of the serotonin receptors induces changes in the expression of genes involved in synaptic plasticity, making these receptors good candidates for a learning-based process such as response to psychotherapy (66). The production of serotonin occurs in a two-step process, in which tryptophan hydroxylase is the rate-limiting enzyme (68). *TPH2* encodes a brain-specific isoform of this protein and as such has a central role in the action of serotonin in the brain (69).

There is little evidence to date for an effect of genetic variants in the wider serotonergic system on the outcome of psychotherapy. Two studies have explored the role of the rs6295 variant in *HTR1A* in different therapies, but both found no association (51, 53). Further variants have been studied in *HTR2A*, with two associations with response reported (39, 52). However, there have been no direct replication attempts for either of these findings, and a later study investigating the association between rs7997012-G and improved treatment response found no effect (although the differences between the cohorts studied were considerable; Table 26.1; 39, 51). Two studies have investigated variants in *TPH2*, with inconsistent or null findings (23, 45).

26.4 Catecholamine system genes: Catechol-*O*-methyltransferase (*COMT*), dopamine receptor D4 (*DRD4*), dopamine receptor D2 (*DRD2*), dopamine transporter 1 (*SLC6A3/DAT1*), and monoamine oxidase A (*MAOA*)

Beyond the serotonergic system, there has been considerable interest in the role of the catecholamine neurotransmitters (which include dopamine, adrenaline, and noradrenaline) in emotional disorders and their treatment, as these systems are targeted by antidepressant monoamine oxidase inhibitors (70). The dopaminergic system has been of particular interest due to its robust associations with psychiatric disorders, particularly dopamine receptor D2 (the product of *DRD2*), which is the site of action for all known antipsychotic drugs (71).

Multiple genetic variants near catecholamine genes have been studied to assess their effects on the etiology and treatment of emotional disorders, but the functional rs4680 (Val158Met) *COMT* polymorphism has drawn the most attention. COMT catalyzes the inactivation of dopamine, adrenaline, and noradrenaline, a process which has been implicated in psychiatric disorders (72). Several studies have investigated the effect of rs4680 on response to psychological therapies, but results have been unconvincing. Of the four independent studies published to date—the study by Andersson and colleagues (23) extends Hedman and co-workers' study (36)—only one reported a positive result, an association with poorer response in Met-Met homozygotes (43). Given the relatively small sample size of this study, and the lack of replication of this result (including in a genome-wide association study; 29), there is little evidence to support an association of rs4680 with response to psychological therapy.

A single study has expanded the field of therapygenetics into the dopaminergic system, studying variants around *DRD2, DRD4*, and *SLC6A3/DAT* in the context of mindfulness-based CBT for depression (24). No significant associations were identified for variants around *DRD4* or *SLC6A3*, and although a significant relationship between *DRD2* variants and change over time was identified, this was driven by changes in the treatment as usual control group.

The most promising of the catecholamine system genes in therapygenetics is monoamine oxidase A (*MAOA*). A study of the effect of the promoter u-VNTR on CBT response in individuals with panic disorder found a significantly poorer outcome in carriers of the longer, high activity allele (46).

26.5 Other neurotransmitter system genes: Glutamate receptor, ionotropic, *N*-methyl D-aspartate 2B (*GRIN2B*), and glutamate receptor, ionotropic, kainate 4 (*GRIK4*), cholinergic receptor, muscarinic 2 (*CHRM2*), and the opioid receptor, μ 1 (*OPRM1*)

The role of neurotransmitters remains one of the most promising avenues of research into the emotional disorders, providing a potential means by which environmental influences (such as receiving therapy) might result in biological alterations through altering neuronal activity (73). As a consequence, multiple neurotransmitter and receptor systems have been the subject of therapygenetic investigations. Glutamatergic excitation is a common feature of vertebrate nervous systems and within the human brain (74, 75). The glutamate system has been of interest in psychiatric and psychological studies because chemical blockade of glutamate receptors in the treatment of stroke (among other neurological disorders) can lead to behavioral side effects, suggesting a potential role in mental disorders more generally (76). Similarly, interest in the opioid system has been driven by the euphoric effects of opiate drugs and their influence on the brain reward pathway, which is hypothesized to contribute to positive affective experiences (24, 77). Finally, other candidate genes have emerged from a data-driven approach—for example, a genome-wide linkage study implicated variation proximal to *CHRM2* with electrophysiological inhibition in individuals at risk for alcoholism (78).

Therapygenetic investigations into receptor systems have yielded findings with some preliminary interest. Sakolsky and colleagues explored the effect of variants near glutamatergic system genes on CBT response, and, although no association was found with the rs1019385 variant near *GRIN2B*, a significant association was found with rs1954787, located in an intron of *GRIK4* (48). Similarly, variants in *CHRM2* and *OPRM1* were associated with increased positive affect in a separate analysis of depressed individuals undergoing mindfulness-based CBT (24). However, all these findings result from single studies and require independent replication in well-powered samples.

26.6 Neurotrophins: Brain-derived neurotrophic factor (*BDNF*) and nerve growth factor (*NGF*)

Much of the biological theory underlying treatment response concerns neuronal plasticity. One aspect of this is the recruitment and action of neurotransmitter systems at neuronal synapses. Another important contributor may be neurotrophins, chemical guidance cues involved in the construction of new neurons and neuronal networks (79). Studies of protein levels in the brain have suggested decreased levels of such cues in animals displaying anxious and depressive behaviors, particularly for *BDNF* and *NGF* (80, 81).

The neurotrophins have been the subject of several therapygenetic investigations, but such studies have largely focused on rs6265, the Val66Met polymorphism in *BDNF*. This variant has been studied in eight separate investigations but results lack promise. Although two studies identified an association between poorer treatment response and Met-allele count (33–34), null results were seen in seven other investigations, some of which were larger and more powerful (29, 36, 40, 45, 48, 50). Other variation proximal to *BDNF* has not been implicated in response to CBT (24). One study identified an association between the T-allele of the rs6330 variant in *NGF* and response to CBT at follow-up (40). However, this association did not replicate in a genome-wide association study which included the same cohort (29).

26.7 Hypothalamic–pituitary–adrenal axis genes: Glucocorticoid receptor (*NR3C1/GR*), corticotropin-releasing hormone receptor 1 (*CRHR1*), FK506 binding protein 5 (*FKBP5*)

The hypothalamic–pituitary–adrenal (HPA) axis is an important neuroendocrinological system that directs the reaction to acute stress (82). Given that the symptoms of anxiety disorders are a disproportionate or chronic reaction to stress, there is obvious interest in the effects of stress-related biological systems in treatment response for anxiety, and this represents part of the wider theory that cause might inform cure (83–85).

Variation in the genes associated with the HPA axis has been studied in four investigations to date. Despite some positive results, there is only weak evidence for an association with psychological therapy response thus far. Variation in both *NR3C1* and *FKBP5* have been associated with differential treatment response in small samples (55, 56), but these effects have not been replicated in a much larger sample, albeit one with very different characteristics (47). An investigation of the rs7209436–rs110402–rs242924 TAT haplotype in *CRHR1* found that individuals who did not carry the haplotype responded better to interpersonal psychotherapy than to treatment as usual, but this investigation did not assess differences within treatment (28).

26.8 Cyclic AMP-responsive element binding protein 1 (*CREB1*)

Many of the candidate pathways of interest in emotional disorders and their treatment cause the phosphorylation of *CREB1*, including post-synaptic signaling involving dopamine and serotonin receptors (86). Phosphorylation of *CREB1* enables it to act as a transcription factor, resulting in a cascade of altered gene expression. As such, functional variation in *CREB1*, or alterations to the expression of the gene, could have downstream effects on neuronal activity. Only one small study has investigated variation proximal to this gene, finding no association with response to interpersonal counselling (45).

26.9 Endocannabinoids: Cannabinoid receptor 1 (*CNR1*), cannabinoid receptor 2 (*CNR2*), fatty acid amide hydrolase (*FAAH*)

Many hypothesis-driven therapygenetics studies have subscribed to the idea that "cause informs cure" (85). The endocannabinoid system has been implicated in emotional processing, and particularly in the acquisition and loss of fears (87–89).

One study has investigated the effect of genetic variants near endocannabinoid genes on response to CBT (41). This study made use of variation across the *CNR1*, *CNR2*, and *FAAH* genes (covering approximately 30 independent genetic regions), and identified a significant association between the rs806365 variant in *CNR1* and change in symptom severity from post-treatment to a six-month follow-up. Given the exploratory nature of this study, this finding requires replication.

26.10 Genome-wide association studies

To date, two studies have explored the association between variants genome-wide on response to CBT, modeled as the change in clinician-rated severity from pre-treatment to post-treatment and to a six-month follow-up (29, 30). No significant associations with response were identified, suggesting the effect sizes of individual variants on CBT response are likely to be small. In the larger of these studies, nine variants from previously reported candidate gene studies were explicitly tested (*HTR2A*: rs6311, rs6313; *COMT*: rs4680; *NGF*: rs6330; *BDNF*: rs6265, rs7934165, rs1519480, rs11030104, rs1954787), with none reaching nominal significance in the analyses (all $p > 0.05$; 29).

26.11 Future directions

As is apparent from the studies included in the review, the field of therapygenetics has been dominated to date by hypothesis-driven candidate gene studies. These investigations have derived largely from the proposed etiological role of neurotransmitters, corticosteroids, and axon guidance cues in emotional disorders. While these overall hypotheses appear to be well-supported by evidence, particularly from the response of these systems to pharmacotherapy, no individual causal genetic variants have been robustly identified near related genes. This is in concordance with the failure of biological hypotheses to translate into robust genetic findings for psychiatric traits more generally (90–92). Few variants proposed by candidate gene studies have been successfully replicated in hypothesis-neutral genome-wide association studies (GWAS; 93). This lack of replication has resulted in part from the limitations of the candidate gene method, which are apparent in the therapygenetic literature. Specifically, candidate gene studies have tended to assess a limited number of genes, usually without capturing the full range of polymorphisms surrounding those genes, and using cohorts that are underpowered to detect the (most likely very small) effects of individual genetic variants on psychiatric outcomes (94, 95).

There is a lack of attempted replication across the therapygenetic literature. In part, this is due to the relative novelty of the field (the majority of studies have been published in the last five years), but it also reflects the numerous potential variants for hypothesis-driven study and the variety of analytical approaches possible to use. To date, no positive association of an individual genetic variant with response to psychological therapy has been successfully replicated. A possible exception is the Met-allele of rs6265 in BDNF, which has been implicated in poorer response to exposure therapy in two cohorts treated for different disorders; however, this variant was not associated with response in seven other studies (33, 34).

Any predictor must explain a sufficient amount of variance in response in order to be effective. The evidence to date, both from the burgeoning field of therapygenetics and from behavioral genomic studies in general, suggests that individual variants are likely to explain only a very small amount of variance, and that the common genetic component of variance in behavioral traits is likely to be made up of many variants (96). As such, focusing on the role of individual variants seems unlikely to be productive in explaining differential response to psychological therapies.

Instead, the field must turn to the power of genome-wide methods to detect the small effects of individual variants and to understand their combined influence. GWAS can assay millions of variants across the genome. The cost of such studies has decreased considerably since their inception a decade ago such that it is now feasible to genotype the thousands of individuals necessary to detect the small effects of individual variants.

Genome-wide genotype data can also provide insight beyond the discovery of variants. All individuals share some genetic variants. Through identifying thousands of variants across the genome, it is possible to infer how much variation any two individuals share. Approaches such as GREML (genomic-relatedness-matrix restricted maximum likelihood) from the software package GCTA (97) combine this information with phenotype data to estimate the contribution of genetic variants to phenotypic variance (known as the common-variant heritability, Figure 26.1). This estimate is only a portion of the heritability estimated from twin studies because it includes only variance captured by the common variants assessed. However, the ability to assess the relatedness of individuals directly enables estimates of heritability to be made from population cohorts, rather than requiring family-based studies.

The GREML method requires the full genotype data of all participants to be available. This restraint can be lifted by using the summary statistics from existing GWAS analyses to predict heritability, as is implemented in the LDScore method and program (98, 99). Both GREML and LDScore can also be used to calculate the genetic correlation between phenotypes.

Genome-wide genotype data can also be used to predict an outcome (such as response to psychological therapy) from individuals' genotypes. This would have obvious clinical value, provided the prediction was sufficiently accurate. For example, results from an existing GWAS can be used to create polygenic risk scores (100–102). Variants are ranked by their strength of association (p-value), and each risk allele is weighted by the effect size (log odds-ratio or beta) from the previous analysis. A weighted risk score can then be created from the known alleles in each individual, and

Case **A**	Case **A**		
Case **B**	+	Case **B**	
Control **C**	–	–	Control **C**
Control **D**	–	–	+

Figure 26.1 Simplified genomic relatedness matrix. The cases (A and B) are genetically more similar to each other than the average similarity (+), as are the controls (C and D). Cases and controls are genetically dissimilar (–). Genetic similarity is correlated with case-control status in this example, and a heritable component of variance could be calculated.

this can be used to predict likely outcome. The existing study could be of the same phenotype (response to psychological therapy), or a related trait of interest (such as depression case-control status). This approach has already been used effectively in the study of CBT response: individuals scoring highly on a polygenic risk score derived from a study of discordant anxiety between monozygotic twins responded better to individual CBT than to CBT delivered in a group (although replication is required; 103). Although these methods cannot yet predict phenotypes accurately enough for clinical usage, their power is likely to increase with the growing cohort sizes studied by international genomics consortia, and the development of secondary analytical methods, which represents a rapid growth area in genomics (104). Therapygenetics can benefit from such developments if the genomic approach is adopted.

The primary obstacle to developing "therapygenomics" is the need for large sample sizes to overcome the high level of multiple testing inherent in genome-wide investigations. There are a number of factors inherent to the study of response to psychological therapy that need to be overcome in establishing a suitable cohort. Firstly, the measurement and definition of "treatment response" is inconsistent between studies (14). Using multiple modalities and different reporters to measure treatment response produces a lower estimate of treatment success that is more valid than defining response with a single measure (14). Studies that use an intention-to-treat approach (rather than assessing only treatment completers) similarly show more robust results (14). The widespread adoption of a standardized definition and measurement of treatment response, conducted over multiple assessments during and after treatment, would produce a more valid and robust phenotype for genetic assessment, albeit with a considerable increase in investment from therapists and from participants.

The variability of measurement in studies of psychological therapy is one example of a broader issue of heterogeneity in the field. In addition to the obvious differences between different types of psychological therapy, heterogeneity is also present in terms of patient and therapist characteristics. Given the evidence that therapy response is affected not just by factors related to the participant and the therapist, but also by the effects of the therapist–participant relationship, it is arguable that the therapeutic experience may differ between all participants, even within a single, well-controlled trial (105–107). As such, heterogeneity may be an unavoidable feature of studying response to psychological therapies. This does not prevent genetic studies from working, but it is likely to increase the necessary cohort size required to identify genetic effects. This may be ultimately beneficial, as if any predictor is to inform treatment response it must be robust to the variability of real treatment settings (108, 109).

Sufficiently sized cohorts should be achievable through targeting large, nationwide programs such as the IAPT program in the United Kingdom and the NewAccess program in Australia (8, 110). In addition, Internet-delivered CBT has been demonstrated to have comparable efficacy to that delivered clinically, with the benefit of being able to reach considerably larger numbers, and this also represents an opportunity to obtain cohorts of sufficient size (111, 112). Finally, genomic studies can be effectively meta-analyzed (provided core similarities exist between studies), so small studies that lack the power to effect discovery alone can still contribute valuably as part of a broader meta-analytical effort.

The future need not be limited to genomics; for example, the development of large cohorts can also provide valuable insight into the demographic and clinical prediction of treatment response (16). Furthermore, because psychological therapy represents a controlled environmental exposure, it is likely that insight can be gained by examining the effect of this exposure on the genome itself (in the form of epigenetic modifications such as DNA methylation) and on the expression of genes (113). Such studies as have been done to date in this area have largely focused on a limited set of candidate regions, especially on the expression of *FKBP5* and HPA axis genes more generally (47, 56, 114–118). Initial genome-wide studies have been performed, and this approach may yield valuable insight as larger cohorts emerge (30).

Therapygenetics is beginning to come of age. Initial bounds for the most likely small effect sizes of genetic variants have been set, and the wider field of behavioral genetics and epigenetics can provide clear guidance of the path to successful variant discovery. By developing larger cohorts, and moving investigations genome-wide, the field can build on its solid foundations, and begin to contribute to improving the provision of psychological therapy to those in need.

References

1. **Gustavsson A, Svensson M, Jacobi F, Allgulander C, Alonso J, Beghi E**, et al. Cost of disorders of the brain in Europe 2010. Eur Neuropsychopharmacol. 2011;**21**:718–79.
2. **Whiteford HA, Degenhardt L, Rehm J, Baxter AJ, Ferrari AJ, Erskine HE**, et al. Global burden of disease attributable to mental and substance use disorders: Findings from the Global Burden of Disease Study 2010. Lancet. 2013;**382**:1575–86.
3. **Barlow DH, Bullis JR, Comer JS, Ametaj AA.** Evidence-based psychological treatments: An update and a way forward. Annu Rev Clin Psychol. 2013;**9**:1–27.
4. **Beck JS.** Cognitive therapy. In: IB Weiner, E Craighead (eds). The Corsini Encyclopedia of Psychology. **10**:561. Hoboken, NJ, USA:John Wiley & Sons, Inc.;2010.
5. **Baldwin DS, Anderson IM, Nutt DJ, Allgulander C, Bandelow B, den Boer JA**, et al. Evidence-based pharmacological treatment of anxiety disorders, post-traumatic stress disorder and obsessive-compulsive disorder: A revision of the 2005 guidelines from the British Association for Psychopharmacology. J Psychopharmacol. 2014;**28**:403–39.
6. **Bandelow B, Lichte T, Rudolf S, Wiltink J, Beutel ME.** The diagnosis of and treatment recommendations for anxiety disorders. Dtsch Arztebl Int. 2014;**111**:473–80.
7. **Bandelow B, Lichte T, Rudolf S, Wiltink J, Beutel ME.** The German guidelines for the treatment of anxiety disorders. Eur Arch Psychiatry Clin Neurosci. 2015;**265**:363–73.
8. **Clark DM.** Implementing NICE guidelines for the psychological treatment of depression and anxiety disorders: The IAPT experience. Int Rev Psychiatry. 2011;**23**:318–27.
9. **National Collaborating Centre for Mental Health (Great Britain).** 2011. Common Mental Health Disorders: Identification and Pathways to Care. Leicester: RCPsych Publications.
10. **Hofmann SG, Smits JA.** Cognitive-behavioral therapy for adult anxiety disorders: A meta-analysis of randomized placebo-controlled trials. J Clin Psychiatry. 2008;**69**:621–32.
11. **James AC, James G, Cowdrey FA, Soler A, Choke A.** Cognitive behavioural therapy for anxiety disorders in children and adolescents. Cochrane Database Syst Rev. 2013;**6**:CD004690.
12. **Pompoli A, Furukawa TA, Imai H, Tajika A, Efthimiou O, Salanti G.** Psychological therapies for panic disorder with or without agoraphobia in adults: A network meta-analysis. Cochrane Database Syst Rev. 2016;**4**:CD011004.
13. **Shinohara K, Honyashiki M, Imai H, Hunot V, Caldwell DM, Davies P**, et al. Behavioural therapies versus other psychological therapies for depression. Cochrane Database Syst Rev. 2013;**10**:CD008696.
14. **Loerinc AG, Meuret AE, Twohig MP, Rosenfield D, Bluett EJ, Craske MG.** Response rates for CBT for anxiety disorders: Need for standardized criteria. Clin Psychol Rev. 2015;**42**:72–82.
15. **Haby MM, Donnelly M, Corry J, Vos T.** Cognitive behavioural therapy for depression, panic disorder and generalized anxiety disorder: A meta-regression of factors that may predict outcome. ANZ J Psychiatry. 2006;**40**:9–19.
16. **Hudson JL, Keers R, Roberts S, Coleman JRI, Breen G, Arendt K**, et al. Clinical predictors of response to cognitive-behavioral therapy in pediatric anxiety disorders: The Genes for Treatment (GxT) Study. J Am Acad Child Adolesc Psychiatry. 2015;**54**:454–63.
17. **Olatunji BO, Cisler JM, Deacon BJ.** Efficacy of cognitive behavioral therapy for anxiety disorders: A review of meta-analytic findings. Psychiatr Clin North Am. 2010;**33**:557–77.

18. **Flückiger C, Del Re A, Wampold BE, Symonds D, Horvath AO.** How central is the alliance in psychotherapy? A multilevel longitudinal meta-analysis. J Couns Psychol. 2012;**59**:10–17.
19. **Webb CA, DeRubeis RJ, Barber JP.** Therapist adherence/competence and treatment outcome: A meta-analytic review. J Consult Clin Psychol. 2010;**78**:200–11.
20. **Heijmans BT, Mill J.** Commentary: The seven plagues of epigenetic epidemiology. Int J Epidemiol. 2012;**41**:74–8.
21. **Lester KJ, Eley TC.** Therapygenetics: Using genetic markers to predict response to psychological treatment for mood and anxiety disorders. Biol Mood Anxiety Disord. 2013;**3**:1–16.
22. **Lueken U, Zierhut KC, Hahn T, Straube B, Kircher T, Reif A,** et al. Neurobiological markers predicting treatment response in anxiety disorders: A systematic review and implications for clinical application. Neurosci Biobehav Rev. 2016;**66**:143–62.
23. **Andersson E, Rück C, Lavebratt C, Hedman E, Schalling M, Lindefors N,** et al. Genetic polymorphisms in monoamine systems and outcome of cognitive behavior therapy for social anxiety disorder. PLoS One. 2013;**8**:e79015.
24. **Bakker JM, Lieverse R, Menne-Lothmann C, Viechtbauer W, Pishva E, Kenis G,** et al. Therapygenetics in mindfulness-based cognitive therapy: Do genes have an impact on therapy-induced change in real-life positive affective experiences? Transl Psychiatry. 2014;**4**:e384.
25. **Bockting CLH, Mocking RJ, Lok A, Koeter MWJ, Schene AH.** Therapygenetics: The 5HTTLPR as a biomarker for response to psychological therapy? Mol Psychiatry. 2013;**18**:744–5.
26. **Bryant RA, Felmingham KL, Falconer EM, Benito LP, Dobson-Stone C, Pierce KD,** et al. Preliminary evidence of the short allele of the serotonin transporter gene predicting poor response to cognitive behavior therapy in posttraumatic stress disorder. Biol Psychiatry. 2010;**67**:1217–19.
27. **Castellini G, Ricca V, Lelli L, Bagnoli S, Lucenteforte E, Faravelli C,** et al. Association between serotonin transporter gene polymorphism and eating disorders outcome: A 6-year follow-up study. Am J Med Genet B. 2012;**159B**:491–500.
28. **Cicchetti D, Toth SL, Handley ED.** Genetic moderation of interpersonal psychotherapy efficacy for low-income mothers with major depressive disorder: Implications for differential susceptibility. Dev Psycopathol. 2015;**27**:19–35.
29. **Coleman JRI, Lester KJ, Keers R, Roberts S, Curtis C, Arendt K,** et al. Genome-wide association study of response to cognitive behavioral therapy in children with anxiety disorders. B J Psychiatry. 2016;**209**:236–43.
30. **Coleman JRI, Lester KJ, Roberts S, Keers R, Lee, SH, De Jong S,** et al. Separate and combined effects of genetic variants and pre-treatment whole blood gene expression on response to exposure-based cognitive behavioural therapy for anxiety disorders. World J Biol Psychiatry. 2017;**18**:215–26.
31. **Di Bella D, Erzegovesi S, Casolari A, Bassi T, Bellodi L.** Eating disorders, treatment response and 5-HTT gene. Am J Med Genet. 2002;**114**:739.
32. **Eley T, Hudson J, Creswell C, Tropeano M, Lester K, Cooper P,** et al. Therapygenetics: The 5HTTLPR and response to psychological therapy. Mol Psychiatry. 2012;**17**:236–7.
33. **Felmingham KL, Dobson-Stone C, Schofield PR, Quirk GJ, Bryant RA.** The brain-derived neurotrophic factor Val66Met polymorphism predicts response to exposure therapy in posttraumatic stress disorder. Biol Psychiatry. 2013;**73**:1059–63.
34. **Fullana MA, Alonso P, Gratacòs M, Jaurrieta N, Jiménez-Murcia S, Segalàs C,** et al. Variation in the BDNF Val66Met polymorphism and response to cognitive-behavior therapy in obsessive-compulsive disorder. Eur Psychiatry. 2012;**27**:386–90.
35. **Gloster AT, Gerlach AL, Hamm A, Hoefler M, Alpers GW, Kircher T,** et al. 5HTT is associated with the phenotype psychological flexibility: Results from a randomized clinical trial. Eur Arch Psychiatry Clin Neurosci. 2015;**265**:399–406.
36. **Hedman E, Andersson E, Ljótsson B, Andersson G, Andersson E, Schalling M,** et al. Clinical and genetic outcome determinants of Internet- and group-based cognitive behavior therapy for social anxiety disorder. Acta Psychiatr Scand. 2012;**126**:126–36.

37. **Knuts I, Esquivel G, Kenis G, Overbeek T, Leibold N, Goossens L,** et al. Therapygenetics: 5-HTTLPR genotype predicts the response to exposure therapy for agoraphobia. Eur Neuropsychopharmacol. 2014;**24**:1222–8.

38. **Kohen R, Cain KC, Buzaitis A, Johnson V, Becker KJ, Teri L,** et al. Response to psychosocial treatment in poststroke depression is associated with serotonin transporter polymorphisms. Stroke. 2011;**42**:2068–70.

39. **Kotte A, McQuaid JR, Kelsoe J.** Genotypic Predictor of Depression Psychotherapy Treatment Outcome. 62nd Annual Scientific Convention and Meeting of the Society of Biological Psychiatry. San Diego, CA; 2007.

40. **Lester KJ, Hudson JL, Tropeano M, Creswell C, Collier DA, Farmer A,** et al. Neurotrophic gene polymorphisms and response to psychological therapy. Transl Psychiatry. 2012;**2**:e108.

41. **Lester KJ, Coleman JRI, Roberts S, Keers R, Breen G, Bögels S,** et al. Genetic variation in the endocannabinoid system and response to cognitive behavior therapy for child anxiety disorders. Am J Med Genet B. 2017;**174**:144–55.

42. **Lester KJ, Roberts S, Keers R, Coleman JRI, Breen G, Wong CCY,** et al. Non-replication of the association between 5HTTLPR and response to psychological therapy for child anxiety disorders. B J Psychiatry. 2016;**208**:182–8.

43. **Lonsdorf TB, Rück C, Bergström J, Andersson G, Öhman A, Lindefors N,** et al. The COMTval158met polymorphism is associated with symptom relief during exposure-based cognitive-behavioral treatment in panic disorder. BMC Psychiatry. 2010;**10**:99.

44. **Lueken U, Straube B, Wittchen H-U, Konrad C, Ströhle A, Wittmann A,** et al. Therapygenetics: Anterior cingulate cortex-amygdala coupling is associated with 5-HTTLPR and treatment response in panic disorder with agoraphobia. J Neural Transmission. 2015;**122**:135–44.

45. **Matsumoto Y, Fabbri C, Pellegrini S, Porcelli S, Politi P, Bellino S,** et al. Serotonin transporter gene: A new polymorphism may affect response to antidepressant treatments in major depressive disorder. Mol Diagn Ther. 2014;**18**:567–77.

46. **Reif A, Richter J, Straube B, Höfler M, Lueken U, Gloster AT,** et al. MAOA and mechanisms of panic disorder revisited: From bench to molecular psychotherapy. Mol Psychiatry. 2014;**19**:122–8.

47. **Roberts S, Keers R, Lester KJ, Coleman JRI. Breen G, Arendt K,** et al. HPA axis related genes and response to psychological therapies: Genetics and epigenetics. Depress Anxiety. 2015;**32**:861–70.

48. **Sakolsky D, Nurmi E, Birmaher B, March JS, Walkup J, Piacentini J,** et al. Association of GRIK4 with treatment response in the Child/Adolescent Anxiety Multimodal Study (CAMS). American Academy of Child & Adolescent Psychiatry 57th Annual Meeting. New York, NY; 2010.

49. **Sakolsky D, Nurmi E, Birmaher B, March JS, Walkup J, Piacentini J,** et al. Serotonin transporter variation and treatment response in the Child/Adolescent Anxiety Multimodal Study (CAMS). American Academy of Child & Adolescent Psychiatry + Canadian Academy of Child and Adolescent Psychiatry Joint Annual Meeting. Toronto; 2011.

50. **Santacana M, Arias B, Mitjans M, Bonillo A, Montoro M, Rosado S,** et al. Predicting response trajectories during cognitive-behavioural therapy for panic disorder: No association with the *BDNF* gene or childhood maltreatment. PLoS One. 2016;**11**:e0158224.

51. **Serretti A, Fabbri C, Pellegrini S, Porcelli S, Politi P, Bellino S,** et al. No effect of serotoninergic gene variants on response to interpersonal counseling and antidepressants in major depression. Psychiatry Investig. 2013;**10**:180–9.

52. **Steiger H, Joober R, Gauvin L, Bruce KR, Richardson J, Israel M,** et al. Serotonin-system polymorphisms (5-HTTLPR and-1438G/A) and responses of patients with bulimic syndromes to multimodal treatments. J Clin Psychiatry. 2008;**69**:1565–71.

53. **Straube B, Reif A, Richter J, Lueken U, Weber H, Arolt V,** et al. The functional -1019C/G HTR1A polymorphism and mechanisms of fear. Transl Psychiatry. 2014;**4**:e490.

54. **Wang Z, Harrer J, Tuerk P, Acierno R, Hamner M, Timmerman MA,** et al. 5-HTTLPR Influence PTSD Treatment Outcome. 64th Annual Scientific Convention and Meeting of the Society of Biological Psychiatry. Vancouver, Canada; 2009.

55. **Wilker S, Pfeiffer A, Kolassa S, Elbert T, Lingenfelder B, Ovuga E,** et al. The role of FKBP5 genotype in moderating long-term effectiveness of exposure-based psychotherapy for posttraumatic stress disorder. Transl Psychiatry. 2014;**4**:e403.
56. **Yehuda R, Pratchett LC, Elmes MW, Lehrner A, Daskalakis NP, Koch E,** et al. M. Glucocorticoid-related predictors and correlates of post-traumatic stress disorder treatment response in combat veterans. Interface Focus. 2014;4.
57. **Furmark T, Carlbring P, Hammer S, Wahlgren I, Ekselius L, Eriksson E,** et al. Effects of serotonin transporter and tryptophan hydroxylase-2 gene variation on the response to cognitive-behavior therapy in individuals with social anxiety disorder. Biol Psychiatry. 2010;**67**:114S-114S.
58. **Blakely RD, Berson HE, Fremeau Jr. RT, Caron MG, Peek MM, Prince HK,** et al. Cloning and expression of a functional serotonin transporter from rat brain. Nature. 1991;**354**:66–70.
59. **Hoffman BJ, Mezey E, Brownstein MJ.** Cloning of a serotonin transporter affected by antidepressants. Science. 1991;**254**:579–80.
60. **Heils A, Teufel A, Petri S, Stöber G, Riederer P, Bengel D,** et al. Allelic variation of human serotonin transporter gene expression. J Neurochemistry. 1996;**66**:2621–4.
61. **Caspi A, Sugden K, Moffitt TE, Taylor A, Craig IW, Harrington H,** et al. Influence of life stress on depression: Moderation by a polymorphism in the 5-HTT gene. Science. 2003;**301**:386–9.
62. **Sharpley CF, Palanisamy SK, Glyde NS, Dillingham PW, Agnew LL.** An update on the interaction between the serotonin transporter promoter variant (5-HTTLPR), stress and depression, plus an exploration of non-confirming findings. Behav Brain Res. 2014;**273**:89–105.
63. **Nakamura M, Ueno S, Sano A, Tanabe H.** The human serotonin transporter gene linked polymorphism (5-HTTLPR) shows ten novel allelic variants. Mol Psychiatry. 2000;**5**:32–38.
64. **Nichols DE, Nichols CD.** Serotonin receptors. Chem Reviews. 2008;**108**:1614–41.
65. **GTEx Consortium.** The Genotype-Tissue Expression (GTEx) pilot analysis: Multitissue gene regulation in humans. Science. 2015;**348**:648–60.
66. **Nichols CD, Sanders-Bush E.** A single dose of lysergic acid diethylamide influences gene expression patterns within the mammalian brain. Neuropsychopharmacology. 2002;**26**:634–42.
67. **Nichols DE.** Hallucinogens. Pharmacol Ther. 2004;**101**:131–81.
68. **Fitzpatrick PF.** Tetrahydropterin-dependent amino acid hydroxylases. Ann Rev Biochem. 1999;**68**:355–81.
69. **Walther DJ, Peter J-U, Bashammakh S, Hörtnagl H, Voits M, Fink H,** et al. Synthesis of serotonin by a second tryptophan hydroxylase isoform. Science. 2003;**299**:76.
70. **Lachman HM, Papolos DF, Saito T, Yu YM, Szumlanski CL, Weinshilboum RM.** Human catechol-*O*-methyltransferase pharmacogenetics: Description of a functional polymorphism and its potential application to neuropsychiatric disorders. Pharmacogenetics. 1996;**6**:243–50.
71. **Kapur S, Mamo D.** Half a century of antipsychotics and still a central role for dopamine D2 receptors. Prog Neuro-Psychopharmacol Biol Psychiatry. 2003;**27**:1081–90.
72. **Morilak DA, Frazer A.** Antidepressants and brain monoaminergic systems: A dimensional approach to understanding their behavioural effects in depression and anxiety disorders. Int J Neuropsychopharmacology. 2004;**7**:193–218.
73. **Moghaddam B.** Stress activation of glutamate neurotransmission in the prefrontal cortex: Implications for dopamine-associated psychiatric disorders. Biol Psychiatry. 2002;**51**:775–87.
74. **Bar-Peled O, Ben-Hur H, Biegon A, Groner Y, Dewhurst S, Furuta A,** et al. Distribution of glutamate transporter subtypes during human brain development. J Neurochemistry. 1997;**69**: 2571–80.
75. **Meldrum BS.** Glutamate as a neurotransmitter in the brain: Review of physiology and pathology. J Nutrition. 2000;**130**:1007S-1015S.
76. **Cull-Candy S, Brickley S, Farrant M.** NMDA receptor subunits: Diversity, development and disease. Curr Opin Neurobiol. 2001;**11**:327–35.

77. **Van Ree JM, Niesink RJ, Van Wolfswinkel L, Ramsey NF, Kornet MM, Van Furth WR,** et al. Endogenous opioids and reward. Eur J Pharmacology. 2000;**405**:89–101.
78. **Wang JC, Hinrichs AL, Stock H, Budde J, Allen R, Bertelsen S,** et al. Evidence of common and specific genetic effects: Association of the muscarinic acetylcholine receptor M2 (CHRM2) gene with alcohol dependence and major depressive syndrome. Hum Mol Genet. 2004;**13**:1903–11.
79. **Duman RS, Monteggia LM.** A neurotrophic model for stress-related mood disorders. Biol Psychiatry. 2006;**59**:1116–27.
80. **Martinowich K, Manji H, Lu B.** New insights into BDNF function in depression and anxiety. Nat Neurosci. 2007;**10**:1089–93.
81. **Schmidt HD, Duman RS.** The role of neurotrophic factors in adult hippocampal neurogenesis, antidepressant treatments and animal models of depressive-like behavior. Behav Pharmacology. 2007;**18**:391–418.
82. **Gray TS.** Amygdaloid CRF pathways. Role in autonomic, neuroendocrine, and behavioral responses to stress. Ann N Y Acad Sci. 1993;**697**:53–60.
83. **Boyer P.** Do anxiety and depression have a common pathophysiological mechanism? Acta Psychiatr Scand. 2000;**102**:24–9.
84. **Pariante CM, Lightman SL.** The HPA axis in major depression: Classical theories and new developments. Trends Neurosci. 2008;**31**:464–8.
85. **Uher R.** The implications of gene-environment interactions in depression: Will cause inform cure? Mol Psychiatry. 2008;**13**:1070–8.
86. **Mayr B, Montminy M.** Transcriptional regulation by the phosphorylation-dependent factor CREB. Nat Rev Mol Cell Biol. 2001;**2**:599–609.
87. **Gunduz-Cinar O, MacPherson KP, Cinar R, Gamble-George J, Sugden K, Williams B,** et al. Convergent translational evidence of a role for anandamide in amygdala-mediated fear extinction, threat processing and stress-reactivity. Mol Psychiatry. 2013;**18**:813–23.
88. **Kathuria S, Gaetani S, Fegley D, Valiño F, Duranti A, Tontini A,** et al. Modulation of anxiety through blockade of anandamide hydrolysis. Nat Med. 2003;**9**:76–81.
89. **Lafenetre P, Chaouloff F, Marsicano G.** The endocannabinoid system in the processing of anxiety and fear and how CB1 receptors may modulate fear extinction. Pharmacol Res. 2007;**56**:367–81.
90. **Collins AL, Sullivan PF.** Genome-wide association studies in psychiatry: What have we learned? B J Psychiatry. 2013;**202**:1–4.
91. **Krishnan V, Nestler EJ.** Linking molecules to mood: New insight into the biology of depression. Am J Psychiatry. 2010;**167**:1305–20.
92. **Lehner T, Senthil G, Addington AM.** Convergence of advances in genomics, team science, and repositories as drivers of progress in psychiatric genomics. Biol Psychiatry. 2015;**77**:6–14.
93. **Dunn EC, Brown RC, Dai Y, Rosand J, Nugent NR, Amstadter AB,** et al. Genetic determinants of depression: Recent findings and future directions. Harv Rev Psychiatry. 2015;**23**:1–18.
94. **Dick DM, Agrawal A, Keller MC, Adkins A, Aliev F, Monroe S,** et al. Candidate gene-environment interaction research: Reflections and recommendations. Perspect Psychol Sci. 2015;**10**:37–59.
95. **Munafò MR, Zammit S, Flint J.** Practitioner review: A critical perspective on gene–environment interaction models—what impact should they have on clinical perceptions and practice? J Child Psychol Psychiatry. 2014;**55**:1092–101.
96. **Gratten J, Wray NR, Keller MC, Visscher PM.** Large-scale genomics unveils the genetic architecture of psychiatric disorders. Nat Neurosci. 2014;**17**:782–90.
97. **Yang J, Lee SH, Goddard ME, Visscher PM.** GCTA: A tool for genome-wide complex trait analysis. Am J Hum Genet. 2011;**88**:76–82.
98. **Bulik-Sullivan BK, Loh P-R, Finucane HK, Ripke S, Yang J, Schizophrenia Working Group of the Psychiatric Genomics Consortium,** et al. LD Score regression distinguishes confounding from polygenicity in genome-wide association studies. Nat Genet. 2015;**47**:291–5.

99. **Finucane HK, Bulik-Sullivan B, Gusev A, Trynka G, Reshef Y, Loh P-R,** et al. Partitioning heritability by functional annotation using genome-wide association summary statistics. Nat Genet. 2015;**47**:1228–35.
100. **Dudbridge F.** Power and Predictive Accuracy of Polygenic Risk Scores. PLoS Genet. 2013;**9**:e1003348.
101. **International Schizophrenia, Consortium, Purcell SM, Wray NR, Stone JL, Visscher PM, O'Donovan MC,** et al. Common Polygenic Variation Contributes to Risk of Schizophrenia and Bipolar Disorder. Nature. 2009;**460**:748–52.
102. **Wray NR, Lee SH, Mehta D, Vinkhuyzen AAE, Dudbridge F, Middeldorp CM.** Research Review: Polygenic methods and their application to psychiatric traits. J Child Psychol Psychiatry. 2014;**55**:1068–87.
103. **Keers R, Coleman JRI, Lester KJ, Roberts S, Breen G, Thastum M,** et al. A genome-wide test of the differential susceptibility hypothesis reveals a genetic predictor of differential response to psychological treatments for child anxiety disorders. Psychotherapy and Psychosomatics. 2016;**85**:146–58.
104. **Corvin A, Sullivan PF.** What next in schizophrenia genetics for the Psychiatric Genomics Consortium? Schizophrenia Bull. 2016;**42**:538–41.
105. **Ackerman SJ, Hilsenroth MJ.** A review of therapist characteristics and techniques positively impacting the therapeutic alliance. Clin Psychol Rev. 2003;**23**:1–33.
106. **Ilardi SS, Craighead WE.** The role of nonspecific factors in cognitive-behavior therapy for depression. Clin Psychol (New York). 1994;**1**:138–55.
107. **Martin DJ, Garske JP, Davis MK.** Relation of the therapeutic alliance with outcome and other variables: A meta-analytic review. J Consult Clin Psychol. 2000;**68**:438–50.
108. **Eley TC.** The future of therapygenetics: Where will studies predicting psychological treatment response from genomic markers lead? Depress Anxiety. 2014;**31**:617–20.
109. **Paulus MP.** Pragmatism instead of mechanism: A call for impactful biological psychiatry. JAMA Psychiatry. 2015;**72**:631–2.
110. **Cromarty P, Drummond A, Francis T, Watson J, Battersby M.** NewAccess for depression and anxiety: Adapting the UK Improving Access to Psychological Therapies Program across Australia. Australasian Psychiatry. 2016;**24**:489–92.
111. **Barak A, Hen L, Boniel-Nissim M, Shapira NA.** A comprehensive review and a meta-analysis of the effectiveness of Internet-based psychotherapeutic interventions. Journal of Technology in Human Services. 2008;**26**:109–60.
112. **Olthuis JV, Watt MC, Bailey K, Hayden JA, Stewart SH.** Therapist-supported Internet cognitive behavioural therapy for anxiety disorders in adults. Cochrane Database Syst Rev. 2016;**3**:CD011565.
113. **Szyf M.** Epigenetics, a key for unlocking complex CNS disorders? Therapeutic implications. Eur Neuropsychopharmacology. 2015;**25**:682–702.
114. **Kéri S, Szabó C, Kelemen O.** Expression of toll-like receptors in peripheral blood mononuclear cells and response to cognitive-behavioral therapy in major depressive disorder. Brain, Behav Immunity. 2014;**40**:235–43.
115. **Kéri S, Szabó C, Kelemen O.** Blood biomarkers of depression track clinical changes during cognitive-behavioral therapy. J Affect Disord. 2014;**164**:118–22.
116. **Levy-Gigi E, Szabó C, Kelemen O, Kéri S.** Association among clinical response, hippocampal volume, and FKBP5 gene expression in individuals with posttraumatic stress disorder receiving cognitive behavioral therapy. Biol Psychiatry. 2013;**12**:1–8.
117. **Roberts SR, Lester KJ, Hudson JL, Rapee RM, Creswell C, Cooper PJ,** et al. Serotonin transporter methylation and response to cognitive behaviour therapy in children with anxiety disorders. Transl Psychiatry. 2014;**4**:e444.
118. **Yehuda R, Daskalakis NP, Desarnaud F, Makotkine I, Lehrner AL, Koch E,** et al. Epigenetic biomarkers as predictors and correlates of symptom improvement following psychotherapy in combat veterans with PTSD. Front Psychiatry. 2013;**4**:1–14.

Chapter 27

The role of pharmacogenetics in the treatment of depression

Airiss R Chan, Ilona Gorbovskaya, and Daniel J Müller

27.1 Introduction

Antidepressants are the largest class of medications used in the treatment of psychiatric disorders, in particular for depression and anxiety. However, there are several issues surrounding the use of antidepressants, including the high variability in response and side effects among patients, as well as their varied response time. In addition, 60–70% of major depression disorder cases are treatment-resistant, thus complicating the treatment procedures for these patients (1–3). These complexities frequently contribute to a weakened management of the underlying psychiatric condition, and negatively affect patients' quality of life and healthcare costs.

Notably, antidepressant response has been proven to have a genetic basis, where genetic variants can vary substantially across ethnicities (4). Therefore, genetics may hold the key to unlocking personalized medicine, and it paves the way for pharmacogenetics, an emerging field of psychiatric research studying genetic differences in the pharmacokinetic and pharmacodynamic properties among patients. An increased understanding in psychiatric pharmacogenetics can provide physicians with the clinical advantage to overcome the challenges associated with outcomes to antidepressants.

27.2 Current stance in pharmacogenetics of antidepressants

Pharmacogenetics has evolved in response to the increasing need for treatment optimization. Once genetic variants have been identified and validated, there is great potential for clinical use since genetic tests can these days be easily performed in a non-invasive manner, that is by collecting saliva samples or cheek swabs (5, 6). In addition, costs have steadily dropped for these genetic sampling procedures over recent years, making pharmacogenetic testing increasingly more applicable for clinical use.

Most previous studies have mainly focused on identifying common variants and single-nucleotide polymorphisms (SNPs) in genes known to be involved in drug metabolism (pharmacokinetics) and target organs (pharmacodynamics). These studies have been complemented using hypothesis-free approaches such as genome-wide association studies (GWAS), which provide large amounts of information (i.e. typically millions of SNPs per individual) (7). Currently, there are a number of large pharmacogenetic studies (i.e. GENDEP, MARS, STAR*D, PGRN-AMPS, and ISPC)[1] which have used GWAS strategies to examine the metabolism, and the targeted receptors and transporters of antidepressants. While a wide number of polymorphisms associated with antidepressant response have been identified, to date, variants within the cytochrome P450 enzymes,

[1] GENDEP—Genome-Based Therapeutic Drugs for Depression (8); MARS—Munich Antidepressant Response Signature (9); STAR*D—Sequenced Treatment Alternative to Relieve Depression (10–12); PGRN-AMPS—Pharmacogenetics Research Network (13); ISPC—International SSRI Pharmacogenomics Consortium (13).

the serotonin transporter, and, for some TCAs and SNRIs, the norepinephrine transporter have been among the best studied, and have therefore the highest potential for clinical applicability.

27.3 Pharmacokinetics

27.3.1 Cytochrome P450 enzymes

The most prominent variants associated with plasma levels and therapeutic outcomes have been identified in cytochrome P450 (CYP) genes. CYP enzymes represent a superfamily of enzymes which are largely involved in the first phase of metabolism for a wide variety of medications. The most important CYP enzymes involved in most psychiatric medications are CYP2D6 and CYP2C19, with approximately 80% of antidepressants and antipsychotics being metabolized by CYP2D6 (14). CYP2C19 also metabolizes a wide variety of drugs, and plays an important role in the pharmacological activation of tricyclic antidepressants (15, 16). Therefore, early on, differences in CYP genotype have been hypothesized to affect how an individual will respond to antidepressant treatment (17).

Since various alleles and genotypes within CYP2D6 and CYP2C19 result in different metabolizer profiles, four classes of CYP metabolizer profiles have been established: poor (PM), intermediate (IM), extensive (EM), and ultra-rapid metabolizers (UM), although other categories such as activity have been proposed (18) and keep being discussed among experts. The EM/IM/PM/UM metabolizer profile has been used in numerous studies to determine how an individual might react to a drug. Table 27.1 describes the metabolizer profiles for CYP2D6 and CYP2C19.

Notably, in some cases, both the original compound and their metabolites can be active and exert antidepressant effects (e.g. amitriptyline being converted to nortriptyline through CYP2C19 or venlafaxine being converted to desvenlafaxine through CYP2D6) while for prodrugs, the original compounds are typically inactive or substantially less active but will become activated upon hydroxylation provided by CYP enzymes. However, none of the antidepressants reviewed here function as prodrugs, and most metabolites resulting from CYP450 hydroxylation are inactive or the contribution of metabolites to response and side effects remain currently unknown.

Extensive metabolizers are considered to have normal (standard) metabolism, as they mostly have two fully functional alleles. In contrast, poor metabolizers have genetic variants resulting in two non-functional alleles, and are less capable of metabolizing particular medications which are substrates of these CYP enzymes. This has been shown to result in increased serum concentrations, and an increased risk for side effects for selective serotonin reuptake inhibitors (SSRIs) (19). At the other end of the spectrum, ultra-rapid metabolizers have duplicated alleles resulting in increased function which leads to increased metabolism, decreased serum concentrations of the original compound, and increased probability for pharmacotherapy failure (19). Based on these differences, it would seem intuitive for physicians to use patient metabolizer profiles in order to optimize treatment.

CYP2D6 and 2C19 allele frequencies can vary significantly depending on ethnicity and population admixture. For example, the CYP2D6 poor metabolizer status is present in 5–10% of the general "Caucasian"[2] population, but is rarer in non-white populations with an estimated prevalence of 3% (21, 22). In "Caucasians," 7% of the population has been reported to be CYP2D6 poor metabolizers, and 5% of the population are CYP2D6 ultra-rapid metabolizers (14). Quite notably, the CYP2D6 IM phenotype is significantly higher in East Asian than other populations (approximately 39.4%), and the CYP2D6 UM phenotype has also been reported to be significantly higher

[2] While the literature frequently refers to Caucasians for white individuals of European ancestry, the term is misleading (20). Therefore, we advise against using the term and have used quotation marks whenever referring to the group.

Table 27.1 Metabolizer profiles for CYP2D6 and CYP2C19 phenotype.

Gene	Phenotype	Prevalence	Description of Genotype	Examples of Genotypes
CYP2D6	UM	1–2%	Duplications of functional alleles	*1/1xN, *1/2xN, *2/2xN[b]
	EM	77–92%	One of the following: - Two normal functional alleles - Two decreased function alleles - One normal function and one non-functional allele - One normal function allele and one decreased function allele	*1/*1, *1/*2, *1/*4, *1/*5, *1/*9, *1/*41, *2/*2, *41/41
	IM	2–11%	One decreased function and one no function allele	*4/*10, *4/*41, *5/*9
	PM	5–10%	No functional alleles	*3/*4, *4/*4, *5/*5, *5/*6
CYP2C19	UM	5–30%	One of the following: - Two increased function alleles - One normal function allele and one increased function allele	*17/*17, *1/*17
	EM	35–50%	Two normal function alleles	*1/*1
	IM	18–45%	One of the following: - One normal function allele - One increased function allele and one non-functional allele	*1/*2, *1/*3, *2/*17
	PM	2–15%	Two non-functional alleles	*2/*2, *2/*3, *3/*3

Adapted from Hicks JK, Swen JJ, Thorn CF, Sangkuhl K, Kharasch ED, Ellingrod VL, Skaar TC, Müller DJ, Gaedigk A, and Stingl JC., 'Clinical Pharmacogenetics Implementation Consortium guideline for CYP2D6 and CYP2C19 genotypes and dosing of tricyclic antidepressants', *Clinical Pharmacology and Therapeutics*, 93 (5), pp. 402–8. © 2015 American Society for Clinical Pharmacology and Therapeutics, published by John Wiley and Sons.

than other metabolizer profiles in North-east African populations (approximately 28.3%) (23, 24). Table 27.2 shows the allelic frequencies of select CYP polymorphisms among ethnicities.

Early identification of a patient's metabolizer profile can greatly inform treatment as drug dosing can be altered according to the predicted response of a patient. To illustrate further the importance of genetic screening, the following case report describes one such instance where knowledge of the metabolizer status can help to understand previous non-response and occurrence of side effects and inform future antidepressant treatment.

Case report: Ms X

A 24-year-old female patient diagnosed with major depressive disorder (MDD) and mild anxiety symptoms had been treated with three different antidepressants with numerous intolerable side effects. When taking a low dose of sertraline (25–50 mg/day), she experienced ongoing adverse gastric effects such heartburn and gastric reflux, forcing her to discontinue the medication after two months. She was then switched to Cipralex® (escitalopram) at 10 mg/day. While she showed some therapeutic benefits, she continued to experience impaired libido and discontinued this medication after 12 months. She was later titrated on 150 mg/day of Wellbutrin® (bupropion) which she did not tolerate due to cognitive side effects affecting her working memory and feeling "jittery" and medication was discontinued after a few weeks.

Table 27.2 Genotypic variance of select CYP2D6 and CYP2C19 polymorphisms across ethnicities.

Gene	Polymorphism & Metabolic Status	Ethnicity/ Population	Frequency	Treatment Response
CYP2D6	*10 (IM)	Caucasian	0.028	PM or IM may experience more side effects than EM → lower dose or alternate drug may be used. UM may need higher dose for therapeutic plasma concentrations.
		East Asian	0.42	
		Hispanic	0.028	
		African American	0.043	
	*4 (PM)	Caucasian	0.18	
		East Asian	0.0045	
		Hispanic	0.11	
		African American	0.06	
CYP2C19	*2 (PM)	Caucasian	0.14	
		East Asian	0.29	
		Hispanic	0.125	
		African American	0.183	
	*17 (UM)	Caucasian	0.2	
		East Asian	0.027	
		Hispanic	0.117	
		African American	0.194	

Adapted from Reyes-Barron C, Tonarelli S, Delozier A, Briones DF, and Su BB. Pharmacogenetics of antidepressants, a review of significant genetic variants in different populations, *Clinical Depression*, 2(2), pp. 1–0.

Due to the multitude of adverse events associated with antidepressant treatment, she was referred for psychiatric pharmacogenetic testing in attempts to elucidate potential underlying genetic causes. It was revealed that Ms X was an intermediate metabolizer for CYP2D6 (*1/*4), and a poor metabolizer for CYP2C19 (*2/*2), which might explain the occurrence of side effects with sertraline and escitalopram. While other factors may have contributed to her side effects from bupropion, it was advised to choose medications which were not metabolized by CYP2D6 or CYP2C19 such as desvenlafaxine, and to consider additional non-pharmacological interventions such as psychotherapy and exercise. This case exemplifies the potential benefits of pre-emptive genetic testing, as occurrence of side effects could probably have been avoided in this case.

Several internationally renowned expert groups have been founded such as the Clinical Pharmacogenetics Implementation Consortium (CPIC). CPIC is using high-quality standards and puts great efforts to review actionable gene–drug pairs across general medicine including antidepressant medications (25). CPIC guidelines have been published for CYP2D6 and CYP2C19 genotypes to help direct physicians in their approach to selective serotonin reuptake inhibitor (SSRI) and tricyclic antidepressant (TCA) treatment (15, 19, 26). Table 27.3 summarizes

Table 27.3 Therapeutic recommendations for selected SSRIs and CYP2D6 and CYP2C19 metabolizer profile.

SSRI	Gene	Phenotype	Therapeutic Recommendation	Strength of Recommendation
Paroxetine	CYP2D6	UM	Select alternative drug not predominantly metabolized by CYP2D6.	**Strong**
		EM	Initiate therapy with recommended starting dose.	**Strong**
		IM	Initiate therapy with recommended starting dose.	Moderate
		PM	Select alternative drug not predominantly metabolized by CYP2D6, or use 50% reduction of recommended starting dose and titrate to response.	Optional
Citalopram/ Escitalopram	CYP2C19	UM	Select alternative drug not predominantly metabolized by CYP2C19.	Moderate
		EM	Initiate therapy with recommended starting dose.	**Strong**
		IM	Initiate therapy with recommended starting dose.	**Strong**
		PM	Select alternative drug not predominantly metabolized by CYP2C19, or use 50% reduction of recommended starting dose and titrate to response,	Moderate

For updates on CYP2D6 and CYP2C19 Guidelines, see also: <https://cpicpgx.org/guidelines/>.

Adapted from Hicks JK, Swen JJ, Thorn CF, Sangkuhl K, Kharasch ED, Ellingrod VL, Skaar TC, Mü ler DJ, Gaedigk A, and Stingl JC., 'Clinical Pharmacogenetics Implementation Consortium guideline for CYP2D6 and CYP2C19 genotypes and dosing of tricyclic antidepressants', Clinical Pharmacology and Therapeutics, 93 (5), pp. 402–8. © 2015 American Society for Clinical Pharmacology and Therapeutics, published by John Wiley. and Sons.

the CPIC therapeutic recommendations for selected SSRIs based on CYP2D6 and CYP2C19 predicted phenotypes, as well as the strengths of recommendation.

In summary, the CYPs are highly implicated in the pharmacokinetics of many antidepressant drugs and thereby affect response and side effects. As demonstrated by the case report, genetic testing for CYP2D6 and CYP2C19 metabolizer profiles can greatly inform treatment selection, while ameliorating side effects, particularly in patients who have failed with first medications (27). Nonetheless, more research is required to determine the clinical and economic benefits of pre-emptive genetic testing versus treatment as usual and such studies are currently conducted at our center; see <http://www.im-pact.ca and https://clinicaltrials.gov/ct2/show/NCT02466477>.

27.4 Pharmacodynamics

Several genes involved in the pharmacodynamics of antidepressant response have been repeatedly identified across studies. Most interesting findings are highlighted in the following with emphasis on the serotonin and norepinephrine transporter genes.

27.4.1 Serotonin transporter

Most antidepressants such as SSRIs and TCAs exert their effects by increasing the availability of serotonin in the synaptic cleft. The SERT or SLC6A4 gene codes for the serotonin transporter,

Table 27.4 Common SERT (SLC6A4) polymorphisms and their effects.

Polymorphism Name	Change	Notes
5-HTTLPR (rs4795541/indel promoter)	43 or 44 bp deletion in promoter of SLC6A4	Short allele associated with poorer response and remission rate (28, 30).
rs25531	A > G in promoter region	G-allele decreased promoter activity in the L-allele to same extent as the S-allele of the 5-HTTLPR (31).
rs25532 (469C/T)	C > T at nucleotide 469 in promoter	May lower SERT expression by 50–80% (32). Studies regarding individual effects on antidepressant response not yet performed.
rs57098334 (2nd intron VNTR)	17 bp variable-number tandem repeat (VNTR) in intron 2	Short allele associated with improved paroxetine response compared to long-allele (33).
rs28914829 (83C/T)	C > T at nucleotide 83 in intron 7	Associated with improved response to citalopram in subjects who identified as "white" or "not Hispanic" (34).
rs28914832 (1273A/G)	A > G at nucleotide 1273 in exon 9	Associated with many neuropsychiatric disorders, such as obsessive-compulsive disorder; may be associated with resistance to treatment with SSRIs (35, 36)

which is involved in the synaptic reuptake of serotonin. Therefore, differences in the expression or conformation of the serotonin transporter were thought early on to affect the ability of neurons to use serotonin, thereby impacting the efficacy of the antidepressant.

There are six well-known polymorphisms in the SERT gene, the most studied being the 5-HTTLPR polymorphism, located in the promoter region of the gene (1). Current research confirms the importance of the 5-HTTLPR polymorphism and the associated changes in antidepressant response (28, 29). Table 27.4 highlights the most well-studied SERT polymorphisms.

Similar to the CYP polymorphisms, there is substantial genotypic variation across different ethnic populations. Figure 27.1 illustrates the genotypic variance of the 5-HTTLPR polymorphism across four ethnic populations.

There have been several studies conducted on the utility of SERT polymorphisms and antidepressant response, including some comprehensive meta-analyses. One meta-analysis conducted by Serretti and colleagues (30). reviewed the role of the 5-HTTLPR polymorphism in the efficacy of SSRI treatment in depressed patients of European and Asian ancestry. Serretti and colleagues found that individuals from European ancestry who were homozygous for the variant short (S-) allele had a poorer response rate and remission rate in their treatment with SSRIs. In contrast, patients of European ancestry who were homozygous for the long (L-) allele had an improved or normal response and remission rate, and those who were heterozygous had an intermediate response to SSRIs. While some conflicting findings have been reported in smaller studies using Asian populations, discrepant findings might be related to ethnic differences with respect to the lower allele frequencies of the 5-HTTLPR polymorphism or related linkage disequilibrium patterns. Notably, a meta-analysis by Porcelli and colleagues confirmed the association between L-allele and better response in European populations, albeit the odds ratio was rather low (OR = 1.53 C.I. 95% 1.14–2.04, $p = 0.004$) (28).

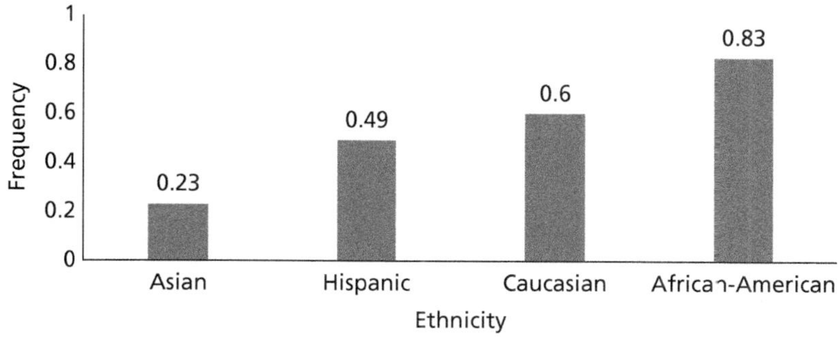

Figure 27.1 Frequency of SERT 5-HTTLPR L-allele across ethnicities[6].

Only a limited number studies have been conducted with the other polymorphisms and the interacting effects are not well understood. Therefore, there is a need for further research in SERT polymorphisms in order to determine the clinical significance of SERT profiling.

27.4.2 **Norepinephrine transporter**

The norepinephrine transporter will be highlighted in this review as another key player in the pharmacodynamics of antidepressant response, in particular for the tricyclic antidepressants and serotonin-norepinephrine reuptake inhibitors (SNRIs). Specific variants of the norepinephrine transporter (NET) gene (*SLC6A2*) have been investigated and shown to affect antidepressant response. The most studied NET polymorphisms are rs2242446 (182T/C) and rs5569 (1287G/A), which were investigated with response of nortriptyline, milnacipran, and venlafaxine. Kim and colleagues investigated the rs5569 polymorphism, and reported that Korean patients who were homozygous for the G-allele had a greatly improved response to nortriptyline compared to patients who were homozygous for the A-allele (37). Other studies have also reported that the G-allele of rs5569 was associated with an improved response to SNRIs and milnacipran (38, 39). As for the rs2242446 marker, a sample of Japanese patients ($N = 80$) found that patients with T-allele had a better response to milnacipran treatment compared to patients who had the C-allele (40). In a small Australian sample including individuals of European ancestry ($N = 38$), Singh and colleagues reported that depressed patients with a history of childhood abuse carrying the C-allele of the rs2242446 marker were more likely to respond to venlafaxine or escitalopram treatment (41).

Furthermore, a recent study by Marshe and colleagues has reported the rs2242446 polymorphism to be associated with venlafaxine response in 350 patients of European and African ancestry with late-life depression (42). While a significant confounding effect related to ancestry could be ruled out, patients with the C/C genotype were found to have a shorter time to remission, greater odds of remission, and a greater percentage change in depressive symptoms when compared to patients with the T/T or T/C genotype. Interestingly, C/C patients also experienced a significant earlier response to venlafaxine. Despite the need for further investigation, these findings affirm the importance of NET polymorphisms on antidepressant response. Although the literature on NET variants and antidepressant response remain inconclusive due to the limited number of studies, these findings point to the strong need to continue research in NET variants in response to antidepressants before it might be considered to be routinely assessed in clinical practice.

27.4.3 **Other genes**

While many other pharmacodynamic genes have been reported to be tentatively associated with antidepressant response, many findings await further validation. Beside the SERT and NET genes which are targeted by SSRIs, TCAs, and SNRIs with great certainty, the effects of other genes remain more speculative at this point of time limiting the plausibility of some findings. Some prominent genes that have also been shown to be involved in antidepressant response include the *ABCB1* gene (43), the *BDNF* gene (44), and the *FKBP5* gene (45, 46). However, most articles conclude that additional research is required to elucidate their roles in antidepressant response. The reader is encouraged to consult recent review articles for further details (4, 47–50).

27.5 **Perspectives**

As treatment of depression remains challenging, research is driven both by the need for new medications and new approaches to existing treatments. While pharmacogenetics holds the promise to improve existing treatment forms and the potential to identify new drug targets substantially, there are several considerations which should be addressed in the future.

One particular problem is the heterogeneity of depression with its various etiologies and clinical subtypes, the nature of which may influence an individual's response to antidepressants. For example, some studies in MDD patients found that individuals who have experienced early-life trauma such as abuse, family distress, disaster, or other traumatic events, were more likely to respond poorly to some specific antidepressant treatments (51), particularly if the trauma was experienced at a critical age range of four to seven years (52). Escitalopram and fluvoxamine response have also been found to be negatively affected by stressful and adverse life events, thus reinforcing the role of environmental stressors in determining the response to antidepressant treatment (53, 54). Additional factors that were reported to influence response and side effects include age, gender, ethnicity, clinical symptoms, comorbidities, and drug–drug interactions (4). Therefore, it would seem highly beneficial to incorporate genetic as well as non-genetic factors into a common future algorithm to predict outcome to antidepressant medications and guide treatment, including non-pharmacological interventions. While there is need for additional research which includes large-scale collaborations (e.g. the Psychiatric Genomics Consortium or CAN-BIND),[3] collection of new samples, and extension of existing samples, the advent of novel methods in biostatistics will enhance successes such as applying complex machine-learning strategies (55).

Another area of molecular research which will be important to investigate are the effects of epigenetics, and RNA modification (epitranscriptomics) on antidepressant response. Epigenetics have only recently starting to be explored in psychiatric pharmacogenetics, with most research investigating its impact on the SERT gene. Domschke and colleagues found that in "Caucasian" patients with MDD, there was a significant decrease in methylation at several CpG islands within the 5-HTTLPR region, which was associated with impaired SSRI response (56). MicroRNAs (miRNAs) have also proven to be modulated in response to treatment, and could therefore be reliable biomarkers in the vigilance of patient response. Lopez and colleagues conducted a randomized placebo-controlled study on depressed patients taking duloxetine, and found that over the treatment course of eight weeks, several miRNAs expression levels were significantly altered (57). Specifically, miR-146a-5p, miR-146b-5p, miR-24-3p, and miR-425-3p were found to be consistent biomarkers of duloxetine response in the peripheral blood, and these miRNAs were also found to be regulators of psychiatrically relevant signaling pathways. In summary, future research

[3] CAN-BIND—Canadian Biomarker Integration Network in Depression.

efforts using epigenetic and epitranscriptomics are likely to complement pharmacogenetic studies and advance clinical utility.

Finally, the benefits and risks of genetic testing will need to be carefully balanced because psychological and ethical concerns have been expressed. For example, treatment may be undermined by the psychological perception of genetic test results; individuals who perceive their results to be positive may experience favorable treatment (i.e. placebo) effects, while among individuals who perceive test results as negative may experience psychological distress. Further concerns have also arisen with potential discrimination issues, as information regarding potential predispositions to poor antidepressant response or vulnerability to depressive behaviors can have potential negative effects with healthcare providers, insurance companies, bank insurance premiums, and the like (58). This emphasizes the need to address and avoid discriminations based on an individual's genetic profiles, as in the Genetic Information Nondiscrimination Act (GINA) in the United States, or the Genetic Non-Discrimination Act in Canada.

27.6 **Conclusion**

Overall, pharmacogenetics has great utility in medical treatment, particularly in psychiatry due to the highly unpredictable nature of antidepressant response among patients. As described in this chapter, several important genes involved in antidepressant metabolism have been identified, where specific gene variants can alter how the individual responds and remits to medication. While the implication of pharmacokinetic gene studies (CYP2D6 and CYP2C19) is well-supported in the literature, pharmacodynamic studies remain more limited and are as yet inconclusive. For the most obvious candidates (SERT and NET), the outlook for their clinical use is promising, however there is a need for further research to confirm the mechanism of their involvement, and to incorporate these genetic variants into future clinical algorithms, including also non-genetic variables. This will ultimately improve prediction on how a given patient responds to treatment, mitigates adverse effects, providing a better quality of life through the course of treatment and saving healthcare costs.

Acknowledgements

DJM is supported by the Canadian Institutes of Health Research (CIHR Operating Grant MOP 142192), the National Institutes of Health (R01MH085801), the Centre for Addiction and Mental Health Foundation (Joanne Murphy Professorship) and the Canadian Biomarker Integration Network for Depression. DJM is a member of the Clinical Pharmacogenomics Implementation Consortium (CPIC), the Pharmacogenomics Research Network (PGRN) and the Pharmacogene Variation (PharmVar) Consortium.

References

1. **Crisafulli C, Fabbri C, Porcelli S, Drago A, Spina E, De Ronchi D, Serretti A.** Pharmacogenetics of antidepressants. Front Pharmacol. 2011;2.
2. **Rush AJ, Trivedi MH, Wisniewski SR, Nierenberg AA, Stewart JW, Warden D**, et al. Acute and longer-term outcomes in depressed outpatients requiring one or several treatment steps: A STAR* D report. Am J Psychiatry. 2006 Nov;**163**(11):1905–17.
3. **Trivedi MH, Rush AJ, Wisniewski SR, Nierenberg AA, Warden D, Ritz L**, et al. Evaluation of outcomes with citalopram for depression using measurement-based care in STAR* D: Implications for clinical practice. Am J Psychiatry. 2006 Jan 1;**163**(1):28–40.
4. **Gvozdic K, J Brandl E, L Taylor D, J Müller DJ.** Genetics and personalized medicine in antidepressant treatment. Curr Pharm Design. 2012 Dec 1;**18**(36):5853–78.

5. **Ortega VE, Meyers DA.** Pharmacogenetics: Implications of race and ethnicity on defining genetic profiles for personalized medicine. J Allergy Clin Immunol. 2014 Jan 31;**133**(1):16–26.
6. **Reyes-Barron C, Tonarelli S, Delozier A, Briones DF, Su BB.** Pharmacogenetics of antidepressants, a review of significant genetic variants in different populations. Clin Depress. 2016;**2**(2):1–0.
7. **García-González J, Tansey KE, Hauser J, Henigsberg N, Maier W, Mors O, Placentino A, Rietschel M, Souery D, Žagar T, Czerski PM.** Pharmacogenetics of antidepressant response: A polygenic approach. Prog Neuro-Psychopharmacol Biol Psychiatry. 2017 Apr 3;**75**:128–34.
8. **Uher R, Huezo-Diaz P, Perroud N, Smith R, Rietschel M, Mors O,** et al. Genetic predictors of response to antidepressants in the GENDEP project. **Pharmacogenomics J.** 2009 Aug 1;**9**(4):225–33.
9. **Hennings JM, Owashi T, Binder EB, Horstmann S, Menke A, Kloiber S,** et al. Clinical characteristics and treatment outcome in a representative sample of depressed inpatients–findings from the Munich Antidepressant Response Signature (MARS) project. J Psychiatric Res. 2009 Jan 31;**43**(3):215–29.
10. **Rush AJ, Fava M, Wisniewski SR, Lavori PW, Trivedi MH, Sackeim HA,** et al. Sequenced treatment alternatives to relieve depression (STAR* D): Rationale and design. Controlled Clinical Trials. 2004 Feb 29;**25**(1):119–42.
11. **Howland RH.** Sequenced Treatment Alternatives to Relieve Depression (STAR* D)—Part 2: Study outcomes. J Psychosoc Nurs Ment Health Serv. 2008 Oct 1;**46**(10):21–4.
12. **Gaynes BN, Rush AJ, Trivedi MH, Wisniewski SR, Spencer D, Fava M.** The STAR* D study: Treating depression in the real world. Cleveland Clinic Journal of Medicine. 2008 Jan;**75**(1):57–66.
13. **Biernacka JM, Sangkuhl K, Jenkins G, Whaley RM, Barman P, Batzler A,** et al. The International SSRI Pharmacogenomics Consortium (ISPC): A genome-wide association study of antidepressant treatment response. Transl Psychiatry. 2015 Apr 1;**5**(4):e553.
14. **Maier W, Zobel A.** Contribution of allelic variations to the phenotype of response to antidepressants and antipsychotics. Eur Arch Psychiatry Clin Neurosci. 2008 Mar 1;**258**(1):12–20.
15. **Hicks JK, Swen JJ, Thorn CF, Sangkuhl K, Kharasch ED, Ellingrod VL,** et al. Clinical Pharmacogenetics Implementation Consortium guideline for CYP2D6 and CYP2C19 genotypes and dosing of tricyclic antidepressants. Clin Pharmacol Therap. 2013 May 1;**93**(5):402–8.
16. **Lee SJ.** Clinical application of CYP2C19 pharmacogenetics toward more personalized medicine. Front Genet. 2013 Feb 1;**3**:318.
17. **Llerena A, Herraíz AG, Cobaleda J, Johansson I, Dahl ML.** Debrisoquin and mephenytoin hydroxylation phenotypes and CYP2D6 genotype in patients treated with neuroleptic and antidepressant agents. Clin Pharmacol Therap. 1993 Dec 1;**54**(6):606–11.
18. **Gaedigk A, Simon SD, Pearce RE, Bradford LD, Kennedy MJ, Leeder JS.** The CYP2D6 activity score: Translating genotype information into a qualitative measure of phenotype. Clin Pharmacol Therap. 2008 Feb 1;**83**(2):234–42.
19. **Hicks JK, Bishop JR, Sangkuhl K, Müller DJ, Ji Y, Leckband SG,** et al. Clinical Pharmacogenetics Implementation Consortium (CPIC) guideline for CYP2D6 and CYP2C19 genotypes and dosing of selective serotonin reuptake inhibitors. Clin Pharmacol Therap. 2015 Aug 1;**98**(2):127–34.
20. **Heinz A, Müller DJ, Krach S, Cabanis M, Kluge UP.** The uncanny return of the race concept. Front Hum Neurosci. 2014;8.
21. **Porcelli S, Drago A, Fabbri C, Gibiino S, Calati R, Serretti A.** Pharmacogenetics of antidepressant response. J Psychiatry Neurosci. 2011 Mar;**36**(2):87.
22. **de Leon J.** The crucial role of the therapeutic window in understanding the clinical relevance of the poor versus the ultrarapid metabolizer phenotypes in subjects taking drugs metabolized by CYP2D6 or CYP2C19. J Clin Psychopharmacol. 2007 Jun 1;**27**(3):241–5.
23. **Gaedigk A, Sangkuhl K, Whirl-Carrillo M, Klein T, Leeder JS.** Prediction of CYP2D6 phenotype from genotype across world populations. Genet Med. 2017 Jan;**19**(1):69.
24. **Caudle KE, Klein TE, Hoffman JM, Müller DJ, Whirl-Carrillo M, Gong L,** et al. Incorporation of pharmacogenomics into routine clinical practice: The Clinical Pharmacogenetics Implementation Consortium (CPIC) guideline development process. Curr Drug Metab. 2014 Feb 1;**15**(2):209–17.

25. **Sistonen J, Sajantila A, Lao O, Corander J, Barbujani G, Fuselli S.** CYP2D6 worldwide genetic variation shows high frequency of altered activity variants and no continental structure. Pharmacogenet Genom. 2007 Feb 1;**17**(2):93–101.

26. **Hicks JK, Sangkuhl K, Swen JJ, Ellingrod VL, Müller DJ, Shimoda K**, et al. Clinical Pharmacogenetics Implementation Consortium guideline (CPIC) for CYP2D6 and CYP2C19 genotypes and dosing of tricyclic antidepressants: 2016 Update. Clin Pharmacol Ther. 2017 Feb 1.

27. **Müller DJ, Kekin I, Kao AC, Brandl EJ.** Towards the implementation of CYP2D6 and CYP2C19 genotypes in clinical practice: Update and report from a pharmacogenetic service clinic. Int Rev Psychiatry. 2013 Oct 1;**25**(5):554–71.

28. **Porcelli S, Fabbri C, Serretti A.** Meta-analysis of serotonin transporter gene promoter polymorphism (5-HTTLPR) association with antidepressant efficacy. Eur Neuropsychopharmacol. 2012 Apr 30;**22**(4):239–58.

29. **Brandl EJ, Walter H.** From pharmacogenetics to imaging pharmacogenetics: Elucidating mechanisms of antidepressant response. Pharmacogenomics. 2017 Jun 22;**18**(10): 927–30.

30. **Serretti A, Kato M, De Ronchi D, Kinoshita T.** Meta-analysis of serotonin transporter gene promoter polymorphism (5-HTTLPR) association with selective serotonin reuptake inhibitor efficacy in depressed patients. Mol Psychiatry. 2007 Mar 1;**12**(3):247–57.

31. **Hu XZ, Lipsky RH, Zhu G, Akhtar LA, Taubman J, Greenberg BD**, et al. Serotonin transporter promoter gain-of-function genotypes are linked to obsessive-compulsive disorder. Am J Hum Genet. 2006 May 31;**78**(5):815–26.

32. **Murphy DL, Fox MA, Timpano KR, Moya PR, Ren-Patterson R, Andrews AM**, et al. How the serotonin story is being rewritten by new gene-based discoveries principally related to SLC6A4, the serotonin transporter gene, which functions to influence all cellular serotonin systems. Neuropharmacology. 2008 Nov 30;**55**(6):932–60.

33. **Bozina N, Mihaljevic Peles A, Sagud M, Bilusic H, Jakovljevic M.** Association study of paroxetine therapeutic response with SERT gene polymorphisms in patients with major depressive disorder. World J Biol Psychiatry. 2008 Jan 1;**9**(3):190–7.

34. **Mrazek DA, Rush AJ, Biernacka JM, O'Kane DJ, Cunningham JM, Wieben ED**, et al. SLC6A4 variation and citalopram response. Am J Med Genet B. 2009 Apr 5;**150**(3):341–51.

35. **Delorme R, Betancur C, Wagner M, Krebs MO, Gorwood P, Pearl P**, et al. Support for the association between the rare functional variant I425V of the serotonin transporter gene and susceptibility to obsessive compulsive disorder. Mol Psychiatry. 2005 Dec 1;**10**(12):1059–61.

36. **Ozaki N, Goldman D, Kaye WH, Plotnicov K, Greenberg BD, Lappalainen J**, et al. Serotonin transporter missense mutation associated with a complex neuropsychiatric phenotype. Mol Psychiatry. 2003 Nov 1;**8**(11):933–6.

37. **Kim H, Lim SW, Kim S, Kim JW, Chang YH, Carroll BJ**, et al. Monoamine transporter gene polymorphisms and antidepressant response in Koreans with late-life depression. JAMA. 2006 Oct 4;**296**(13):1609–18.

38. **O'Leary OF, O'Brien FE, O'Connor RM, Cryan JF.** Drugs, genes and the blues: Pharmacogenetics of the antidepressant response from mouse to man. Pharmacol Biochem Behav. 2014 Aug 31;**123**:55–76.

39. **Singh AB, Bousman CA, Ng CH, Byron K, Berk M.** Psychomotor depressive symptoms may differentially respond to venlafaxine. Int Clin Psychopharmacol. 2013 May 1;**28**(3):121–6.

40. **Yoshida K, Takahashi H, Higuchi H, Kamata M, Ito KI, Sato K**, et al. Prediction of antidepressant response to milnacipran by norepinephrine transporter gene polymorphisms. Am J Psychiatry. 2004 Sep 1; **161**(9):1575–80.

41. **Singh AB, Bousman CA, Ng CH, Byron K, Berk M.** Effects of persisting emotional impact from child abuse and norepinephrine transporter genetic variation on antidepressant efficacy in major depression: A pilot study. Clin Psychopharmacol Neurosci. 2015 Apr;**13**(1):53.

42. **Marshe VS, Maciukiewicz M, Rej S, Tiwari AK, Sibille E, Blumberger DM**, et al. Norepinephrine transporter gene variants and remission from depression with venlafaxine treatment in older adults. Am J Psychiatry. 2017 Jan 10;**174**(5):468–75.

43. **Breitenstein B, Brückl TM, Ising M, Müller-Myhsok B, Holsboer F, Czamara D.** ABCB1 gene variants and antidepressant treatment outcome: A meta-analysis. Am J Med Genet B. 2015 Jun 1;**168**(4):274–83.
44. **Tsai SJ, Hong CJ, Liou YJ.** Brain-derived neurotrophic factor and antidepressant action: Another piece of evidence from pharmacogenetics. Pharmacogenomics. 2008 Sep 9;**9**(9): 1353–8
45. **Sarginson JE, Lazzeroni LC, Ryan HS, Schatzberg AF, Murphy GM.** (2010). FKBP5 polymorphisms and antidepressant response in geriatric depression. Pharmacogenet Genomics. 2010 Aug 1;**20**(8):467–75.
46. **Binder EB, Salyakina D, Lichtner P, Wochnik GM, Ising M, Pütz B,** et al. Polymorphisms in FKBP5 are associated with increased recurrence of depressive episodes and rapid response to antidepressant treatment. Nat Genet. 2004 Dec 1;**36**(12):1319–25.
47. **Serretti A.** Genetics and pharmacogenetics of mood disorders. Psychiatria Polska. 2017 Apr 30;**51**(2):197.
48. **Fabbri C, Serretti A.** Pharmacogenetics of major depressive disorder: Top genes and pathways toward clinical applications. Curr Psychiatry Rep. 2015 Jul 1;**17**(7):50.
49. **Niitsu T, Fabbri C, Bentini F, Serretti A.** Pharmacogenetics in major depression: A comprehensive meta-analysis. Prog Neuro-Psychopharmacol Biol Psychiatry. 2013 Aug 1;**45**:183–94.
50. **Kato M, Serretti A.** Review and meta-analysis of antidepressant pharmacogenetic findings in major depressive disorder. Mol Psychiatry. 2010 May 1;**15**(5):473–500.
51. **Quilty LC, Marshe V, Lobo DS, Harkness KL, Müller DJ, Bagby RM.** Childhood abuse history in depression predicts better response to antidepressants with higher serotonin transporter affinity: A pilot investigation. Neuropsychobiology. 2016;**74**(2):78–83.
52. **Williams LM, Debattista C, Duchemin AM, Schatzberg AF, Nemeroff CB.** Childhood trauma predicts antidepressant response in adults with major depression: Data from the randomized international study to predict optimized treatment for depression. Transl Psychiatry. 2016 May 1;**6**(5):e799.
53. **Keers R, Uher R, Huezo-Diaz P, Smith R, Jaffee S, Rietschel M,** et al. Interaction between serotonin transporter gene variants and life events predicts response to antidepressants in the GENDEP project. **Pharmacogenomics J.** 2011 Apr 1;**11**(2):138–45.
54. **Mandelli L, Marino E, Pirovano A, Calati R, Zanardi R, Colombo C,** et al. Interaction between SERTPR and stressful life events on response to antidepressant treatment. Eur Neuropsychopharmacol. 2009 Jan 31;**19**(1):64–7.
55. **Maciukiewicz M, Marshe VS, Hauschild AC, Foster JA, Rotzinger S, Kennedy JL,** et al. GWAS-based machine learning approach to predict duloxetine response in major depressive disorder. J Psychiatric Res. 2018;**99**:62–8.
56. **Domschke K, Tidow N, Schwarte K, Deckert J, Lesch KP, Arolt V,** et al. Serotonin transporter gene hypomethylation predicts impaired antidepressant treatment response. Int J Neuropsychopharmacology. 2014 Aug 1;**17**(8):1167–76.
57. **Lopez JP, Fiori LM, Cruceanu C, Lin R, Labonte B, Cates HM,** et al. MicroRNAs 146a/b-5 and 425-3p and 24–3p are markers of antidepressant response and regulate MAPK/Wnt-system genes. Nature Communications. 2017 May 22;**8**:15497.
58. **Rasmussen-Torvik LJ, McAlpine DD.** Genetic screening for SSRI drug response among those with major depression: Great promise and unseen perils. Depress Anxiety. 2007 Jan 1;**24**(5):350–7.

Index

Figures, tables and boxes are indicated by an italic *f*, *t* or *b* following the page number